T0418952

ROUTLEDGE COMPANION TO CYCLING

Routledge Companion to Cycling presents a comprehensive overview of an artefact that throughout the modern era has been a bellwether indicator of many major social, economic and environmental trends that have permeated society.

The volume synthesizes a rapidly growing body of research on the bicycle, its past and present uses, its technological evolution, its use in diverse geographical settings, its aesthetics and its deployment in art and literature. From its origins in early modern carriage technology in Germany, it has generated what is now a vast, multi-disciplinary literature encompassing a wide range of issues in countries throughout the world.

Glen Norcliffe is an Emeritus Professor of Geography at York University, Canada.

Una Brogan is a translator and independent researcher from Northern Ireland, UK.

Peter Cox is Professor of Sociology at the University of Chester, UK.

Boyang Gao is Professor of Urban and Real Estate Management, Central University of Finance and Economics, Beijing.

Tony Hadland is a Chartered Building Surveyor and historian living in Oxfordshire, UK.

Sheila Hanlon is a historian specializing in the history of women's cycling. She works with a number of organizations such as Cycling UK and the Cycling History Education Trust.

Tim Jones is Reader in Urban Mobility in the School of the Built Environment at Oxford Brookes University, UK.

Nicholas Oddy is Head of the Department of Design History & Theory at Glasgow School of Art, UK.

Luis Vivanco is Professor of Anthropology and Director of the Humanities Center at the University of Vermont, USA.

ROUTLEDGE COMPANION TO CYCLING

Edited by

Glen Norcliffe, Una Brogan, Peter Cox, Boyang Gao, Tony Hadland, Sheila Hanlon, Tim Jones, Nicholas Oddy and Luis Vivanco

LONDON AND NEW YORK

Cover image: Greg Curnoe, Mariposa 10 Speed No. 2, 1973.
Watercolour over graphite on wove paper, 101.1 × 181.4 cm.
National Gallery of Canada, Ottawa

First published 2022
by Routledge
4 Park Square, Milton Park, Abingdon, Oxon OX14 4RN

and by Routledge
605 Third Avenue, New York, NY 10158

Routledge is an imprint of the Taylor & Francis Group, an informa business

British Library Cataloguing-in-Publication Data
A catalogue record for this book is available from the British Library

Library of Congress Cataloging-in-Publication Data
Names: Norcliffe, G. B., editor.
Title: Routledge companion to cycling / Edited by Glen Norcliffe, Una Brogan, Peter Cox, Boyang Gao, Tony Hadland, Sheila Hanlon, Tim Jones,
Nicholas Oddy, and Luis Vivanco.
Description: Abingdon, Oxon ; New York, NY : Routledge, 2022. | Includes
bibliographical references and index.
Identifiers: LCCN 2022007864 (print) | LCCN 2022007865 (ebook) | ISBN
9780367683993 (hardback) | ISBN 9780367695088 (paperback) | ISBN 9781003142041 (ebook)
Subjects: LCSH: Cycling--History. | Cycling--Social aspects. | Cycling--Economic aspects. | Bicycles--History.
Classification: LCC GV1040.5 .R68 2022 (print) | LCC GV1040.5 (ebook) |
DDC 796.6--dc23
LC record available at https://lccn.loc.gov/2022007864
LC ebook record available at https://lccn.loc.gov/2022007865

ISBN: 978-0-367-68399-3 (hbk)
ISBN: 978-0-367-69508-8 (pbk)
ISBN: 978-1-003-14204-1 (ebk)

DOI: 10.4324/9781003142041

Typeset in Bembo
by SPi Technologies India Pvt Ltd (Straive)

CONTENTS

List of figures xi
List of tables xiv
About the contributors xv

An introduction to the companion to cycling 1
Glen Norcliffe

Part I
Cycling and society: an introduction **11**
Peter Cox

1 Theorizing cycling 15
Peter Cox

2 Cycling and gender: past, present and paths ahead 24
Jennifer Bonham and Kat Jungnickel

3 The precarious work of platform cycle delivery workers 33
Cosmin Popan and Esther Anaya-Boig

4 The sociality of cycling 42
Simon Batterbury and Alejandro Manga

Vignette A Black cyclists matter: Major Taylor – Au Parc des Princes 1901 52
Andrew Ritchie

5 Programs for cycling inclusion 55
Angela van der Kloof

6 The potential of "bike-like" vehicles to provide big wins for climate change, safety and justice 65
Kevin J. Krizek

7 Mobility, freedom and self-determination: the benefits (and barriers) to disabled people cycling 73
Kay Inckle

Part II
Cycle technology: an introduction **83**
Tony Hadland

8 Configuration of cycles 87
Hans-Erhard Lessing

9 Frames and materials 97
David Minter

10 Wheels and shock absorption 108
Tony Hadland

11 Transmission and brakes 118
Dan Farrell

12 Passenger carrying 128
David Henshaw

Vignette B Micromobility in Rwanda 137
Hilary Angus

13 Cycling technologies and disability 140
Ron Buliung, Annika Kruse, Glen Norcliffe and John Radford

Part III
The cycling economy: an introduction **149**
Glen Norcliffe

14 The global bicycle industry 153
Glen Norcliffe and Boyang Gao

15 The value chains and production clusters of Taiwan's bicycle industry 163
Cheng-Mei Tung

16 Bicycle trade shows as transactional spaces 172
Michael Andreae and Glen Norcliffe

17 Retailing bicycles 181
Jay Townley, Bradley Hughes and Michael Fritz

18 On the shoulders of giant: Cluster innovation and entrepreneurship in the Taiwanese bicycle industry 192
Yu-Chun Lin

19 Street trades and work cycles 201
Claudio Sarmiento-Casas

Vignette C Mobile cycle repairing in Beijing 211
Tian Yun

Part IV
Urban cycling: an introduction **215**
Sheila Hanlon

20 Cycling infrastructure: planning cycle networks 219
John Parkin

Vignette D Cycling infrastructure in Lund, Sweden 229
Till Koglin

21 Situating the mobility fix of contemporary urban cycling policy 232
Justin Spinney

22 Making Space for Cycling 241
Paola Castañeda and Sergio Montero Munoz

Vignette E B2W Indonesia and the re-cycling of Jakarta 250
Purwanto Setiadi

23 Shared micromobility: policy, practices, and emerging futures 254
Susan Shaheen, Adam Cohen and Jacquelyn Broader

24 E-bikes: expanding the practice of cycling? 263
Dimitri Marincek and Patrick Rérat

25 Cycling safety as mobility justice 272
Léa Ravensbergen, Ron Buliung and Ahmed El-Geneidy

Part V
Sport, health and lifestyle: an introduction **283**
Tim Jones

26 Amateur sport cycling: the rise of the MAMIL 287
Tim Jones

27 Professional road cycling 298
Daam Van Reeth

Vignette F In the peloton 307
Michael Barry

28 Off-road cycling 310
Karen McCormack and Ben Osborn

29 Track cycling 318
Michael Jordan

Vignette G Keirin culture 327
Keizo Kobayashi

30 Health benefits of cycling 332
Adrian Bauman, Sylvia Titze and Pekka Oja

31 Doping in cycling: past, present and future trends 345
Charlotte Smith

Part VI
Places of cycling: an introduction **355**
Luis Vivanco

Vignette H Early Cycling in the Bois de Boulogne, Paris 359
David V. Herlihy

32 Cycling's symphony of place 363
Robert McCullough

Vignette I Constructing peaceful places through bicycles 373
Jeanette Steinmann, Mitchell McSweeney, Lyndsay Hayhurst, and Brian Wilson

33 In quest of adventures 376
Duncan Jamieson

Vignette J Winter cycling: Montreal's four-season bicycle network 385
Bartek Komorowski and Kevin Manaugh

34 The Africanized bicycle 388
Hans Peter Hahn

Vignette K Wheels of Fire: women cycling in the Middle East 396
Alon Raab

35 Cycling in Indian cities: between everyday cyclists and affluent cyclists 399
Rutul Joshi and Jacob Baby

36 The rise of the *Kingdom of Bicycles* 409
Xu Tao

37 Copenhagen is a good place to bike: but it could be better... 419
Malene Freudendal-Pedersen

Vignette L Beach Road, Melbourne 427
Charlie Farren

38 Bogotá: perspectives on the "World bike capital" 430
Luis Vivanco

Part VII
The visual culture of cycling **439**
Nicholas Oddy

39 The machine aesthetic: the visual identity of the bicycle and its representation in advertising and artefacts 443
Nicholas Oddy

40 Dressed to ride 453
Emma Hilborn

41 Cycle posters of the *Belle Époque* 462
Nadine Besse

42 Art and the cycle 472
Scotford Lawrence

Vignette M The space between 481
Hilary Norcliffe

43 Cycling and cinema: revolutionary films 484
Bruce Bennett

Part VIII
Cycling in literature: an introduction 493
Una Brogan

44 The bicycle and the creative pursuit in French literature 497
Edward Nye

45 The liberating bicycle in literature 505
Jeremy Withers

46 Cycling humor in turn-of-the-century literature 514
Una Brogan

47 On bards on bicycles: the art of cycling poetry 521
Justin Daniel Belmont

48 "The stutter of the world beneath you": the literature of cycle travel 533
Dave Buchanan

Index 542

FIGURES

4.1 The Récup'R workshop, in Bordeaux, France (February 2020, before relocation). Source: Simon Batterbury. 46
4.2 Atelier Pignon sur Rue, Ambilly, France. Expert women mechanics teaching a (male) workshop employee a specific repair (December 2020). Source: Alejandro Manga. 47
4.3 The first Ciclovía to take place after the first COVID-19 Lockdown, Bogotá, Colombia (September 2020). Source: Alejandro Manga. 48
A.1 Major Taylor at the Buffalo Velodrome during his comeback in 1908. Source: Author's collection. 53
A.2 Major Taylor and Edmond Jacquelin enter the Parc des Princes followed by press photographers prior to their first race 16th May 1901. Source: Author's collection. 54
5.1 The cycling participation ladder elaborated on physical cycling skills and traffic skills. Source: Author. 57
8.1 Karl von Drais's Draisine of 1817. Illustration by the late Joachim Lessing. Reproduced by permission of Hans-Erhard Lessing. 88
9.1 Five types of force that can act on a structural member. (copyright Tony Hadland). 97
9.2 Clockwise from top left – diamond frame, cross frame, F-frame, short wheelbase recumbent, space frame, monocoque. (copyright Tony Hadland). 100
9.3 Structural members of a typical diamond frame. (copyright Tony Hadland). 102
10.1 Front hub incorporating drum brake, supported by a monoblade for easy tire removal. (copyright of Tony Hadland). 110
11.1 Derailleur transmission. The classical derailleur gear transmission with double chainrings. (author's photo). 120
11.2 Rim brakes. Types of rim brakes, clockwise from top left: Centrepull, cantilever, direct pull 'V-brake', dual-pivot sidepull. (author's photo). 124
12.1 Carrying a six-week old baby in a single-seat trailer. (photo copyright of David Henshaw). 132
B.1 A project Rwanda coffee bike. (author's photo). 138
17.1 Valley Cyclery, Van Nuys, California, 1963. (author's collection). 186
17.2 A New Age Bike Shop Showroom. (author's collection). 190
18.1 Taiwan 27-year Bicycle Export History. (Source: 2021 Taiwan Bicycle Source). 195

19.1 A group of cargo tricycles selling food snacks in a public park in Mexico City. (photograph by the author). 205
19.2 A coffee stand built on top of a (likely unmovable) cargo tricycle, outside a Bohemian public market. (photograph by the author). 206
C.1 Li's bicycle repair stall in Beijing's Chaoyang District, North-East of the City centre. (Source: author's collection). 212
20.1 Priority junctions on cycleways needs appropriate signing and marking. (author's photo). 223
20.2 Cycle optimized traffic signal-controlled junction, Trinity Street, Bolton, UK. Transport for Greater Manchester (2021). Bolton opens the UK's second pioneering CYCLOPS junction with nine more set to be delivered in 2021 (News release, Transport for Greater Manchester). Available at https://news.tfgm.com/news/bolton-opens-the-uks-second-pioneering-cyclops-junction-with-nine-more-set-to-be-delivered-in-2021, accessed on 12th May 2021. 225
20.3 Innovative roundabout design in Zwolle, Netherlands. (author's photo). 226
D.1 Lund, Sweden. Source: Lund Municipality, 2020. 230
D.2 Cycling infrastructure beside LRT line in Lund. Source: Photo by author. 231
E.1 A reggae group bicycle to a performance in central Jakarta. (author's photo). 252
23.1 Spin's three-wheeled scooter. (Source: Ford Media Center, 2021). 260
24.1 Restorative cycling trajectory. (Copyright Marincek & Rérat, 2020). 268
24.2 Resilient cycling trajectory. (Copyright Marincek & Rérat, 2020). 268
25.1 Cycling safety as the outcome of intersectional power relations. (Source: author). 278
29.1 Track racing at the velodrome in Cali, Columbia, 2014. Source: https://en.wikipedia.org/wiki/Track_cycling 320
G.1 Keirin race: Goal sprint. (Copyright JKA with permission). 329
30.1 Health benefits of cycling. (Source: authors). 333
H.1 A velocipede race at the Pré Catelan, May 1868. (Source: L'Univers Illustré, 6 June 1868, p.348). 361
32.1 Photographic Outfits. *Outing* 7 (October 1885): 129. (Author's collection). 366
32.2 "By the Roadside," a wood engraving by H. E. Sylvester after a painting or drawing by F. Childe Hassam for The Wheelman (September 1883). (Author's collection). 368
32.3 Scottsville Sidepath, Monroe County, New York. Photograph by Cline Rogers, 1899. Courtesy Rochester (N.Y.) Public Library, Local History and Genealogy Division. 369
32.4 LAW Pennsylvania Division, L.A.W., Road Book of 1893, Tabulated Route 60. (Author's collection). 370
33.1 George Burston and Harry Stokes. (Courtesy of John Weiss). 379
33.2 Alison Stone reaches the summit of the Simplon Pass, 2014. (Courtesy of John Weiss). 383
J.1 Conventional (left) and parking-protected (right) bicycle lane configurations. (authors' collection). 387
36.1 Bicycle production in China 1949–1978. Source: Data provided by the Chinese Bicycle Industry Technology and Information Office. 412
36.2 People riding bicycles across a railway bridge in Beijing, 1991, as the Kingdom of Bicycles approaches its end (Photo by Wang Wenlan 王文澜 : see https://www.pinlue.com/article/2017/05/2221/381926780814.html). 414

36.3 China – bicycle exports 1953–2020. Source: Data provided by China Bicycle Association. 417
37.1 A curbed bike lane in Copenhagen. (Photo – Jesper Pagh with permission). 420
L.1 Beach Road Melbourne, Australia, Sunday morning. (author's collection). 427
39.1 Hampden/Barber poster by Burch (1895). Single colour stone litho, 22½″ × 35″ printed by Macbride & Macintyre, Glasgow. 444
40.1 A cartoon satirizing women's cycling clothing. (Source: *Punch* 4 June 1898, p. 258). 457
40.2 A well-dressed cycling woman. (Source: Svensk Damtidnings modebilaga 1901). 459
41.1 Georges Richard, 1897. Dessinateur Eugène Grasset. Imprimerie de Vaugirard, Paris. 116 × 153 cm (Collection du musée d'Art et d'Industrie de la ville de Saint-Etienne: Cliché Yves Bresson). 466
41.2 Déesse, vers 1895. Dessinateur PAL. Imprimerie Paul Dupon, Paris. 149 × 111 cm (Collection du musée d'Art et d'Industrie de la ville de Saint-Etienne: Cliché Yves Bresson). 467
41.3 Whitworth, 1897. Dessinateur Jules Alexandre Grün. Imprimerie Bourgerie et cie, Paris. 117 × 84,5 cm (Collection du musée d'Art et d'Industrie de la ville de Saint-Etienne: Cliché Yves Bresson). 468
41.4 Clément Paris, 1892. Dessinateur inconnu. Imprimerie Kossuth, Paris. 150 × 99.5 cm (Collection du musée d'Art et d'Industrie de la ville de Saint-Etienne: Cliché Yves Bresson). 470
42.1 Artist unknown. Exercising a hobby from Wales to Hertford. Sidebethem print 1819. 473
42.2 Joseph Roux. Course de Vélocipèdes. Gray 1869. Oil on Canvas. (Courtesy of National Fietsmuseum Velorama. Netherlands). 475
42.3 Frank Patterson. Line drawing. My Favourite County. 1917. 477
42.4 Claude le Boul. Gouache. The Champion Eddy Merckx. 1987. 479
M.1 Forever bicycles: Ai Weiwei (2015–2016) (National Gallery of Victoria, Melbourne). With the artist's permission. 482
48.1 Joseph Pennell, "The Hospice on the Grimsel." From *Over the Alps on a Bicycle* (1898) by Elizabeth Robins Pennell. 534
48.2 "Kuklos" or W. Fitzwater Wray (1931). Photographer unknown. Credit: The Cycling Photographica Collection of Lorne Shields. 537

TABLES

14.1	Bicycle production in China by province, 1985–2020 (excluding e-bikes)	155
14.2	E-bike production in China by province, 2016–2020	155
14.3	Estimated bicycle production – major countries c.2020	157
17.1	Estimated US Bicycle Market Retail Sales by Channel of Distribution, 2019	184
17.2	Estimated French Bicycle Market Retail Sales by Channel of Distribution 2019	184
17.3	US bicycle Market Estimated Direct Effect Retail Dollars	185
17.4	US Independent Bicycle Dealers Engaged in e-commerce	188
24.1	E-bike sales per country (in thousands of units), 2014 to 2019	265
25.1	Approaches to studying cycling safety	273
27.1	The competition structure of professional road cycling in 2021	302
27.2	Key financial data for professional road cycling (1990–2020, in nominal euros)	303
31.1	Popular doping methods in cycling compared	346

ABOUT THE CONTRIBUTORS

Esther Anaya-Boig has been a cycling advocate and consultant for almost twenty years. She has been recently awarded a PhD in Environmental Policy Research at Imperial College London for a thesis that combines psychology, transport, and health. Her research focuses on the evaluation of cycling policies and how to make cycling more inclusive and accessible for everyone.

Michael Andreae applied performance theory in his master's thesis to explore why people congregate at bicycle trade shows. He is now a computer scientist and systems librarian at the University of Toronto, where he continues to explore how people share information using technology.

Hilary Angus is a bicycle advocate, writer, and the former editor of *Momentum Mag*. Her interest in bicycles was born out of a love for responsible urbanism, which has now led her to a fledgling career in urban farming.

Jacob Baby is an urban planner and researcher. He teaches at the Graduate Foundation Studio at CEPT University in Ahmedabad. His research focus includes urban housing, transit-oriented development and public policy in Indian cities. He is keenly interested in walking and cycling being mainstreamed as part of the urban development planning process.

Michael Barry was a professional cyclist from 1998 to 2012. He represented Canada at three Olympic Games, two Commonwealth Games and rode for the most famous teams of his age, including Team Sky, Columbia – High Road, T-Mobile, Discovery Channel, and the US Postal Service team. He also competed in the Tour de France, the Vuelta a Espana and the Giro d'Italia. His has written four books and now builds Mariposa Bicycles in Toronto.

Simon Batterbury is Associate Professor of Environmental Studies at the University of Melbourne, Australia and Visiting Professor and former Chair at Lancaster University, UK. His work is on political ecology and environmental justice, working in West Africa, South East Asia and the francophone Pacific over 28 years. More recently he has researched and volunteers in community bike workshops in the Western world, leading to a co-authored volume in progress.

Adrian Bauman is Emeritus Professor of Public Health at Sydney University, Australia. He is a physical activity epidemiologist and population health researcher, with a personal and professional interest in population-level cycling research.

Justin Daniel Belmont studied at Tufts, Oxford, and Columbia. He started his career as an editorial assistant at *Bicycling* magazine. He is a lifelong writer, poet, scholar and cyclist, an entrepreneur, and the editor of *The Art of Bicycling: A Treasury of Poems*.

Bruce Bennett is Senior Lecturer in Film Studies at Lancaster University in the UK. His research focuses on transnational cinemas, film history, pedagogy and media technology. His publications include *Cycling and Cinema* (2019), and *The Cinema of Michael Winterbottom: Borders, Intimacy, Terror* (2014).

Nadine Besse is an historian, museologist and former Director of the Museum of Art and Industry of Saint-Etienne, France. She is now retired, after 30 years of research, exhibitions, scientific catalogs and the development of museum collections, including cycles and posters. She is currently writing a book on Julien Faure, a family business of ribbons and silks. She was also a co-founder of the International Cycle History Conference.

Jennifer Bonham has a background in human geography and is an adjunct Senior Research Fellow at the University of South Australia. Her research expertise includes mobility, cycling and sustainable neighborhoods. Jennifer led the multi-partner, Australia Research Council-funded project *Cycle Aware: Driving with Bikes* and her publications include the co-edited volume *Cycling Futures*. She is a councillor for the City of Unley, where cycling is a policy priority.

Jacquelyn Broader is a researcher with the University of California, Berkeley's Transportation Sustainability Research Center (TSRC). Her research has focused on increasing accessibility for people with disabilities, shared mobility, and leveraging the sharing economy for disaster evacuations.

Una Brogan is a translator and independent researcher from Northern Ireland. Her PhD, entitled "Bicycles in Literature: the Alternative Modernities of Human-Powered Locomotion in Britain and France, 1880–1920", was completed at Université Paris 7-Diderot in 2016 and will soon be published as a book by Edinburgh University Press. She is also a cycling and environmental activist, having participated in organizing critical masses and DIY bike workshops in Paris.

Dave Buchanan teaches English at MacEwan University in Edmonton, Alberta. He edited *A Canterbury Pilgrimage/An Italian Pilgrimage* by Elizabeth Robins Pennell (2015), and is co-editor of *Elizabeth Robins Pennell: Critical Essays* (2021).

Ron Buliung is a Professor of Geography in the Department of Geography, Geomatics and Environment at the University of Toronto, Mississauga. For many years, he taught a graduate seminar on the history and geography of cycles and cycling. He is a parent of a child with disability, and his research program is now almost entirely devoted to disability studies.

Paola Castañeda is a historian, geographer, and a doctoral candidate at the University of Oxford's School of Geography and the Environment and Transport Studies Unit. She is currently a visiting researcher at the Universidad de los Andes in Bogotá, Colombia. Working from the in-between space between cycling activism and academia, Paola's research examines cycling mobilizations in Latin America and the intersection of the politics of cycling with social concerns and mobility justice.

Adam Cohen is a Senior Research Manager at the Transportation Sustainability Research Center at the University of California, Berkeley. His research has focused on innovative mobility strategies, including shared mobility, last-mile delivery, automated vehicles, advanced air mobility, smart cities, and other emerging technologies.

Peter Cox is a Professor of Sociology at the University of Chester, UK. He is a founding member of the Cycling and Society Research Group and current chair of the Scientists for Cycling academic network for the European Cyclists' Federation.

Ahmed El-Geneidy is a professor at McGill University's School of Urban Planning and leader of the TRAM (Transportation Research At McGill) research group. He conducts research in land use and transport planning, public transport operations and planning, travel behavior analysis concentrating on the use of motorized and non-motorized modes of transport and their impacts on health and well-being.

Dan Farrell has worked in bicycle manufacturing in the UK for over twenty-five years. Educated in Industrial Design and Design Management, he is a Chartered Engineer and a Chartered Technological Product Designer. An accomplished ultra-distance cyclist, Dan is currently the Chair of the British Standard Institution's Cycles committee. He is also a Fellow, and Trustee, of the Institution of Engineering Designers.

Charlie Farren is a lifelong bike rider with an extensive history of bicycle advocacy and organization of large-scale bicycle events in Australia. She is currently the custodian of a world-class early bicycle collection established by her late husband and she has co-authored a book on the collection.

Malene Freudendal-Pedersen is Professor in Urban Planning at Aalborg University, Denmark. She has an interdisciplinary background linking sociology, geography, urban planning, and the sociology of technology. Her research has been strongly inspired by the mobilities turn and focuses on the interrelation between spatial and digital mobilities and its impacts on everyday life communities, societies, and cities. She has been co-organizing the international Cosmobilities Network linking mobilities researchers in Europe and beyond.

Michael Fritz is a motivated executive professional with more than 45 years of global experience as a CEO and in engineering, engineering management and product development environments.

Boyang Gao is Professor of Urban and Real Estate Management, Central University of Finance and Economics, Beijing. She has researched activities at many cycle industry factories in China, and published papers on them in Mandarin and English.

Tony Hadland is a Chartered Building Surveyor and historian living in Oxfordshire, UK. He is the author or co-author of nine books on cycle history and technology and the contributor to many other books on cycle history and technology.

Hans Peter Hahn is professor for Cultural Anthropology at Goethe University, Frankfurt, Germany. He specializes on West Africa, consumption, and the impact of globalization on local societies. His recent research addresses the materiality of mobility which includes, among others, mobile phones and bicycles.

Sheila Hanlon is a historian specializing in the history of women's cycling. She completed her PhD at York University, Toronto and held a Vera Douie Research Fellowship. Her research has been featured on BBC radio and TV. She works with a number of organizations, such as Cycling UK and the Cycling History Education Trust.

Lyndsay Hayhurst is an Assistant Professor in the School of Kinesiology and Health Science at York University in Toronto. She is a co-author of *Sport, Gender and Development*, and is the principal investigator on the SSHRC-funded *Bicycles for Development* research team.

David Henshaw is a writer, graphic designer and engineer. He has authored and co-authored several transport books, primarily on railways, electric bikes, and folding bikes: his book *Brompton Bicycle* is now on its third edition and fourth reprint. Since 1996, David has published the *A to B Magazine*, covering a broad range of 'green-tinged' environmentally-benign transport and power-generation topics.

David V. Herlihy is a historian and freelance writer specialized in bicycle and motorcycle history. He is the author of *Bicycle: The History* and *The Lost Cyclist*, and numerous articles. He has conducted seminal research on the origins of the bicycle. He has also curated two major exhibitions, *The Bicycle Takes Off* (Lockwood-Mathews Museum of Norwalk, CT 2000–2002) and *Round Trip* (UCLA's Fowler Art Museum 2014–2015.)

Emma Hilborn is a historian and researcher at Lund University, Sweden. Her research interests focus on cultural history, gender history and labor history, primarily during the early 20th century. She is currently studying the feminization of weight-loss dieting in England and Scandinavia at the turn of the century.

Bradley Hughes is a global business leader with more than 45 years of experience who has presided over companies as a CEO and COO and produced triple digit cost reduction and profit results.

Kay Inckle is a social scientist with expertise in disability and mental health research. She has held a number of academic positions, but ultimately left higher education because it is such a hostile environment for disabled women. She currently works as an independent scholar/researcher, writes fiction and teaches Pilates. She is also an avid hand cyclist, swimmer and environmentalist and she has been a candidate for the Green Party in recent local and national elections.

Duncan Jamieson is Professor of History at Ashland University, Ohio, USA. His research interest is in long-distance cycle travel. In addition to *The Self-Propelled Voyager* (2015), he has published several articles on bicycles and tricycles. He maintains a bibliography, *Bicycle Travel and Touring Resources*. When not writing or teaching, he enjoys a relaxing ride on his recumbent.

Tim Jones is Reader in Urban Mobility in the School of the Built Environment at Oxford Brookes University, UK. His research interests focus on how urban environments can be (re) configured to support and promote sustainable and healthy urban mobility, particularly walking and cycling.

Michael Jordan is a cycling coach specializing in track cycling for professional Japanese and Korean keirin competitors, aspiring national representatives, keen amateurs and former military personnel riding for physical and mental wellbeing. He studied Performance Analysis at Australian Catholic University and is an honorary fellow at Queensland University of Technology's school of Exercise and Nutrition Sciences.

Rutul Joshi is an Associate Professor at the Faculty of Planning, CEPT University, Ahmedabad, India. His research interests and publications focus on sustainable mobility, transit-oriented development and reforming urban planning practices. Rutul also writes occasionally in the newspapers and popular media on urban mobility and civic issues.

Kat Jungnickel is a sociologist at Goldsmiths, University of London. Her research explores mobilities, gender, cultures of invention and inventive methods. She is Co-Director of the *Methods Lab* and PI on the ERC funded *Politics of Patents: Re-Imagining Citizenship via Clothing Inventions 1820–2020*. Recent publications include *Transmissions: Critical Tactics for Making and Communication Research* (2020) and *Bikes & Bloomers: Victorian Women & Their Extraordinary Cyclewear* (2018).

Angela van der Kloof is an international strategic advisor in cycling mobility, working at Mobycon. She advises clients in the fields of cycling, safe neighborhoods and school zones and traffic safety education. She has a focus on giving voice to groups in society that are traditionally forgotten in mobility planning. She is one of the founders of *Women In Cycling*, an initiative aiming to help women to get more visibility, impact and leading seats in the cycling sector.

Keizo Kobayashi studied technical history at Ecole Pratique des Hautes Etudes, Paris, and published two books: *Pour une bibliographie du cyclisme* (1984) and *Histoire du vélocipède de Drais à Michaux, 1817–1870, mythes et réalités* (1993).

Till Koglin is a Senior Lecturer in transport and mobility planning at Lund University, Sweden. His research mainly deals with mobility, transport and urban planning, with a strong focus on the marginalization of cycling in urban space. Within his research, he combines critical theory with transport and mobility in order to shed light on injustice issues within transport planning and transport systems.

Bartek Komorowski is an urban planner working in the field of active mobility. He is currently a team leader at the City of Montreal's Urban Planning and Mobility Department. Prior to joining the City, he was a Project Leader in the Research and Consulting department at Vélo Québec. He is a co-author of Vélo Québec's recently-released design manual, *Aménager pour les piétons et les cyclistes*. Bartek is a board member of the Winter Cycling Federation, which organizes the annual international Winter Cycling Congress.

Kevin J. Krizek is Professor of Environmental Design at the University of Colorado Boulder and senior advisor in Washington DC to the Under Secretary of the Environment, Energy and Equity. He has developed informed insights into solving one of the world's most pressing problems – how to reverse the automobile-focused nature of our environments. He has authored or edited several books: *Advanced Introduction to Urban Transport Planning*, *Metropolitan Transport and Land Use* and *End of Traffic and the Future of Access*.

Annika Kruse is a Post-Doctoral Researcher at the Institute of Human Movement Science, Sport and Health of the University of Graz, Austria. She has investigated the effects of training and treatments, especially on muscle-tendon properties and function as well as functional performance, in individuals with cerebral palsy.

Scotford Lawrence is a former racing cyclist, who rode as an *indépendent* in France in the early 1950s. He studied art history at Leeds University and is a member of the Barber Institute of Fine Arts, Birmingham, where he presented his master's thesis on *The Art of the Cycle*. He has researched, written and translated extensively on cycle art and history and is historical advisor to the National Cycle Museum. He lives in Herefordshire, England.

Hans-Erhard Lessing studied physics in Stuttgart, graduated in Berlin, researched at IBM Labs San Jose, was Professor at the University of Ulm and Curator at Technoseum Mannheim and ZKM Karlsruhe. He wrote biographies of Karl Drais and Robert Bosch, and co-authored, with Tony Hadland, the book *Bicycle Design – An Illustrated History*.

Yu-Chun Lin is Associate Professor at the Department of Social Development, National Ping Tung University, Taiwan. His research interests are in economic geography and industrial region research, especially working on the triple topics of industrial clusters, innovation networks and global production networks.

Kevin Manaugh is an Associate Professor jointly appointed in the Department of Geography and Bieler School of Environment at McGill University in Montreal. His research examines on the equity and justice dimensions of transportation infrastructure and policy with a focus on disparities in access to safe and comfortable places to walk and cycle.

Alejandro Manga is a dual degree PhD candidate in the programs of Communication, Culture, and Media at Drexel University in Philadelphia and Urban and Regional Planning at the University Gustave Eiffel in Paris, France. His research focuses on the policies and practices mobilized from the bikespace, how they shape the production of space, and the role of advocacy groups. He is also a board member of L'Heureux Cyclage, the network of French DIY bike co-ops.

Dimitri Marincek is a PhD student at the University of Lausanne, Switzerland, where he is also a member of the Observatory for cycling and active mobilities (OUVEMA). He is working on electrically-assisted bicycles (e-bikes) and their users. His research is inspired by biographical approaches to mobility and the concept of a lifelong cycling trajectory.

Karen McCormack is Dorothy Reed Williams Professor of Sociology and Associate Provost for Academic Administration and Faculty Affairs at Wheaton College, Massachusetts.

Her research on mountain biking has been published in the *Sociology of Sport Journal* and *Information, Communication, and Society*. Karen credits mountain biking with solidifying her love of the wilderness and forcing her to be brave.

Robert McCullough is Professor of Historic Preservation at the University of Vermont. He lives in Montpelier and writes about American landscapes and built environments. His most recent book is *Old Wheelways: Traces of Bicycle History on the Land* (2015), and he is currently exploring bicycle manufacturers' contributions to industrial architecture.

Mitchell McSweeney is a PhD candidate in the School of Kinesiology and Health Science at York University in Toronto. His research focuses on sport for development and peace, social entrepreneurship, livelihoods, postcolonialism, and institutional theory.

David Minter is a lifelong cyclist. David spent a decade racing solos and tandems on the velodrome and road, then racing MTBs in the late 1980s and high-wheelers in the 1990s. He rode recumbents and small-wheelers and began riding long brevets over a quarter-century ago. After stints as the chief mechanic at bike shops and for cycle racing teams, he became a structural engineer before veering into materials technology, mostly in civil infrastructure.

Sergio Montero Munoz is Associate Professor at the Universidad de Los Andes in Bogotá, Colombia. He is based at CIDER, an interdisciplinary research center of development studies. His interests are in urban planning and governance, comparative global urbanism, and local and regional economic development, with an emphasis in Latin American cities and regions. He holds a PhD in City and Regional Planning from UC Berkeley.

Hilary Norcliffe is an Adjunct Professor in the College of the Arts, California State University, Long Beach, USA. She is a mixed-media artist specializing in drawing and found object manipulations, and an author of children's books.

Glen Norcliffe is an Emeritus Professor of Geography at York University, Canada. He is a researcher into industry, development, work, technology and trade and author of *The Ride to Modernity* and *Critical Geographies of Cycling* as well as many articles related to cycling, past and present.

Edward Nye is Associate Professor of French at the University of Oxford, UK. His research interests lie in the eighteenth and nineteenth centuries, in theatre, mime, dance, and Deaf culture. He is the author of a scholarly anthology of literature on cycling, *A Bicyclette* (2000, 2013).

Nicholas Oddy is Head of Department of Design History & Theory at Glasgow School of Art, UK. After being awarded post-graduate diploma in Edinburgh, Nicholas jumped into a master's in design history at the Victoria & Albert Museum/Royal College of Art Design History. A regular presenter at the International Cycle History Conference, his focus is late 19th- and early 20th-century quantity produced items, particularly those relating to cycling and cycling culture generally.

Pekka Oja is Scientific Director (retired) of the UKK Institute for Health Promotion Research, Tampere, Finland. With increasing age, he engages in less cycling and more utilitarian physical activity.

Ben Osborn is a senior physics major at Wheaton College, Massachusetts (USA). He has been mountain biking since he was 7, began a middle school bike club to teach others how to ride, and spent his teenage summers counseling at a mountain bike camp in Vermont.

John Parkin is Professor of Transport Engineering at the University of the West of England, Bristol. He has experience in all stages of transport, from policy formulation, modelling and forecasting to operational analysis and economic appraisal, design and construction, and evaluation. He leads teams undertaking quantitative and qualitative empirical research and evaluations of sustainable transport.

Cosmin Popan is a sociologist and a Leverhulme Early Career Fellow based at Manchester Metropolitan University. His research investigates the gig economy, platform-based food deliveries and their reconfiguration of urban spaces. He is the author of the book monograph *Bicycle Utopias: Imagining Fast and Slow Cycling Futures* (2019).

Alon Raab is a native of Jerusalem and a retired Professor of Religious Studies (UC Davis) whose true religions are football and bicycles. He is the author of the entry on the Middle East for *The Oxford Handbook of World Sports* and co-hosts the KBOO Radio (in Portland, Oregon) Bike Show. A daily cyclist since age 4, he hopes to continue cycling in this world and all future ones.

John Radford is a Professor Emeritus at York University, Toronto, Canada. His research interests are in spatial justice and historical geographies and political economies of disability.

Léa Ravensbergen is a postdoctoral fellow at McGill University's School of Urban Planning. Her mixed-methods transport research is driven by an overarching concern for centering equity in the creation of healthy and sustainable cities. Engaging with feminist geographies and mobilities studies, her work lies at the intersection of sustainable mobilities (city cycling, public transport, rail), and identity (gender, aging, class).

Daam Van Reeth is professor at the KU Leuven (Belgium) at the Faculty of Economics and Business, campus Brussels and Kortrijk. He teaches courses in micro- and macro-economics. His research interests relate to the economics of sport, with a special focus on professional road cycling, media attention for sport and gender balance in sport. He is co-author and editor of *The Economics of Professional Road Cycling*.

Patrick Rérat is Professor of Geography of Mobilities of the University of Lausanne, Switzerland, where he is also the co-director of the Observatory for cycling and active mobilities (OUVEMA). His research focuses on the practices of utility cycling and the politics of velomobility. His latest book is entitled *Cycling to Work: An analysis of the Practice of Utility Cycling*.

Andrew Ritchie is an author, photojournalist and editor of *The Boneshaker* living in Cornwall, UK, and formerly in El Cerrito, CA. He is the author of *King of the Road* (1975), *Major Taylor* (1988, 2010) and *Quest for Speed* (2011), all social and technological histories of the development of the bicycle. [Andrew died in August 2021, shortly after writing his vignette.]

Claudio Sarmiento-Casas is an architect and a doctoral student at the Department of Geography and Planning at the University of Toronto. His research on *biciofícios* lies at the intersection between contested street politics and everyday informal cycling. He works as an urban consultant, researcher, and lecturer in Mexico.

Purwanto Setiadi is a freelance journalist and passionate bicycle user based in Depok, one of the satellite cities adjacent to Jakarta, Indonesia. Cycling has been his passion since late 2014, three years before he retired from his job as managing editor of *Tempo* magazine. He decided to involve himself in bicycle advocacy in his country because he believes bicycles are the key to better, liveable cities.

Susan Shaheen is a Professor in the Department of Civil Engineering and Co-Director of the Transportation Sustainability Research Center at the University of California, Berkeley. She was the founding chair of the subcommittee for Shared-Use Vehicle Public Transport Systems of TRB. She also is a member of the Mobile Source Technical Review Subcommittee to the US Environmental Protection Agency's Clean Air Act Advisory Committee and a former member of the ITS Program Advisory Committee of the US Department of Transportation.

Charlotte Smith is an Associate Professor in Management and Organisation at the University of Leicester, UK. Her research interests broadly concern performance enhancement and its regulation in sport and other forms of work. Particular interest in doping in cycling and potential newer forms of enhancement such as 'mechanical doping'. Charlotte is also a long-distance age-group triathlete for Great Britain, though cycling is her favourite!

Justin Spinney is Senior Lecturer in Human Geography at Cardiff University. His research and teaching focuses on the intersections of mobility, technology, embodiment and political economy. His current research interests are the political economy of sustainable mobilities; the role of virtual mobilities in reshaping the public sphere; mobility inequalities and social justice; and methodologies for enhancing citizen participation.

Jeanette Steinmann is a PhD student at the University of British Columbia. She explores topics related to cycling (in)equity and social justice. She is a member of the SSHRC-funded *Bicycles for Development* research team. She is also a bike mechanic.

Sylvia Titze is Professor of Physical Activity and Public Health, University of Graz, Austria. Her research is focused on the promotion of regular physical activity. Throughout the year she enjoys cycling to work.

Jay Townley is a Resident Futurist and Founding Partner, Human Powered Solutions, a unique consultancy providing advice and services based on over 250 years of hands-on experience in human mobility.

Cheng-Mei Tung is an Assistant Professor in the Department of International Business at Feng Chia University, Taiwan. Her research and teaching focuses on industry competition, innovation and entrepreneurship and marketing management.

Luis Vivanco is a Professor of Anthropology and Director of the Humanities Center at the University of Vermont, USA. Luis is a cultural anthropologist whose work focuses on the cultural politics of environmentalism, sustainability, and urban bicycle activism.

Brian Wilson is Professor in the School of Kinesiology at the University of British Columbia, and Director of UBC's Centre for Sport and Sustainability. He is the author of *Sport & Peace*, and the editor of *Sport and the Environment* and *Sport and Physical Culture in Canadian Society*. He is a member of the SSHRC-funded *Bicycles for Development* research team.

Jeremy Withers is an associate professor of English at Iowa State University. His publications include *The War of the Wheels: H.G. Wells and the Bicycle* (2017) and *Futuristic Cars and Space Bicycles: Contesting the Road in American Science Fiction* (2020).

Xu Tao is Associate Professor of the History Institute at the Shanghai Academy of Social Sciences, and Deputy Secretary-General of the Shanghai Association of History. He has researched the bicycle in China for nearly 20 years, and published a monograph entitled *A History of the Bicycle and Modern China*.

Tian Yun is a doctoral candidate at the Institute of Ethnology and Anthropology of the Chinese Academy of Social Sciences. Her research focuses on multi-ethnic communities, ethnic coexistence and multiculturalism.

AN INTRODUCTION TO THE COMPANION TO CYCLING

Glen Norcliffe

The academic sub-field of cycling studies has recently emerged from the shadows as researchers in fields as diverse as biochemistry and poetry have found common cause in these unpretentious machines and the aficionados who ride them. The publication of major research articles and books challenging the *status quo* in transportation, and demonstrating an earth- and people-friendly alternative, have gained even greater significance as the recent pandemic made cycling a safe form of commuting, shopping, visiting and exercising. The Paris Accords of 2015, and the annual United Nations Climate Change COP Conferences, have identified fairly ambitious long-term targets, while sidelining the immediate progress that could be achieved by creating physical and policy spaces for cycles and cyclists. This reappraisal of the current social, political and cultural significance of the bicycle has been accompanied by a reassessment of its significance in the past, in diverse places, from diverse disciplinary perspectives. The collection of essays presented in this book illustrate many of these exciting developments that collectively are nurturing this new sub-discipline.

There is no universal understanding of a bicycle. Its function is contingent upon the perspective adopted, and its position in space and time, culture and politics, so it may be understood in many ways. It has been described as having various attributes, some dualistic: it has been a tool for peaceful conviviality and an instrument of war, terrorism and torture; a workhorse and a source of relaxation and pleasure; a sophisticated technology and mobility simplified; the key to velomobility and a cause of traffic congestion; an agent for development and a platform for labor exploitation; a definer of both safe and unsafe places; and an instrument promoting racial, class and gender harmony, but sometimes confrontation. It has also served as an expression of culture and identity; a symbol with its own semiotics; an inspiration for creativity; an investment; democracy on wheels; a training engine; a piece of environmental detritus; a valued player in the Anthropocene; a garden ornament; and a quirk of history. The list could go on for pages because the bicycle is such a versatile artefact. Recognition of this versatility has led to cycling becoming an object of serious academic study in a number of disciplines during the past two decades. This collection of essays provides a broad understanding of the scope of this emerging field.

The transformative quality of the bicycle complements the recent growth of interest in interdisciplinary studies, helping this adaptable machine to become a focal point of academic interest. Several developments have influenced this evolution. First, quality of daily life (as opposed

DOI: 10.4324/9781003142041-1

to net income) has become important to people living in affluent consumer societies, with an emphasis on a productive engagement with nature, with the economy, with society, and with politics (Nussbaum and Sen 1993). Research shows that after family incomes reach a comfort level, further improvement in the quality of life results mainly from increasing appreciation of non-monetary considerations, including the dangers of sedentarism, the value of exercise and good health, social engagement in earth-friendly activities, inclusive democracy, comprehensive social support, and engagement with literature, music and the arts, all requiring an interdisciplinary perspective. The World Happiness Report for 2018 found that EU countries with the highest bicycle use are also among the happiest overall (https://thecityfix.com/blog/cyclists-mean-happier-city-yes-no-dario-hidalgo/).

The second development that has raised academic interest in the bicycle results from the priority given to slowing growth of, and reversing the human footprint on planet Earth, with the bicycle repeatedly mentioned as having instrumental powers to lighten human impacts. Human-powered vehicles (HPVs) are playing a significant role, resulting from the development of new cycle technologies, by improvements to cycling infrastructure, by advancing cycling justice, and by creating bikespaces offering an alternative to motorized transport for short trips (which most trips are). Information technology is giving the bicycle more agency in the networks combining humans with cycles, including sensors that automatically select the optimal gear or use continuous gearing, autonomous bicycles that can navigate city streets, and GPS sensors that locate the nearest dockless bicycle.

The third important development that has raised the academic profile of the bicycle results from increased scepticism of the neoliberal mantra stressing privatization, lower taxes, minimal social services, exploitative resource-based development and attacks on labor institutions. The results are not environmentally sustainable; they are socially destructive as wealth becomes highly concentrated even as wages stagnate, and they give energy to populist movements. The new economics stresses cooperative management of production, participatory local democracy, nature as a valued commodity, and people-friendly urban design in which pedestrians and cyclists flourish.

In order to capture the numerous interdisciplinary manifestations of this multipurpose artifact, this book presents 48 chapters and 13 vignettes written by leading researchers in the field of cycling. All of the authors are cyclists who, despite its technical simplicity, view the machine as one of the more useful inventions yet conceived. It is human-powered. A person with a few essential skills can maintain it. Either new or "used," it is affordable for most people on this planet. It enables low-cost mobility for millions of people, worldwide. And with care it will function for up to 100 years. The authors cover many aspects of cycling and the cycles themselves, indeed the diversity of the contributions has led us to organize the chapters into eight main parts that provide the book with a structure.

The **first part** inquires into the associative interactions between society and the bicycle, viewing the bicycle not just as an assemblage of metal, rubber and plastic, but also as the means to engage in cycling as a socially formed and forming activity. The innovative opening chapter on *theorizing cycling*, by the part editor, Peter Cox, provides the framework that the subsequent chapters elaborate. Use of this artefact depends upon the user, the groups with whom the user has social relations, and broader trends in society at large. These relations intersect in the following chapter examining gender (and patriarchy), which has permeated cycling since its origins in the nineteenth century. Also fraught are the precarious jobs of delivery workers employed by platform companies, which do not treat them as employees. Socialization emerges as a key theme in the next chapter, which stresses social movements as key to generating demand for a better bikespace and a culture of cycling. The vignette that follows describes a turning point for Black athletes, when Major Taylor successfully challenged the reigning world champion in Paris in 1901. Cycling justice is the theme of Chapter 5 proposing a cycling inclusion ladder as a means to making cycling a more inclusive activity. The impossibility of continued automobility and the necessity of sustainable alternatives, especially cycling for shorter trips, is stressed in the penultimate chapter, while the final chapter in this part calls for the incorporation of persons with disability into the cycling mainstream.

Part II, edited by long-time researcher into cycling technology, Tony Hadland, literally addresses the nuts and bolts of the bicycles themselves. Chapters emphasize that the bicycle is not only a social construct, but also a technical one that has evolved rapidly at several points in recent history. The part opens by considering the evolution of the configuration of the bicycle as a whole. This is followed by chapters that disassemble the machine by examining, in turn: the frames and the materials used to make them; the wheels that keep them rolling and the technologies that absorb some of the shocks that all cycles encounter; the transmission that gets a bicycle going, and the brakes that stop it; and the ability of bicycles to carry luggage and passengers. A vignette describes the cargo bicycles designed specifically to transport coffee in Rwanda. The final chapter in this part examines technologies designed to help disabled persons to cycle.

Part III addresses the bicycle trade. The bicycle is a successful innovation that has created a major world industry, the subject of this part's opening chapter authored by Boyang Gao and myself. Because makers and users often live on opposite sides of the world, supply chains have evolved to connect widely separated producers and consumers. The following chapter examines the site where makers and sellers commonly connect, namely the bicycle trade show. At the selling end, traditional forms of retailing have been squeezed out by competition from mass-sellers, specialized sport retail chains, online sales and the novelty of mobile curated collections. The making and selling of cycles are brought together in the following chapter on the world's largest bicycle maker, Giant Cycles of Taiwan, which now also has a global presence as a retailer. From the biggest, the final chapter in Part III turns to the smallest – the micro-economies that use bicycles as a platform for business, especially in the Global South, but also in the rest of the world, followed by an illustrative vignette on a mobile bike shop in Beijing.

Tackling urban cycling in **Part IV**, editor Sheila Hanlon has placed urban design and policy front and center by examining the tensions between cyclists and other users of city roads, and the planners and policy makers who try to resolve disputes between interest groups. Providing bicycle-friendly infrastructure and cycleways is clearly a pressing issue, illustrated by a vignette on infrastructure in Lund, Sweden. The following chapter, in contrast, cautions that current policy can be viewed as a political fix that compromises social justice with a neoliberal growth agenda, and pays shallow attention to the fundamental environmental issues facing the planet. And a key element is space, not just in the material sense, but also in the Lefebvrian sense with imagined spaces, policy spaces and activist spaces all contributing to the inclusiveness of cycling. The vignette that follows returns to the physical spaces required for the bike-to-work campaign in Jakarta. Chapters on bike-share programs and e-bikes capture major elements of the contemporary urban scene, while the closing chapter raises important questions of risk, safety and security for the city rider.

Sport, health and lifestyles are the connected themes developed in **Part V**, edited by Tim Jones. A paired contrast is found in the opening chapters on amateur and professional road cycling, the latter being organized to resemble an industry, the former attracting many enthusiastic amateurs dressed to look like professionals. A vignette on the experience of riding in the peloton in professional road racing links these two opening chapters. Cycling in two other settings follow: off-road cycling in various forms – mountain bikes, gravel riding, cyclocross and several more; and track racing (with a vignette on keirin races). Health, injuries and (one hopes) recoveries are considered next, these being prominent themes in a world where the lifespans of the most sedentary citizens are starting to decline. The closing chapter of Part V treats the sensitive issue of doping.

In **Part VI** Luis Vivanco takes the perspective of a cultural anthropologist *cum* geographer by focusing on significant *places* of cycling, some of them being iconic, and others mundane. To set the scene, a vignette takes the reader to the Bois de Boulogne in Paris in 1867, the place where the cycling age was effectively launched. The opening chapter examines the view from the handlebars, as seen by landseërs of the late nineteenth-century landscape, followed by a vignette advocating the bicycle as a place of peace. The following chapter examines tourists who, naturally, seek evocative places, often combining a cycle tour with adventure and the exploration of places unknown, followed by a vignette on the special case of winter riding. The next five chapters examine cycling on four different continents: West Africa, where Africanized bicycles occupy a liminal space between African and European culture; India, where they are imbued with colonial and postcolonial meanings; China, which under Mao was indubitably the Kingdom of bicycles; Denmark, which prides itself as one of the world's most cycle-friendly countries; a vignette on Beach Road in Melbourne, and a closing chapter on Bogota in Colombia, where *La Ciclovía* took shape.

In **Part VII**, the culture and art surrounding the bicycle is scrutinized. The editor, Nicholas Oddy, delves into the visual culture of the machine itself, suggesting three distinct machine aesthetics – the roadster, the lightweight and the fat tire. Parallel issues are taken up in the following chapter when applied to the cyclist herself or himself: how does one dress to ride, in fashion, in everyday outfits, or in clothing specifically designed for the ride? In the *belle époque* era, bicycle promoters engaged a number of France's most talented artists to publicise cycles via posters, leaving behind some of the most creative commercial art ever seen. In contrast, as shown in the following chapter, in the salons of France's *beaux arts* the bicycle was not a popular subject with only a few interesting images handed down for posterity. But the vignette that follows shows that bicycles have caught the attention of the noted installation artist, Ai Weiwei. In the cinema, the bicycle has caught on as a recurring theme not merely as part of the setting, but as central to the storyline.

Part VIII, edited by Una Brogan, turns to the bicycle in literature. The opening chapter on riding and writing presents a synopsis of selected authors who have introduced cycling in their works, with the following chapter taking up one aspect, namely stories about the way cycles liberate their riders and grant a particular type of mobility. The third chapter takes the reader into the realm of humor, which, in some instances, verges on the absurd, with the bicycle offering rich opportunities to amuse. A number of poets have addressed cycling in their musings, often in fanciful flights of language to which the cycle lends itself. Finally, literary tourism has long been a fertile field occupied by pilgrims, ramblers and adventurers.

Cycling and its discontents

Bicycle advocates tend to accept the virtues of cycling uncritically, and then re-enforce their values by socializing with friends who share similar values. Negative opinions about cycling are usually dismissed as uninformed. Yet the reality is that, even in this era of rising environmental awareness, the bicycle is traduced by many citizens who view it either as unimportant, as a nuisance that should be relegated to the status of children's toy, or even as an instrument for capitalist exploitation. In many jurisdictions, politicians and policy makers willingly dismiss arguments that cycles are beneficial, and continue to marginalize this form of transportation in favor of motorized vehicles. For these reasons, I will summarize some major criticisms levelled at cycling, and respond to the critics. They deserve our attention, and an informed response. I have chosen the critique cycling has received from one of its more strident critics, Lawrence Solomon. Solomon's article appeared in Toronto's *Financial Post* on December 1, 2017 (see https://financialpost.com/opinion/lawrence-solomon-ban-the-bike-how-cities-made-a-huge-mistake-in-promoting-cycling). Solomon has seven main criticisms, to which a few other common censorious remarks are added.

1 *Bike lanes represent an inefficient use of road space, lowering total utility.*
By definition, all users consume road space, be they pedestrians on sidepaths, motorists in truck lanes, or geese being driven on medieval droving roads to market in the nearest town. For millennia, roads have been a shared space, with the motorist a newcomer given access early in the twentieth century to the Good Roads already created due to pressure in the late nineteenth century from farmers and cyclists. Approximately eight bicycles fit in the space occupied by one motor vehicle when it is in motion. Hence a motor vehicle on an urban street is more efficient if it moves more than eight times quicker than a bicycle, or transports more than eight passengers. But data presented by Trigg (2015) indicate that on arterial roads in the city, cars average at best 18–19 mph (~30 km/h), in New York it is closer to 15 mph (25 km/h) and in Beijing 7.5 mph (12.5 km/h). In cities such as Bangkok and New Delhi it is often faster to walk than to drive! Bicycles average 9–10 mph (15 km/h) or close to the speed of cars that are not on urban expressways. Since surveys show the average number of passengers per car is 1.59, in crude aggregate numbers, a unit of road space with

1 car @ 25 km/h × 1.59 persons = delivers 39.75 persons
8 bicycles @ 15 km/h × 1 person = deliver 120 persons.

So bicycles use road space about three times as efficiently as motor vehicles. However, Will et al. (2020) find that in some cases the figure for bicycles can rise as high as 20 times more space-efficient.

2 *Bikes have higher negative externalities than motorized road users.*
The negative externalities produced by motor vehicles include nitrogen oxide, carbon monoxide, sulfur oxides, volatile organic compounds and particulates. For bicycles, the externalities are minimal (zero tailpipe emissions), although e-bikes do have externalities associated with generating electricity to re-charge batteries. In addition, the "pass-by" noise generated by cars, motorbikes and trucks is normally in the 67 to 75 decibel range (Schreurs et al. 2011), whereas bicycles register below 40 dB, which is the level of background noise in a residential area.

3 *Bike lanes are a drain on the public purse.*
All roadways, including sidewalks, are a drain on the public purse in that they have to be built and maintained, but the cost of constructing bike trails is approximately about one-ninth per kilometer that of urban roads, and annual maintenance costs are substantially less (Collier 2011). Moreover, when London UK introduced a congestion charge in 2003 of $15.40 on each weekday, transit use rose 60%, and bicycle use 90%, while vehicle traffic and pollution counts dropped 25%. So the reverse of what Solomon asserts applies in practice – bike lanes are much less of a drain on public accounts and more cost-efficient than roads built for vehicles.

4 *Bike lanes make cars idle and increase pollution.*
The main cause of cars idling and creating urban traffic congestion is other cars. Cities with the most cars, such as Mumbai, Shanghai, Jakarta and Sao Paolo, also have the worst traffic jams and the slowest average speeds for cars. Adding bike lanes increases traffic flow because they are separated from cars and keep moving in heavy traffic.

5 *Bicycles cause accidents and kill pedestrians.*
All road users, even pedestrians, cause accidents when they are careless or behave aggressively. It is rarely mechanical failure by the vehicle itself, but the driver/rider/pedestrian that is responsible for an accident. So yes, scofflaw cyclists are a menace when they ignore stop signs and red lights, dart between vehicles, and ride fast along sidewalks. But so, too, are scofflaw motorists. In the UK about 400 pedestrians are killed annually, over 99% by motorists and less than 1% by cyclists. The differences relate to weight and speed. "A 1,000 kg car moving at 22 mph will have 50 kJ of energy; a 15 kg bike with a 70 kg rider at the same speed has less than one-tenth of that" (Laker 2018). Also, commuter cyclists tend to ride at about 15 km/h, while motorists travel at 25 or more km/h if traffic conditions permit it, so motorists generally travel faster which increases the force of impact. Motorists kill nearly 100-fold the number of pedestrians.

6 *Motorists user-pay, whereas cyclists don't pay.*
Motor vehicles require licences, so they generate revenues used to offset some of the costs of road construction and maintenance. In most jurisdictions, bicycles do not require licences, and contribute no licence fees. Note, however, that road licences only cover a fraction of the costs of construction and maintenance, and most roads in urban areas are toll-free, so motorists don't pay a fee to use them. But a very different picture emerges when environmental costs and carbon pricing is factored in. Even with a low carbon tax of $100 per metric ton of carbon dioxide, motorists cover only a small fraction of the environmental costs of running their internal combustion vehicles so they are not covering their full user-costs (the transport sector accounts for 28% of US greenhouse gas emissions). Bicycles have negligible greenhouse gas emission and incur minimal environmental costs.

7 *Cyclists further the gentrification of inner cities.*
This may be true, although separating cause and effect is not easy – it is equally possible that gentrification promotes cycling in the city. Either way, the changing location of work within the city inverted the geography of the early-to-mid-twentieth century. Office work, financial services, multi-media, the creative arts, tourism and recreation now cluster in the central city, while manufacturing, warehousing and distribution have long deserted the areas surrounding the central business district for the outer suburbs. When home and work are located in close proximity, cycling becomes a viable means of commuting. From a transport perspective, increased velomobility due to gentrification is a plus.

8 *Distracted and scofflaw cycling causes accidents.*
Reckless motorists, scofflaw cyclists and jaywalkers are all equally guilty of negligent and distracted behavior. All such lawless acts are wrong. It is worth noting – not as an excuse but as a contributing factor – that motorists and pedestrians have their own road space, whereas cyclists often do not.

9 *Bicycles have become a disposable consumer good creating ugly piles of rusting metal.*
Because bicycles in high-income countries are relatively inexpensive to buy, little care is taken of them. Landfills have piles of discarded bicycles that, with a little care, can be put back in working order. Fortunately, in some places this is happening, with bicycles repaired by volunteers and then re-cycled to those with limited disposable income, or passed on by charities to low-income countries to improve the mobility of those who

cannot afford public transportation or a bicycle. In China, station-less bikesharing schemes based on smartphones, such as Ofo, Mobike and Bluegogo, have led to an increase in urban cycling (Wang, Huang and Dunford 2019). But there has been a cost: over-supply, wilful damage to bicycles and hyper-competition has resulted in massive piles of abandoned and broken dockless bikes, as well as smaller piles on street corners in many Western cities where they have also been introduced and trashed. A return to docking bicycles may be the solution.

10 *New bicycle paths cut through green spaces and damage nature.*
The bicycle is not always a green paragon. Cases are reported of bikeways being cut through bush and meadows, destroying ground-nesting birds' habitats and felling mature trees. This is regrettable, and such impacts should obviously be minimized. But new road-ways and railways typically have far greater impact on the environment not only because they are wider, but also because their base is dug much deeper.

The conclusion to be drawn is that, despite some locally negative outcomes, bicycles can be compared favorably with motor vehicles in term of space consumption, cost for the public purse, and a range of environmental and urban impacts.

Non-conformist uses of the bicycle

Everyone has seen cycles used for commuting, for shopping, for exercise and for other forms of mobility. These uses are examined in depth in the chapters that follow. Not discussed are obscure uses, some alarmingly destructive, so as to avoid giving the reader the impression that they are simply harmless riding machines, a few of the alternate uses to which the bicycle has been put to will be noted.

For over a century bicycles have been used in wars, and, more recently, for acts of terror-ism. In two world wars, bicycles carried messages, moved troops and ammunition with stealth, served as ambulances to carry the wounded and transported infantry (Fitzpatrick 2011). In World War II, the Japanese army successfully and swiftly invaded former European colonies in South-East Asia with their highly mobile bicycle corps, a tactic subsequently copied by Viet Nam in its anti-colonial war with France, and the Viet Cong, when their 64,000 load-bearing pack-bikes moved largely undetected along jungle paths to defeat American forces (Cheney 2017). On March 6, 2021 *Agence France-Presse* reported that a rickshaw loaded with explosives was detonated by al-Shabaab Islamists outside a popular restaurant in Mogadishu, with 20 people killed and 30 others wounded. This was a case where the "below the radar" character of the weaponized bicycle led to it being infiltrated unnoticed into the city center. These acts of war and terrorism result from the lethal potentials of the bicycle being under-rated by higher authorities.

More commonly, the bicycle is an instrument of peace. Inner peace can be found on a leisurely solo ride, and groups find cycling a convivial activity promoting social cohesion (Illich 1973). On one recent occasion, it has physically promoted peace: after motorcycles were banned in Maiduguri, north-east Nigeria (the birthplace of Boka Haram) in 2011 due to their violent attacks, bicycles became popular and peace returned to the city, triggering an economic boom (https://www.bbc.com/news/av/world-africa-37045314 accessed July 15, 2021). Vignette I explains how the bicycle's developmental role can promote peace.

Bicycles have acquired symbolic meanings not directly related to riding them. Old bicycles have become kitsch garden ornaments, a posy of flowers in their baskets. The semiotics of the bicycle often interpret it as a symbol of a past age, redolent of a person's youth, with (*pace*

Walter Benjamin) a lack of critical distance between the observer and the observed. This use of the bicycle carries over to sentimental greeting cards, pot-boiler book covers, comic books and street scenes in film and television. In Britain, the ordinary (penny farthing) bicycle is widely used to signify an antique shop, such is its symbolic power. In up-market clothing stores, bicycles are used in window-dressing to connote fashionable outdoor styles and a sense of freedom. Some view bicycles as sculptures in their own right or as components of a sculpture, the most widely known being Pablo Picasso's bull's head, made of a bicycle saddle and handlebars, and Ai Weiwei's bicycle sculptures discussed in Vignette M. Some of his bicycle installations are large enough to decorate entire urban neighborhoods.

One of the less obvious uses of the bicycle is as a therapeutic treatment for anxiety and depression, attributable to the beneficial sensory experiences of cycling, and the sense of freedom it produces, particularly, as Ross (2021) notes, for women. It activates the body's proprioceptive system as we balance, pedal and steer, there is continuous visual inspiration, while stimulation of the tactile senses – sun on the skin, wind in your face – all engender a sense of escape from the cares of the world. A bicycle's benefits extend to group therapy, as evinced by meetings of Indonesia's KOSTI (komunitas sepeda Indonesia) which holds annual gatherings with several thousand participants riding colonial-era bicycles wearing remarkable costumes in an uplifting display of cycling designed to alleviate a sense of anxiety and promote a sense of community among participants, very few of whom are prosperous.

There are major contradictions in the way bicycles are used, often by *othering* the non-user of the bicycle. Both in fiction and in real life, they help to build social networks yet have also served in acts of brutality. There is an incongruity between designing bicycles and tricycles to increase the mobility of persons with disability, and *No Fear* mountain bikers riding down precipitous mountainsides at significant risk to life and limb. Sturdy cargo bikes carry children to school in the Global North and farm produce to market in the Global South, whereas minimalists strip their fixie bicycles down to bare bones unable to carry anything but their rider. And while radical environmentalists view motorized transport as a disaster, many leisure cyclists happily drive their SUV with trail bikes hanging from a rack to a nearby bicycle trail.

In recent years, the bicycle has frequently been proclaimed not only as a companionable machine but also as an Earth-friendly device offering unknown possibilities around the corner. Although there are critics who traduce the bicycle and speaking disparagingly both of the machine itself and of its use by riders, the following chapters will demonstrate that the bicycle is one of the most useful artefacts ever invented. It is compellingly egalitarian, overwhelmingly a source of pleasure or convenience, a stimulus to travel, art and literature, rarely boring, and mostly an earth-friendly means of functional mobility. The interdisciplinary nature of cycle studies captures this versatility. In the words of the late Archbishop Desmond Tutu:

> Give a man a **fish** and feed him for a day. Teach a man to **fish** and feed him for a lifetime. Teach a man to cycle and he will realize **fishing is stupid and boring**.

References

Cheney, W. (2017) "How the bicycle won the Vietnam war." *Campfire Cycling*, January 30. https://www.campfirecycling.com/blog/2017/01/30/how-the-bicycle-won-the-vietnam-war

Collier, M. (2011) http://www2.hamilton.ca/NR/rdonlyres/A65D0CD6-42F2-47BF-AE4D-86CDE79C1583/0/RoadPricinginNorthAmericaandOtherJurisdictions.pdf

Fitzpatrick, J. (2011) *The Bicycle in Wartime: An Illustrated History* (Kilcoy, Australia: Star Hill Studio).

Helliwell, J., Layard, R., & Sachs, J. (2016). *World Happiness Report 2016, Update* (Vol. I) (New York: Sustainable Development Solutions Network).

Illich, I. (1973) *Tools for Conviviality* (New York: Harper and Row).
Laker, L. (2018) "Killer cyclists? Let's not forget the real threat on our roads." *The Guardian*, March 8. https://www.theguardian.com/commentisfree/2018/mar/08/killer-cyclists-roads-bikes-pedestrian-collision-deaths-britain
Nussbaum, M. & Sen, A. ed. (1993) *The Quality of Life* (Oxford: Clarendon Press).
Ross, H. (2021) *Revolutions: How Women Changed the World on Two Wheels* (London: Weidenfeld and Nicolson).
Schreurs, E., Brown, L., & Tomerini, D. (2011) "Maximum pass-by noise levels from vehicles in real road traffic streams: comparison to modeled levels and measurement protocol issues." *Inter Noise 2011, Osaka Japan*. https://core.ac.uk/download/pdf/143851566.pdf
Trigg, T. (2015) "Cities where it's faster to walk than drive." *Scientific American*, May 16. https://www.researchgate.net/publication/238594974_Daily_Variability_of_Motor_Vehicle_Emissions_Derived_from_Traffic_Counter_Data/figures?lo=1 http://www.diva-portal.org/smash/get/diva2:11495/FULLTEXT01.pdf
Wang, J., Huang, J., & Dunford, M. (2019) "Rethinking the utility of public bicycles: the development and challenges of station-less bike sharing in China." *Sustainability*, 11, 1539. doi:10.3390/su11061539.
Will, M.-E., Munshi, T., & Cornet, Y. (2020) "Measuring road space consumption by transport modes: toward a standard spatial efficiency assessment method and an application to the development scenarios of Rajkot City, India." *The Journal of Transport and Land-Use*, 13(1) 651–669. doi:10.5198/jtlu.2020.1526.

PART I

Cycling and society

An introduction

Peter Cox

Thinking about an introduction to this part on cycling and society, I am taken back to the genesis of the book of the same name (Horton, Rosen and Cox 2007). Some speculative correspondence between a group of people who had previously known each other through cycle campaign networks and were now working in different fields of social science wondered whether there might be any mileage in getting together to make a conscious and concerted connection between cycling and society. A symposium, a mailing list and, eventually, a book followed, deliberately trying to sketch connections between what we collectively knew of existing academic and enthusiast work on cycling and ideas and problems familiar elsewhere in our studies.

Interdisciplinary in intention, the organization of a network for social scientific cycling studies, the Cycling and Society Research Group (CSRG), took advantage of several developments in social science research that provided opportunity, legitimation and frameworks for study. First, sociological recognition of the importance and relevance studies on everyday and mundane activity. An explicitly social-science-based cycling research network opened the possibility of studies going beyond either the existing remits of historical analysis or transport studies. Pioneering work was already in progress in the fields of the study of technology (Bijker 1995). Another important prompt was the work of Albert de la Bruhèze and Veraart (1999) who, by producing a comparative study of the fortunes of cycling in a number of European cities in the twentieth century, gave shape to a whole line of enquiry. Differential fortunes of cycle technologies for transport were not explicable by reference to technology alone. Where instead might explanation lie?

A second important development was scholarship focused on mobilities research, formalized in 2004 with the Centre for Mobilities Studies (CeMoRe) at Lancaster University (UK). Dissatisfied with conventional models based on the idea of societies as fundamentally stable objects that occasionally move (or are moved) from one state to another, mobilities research started from the idea of a world in constant flux. Mobility – the constant shift and transformation of the world – is put at the centre of study, instead of the lingering nineteenth-century focus on social statics. This paradigm shift reframes the study of physical mobilities. Studies of transport, previously neglected in sociology, became important. Processes of movement, how movement creates mobile subjects, how some are rendered less- or immobile and the effects of power relations between those moved and not moved, all become areas for research

DOI: 10.4324/9781003142041-2

and study. Cycling became an interesting case in mobilities study, not merely for its contribution to debates on sustainable transport or liveable cities (although these remain important) but as a gateway to other discussions such as kinaesthetics or the politics of space and urban design. How power is manifest in public space and the consequences of social justice, inclusion and exclusion have all become major foci of cycling research and allied mobilities research, recognized in ongoing work on mobility justice (Hoffman 2016; Golub et al. 2016; Sheller 2018; Cook and Butz 2019). These topics provide a clear basis for the subsequent chapters in this part in Cycling and Society.

Engaging with mobilities research, academic studies on cycling were released from transport studies concern with the modal split and the management of traffic as an adjunct of policy and opened to a wider social science analysis that did not necessarily have to serve a pre-existent agenda. Early studies such as Smith (1972) had explored the social history of cycling, but the explanations proffered largely explained cycling as a function of other social factors. That is, cycling was either a secondary product, to be explained either as an outcome of other social and technological processes or, conversely, a factor *sui generis*. Cycling is contradictorily depicted as having its own social power, for example, as a force of liberation for women or for working classes but in all other aspects its fate was a reflection of other social changes. These analyses are not invalid but lack the capacity to present a more nuanced picture that might, for example, understand the complexities of gender equalities being entangled with other areas of social difference and inequalities in class and race as explored in the chapters that follow. Similarly, deterministic portrayal of cycling activities lacks any space to explore the interplay of forces that could interact with its place in emancipatory politics. Overall, both approaches are problematic inasmuch as they homogenize cycling to be a 'thing': recognizable and measurable. Instead, applying the approaches of critical social theory to the subject of cycling opens up the complex and contradictory identities and processes bound up. The introduction to *Cycling and Society* (Horton et al. 2007) started out by raising a series of divergent images, all of which could be subsumed under the single title 'cycling', but which demand multiple dimensions of analysis using different disciplines and research methods as found in this volume.

Predictably, researching cycling attracted and continues to attract those who were already interested in cycling as a phenomenon. While this may seem remarkably self-evident, in practice connected those with interests in cycling that extended beyond academia, most notably those connected through networks of advocacy and bike activism. After all, academics pursue topics that interest them within disciplines in which they are invested. Prior to the widespread availability of research funding, enthusiasm could be the only incentive for social science research into cycling. Not unexpectedly, this tied social science research in cycling to the broader world of campaigning networks, already engaged in international connections.

Studies addressing cycling as a social phenomenon naturally and obviously draw from sociological framings of social relationships and social problems. However, rather than divide the chapters in this part up into the obvious thematic headings as if presenting an undergraduate introductory volume, the topic headings and approaches here highlight instead the breadth of styles and contribution in current work on cycling and society. Opening with a focus on theory, Cox considers the purpose of thinking abstractly and the legacy of different approaches that are visible in social studies across a number of disciplines. The subsequent chapters range widely in topic and approach. Bonham and Jungnickel address issues of gender and cycling but move us away from simplistic accounts of women's cycling to explore how it is that gendered identities are produced and how this production is entwined with cycling, how cycles and cycling practices become part of the (re)production of gender. In highlighting selected aspects of the social dimensions of cycling, Batterbury and Manga use the term "bikespace" to describe the way that cycling creates new forms of interaction, often intensified. These need not rely on actual movement on the bike but are also visible in the networks created around repair and replenishment of both cycles and riders. Contrasting issues of injustice and inclusion take up the next chapters. Krizek highlights how cycling is not immune from the structural inequalities of wider society. Conversely, van der Kloof daws our attention towards the huge diversity of programmes and activities using cycling as a means to overcome exclusions, or that serve to make cycling provision itself more inclusive. Importantly, this chapter also highlights the location of much knowledge of cycling as existing beyond academia. This theme is also taken up in Inckle's chapter, in which practical everyday experiences are brought to the fore. She ends by reiterating the potential that cycling has to be beneficial, but also the distance that current provisions still have to go in most locations.

References

Albert de la Bruhèze, A.A., and Veraart, F.C.A. 1999. *Fietsverkeer in praktijk en beleid in de twintigste eeuw*. Den Haag: Ministerie vanVerkeer en Waterstaat/Stichting Historie der Techniek.

Bijker, W. 1995. *Of Bicycles, Bakelite and Bulbs: Toward a Theory of Socio-Technical Change*. Cambridge, MA: MIT Press.

Cook, N., and Butz, D. 2019. *Mobilities, Mobility Justice and Social Justice*. Abingdon: Routledge.

Golub, A., Hoffman, M.L., Lugo, A., and Sandoval, G.F. (eds.) 2016. *Bicycle Justice and Urban Transformation. Biking for all?* London: Routledge.

Hoffman, M.L. 2016. *Bike Lanes Are White Lanes: Bicycle Advocacy and Urban Planning*. Lincoln: University of Nebraska Press.

Horton, D., Rosen, P., and Cox, P. (eds.) 2007. *Cycling and Society*. Abingdon: Routledge.

Sheller, M. 2018. *Mobility Justice*. London: Verso.

Smith, R.A. 1972. *A Social History of the Bicycle: Its Early Life and Times in America*. New York: American Heritage Press.

1
THEORIZING CYCLING

Peter Cox

To talk about theory in relation to an activity that appears to be simple, straightforward and self-evident might, at first, seem an odd choice. Surely, cycling simply requires a cycle, someone to ride it, and a space in which the riding can take place? However, when this simple equation is unpacked, it rapidly becomes obvious that this single description covers a multitude of activities. Cycling has a global history. It involves, as this volume demonstrates, a huge variety of stories, perspectives, participants and meanings. A lifetime could be spent in recording the multiple stories and events involved. The emergence of cycling studies as a distinct area of academic and popular research has shown that it also involves different ways of looking at the phenomenon and some radically contradictory ways of thinking about its current significance, its future, and even its past. Acts of interpretation inevitably and ineluctably engage us in the process of theorizing, of creating ways of knowing the events and processes under scrutiny.

Much of the rise in cycling studies in the last two decades has been coupled with a simultaneous interest in theory in order to understand better why cycling phenomenon occur. Theories of change, for example, cast light on why certain societies' attitudes toward cycling have developed differently. Explanatory accounts differ, drawing on a range of disciplines and analytical frameworks. Apparently contradictory at a surface level, multiple explanations develop multilayered explanations to overcome simple linear narratives struggling with complex realities. Further, they allow us to compare the relative importance of given factors in particular locations. Transport studies and cycle historians have naturally paid attention to these questions: the focus here is on the emergence of studies on, and theorization of, cycling in the social sciences.

Why theorize?

The point of social theory is to go beyond immediate observation and to make sense of the observable. Social theories provide means of interpretation to answer the "so what" dimension that arises whenever we are confronted by information. They go beyond data collection and ask, "why is this so?,", "what is significant about it being so?," allowing social scientists not only to explore the presence of different variables, but also to assess their relative significance in relation to other processes. Theoretical models reveal underlying processes. They allow us to move beyond the particulars of the immediate to show how a given example might be generalized.

DOI: 10.4324/9781003142041-3

Beyond this function, at a more abstract level, social theory can have a more philosophical role to play. It can be used in speculative mode to suggest (or posit) a novel way of looking at phenomena. This may include underlying processes not obvious from data alone; for example, contrasting monetarist and Marxist interpretations of bikesharing. Theory allows the building of hypotheses that can be tested to determine whether models proposed have validity, and, if so, what the reach of that validity is, even though the subject matter might be entirely abstract. Speculative thought provides further insights to show other ways of seeing events; explanations that can cast light on other allied, but not similar phenomena. Social scientists of all persuasions can bring pre-existing ways of theorizing sociality, social relations and social phenomena to bear upon cycling. The relation between cycling and social theory has been a fruitful one.

This chapter explores some of the ways in which theorizing cycling has developed and examines some of the key ideas. It also shows some of the functions that they have had in cycling studies, and how abstract conceptualizations may serve practical purposes. Links between theory, research, and policy are not direct. Testing theory and exploring ideas requires research and cycling studies have been notable in utilizing novel methods, including, for example, digital methods to explore experiences and perceptions (see, for example Brown and Spinney 2010, Spinney 2011).

Studies of cycling and society reflect the divergent trends in social science research. On the one hand, studies record (apparently) straightforward empirical measurement, telling us who does (or did) what, and where. Measurements of the identities of cyclists (according to factors such as gender, age, ethnicity or other social variables), of ride or trip lengths and purpose, of frequency, or of modal shift, all serve to enable basic understanding of what is going on. The "why?" needed to explain distributions requires different forms of analysis. Data collection and collation, recording, chronicling, and statistical analysis preserve information on what happens, and what relation it bears to other data. In transport studies, we might, for example, explore user information to determine modal shift or rises and declines in journey frequency or length, or even journey purposes. We can measure who rides, activity distribution and how it relates to social inclusion or exclusion, to the maintenance and reinforcement of social privilege, or to its challenge in pursuit of more just and sustainable societies. Studying cycle sport we might explore participation rates, distributions by gender and the interaction between amateur sports organizations, participation and physical and mental health. As we see in other chapters in the volume, these correlations are vital dimensions of the bigger pictures of cycling studies. However, to go beyond the measurement inevitably engages us with social theory.

Data collection and empirical work is most obviously valued in the application of research to public policy and has led to a rapidly growing body of literature (Pucher and Buehler 2017) as this volume shows. The form, mode and techniques of analysis to interpret data, depend on underpinning theories of how the world is and how it works. Even before the analysis, the very things researched and the questions that seek to be answered depend on underlying presuppositions not only about how things are, but also often on desires and expectations of how things could be, or how things ought to be. These normative questions are, as Oosterhuis (2014, 2016) has observed, very visible in cycling studies as they have emerged over the past twenty years. Persons electing to study cycling through an academic lens frequently reflect prior interests in cycling promotion outside academia.

Whether related to sport, transport, leisure or health, data on cycling and its comparison with other data sets allow deeper investigation into the nature of these relationships. Statistical analysis allows us insight into the relative significance of data correlations and to propose which relations might be causal, and what further research agendas are required to allow us to demonstrate the direction of effect between two (or more) variables. These ways of thinking about researching cycling are the keystones of research design and shape cycling studies.

There is, however, a second dimension of social science research that on immediate inspection might appear a little more esoteric. It might even be assumed to be less valuable in terms of its impact on policy and practice. As we shall see, this assessment is not entirely accurate. It addresses the logical consequence of data analysis by focusing on the 'so what?' factor. It asks why something matters (or not), why should it be of concern or worth further reflection? What might information be useful for?

Knowledge and theory

While investigations of cycling may appear to be self-evident explorations of fact, meanings and interpretations are subject to analysis and construction. As shown by recent work on knowledge production in other areas of social theory, knowledge is never neutral. "Who produces knowledge, what is produced and what is 'left out' are central questions of enquiry within the politics of knowledge" (Jansen 2019: 2). Consequently, cycling research, generating knowledge about cycling has its own politics, especially in relation to what is, and what is not, researched. Different theoretical perspectives shape the form and direction of research. Distinct positions require and pose different sorts of research questions. In cycling studies, as Johnson and Bonham (2012) succinctly point out, a primary divide has emerged between realist and non-realist perspectives, reflecting different basic starting points in general philosophy. Non-realist analyses can be further divided between constructivist and social constructionist approaches. This is necessarily a simplification of a complex set of arguments and the categories are far from impermeable, but as a way of understanding the different directions and research questions taken in cycling studies, it is a useful guide.

Realist approaches, more properly, realist ontologies, view reality as existing independent of the individual. Objective knowledge of that reality is possible. This positive knowledge can then be used to inform decision-making, either by individuals or in policy. These positions provide the foundation for the majority of cycling studies, especially as related to health and to most policy analysis. Focusing on measurable problems and factors, a broad range of tools for research, data gathering, and analysis can be brought to bear from the conventions of existing and familiar social science practice. Critical realist approaches (typified by Melia 2016, 2020) present a valuable departure from straightforward realism, rejecting the positivism sometimes associated with realist ontologies. Applied as a way of understanding policy, particularly cycling polices, they show how deep and often-hidden structures affect decision-making in ways not easily accounted for in straightforward measurement.

Constructivist positions, by contrast, suggest that people are "born into an already interpreted world, they and their interpretations of the world are necessarily shaped by socially available understandings" (Bonham and Johnson 2012: 2). Such positions push toward explorations of how people make sense of the world, how meaning is constructed and how events are interpreted. For cycling research, constructivist approaches allow us to interrogate how certain images and ideas around cycling emerge (Aldred 2010). Exploring worlds of mutable meanings does not preclude policy engagement. Instead, it can open up new ways of thinking about cycling policy, not as an arena of pre-determined optimal solutions but of conflict. Rather than searching for best-practice solutions, questions posed from this perspective recognise and consider how different solutions might address different ways of understanding what cycling is for and for whom (Hoffman and Lugo 2014; Pugh 2019). Again, constructivist analysis has a broad base and legitimacy in current social science, although its precepts are perhaps less self-evident than realist ontologies.

A second line of non-realist approach is in social constructionism, which foregrounds concern with cyclists and cycling as products of different sets of relations. In other words, a cyclist may be literally described as a person on a bicycle (physical reality is not being denied), but what that label 'cyclist' means for the person cycling and for any number of different onlookers, may be very different and mutable things, often contradictory, changing over time, and contested. From a constructionist perspective, it is also legitimate to ask 'when is a cycle a cycle?'; in other words, to deconstruct the very category (Cox and Van de Walle 2007). How many wheels? How is it propelled? Is an e-bike still a bike? At what level of power augmentation can it still be considered as such (conceptually rather than legally)? Constructionist approaches require unpacking the categories of objects and activities under examination, not assuming that there are pre-existent, uncontested shared understandings of cycling practices and persons (Cox 2019).

A notable example of this approach in practice is Horton's (2007) paper on the social construction of fear. Without denying people's very real fears about cycling, he demonstrated how safety campaigns and regulations, ostensibly intended to allay fears, serve instead to create a field of images, narratives and conversations (that is, a discourse), in which cycling is depicted as an intrinsically unsafe activity. Through this discourse, the language of unsafety becomes embedded (inherent) in the public image of cycling. Thus, individual apprehension is disconnected from risk and fed by the discourse. Rational analysis of fear as a response to risk becomes impossible in such circumstances.

Constructionist and constructivist approaches to thinking about cycling have led to greater dialogue with other areas of contemporary social theory. There are geographical variations in the prominence with which these positions appear in cycling studies, reflecting localized traditions in wider social science research, and marking different academic disciplinary sympathies or hostilities. While the debates over fundamental positions and theories of knowledge may seem obscure and arcane, even irrelevant, by entering these debates in wider social science research and thought, new directions in cycling studies have emerged.

Varieties of thinking and theorizing

Directions in which cycling can be theorized fall into a number of interrelated but distinct streams. Different disciplines, as this volume illustrates, bring different concerns and methodological norms with implicit theoretical presumptions. That is, different ways of looking are rooted in different explanatory models. Historical accounts might seek explanation for the changing fortunes of cycling in business organization and manufacturing histories (Epperson 2000), in the legal regulation of cycling (Longhurst 2015), or the agency of cyclists themselves (Reid 2017). Each is enriched by theoretical explanations grounded in disciplinary sensitivities. For example, examining the destructive impact of colonialism, Boal (2001) also noted the relationship between crop failures caused by volcanic eruption and Karl von Drais' construction of his laufmaschine. While controversial, his analysis provides a different way of thinking through events. Similarly, Kat Jungnickel's (2018) use of feminist theory to re-examine the use of patent innovations among Victorian cyclists allows us not just to record and reflect on their actions, but also to rethink the agency of early women cyclists and what this has to contribute to wider narratives of gender and social roles (see Chapter 2).

Social theorizing on cycling loosely divides between approaches concerned with the social construction of technology and those that place cycles and cycling within broader social analysis. Theories of technology deal with the machinery and mechanics of bicycles and tricycles and expand this to include users and geographies (Norcliffe 2009). Cycles have to be ridden and so further questions need to be asked in relations to who rides and why? What factors and processes ensure the distributions of technologies, their acceptability (or not) to different groups of users? Explanations are sought for the different distributions of cycling across geographical territories and between social groups divided by gender, class, race and other markers of distinction within the same location.

Cycling as a sociotechnology

In their ground-breaking work on the social construction of technological systems [SCOT] (1987), Bijker, Hughes and Pinch demonstrated that technologies do not develop or progress purely in response to technical imperatives, nor necessarily in a rational manner or direction. Rather, by considering the relationships between technologies and society, they showed how technological development is the outcome of a complex interaction of forces, often unexpected and rarely rational. Technological artefacts – sociotechnologies – are not comprehensible without consideration of their social origins. This means that sociotechnologies not only reflect but embody and, at least in part, reproduce the divisions, stratifications, and exclusions of the societies in which they are formed (Leonard 2003; Norcliffe 2009). Bijker (1995) examined the transition from high bicycle to safety bicycle as one of three case studies to show how this works. While his depiction of the details of historical events was sharply criticized by specialist cycle historians, the basic argument remained sound (Shrivastava 2005).

SCOT analyses show the role of the social in technological development. Initially, technologies can be understood in many ways. Through familiarity and common practice in the actions of users, flexibility of interpretation gives way to obduracy: further innovation and redefinition become difficult. Examining changing patterns in the 20th-century cycle industry, especially the flurry of innovation and change occurring from the 1970s with the development of BMX and mountain biking, Rosen (1993, 2002) investigated how previously obdurate conditions may once again become fluid through the impact of broader social changes.

Oudshorn and Pinch (2003) further argued that technologies are effectively co-constructed by their users. In use, artefacts become repurposed in often-unexpected ways. This renewed emphasis on users also helps highlight how innovation in cycling since the 1960s has largely come from outside the cycle industry. Where users pioneer new ideas and new uses, industry follows, not always successfully (Stoffers 2016). Significant developments in cycle technologies of sustainable transport, for new uses and new users, are actually produced as users become manufacturers (Cox and Rzewnicki, 2015; see also Chapter 9).

Seeing cycles as technological artefacts inseparable from the societies in which they are used and developed, situates cycle use and users within broader social contexts. For example, gendered conventional bicycle designs embody particular expectations of gendered social roles and behaviours. Beyond cycle technology, we need also to examine use and users, ways in which cycling behaviours and practices are mixed up with other concerns about social equity and inequality. This requires engagement with other dimensions of social analysis.

Cycling as a cyborg activity

Haraway's' influential 1985 essay "A Cyborg Manifesto: Science, Technology, and Socialist-feminism" made a constructionist approach to feminist thought widespread. Lupton (1999) applied her image of the cyborg to illuminate the transformations of vehicular road users.

Even without engaging Haraway's wider philosophico-political project of boundary dissolution, Butryn and Masucci (2003) used the cyborg to explore how cyclists emerge as an interplay of human body and technology. The basic imagery is not new, as Brogan (2016) points out, but the cyborg imagery and its constructionist basis provided a means by which to develop further thought on the ways in which the persons, technologies and spaces are connected in the act of cycling.

Still thinking about the interactions of humans and machines, Akrich and Latour (1992) and Latour (2005) explored the ways in which technologies (such as cycles) are bound up in networks of action (hence Actor–Network Theory, or ANT). Technologies are not inert objects within these networks: differences in technology afford (and, conversely, constrain) different possibilities of action for different users. While not having wilful agency in the same way as human participants, technologies are 'actants' not neutral objects. Applied to cycling, this opens up new ways of thinking, for example, about how cycle use can shape cities and social relations of spaces within them; or the historic effects of cycle mobility to open opportunities for autonomous travel in sections of the community previously denied through class, gender or race. Cycling is also almost always a public act, so how these actions are performed also matters. Without theory, we may observe patterns of change but be unable to use those observations.

Deleuze and Guattari's (1988) concept of an assemblage allows further progression. Machines and humans are intertwined, but these relations take place in a broader network of connections that give meaning to the technologies and their usage. Cycles (when used in mobility and not just as items of display) are meaningless without the spaces in which they are used. Just as different cycle designs provide different opportunities of action (to different users), different spaces of use afford different possibilities to various combinations of user/machine (Norcliffe 2009). Cycling is thus a hybrid assemblage of cycle, user and space, and complex interactions between these (Cox 2019). One obvious implication of this is that policy needs to take account of the multiple 'cycling*s*' that emerge from different combinations of elements.

Cycling as a social practice

Social practice theory (Shove, Pantzar and Watson 2012; Watson 2012), uses cycling as an illustrative example. Social practices are those actions in society undertaken by many people, often on a mundane basis (cycling, doing the laundry, brushing one's teeth, recycling). However, rather than thinking about social practices as the sum of individual actions, Shove and her colleagues have examined how they can be depicted as the conjunctions of materials (physical things involved), competencies (skills and knowledge of participants) and meanings (values and discourses attached to the action). Social practices therefore take on an existence beyond the participation of individuals: the relationship between action and practice is inverted. Social practices can be said to recruit people, to offer potential participants rewards (real and/or symbolic) or, conversely, to dissuade participation by constructing barriers, depending on the arrangements of the elements. Further, social practices do not exist in isolation, they are interlinked with one another. Thinking through cycling in this way exposes cycling practices as more than the sum of individual behavioural choices (Spotswood 2016). Infrastructures and machines, the skills and knowledge required to use them, and the acquisitions processes, social reputation, all interact to shape cycling practices. This has clear implications for policy thinking (Shove 2011, 2015), especially considering changes required for greater sustainability: a situation in which cycling has much to offer (Parkin 2012).

Automobility and vélomobility

Perhaps the most important contribution to theorizing cycling and society (from the perspective of this writer's particular concern with cycling and sustainability), are those ways of thinking prompted by Urry's work on *mobilities*, specifically his analysis of *automobility*. (Urry 2005). His innovation was to move from examining the car either as an object or through its use, and to adopt a systems analysis of the sets of social, economic and political relations that are bound up with car use. As Cass and Manderscheid (2019) summarize, beyond the car lies a tangle of relations comprising "its production, fuel and infrastructure industries, the policies that create automobile landscapes that separate work, residence and other activities in space, as well as the discursive and cultural association of cars with freedom and autonomy. Modern lifestyles that archetypally centre on one-family-houses in suburbia, shopping centres and leisure facilities on the edge of cities represent the ideal of the 'good life' under automobility." Urry also notes the way this system also serves as the 20th century's archetype of a growth-oriented carbon economy, in what he later called carbon capitalism (Urry 2011).

Although not directly concerned with cycling, the system of automobility Urry describes has nevertheless provided the context for most of cycling's history. Even when cycling was numerically greater, political decision-making and planning was designed to favor the car, and popular discourse held motoring as the most desirable from of personal mobility (Carter 2021). In the light of dramatic changes in carbon emissions required to address the climate crisis (WMO 2020), especially in the transport sector, cycling for transport has the potential to make significant contributions to future low-carbon transport scenarios. As the systemic analysis shows, however, the problem of automobility is not just an issue of the internal combustion engine. Electrification of the auto fleet will not address the energy-intensive requirements of the system. Consequently, authors such as Koglin (2013) have adopted vélomobility as a term to explore the parameters and possibilities of a system of mobility predicated not on private motoring but on human scaled mobilities based on active travel forms (see Chapter 21).

Conclusions

This is far from a comprehensive analysis of theoretical perspectives on cycling. Instead, it suggests social theory as necessary to comprehend cycling and society. Cycling activities are not separated from the rest of the social world. As we address the interaction of cycling with issues of work, of gender relations and roles, inequalities of class and race or of the social exclusion of people with impairments (and the ways in which these may be alleviated or exacerbated by the ways in which we think about and act on cycling), we are encountering issues of knowledge. Here we come back to the quotation from Jansen: that knowledge is always political – whose knowledge counts, and who is and is not included as having valid knowledge, is a political decision. Theorizing cycling, the choice of ontological positions made in designing research and writing about cycling matters. There is always a temptation in social science research to identify problems and immediately focus on ways of solving them. Cycling researchers are not immune from this. However, by reflecting on the underlying bases of thinking and the ways in which these have the potential to define problems in particular ways, this chapter is an attempt to assist better research design.

References

Albert de la Bruhèze, A.A. and Veraart, F.C.A. 1999. *Fietsverkeer in praktijk en beleid in de twintigste eeuw*. Den Haag: Ministerie van Verkeer en Waterstaat/Stichting Historie der Techniek.

Aldred, R. 2010. 'On the Outside': Constructing Cycling Citizenship, *Social & Cultural Geography*, 11:1, 35–52.

Bijker, W. E., Hughes, T.P., and Pinch, T.J. (eds.) 1987. *The Social Construction of Technological Systems: New Directions in the Sociology and History of Technology*. Cambridge, MA: MIT.

Bijker, W. 1995. *Of Bicycles, Bakelite and Bulbs: Toward a Theory of Socio-Technical Change*. Cambridge, MA: MIT Press.

Boal, I.A. 2001. Towards a world history of cycling, *Cycle History*, 11, 16–22.

Brogan, Una. 2016. Two-wheeled sensibility: Sensory engagement with place in British, American and French cycling narratives, 1880–1914, *Cahiers Victoriens & Édouardiens*, 83 (83 Printemps), 8.

Brown, K. and Spinney, J. 2010. Catching a glimpse: The value of video in evoking, understanding and representing the practice of cycling. In B. Fincham, M. McGuinness, and L. Murray (eds.) *Mobile Methodologies*. Farnham: Ashgate, pp. 130–151.

Butryn, T.M. and Masucci, M.A. 2003. It's not about the book: A cyborg counternarrative of Lance Armstrong, *Journal of Sport and Social Issues*, 27:2, 124–144.

Carter, N. 2021. *Cycling and the British*. London: Bloomsbury.

Cass, N. and Manderscheid, K. 2019. Call for papers for Paper for "Shapes of Post-growth Societies", at the *Final Conference of the DFG Research Group 'Landnahme, Acceleration, Activation' and the 2nd Regional Conference of the German Sociological Association at the Friedrich Schiller University Jena*, September 23 to 27, 2019.

Cook, N. and Butz, D. (eds.) 2019. *Mobilities, Justice and Social Justice*. Abingdon: Routledge.

Cox, P. 2019. *Cycling: A Sociology of Vélomobility*. Abingdon: Routledge.

Cox, P. and Van De Walle, F., 2007. Bicycles don't evolve: Velomobiles and the modelling of transport technologies. In D. Horton, P. Rosen, and P. Cox (eds.) *Cycling and Society*. Abingdon: Routledge, pp. 113–131.

Deleuze, G. and Guattari, F. 1988. *A Thousand Plateaus: Capitalism and Schizophrenia*. [Tr. B. Massumi]. London: The Athlone Press.

Epperson, B. 2000. Failed colossus: Strategic error at the Pope Manufacturing Company, 1878–1900, *Technology and Culture*, 41:2, 300–320.

Golub, A., Hoffman, M.L., Lugo, A., and Sandoval, G.F. (eds.) 2016. *Bicycle Justice and Urban Transformation. Biking for All?* London: Routledge.

Haraway, D. 1991/1985. A cyborg manifesto: Science, technology, and socialist-feminism in the late twentieth century. In Haraway, D. (ed.) *Simians, Cyborgs and Women: The Reinvention of Nature*. London: Routledge.

Hoffman, M.L. 2016. *Bike Lanes are White Lanes: Bicycle Advocacy and Urban Planning*. Lincoln: University of Nebraska Press.

Hoffman, M.L. and Lugo, A. 2014. Who is 'World Class'? Transportation Justice and Bicycle Policy, *Urbanities*, 4:1, 45–61.

Horton, D. 2007. Fear of cycling. In D. Horton, P. Rosen, and P. Cox (eds.) *Cycling and Society*. Abingdon: Routledge, pp. 133–152.

Jansen, J. (ed.) 2019. *Decolonisation in Universities. The Politics of Knowledge*. Johannesburg: Wits University Press.

Jungnickel, K. 2018. *Bikes and Bloomers*. London: Goldsmiths' Press.

Koglin, T. 2013. *Vélomobility – A Critical Analysis of Planning and Space*. PhD thesis, Lund University.

Koglin, T. 2015. Organisation does matter – planning for cycling in Stockholm and Copenhagen, *Transport Policy*, 39, 55–62.

Leonard, E. B. 2003. *Women, Technology and the Myth of Progress*. Upper Saddle River, NJ: Prentice Hall.

Longhurst, J. 2015. *Bike Battles: A History of Sharing the American Road*. Seattle, WA: University of Washington Press.

Lupton, D. 1999. Monsters in metal cocoons: 'Road Rage' and cyborg bodies, *Body & Society*, 5:1, 57–72.

Melia, S. 2016. Evaluating the impact of policy: The built environment and travel behaviour. In Fiona Spotswood (ed.) *Beyond Behaviour Change Key Issues, Interdisciplinary Approaches and Future Directions*, Bristol: Policy Press, pp. 89–112.

Melia, S. 2020. Learning critical realist research by example: Political decision-making in transport. *Journal of critical realism*, 19: 3, 285–303.

Norcliffe, G. 2009. G-COT: The geographical construction of technology, *Science, Technology and Human Values*, 34:4, 449–475.

Oosterhuis, H. 2014. Bicycle research between bicycle policies and bicycle culture, *Mobility in History*, 5, 20–36.

Oosterhuis, H. 2016. Cycling, modernity and national culture, *Social History*, 41:3, 233–248.
Oudshorn, N. and Pinch, T. (eds.) 2003. *How Users Matter: The Co-Construction of Users and Technology*. Boston, MA: MIT.
Parkin, J. (ed.) 2012. *Cycling and Sustainability*. Bingley: Emerald.
Pucher, J. and Buehler, R. 2017. Cycling towards a more sustainable transport future, *Transport Reviews*, 37:6, 689–694.
Pugh, N. 2019. *Orrery for landscape, sinew and serendipity. Presentation and performance at Cycling and society symposium*, University of Chester, 2–3/09/2019. http://www.cyclingandsociety.org/2019-symposium-chester/
Reid, C. 2017. *Bike Boom: The Unexpected Resurgence of Cycling*. Washington: Island Press.
Rosen, P. 1993. The social construction of mountain bikes: Technology and postmodernity in the cycle industry, *Social Studies of Science*, 23:3, 479–513.
Rosen, P. 2002. *Framing Production: Technology, Culture and Change in the British Bicycle Industry*. Cambridge, MA: MIT Press.
Sheller, M. 2018. *Mobility Justice*. London: Verso.
Shove, E., Pantzar, M., and Watson, M., 2012. *The Dynamics of Social Practice: Everyday Life and How It Changes*. London: Sage.
Shove, E. 2011. How the social sciences can help climate change policy. Transcript of the extraordinary lecture at British Library, 17 January.
Shove, E. 2015. Linking low carbon policy and social practice. In Y. Strengers and C. Maller (eds.) *Social Practices, Intervention and Sustainability: Beyond Behaviour Change*. Abingdon: Routledge, pp. 31–44.
Shrivastava, P. 2005. Towards a socio-technical history of bicycles. In *Proceedings of the 15th International Cycle History Conference, Vienna, Austria September 2004*. Bicycle Books, San Francisco, pp. 9–24.
Smith, R.A. 1972. *A Social History of the Bicycle: Its Early Life and Times in America*. New York: American Heritage Press.
Spinney, J. 2011. A chance to catch a breath: Using mobile video ethnography in cycling research, *Mobilities*, 6:2, 161–182.
Spotswood, F. (ed.) 2016. *Beyond Behaviour Change. Key Issues, Interdisciplinary Approaches and Future Directions*. Bristol: Policy Press.
Stoffers, M. 2016. The politics of bicycle innovation: Comparing the American and Dutch Human-Powered vehicle movements, 1970s–Present. In R. Oldenziel and H. Trischler (eds.) *Cycling and Recycling: Histories of Sustainable Practices*. Oxford: Berghahn.
Urry, J. 2005. The system of automobility, *Theory, Culture & Society*, 21:4/5, 25–39.
Urry, J. 2011. *Climate Change and Society*. Cambridge: Polity.
Watson, M. 2012. How theories of practice can inform transition to a decarbonised transport system, *Journal of Transport Geography*, 24, 488–496.
World Meteorological Organization (WMO) 2020. *United in Science 2020. A multi-organization high-level compilation of the latest climate science information*. Available electronically at www.public.wmo.int/en/resources/united in science.

2

CYCLING AND GENDER

Past, present and paths ahead

Jennifer Bonham and Kat Jungnickel

Introduction

Cycling's possibilities and also its problems are not equally distributed or experienced. Discussions, reflections and questions of gender have formed an integral part of cycling inquiry for more than a century or for as long as the bicycle has been popularized, practiced and pilloried. The who, along with the what, where, when and why of cycling, has been a primary topic of debate. Since the 1990s, this debate has often started with the observation that in low-cycling countries, including Australia, Brazil, Canada, Chile, the UK and the USA, men are much more likely to cycle than women. By contrast, women in high-cycling countries like The Netherlands and Denmark are as, if not more, likely to cycle than men (e.g. Aldred et al. 2017). These national differences, along with historical accounts of women cycling (e.g. Jungnickel 2018), counter arguments that women have a 'natural' aversion to riding a bike (Garrard, Handy & Dill 2012). Much of the gender and cycling literature is concerned with examining how and why cycling is or isn't available to diverse populations and the barriers in place that impede its uptake.

This chapter focuses on why intersections of gender and cycling have long been, and continue to be, critical subjects attracting the attention of interdisciplinary and international writers, practitioners and researchers. This covers work in fields such as sociology, geography, cultural studies, and history amongst others attempting to diversify and broaden the spectrums of cycling. While it is not possible to delve deeply into all of these rich debates, in the following we present the "state of the art" overview of gender and cycling to provide a map of sorts.

We start with a discussion of gender and gendering practices, review historical underpinnings, consider present-day themes in everyday cycling (understood as cycling to a destination such as work, shops, education, social and other activities), and conclude with some of the future challenges to broadening cycling reach.

From gender to gendering

Gender is used in cycling literature in various ways. At times it refers to a person's biological sex as one attribute among many. In other instances, it refers to the sets of attributes acquired by biologically sexed bodies (women and men) in the process of socialization. The biological

 DOI: 10.4324/9781003142041-4

body is presumed to pre-exist and serve as the material basis for the socially constructed gender norms anticipated of it.

Alternatively, gender refers to the qualities, characteristics, behaviours and functions constituted as biologically sexed bodies (Butler 1990). In this understanding, biological sex and gender do not align with a nature/culture divide but are co-constituted within practices (research, policymaking, legal proceedings, industrial design etc.) referred to as gendering. These 'gendering practices' produce 'women' and 'men' as specific kinds of unequal political subjects (Bacchi 2017) and operate toward regulating populations (Butler 1990). However, because 'women' and 'men' are produced within practices (including a variety of scientific practices, Mol 2015) their production is ongoing, incomplete and open to change. Bringing this conceptualization of 'gendering' into cycling (Bonham & Bacchi 2017) responds, in part, to calls for a more detailed and systematic examination of the relation between gender and mobility (Ravensbergen, Buliung & Laliberté 2019).

Recent scholarship is providing insights into how gender identities are produced and how they are claimed, consolidated, contested and reconfigured through entanglements both on and off the bicycle. The proliferation of identities – trans, queer, non-binary – demonstrates the possibilities for reconstituting gender. These identities are being produced in cycling literature with researchers enabling respondents to claim different gender statuses. Importantly, we expect research on gender and cycling to open up new ways of doing bicycling beyond the dichotomous female/risk-averse/slow/defensive and male/risk-taking/fast/aggressive that often populates the cycling literature. Our review of the current state of gender and cycling research points to opportunities for disruption as a way of re-shaping bicycling and making it available to diverse populations.

We begin with an overview of cycling in the late 19th century where research has focused on how mobility reconfigured gender in Australia, Europe, New Zealand and North America.

Recovering cycling histories

One of the primary reasons cycling is intertwined with gender relates to the social and cultural contexts of its invention. In 1895, Frances E. Willard wrote *A Wheel Within a Wheel* during the cycling craze that swept England. She took up cycling at the age of 53 and evangelized bike riding for people of maturing years and for women in general. Willard believed bicycling held the power to mitigate some of the many restrictions on Victorian women's freedom of movement in relation to clothing, social lives, physical health and political expression. To Willard, the bicycle was a 'new implement of power' that held 'special value for women' and would 'help women to a wider world' (1895, p. 73).

Women have long been enthusiastic cyclists, yet there are far fewer records documenting their interest and achievements. From the advent of popular cycling in the 1890s to today, this blindness lingers. Researchers are recovering the history of women's engagement with cycling and using it to argue for the importance of representational equality and diversity. At the end of the 19th century, women embraced bicycle racing (Simpson 2007), were early adopters of popular 'Bike Portraits', a fusion of new technologies (Kinsey 2012), and inventors of radical new forms of early cycle clothing (Jungnickel 2015, 2018: see, also, Chapter 40). Then, through to the mid-20th century women took up endurance cycling (Bootcov 2019) and utility cycling as enthusiastically as men (Carstensen & Ebert 2012). Remarkably, this engagement persisted despite many social, political, physical and institutional barriers to women's cycling.

Bicycling provided a means for women to challenge a life anchored to the domestic sphere. It gave middle- and upper-class women especially 'a taste of independence... and physical freedoms' (The Sketch 1896: 311). Changes in bicycle and clothing designs afforded opportunities for women to take up cycling in different ways. In contrast to many accepted accounts, women also drove many of these shifts, as keen consumers and also as actively engaged inventors.

'What' a person cycled, and cycled in, was shaped and regulated in terms of gender. Victorian society was highly differentiated by class, race and gender. How a person spoke, what they wore and the places they inhabited determined their course in life and how they were treated. By the mid-1890s, bicycles with a lowered top-tube featured in bicycle manuals and marketing materials as 'ladies' safety bicycles (Bonham, Bacchi & Wanner 2015). In creating the ladies' safety bicycle, engineering, design, metals, clothing, biology and class were brought together in a gendering practice. Differentiating safety bikes into ladies' (and necessarily men's) 'safeties' – rather than 'safety with'/'safety without' a top-tube – produced bicycles as gendered objects (ibid.). Gender was being co-constituted with the vehicle itself. Ladies' bicycles both consolidated femininity as a performance of 'modesty' and began reconfiguring it by opening the possibility of more vigorous exercise and greater travel distances.

Bicycle designs were also a response to what became known as the 'dress problem'. While men's clothes were more oriented toward physical activity, women's were not. Middle- and upper-class women's fashions, with tightly bound and heavy cumbersome layers, were perilously problematic on bicycles. The Rational Dress movement recognised the bicycle boom sweeping the nation as another way to continue their campaigning for lighter, looser layers to enable women to lead more active lives. Some pioneering women even patented radical new forms of cycle wear (Jungnickel 2015, 2018). However, while shorter skirts and bloomers or knickerbockers were safer and more comfortable to cycle in, they potentially exposed wearers to harassment from shocked onlookers who viewed wearers as masculine and threatening the status quo.

One reason cycling has been so political through the centuries is related to the public spaces it inhabited. Located outside, often in highly populated streets and parks, meant cycling garnered much more attention than other popular sports such as horse riding and swimming. Cycling was far less controlled by sporting's rules and young women could occasionally 'lose' their chaperones and experience personal private time. This was problematic in Victorian society, as upper-class women appearing unaccompanied in public were vulnerable to social disgrace. In contrast, swimming was considered acceptable for women because bodies were concealed within the confines of swimming enclosures.

'How' a person cycled was shaped by the bicycle they had access to, the advice available and the accumulated development of their capacities. Like bikes and clothing, advice on riding a bike was gendered, thereby making cycling available to women at the same time as altering the content of gender categories. Cycling catalyzed new periodicals and newspapers, which provided a visual imaginary and encouragement to cycle. While men's media were often oriented to racing, speed records and new technologies, women's titles implored riders to maintain ladylike decorum while cycling.

Highwheel Bicycles were first available to men and they established clubs to support their cycling participation. Cycling was one of many Victorian sports considered the 'natural domain of men and that to be good at them was to be essentially "masculine"' (Hargreaves 1994: 43). Men's cycling progressed rapidly, and attracted much media attention, because the advancements made in racing trickled down into ordinary men's cycling cultures. This meant it was more fitting for men to exert themselves on bicycles and also enter in bicycle retailers and claim expert consumer identities.

Club activities included staging bike races and fostering bicycle knowledge, but perhaps more importantly providing advice on how to conduct oneself on the roads (Mackintosh and

Norcliffe 2007: 153). Clubs rejected the practice of 'scorching' (riding recklessly fast) that involved young men and has been regarded as an assertion of masculinity especially with the 'domestication' of cycling by women (2007: 163). This point on domestication deserves further research attention. It flags how qualities constituted as 'feminine' might have modified the performance of cycling, thereby ensuring streets continued to be available for leisure, recreation and, eventually, utility cycling.

Women's engagement with bicycles continued into the twentieth century. Female athletes rode the post-war bicycle boom of 1930s Australia, with women's clubs emerging to focus on endurance racing (Bootcov 2019). Despite discrimination and resistance, women persisted and, in the process, challenged 'fears of "fast women", and the notion that strenuous exercise had deleterious consequences' (2019: 1448).

Recovering bicycling histories demonstrates that women's 'lack' of participation in cycling is a relatively recent phenomenon. While past practices suggest future potential, these histories focus on leisure, recreation and sport cycling in countries of the global north. More work from South America, Africa and Asia will enrich these stories. We also need to flesh out histories of gender and everyday cycling. While much has been written about the highs and lows of everyday cycling through the 20th century more attention on gender is needed. Many of the themes flagged in historical research provide insights into the gendering of everyday cycling in the present day.

Gendering cycling today

Over the past 30 years, activists, researchers and policymakers have promoted bicycling as a healthy and sustainable mode of transport (Chapter 30). Gender has been a key theme in developing the discourse on everyday cycling with women (and girls) consistently reported as less likely to ride than men (<30% and >70% respectively). Examining differences in cycling rates and recommending policies to facilitate participation is important in highlighting inequalities and it is also risky when women are produced as 'lacking' – lacking assertiveness, lacking speed, lacking skill, lacking knowledge, lacking time and lacking courage (being fearful or risk-averse). Research into *how* women engage with bicycling can challenge the concept of 'lack' and provide new ways of thinking about both gender and cycling. The following discussion is organized around conventional themes of 'challenges to cycling' and 'enabling cycling'. However, we have paid attention to gender attributions and how attributes are claimed, consolidated, contested or reconfigured by both researchers and their research subjects.

Challenges to cycling

Riding a bike in public space can be as fraught today as in the late 19th century. Cultural heritage and local cultural context – what cycling means and how it is 'done' – influences engagement with cycling (Aldred & Jungnickel 2014). As cyclists are in full public view, they can be observed, objectified and judged in that local context. At the heart of this surveillance is the regulation of presence, appearance and conduct in public space. Four issues related to 'public scrutiny' and self-regulation are often raised in cycling studies: physical exercise, maturity, sociality and personal safety.

Physical exercise, exertion and associated bodily secretions are often given as reasons for girls and women not cycling. This concern has been widely discussed in the gender and health literatures with representations of femininity and masculinity in popular culture often called out as the problem. However, we have barely considered how norms in present day preventive health and related literatures, such as 'women walk and men ride bikes', encourage people identifying as girls/women and boys/men to regulate their mobility. We laugh at 19th-century health warnings about women cycling yet Davara Bennet (2017) demonstrates how ambiguous medical advice given to pregnant women can regulate their cycling.

In low-cycling countries, physical activity and the bicycle itself have been linked to childhood and immaturity (Frater & Kingham 2018). Migrants in Toronto reported bike riding in their home countries was acceptable for young boys but it was unacceptable in adulthood (Ravensbergen 2020). Similarly, boys and men who cycle have been infantilized in US popular culture (Furness 2010) and Australian anti-drink-driving advertising (Nielsen & Bonham 2015). This representation resonates with views among some teenage girls in New Zealand that cycling is not 'cool' and adults do not ride bicycles. We return to this below.

The third aspect of self-regulation is sociality. Frater and Kingham (2018) reported girls feared rejection as friends would laugh at them and the bicycle was an awkward object when accompanying friends who chose to walk (the 'norm' for women). Yet for boys the bicycle served as a vehicle for social engagement which resonates with the participation of men in fitness bunch rides. By contrast, girls and boys in The Netherlands appreciated the social interaction provided by cycling (Frater & Kingham 2020). Exploring this link between gender, cycling, sociality and local context could open up new ways to make cycling available to teenagers.

Concern for personal safety, such as being sexually assaulted, mugged or harassed, is often cited by women as a reason for not cycling or limiting their cycling to particular times and certain places (Ravensbergen, Buliung, & Laliberté 2020). Targeting women via sexual assault and harassment produces public space as masculine and fosters women's self-regulation. However, in some contexts, bikes are preferred at night as women can travel more quickly (Montoya-Robledoa & Escovar-Álvarez 2020). Women cyclists in Bogotá, Columbia, reported property theft as their biggest concern with some running red lights to avoid or minimize targeting by thieves (Montoya-Robledoa et al. 2020). But harassment and threats to personal safety in Bogotá also came from male cyclists who challenged women's competence and admonished them for how they dressed (Chapter 38). The issue of harassment significantly shapes the journeys of minority women (and men) who are as concerned about the conduct of law enforcement officers as other citizens (Lubitow, Tompkins & Feldman 2019). While cycling in these contexts is fraught, it participates in contesting the masculinization (and, in some contexts, whiteness) of public space.

Related to public space is the issue of motor vehicle traffic. Traffic and potential for traffic-related injuries are frequently given as the reason women don't cycle or are selective about where and when they ride. This concern is explained as the different 'risk' tolerance of men and women. Rather than risk aversion, some are 'fed up' with the hypervigilance required to safely negotiate traffic (Bonham & Wilson 2012). Women are characterized as either 'naturally' more risk-averse than men or socialized as potential mothers and carers into risk aversion (see Garrard et al. 2012). Alternatively, traffic-related concerns have recently been explained in terms of differences in childhood cycling. Based on their Canadian research, Sersli et al. (2021) argue when girls are forbidden or discouraged from cycling, they do not have an opportunity to develop the necessary skills, knowledge and capacities. However, as they acquire knowledge and practice cycling in their new country, they became more confident in negotiating road environments (ibid.).

In the 19th century women celebrated the bicycle for the escape it offered from the domestic sphere. In the 21st century heterosexual households, women's greater share of domestic and carer responsibilities account for their lower rates of cycling (Emond, Tang & Handy 2009).

These complex journeys often requiring trip chaining, accompanying others (children, elderly parents, relatives with disabilities), and carrying goods, all of which make riding a difficult option, particularly in car-oriented cities. By contrast, research from The Netherlands comparing women with (n = 20) and without (n = 17) children reported all respondents felt very positive about cycling. The analysis noted variations in when and where women cycled and only marginal differences in distances travelled (Eyer & Ferreira 2015). Unlike many of their counterparts in low-cycling countries, having children did not spell the end of cycling for mothers.

Partners that share household and caring tasks more equitably can enable each partner to ride. Bonham and Wilson (2012) noted that for some women, carer responsibilities provided an opportunity for women to cycle. Accompanying children to school by bike was relationship building and allowed mothers to model personal health and environmental sustainability. We need more research on how men perform the journey-to-childcare and/or the journey-to-school with their children. Research from Bogotá on father's accompanying children suggests the persistence of 'toxic masculinity' with men performing aggressive, fast, risk-taking riding (Montoya-Robledoa et al. 2020). Research across different countries and cultures will provide alternative productions of masculinity.

Enabling participation

Provision of infrastructure and developing cycling skills and capacities are two key policy recommendations aimed at increasing women's cycling participation and reducing the disparities between women and men (Chapter 22). Surveys, interviews and observation studies consistently report women prefer or are more often observed using separated cycling facilities (Aldred et al. 2017). This raises questions about the very possibility of designing, constructing and regulating streets and roads that do not ensure the safety of *all* users, including cyclists. We might ask: what knowledge (data collection, travel surveys) has informed the exclusion of cyclists?; does catering to cyclists constitute them as a 'special needs' group that can be ignored in tight budgets?; and do infrastructure 'preference' studies participate in gendering cycling spaces? With the latter question, we might add: what are the effects of this gendering practice?

Over the past decade, attention has turned to developing cycling knowledge, skills and capacities that address people's (but especially women's) 'lack' and, arguably, prepare them for 'fitting into' existing conditions. These courses range from checking and maintaining bike components and learning about bike-handling to moderate or advanced riding skills.

School-based bicycle education and training is widespread in The Netherlands and Denmark, but it is variable in most other countries. Courses are usually conducted by private companies or bike advocacy organizations with the latter also catering to adults. There have been surprisingly few evaluations of the impact of these programs on increasing participation in cycling (Sersli et al. 2019). Importantly in the current context, Transport for London found the disparity in cycling rates of women and men reduced after training. However, studies among migrant women have been mixed (van der Kloof, Bastiaanssen & Martens 2014). Many women develop the skills necessary to riding independently; whether they actually commence cycling is another matter, especially if they have children.

Taking a different approach, alley cat races initiated by the women's collective Carishina en Bici (Bad Housewives that Bicycle) in Quito, Ecuador, develop cycling knowledge, skills and capacities (Gamble 2019). Deep play is used strategically to shift the physical and emotional experience of cycling in public space. In place of violence and aggression, alley cat creates public space as entertaining, playful and safe. It produces women as funny, fun loving, courageous and determined, although these qualities are not emphasized by Gamble. In Gamble's analysis, women continue to be constituted as caring and relationship-oriented but, importantly, these characteristics become a positive attribute of bicycling. Rather than women being problematized for their risk aversion, lack of speed and lack of assertiveness, cycling is constituted as a site of care and joy.

Working on the cycling body via formal programs, informal groups or individually has been critically examined for the production of gender itself. Using Butler's theorization of performativity, Ravensbergen (2020) has examined the translation of gender performance into cycling. Based on interviews with mainly migrant women and men completing a cycle training program, she reports on how participants regulate themselves toward gender norms when riding a bike. As noted above, bicycling in itself contests gender norms of some cultures. Participants' selection of clothing, times and places of travel, and practices of the journey begin to reconfigure gender performances.

Indeed, this links to the discussion of violence and aggression as performances of 'toxic masculinity' (Montoya-Robledoa et al. 2020) and raises questions as to how this version of masculinity became possible and what other versions are being produced within and beyond the academic literature. Addressing the first question, we can look more closely at the cultural context in which cycling takes place. In Australia and the USA, adulthood has become closely associated with getting a driver's licence. Representing the driver licensing process as a 'rite of passage' demonstrates the significance of driving in these societies, but how it has become a 'rite of passage' is rarely interrogated (see Nielsen & Bonham 2015). Driving replaces rather than adds to mobility options so that men, in particular, who continue to ride can be characterized as 'lacking'.

How then can men who ride bicycles in car-oriented cities perform masculinity? The behaviour directed at female cyclists in Quito and Bogota is an assertion that a certain type of physical strength and willingness to fight are qualities necessary to cycling in these cities. Only strong, aggressive men (not women) can cycle. Alternatively, in Australia and North America performances of masculinity involve fast speeds, risk taking and competition. Barrie et al. (2019) examine how sport cyclists located within virtual, material and social networks work on themselves to foster these 'masculine' attributes. Sersli et al. (2021) argue that John Forester's 'vehicular cycling' (cyclists riding like they are driving a car) is a performance of masculinity. We might ask whether 'vehicular cycling' made bike riding available to men at a time when driving a car was entrenched as the transport norm. Further, in constituting speed, competition and risk taking as masculine, men are encouraged to ride or speak of their riding in this way (Steinbach et al. 2011). Research by Barrie et al. (2019) directs attention toward men doing cycling differently, thereby reconfiguring or providing alternative masculinities.

It is in the enactment of networks of relations that bikes, bicycling and bicyclists are gendered. Tracing these relations, we can see how gendering happens and how it can be disrupted. We previously noted, entanglements that produced the 'ladies safety bicycle' in the 19th century. Today, bicycles without a top-tube are differentiated by the action in mounting/dismounting (step-through) or the places they are likely to be ridden (town bikes) rather than the body of the rider. But 'road' bikes, previously referred to as men's bikes, are being gendered in new ways. 'Women's (and consequently men's) specific' road bikes are produced within relations of engineering, biomechanics, anatomy, materials and design. Height, reach and hand size (to name a few) are used in differentiating both people and bikes. This new gendering practice produces 'women' as engaged in vigorous physical exercise. It has implications for bodies within and between populations that don't fit 'women' and 'men' constituted in this way (Bonham, Bacchi & Wanner 2015).

Conclusion

While many things have changed since the first cycle boom swept much of the world at the turn of the last century, a number of challenges remain remarkably similar. This chapter has attempted to map intersections of gender and cycling from the 1890s through to today and into the near future. We approached this ambitious task by marking how gender is claimed, consolidated, contested and reconfigured in the process of interacting with the bicycle. We journeyed through the why, what, where, and how of historic cycling and more contemporary discussions of cycling cultures around themes of 'challenging' and 'enabling' cycling. Throughout, we reiterated the potential for further research, inviting deeper investigation of ideas.

References

Aldred, R., Elliott, B., Woodcock, J., Goodman, A. (2017). Cycling provision separated from motor traffic: A systematic review exploring whether stated preferences vary by gender and age. *Transport Reviews*, 37(1), 29–55.

Aldred, R., Jungnickel, K. (2014). Why culture matters for transport policy: The case of cycling in the UK. *Journal of Transport Geography*, 34, 78–87.

Bacchi, C. (2017). Policies as gendering practices: Re-viewing categorical distinctions. *Journal of Women, Politics and Policy*, 38(1), 20–41.

Barrie, L., Waitt, G., Brennan-Horley, C. (2019). Cycling assemblages, self-tracking digital technologies and negotiating gendered subjectivities of road cyclists on-the-move. *Leisure Sciences*, 41(1–2), 108–126.

Bennet, D.L. (2017). Bumps and bicycles: Women's experience of cycle-commuting during pregnancy. *Journal of Transport and Health*, 6, 439–451.

Bonham, J., Bacchi, C. (2017). Cycling "subjects" in ongoing-formation: The politics of interviews and interview analysis. *Journal of Sociology*, 53(3), 687–703.

Bonham, J., Bacchi, C., Wanner, T. (2015). Gender and cycling: Gendering cycling subjects and forming bikes, practices and spaces as gendered objects. In J. Bonham & M. Johnson (eds.) Cycling Futures, University of Adelaide Press, Adelaide, pp. 179–201.

Bonham, J., Wilson, A. (2012). Women cycling through the life course: An Australian case study. In J. Parkin (ed.) *Cycling and Sustainability* (*Transport and Sustainability, Vol. 1*), Bingley, UK: Emerald. pp. 59–81.

Bootcov, M. (2019). Australian female endurance cyclists of the 1930s and the commercialization of their athletic femininity. *The International Journal of the History of Sport*, 36(15–16), 1433–1456.

Butler, J. (1990). *Gender Trouble*. New York: Routledge.

Carstensen, T., Ebert, A.K. (2012). Cycling cultures in Northern Europe. In J. Parkin (ed.) *Cycling and Sustainability*, Bingley, UK: Emerald. pp. 23–58.

Emond, C., Tang, W., Handy, S. (2009). Explaining gender difference in bicycling behaviour. *Transportation Research Record*, 2125, 16–25.

Eyer, A., Ferreira, A. (2015). Taking the tyke on a bike: Mothers' and childless women's space-time geographies in Amsterdam compared. *Environment and Planning A*, 47, 691–708.

Frater, J., Kingham, S. (2018). Gender equity in health and the influence of intrapersonal factors on adolescent girls' decisions to bicycle to school. *Journal of Transport Geography*, 71, 130–138.

Frater, J., Kingham, S. (2020). Adolescents and bicycling to school: Does behaviour setting/place make a difference? *Journal of Transport Geography*, 85, 102724.

Furness, Z. (2010). *One Less Car: Bicycling and the Politics of Automobility*. Philadelphia, PA: Temple University Press.

Gamble, J. (2019). Playing with infrastructure like a Carishina: Feminist cycling in an era of democratic politics. *Antipode*, 51(4), 1166–1184.

Garrard, J., Handy, S., Dill, J. (2012). Women and cycling. In J. Pucher & R. Buehler (eds.) *City Cycling*, Cambridge, MA: MIT Press. pp. 211–234.

Hargreaves, J. (1994). *Sporting Females: Critical Issues in the History and Sociology of Women's Sports*. London and New York: Routledge, p. 43.

Jungnickel, K. (2015). "One needs to be very brave to stand all that": Cycling, rational dress and the struggle for citizenship in late nineteenth century Britain. *Geoforum*, 64, 362–371.

Jungnickel, K. (2018). *Bikes & Bloomers: Victorian Women and their Extraordinary Cycle Wear*. Goldsmiths Press.

Kinsey, F. (2012). Reading photographic portraits of australian women cyclists in the 1890s: From costume and cycle choices to constructions of feminine identity. *The History of Sport*, 28(8–9), 1121–1137.

Lubitow, A., Tompkins, K., Feldman, M. (2019). Sustainable cycling for all? Race and gender-based bicycling inequalities in Portland. Oregon. *City & Community*, 18(4), 1181–1202.

Mackintosh, PG, Norcliffe, G. (2007). Men, women and the bicycle: Gender and social geography of cycling in the late-nineteenth century. In D. Horton, P. Rosen, & P. Cox (Eds.) *Cycling and Society*, Aldershot: Ashgate. pp. 153–177.

Mol, A. (2015). Who knows what a woman is … On the differences and the relations between the sciences. *Medicine Anthropology Theory*, 2(1), 57–75.

Montoya-Robledoa, V., Calerob, L., Carvajala, L., Molinac, D., Pipicanod, W., Peñad, A., Pipicanod, C., Valderramae, J., Fernándezd, M., Porrasd, I., Ariasd, N., Miranda, L. (2020). Gender stereotypes affecting active mobility of care in Bogotá. *Transportation Research Part D*, 86, 102470.

Montoya-Robledoa, V., Escovar-Álvarez, G. (2020). Domestic workers' commutes in Bogotá: Transportation, gender and social exclusion. *Transport Research Part A*, 139, 400–411.

Nielsen, R., Bonham J. (2015). More than a message: Producing cyclists through public safety advertising campaigns. In J. Bonham & M. Johnson (eds.) *Cycling Futures*, Adelaide: University of Adelaide Press. pp. 229–250.

Ravensbergen, L. (2020). 'I wouldn't take the risk of the attention, you know? Just a lone girl biking': Examining the gendered and classed embodied experiences of cycling, *Social & Cultural Geography*, on-line. doi: 10.1080/14649365.2020.1806344.

Ravensbergen, L., Buliung, R., Laliberté, N. (2019). Toward feminist geographies of cycling. *Geography Compass*, 13(7), e12461.

Ravensbergen, L., Buliung, R., Laliberté, N. (2020). Fear of cycling: Social, spatial, and temporal dimensions. *Journal of Transport Geography*, 87, 102813.

Sersli, S., DeVries, D., Gislason, M., Scott, N., Winters, M. (2019). Changes in bicycling frequency in children and adults after bicycle skills training: A scoping review. *Transportation Research Part A: Policy and Practice*, 123, 170–187.

Sersli, S., Gislason, M., Scott, N., Winters, M. (2021). Easy as riding a bike? Bicycling competence as (re)learning to negotiate space, *Qualitative Research in Sport, Exercise and Health*. doi: 10.1080/2159676X.2021.1888153.

Simpson, C. (2007). Capitalising on curiosity: Women's professional cycle racing in the late nineteenth century. In D. Horton, P. Rosen, & P. Cox (Eds.) *Cycling and Society*, Aldershot: Ashgate. pp. 47–65.

Steinbach, R., Green, J., Datta, J., Edwards, P. (2011). Cycling and the city: A case study of how gendered, ethnic and class identities can shape healthy transport choices. *Social Science & Medicine*, 72(7), 1123–1130.

The Sketch. (1896). *Society on Wheels*, 11 March, p. 311.

van der Kloof, A., Bastiaanssen, J., Martens, K. (2014). Bicycle lessons, activity participation and empowerment. *Case Studies on Transport Policy*, 2(2), 89–95.

3

THE PRECARIOUS WORK OF PLATFORM CYCLE DELIVERY WORKERS

Cosmin Popan and Esther Anaya-Boig

Introduction: from cycle work to gig work. A brief history

'As easy as riding a bicycle'. While this is easier said than done, the aphorism remains a powerful claim that cycling is a straightforward skill, which produces normative assumptions regarding individual body capabilities. This overoptimistic assertion may also imply that the bicycle is the most basic vehicle to navigate contemporary urban agglomerations, often associated with a desired "sustainability", while often disregarding the overwhelming hostility of automobile-dominated road environments. Similar postulations are also found in contemporary discourses about work, where flexibility and easy access have come to dominate the narrative of the last four decades. Neoliberalism has dramatically reconfigured jobs away from the security of decent full employment, recasting them as malleable while at the same time obscuring the increasingly precarious nature of work. The gig economy, a labor market characterized by independent contacting that takes place via and on digital platforms, represents the most recent attempt to legitimize this flexibility as both normal and desirable.

These strong beliefs that cycling and entrepreneurship are accessible to everyone have come together to explain the popularity that platform food deliveries such as Deliveroo, Uber Eats or Glovo have gained in recent years in urban environments across the world. One only needs a functional bicycle, undertake a summary selection process, install an app on one's smartphone, and one can become 'one's own boss'. The bicycle is seen, within this logic, both as a basic requirement and as a simple working tool that anyone can ride and afford, which facilitates access to a flexible job. Yet the reality is quite different. Contrary to the mainstream appraisal of cycling as empowering and liberating and of the gig economy as flexible and entrepreneurial, we observe the opposite situation. Namely, we argue that platform work adds an extra level of precarity to the already precarious practice of riding a bicycle in the city. And gender, ethnicity and migrant status all further compromise the already fragile living and working conditions of these workers (Chapter 19). We echo here as well the concept of intersectionality coined by Kimberlé Crenshaw (1989) to account for how cycling, race, class, gender, and other individual characteristics "intersect" with one another and overlap. In doing so, Crenshaw addresses not merely questions of identity and representation, but instead tackles deep structural and systemic questions about discrimination and inequality.

DOI: 10.4324/9781003142041-5

Research has focused on cycling for transport, with a particular attention to commuting. Interest in work amongst cycling researchers was tangential: it focused on cycling as means to access work rather than work *per se*. The notable exception was the case of the literature on cycle messengers from the 2000s (Kidder 2005; Fincham 2006, 2007, 2008). Authors of cycle messenger literature focused on subcultural identities and lifestyles, while actual working conditions and labor struggles were not an issue.

Although the tasks that platform cycle workers perform might seem similar to that of their predecessors, advances of the neoliberal economy and information and communications technology (ICT) in the last three decades have given rise to platform work. The bicycle messengers of 2000s were a romanticized dying breed, as they were essentially carrying physical objects such as legal papers, video tapes and DVDs, which were soon turned into virtual items thanks to advances in technology (Day 2015). By contrast, platform cycle couriers are either celebrated or feared to represent the future of work as the gig economy is seen as the laboratory of platform capitalism, where new techniques of management, control, exploitation and extraction of profit are tested and refined (Cant 2020; Popan 2021). During the SARS-CoV-2 pandemic, the need to deliver food to the locked-down and quarantined world population, and the growth of unemployment, has contributed to an increase of platform cycle couriers; essential but unprotected workers.

In this chapter, we contribute to bridging the gap between cycling research and work research by unpacking the implications of gig workers' use of bicycles as a collection of levels of intersectional precarity. In doing so, we aim to connect the burgeoning domain of cycling studies with some of the most pressing issues impacting contemporary societies: neoliberal rationalities, platform capitalism and the future of work.

Throughout history, the bicycle has found itself at the intersection of work and play. On the one hand, it is intrinsically linked to the rise of capitalism, mass production and consumption. It has mobilized, for most of the previous century, a working class commuting to and from factories while at the same time being used on a large scale to transport a great variety of things by a similar great variety of urban tradesmen and service workers. On the other hand, from its very inception in the late 1800s, the bicycle was first a bourgeois pastime and an object of conspicuous consumption before it eventually reached the masses. After half a century of being taken seriously, from the 1950s onwards the bicycle became, at least in the Western world, a leisure object, a child's toy or a sports machine. In the age of mass automobility, reliance on cycles for everyday commuting and transport diminished drastically.

The emergent gig economy has taken advantage of the bicycle and its versatile role for both work and play. Thanks to platforms such as Deliveroo, Uber Eats, Foodora, Glovo or Rappi, which operate today across all continents, cycling as a job has re-entered public attention. In most cases, riders are not considered employees, but freelance workers performing "gigs". In the same way that the bicycle both is and is not work, the gig economy is and is not employment.

The bicycle is essential to the prosperity of these platforms; it provides affordable access to a job to anyone who knows how to use it, as well as easily mobilizing a pool of cheap labor for transnational capital. During their early stages, when entering a new market, platforms initially cover central urban areas, offer better wages and welcome as many workers as possible. As platforms expand and progress in a specific urban environment, delivery distances may increase and companies start rewarding, through algorithm management, those using motorized vehicles as they can travel far-reaching destinations faster and deliver more meals in one ride. However, the algorithms disadvantage those who cannot afford a motorized vehicle or are not available to work in certain areas, creating inequalities amongst workers. Additionally, neoliberal platform companies have been taking advantage of the increase of people in need of jobs during the pandemic, by lowering the wages and terminating contracts unscrupulously.

This chapter adds to a growing body of literature focusing on the precarious working conditions of food couriers in the gig economy. We aim to expand these debates by focusing on the role of cycling in the gig economy, accounting for the ways in which issues of gender, ethnicity and migration status amplify the job and income insecurities of cycle couriers.

Conceptualizing precarity from an intersectional perspective. From the right to a livelihood to the right to the road

The "gig economy" is driven by the "lean platform economy", developed after the financial crisis of 2007–2008 and relies heavily on venture capital for its rapid growth. It appears as "an outlet for surplus capital in an era of ultra-low interest rates and dire investment opportunities rather than the vanguard destined to revive capitalism" (Srnicek 2016: 91). The term "gig economy" describes a labor market characterized by the prevalence of short-term insecure work as opposed to permanent jobs, and has generated heated debates in recent years with the development of a plethora of digital platforms intermediating not only food delivery services, but also transportation (Uber), hosteling (Airbnb), pet, children and elderly care, cleaning, chores and many others.

Woodcock and Graham (2020) identify a set of factors contributing to the emergence of platform work world-wide. Amongst them are the ubiquity of mass connectivity and cheap technology, but also the changing socio-economic landscape of the last forty years which has led to state deregulation of work and the weakening of employment protection (2020: 23–38). The desire for flexibility for and from workers is equally important here, with workers wanting to get away from the rigid 9-to-5 jobs, while the companies requiring work on demand and rewarding those who can accommodate their requests.

In Latin America, the emergence of the gig economy must be understood within a context in which historical and structural forms of oppression (colonial and related to the international, sexual and racial labor division) operate in addition to the neoliberal ones (Chapters 19 and 38). These historical and structural forms of oppression generate the ideal conditions for neoliberal companies to flourish (Hidalgo Cordero and Salazar Daza 2020). Colonial forms of exploitation continue to this day with most of the platform companies operating in Latin American countries originating and/or being funded by venture capital from the "Global North" (see, for example, Azevedo and Mascarenhas 2019). Platforms can profit from a weakened labor market hiring cheap human resources with promises of false "entrepreneurship", which intensifies the extractive methods of capitalism from people who are most in need. Additionally, lower internet penetration rates and internet data traffic are partly responsible for slightly fewer South American workers involved in the gig economy (Grigera 2020). The number of gig workers in the Global South is estimated between 30 and 40 million, representing around 1.5% of the global South workforce, with around 2 million gig workers based in Latin America (Heeks 2019). Labor markets in Latin America are often informal, precarious and exploitative.

The appeal of such gig jobs amongst cyclists is due, at least partly, to the flexibility enabled by the bicycle itself through its aforementioned ambiguous relation to work and play. As Bennett (2019: 56) observes, 'although it is firmly associated with employment, it also promises escape from work'. This is an important issue to consider if we aim to understand the appeal of cycle deliveries amongst many such workers. Previously, authors have pointed to the conundrum posed by the triad cycling–work–play when highlighting the difficulty to promote cycling as utility transport due to its problematic associations with leisure and pleasure (Aldred 2015).

In our case, the playfulness associated with cycling is compounded by gamification strategies deployed by digital platforms to recruit, exploit and keep captive users and workers.

Having investigated the lived experiences of food cycle couriers in the gig economy in three different contexts, we aim to challenge the prevalent celebratory and emancipatory narrative of the gig economy. In doing so, we follow Anna Tsing's (2015: 3) astute observation that 'the irony of our times [...] is that everyone depends on capitalism but almost no one has what we used to call a "regular job"'. The anthropologist argues, in consequence, for 'thinking with precarity' as a means to engage with critical lines of inquiry not only into the recent transformations in work and employment, but, more generally, into people's livelihoods. Tsing proposes thus 'an appreciation of current precarity as an earthwide condition [which] allows us to notice this – the situation of our world' (2015: 4).

What we are left with is more flexibility for some workers than for others within an otherwise generalized state of exploitation. Some of these couriers are doing deliveries as a side job and are able to maintain relative control over how much they work, when and even where. Others, relying on platforms for a living, are deprived of any control they have over when and for how long they need to be out on the road with their apps turned on. Platforms, for their part, are keen to emphasize and promote the fact that the majority of gigs are side jobs, when this could not be further from the truth. A notable gap between rhetoric and reality characterizes platform discourses (Fairwork 2020). For example, the language used by platforms to describe their relationship with the workers is carefully picked up to displace their responsibility: talking about "disconnection" instead of officially dismissing a worker; or calling the workers "partners", to avoid legal obligations. Platforms' manipulation of language can create frictions with policies and regulations. These frictions might end up protecting workers or, conversely, benefitting the companies.

We aim to expand the notion of precarity beyond the rather narrow confines of economic insecurity, which results primarily from labor market experiences (Burridge and Gill 2017: 26), and starts with the precarity of cyclists as road users. While prevalent uses and understandings of the term "precarious" are linked with anxiety about raw unemployment, there is a more pervasive sense of insecurity and uncertainty suggested by an adjective that has come to describe a deteriorated life condition distributed across regions and social classes (Ferguson and Li 2018). We argue that the precarity of cycle food couriers is intersectional, in the sense that issues of gender, race and migration status further impact on what it means to be precarious.

Precarity has also been applied to the type of entitlement that cyclists have as road users (Egan and Philbin 2021). The three properties that, according to Egan, make the conditions for this precarious entitlement – insecure space, spatial disregard and police neglect – are exacerbated for cycling food couriers. The spaces to which cyclists are entitled and, at times, obliged to use expose them to danger and threats and are perceived as insecure. Due to the rushed nature of their work, couriers are also forced to navigate these spaces fast while carrying bulky bags that make cycling less comfortable. Furthermore, since these cycle spaces are not designed for the use of couriers, other cyclists may feel unsafe sharing already precarious spaces with faster couriers who often overtake them in inadequate spaces such as narrow cycle lanes. This connects with the aforementioned spatial disregard, which refers both to a disregard for a cyclists' space as well as an inconsideration of cyclists within a particular space. The risk of collision with other road users amongst couriers is undoubtedly amplified by the fact that they spend a good part of their working time on the roads.

Finally, the protection of cyclists is neglected. Egan talks about policing, in the sense of a lack of punishment for those creating insecurity or invading cycling space (e.g. parking on cycle lanes, close passes). In the context of cycle couriers, this lack of protection includes other institutions failing to protect the gig cycle workers who cannot defend themselves against harassment or access adequate accident insurance. Additionally, their physical integrity and health have been further threatened by the inadequate hygienic conditions endured during the Covid-19 pandemic.

This chapter draws on ongoing research investigating working conditions amongst platform-based cycle food couriers undertaken in the UK, Spain and several countries in Latin America (Mexico, Colombia, Argentina and Chile). The authors have conducted ethnographic work consisting of participant observation and 15 in-depth interviews in Manchester (involving 13 migrant workers and 4 women) and an analysis of recorded interviews and webinars involving 7 women in unions in Spain, Mexico, Colombia, Argentina and Chile, and a cooperative in Spain. These have been supplemented by data produced by riders, unions and similar collective associations gathered from social media platforms, webinars as well as media representations. Broader contextual data was gathered from existing grey literature and online media outlets.

Gender

The platform economy can provide a source of income for marginalized groups, such as low-skilled or untrained women. For them, and for women who perceive their main responsibility to be unpaid care work, this could be a gateway into the labor market as it offers them flexibility and task variety. However, the transport and delivery sector has been historically male-dominated, and transport-related platforms are not an exception. For example, in Spain only 13% of food delivery riders are women (Adigital 2020), in the UK, 6% of all Deliveroo riders are women (Dupont et al. 2018) and in Argentina they are below 5% (Madariaga et al. 2019). In cities with a low uptake of cycling like the ones analyzed in this chapter, women are underrepresented in cycling mobility; 33% in Manchester (United Kingdom), 35% in Barcelona (Spain), 38% in Mexico City (Mexico), 30% in Rosario (Argentina) 23% Bogotá (Colombia), 17% Santiago de Chile (Chile). More generally, women cyclists are perceived to be less competent because their bodies are not considered to be 'proper cycling bodies' (Aldred 2013). As a minority in this working environment, most women develop strategies of self-protection against hostility, sexual harassment and neglect, such as avoiding socialization in the workplace, which in turn has a negative impact in skill development and earnings. Women riders try to find quiet spaces to be on their own or even cover their face to avoid gender assumptions or direct interaction. Women miss opportunities of getting advice from more experienced male colleagues that could help them be more efficient at their jobs. This can lead to making beginner mistakes while using the apps and consequently getting lower app ratings. A second strategy is for women to come together by organizing online gatherings and social media networks. Women riders in Argentina, Colombia, and Mexico, for example, participated in a webinar with the meaningful title "Platforms don't take care of us, our female colleagues do" (*Las plataformas no me cuidan, me cuidan mis compañeras* 2020).

Doing cycle deliveries exposes workers to public spaces for long periods of time. For women, this means constantly experiencing fears and feelings of vulnerability, which makes them feel unsafe, influences their movement and leads to an avoidance of public exposure. This suggests that, for some women, engaging in cycle delivery work means forcing themselves into hypermobility and having to negotiate it by, for example, choosing to work in familiar and/or local areas.

In the context of cycle platform deliveries, gender inequalities have health, safety, security and economic impacts on women. For example, access to toilets is key to women's health; these are basic hygiene requirements during menstruation (approximately one-fifth of the time for a woman in reproductive age) and pregnancy, and not being able to urinate for long periods of time increases dramatically the risk of urinary infections. It is a basic and frequent need for women to use toilets, which at the same time can expose them to more harassment and abuse, when they need to request restaurants to enter their premises. The Covid-19 pandemic has further limited access for women to these spaces, which have been partially or totally closed, or made riskier to use due to compromised hygiene conditions.

There are features of platform delivery work that expose women to higher safety and security risks, which often require them to go into buildings or to share their personal contact details (full names, phone numbers) that can be misused to stalk and harass them. Protection strategies are a matter of goodwill from clients (e.g. meeting the women riders outside their buildings to collect their food) or depend on women networks (e.g. sharing information about open, clean toilets through social networks). Adding an intersectional approach to Egan's precarious entitlement to the road, women cyclists seem to have a higher exposure to road unsafety. Women report more near misses per hour and per distance unit (mile), with 50% more close passes per distance unit than men (Aldred and Crosweller 2015). We can expect women riders being more exposed to a higher crash risk while doing the same job.

The platform economy not only reproduces the already existing disadvantages for women in the workplace, but also exacerbates them, especially in vulnerable situations and/or intersectional contexts. Women with caring responsibilities who find themselves in situations of financial insecurity or poverty are profoundly affected by the consequences of precarious platform work. Unionized women in Mexico, Argentina and Colombia call the relationship between these vulnerable women riders and the companies "predatory" and "violent", referring to this oppressive source of power as an abusive male (*Las plataformas no me cuidan, me cuidan mis compañeras* 2020).

Women are a minority in delivery platforms, but they have been at the forefront of resistance strategies. Whether it is unionization, peer support groups (e.g. WhatsApp groups), labor movements or cyber activism, women have been involved in all kinds of organized forms of resistance. For example, there are women amongst the co-founders of the organization for the rights of riders "Riders X Derechos" (Riders for rights) in Spain. Women-specific groups have also been created within unions, e.g. the IWGB Women's project in the UK or "Ni una repartidora menos" (Not a woman delivery worker less) as part of the Mexican organization "Ni un repartidor menos" (Not a delivery worker less). Alternatives to platform work in the shape of cooperatives like "Mensakas" in Barcelona are also co-founded by women and have a clear feminist perspective, paying women 5% higher wages to try to compensate for the gender pay gap in the sector. Accordingly, the color of their visual identity is violet.

Ethnicity and migrant status

There is a lack of comprehensive or comparative public data on the precise number of ethnic minority and migrant workers in urban gig economies. The few existing statistics show a high volume of such workers undertaking these jobs, providing an 'infrastructural role' for these platforms, 'one that is as vitally important to their business model's viability as the steady influx of investment capital' (van Doorn et al. 2020: 2).

Amongst food couriers, ethnic minority and migrant workers represent the majority in many European countries. In our case studies, an overrepresentation of migrants and minority ethnic groups can be observed as well. The majority of cycle couriers in British cities have a migrant background and often rely on these jobs as their main source of income. This reality echoes existing research which shows that migrants and ethnic minorities living in the UK are disproportionately represented in precarious jobs, are exposed to higher risks of exploitation and are often willing to tolerate poorer working conditions. The situation is similar in Spain, where 64% of riders are from Latin America, 28% from Spain and the rest from the EU and other countries. In Argentina, they represent an even bigger fraction of the workers, as 65% and 84% of Glovo and Rappi riders, respectively, are migrants.

One's ethnicity can negatively impact access to work. Algorithmic bias is one common experience amongst ethnic minority populations and increasingly visible in the workplace, where automated processes discriminate during hiring and job allocation processes. App renting also represents a prevalent phenomenon amongst couriers with an undocumented migrant status, who pay an average of £50 per week to rent an account in the UK. Renting occurs either because couriers have to wait weeks and months in a row to have their account activated, but also because of their undocumented migrant status. The exploitative nature of this job outsourcing can be compounded by inflammatory anti-immigration headings in the right-wing media.

The migrant and transient status of many app-based food-delivery workers often prevents them as well from expressing collective agency, either because of the aforementioned undocumented status or because they are simply less aware of their rights. The invisibility of app-renters is perceived as an advantage for those who lack the documentation to access formal jobs, but it silences them and leaves them completely unprotected against any risk. The lethal collision suffered by a Nepalese cyclist while renting the Glovo app in Barcelona is illustrative of these risks (Newsdesk/ACN 2019). Within this context, the calls to reclassify cycle couriers as employees rather than self-employed could be interpreted as a barrier for undocumented migrant workers to access these jobs. Exploitative as they are, these jobs nevertheless offer a lifeline to many workers; if these companies are demanded to reclassify their workforce, it is without doubt that this will result in a purge of undocumented migrants from the platform (van Doorn et al. 2020). We argue nevertheless that what these migrant workers ultimately need are systemic policies and procedures that allow them to access a dignified job in the first place.

Concluding reflections

The public and academic debates of the last few years on the nature of the gig economy have focused on the 'self-employed' without much consideration on who this 'self' is. While examining the phenomenon, the critics of platform work, with few exceptions, did not explore in great depth the implications of the use of cycles for work, and the socio-demographic characteristics of these workers, let alone their historical and cultural backgrounds. This chapter is a first step to address these omissions and show the intersectional layers of complexity to what is already regarded as precarious work, which start by the intrinsic precarity of cycling mobility. We have demonstrated that, amongst cycle food couriers, there is a minority of women as well as a majority of migrants whose work experiences are not only hidden from view under a generic 'self-employed' category, but, most importantly, both these under and overrepresented populations are indicative of additional discrimination and exploitation processes at work in the gig economy.

The fact that platform companies do not regularly use images of women and ethnic minorities in their promotional materials is indicative of an effort to conceal and even deny their individual work experiences (Vyas 2020). This chapter has begun to unpack these experiences, showing that women riders suffer multiple levels of oppression from the platforms, their workmates, their clients, and the people they are constantly exposed to while on the road. For ethnic minorities and migrants, on the other hand, the platform work experience is aggravated by algorithmic discrimination, while their undocumented status forces them to rent accounts and remain invisible.

We need situated knowledge in order to tackle these global issues, since both the operation of these companies and the associated worker struggles are global phenomena, albeit with local specificities. We need to draw on research produced in local contexts and process it collaboratively. The perspective from Latin American countries shows that colonial dynamics are still playing a remarkable role in both how oppression and resistance occur. Colleagues from that region make it clear that they want to recover technological sovereignty and manage it according to their local culture, proposing a conservational approach to production, respectful of life and nature, rather than a neoliberal exploitative one. Challenging neoliberalism and conquering technology could work as a way to resist the history of conquest that Latin America has been subjected to (Hidalgo Cordero and Salazar Daza 2020).

Platform delivery work is a fragmented and isolated activity, which makes it difficult for workers to interact and organize. Additionally, companies actively discourage and block any possibilities for communication and networking. Despite this, and the extreme vulnerabilities faced by gig workers, there is room for creative resistance strategies that go beyond traditional forms of collective organization and happen in the streets or on social media.

Acknowledgement

Cosmin Popan's work was supported by the Leverhulme Trust [grant number ECF-2020-523].

Bibliography

Adigital, 2020. *Importancia económica de las plataformas digitales de delivery y perfil de los repartidores*. Spain: Adigital.

Aldred, R., 2013. Incompetent or Too Competent? Negotiating Everyday Cycling Identities in a Motor Dominated Society. *Mobilities*, 8 (2), 252–271.

Aldred, R., 2015. A Matter of Utility? Rationalising Cycling, Cycling Rationalities. *Mobilities*, 10 (5), 686–705.

Aldred, R. and Crosweller, S., 2015. Investigating the Rates and Impacts of Near Misses and Related Incidents among UK Cyclists. *Journal of Transport & Health*, 2 (3), 379–393.

Azevedo, M.A. and Mascarenhas, N., 2019. Colombian On-Demand Delivery Unicorn Rappi Raises $1B From SoftBank [online]. *Crunchbase News*. Available from: https://news.crunchbase.com/news/colombian-unicorn-rappi-reportedly-raising-1b-from-softbank/ [Accessed 4 March 2021].

Bennett, B., 2019. *Cycling and Cinema*. London: Goldsmiths Press.

Burridge, A. and Gill, N., 2017. Conveyor-Belt Justice: Precarity, Access to Justice, and Uneven Geographies of Legal Aid in UK Asylum Appeals. *Antipode*, 49 (1), 23–42.

Cant, C., 2020. *Riding for Deliveroo: Resistance in the New Economy*. Cambridge: Polity.

Crenshaw, K., 1989. Demarginalizing the Intersection of Race and Sex: A Black Feminist Critique of Antidiscrimination Doctrine, Feminist Theory and Antiracist Politics. *The University of Chicago Legal Forum*, 1989, 139–167.

Day, J., 2015. *Cyclogeography: Journeys of a London Bicycle Courier*. Honiton, Devon: Notting Hill Editions.

van Doorn, N., Ferrari, F., and Graham, M., 2020. *Migration and Migrant Labor in the Gig Economy: An Intervention*. Social Science Research Network. Doi: 10.2139/ssrn.3622589.

Dupont, J., Hughes, S., Wolf, R., and Wride, S., 2018. *Freedom and Flexibility. The relationship Deliveroo Riders Have with the Labor Market*. London: Public First.

Egan, R. and Philbin, M., 2021. Precarious Entitlement to Public Space & Utility Cycling in Dublin. *Mobilities*, 16 (4), 509–523.

Fairwork, 2020. *The Gig Economy and Covid-19: Fairwork Report on Platform Policies*. Oxford, UK.

Ferguson, J. and Li, T.M., 2018. Beyond the "Proper Job": Political-economic Analysis after the Century of Laboring Man. https://media.africaportal.org/documents/WP_51_Beyond_the_proper_job_12_Apr_2tl2_FINAL.pdf

Fincham, B., 2006. Bicycle Messengers and the Road to Freedom. *The Sociological Review*, 54 (s1), 208–222.

Fincham, B., 2007. Bicycle Messengers: Image, Identity and Community. In P. Rosen, D. Horton, and P. Cox, eds. *Cycling and Society*. Aldershot: Ashgate, 179–195.

Fincham, B., 2008. Balance is Everything: Bicycle Messengers, Work and Leisure. *Sociology*, 42 (4), 618–634.

Grigera, J., 2020. Futures of Work in Latin America: Between Technological Innovation and Crisis. https://unesdoc.unesco.org/ark:/48223/pf0000374436

Heeks, R., 2019. How Many Platform Workers Are There in the Global South? *ICTs for Development*.

Hidalgo Cordero, K. and Salazar Daza, C., eds., 2020. *Precarización Laboral en Plataformas Digitales. Una Lectura desde América Latina*. Quito, Ecuador: Friedrich-Ebert-Stiftung Ecuador FES-ILDIS.

Huws, U., Spencer, N., Syrdal, D., and Holts, K., 2017. *Work in the European Gig Economy. Research Results from the UK, Sweden, Germany, Austria, the Netherlands, Switzerland and Italy*. Brussels: FEPS, UNI Europa, University of Hertfordshire.

Kidder, J.L., 2005. Style and Action: A Decoding of Bike Messenger Symbols. *Journal of Contemporary Ethnography*, 34 (3), 344–367.

Las plataformas no me cuidan, me cuidan mis compañeras, 2020.

Madariaga, J., Buenadicha, C., Molina, E., and Ernst, C., 2019. *Economía de plataformas y empleo: ¿Cómo es trabajar para una app en Argentina?* Inter-American Development Bank.

Manyika, J., Lund, S., Bughin, J., Robinson, K., Mischke, J., and Mahajan, D., 2016. *Independent Work: Choice, Necessity, and the Gig Economy*. McKinsey Global Institute.

Newsdesk / ACN, 2019. Trade Union Demands Investigation into Glovo. *Spain in English*, 1 June.

Popan, C., 2021. Algorithmic Governance in the Gig Economy: Entrepreneurialism and Solidarity Amongst Food Delivery Workers. *In*: D. Zuev, K. Psarikidou, and C. Popan, eds. *Cycling Societies: Innovations, Inequalities and Governance*. Oxford, UK: Routledge, 239–257.

Srnicek, N., 2016. *Platform Capitalism*. Cambridge, UK; Malden, MA: Polity Press.

Tsing, A., 2015. *The Mushroom at the End of the World. On the Possibility of Life in Capitalist Ruins*. Oxford: Princeton University Press.

Vyas, N., 2020. 'Gender Inequality- Now Available on Digital Platform': An Interplay between Gender Equality and the Gig Economy in the European Union. *European Labor Law Journal*, Doi: 10.1177/2031952520953856.

Woodcock, J. and Graham, M., 2020. *The Gig Economy: A Critical Introduction*. 1st edition. Cambridge; Medford, MA: Polity.

4
THE SOCIALITY OF CYCLING

Simon Batterbury and Alejandro Manga

Introduction

As Peter Cox notes, "... cycling as primary mobility has shaped and still shapes significant numbers of social lives" (2019: 3). Cycling is a social activity as well as a material practice (Popan 2019), and therefore involved in the production of space (Castañeda 2020). The act of cycling fulfils basic requirements – to reach a destination, to travel through the natural and built environment, or to maintain health and fitness. But like the two other major forms of active travel, running and walking, cycling generates and is embedded in various types of social institutions, fashions and trends, and community spaces and campaigns. These include clubs, societies and interest groups dedicated to cycling in all its forms – from the *flâneurs* of the 1890s (Mackintosh & Norcliffe 2006) to contemporary urban "fixie crews" and youth groups (Azzarello et al. 2016), and from the lowriders of Los Angeles to the competitive bike racing fraternity and their teams and organizations (Mundler & Rérat 2018). The material presence of a bike also has symbolic importance, as an object of desire, as a material possession, and as an expression of good technical design (Nurse 2021). For some, bikes are a favored form of transport because they are a harbinger of a low-carbon economy, in which low-impact and active transport modes will become commonplace once again. Cycling has its own social politics – commonly, as a celebration and manifestation of resistance to the culture of automobility and the domination of the streetscape by polluting vehicles.

In this chapter, we explore selected aspects of the social dimensions of cycling. We show that in the many debates about how to reclaim space for bikes and increase ridership, the "sociality" of cycling and the pleasure of riding a bike also has political ramifications. Increasing the modal share of bikes means investing in much more than bicycle infrastructure like bike lanes, road and intersection treatments and bike parking (described in this Companion, Chapter 20). Such transport engineering "hardware" may "attract" more cyclists to the roads, as many bike-friendly municipal planners hope (Pucher and Buehler 2008). Some even argue that we know everything that is needed to promote cycling from this planning-centric perspective (Nello-Deakin 2020). We argue that it is the culture of mobility, boosted by social movements and organizations that create "demand" for cycling, through socialization, enlisting new cyclists, lobbying, and bike-friendly actions. These form the "software" that complements, and builds, a cycling culture. We term this the "bikespace."

DOI: 10.4324/9781003142041-6

The bikespace and the importance of demand for cycling

"Bikespace" is something produced by various communities of cyclists, as well as by bike phenomena and social movements of different forms. These spaces are not just physically anchored ones like velodromes, bike stores or networks of cycle lanes, but also include spaces that are formed through the practice of cycling – in movement – like Critical Masses and social rides. We argue that through the interaction of "kinopolitical constellations" within the bikespace, enhanced practices and socialities emerge that can be used to boost the demand for cycling. Kinopolitical constellations are mobile and have scale. They collapse memories into the present: one of the authors grew up in Bogotá in a time of drug trafficking and violence. He started riding from his house in the suburbs to his school on the newly minted cycle paths, and cycling made him proud of his city, and made him a cycling scholar. This emotional attachment to the bike is always present. Bikes have meanings, bringing a multiplicity of perspectives related to this one object. In the aggregate form, in conjunction with infrastructures and social relations, they form a system of vélomobility (Cox 2019), what the French call, a *système vélo*, what we here consider a local constellation. We also show that while bikespaces are local and rooted in place, the practices and socialities that emerge from them show some similarities across the world, existing in different locations as forms of translocal assemblages (McFarlane 2009).

The idea of a bikespace is based on the representations of space defined by Lefebvre in *La Production de L'espace* (Lefebvre 1974) and by Soja in *Thirdspace* (1996). From their perspective, space is socially constructed. Lefebvre considers that space is defined by a spatial triad: (1) "*l'espace perçu*" is the space we perceive through our senses, supporting the materiality of its elements; (2) "*l'espace conçu*," the conceived space, and its production is intertwined with social conventions. It is an abstract form that is perceived but not lived, dominant among planners and professionals; (3) "*l'espace de representation*" is a representational space of symbols, a semiotic space that includes art and literature, but also the impression of being free while riding. Each one of these forms is related to the production of knowledge, to producing material things and spaces, and the production of meaning (see Chapter 22).

The bikespace combines these three elements as a form of assemblage, or mobility praxis. Although more complex definitions exist, we consider it to be a mobile social field in which bicycles are a focus for human agency, knowledge, institutions and, in places, also attain symbolic power. Practicing bike mobilities means the bikespace forms around us and is potentially transformed by our presence as it interacts with other assemblages. We (re)make space by moving through complex assemblages of mobile knowledge, actions and power. Creating positive bikespaces is integral to mobility justice (Sheller 2018) and to supporting the mobile commons (Nikolaeva et al. 2019).

Bicycle riders, activists, and movements perceive and react to elements of the bikespace, and express it in ways we detail in the next section. Doreen Massey has a useful expression of this interaction: the "articulated moments in networks of social relations and understandings." These allow "a sense of place which is extraverted, which includes a consciousness of its links with the wider world, which integrates in a positive way the global and the local" (Massey 1993: 67). Here, we are particularly interested in constellations of *vélomobilty*, a concept akin to the bikespace and following Peter Cox "… the assemblage of rider, machine, and space and the systemic relations of society, economy, polity and history within which they are performed" (Cox 2019: 27).

Bikespaces take into account the many layers that are often forgotten when thinking about, promoting and researching cycling. In particular, participation within the bikespace is informed by different positions: by race (Lugo 2013; Hoffmann & Lugo 2014; Hoffmann 2016); by gender (Abord de Chatillon 2020); by ability or age (Goodman & Aldred 2018; Inckle 2019; Chapter 7); by passion; by the importance of mobile autonomy; and by the value that some cyclists place on low-carbon transport, and even on degrowth (Rigal 2020). Such identities often overlap and intersect. Each individual who enters the bikespace does so with their own kinopolitical constellation, and this can transform over time: for example, after discovering cycling, then beginning to commute by bike eventually leads to participation in cyclist organizations or networks. Castañeda alerts us to the playfulness and the empowerment that arises from participation in Bogotá's Critical Mass in Colombia (Castañeda 2020; Vivanco Chapter 38 in this volume). Popan describes the different types of sociabilities that arise while riding together and conversing, belonging to a club, or by participating in a carnival as people do when riding together in Critical Mass events (Popan 2019; Nurse 2021).

Examples of enhanced sociality

Cycling has different meanings depending on class, space, time and culture – it has "rolling signification" that changes as society has passed through different epochs, and socioeconomic and cultural spaces (Hoffmann 2016). Héran describes how cyclists and their spatial attachment have changed over the decades in Europe: from the wheelmen of the late nineteenth century who lobbied for the expansions of roads from central urban locations to the countryside, to the working class who cycled from factories to their suburbs in the mid-twentieth century, to the "Bobos" of the early 2000s who rode in central urban locations and adopted alternative lifestyles (Héran 2015). In high mode-share cities like Copenhagen, the bike itself does not stand out, since it is a quotidian and ubiquitous object (Chapter 37). Nonetheless, the city's bike culture has been represented globally and is emblematic (Colville-Andersen 2018). Jordan (2013) tells a similar story for Amsterdam, showing how bikes defined the city over more than a century through war and peace, and survived attacks on access to key routes and spaces by pro-car planners at different historical moments.

The role of culture and identity cannot be overstated. Recent decades have seen affluent, environmentally conscious publics adopt cycling as a marker of progressivism and "wokeness," what Hoffmann and Lugo termed "creative class bait" (Hoffmann & Lugo 2014: 45). Examples in the West include largely white and affluent urban advocates who lobby and campaign for better infrastructure and bike lanes, thereby producing spatial relations in ways that fit their favored image of an urban streetscape (Stehlin 2019). An archetype would be the recent college graduate who lives in a central location in a moderate to high-density city like Portland, Oregon, USA (as represented in TV shows like *Portlandia*). By contrast, in Mexico and parts of the US, stereotypical cycle commuters are low-income men who have no access to motor vehicles (Guerra et al. 2020) – a group inappropriately labeled as "invisible" cyclists (Barajas 2016). Only 2.2% of the wealthiest quintile of Mexican workers commute by bike (Guerra et al. 2020: 8). In Ouagadougou, Burkina Faso, dubbed Africa's "city of two wheels," bicycles and mopeds still predominate after decades of urban expansion and roadbuilding, forming a unique roadside culture (Ouedraogo 1986). Culture, class, intersectional identities and the meaning of cycling vary within and between urban populations, and according to wealth and access to other transport modes.

The French city of Grenoble is the most bikeable in the country for a city of its size (Fédération des usagers de la Bicyclette 2019). This is due to decade-long investments in infrastructure and, more recently, an innovative long-term bike hire scheme (Métrovélo), as well as the presence of a very lively network of community organizations (Rigal 2020).

Yet some residents still feel disenfranchised and excluded from spaces that provide access to cheap and affordable mobility. They include recent immigrants working as food couriers, and some Muslim youth and women (Vietinghoff 2021; pers. obs. 2020). Vietinghoff argues racism exists in this bicycle culture, but is concealed by France's official statistics in which religious background or ethnicity cannot be gathered by law. In New York, restrictions have been placed on food delivery workers who use e-bikes, many of them migrants, while preserving access for publics with more privileged identities (Lee 2018). Race is erased when discussing policies that could inform and improve the demand for cycling in the French-speaking context, while class and racial inequalities are perpetuated in the US (Lugo 2013).

Over forty years in Santiago, Chile, bike movements have promoted cycle inclusion while shaping participatory processes in the city (Sagaris 2015). These movements have: (1) helped shape urban policy; (2) encouraged behavioral change; and (3) promoted what Sagaris calls the cycling economy. Political engagement by cycling movements and the spaces they provide have helped to teach people to ride and learn repair skills, and have provided spaces of sociality that help to build community bonds. Following the September 2017 earthquake in Mexico City, "bikepilling" volunteer crews were organized by the community to bring supplies to collection points across the city as part of the emergency response, and to tend to the wounded and other victims (Cabezas 2017). Barajas (2020) found Spanish-speaking migrants in San Francisco's Bay Area participate in the promotion of a cycling economy. They have lobbied for the development of a bike system that goes beyond hard infrastructure investments, to include services such as bike repairs, shared mobility services, and bike rentals.

Bike cultures have also benefited from short-term changes. During the global Covid-19 pandemic from 2019, cycling has received support, particularly given the greater risks of infection on public transport and the need to maintain fitness during lockdowns. In France and the UK, subsidies were available for bike checkups, reconditioning and repairs (D'Halluin 2020). There was a lack of new bikes to satisfy the demand induced by the pandemic, favoring repair and re-use. Under the French *Coup de Pouce Vélo*, 900,000 bikes were refurbished by November 2020, some by community bike workshops (*ateliers vélo*), and 500 workplaces were supported through government assistance totaling 80 million euros (Fédération des Usagers de la Bicyclette 2021). Temporary "coronapistes" (covid tracks) emerged in France – some of which have turned permanent under public pressure (c. 2020).

Community bike workshops

Community bike workshops, "bike kitchens," or *ateliers vélo* are important bikespaces in cities. In all of them, donated or scavenged bicycles are repaired, and parts are re-used (Figure 4.1). There are at least two types of workshops: (1) those that fix bikes for others – often for disadvantaged populations, with a few even exporting bikes to the Global South, like Working Bikes in Chicago (https://workingbikes.org); (2) those where tools, parts and bike stands are on offer for anybody to use, assisted by workshop volunteers and sometimes by salaried mechanics. The latter mode contributes to *vélonomie*, enabling an autonomous cyclist to ride safely and to maintain a bicycle. Workshops perform social and material functions, and they have grown in numbers, particularly since the 1990s. They can help little-used and abandoned bikes get back on the road much more cheaply than most bike shops, and they can interrupt the waste stream, re-using bikes and parts, for example those collected by authorities, or cleared from storage. They are, therefore, agents in the circular economy. Some secondhand bikes may be sold to support the workshop (to meet its ongoing costs and pay for often-precarious premises), but their ethos is largely not-for-profit (Batterbury and Dant 2019). Workshops exist in the

Figure 4.1 The Récup'R workshop, in Bordeaux, France (February 2020, before relocation).
Source: Simon Batterbury.

liminal space of the alternative economy: they vary from highly organized and rule-driven, to cooperatively managed along consensus or anarchist principles (Batterbury and Dant 2019).

In some countries, particularly in France, workshops have become a social movement, with structure and collective purpose. Bike!Bike! is an annual get-together for community workshops in the Americas (https://en.bikebike.org). L'Heureux Cyclage (LHC, https://www.heureux-cyclage.org) is the national network that includes most French DIY bike coops. Its activities include cycle promotion, by lobbying local and regional governments; guidance for community organizers to develop new bike coops, and popular education programs that teach how to ride bikes and to repair them (LHC 2019). LHC's annual Panorama reports register the growth of workshops, and detail their premises, volume of bikes fixed, personnel and staffing. There were at least 250 workshops registered in France serving 110,000 people in 2019 (LHC 2019: 3). LHC also works with other organizations that promote the circular economy, and a slower way of life based on principles that value justice, community, and self-reliance.

A few studies illustrate the important social dimensions of community bike workshops (Batterbury & Vandermeersch 2016: Rigal 2020). Rigal shows how DIY bicycle workshops become spaces of resocialization, where members (like the authors) can become bike advocates. Bike activists can help to shape policies that give form to the territories they traverse at the local level (Rigal 2020). The P'tit Vélo dans la Tête workshop in Grenoble, one of France's oldest, helped to shape and implement the Métrovélo program, which is managed by metropolitan authorities (pers. obvs. 2020). In cities like Lyon and Melbourne, struggle for the rights of women to work in bikespaces as mechanics, enthusiasts and volunteers has challenged a predominantly male-dominated space, and has involved new organizations, actions, and collaboration (Figure 4.2; Abord de Chatillon 2020).

Figure 4.2 Atelier Pignon sur Rue, Ambilly, France. Expert women mechanics teaching a (male) workshop employee a specific repair (December 2020).

Source: Alejandro Manga.

Critical mass/vélorution and bike events

Critical Masses are social rides where cyclists take over city streets, congregating and then cycling slowly in large numbers, under the slogan "Critical Mass isn't BLOCKING traffic – We ARE traffic" (D'Andrade 2021). From its countercultural origins in the city of San Francisco bike messenger culture during the 1990s, by the mid-2000s it had spread to over 200 cities around the world (Furness 2007: 300). The term in the French-speaking world is *Vélorution* – referring to the major symbolic challenge of tackling automobility as a hegemonic system (Carlsson 2002; Rosen 2004; Furness 2007). In Bogotá, Colombia, the *Ciclovía* involves the closure of 117 km of arterial roads so that people can walk, exercise and cycle on a Sunday, and many other cities have adapted the practice (Figure 4.3; Montero 2017) (see Chapter 38). Significant bikespaces have emerged, and cycling policies have been shaped by these convivial mass cycling events.

For instance, during a Critical Mass in Porto Alegre, Brazil in 2011, an angry car driver crashed into the crowd and injured at least a dozen people (G1 2011). One year later, the first *Foro Mundial de la Bicicleta* was organized as a memorial to the occasion. Some 2,000 riders

Figure 4.3 The first Ciclovía to take place after the first COVID-19 Lockdown, Bogotá, Colombia (September 2020).

Source: Alejandro Manga.

rode a commemorative ride as part of the Foro (Belotto et al. 2014). From its countercultural roots, the Foro has grown to become the biggest cycling event in Latin America. Government officials, academics and planning practitioners have participated in the yearly event, including former New York transportation commissioner Janette Sadik-Khan, academic Peter Cox, and the pro-cycling former Mayor of Bogotá, Enrique Peñalosa (Navarrete 2015). The 2015 event was conducted in Medellín, Colombia and had more than 6,500 participants (Morales and Pareja 2015). Events like these also enhance socialities, and new kinopolitical constellations emerge.

Bike collectives and Critical Mass have been part of larger social protest movements, for example the Chilean *"el estallido social"* protests following an increase of transit fares in Santiago in October 2019 (Trejo 2020). These have spread to campaigns to defend women's rights and against street harassment; against car-based violence; and in favor of a new constitution (Figueroa 2020). This *Revolución Ciclista Plurinacional* has linked diverse social movements, persisting right through the COVID-19 pandemic when mobility was severely constrained (Schüller 2020). In Los Angeles, Adonia Lugo describes the longstanding campaign to increase adoption of cycling, including organizing the first CicLAvia in 2010. These social rides and celebrations of mobility are still taking place today (Lugo 2013; https://www.ciclavia.org). Important networks and new practices are created through such social interactions.

Changing culture is key to altering mobility practices toward more active travel, and, sometimes, to the production of new public space, building on cycling networks. A "commoning of mobility" agenda goes beyond individual bikespaces and campaigns: it involves "... a process that encompasses governance shifts to more communal and democratic forms while also seeking to move beyond small-scale, niche interventions and projects... rethinking the value, meaning, and practice of mobility as a step towards reconfiguring societal mobility regimes in more equitable and environmentally sustainable ways" (Nikolaeva et al. 2019: 353). The governance and control of urban movement and mobilities remain highly unequal and commercialized, and movements to reclaim *vélomobility* are both social and political (Sheller 2018).

A just urban mobility transition begins on the streets and in bikespaces, and less so with the "displacement of poorer households under neoliberal urbanization" from real estate investment and gentrification, in which bike lanes and road treatments are sometimes included (Temenos et al. 2017: 117).

Conclusion

We have made three points in this brief chapter. **Firstly**, we have shown that bicycles form part of an assemblage of social and material processes, together forming the bikespace. This is an all-encompassing term that nonetheless describes the particular constellation of forces that are anchored by cycling, social relations, and mobility spaces. The bikespace and the conviviality and social relationships that bikes enable, can be a window into broader societal issues that involve mobility justice (Sheller 2018).

Secondly, the contemporary desire by urban planners and some political actors to increase cycling mode share, particularly in cities, as part of a drive toward sustainable mobilities and lower carbon forms of transport, is unlikely to succeed without cultural change (Chapter 6). It also requires recognition of diversity and the social life of the bike and its users, complementing and feeding the material changes to the supply of cycling infrastructure described in Chapter 20. Demand-side variables, or the cycling "software" we describe above, are essential tools for engineers, urban planners, educators, and transportation professionals.

Thirdly, we have provided examples of some of the cultural spaces where sociality thrives. Within mobile bikespaces like Critical Mass rides and *ciclovías*, different publics mix together, from politically motivated participants to inadvertent observers, swept up in a collective experience. While cycling generates a degree of conviviality, particularly in the growing phenomenon of bicycle workshops and the sudden interest in bikes bought about by the COVID-19 pandemic, it also takes the form of protest. People are prepared to join together to struggle and fight for safer and less polluted roads, better access to public space, and to battle the culture of automobility that still dominates most western cities and across the global South.

The remaining question for those interested in this heady mix of bicycle assemblage, planning and protest is how best to bring about the sustainable mobility transitions that cities deserve and need, while involving those excluded from urban and transport planning.

References

Abord de Chatillon M. (2020) Feminine Vélonomy: Characterising Women's Experiences of Bicycle Repair and Maintenance within Patriarchal Contexts. In D. Zuev, K. Psarikidou, & C. Popan (eds) *Cycling Societies: Emerging Innovations, Inequalities and Governance*. London: Routledge, 137–155.

Azzarello, P., Pirone, J., & Mattheis, A. (2016) Collectively Subverting the Status Quo at the Youth Bike Summit. In A. Golub, M.L. Hoffmann, A.E. Lugo, & G.F. Sandoval (eds) *Bicycle Justice and Urban Transformation: Biking for all?* London: Routledge, 231–248.

Barajas, J.M. (2016) *Making Invisible Riders Visible: Motivations for Bicycling and Public Transit Use among Latino Immigrants*. Doctoral dissertation. Berkeley: UC Berkeley.

Barajas, J.M. (2020) Supplemental Infrastructure: How Community Networks and Immigrant Identity Influence Cycling. *Transportation*, 47(3), 1251–1274.

Batterbury, S.P.J., & Dant, T. (2019) The Imperative of Repair: Fixing Bikes, for Free. In F. Martinez, & P. Laviolette (eds) *Repair, Brokenness, Breakthrough: Ethnographic Responses*. New York: Berghahn, 249–266.

Batterbury S.P.J., & Vandermeersch, I. (2016) Bicycle Justice: Community Bicycle Workshops and "Invisible Cyclists" in Brussels. In A. Golub, M.L. Hoffmann, A.E. Lugo, & G.F. Sandoval (eds) *Bicycle Justice and Urban Transformation: Biking for all?* London: Routledge, 189–202.

Belotto, J.C., Nakamori, S., Nataraj, G., & Patricio, L.C. (2014) *A Cidade em Equilíbrio: Contribuições Teóricas ao 3º. Fórum Mundial da Bicicleta-Curitiba 2014*. Curitiba: UFPR.

C., J.-P. (2020) Métropole de Lyon. Conserver les coronapistes? Trois Grands Lyonnais sur quatre sont pour, selon un sondage. [online] *Leprogres.fr*. Available at: https://www.leprogres.fr/economie/2020/09/11/conserver-les-coronapistes-trois-grands-lyonnais-sur-quatre-sont-pour-selon-un-sondage [Accessed 17 March 2021].

Cabezas, D. (2017) La bicicleta, clave tras el terremoto de Ciudad de México. [Online] *Ciclosfera.com*. Available at: https://ciclosfera.com/a/terremoto-ciudad-mexico-bicicleta [accessed 7 May 2021].

Carlsson, C. (ed). (2002) *Critical Mass: Bicycling's Defiant Celebration*. San Francisco: AK Press.

Castañeda, P. (2020) From the Right to Mobility to the Right to the Mobile City: Playfulness and Mobilities in Bogotá's Cycling Activism. *Antipode*, 52(1), 58–77.

Colville-Andersen, M. (2018) *Copenhagenize – The Definitive Guide to Global Bicycle Urbanism*. Washington, DC: Island Press.

Cox, P. (2019) *Cycling: A Sociology of Vélomobility*. London: Routledge.

D'Andrade, H. (2021) We are Traffic. [online] *Scorcher.org*. Available at https://www.scorcher.org/cmhistory/traffic.html [Accessed 17 March 2021].

D'Halluin, J. (2020) Vé-lobbying. *Revue Projet*, 5, 14–15.

Fédération des usagers de la Bicyclette (2019) Parlons Vélo: Baromètre des Villes Cyclables. https://www.parlons-velo.fr/barometre-des-villes-cyclables

Fédération des usagers de la Bicyclette (2021) Alvéole Coup de Pouce Vélo. [online] Available at: https://www.coupdepoucevelo.fr/auth/programme [Accessed 17 March 2021].

Figueroa, N. (2020) Revolución feminista en bicicleta: Ciclomarchas contra el acoso callejero y nuevas estrategias de acompañamiento. [online] *El Desconcierto.cl*. Available at: https://www.eldesconcierto.cl/nacional/2020/12/14/mujeres-en-ciclomarchas-vivimos-inseguridad-en-eventos-mixtos-a-causa-del-acoso-de-hombres.html [Accessed 17 March 2021].

Freudendal-Pedersen, M. (2020) Sustainable Urban Futures from Transportation and Planning to Networked Urban Mobilities. *Transportation Research Part D: Transport and Environment*, 82, 102310.

Furness, Z. (2007) Critical Mass, Urban Space and Velomobility. *Mobilities*, 2(2), 299–319.

G1. (2011) Grupo de ciclistas é atropelado em Porto Alegre. [online] Available at: http://g1.globo.com/brasil/noticia/2011/02/grupo-de-ciclistas-e-atropelado-em-porto-alegre.html [Accessed 17 March 2021].

Goodman, A., & Aldred, R. (2018) Inequalities in Utility and Leisure Cycling in England, and Variation by Local Cycling Prevalence. *Transportation Research Part F: Traffic Psychology and Behaviour*, 56, 381–391.

Guerra, E., Zhang, H., Hassall, L., Wang, J., & Cheyette, A. (2020) Who Cycles To Work and Where? A Comparative Multilevel Analysis of Urban Commuters in the US and Mexico. *Transportation Research Part D: Transport and Environment*, 87, 102554.

Héran, F. (2015) *Le Retour de la Bicyclette: Une Histoire des Déplacements Urbains en Europe, de 1817 à 2050*. Paris: La Découverte.

Hoffmann, M.L. (2016) *Bike Lanes are White Lanes: Bicycle Advocacy and Urban Planning*. Lincoln: University of Nebraska Press.

Hoffmann, M.L., & Lugo, A. (2014) Who is 'World Class'? Transportation Justice and Bicycle Policy. *Urbanities*, 4(1), 45–61.

Inckle, K. (2019) Disabled Cyclists and the Deficit Model of Disability. *Disability Studies Quarterly*, 39(4). doi:10.18061/dsq.v39i4.6513.

Jordan, P. (2013) *In the City of Bikes: The Story of the Amsterdam Cyclist*. London: Harper.

L'Heureux Cyclage. (2019) *Panorama des Ateliers Vélo Participatifs et Solidaires*. [online]. https://www.heureux-cyclage.org/panorama-2019-des-ateliers-velo-en.html?lang=fr [Accessed 24 February 2021].

Lee, D.J. (2018) *Delivering Justice: Food Delivery Cyclists in New York City*. PhD Dissertation. New York: CUNY.

Lefebvre, H. (1974) *La production de l'espace*. Paris: Ed. Anthropos.

Lugo, A.E. (2013) CicLAvia and Human Infrastructure in Los Angeles: Ethnographic Experiments in Equitable Bike Planning. *Journal of Transport Geography*, 30, 202–207.

Mackintosh, P. G., & Norcliffe, G. (2006) Flâneurie on Bicycles: Acquiescence to Women in Public in the 1890s. *Canadian Geographer/Le Géographe canadien*, 50(1), 17–37.

Massey, D. (1993) Power, Geometry and a Progressive Sense of Place. In Bird J., Curtis B., Putnam T., & Tickner L. (eds) *Mapping the Futures: Local Cultures, Global Change*. London: Routledge, 59–69.

McFarlane, C. (2009) Translocal Assemblages: Space, Power and Social Movements. *Geoforum*, 40(4), 561–567.

Montero, S. (2017) Worlding Bogotá's Ciclovía: From Urban Experiment to International "Best Practice". *Latin American Perspectives*, 44(2), 111–131.

Morales, P., & Pareja, D. (2015) Medellín muestra apuesta por la bici en Foro Mundial de la Bicicleta. [online] *El Tiempo*. Available at: https://www.eltiempo.com/archivo/documento/CMS-15303015 [Accessed 17 March 2021].
Mundler, M., & Rérat, P. (2018) Le Vélo Comme Outil D'empowerment: Les Impacts des Cours de Vélo Pour Adultes sur les Pratiques Socio-Spatiales. *Les Cahiers Scientifiques du Transport*, 73, 139–160.
Navarrete, A.M. (2015) Pedaleando hacia el Foro Mundial de la Bicicleta de Medellín. [online] *Ciclosfera.com*. Available at: https://ciclosfera.com/a/pedaleando-hacia-el-foro-mundial-de-la-bicicleta-de-medellin [Accessed 17 March 2021].
Nello-Deakin, S. (2020) Environmental Determinants of Cycling: Not Seeing the Forest for the Trees? *Journal of Transport Geography*, 85, 102704.
Nikolaeva, A., Adey, P., Cresswell, T., Lee, J. Y., Nóvoa, A., & Temenos, C. (2019) Commoning Mobility: Towards a New Politics of Mobility Transitions. *Transactions of the Institute of British Geographers*, 44(2), 346–360.
Nurse, S. (2021) *Cycle Zoo: Bikes for the 21st Century*. Melbourne: Silverbird Publishing.
Ouedraogo, I. (dir.) (1986) *Ouagadougou, Ouaga Deux Roues* (film). Ouagadougou: IDHEC. Available at: https://www.youtube.com/watch?v=Dt_Y_Ucz8F4 [Accessed 5 May 2021].
Popan, C. (2019) *Bicycle Utopias: Imagining Fast and Slow Cycling Futures*. London: Routledge.
Pucher, J., & Buehler, R. (2008) Making Cycling Irresistible: Lessons from The Netherlands, Denmark and Germany. *Transport Reviews*, 28(4), 495–528.
Rigal, A. (2020) *Changer la Vie dans un Atelier d'Autoréparation de Vélo*. Forum Vies Mobiles.
Rosen, P. (2004) Up the Velorution! In Eglash R., Croissant J.L., Di Chiro G., & Fouché R. (eds) *Appropriating Technology: Vernacular Science and Social Power*. Minneapolis: University of Minnesota Press, 365–390.
Sagaris L. (2015) Lessons from 40 years of planning for Cycle-Inclusion: Reflections from Santiago, Chile. *Natural Resources Forum*, 39, 64–81.
Schüller, P. (2020) Masiva cicletada en Santiago a favor del Apruebo y contra la muerte de pedaleros – La Nación. [online] *La Nación*. Available at: http://www.lanacion.cl/masiva-cicletada-en-santiago-a-favor-del-apruebo-y-contra-la-muerte-de-pedaleros/ [Accessed 17 March 2021].
Sheller, M. (2018) *Mobility Justice: The Politics of Movement in an Age of Extremes*. London: Verso.
Soja, E.W. (1996) *Thirdspace: Journeys to Los Angeles and Other Real-and-Imagined Places*. Oxford: Blackwell.
Stehlin, J.G. (2019) *Cyclescapes of the Unequal City: Bicycle Infrastructure and Uneven Development*. Minneapolis: University of Minnesota Press.
Temenos, C., Nikolaeva, A., Schwanen, T., Cresswell, T., Sengers, F., Watson, M., & Sheller, M. (2017) Theorizing Mobility Transitions: An Interdisciplinary Conversation. *Transfers*, 7(1), 113–129.
Trejo, C. (2020) Las protestas en Chile hacen la revolución en dos ruedas. [online] *Sputniknews.com*. Available at: https://mundo.sputniknews.com/20200117/las-protestas-en-chile-hacen-la-revolucion-en-dos-ruedas-1090158879.html [Accessed 17 March 2021].
Vietinghoff, C. (2021) An Intersectional Analysis of Barriers to Cycling for Marginalized Communities in a Cycling-Friendly French City. *Journal of Transport Geography*, 91, 102967.

Vignette A
BLACK CYCLISTS MATTER
Major Taylor – Au Parc des Princes 1901

Andrew Ritchie

The background to the photos of Major Taylor (Figure A.1) and Taylor with Edmond Jacquelin during their well-advertised races in Paris in May 1901 (Figure A.2) was a crescendo of publicity, orchestrated by managers, publicists and company promoters. Through 1899 to 1901 and after, the American and French press was constantly discussing whether Taylor, as a world champion, would come from America to meet the Europeans. Photos of him as American national champion were published close to those of the French champion.

When I first started digging into Taylor's visit to France in the Parisian papers in 1899–1901, the coverage of a newly dominant sport had become huge. I believe no black athlete had ever occupied such media attention (Ritchie 2010, Chapters 9 and 10). Evaluation of the qualities of the two champions was conflated with national and especially racial identities.

Taylor's race had made him the object of ferocious racial opposition within the sport – the League of American Wheelmen had in fact declared a ban against black members, all of which he faced with remarkable dignity. Since he had entered the sport as a professional in 1896, Taylor had been the object of fierce, often physical, opposition, frequently baulked and physically confronted on the track. Who would win in Paris, the now-famous black man or the established white champion? What would the result say about the relative qualities of black and white athletes, and therefore of the fight between the races? There had been very little testing of this question.

The races took place in the *Parc des Princes*, a velodrome-cum-stadium created in 1897 where many important French sporting events took place. The crowd of about 20,000 at the first of the races on 16 May 1901 showed the extremely high level of interest in the question of cyclists in actual competition – which makes these photos published in a prominent sporting/social newspaper especially significant.

The use of the photo of the two rivals pursued by a flock of press photographers as a cover in *La Vie Illustrée* – a nationally known weekly sports and news outlet – expresses a lot about this event as a "moment" of cycling history – a 19th-century "black lives matter" moment if ever there was one. The photo was taken by Jules Beau (see Figure A.2), while other pioneer Parisian news and sport photographers can be seen in the background also publicizing the moment.

DOI: 10.4324/9781003142041-7

Figure A.1 Major Taylor at the Buffalo Velodrome during his comeback in 1908.

Source: Author's collection.

A Google search will show the importance of "the photographic moment" not just for bicycle racing, but for all the other emerging team and individual sports, with frequent competition making it possible to earn substantial amounts in the illustrated press. Actually, cycling was a test case of the expansion of this mode of commercial photography.

The publication of the first edition of my book (Ritchie 1988), had a major impact on awareness of Major Taylor as an important figure in cycling and in the history of US sport, leading to discussion of a film biography. As a youngster (aged 13/14/15) he stood alone, and even after becoming a functioning professional (post-1896) he endured constant harassment, racism and physical interference. He was attacked physically on the track many times. Since the publication of my book, several other authors – recognizing the importance of Major Taylor's achievements – have followed suit. A century later his achievements are being widely recognized with, for instance, over 30 Major Taylor cycling clubs founded in the USA, and many talented black racers taking to the saddle.

Note: A Google search for Jules Beau will yield a biography-level website. He is a recently well-researched early news photographer. The *Bibliothèque Nationale de France* has an exhaustive collection of Beau's original work bound into albums, many of which gave information on Taylor's career. It was through familiarity with this material that I have come to think of Beau as THE 'pioneer' sports photographer.

Figure A.2 Major Taylor and Edmond Jacquelin enter the Parc des Princes followed by press photographers prior to their first race 16th May 1901.

Source: Author's collection.

Bibliography

Ritchie, A. (1988) *Major Taylor: The Fastest Bicycle Rider in the World* (San Francisco, CA: Van Der Plas Publications).

Ritchie, A. (2010) *Major Taylor: The Fastest Bicycle Rider in the World*, 2nd ed. (San Francisco, CA: Van Der Plas Publications).

5

PROGRAMS FOR CYCLING INCLUSION

Angela van der Kloof

Cycling presents a strange paradox. While it theoretically offers a cheap and easily accessible mode of transport, in practice the cycling mobility system is not readily accessible to everyone. Fortunately, many initiatives worldwide work toward an inclusive cycling mobility system. This chapter first explains why cycling inclusion is important in the understanding of cycling and society. Then it presents a framework through which programs and initiatives working on cycling inclusion described in the third part can be understood.

Why cycling inclusion matters

At a personal level, it is important for individuals to have access to cycling as a healthy and inexpensive way of moving around. At a societal level, cycling has moved from an alternative to the automobile, toward an intrinsically important and serious mode of transport. Increasingly, cities and countries in the global North as well as in the South have started to invest resources in cycling planning, policies, physical infrastructure, services, and promotion. The Covid-19 pandemic has accelerated this process. When public funds are made available to invest in cycling mobility, it is important to develop this system in an inclusive way. Working toward an accessible cycling mobility system means that on a societal level, resources spent should provide the highest possible overall benefit.

To understand the term cycling *inclusion*, we need to be aware of processes of cycling *exclusion*. The starting point here is, that cycling is not *intrinsically* inclusive, despite claims easily made by cycling advocates and policy makers who, in their enthusiasm, are unaware of the paradox above. It is challenging for advocates and professionals to comprehend the challenges facing people who experience barriers preventing them from cycling. Many have ridden since they were children and have always had a bicycle at their disposal, so find it difficult to imagine not being able to ride a bicycle or having to prioritize buying food for the family over acquiring a cycle. Advocates with access to the cycling system may find joy, even identity, in it, but it is important to understand that the same system has exclusionary facets, invisible to those not affected.

We need, therefore, to ask questions to groups that generally experience issues accessing public goods and, specifically, transport. What are the barriers facing particular groups when they think of taking up cycling? What barriers make them quit cycling? What barriers make

DOI: 10.4324/9781003142041-8

people think that cycling is not for them in the first place? Going deeper, to which social processes are these barriers related? How can they be overcome, and who are the best actors to make this happen?

The good news is, that these questions are not new: many initiatives and programs deal with specific barriers for specific target groups. In many cases, these arise from grassroots action, based on the identification of concrete needs and ideas or experiments on how to meet them. Before delving into examples, I offer a framework to understand different kinds of barriers that groups can experience, and the relationships between those kinds of barriers. It is designed to assist thinking about different types of barriers surrounding cycling as a mode of transport, for daily trips as well as for recreational purposes.

The cycling participation ladder

We cannot assume that cycling is easy or self-evident for everyone. There are, for example, people with health problems, or those who do not know the possibilities of new types of bicycles or cannot afford them (see Chapter 13). There are adults who have never cycled before, or who have lost familiarity since driving and feel uncomfortable on a bicycle. Although cycling can be a relatively cheap mode of transport, there are people living under precarious conditions for whom bicycle ownership and repair is unaffordable. For others, cycling is simply incomprehensible as a personal choice.

To make cycling attractive and achievable for all these groups, requires approaches that differ from more general cycling promotion campaigns. Positive messaging about cycling is by itself inadequate. Alongside good basic information, concrete opportunities to learn new skills and opportunities to overcome mental blocks are required. The *cycling participation ladder* is an instrument used to interpret the myriad of initiatives to make the cycling mobility system more inclusive. To do this, it divides similar types of barriers that people experience into three groups. Simultaneously, it scores the extent to which people experience that type of barrier to place them on a ladder, offering a visualization of the possibility of small progressive steps alongside the level of progress in different types of cycling environments.

The theoretical basis for the types of barriers used in our cycling participation ladder stems from Kaufmann's discussion of motility (Kaufmann et al. 2004; Flamm and Kaufmann 2006). Motility is used to discuss mobility potential in relation to the activity locations that people can potentially reach. The motility of a person is the potential mobility this person has, determined by characteristics of the geographical environment and the social context. A persons' motility consists of three interrelated elements: competences, access and embrace. (Note that, originally, Kaufmann uses the term *appropriation*. This is less useful in the context of practical interaction, hence the substitution of *embrace*.)

Level of *access* to the cycling system is the first element. It deals, for example, with access to physical infrastructure fit for cycling like cycle lanes, low speed neighbourhoods, and well-lit streets (Chapter 20) and access to a cycle that fits the needs of the user and the type of trip – size and model of the cycle, e-assist or not, child seats and panniers etc. (Chapters 8, 9 and 12). The second element is *competences*; this is the level of knowledge and skills that potential users have of the cycling system. It includes physical cycling skills, traffic knowledge, knowledge of the environment and the ability to remember routes and find one's way. The third and last element is *embrace*; the degree to which the potential users have made the cycling system mentally their own. It means that a person or group can think "Cycling suits people like me" and have concrete reasons and ideas when and where to use the cycle. In short, capacity to choose cycling for transportation requires three elements: ability to cycle in local traffic; availability of a suitable, working bicycle; and the skills to use it. Absence or poor development of any one of these elements limits the motility (the potential mobility) of this (group of) person(s), making the cycling system less inclusive.

Scoring for the different barriers is then applied to the ladder. A person's score on the ladder can change. Over time, after having gone through an intervention or experience, the same person can score on a higher (or lower) rung for the same element. The higher on the ladder a person stands, the more included they are in the cycling system. And the fewer people at the bottom of the ladder, the more inclusive the cycling system.

To make the idea of the ladder even more concrete, we zoom in to the element of competences and, more specifically, physical cycling skills and traffic skills. Someone who is at the bottom of the cycling participation ladder cannot cycle at all. Someone who is at the top of the ladder can cycle confidently in daily traffic. In between there are several steps of becoming more skilled, knowledgeable, and confident and each step is a new rung on the cycling participation ladder. Figure 5.1 elaborates this idea.

Each successive intermediate step in Figure 5.1 'can cycle on a closed terrain', 'can cycle in a park' and 'can cycle in a quiet street', requires increasingly stronger skills and knowledge (both tacit and codified). Recruits can work by themselves, with the help of family and friends, or enrol in a variety of programs to climb the ladder. The other two essential elements of cycling mobility (*access* to the cycling system and *embrace* of the cycle) also have intermediate steps. The ladder offers a constructive way of understanding how groups of people, for whom the use of the cycle is less obvious, are supported and engaged by grassroots initiatives and more formal programs.

The metaphor of the ladder is not intended to give the impression that programs are either only working one element at a time, or that this should be done separately. The three elements are interrelated. While an initiative may be intended to focus on one or two of the elements, there will always be an effect on all three elements.

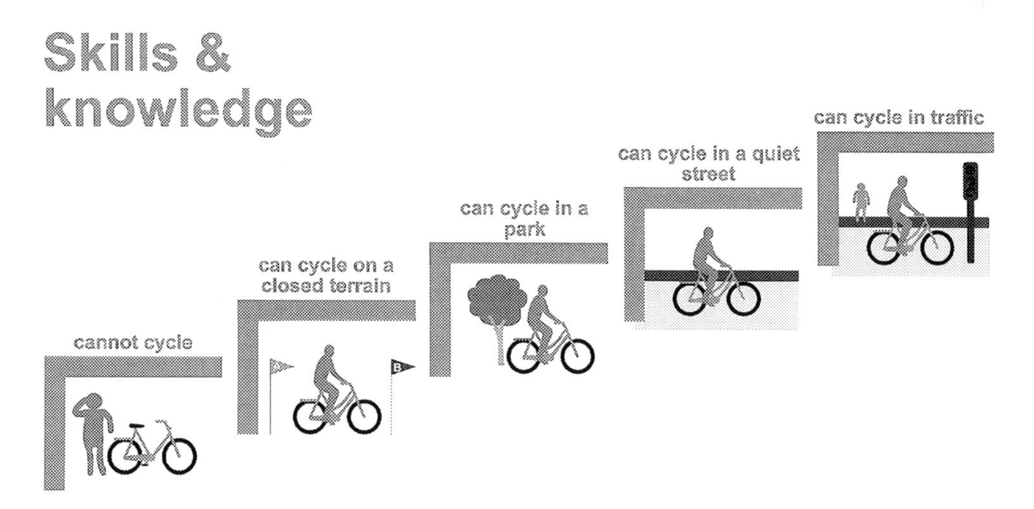

Figure 5.1 The cycling participation ladder elaborated on physical cycling skills and traffic skills.

Source: Author.

Examples of initiatives and programs working toward cycling inclusion

A myriad of initiatives and formal and informal programs aim at promoting an inclusive cycling system and culture. The participation ladder allows us to group them according to type of activity, allowing presentation of a variety of examples from different locations without trying to produce an exhaustive list. It is simply impossible to make a complete list of the hundreds of (hyper)local initiatives and programs around the world. Such a database would need a weekly update: constant changes result from fluctuations in resources and opportunities. We therefore remind the reader to keep in mind that this overview is meant as a stepping stone that showcases a range of possibilities of initiatives.

Since we use the cycling participation ladder, the first three main groups of examples represent the three elements that together determine the cycling motility of a person or group of persons: access, competences, embrace.

Access

The level of access to the cycling system deals with access to different types of resources. Each represents another rung on the cycling participation ladder. The first *access* rung is to physical infrastructure intended for cycling for everyone. A dedicated network connecting origins and destinations meaningful for a broad group of inhabitants providing physically separated cycle lanes and low speed (and low traffic volume) streets in combination with safely designed junctions, tunnels and bridges, forms a good base level for access to cycling for many. The following examples show how cities and towns begin to gradually rebuild their physical and social cycling infrastructures to address the needs of a broader population.

Numerous local initiatives address specific safety concerns, in a mix of traffic safety and personal security. For example, streets in front of a school can have temporary restrictions on motorized traffic at school drop-off and pick-up times (schoolstreets.org.uk). They provide children and parents with the opportunity to arrive at the school gate without the usual traffic chaos caused by parked cars and through traffic. In the rural municipality of Halderberge in the Netherlands, youngsters often need to cycle to secondary school over long distances (>10 km). To offer them extra security a network of voluntary safe havens has been established (*Veilig Honk*, halderberge.nl/project-veilig-honk). At each designated location a youngster can ring the doorbell and ask for help. A third kind of intervention addresses year-round safe and secure cycling. Snow or ice on routes presents an obstacle to cycling, while darkness discourages many people from riding if it is not dealt with properly. The Finnish city of Oulu is exemplary for winter maintenance of the cycling network (Beaulé and Evans 2020) while the Dutch city of Zoetermeer has developed a Night Cycling Network, offering a network of well-lit routes throughout the city (Mobycon 2016; nachtnetfiets.nl; compare Lucia 2020). The Fietsersbond (Cyclists' Union) in Utrecht commissioned a research project (*Meefietslijn*, The Ride-Along Phone Line) to understand and seek possible solutions to address the feeling of insecurity felt by people riding in the evening (Brink 2020).

The second material dimension of *access* is the cycle itself. More specifically, a cycle that fits the needs of the user and the type of trip. Bike share programs generally aim at giving access to cycles in certain locations (see Chapter 23). Several initiatives aim to make bikeshare more accessible for groups that face more (persistent) barriers. One such an example is offered by the Community-Based Bike Share System executed by *Equiticity* in Chicago, US, which targets Afro-American communities (Equiticity.org). Another example is *Bikes for All* in Glasgow

(COMOUK 2021). This is the UK's first social equity bike share project, combining reduced rate access with group rides, city navigation, road skills sessions and family rides. Most bikeshare systems are not designed with families and (young) children in mind and lack the options to rent children's bicycles or attach accessories like child seats, trailers, or panniers. An exception is *Mini Bicicletar* found in Fortaleza, Brazil, a bikeshare system for young children (bicicletar.com.br). Another example can be found in Belgium where several *Fietsbieb* (children's bicycle library) initiatives offer a long-term lending system that adapts the sizes of the cycle throughout the childhood years. This service is typically offered for a low fee to low-income families (Fietsberaad 2015). An example of a bikeshare scheme of adaptive bicycles aimed at people with a variety of physical disabilities is *Biketown* in Portland, US (S. Cox 2018) (see also Chapter 13).

Carry me Bikes in London, UK (carryme.org.uk) lets families try different kinds of cycles. A Covid-19 response from the French national government, the *Coup de Pouce Vélo* (Bike Boost) offered a voucher worth up to 50 euros per person toward the repair of a cycle (Coulon 2020; Ortar and Adam 2021). Other programs specifically provide for the elderly, offering an opportunity to ride types of cycles that may suit their bodily needs better than regular cycles. Examples are the Dutch initiative *Van 2 naar 3* (From 2 to 3), an initiative that introduces people to the tricycle (Fietsersbond n.d.) and *Cycling without Age*, which started in Denmark and is now represented in 51 countries, offering trishaw tours for the elderly (cyclingwithoutage.org). Lastly, numerous projects provide low-income families or individuals with a (re-cycled) bicycle. Examples include *The Bike Project* in London, UK (thebikeproject.co.uk), *Coop-Africa* in Kenya and Uganda (coop-africa.org) and the *Bicycling Empowerment Network* (BEN) in South Africa and Namibia (benbikes.org.za, bennamibia.org). Similarly, the *Give a Bike Foundation* in Sudan offers bicycles to people; in this case in combination with bicycle lessons for women (giveabike.nl/project-sudan). The focus of these projects is to give targeted groups access to a cycle through personal ownership. This category also includes numerous initiatives and organizations that focus on access to cycles for children (van der Kloof and Kurz 2021), such as the Dutch Kinderfietsenplan initiated by the ANWB (the Royal Dutch Touring Club) (anwb.nl), Aromeiazero in Sao Paulo, Brazil (aromeiazero.org.br), Dr Cranky's in Melbourne, Australia (drcrankys.com.au), and the Cycle Program in Bihar, India (Muralidharan and Prakash 2017).

Competence

The second element of cycling motility is the level of competence, based on the knowledge and skills that potential users have of the cycling system. One type of knowledge needed deals with the cycle itself. Initiatives to build technical and practical knowledge of cycles are often aimed at learning or providing maintenance and repair. Examples are *Solicycle* in Ile-de-France (solicycle.org) and innumerable *Bike Kitchens* worldwide that run workshops and provide people with the tools to start maintaining and repairing their cycles (bicyclekitchen.org, velokitchen-dortmund.de; see also Zapata Campos and Zapata 2017; Batterbury and Dant 2019; Lin 2019). Several Bike Kitchens have specific women's hours (such as those at the *London Bike Kitchen* lbk.org.uk/pages/women-and-gender-variant-wag-night), in other places a repair initiative may be initiated by and focused on women only, or with a Queer focus like *Dynamonik* in Anderlecht, Belgium (facebook.com/AtelierDynamonik). Organizations like *The Wrench* (thewrench.ca) in Winnipeg, Canada, take it a step further and help youngsters build their own cycles from recycled parts.

Other competences cover physical cycling skills, combined with traffic knowledge. Bicycle lessons for adults have existed since the invention of the bicycle. In the early years cycling schools were visited by people who could afford to learn new skills for leisure in their spare time. Since the 1980s local initiatives have started to offer lessons for groups that experience mobility poverty, like migrant and refugee women in the Netherlands (van der Kloof and Bek 2019). Many similar examples for women or migrants and refugees can be found in cities around the world, including *Macleta* in Chili (facebook.com/Macletas, Sagaris 2009), *Bike Bridge* in Freiburg, Germany (bikebridge.org), and *Cycle Sisters* in London, UK (cyclesisters.org.uk). This latter initiative combines training sessions ("inspiring and enabling Muslim women to cycle") with group rides as a "supportive, sociable, safe space to develop your cycling skills, explore your local area and make new friends!," a combination also seen at *Bikeygees* in Berlin, Germany (bikeygees.org).

A range of initiatives and programs build children's skills and knowledge and ensure that they do not depend solely on the enthusiasm, time and skills of their parents.

The *Cornerstones* programme in Washington, DC (Austermuhle 2015) is an example of schools-based training. Physical education and health teachers are taught cycle training so they can deliver group lessons in school. In the UK, Bikeability, (bikeability.org) uses external trainers to provide guest lessons in schools. Teaching cycling on school facilities is best delivered using materials props such as traffic cones and signs. Initiatives, like the *Mobiel Verkeerspark* (Mobile Traffic Park) in Belgium (verkeeropschool.be), offer these to schools for a dedicated period. Outside schools, organizations focused on cycling sports offer cycling skills training to children. The *BMX Fietsschool* in Rotterdam, the Netherlands (https://dansh.nl/bmx-fietsschool), is not aimed at BMX sport *per se*, but at increasing children's independence and self-sufficiency in traffic in neighborhoods where there is no strong cycling culture.

Once children know the basics, they would ideally have parents who can ride with them to create opportunities to build experience and, through that, confidence and understanding of expected behavior in traffic. Where this is not the case, *Cycle Bus* initiatives (aka cycle train, bike bus, and bike train) group children together to cycle at specific times (see, for example, greenschoolsireland.org/travel/cyclebusnetwork, galwaycyclebus.weebly.com). These are typically organized by and for parents of schoolchildren who want to drop off and pick up the children by cycle. Children join or leave the cycle bus at various points along the route.

Parents of children not old enough to cycle may lack knowledge of solutions to transport them. The German website *Radfahren mit Baby* (Cycling with Baby, radfahren-mit-baby.de) was developed as part of a research project on cycling during pregnancy and with a baby. It offers information to reduce and eliminate fears of family cycling with babies and young children.

The last target group for *competence* interventions are the elderly. Training may not seem that necessary for those who have cycled their whole life, but *Cycle On*, a large-scale project in the Netherlands, exists to help make elderly cyclists aware of traffic safety issues linked to aging (doortrappen.nl). These range from adjustments to the cycle itself (lowering the saddle, adding a mirror, changing to a step-through cycle), to riding style and familiarity with new types of infrastructure. An important key in this program is that learning new cycling safety facts and personal strategies relevant to the older rider are consciously combined with *embrace*, the third element of motility. Instead of stressing the dangers of cycling, messages highlight the health and social benefits of an active lifestyle.

Embrace

Embrace is about making the cycling system yours, mastering cycling, feeling positive about it, knowing that cycling fits with your identity and image and having reasons to ride and destinations to ride to. Many examples already mentioned contain aspects that deal with this element. Learning physical cycling skills, or bike mechanics can go hand in hand with showing that these really belong to the learner. Yet programs can still take an exclusive approach, consciously or unconsciously, by, for example, only rewarding the best-performing learners, having a lack of diversity in their staff, or not paying enough attention to those who need more time for an adapted approach for learning.

It may also be that a non-inclusive image of cycling is promoted. Here we stumble into the issue that cycling (a vulnerable and often-stigmatized activity) intersects with gender, age, race, migration background, and ability, addressed in Chapters 2, 4, 5 and 7. Current images of cycling in a certain location may be one of the *causes* for people not to cycle: they either cannot or do not desire to fit to the projected images and associated identities. Programs with a focus on the embrace-element to create an inclusive cycling system typically address current images of cycling and work to find ways to change the status quo of what cycling *is* and what it *means* to people. Working on an inclusive cycling system involves finding answers provided by groups in society excluded from access to public goods. In other words, where elements of *competence* and *access* tend to work on 'fixing people' (removing personal barriers to cycling) the embrace-element typically contains a 'fixing the system' approach. It is about fixing the system from the perspective of specific groups currently excluded from cycling, because the dominant image of cycling is a white, young, well-off, able-bodied male activity. The following examples highlight initiatives or organizations that address dominant images from either a specific or an intersectional perspective.

Many initiatives address the idea of cycling as a masculine activity. Some focus on girls and young women to encourage them to take up cycling or keep on riding in their teens. In 2009 the *Beauty and the Bike* project asked young women in both Darlington, UK, and Bremen, Germany, why so many of them didn't cycle (Wupperman and Grassick 2009; Newcastle Cycling Campaign 2018). In Darlington, cycling being "not cool" was a barrier, yet in Bremen cycling was perceived as cool and easy. Consequently, they gave the women in Darlington Dutch-style bikes for a token fee of £1 per week and an exchange between the women in the two cities was organized that was documented to affect an attitude change. In Ireland, the national Green-Schools Travel campaign *#andshecycles* supports girls to get back on their bikes, at the same time addressing reasons why the girls are reluctant to cycle (greenschoolsireland.org/andshecycles-campaign). *Black Girls Do Bike*, in Verona PA, aims to grow and support a community of women of color who share a passion for cycling. They "actively share positive images of ladies and their bikes to affirm the truth that black girls do indeed bike!" (blackgirlsdobike.org). Other initiatives put women center stage. In Barcelona, Spain, data on the barriers faced by women were collected and a manifesto for a feminist cycling city drawn up, signed and handed over to local government (Col·lectiu Punt 6 2020, sindominio.net/bicifeminista). Another Spanish collective initiative is *Cicliques*, comprised of women, lesbians and trans women who unite feminism and cycling, and create safe spaces to promote the bicycle as a tool for social transformation (cicliques.net). In the *Women on Bicycle* Initiative (Turkey), rides were combined with story-sharing and support (Cojocaru 2019) (Vignette K). It resulted in the documentary *My Bicycle, My City*, telling the personal stories of women who own a bicycle and use it in their daily lives. The *Fancy Women Bike Ride* originated as a one-off, World Car Free Day initiative in Izmir, Turkey, to demand space for inclusive cycling (suslukadinlarbisikletturu.com). It has since grown to an annual, global women-only cycling event. Similarly, *Kidical Mass* originated in the US and is now held in many cities in North America and Europe (facebook.com/KidicalMass, kinderaufsrad.org).

As the *Kidical Mass* shows, gender issues are not the only excluding factor. BikEquity (bikequity.org), for example, focuses on people of color (Bhargaw 2021). The *Community Bike Hub* in Canada addresses the image of cycling as an urban means of transport and brings it into rural communities (Ledsham and Verlinden 2019), and *Wheels for Wellbeing* in the UK aims to remove barriers to cycling for disabled children and adults (wheelsforwellbeing.org.uk).

Initiatives not only address the dominant and excluding image of cycling, but also develop strategies for change within the organizations and the professional field. This is about changing routines and offering new ways of organizing. In Berlin, Germany, there is a women's network as part of the city's cyclists' union (ADFC-Berlin.de). The initiator wondered why there are so few women working (professionally and voluntarily) in the organization and what could be done about it. She set up a network that meets regularly and organizes events. Something similar happened on an international scale, where *Women in Cycling* has been set up as an initiative with the vision for a diverse, inclusive cycling sector that provides equal opportunities and contributes to achieving cycling's full potential (cyclingindustries.com/wic). In Spain, the Cyclists Union Conbici, created an *equity manual* for their member cycling associations (Conbici n.d.) and in the US, the League of American Bicyclists works on their Equity initiative (Lugo, Murphy and Szczepanski 2015).

Conclusion

Myriad formal and informal programs aim at an inclusive cycling system and culture. The diversity of programs shows that there are multiple exclusionary processes that need to be tackled for cycling to be a truly inclusive activity. What the initiatives have in common, is that they move away from the assumptions that everyone can already use a cycle, or that access to cycling is easy and that cycling is an attractive activity for everyone. Either by providing people with skills, access, and motivation, or by working on structural changes to the system and using the cycle as a tool for social change, they address gaps in current provision. That so many are autonomous, self-organized initiatives is important. They show that, apart from building physical cycling infrastructure for those who already cycle, there is a need to provide physical cycling infrastructure attractive to those who do not currently cycle, as well as a need to work in the technical and social cycling infrastructure needed to make it a welcoming and achievable activity for everyone.

The relationship between these programs for change and academic cycling studies is an interesting one. Since the emergence of studies in cycling and society as a recognisable phenomenon, there has been considerable cross-over engagement: many of those engaged in academic analysis have a background in, or connection with, forms of advocacy. Yet the programs discussed here are visible mainly in the grey literature. Consequently, many of the tools of analysis as shown here are developed at the interface of academia and advocacy. Adaptive, rapidly changing initiatives, capable of quick response to alterations in local conditions, are a difficult topic for academic analysis, often working to a much more glacial timescale. Without recognition of these initiatives, little understanding can be made of the broader patterns of citizen engagement and agency. Whereas conventional models of analysis are accustomed to imaging social change through institutional modes, the programmes presented here demonstrate ways in which institutions of governance and administration are rarely initiators of change but more often brought in as partners to follow grassroots actions and agency.

References

Note: Where intext reference is to a generic organizational website, no separate listing is necessary here. Specific sub-pages and projects are listed normally

Austermuhle, M., 2015. D.C. Looks To 'Cornerstones' To Bring Education Equity To Public Schools. WAMU American University Radio https://wamu.org/story/15/08/24/dc_looks_to_cornerstones_to_bring_education_equity_to_public_schools/ (Accessed 23/08/2021).

Batterbury, S.P.J., and Dant, T., 2019. The Imperative of Repair: Fixing Bikes – for Free. In P. Laviolette and F. Martinez (eds.) *Repair, Breakages, Breakthroughs: Ethnographic Responses*, pp. 249–266, Berghahn.

Beaulé, C.I., and Evans, P., 2020. Living in the Near North: Insights from Fennoscandia, Japan and Canada. In *Relate North: Tradition and Innovation in Art and Design Education*, pp. 140–161, InSEA Publications. doi:10.24981/2020-7-6.

Bhargaw, S., 2021. Two Madison Nonprofits Work to Create More Inclusive Cycling Community. *NBC15*. https://www.nbc15.com/2021/07/25/two-madison-nonprofits-work-create-more-inclusive-cycling-community/

Brink, S., 2020. De Meefietslijn Hulplijn voor fietsen in het donker. https://www.fietsersbond.nl/nieuws/de-meefietslijn/ (Accessed 20/08/2021).

Cojocaru, A., 2019. *Women on Bicycle Initiative in Turkey*. https://www.worldcyclingalliance.org/advocacy/women-on-bicycle-initiative-in-turkey/ (Accessed 23/08/2021).

Col·lectiu Punt 6, 2020. *Mujeres y Personas No Binarias en Bici. Estudio de Movilidad Ciclista en Barcelona desde una perspectiva feminista. Con el apoyo del Ajuntament de Barcelona "Convocatòria de subvencions projectes de districte i ciutat 2019"*. http://www.punt6.org/wp-content/uploads/2020/07/Informe-Final-Dones-Bici_-Castella%CC%80.pdf

COMOUK, 2021. Supporting Shared Transport. https://como.org.uk/project/bikes-for-all/ (Accessed 20/08/2021).

Conbici, n.d. Guia Asociaciones y Equidad. https://conbici.org/noticias/guia-asociaciones-y-equidad (Accessed 23/08/2021).

Coulon, J., 2020. France Wants More People to Bike. So the Country Is Paying Them (…Kind Of). *Bicycling* May 4, 2020. https://www.bicycling.com/news/a32369097/france-bike-repairs-50-euros/ (Accessed 20/08/2021).

Cox, S., 2018. Portland's Adaptive BIKETOWN Shares Pilot Results Better Bikeshare Partnership. https://betterbikeshare.org/2018/01/05/infographic-portlands-adaptive-biketown-shares-pilot-results/ (Accessed 23/08/2021).

Fietsberaad, 2015. Kinderfietsdeelsystemen (fiche). https://fietsberaad.be/wp-content/uploads/Fiche_Kinderfietsdeelsystemen.pdf (Accessed 20/08/2021).

Fietsersbond, n.d. De Fietsersbond en de driewielfiets. https://www.fietsersbond.nl/de-fiets/fietssoorten/driewielfiets/project-van-2-naar-3/ (Accessed 20/08/2021).

Flamm, M., and Kaufmann, V., 2006. Operationalising the Concept of Motility: A Qualitative Study. *Mobilities* 1(10), 167–189.

Kaufmann, V., Bergman, M.M., and Joye, D., 2004. Motility: Mobility as Capital. *International Journal of Urban and Regional Research* 28(4), 745–756.

Ledsham, T., and Verlinden, Y., 2019. *Building Bike Culture Beyond Downtown: A Guide to Suburban Community Bike Hubs*. Toronto: The Centre for Active Transportation at Clean Air Partnership. https://metcalffoundation.com/publication/building-bike-culture-beyond-downtown-a-guide-to-suburban-community-bike-hubs/

Lin, J., 2019. *Taking Back the Boulevard: Art, Activism, and Gentrification in Los Angeles*. New York: NYU Press.

Lucia, 2020. Hamburg, Borough of Altona. Public Lighting for the Liveable Public Pathway. "Elbewanderweg". https://lucia-project.eu/pilot-sites/hamburg-germany/ (Accessed 20/08/2021).

Lugo, A., Murphy, E., and Szczepanski, C., 2015. *The New Movement: Bike Equity Today*. Washington, DC: League of American Bicyclists. https://bikeleague.org/sites/default/files/The_New_Movement_Report_Web.pdf

Mobycon, 2016. Night Cycling Network, Municipality of Zoetermeer, The Netherlands. August 16, 2016. https://www.youtube.com/watch?v=gQ73EAf3fJg (Accessed 20/08/2021).

Muralidharan, K., and Prakash, N., 2017. Cycling to School: Increasing Secondary School Enrollment for Girls in India. *American Economic Journal: Applied Economics* 9(3), 321–350.

Newcastle Cycling Campaign, 2018. Beauty and the Bike – What Happened Next? newcycling.org/beauty-and-the-bike-what-happened-next

Ortar, N., and Adam, N., (eds) 2021. *Becoming Urban Cyclists. From Socialization to Skills*. Chester: University of Chester Press.

Sagaris, L., 2009. Living City: Community Mobilization to Build Active Transport Policies and Programs in Santiago, Chile. *Journal of Field Actions, Field Action Science Reports* 2, 41–48.

van der Kloof, A., and Bek, P., 2019. Niet-westerde migrantenvrouwen pakten zelf de fiets: Mobiliteitsarmoede in historisch perspectief. *Verkeerskunde* 2019, 2.

van der Kloof, A., and Kurz, S., 2021. *International Inspiration for the "A Bicycle for Every Child" Approach*. Mobycon for Tour de Force, in partnership with the Dutch Cycling Embassy.

Wupperman, B., and Grassick, R., 2009. *Beauty and the Bike*. Darlington: Darlington Media Group.

Zapata Campos M.J., and Zapata, P., 2017. Infiltrating Citizen-Driven Initiatives for Sustainability. *Environmental Politics* 26(6), 1055–1078.

6

THE POTENTIAL OF "BIKE-LIKE" VEHICLES TO PROVIDE BIG WINS FOR CLIMATE CHANGE, SAFETY AND JUSTICE

Kevin J. Krizek

Urban living and car travel

Urban living is the norm for more than half of humanity (Perry et al. 2021). Even in one of the most spread-out countries, the United States, three-quarters of the population reside in cities and suburbs. Residents in these locations share a common denominator: proximity to destinations. Many forms of transport are available to satisfy the trips that connect origins to destinations.

For the past century, public policies in most countries have steered residents to use cars, even for short trips. Reasons behind the automobile orientation for these trips run deep (Mackett 2003). Traveler decisions are moderated by reams of policies (e.g., traffic, fiscal, and parking), design codes, planning practices and culture – all of which support car use (Shill 2020). Now, every other time someone gets in a car, at least in the US, they travel fewer than four miles (Krizek & McGuckin 2019). In England, more than half of all trips between two and three kilometers are made by car (House of Commons Transport Committee 2019). Even in the most heavily auto-oriented communities, a sizable proportion of urban trip making is both short and by auto, a fact that applies to many countries.

The cumulative impact of transport planning practices to support cars carries many costs. The costs of auto-dependence have long been recognized by transport professionals and have become topics of concern for decision-makers and the public. For example, annually, across the globe, more than one million people die in automobile crashes. Emissions from cars are a leading contributor to greenhouse gases. Non-exhaust emissions from cars, such as the cancer-causing particulate matter from tire wear and brake pads (Timmers & Achten 2016) contribute to more than 50,000 premature deaths a year (Caiazzo et al. 2013). And disenfranchised populations that do not have a car or convenient transit suffer a disproportionate burden of basic mobility that limits their access to jobs, groceries, and necessities.

Which of these costs will continue to be borne by future generations? In a 21st-century city, ideally, the extensive costs borne by past practices will be minimized. We know that future

DOI: 10.4324/9781003142041-9

transport systems will differ from current features. Innovation will lead to transport systems that are cleaner, greener, safer, and more affordable. Modified or new transport networks will move people across space, and altered land-use patterns will distribute people, goods, and services across space. Transport systems will morph to a future state. Which changes are certain, likely or desirable when compared against the tripartite barometer of climate, safety and justice?

This future state, ideally, will allow residents to get around using quiet vehicles, emit as little carbon as possible, and be affordable and convenient for most users. The network to support the use of such vehicles will connect people to the destinations they want to go to. Many experts and advocates, though certainly not all, suggest that vehicles will operate at speeds that allow space to be shared between many users doing many things and be space-efficient in their own right.

In communities everywhere, the first step is to recognize the startling differences between what exists and what is desired. A second step lies in identifying ways to bridge that divide. And third and most importantly, we need to frame the solution in terms of practicality and timing: What is the feasibility of change and how quickly can results be realized?

There is a sense of urgency that underscores these matters. The 2021 United Nations climate change report (IPCC 2021) signals the need to do something different and to do it quickly. It suggests a death knell, specifically, for internal combustion cars. The lack of progress made on transport-related matters, as reported between iterations of the 2013 and 2021 IPCC documents, doesn't bode well for the effectiveness of past progress.

Furthermore, astute readers of the report would conclude, from a transport perspective, that most cities in most countries are beyond a point of meaningfully repairing past woes to satisfy pressing climate goals. Marginal adjustments to existing transport systems – increasing ridesharing, walkable cities, telepresence or bus rapid transit – have been insufficient and are part of why we are facing a global crisis. Even a tripling of current transit rates would yield fewer car trip reductions than are currently being achieved by telecommuting.

Look at the electrification of vehicular fleets. In 2020, only one in 50 new cars were fully electric. The average age of a car in North America is 11 years old (Electric Vehicles 2020). If all new cars that were sold tomorrow were 100% electric, most forecasts suggest a full transition could extend well beyond 2050 (recognizing, of course, that vehicular turnover could be considerably quickened through initiatives being used, for example, in China or Japan). And, as a policy solution, electric vehicles have not yet addressed safety, who has access, or livability – open questions that will extend the timeframe of dependency on fossil-fuel-burning vehicles.

Public transit could, assuming they are used at close to full capacity, dramatically reduce emissions per distance traveled mile. Yet, at least in the US, transit carries less than 5% of trips to work – and fewer for all purposes – because so few people live close to the high-quality transit needed to get them to where they need to go. And transit's marginal strength lies in longer trips (i.e., more than four miles), which are more variable and fewer in number than the typical daily run of travel-making. As a remedy, relying on transit is costly, especially for rail. Appraisal periods to install new systems can take years as planning, design and buildout can extend a decade – and even longer for concomitant and supportive land-use changes to follow suit.

Viewing any remedy through the prism of practicality and timing suggests that effective pathways would prize the following: (1) simple and effective vehicles that have the fewest negative consequences; (2) vehicles that have the greatest flexibility to satisfy the spatial demands of the public; and (3) vehicles that would be supported by a network that could be furnished quickly.

Furthermore, any remedy would require overcoming the forces that are unaccustomed to quick change. Transport, being a historically slow-moving industry, is sclerotic in so many dimensions. The materials used for transport infrastructure (e.g., concrete, steel) are among the most durable. Think of how the layout of cities and their streets, once formed, last centuries and how many Roman road alignments still survive in Europe. The industry is also its own worst enemy, in a sense. The standards and regulations that it has developed and upheld for decades are enormously resistant to innovation and change.

Emerging solutions would need to be available to users across urban regions, offer multiple modes of travel, and require tens of millions of vehicles. Human behavior and attitudes regarding transport have been firmly ingrained for decades and would need to be adapted. Any progress would be moderated by strongly entrenched public and private interests and policies, requiring revamped transport networks to enact alternatives. Which actions, from government or other, could quickly and feasibly meet these challenges?

We can draw two conclusions from what we know about the challenges we face: First, the status quo of urban mobility must yield to alternative systems that reflect both realities and priorities for 21st-century transport planning. Second, any changes need to be both practical and timely – we need bold steps, quick actions, big wins and sharp results. This means doing more with what already exists.

Decision-makers in this realm are well-advised to understand the valuable role that vehicle innovation play in advancing such discussions. Owing to rapid technological innovation (e.g., electrification, robotics, materials), there are many means of travel (other than cars) that are available today, especially in urban areas. The bicycle, one of the oldest, best-known forms of human locomotion, is a baseline of existing success that can inspire the broader roll-out of other innovative vehicles. Anything smaller than 40 inches wide and 100 inches long that can move a person fits into this new constellation of vehicles but generally, we are talking about bicycles, electric bikes (known as e-bikes), scooters, boosted skateboards or electric three- or four-wheeled variants of a bike (or car) – all of which provide increased stability and space to carry goods (Chapters 23 and 24).

These new forms are already providing more people with mobility across a broader spectrum of physical abilities. They also run the gamut of affordability. They can be adapted to provide comfort or protection from the weather if desired by the consumer. All of these adjustments could be made to allow increased safety relative to cars.

As smaller vehicles become available, their success will hinge on two related factors. First, we need to get more micromobility vehicles into the hands of more people. Doing this will require incentivizing ownership or making access more convenient across the population. Greater ownership and use will drive demand for private sector production of these modes.

Second, and more germane to this chapter, the public sector must build an urban network to serve these vehicles and their users. Honing in on required characteristics for bicycles can be used as a bellwether test for the type of infrastructure decisions can be made. Building a new network using existing streets and enabling innovative uses of street space, will provide a low-cost, high-impact means to meet demands for smaller vehicles.

Design factors

In advancing infrastructure planning principles for micromobility, transport planners will need to make a case for what the optimal characteristics are of these vehicles. From that, we can set new rules for the size of vehicles, speed, sound and other factors. Ultimately, vehicle design can adapt to principles of good city building, as opposed to the other way around.

Safety and speed

Deaths on streets, at least in the US, are at an all-time high. (This was true even despite the reduction in car traffic that occurred during the pandemic.) Countless factors moderate traffic safety, including vehicle size, driver negligence, and user error, but one factor stands out: speed. Excessively heavy and fast-moving objects, namely cars, underscore the unforgiving nature of most urban transport systems. The odds for different types of injury relative to vehicular speed is well-traveled knowledge (Aldred et al. 2018). Among analysts, a consensus is gaining that travel speeds of roughly 20 mph (30 km/h) yield an appropriate compromise balancing safety with matters of economic vitality, which is close to the upper bound for how fast standard bicycles can comfortably roll.

The tradeoffs, however, for how 30 km/h travel interacts with urban distances and travel budgets are, for most residents, difficult to grasp. Windshield bias and motorization dominate perceptions of urban travel. Assume a three-mile trip across town. In a car, using stretches of arterials (45 mph) or collectors (35 mph), and accounting for intersections, congestion and parking, it is reasonable to assume a 27 mph average speed and therefore a 13-minute round-trip (3/27*60*2). For the same trip, five additional minutes would be required to use a smaller device average averaging 20 mph. Relative to a typical intra-urban, four- or five-mile car trip, the alternative of using new smaller electric-assisted vehicles could even be time-competitive, especially when parking issues are considered.

Human-scaled vehicles

In terms of how urban space is used, bicycles are built and designed around human dimensions. In 1490, Leonardo Da Vinci described the Vitruvian man. It represents a concept that relates humans to nature and the workings of the universe. Da Vinci asserted that everything that was built should be constructed with the human's perspective in mind. Cities were quite different back then, and it is unclear where Leonardo would stand on current transport discussions, though he did produce hundreds of sketches and written materials about human mechanical flight and the architecture of cities that were centered on human-scaled environments.

New micromobility devices, whether their users are sitting or standing up or whether they are designed for one person or two, center on the human scale. Looking at efficient use of roadways, in the 1970s Austrian civil engineer Herman Knoflacher highlighted the over-consuming nature of cars in terms of how much space they require. He constructed a wood-frame outline of a car and allowed a pedestrian to hang it over their shoulders, showing the excess space someone driving alone consumes. The ride-hailing company Uber riffed the idea to help make the point that sharing cars is space-efficient in a 2017 advertising campaign (Uber 2017). To humanize the emotional toll of the devastating crush of traffic, the video shows drivers in extremely large cardboard boxes, not automobiles, ending with the city being overrun by boxes (https://www.youtube.com/watch?v=Wa1WSf1BlaQ accessed 13 September 2021).

Energy efficiency

Fifty years ago, *Scientific American* showcased the cost of energy to transport one unit of body weight for a unit of distance for different species (Wilson 1973). The article drew on the work of Vance Tucker, who studied comparative physiology, particularly the energetics of locomotion, and the interactions between an organism's natural environment and its respiratory and circulatory systems. In his analysis of the aerodynamics and energetics of avian flight, mammals, reptiles, amphibians and insects, Tucker found that larger birds are about as economical as small propeller planes or jet fighters and are better than helicopters. Jet transport is more economical than any bird yet measured, but an extrapolation of the line for fliers indicates that a swan or goose weighing 10 kg might do just as well. Larger walking and running mammals do about as well as jet transport or automobiles.

A human on a bike ranks first in efficiency among traveling animals and machines in terms of energy required to move a certain distance as a function of body weight. The rate of energy consumption for a bicyclist (about 0.15 calories per gram per kilometer) is approximately a fifth of that for an unaided walking man (about 75 calories per gram per kilometer). A bicycle moves five times faster than walking and goes three times as far on the same amount of caloric energy. As far as overall efficiency, bicycling outshines cars and pedestrians and even birds and planes.

Bicycles are the template

Bicycles are a known entity in transport research and real-world implementation. What we know now about bicycle safety, speed, accessibility, and efficiency can inform the entire effort to move quickly to a micromobility-driven transport network.

As the matter of access in transport, sustainability, and urban economics is increasingly recognized, bicycles and micromobility have a strong role to play because the positive correlation between access and land value originates from the tradeoff between time (transport cost) and space (the price of land) (Vickerman 2008). In this light, bicycles could and should be used as a bellwether test to record progress toward any goal that is pursued. We know that people should have choices on how they get around and that choices need to be safe and green. The next question revolves around providing the needed infrastructure to support those choices.

Cumulatively, the ability to satisfy the above criteria – speeds of roughly 20 mph, space and energy efficiency and equality – are precisely key characteristics of bicycles. Recent transport innovations, which vary widely by geography and culture, can rally around bicycling – writ large – to address an urgent situation. Other small vehicles, such as electric bicycles, scooters, modified golf carts, or small single-seater cars that share characteristics of bicycles, are also being developed; using these devices could provide mobility needs during inclement weather or for the elderly and those with disabilities. All such variations can coexist with other street users.

Despite persistent calls, transport planning practices counter constant challenges in providing for smaller, lighter, short-trip vehicles. It is difficult to envision meaningful alternatives. Most residents simply cannot imagine what could replace the car and how any substitute could meaningfully scale. Politicians, in response, double down on a dinosaur of a transport program that carries excessive costs, including a massive repair backlog.

Toward a solution

To see an alternative future realized, vehicle innovation needs to be incentivized. That endeavor is best suited for stronger marriages between economic development initiatives and the private sector. Sprouting a stronger market for their use lies in the success of making a network for their use.

Calls to alter streets away from being centered on automobiles are hardly new. Yet any future needs to consider the current structure for how rules oriented toward the same outcomes rule the existing system. Decision-makers need to guide new actions that reduce the uncertainty about where these vehicles could be employed on existing streets. They need reassurances that changed processes will work. Experimenting with streets to create new expectations about how these spaces are conceived of – and used – is essential and can provide quick gains (Glaser & Krizek 2021; King & Krizek 2021).

During 2020, in cities across the world, through a seemingly overnight metamorphosis, hundreds of streets were reoriented to prioritize walking, gathering safely, and accommodating smaller vehicles. This action prompts the question: what is stopping cities from using land area currently devoted to streets for public benefits that transcend standard car travel? What opportunities lie for using existing street space to grow the transport networks to accommodate electric bicycles, micromobility devices, wheelchair users, delivery persons, and mobile food vendors alongside standard bicycles?

Beginnings of a research and policy agenda

Drawing on evidence of the potential for use will determine where building out new street networks will produce the quickest gains. This requires employing what we know from decades of bicycle research and from recent anecdotal research about repurposing streets. In addition, it necessitates supporting a rapid transition to micromobility. Federal-level governments, too, have a strong role, as dependent local municipalities often lack the resources to initiate projects by themselves.

Location of improvement is key. Many strides towards this end already exist in Europe. For example, the *Propensity to Cycle Tool* in the UK provides data to prescribe where additional bicycle facilities are needed. Progress is being made in this direction in the Netherlands, with the development of an *Integrated Mobility Analysis* (IMA) which includes accessibility indicators (the number of destinations that can be reached from a specific location within a given travel time for different modes). Explaining further how these tools and others could be used to decide which streets to prioritize would go a long way.

At the local level, council members will need to make hard decisions about which types of vehicle will be allowed to use which streets. Transport researchers and professionals must provide the research and knowledge on all aspects of micromobility to support these decision-makers. Using bicycle lanes for all sorts of small vehicles may be feasible in special circumstances, but it may not be acceptable for many bicyclists if it is adopted as a general policy. Just as automobiles and trucks threaten bikes, in turn, bikes threaten other users. Co-mingling may not be appropriate when bicycle use or other traffic volumes are high or where bicycle lanes or paths are narrow.

Inevitably, for most communities, a new regulatory transition task force is warranted. Existing city codes and zoning and development regulations must be revised to include both land-use standards and facilities standards and requirements for new modes. Under the pandemic, most street changes have been enacted under declared emergency response orders. How such changes would play out in the long term remains to be seen. Revolutionary initiatives would require waiving the stifling restrictions imposed by existing codes, such as in the US, the Manual of Uniform Traffic Control Devices and other stipulations as prescribed by the proverbial "Green Book" as issued by the American Association of State and Highway Transportation Officials (AASHTO). These outdated legislative procedures hamper change, reinforcing "an automobile-dominated regime that restricts innovation, learning and change" (Glaser, Krizek, & King 2020). Desperately, future research is needed to support innovation and accelerate transformation in how future streets would be managed to harness the power of local government to strategically experiment with new ways that space could be used.

Changing the size and character of vehicles so that smaller vehicles can gain prerogative is no small feat. The public would need to be brought along in articulating short-, medium-, and long-term actions. Actions here would be community-specific as the culture of each needs to be met where they are at on the broader transition, in general, and on accepting reasonable alternatives to standard cars. For example, a safety task force will be required to mediate information and action about impending crashes that might result from such a dramatic transition.

Done successfully, traffic lights will need to be reversed, road signs changed, intersections redesigned, lines on the road repainted, buses modified, and bus stops moved. Massive public relations campaigns will be needed to bring public opinion along and to inform people about how changes might be implemented. The task is roughly analogous to "Högertrafikomläggningen" ("The right-hand traffic reorganisation") put forth by the Swedish government on September 3, 1967, in which the traffic in the country switched from driving on the left-hand side of the road to the right.

A concept typically considered to be outside the range of acceptable outcomes – what political economists refer to as the 'Overton window' – has recently shifted, thereby expanding what is possible. City leaders, supported by transport industry representatives, planners and designers, are well-suited to solidifying futures based on the potential of smaller vehicles. By not changing streets to noticeably usher in smaller vehicles our efforts will likely just result in pushing on a string.

Urban transport planning warrants a complete reboot. The time is now to leverage what we have – our streets, our existing urban forms – to prioritize smaller vehicles. We know what will likely be the result if we keep doing the same thing. Failing to take action, cities will have missed an opportunity to make a big change when it is most needed. To quickly address greenhouse gas emissions, public safety, livability and infrastructure maintenance, a remedy with quick changes stands before us. Urban mobility has seen dramatic innovations over the past decade that can help solve many pressing problems. Prizing more vehicles with bicycle-like characteristics provides a low-cost solution with great payback to cities.

References

Aldred, R., Goodman, A., Gulliver, J., & Woodcock, J. (2018). Cycling injury risk in London: A case-control study exploring the impact of cycle volumes, motor vehicle volumes, and road characteristics including speed limits. *Accident Analysis & Prevention*, 117, 75–84.

Caiazzo, F. et al. (2013–11). Air pollution and early deaths in the United States. Part I: Quantifying the impact of major sectors in 2005. *Atmospheric Environment*, 79, 198–208.

Electric Vehicles, (2020). https://www.iea.org/reports/global-ev-outlook-2020

Glaser, M., & Krizek, K. J. (2021). Can street-focused emergency response measures trigger a transition to new transport systems? Exploring evidence and lessons from 55 US cities. *Transport Policy*, 103, 146–155.

Glaser, Meredith, Krizek, Kevin J., & King, David. (2020). Accelerating reform to govern streets in support of human-scaled accessibility. *Transportation Research Interdisciplinary Perspectives*, 7, September, 100199. https://doi.org/10.1016/j.trip.2020.100199

House of Commons Transport Committee, Active travel: Increasing levels of walking and cycling in England, July 2019.

IPCC, 2021: Climate Change 2021: The Physical Science Basis. Contribution of Working Group I to the Sixth Assessment Report of the Intergovernmental Panel on Climate Change [Masson-Delmotte, V., P. Zhai, A. Pirani, S. L. Connors, C. Péan, S. Berger, N. Caud, Y. Chen, L. Goldfarb, M. I. Gomis, M. Huang, K. Leitzell, E. Lonnoy, J. B. R. Matthews, T. K. Maycock, T. Waterfield, O. Yelekçi, R. Yu and B. Zhou (eds.)]. Cambridge University Press. In press.

King, David A., & Krizek, Kevin J. (2021). Visioning transport futures through windows of opportunity: Changing streets and human-scaled networks. *Town Planning Review*, 92(2), 157–163.

Krizek, K. J., & McGuckin, N. (2019). Shedding NHTS light on the use of little vehicles' in urban areas. *Transport Findings*. https://doi.org/10.32866/10777

Mackett, R. L. (2003). Why do people use their cars for short trips? *Transportation*, 30(3), 329–349.

Perry, Gad, Stone, Lesley A., & Obaid, Obaidullah. (2021). Adapting U.S. foreign assistance for a rapidly urbanizing world. *Science & Diplomacy*, 10(2). https://www.sciencediplomacy.org/article/2021/adapting-us-foreign-assistance-for-rapidly-urbanizingworld

Shill, G. H. (2020). Should law subsidize driving? *NYUL Review*, *95*, 498.

Timmers, V.R.J.H., & Achten, Peter A.J. (2016–06). Non-exhaust PM emissions from electric vehicles. *Atmospheric Environment*, 134, 10–17.

Uber (2017). https://www.youtube.com/watch?v=oNyq2_92H0Y

Vickerman, R. (2008). Transit investment and economic development. *Research in Transportation Economics*, 23(1), 107–115.

Wilson, S. S. (1973). Bicycle technology. *Scientific American*. 228(3), 81–91.

7

MOBILITY, FREEDOM AND SELF-DETERMINATION

The benefits (and barriers) to disabled people cycling

Kay Inckle

Introduction

In the UK the stereotypical cyclist is often referred to as a MAMIL – a middle aged [white] man in Lycra (Aldred et al. 2016; Chapter 26). Likewise, the symbol for a cycle or cycle facilities depicts a standard two-wheeled bicycle (Hickman 2016). Older people, people of colour and women rarely feature in cycling imagery, and disabled people of any demographic are notable only by their complete absence.[1] This creates a social, cultural and policy context where cycling and disability are assumed to be diametrically opposed. The notion of a disabled cyclist becomes only the rarest of "super-crips" or some kind of disability cheat. Nonetheless, cycling is easier than walking or wheelchair propulsion for many disabled people (Andrews et al. 2018; Arnet et al. 2016; Inckle 2019; van Drongelen et al. 2009; WfW 2016) and cycling also provides mobility, exercise, autonomy along with the attendant physical and mental health benefits that disabled people often struggle to access elsewhere (Inckle 2019, 2020a).

In this chapter I explore the experiences of disabled cyclists and highlight that cycling is key to mobility, exercise and health for disabled people. I draw attention to the barriers that disabled cyclists encounter – be that whether they use a standard two-wheeled bicycle or a non-standard cycle such as a recumbent, trike or handcycle (Chapter 13). I draw on data from a small-scale qualitative research project which I conducted in partnership with Wheels for Wellbeing (WfW), a London (UK)-based inclusive cycling provider and campaign group. The project was grounded in disability research ethics, a rights-based approach which ensures that all aspects of the research are devised by and for the benefit of disabled people (Barnes 2003; Kitchen 2000; Payne et al. 2016). Everyone involved in this project: the partner organisation, the researcher (me),[2] and the research participants, all identified as disabled people.

There were seven participants, three female and four male, and their ages ranged from 31–64 years. Of the seven, six were white European, five lived in the south-east of England and six currently or previously worked in white-collar jobs. All identified as having a physical disability or impairment or a mobility impairment, and all had the capacity to consent.[3] Four participants had been disabled since birth/early infancy and three (all male) had acquired their disability in adulthood. Three rode standard bicycles, three rode non-standard cycles (recumbent, handcycle, trike) and one rode a combination of standard and non-standard cycles. Their cycling experience ranged from ten to more than fifty years.

DOI: 10.4324/9781003142041-10

I interviewed participants either face-to-face or using Skype, and all interviews were audio recorded, transcribed verbatim and then thematically analysed in a two-stage process (see Inckle 2020b). The themes explored in this chapter highlight the participants' experiences of cycling as mobility, the physical and mental health impacts of cycling, and the barriers to cycling that they encountered. Significant among these barriers were the knowledge deficits of health professionals and wider social attitudes towards disabled people in general, and disabled cyclists in particular.

Cycling, mobility & health

Cycling as mobility

All of the research participants described cycling as essential to their mobility, it allowed them to travel distances and participate in activities that would otherwise be inaccessible to them. For, example Rob, who was unable to walk more than 50 metres, had completed a 110 kilometre group bicycle ride just prior to my interview with him. He said:

> I do a *lot* of cycling, actually, despite the fact that I have real problems walking any distance. I do cycle a lot and I cycle quite long distances… So I have *no* problems with cycling as such, so I *just love* cycling, it's a *fantastic* way for me to get around, otherwise I wouldn't be able to.

For Hélen, a wheelchair-user, her handcycle enabled her to leave her car behind and travel in ways which connected her with her local environment and community and enabled her to fulfil her own green aspirations. Without the handcycle, her mobility was reduced to car use.

> I would never by choice wheel somewhere [e.g. self-propel wheelchair] unless it's *literally round the corner* because that is not a pleasant situation to be in, because pavements are so rubbish, lack of cut kerbs, all of that rubbish.

Disabled people are often forced to rely on private cars for their mobility because of the lack of accessibility of public transport (Maynard 2009). Whilst rudimentary access is provided on some routes, public transport, and public attitudes, are not designed to facilitate disabled people commuting at peak times. Cycling can therefore provide essential transport, as Paul explained:

> If I cycle to work it means I don't have to use public transport which *can* be a bit difficult: getting on and off a bus, you know, and on the tube. I have found that difficult because obviously you have got lots of stairs and escalators and, you know I'm ok, but in the rush hour it's completely packed and I have actually fallen over in the tube before… So in a way it's easier than public transport cos it's just me and my bike and I can go from A to B, and I don't have to worry about getting on a bus or anything like that so it just makes things easier.

The predictability and safety of cycling compared to other forms of transport is a crucial and often-overlooked dimension of cycling for disabled people. Cycling can be safer as well as easier than walking or wheelchair use. Over the long term, propulsion of a manual wheelchair is known to be detrimental to shoulder joints in ways that handcycling is not (Arnet et al.

2016; van Drongelen et al. 2009). And for participants who use mobility aids to walk, cycling a three-wheeler such as a recumbent or trike can feel much less risky. Rosie, who had fallen and broken a leg while walking with her crutches, explained that:

> Crutches I would be more anxious about moving around, I'm worried I'll fall or something and then on the bike you don't really [worry] because you know you are not going to fall.

Cycling provides an essential form of mobility for people with disabilities, it creates opportunities for social participation and sustainable living and may also feel safer than other forms of mobility. Cycling also provides access to exercise and physical activity which has positive impacts on mental and physical health.

Cycling and health

All of the research participants described cycling as their main form of exercise and for a number of them it was their only physical activity. Nasia described how cycling was both functional, her "commute", as well as health-promoting:

> It's really good for my health, it keeps my weight down, I do it because it *does* keep my weight down, and it is constant exercise… It's my commute but it's good for my exercise.

Prior to his disability Michael had lived a very active lifestyle which included mountain biking and fell running. Following his injury cycling became his essential exercise:

> The other most normal thing that I do, the highest frequency, would be leisure riding on the recumbent. And that's really my kind of gym session, my opportunity to get a proper work out, just spend a good hour or two getting the heart rate up, getting the breathing rate up, and, yeah, cos taking the kids to school [on a family trike], as much fun as it is, I don't really get that much of a sweat on!

Disabled people often encounter significant infrastructural and attitudinal barriers to exercise (Brittain 2004; Mulligan et al. 2017). This is particularly so for cardio-vascular activity which is essential for long-term health, especially after the onset of a disability. Eric described how his discovery that he could cycle a trike not only helped to limit the impact of his stroke but also impacted another long-term health condition.

> The other thing, with this stroke I'm not able to walk far *at all*, and I'm very slow walking, so there is no exercise there and the thing is they say to people if you are not getting enough exercise it makes you worse and worse and worse… It [cargo trike] means that I *am* able to do a full shop, and I am able to get exercise. I am getting a lot of exercise. I've got asthma as well and they said, the asthma nurse said, my lung capacity was improving, it's better than it had been in previous years, which is what you want.

Exercise is also essential for mental health, and participants described this in terms of the "endorphin rush" (Rob) brought on by cycling and "how *good* that felt" (Hélen). For participants who became disabled as adults, cycling was crucial for positive adjustment to their new identity and life circumstances. Paul highlighted that:

> To my mind you can go one of two ways if you have an accident – obviously for people who are already disabled they have grown up with that – but when something *happens* to you like this you can go one of two ways: you can sit in a chair and go, "Oh my goodness this is terrible, what am I going to do?!" Or you can try and get on with life, and I think that if you can stay positive and do what you could do *before* then that certainly [helps]. I think if I hadn't had *that* [cycling], that would have been a very different story *for me*.

Michael was prescribed psychiatric and pain relief medication to help him to cope with the physical and mental trauma of his accident. However, after he took up cycling a recumbent he ceased using both. He described cycling as "better than counselling."

Participants also highlighted the impacts of cycling on positive wellbeing more broadly; they spoke about "freedom" (Eric and Rosie), "independence" (Michael and Nasia) and "self-determination" (Hélen). The opportunity to be free, independent and self-determined is often rare for disabled people who encounter multiple barriers to autonomy and mobility as Nasia explained:

> It's the *independence*, which I haven't said. I think that's the other thing, you're not *relying* on transport you are relying on your own transport: you are relying on *yourself* and I think that is a big boost to anybody with a disability. You rely on yourself, and wherever you are when you are on your bicycle, you rely on yourself. And wherever you are you can dictate what you do, no-one's dictating to you what you can do.

Cycling also offers people with physical disabilities access to nature which is crucial to mental health and wellbeing (Barton & Pretty 2010). The physical and mental health benefits of cycling are especially important given that disabled people have the worst health outcomes of any population group and encounter the greatest barriers to health promoting activity (Mulligan et al. 2017; Reichard et al. 2011). However, notwithstanding the indisputable benefits that cycling provides there are significant barriers which prevent disabled people's access to and uptake of cycling.

Barriers to cycling

Cycling infrastructure, like transport infrastructure more broadly, is largely designed by able-bodied people to prioritise the needs of able-bodied people (Chapter 20). This means that cycle paths and facilities such as entrance gates, bollards, cycle parking and storage are often inaccessible to non-standard cycles and disabled cyclists. Moreover, poor-quality road surfaces such as potholes and camber can also cause problems for those who cycle on three (or more) wheels (WfW 2016, 2020). The cost of non-standard cycles is prohibitive, especially in the context of the disability employment and pay gaps which result in more than half of the disabled population living in poverty in the UK (EHRC 2017). Coupled with starting prices of around £3,000 for a recumbent or handcycle, and no funding or loan schemes, the majority of disabled people are priced out of the market.

Disabled people who can ride a standard two-wheeled cycle also encounter significant barriers. These can include cycle parking and storage facilities that are accessed via steps or located too far from the destination point. Pedestrianised areas and public transport hubs which prohibit cycling exclude those who use their bike as a mobility aid and cannot otherwise access the space. My research participants highlighted all these issues, which are increasingly documented in cycle design guidance (e.g. WfW 2016, 2020). However, in this chapter I focus on two other themes which relate to wider knowledge and attitudinal barriers toward disabled people who cycle. These barriers are particularly significant because it seems that they also underpin the policy and practice which excludes disabled cyclists (see Inckle 2019, 2020a).

Health professionals

One of the most common barriers to cycling that the participants experienced was the lack of knowledge among health professionals. This meant that participants only discovered cycling of their own volition, by chance via own resourcefulness and/or creativity. This was true for Rob for whom cycling was his primary day-to-day mobility and exercise:

> No, no, never, no, no, no. It's definitely something that *I* discovered for myself and it kind of evolved in a way, it just [pause], yeah, nobody ever said that cycling might be an option for you, it's just something which I *found, myself.*

This knowledge barrier persists even when disabled people are given static (hand and/or leg) cycle machines to use as part of their physiotherapy as Hélen recounted:

> There was a period of time where I got quite a bit of support through the neuro-physiotherapy gym at [name] and they have got a static bike that they put you in front of to get your legs, it's legs *and* it does have arm things like that [gesturing handcycling]. And that was brilliant, that was really good, I could get my legs moving using muscles that I don't normally use, but I saw it as physio. Nobody then said 'Oh how about looking into doing it as a mode of transport', or as fun, or as exercise.

Rosie was similarly instructed on a static cycle machine during physiotherapy without being informed that a trike or recumbent could provide her with mobility as well as exercise, "No, no, only on the exercise bike, but not ever – I didn't know, I had *no* clue whatsoever that there was inclusive cycling." Likewise, Paul was perplexed that his post-injury physiotherapy only focused on him learning to walk using his prosthetic and not relearning any of the other activities he had previously enjoyed, including cycling.

> After my accident I had physiotherapy and that was about getting me walking again. Basically, it was to teach me to walk again. But why not have a little bike and people can try out riding in the rehabilitation area? It wouldn't be very much to have just have a little bike that they could try. It doesn't have to be a big one, just a small-wheeled one, that would be easy to step on to so that people could get their confidence back in the same way that I got my confidence back walking with the physiotherapist. I think that would be quite a good idea.

Michael was similarly left to discover post-injury cycling by himself, despite his key health professional being a keen cyclist and also aware that Michael had been an avid mountain biker before his accident. Eric conducted his own research with health professionals outside the UK in order to learn about his cycling options: "When I first had my stroke I did a lot of research on email and had email conversations with Danish neuro-physiotherapists and they said, 'Oh yes definitely go for a cargo trike, it's what you do'."

This deficit in knowledge and practice may have significant impact on the uptake of cycling among disabled people, denying them a source of mobility and health-promoting exercise. It also means that there is no official recognition among state-accredited health and social service providers that disabled people can cycle. This can reinforce assumptions that any level of physical activity is a contraindication to disability, which has hugely damaging implications for welfare support claimants and the health of disabled people (Brown & Pappous 2018). Such beliefs are also integral to wider perceptions about the abilities of disabled people and are evident in widespread social attitudinal barriers towards disabled cyclists.

Social attitudes – and their impact

Regardless of the type of cycle that the participants used, all encountered attitudinal barriers toward them as disabled people. For participants with a visible disability who used a non-standard cycle, their visibility attracted attention which often had unclear motivations. Rosie described an encounter during an inclusive cycling session which had initially alarmed her: "There was a guy actually in the park, in the park where we did it the other day, who was just like shouting over at me and I was like [concerned], 'Oh what does he want?'" It transpired that the man was merely curious, but given the high levels of street harassment and hate crime that disabled people encounter in the UK (EHRH 2017) this kind of attention can be unnerving.[4]

Michael, who often cycled with his young children using a family trike, experienced direct ableist abuse whilst cycling alone with his recumbent in his locality:

> I have had cause to speak to the police just once about getting shouted at by some builders. And the reason I went and spoke to the police is I don't want my kids to have to put up with this or see this kind of behaviour.

However, unnerved by the police response, Michael decided to withdraw the formal complaint and deal with the matter himself:

> I did ring the building company and say 'look, you know, I appreciate you've got lads working there, but just tell them to keep their opinions to themselves and I don't really care,' erm, and then my window got put through.

Cyclists whose disability is relatively invisible and who ride a standard two-wheeled bicycle encounter a different set of attitudinal barriers. In these instances the perception that someone is either disabled or a cyclist means there is no recognition that for some people a bike is a mobility aid in the same way that a walker or mobility scooter might be for others. Nasia explained, "For *me*, it's a mode of transport, it's *my* mode of transport, *you've* [e.g. Kay/author] got a wheelchair that's your mode of transport mine is a cycle." Rob was similarly frustrated that people could not understand that his bicycle was essential to his mobility:

> People just see a *bicycle*, and that's all that they see, despite the fact that they can obviously see that I have problems walking, they will still ask me to get off and walk with it in various places, or they won't allow me into, for example, parks.

Nasia finds it very difficult to walk alongside her bike. So, when she enters pedestrianised areas, she partially dismounts and scoots along with one leg in order to attempt to comply with the restrictions whilst remaining mobile. Nonetheless she still encounters hostility:

> I don't get off the bike, I just scoot, I just use one leg and I'm on a pedestrian area and everyone just looks at you like, "Get off your bike" and I'm [thinking] like, "Well it's actually easier for me to just do that rather than to get off the bike and push it". You do get people looking at you going, "Why are you doing that?", you know, you just *take* it, if people say something I just say "I can't walk" or "I find it difficult to walk", and just carry on.

Eric also described the constant negotiations that he has to undertake in order to be able to use his cycle as a mobility aid:

> You don't want to *have* them, I don't want to have to go round being a preacher or trying to convert people, you don't *want* it, you are hoping for some kind of basic standard there of civilisation that that's how you do things, but that's right, it isn't there... I do it reluctantly because it *has* to be done, you know?

These attitudes create significant social and emotional burdens for disabled people when they use their cycle. The freedom, mobility, exercise and independence that cycling provides is tempered with an emotional burden that impacts on the overall experience, as Rob described:

> There is that sense *all the time* that you're thinking, "How are people looking at me here?"' Cos I'm quite sensitive, I like to think that I'm not irritating people, but I think I find that quite often *I am* when I'm on my bike because of *their* expectations and *their* preconceptions of what I should be doing. Yeah it is, it's *wearing*, it is quite wearing to have to think about this every time I go out.

Eric described having to consciously present a façade in order to manage these kinds of encounters: "I put my war paint on, I put my theatrical stuff on, you don't want to be doing that all the time, you know." Encounters such as these have a lasting impact. Nasia highlighted the lingering emotional effects of being constantly challenged for using her bike, even when scooting slowly along with it:

> I sort of get *annoyed* and sort of think about it for a while and I think "Well, actually, no I can, I will just carry on". You get angry, you know, when anyone says it that you can't do something, you *explain* it to them that this what I'm going to do, but I just carry on. I do carry on, but you think about it, that's what I do, I think about it more than I should.

These attitudinal and knowledge barriers mean that disabled people are often prevented from accessing public spaces if they have limited walking ability but are able to cycle. Conversely, while those who use non-standard cycles may be more easily identified as disabled people and therefore encounter less attitudinal barriers to accessing public spaces/pedestrianised areas, they are still subject to unwanted and/or hostile attention and encounter significant infrastructural barriers.

Conclusion

Cycling can provide a key form of mobility, exercise and social participation for disabled people along with the associated improved mental and physical health outcomes. However, disabled people encounter many barriers to cycling and cycling uptake among disabled people who can cycle is low. Increased knowledge across health and transport professionals and policymakers would go some way to addressing the knowledge deficit and could concurrently also increase the likelihood of accessible cycle infrastructure. Greater public awareness that disabled people can cycle and widespread knowledge about the variety of cycles available would make a difference. Likewise, a thorough understanding among professionals and the public that for many disabled people a cycle is an essential mobility aid would be transformative to disabled people's experience. One means to achieve this is to develop a "blue badge" scheme for cycles similar to that which already operates for car drivers – a facility that WfW have been campaigning for. Finally, given the cost of non-standard cycles, funding and hire schemes would be an important route for many disabled people to access cycling, along with nationwide, widespread provision of inclusive cycling sessions where people can try out different types of cycle and be trained so that they are confident using them. Cycling uptake is not merely important for the mobility, health and wellbeing of disabled people; given the climate crisis we are facing as well as the global pandemic, it also has implications for the health of the whole planet.

Notes

1 A notable exception is Lambeth council (London, UK) who in 2020 introduced a "mobility lane" – for cycles, scooters and mobility scooters – which included a painted symbol of a female handcyclist alongside the usual bicycle image.
2 I am also an avid handcyclist.
3 It is important to highlight that all the participants identified as having a physical disability (rather than sensory, intellectual or mental health). This is significant because the barriers that different disability groups encounter to mobility transport and cycling are very different. For people with physical disabilities mobility, transport and exercise can be completely inaccessible and they may require non-standard cycles and cycle facilities.
All names of participants quoted are pseudonyms, and emphasis in the text reflects participants' original speech.
4 I also encounter constant unwanted attention as a handcyclist, and whilst a lot of it is potentially well meaning, being shouted at, pointed and marvelled at on a daily basis is very wearing, especially when I just want to do my shopping or go home from work.

References

Aldred, R., Woodcock, J., Goodman, A. (2016) "Does more cycling mean more diversity in cycling?" *Transport Reviews* 36(1), 28–44.
Andrews, N., Clement, I., Aldred, R. (2018) "Invisible cyclists? Disabled people and cycle planning – A case study of London," *Journal of Transport & Health* 8, 146–156.
Arnet, U., Hinrichs, T., Lay, V., Bertschy, S., Friel, H., Brinkhof, M.W.G. (2016) "Determinants of handbike use in persons with spinal cord injury: Results of a community survey in Switzerland," *Disability & Rehabilitation* 38(1), 81–86.
Barnes, C. (2003) "What a difference a decade makes: Reflections on doing 'emancipatory' disability research," *Disability & Society* 18(1), 3–17.
Barton, J., Pretty, J. (2010) "What is the best dose of nature and green exercise for improving mental health? A multi-study analysis," *Environmental Science & Technology* 44(10), 3947–3955.
Brittain, I. (2004) "Perceptions of disability and their impact upon involvement in sport for people with disabilities at all levels," *Journal of Sport & Social Issues* 28(4), 429–452.

Brown, C., Pappous, A. (2018) ""The legacy element …it just felt more woolly": Exploring the reasons for the decline in people with disabilities' sport participation in England 5 years after the London 2012 paralympic games," *Journal of Sport and Social Issues* 42(5), 343–368.

Equality and Human Rights Commission (EHRC). (2017) *Being disabled in Britain: A journey less equal*, London: EHRC.

Hickman, K. (2016) "Disabled cyclists in England: Imagery in policy and design," *Urban Design and Planning, Institute of Civil Engineers* 169(3), 129–137.

Inckle, K. (2019) "Disabled cyclists and the deficit model of disability," *Disability Studies Quarterly* 39(4). 10.18061/dsq.v39i4.6513.

Inckle, K. (2020a) "Disability, cycling and health: Impacts and (missed) opportunities in public health," *Scandinavian Journal of Disability Research* 22(1), 417–427.

Inckle, K. (2020b) "Poetry in motion: Qualitative analysis, I-poems and disabled cyclists," *Methodological Innovations* 13(2), 1–13.

Kitchen, R. (2000) "The researched opinions on research: Disabled people and disability research," *Disability & Society* 15(1), 25–47.

Maynard, A. (2009) "Can measuring the benefits of accessible transport enable a seamless journey?" *Journal of Transport and Land Use* 2(2), 21–30.

Mulligan, H., Motohide M., Nichols-Dunsmuir, A. (2017) "Multiple perspectives on accessibility to physical activity for people with long-term mobility impairment," *Scandinavian Journal of Disability Research* 19(4), 295–306.

Payne, D.A., Hickey, H., Nelson, A., Rees, K., Bollinger, H., Hartey, S. (2016) "Physically disabled women and sexual identity: A photo voice study," *Disability & Society* 31(8), 1030–1049.

Reichard, A., Stolzle, H., Fox, M.H. (2011) "Health disparities among adults with physical disabilities or cognitive limitations compared to individuals with no disabilities in the United States" *Disability & Health Journal* 4, 59–67.

van Drongelen, S., Arnet, U., van der Woude, L., Veeger, D. (2009) "Is synchronous hand-cycling less straining than hand-rim wheelchair propulsion?" In *XXII Congress of the International Society of Biomechanics Proceedings* 413, 5–9.

Wheels for Wellbeing (WfW). (2016) *Beyond the bicycle: A manifesto for an inclusive cycling policy*, London: WfW.

Wheels for Wellbeing (WfW). (2020) *A guide to inclusive cycling*, 4th Edition, London: WfW.

PART II

Cycle technology

An introduction

Tony Hadland

This part of the *Routledge Companion to Cycling* examines various aspects of cycle technology. It provides an overview of current options, and thoughts on future developments, with brief historical context provided where appropriate. The contributing authors are from Australia, Austria, Canada, Germany and the United Kingdom.

Why is cycle technology important?

Cycle technology is important partly because its development demonstrates technical progress. The development of practical and affordable bicycles has had a profound effect in many countries and cultures worldwide, both as a means of transport and also for sport and leisure. Advances in cycle technology have thereby contributed to social change.

Technological progress has allowed cycles to compete successfully with other forms of transport, which is especially significant as environmental issues arise. Environmentally-friendly transport policy is necessary, but cannot succeed without good technology. Improved technology allows cycles to function better: to translate the rider's energy input into motion more efficiently, in greater safety, more comfortably, more reliably and more enjoyably. It also facilitates use by a wider clientele: children, the elderly, parents doing the shopping, and people with disabilities.

When, as in Copenhagen, transport infrastructure, policy and cycle technology come together, cycling becomes the norm across a wide range of demographic groups, because it is easier and more convenient than other options. But this is not possible without appropriate technology that allows the journey to be made by people of all ages and abilities reliably, safely and in ordinary clothes. Two of our chapters are of particular relevance in this context: "Cycling Technologies and Disability" by Ron Buliung, Annika Kruse, Glen Norcliffe and John Radford raises important issues which are commonly given less attention than they deserve. David Henshaw's chapter "Luggage and Passenger Carrying" focuses particularly on cycle technology for conveying children.

DOI: 10.4324/9781003142041-11

Evolution of cycle technology

As Hans-Erhard Lessing points out in his chapter "Configuration of Cycles", the history of technology is evolutionary. Cycle technology can be traced back to the early 19th century for two-wheelers and well into the 18th century for human-powered machines with three or four wheels. This evolution fed on existing technologies and contributed to evolving ones. In its early days cycle technology adopted and adapted the techniques of wheelwrights, blacksmiths, carpenters, joiners and leather workers. Later, as cycles became more intricately engineered, there were borrowings of a more delicate nature, such as the adoption of the clockmaker's fusée chain to link hub gears to their control mechanisms. In the late 19th century, cycle technology made important developments in the fields of change-speed gearing and differential gears, which contributed to the development of motorcycles and automobiles.

Advances in cycle technology stimulated developments in the use of materials, such as vulcanized rubber, lightweight steel tubing, pressed steel and aluminum. It is significant that the Wright Brothers were bicycle mechanics: by the time they started experimenting with airplanes, bicycle technology was at the leading edge of strong lightweight construction and efficient energy transmission.

Technological dead ends

Cycle technology is subject to various influences, some of which limit or curtail the widespread adoption of innovative ideas and improvements. In his chapter, Lessing highlights design restrictions imposed by the Union Cycliste Internationale (UCI) with "the objective of preserving the culture and image of the bicycle as an historical fact". The UCI is no stranger to controversy concerning technical developments. The widespread adoption of recumbent cycles received a huge setback in the 1930s by a UCI ruling, rumored to have been encouraged by the French cycle industry.

But a technological dead end may be reached for reasons other than rules and regulations. It may be because a different solution prevails, through a change in public perception or fashion. The high-wheelers of the 1880s were refined and efficient machines, mainly used by relatively wealthy and athletic young men; whereas other cyclists of that time (who also needed to be quite rich) mostly used tricycles – safer and more user-friendly but bulky and heavier. High-wheelers and tricycles were rapidly superseded by the safety bicycle, which was safer for a wider range of users and more compact, therefore easier to store. Compared to a tricycle, the safety bicycle was significantly cheaper, and it was easier to propel, once the rider had mastered the art, and overcome the fear, of balancing on two wheels.

The influence of fashion

Fashion, peer group pressure and popular perceptions of what is "cool" or "uncool", may all play a major part in the adoption or rejection of aspects of cycle technology. This can apply to almost anything, from the general configuration of the cycle to the individual components on it. Examples include the frame material, whether or not the frame has a top tube, the style of handlebar, and the type of gearing. These perceptions, preferences and prejudices vary over time and from country to country. There are often sexist and ageist undertones: certain frame configurations, wheel sizes, handlebar styles, and even the use of fenders (mudguards) being considered by some as only for girls, gays, adolescents, elderly spinsters or middle-aged men in Lycra.

David Minter's chapter, "Frames and Materials," covers, among other things, the materials from which cycle frames may be built. A major factor leading to the Raleigh cycle company ceasing frame production in the United Kingdom was a relatively sudden shift in customer preference away from steel frames and towards aluminum. Raleigh had succeeded in considerably rationalizing and modernizing their main production facility to make automatically welded steel frames very efficiently and with very little waste. But facing the unexpected demand for aluminum frames, the company could not afford to re-equip and instead commissioned production from overseas suppliers. There was no significant technological advantage in these aluminum frames: they weighed about the same as comparable steel ones. But customers liked the look of aluminum frames and the word on the street was that aluminum was better.

The influence of cycling fashion can lead to the rejection of superior or more appropriate technology. But in the case of the mountain bike (MTB), an apparently irrational fashion led to major improvements in the design of cycles and componentry. Whilst it was easy to understand why an MTB would be the bike of choice for hurtling down a Californian mountain, major cycle makers were initially bemused as to why anyone in a European or North American city would want one. MTBs nonetheless became so popular from the mid-1980s onwards, that they soon dominated the market. The benefits included huge improvements to gears and brakes, and refinements to aspects of frame design. MTB tires coped better with potholed city streets, and flat handlebars were ergonomically superior for city use. Over time, aspects of MTB technology have crossed over to other styles of cycle, ranging from top quality road bikes to inexpensive utility cycles.

Technology push versus market pull

Sometimes worthwhile developments in cycle technology appear at a time when there is no market for them, or where the cost of manufacture is prohibitive, or suitable materials are unavailable. Nonetheless, many years later such innovations may gain widespread adoption. For example, Dan Farrell's chapter "Transmission and Brakes" deals, *inter alia*, with disc brakes for cycles. These can be traced back to 1894 but it took a century for them to become commonplace.

Cycle technology comes in many forms: some are high profile and catch the public imagination, such as Boardman's carbon fiber bike in the Barcelona Olympics; others are low-key and taken for granted, such as the automatic freewheel, which in various forms rapidly became the norm in the early years of the 20th century. The introduction of the freewheel depended on the development of improved brakes to compensate for the lack of fixed-wheel drive. In this case, technology push in the areas of braking and transmission coincided fortuitously with market pull. Technological advances have contributed hugely to the improved cycling experience of countless millions of cyclists, past and present, and thus to the popularity of cycling, with all the social, economic and cultural consequences that arise from it. We hope that the chapters in this part will help readers appreciate key elements of cycle technology and heighten awareness of the possibilities it offers.

Chapters in this part

The first chapter, by Hans-Erhard Lessing, deals with the many possible configurations of cycles, some of which may be unfamiliar to the reader. A cycle may have one, two, three or four wheels, arranged in various ways. It may have one, two, three or more riders, who may be sitting upright, recumbent, semi-recumbent or even prone. Some may be propelling the vehicle while others are passive passengers. By way of historical context, Lessing also covers the 1817 draisine, the machine that proved the counter-intuitive idea that a single-track steerable two-wheeler could easily be balanced, enabling a rider to travel considerably further and faster than a pedestrian.

In the second chapter, David Minter discusses cycle frames. He provides a clear, non-mathematical introduction to the forces a cycle frame has to withstand and relates this to the various designs adopted by cycle makers. Apart from the familiar diamond frame, there are many other types that have their own advantages and disadvantages. Minter also discusses frame geometry and the various materials from which frames are made.

A few cycles have one wheel, most commonly they have two and some have more. The very word "cycle" derives from *kuklos*, the Greek word for a circle, the shape of a wheel. In the late 19th century, the wheel became symbolic of cycling, as in the badge of the UK's Cyclists' Touring Club; and cyclists were sometimes known as wheelmen or wheelers. In the third chapter, Tony Hadland provides an overview of current wheel technology. He covers hubs, including those incorporating planetary gears, multiple freewheels, brakes and dynamos, and discusses rims, spokes and tires. The most common forms of wheel suspension are also described, including those found on mountain bikes, road bikes and small-wheelers.

The link between transmission and braking lies in motion; the former transmitting the effort of the rider into forward progress, the latter in arresting it. Although the inventor of the draisine incorporated a rudimentary brake, much of the slowing of such machines was by means of the rider's feet. Forward motion required those same feet to transmit the rider's energy to the ground. Fortunately, this long ceased to be the case. In the fourth chapter, Dan Farrell deals with primary drive, including chains, belts and shafts; various forms of change-speed gearing, such as derailleurs, internally-geared hubs and bottom bracket gears; and rim brakes, drum brakes and disc brakes.

From the earliest days of the draisine there was an appreciation that the machine could carry not only a rider but also luggage. Each wheel could have a pannier on either side and a luggage rack over it, and additional bags could be attached to the handlebar and saddle. The situation is very similar today, the main differences being in the materials from which the bags are made and in prevailing fashions. But cycles can also carry human passengers and in the fifth chapter David Henshaw explores the topic, with particular emphasis on child carrying. In so doing he considers child seats, trailer bikes, child trailers, rickshaws and electric assistance.

The final chapter is by Ron Buliung, Annika Kruse, Glen Norcliffe and John Radford. They ask why disability and cycling has received little attention in planning practice, despite the relationship between cycling and disability being centuries-old. They question the neglect in professional planning of the relationship between cycling and supportive infrastructures and they examine the cycling technologies developed for persons with disability.

8
CONFIGURATION OF CYCLES

Hans-Erhard Lessing

Defining a cycle

A useful definition of the machine called a bicycle, bike or wheel has been provided by the *United Nations Convention on Road Traffic* of 1968 using the broader term "cycle":

> "Cycle" means any vehicle which has at least two wheels and is propelled solely by the muscular energy of the person on that vehicle, in particular by means of pedals or hand-cranks.
>
> *(UN 1968: 3)*

This definition encompasses virtually anything from the earliest two-wheeler to present-day human-powered vehicles, except unicycles.

It turns out to be also particularly useful as it puts an end to discussions about the concept of the "first true bicycle". This concept completely ignores the evolutionary character of the history of technology. Since the term "bicycle" first appeared on front-cranked machines and high-wheelers from 1868 onward, the "first true bicycle" is claimed to date from 1868. But nobody at that time yelled: "Look, there comes the first true bicycle", and today's bicycle appeared today. The French designation "la bicyclette" dates from 1886, "le vélo(cipède)" in 1818, and the Polish designation "Rower" in 1885 – offering a fertile soil for further jingoist firsts. Today bicycle historians are unanimous that cycling, defined as balancing seated upon two wheels in line, was invented in 1817 by Karl von Drais (Figure 8.1) (ICHC 1990–2019). The German terms "fahrrad" for cycle and "radfahren" for cycling were introduced by language patriots as late as 1885. Curiously, the latter originally meant a carousel of children's sledges, connected by ropes to a central turnstile posted on ice, that workers turned around in return for a penny. The UN definition also includes roller skates and scooters, the latter having been invented in Wroclaw in 1817 (Hadland & Lessing 2014: 29).

Since cycles are ridden on roads, there are legal restrictions and tighter definitions within national regulations. In the United States, the Code of Federal Regulations defines the bicycle as *a two-wheeled vehicle having a rear drive wheel solely human-powered* (US 1978: §1512.2). But the most restrictive definition stems from the Union Cycliste Internationale (UCI), the

DOI: 10.4324/9781003142041-12

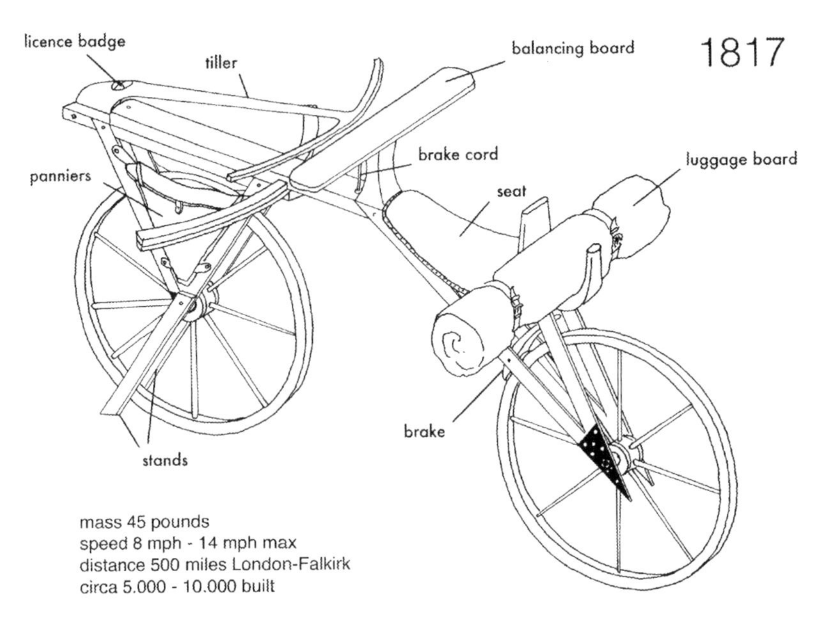

Figure 8.1 Karl von Drais's Draisine of 1817. Illustration by the late Joachim Lessing. Reproduced by permission of Hans-Erhard Lessing.

organization ruling cycle sports, based at Aigle, Switzerland. Only a few sentences shall be extracted from their lengthy charter limiting the racing machine down to the millimeter:

> The bicycle is a vehicle with two wheels of equal diameter. The front wheel shall be steerable; the rear wheel shall be driven through a system comprising pedals and a chain.
>
> The weight of the bicycle cannot be less than 6.8 kilograms.
>
> The frame and forks must be able to fit entirely within the template formed by seven rectangular boxes of 80 mm width. [Each of the notional 80 × 80 mm boxes of different lengths encloses an element of the frame, such as the top tube, fork, etc.]
>
> This rule results from the Lugano Charter [1998] and has the objective of preserving the culture and image of the bicycle as an historical fact.
>
> *(UCI 2018)*

This last statement helps explain why the sport and everyday bicycles haven't changed for a hundred years (Schmitz 1990).

The rider of the cycle is a cyclist, and the UN Convention of 1968 was kind enough to elevate him/her to the ranks of an operator of the cycle. This rehabilitation was overdue since the German Road Construction Guidelines of 1956 categorized cyclists as follows:

> Cyclists, mopeds and pedestrians on the roadway or on the hard shoulder are to be regarded as sideways relocatable obstacles which reduce the width of the roadway and the obstacle-free width existing next to the roadway.
>
> *(Lessing 2017: 239)*

Contrast this with the sanguine definition of Italian anthropologist Paolo Mantegazza in 1896 (Gentile 1896).

> Cycling is the triumph of human thought over the inertia of matter. Two wheels that barely touch the ground, that carry you far away as if on wings at a dizzying, exhilarating speed, without the cruel sweat of whipped draft animals, without the hated noise of fuming machines – a miracle of balance, of lightness, of simplicity – a maximum of power and a minimum of friction – a miracle of speed and elegance – man who wants to become an angel and no longer touches the earth – Mercury, which has risen from his ancient Hellenic tomb and appears alive before us – that is the modern cyclist.

Cycling as machine sports

The present-day configuration of everyday cycles has been determined by history, cycle-racing officials and the bicycle sector. Since the earliest velocipedes were meant to replace saddle horses, horse racing was soon transferred into velocipede and hobby-horse racing as side events at British horse tracks from 1819 onward. The earliest organized velocipede race took place "in silence" at Munich in 1829. The winner averaged 14 mph for half an hour and received six thalers and a flag signed by the Bavarian king (Hadland & Lessing 2014: 18).

Club cycle races began in France in 1868 (Vignette H). Cycling sports secured the survival of bicycle manufacturers after the Franco-Prussian war, when France was in crisis, and British enthusiasts kept their high-wheeler manufacturers going. Somehow the United States avoided the agonizing multi-stage races like the Giro d'Italia and the Tour de France that became prime media events. The League of American Wheelmen insisted early on that stressful bicycle races send the wrong message about the ease of civil cycling, as expressed by Michigan's chief consul in 1888:

> There is no more sense in the LAW running bicycle races than the poultry association staging cock fights or the dairy association, bull fights.
>
> *(Epperson 2010: 93)*

Machine sports impose a disastrous restrictive logic that contrasts with the problem-solving capabilities of engineers, as seen in an example from rowing sports of the 1980s (Miller 2000). A similar conflict arose in the 1930s at the Union Cycliste Internationale. The phenomenon behind is best explained via the logo of Levi's blue jeans, with two horses trying to tear the jeans apart. One horse alone could achieve the same, if the other side were tied to a tree. What is true for pulling apart also holds for pressing apart, as realized by Charles Challand at Geneva, who patented the earliest recumbent cycle in 1895. By changing an operator's position and adding a firm backrest (analogous to a tree) instead of pulling by the arms on the handlebar (analogous to a second horse), the force on the pedals is nearly doubled. This principle was rediscovered by Georges Mochet, a successful French manufacturer of four-wheeled cycle-cars for commuting and leisure under the brand name Velocar since 1920. He patented a two-wheeled recumbent called the Velo Velocar that had the operator sitting low and provided with a backrest. Needless to say, test runs by a racing cyclist broke all track records of the day. After heated debates in the cycling press, the UCI excluded the Velo Velocar from standard races, notwithstanding that it had complied with UCI regulations thus far. They annulled its records in 1934, a decision criticized as "a wasted chance" and "a disaster for cycle technology" (Schmitz 2010: 67). Influence on this ruling by the French cycle industry was rumored

To make the personal performance of racers comparable machine sports organizations needed to standardize the configuration of machines. This is reasonable enough, but they should also evolve their rules to reflect the state of the art, in the same way as the Formula One World Championships regulated by the FIA organization. Trying to keep the machines simple, the UCI was extremely reluctant to adopt new cycle technology. Every improvement was termed an "unfair advantage," e.g., the derailleur gear that French tourers used since 1900 was forbidden in races until finally accepted by the UCI in 1936. The statement that cycle races help to improve the everyday bicycle appears mere eyewash in view of this. Commuters want as many unfair advantages as they can get (Hadland & Lessing 2014: xii).

The bicycle sector

Comparing the bicycle market with the car market can be misleading. The automobile is a sellers' market where each manufacturer has contracts with its dealers, who have to sell and repair that brand exclusively, but have no influence on the design of the cars. Car manufacturers do market research to find out what the operators of their cars want. Only one exception is known where the dealers finally achieved the omission of car bumpers in Germany in order to gain more sheet-metal repairs.

The bicycle market is generally a buyers' market like those of most other consumer products (the exception being Giant – Chapter 18). But the working of the bicycle market differs among continents and countries. In Europe many bicycle dealers are ex-racers, and their ideal of a cycle operator clearly is represented by themselves in their glorious younger years. Even the city cyclist is imagined as a bachelor hastening at utmost speed through town. In France in 2019, for example, independent dealers accounted for 55% of bicycle sales, by value (Union Sport & Cycle 2020: Chapter 17). These independent dealers order what they think is right for their customers at trade fairs, mainly machines for sports or lookalikes. For their bread-and-butter business they also order ladies' and children's bicycles with less fervor. In contrast, in North America the great majority of sales are made by mass sellers (big boxes): they accounted for 74% of unit sales (but only 29% by value as they concentrate on low end and child bicycle sales). Sporting chains accounted for 5% of sales, while independent dealers accounted for just 12% of the units sold (but 49% by value). The 3% of units sold on-line in 2019 is estimated to have doubled during the Coronavirus pandemic of 2020 as Americans took to indoor and outdoor cycling in large numbers. The configuration of branded bikes sold by big boxes, chains and on-line is usually negotiated between the brand owners and the makers (mostly in China) (see Chapters 14 and 17). In Asia the picture is brighter because the world's largest bicycle maker, Giant of Taiwan, has established over the past 20 years a network of 12,000 retail outlets in 50 countries, especially in Asia (see Chapter 18). Giant's arrangement resembles a sellers' market, much like the car dealers.

The fatal misconception in public and city planners' opinion was and still is that the status of the bicycle is the inevitable result of a functioning sellers' market like that of the car whereas, depressingly, in terms of design and safety it is mainly an unregulated buyers' market following buyers' tastes. In the boom following the oil crisis, industrial designers had to learn the bitter lesson that their proposals for more sensible cycles for city dwellers had no chance in the face of dealer power. Help for the city bicycle has to come from the transport and environment departments of the governments, by funding design competitions and start-up companies, but through more specific regulations, too. Also bring back single-brand manufacturers and shops to Europe and the US!

Gender

Cycling has its origins in skating on ice, then mostly by students, as was recorded by German inventor Karl von Drais. Women were also present on ice, but there was a difference. While British women could skate, it was socially unacceptable for French and German women: in the event of a fall, their ankles or even more of their legs would be exposed. Therefore women sat on chair sledges and were pushed around on the ice by the students. Similarly, Drais provided three-wheeled velocipedes with a ladies' seat between the front wheels, whereas British builder Denis Johnson built a ladies' two-wheeler with deep instep and a belly support. The high-wheeler (aka ordinary) of the 1870s was a no-go for the ladies, but the costly high-wheeled tricycles had ample space for long skirts.

The diamond-frame bicycles of the 1890s promised easy handling, but initially required straddling the top tube. Puffy trousers from the agenda of feminist Amelia Bloomer and divided skirts were necessary, until the triangulated frame of Charles Brown with low step-thru finally eased cycling for ladies in long skirts. Women in trousers on bicycles have been a common view only since the 1920s. American bicycle magnate, Albert Pope, had favored chainless bicycles, driven by a shaft and beveled gear encapsulated within the right chain stay, in the late 1890s to avoid soiled clothing, but these were costly and suffered from broken teeth of their gears on bumpy roads.

Cycling science

Contrary to common belief the bicycle is a most complicated subject of physical mechanics. Arend Schwab's group at Delft University has set up a computer model with 25 parameters that describes cycle behavior exactly, in the process destroying many a cycling myth. For instance, the rule "always-steer-into-the-undesired-fall" relies on the centrifugal effect and not on shifting support points under the center of gravity for speeds above 0.4 mph (van Dijk 2007). This rule was stated long ago by Drais. Luckily, the basic laws of cycle physics and their influence on cycle configuration can be understood using calculus to deal with simple proportionality.

Friction

Coulomb's global law says that the frictional force opposing the moving force onto a body is proportional to the force acting vertically onto this, usually its weight:

friction equals coefficient times weight

the coefficient for a bicycle amounting to circa one percent. This explains the advantage of the bicycle or of any vehicle when, instead of holding your body upright and walking, you put it to rest on a stable cart and – forgetting about pedaling and balancing – pull along a string fastened to a distant post. If your pulling force equals the opposing friction force, you cruise at constant speed. Coulomb's law can't help to decide about the optimal number of wheels, e.g., if the weight is distributed equally on four wheels, each contributes a quarter of friction, which taken four times adds up to the total friction again; and the same holds for three or two wheels. Hence this law would not show an advantage of the two-wheeler.

Inertia

If you pull harder on that string, you will accelerate according to Newton's law (wherein to use mass instead of weight)

acceleration equals force divided by mass

Accordingly, the heavier the bicycle plus rider, the slower to accelerate. Force here means the force of the rear wheel onto level ground, which is the pedal thrust diminished by the leverage of crank length, chainring, sprocket and rear-wheel radii, so that even a child can hold the cyclist back from starting. In addition, friction reduces the driving force even more. The wheels' rotation also has to be sped up against their rotational inertia, usually calculated as the mass of their rim plus tire divided by their radius squared. Taking this into account the bicycle's linear acceleration is hindered by an

effective mass being MASS plus four times rim/tire mass

where MASS means bicycle plus rider without rims and tires. This explains why racers try to use as light rims and tires as possible (Hadland & Lessing 2014: 66).

Tire friction

Tire friction is the number two impediment in cycling after air drag. Coulomb's coefficient includes bearing friction and rolling friction of tires. This is where engineering begins and calculus doesn't help anymore, since that coefficient hides all true influences of the real situation (Wilson & Schmidt 2020: chapter 6). Part of the driving force goes into kneading the tires, while they roll, heating the tires and their surroundings, and thus contributing to rolling resistance. With new tire types frequently being introduced, measurements of their rolling resistance can hardly keep up and aren't usually published.

Air drag

Another culprit reducing the driving force is air drag impeding racers and calculated as

drag equals half c_D times air density times cross-sectional area times speed squared

where c_D is the drag coefficient. Hence crouching on the handlebars reduces the cyclist's frontal area from, say, 5.9–3.9 ft^2 and thus drag to 66%. For the peaceful city cyclist 10 mph is enough, but the racer wants 40 mph, which results in a 16-fold air drag. For the city cyclist air drag is not the dominant resistance, so he can ride in the upright position, which is the most efficient one for long-time power, as has been demonstrated by pedal-powered aircraft, where streamlined cabins take care of air drag.

How many wheels? How many operators?

Unicycles

Richard Hemmings, a mechanic from Dubuque, Iowa, obtained the earliest patent in 1869, during the heyday of the cranked two-wheeler from France. Unicycles have the rider sitting above, while monocycles within the wheel. In unicycles, the advantages of minimization are cancelled out by the need for double balancing – to and fro plus left or right. Nevertheless, balancing and cranking can be learned, and there is a subculture comprising young people, sports and circus artists (Unicyclopedia 2020).

Dicycles

Two large wheels side by side, supporting the rider in between and enabling him or her to scoot along, appeared in England in 1820. Cruising speeds of 12 mph or a top speed of 20 mph over one mile were claimed. Carl Otto's dicycle, produced in Birmingham in the 1880s, offered a safer alternative to dangerous high-wheelers and costlier tricycles. Its seat plus pedal-crankshaft mount rotated on the wheels' axle, and its steering worked by slowing one wheel or the other. Ladies preferred Otto's machines, because these could be mounted easily when wearing a long skirt.

Bicycles

According to a theoretical model, placing two firm wheels in line instead of side by side, reduces rolling resistance to 89% on a hard sandy road. (Bekker 1956: 201). A real measurement under different conditions gave a lower figure of 77%. This, and the reduced rotational inertia of only two wheels, as compared to three or four, are the main reasons for the two-wheeler's superiority over multi-wheeled vehicles of the same weight. Stability to and fro was provided in Drais's two-wheeler, yet most contemporaries abhorred mounting a vehicle in merely labile equilibrium – except students and mechanically-minded persons. Present-day cyclists learn to balance intuitively as a child. But Drais thought it was impossible to offer a mechanical drive while balancing with one's feet off the ground. Actually, he had already experimented with a foot-operated crankshaft between the rear wheels of his earlier four-wheeler, but this was stable of course. By placing the front-wheel axle six inches behind its pivot, Drais introduced the caster to vehicles and revolutionized human-powered vehicular transport (Bulliet 2015: 210). Another innovation at that time was the friction brake on the rear wheel, operated by a string ending in a ring which Drais put on one finger. Carriages adopted such brakes rather late, copying from railroad wagons in the 1840s. Amazingly, many features of today's bicycles were envisioned by Drais, like equal-sized wheels of 27-inch diameter, attachable stands for parking, carrier, panniers, and oiled bearings. With serendipity, evolution could have avoided 70 years of detour thru front cranks and high wheels and continued directly with safety bicycles like the 1888 model of the British Rover.

Tandem

Drais imagined a lengthy velocipede with two seats in line, later called a *tandem*. The word *tandem* in Latin means "finally" or "at length". After an 1800 tax increase on prestigious four-wheeled coaches in London, carriage builder Tilbury invented a light two-wheeled (side by side) cabriolet plus horse which he called "tantum" meaning "only" or "alone" – a pun referring to the horse being reduced to one and the lower taxes. Subsequently, the hunters' cabriolet with two horses in line became fashionable, since it demonstrated that one could afford more than one horse, and the term "tantum" transferred to that one. Written phonetically as "tandam" or "tandem" the word became a synonym for "two in line", whereas the original "tantum" was re-named a "Tilbury" (Lessing 2014). Tandem bicycles had their heyday in France in the 1930s when young couples used them before and after weddings.

Races were run because of the promise of higher speed with two operators pedaling. And in paced races they could act as a windshield for the following single racer, who therefore could achieve higher speeds.

Riding a tandem needs cooperation between the captain on the handlebars and the stoker sitting behind. The stoker should refrain from any balancing lest he/she counteracts the captain's balancing which could result in a fall. Because of their extension and weight, tandems are cumbersome to use in the city, so they are used mainly for touring, except for the Italian short tandem Bi-Bici.

Sociable

Seating two operators side-by-side on two wheels began with James Starley's Sociable Ariel in 1870. The doubled frontal area meant doubled air drag, and starting off needed good cooperation between the operators. On narrow bike lanes, these machines were hard to overtake.

Triplet

A two-wheeler seating three operators in line holds the promise of still more speed, and the Rudge Triplet of 1892 achieved period records. Paced races used it as the stayer machine, succeeded by quadruplets, quintuplets, sextuplets and finally a motorcycle (such as in Keirin races).

Tricycle

Tricycles are heavier than bicycles and have about 10% more rolling resistance plus more rotational inertia due to the third wheel. This is the penalty paid for stability. To spare anxious souls balancing on two wheels, the tricycle was propagated 200 years ago. Today, unfortunately, it has the image of being useful only for aged or disabled people. Whereas around 1880s there were numerous tricycle configurations with high wheels, today the front-steerer/rear-driver predominates, i.e. with one steered wheel in front and two chain-driven rear wheels, usually of the standard size. Like any car, it benefits from a differential to apportion power transmission between the driven wheels when turning. American magnate Pope built another standard tricycle configuration, the Columbia "Surprise", with two front wheels with Ackermann steering and one chain-driven rear wheel. This spared the costly differential and enabled an important feature: the spacing between the front wheels could be reduced by hand, e.g. to pass narrow doorways. This would be the ideal cycle for city dwellers who hadn't learned cycling in younger years and owned no garage. Some tricycles have a lockable parking brake that makes mounting easy. Operating a tricycle is different from riding a bicycle, which is why riders of motorcycles with sidecars usually need an extended driver's license. A rider who comes from bicycling has to learn leaning his body rather than the tricycle into a curve, otherwise it would tip over.

Quadricycles

In spite of even more rotational inertia due to the fourth wheel, quadricycles found their niche around 1890 when Rudge of England and Elliott of America built theirs, the former with Ackermann steering and the latter with a child's seat. After World War 1 there was a renaissance of quadricycles called cycle cars that were set lower. Since the oil crisis in the 1970s, human-power designers have built what they call "quadracycles," streamlined recumbents for weather-protected every-day use.

Recumbents

Placing the operator in a leaned-back position raises the thrust onto the pedals, lowers the frontal area (and therefore air drag) and reduces the drop height and risk of bone fractures. After dissatisfaction with the UCI's ideology led to the foundation of today's International Human-Powered Vehicle Association (IHPVA) in 1976, there was an outburst of creativity, building all possible configurations but set lower. Since the legs pedaled horizontally, new cloth seats in the style of a director's chair could replace the usual saddle. Direct drive with pedal cranks on the front wheel was ergonomically obvious, but unfortunately an affordable, mass-produced hub gear with cranks and pedals was, and remains, unavailable. So, either the recumbent had to be extended and the pedal crank placed behind the front wheel (long-wheelbase); or one made the front wheel small and placed the pedal crank above it (short-wheelbase). Alas the short recumbents, in particular, are difficult to start. Treading down the right pedal one has then to lift the outstretched left leg up to nearly eye level and hit the left pedal, which rarely succeeds before the machine loses speed and tumbles. A common remedy is to lean against a wall or post, bring both feet onto the pedals, and then pedal away. Because of this most recumbents have found wider acceptance as tricycles that are more stable.

A front-wheel pedaled hub gear would be needed for recumbents and other cycles, such as upright convertible tricycles, so that a single rear wheel could be switched for a two-wheeled one with a shopping basket by snap-on action, as launched by Australian designer Paul Cockburn. Hub or crank-case gears with cranks and pedals have been devised by Kervelo, France, with up to 12 speeds, but still at high, short-series cost.

Aerodynamics: streamliners, velomobiles and fairings

Students of engineering and autodidacts have created streamlined HPVs with low c_D and small frontal area surpassing the 80 mph speed limit – called either "streamliners" (2-wheeled) or "velomobiles" (3- or 4-wheeled). For the upright city cyclist this can't be done easily as side gusts impact a large area to unbalance the rider. Nevertheless at least a rounded front shield could improve c_D and protect from rain and splash. Such a vertical transparent windshield rising up to eye level was offered as an add-on in France by 1893. Forbidden by the UCI, more add-on roll-up windshields, as well as bubbles for the crouched cyclist and for recumbents, branded Zzipper, have been manufactured by Glen Brown since the mid-1970s.

The BMW motor scooter C1 with its enclosed safety cell brought a new idea to two-wheeled transport, and needed no helmet due to fastened seatbelts. A similar pedal or e-bike, but without roof or windshield, was proposed by Crispin Sinclair in 2016. The C1 was discontinued, but will be revived by Govecs, Poland, who plan an electric hybrid between the C1 and three-wheeled Piaggio MP3.

Future city cycles

Only by making cycling much more convenient than at present, is there hope to attract more citizens to mount the steed. Ladies on e-bikes certainly would love to have a cloth seat like on recumbents that wouldn't crumple their skirts. Motor scooters present a model for more convenience. Their platform makes mounting easier than on a swinging pedal crank, and operators can offer a lift to any passenger. An expired German patent on a pedaled "Fahrroller" (bikooter) by Hans-Georg Ruffer enabling this merely needs to be put into production at last. More generally, the eternal risk and servicing of punctures should be tackled. Designer Mike

Burrows has long advocated single-sided wheel mounting, on monoblades and structural chain cases, to enable easy repair or replacement of tires. (Burrows & Hadland 2014: 149). Designer bicycles like the "Strida" or the folding "Gocycle" already sport these interchangeable wheels. A component manufacturer is needed to launch such a novel standardized power-transmission kit upon which manufacturers could build their bicycles and e-bikes. The present e-bike boom holds the promise to surmount the UCI blockade in sector categories and to generate convenient new features from which pedal cycles should also benefit.

References

Bekker, M. G. (1956) *Theory of Land Locomotion*, Ann Arbor: University of Michigan Press.

Bulliet, R. W. (2015) *The Wheel – Inventions and Reinventions*, New York: Columbia University Press

Burrows, M. and Hadland, T. (2014) *Bicycle Design – Towards the Perfect Machine*, Haddenham: Snowbooks.

Epperson, B. D. (2010) *Peddling Bicycles to America*, Jefferson: MacFarland.

Hadland, T. and Lessing, H.-E. (2014) *Bicycle Design – An Illustrated History*, Cambridge, MA: MIT Press.

International Cycle History Conference (ICHC) (1990–2019) Cycle History Volumes 1 to 30. http://www.ichc.biz/

Lessing, H.-E. (2014) "Ein Etym, das Arno Schmidt nicht knacken konnte," Frankfurter Allgemeine Zeitung of 19.02.2014.

Miller, B. (2000) "The development of rowing equipment." http://www.rowinghistory.net/resources/equipment

Schmitz. A (1990) "Why your bicycle hasn't changed for 106 years," *Cycling Science* 2, June: 3–9.

Schmitz, A. (2010) *Cyclists Cycling Cycles & Cycle Parts*, Faringdon: Tony Hadland.

UCI (2018) Technical Regulation, Article 1.3.007-020. https://www.uci.org/docs/default-source/equipment/clarificationguideoftheucitechnicalregulation-2018-05-02-eng_english.pdf

UN (1968) Code of Road Regulations, Vienna, Article I, p. 3. https://treaties.un.org/doc/Treaties/1977/05/1977052400-13 AM/Ch_XI_B_19.pdf

Unicyclopedia (2020) http://en.wikibooks.org/wiki/The_Unicyclopedia

Union Sport & Cycles (2020) https://www.unionsportcycle.com/cycle-mobilite/le-marche-du-cycle-en-france

US (1978) 16 CFR § 1512.2. https://www.law.cornell.edu/cfr/text/16/1512.2

Van Dijk, T. (2007) "Bicycles made to measure," Delft Outlook, July, 7–11.

Wilson, D. G. and Schmidt, T. (2020) *Bicycling Science*, 4th edition, Cambridge, MA: MIT Press.

9
FRAMES AND MATERIALS

David Minter

The frame and fork comprise the main structure of a typical cycle, connecting the wheels, other components and component assemblies and transferring forces between them. The almost universal two-bladed fork provides a pivoting support for the axle of a steerable wheel. Instead of a fork, a monoblade or kingpin may be used to provide single-sided pivoting wheel support. Monoblades have been used to improve the aerodynamics of racing bikes, for ease of tire changing, and to reduce the folded size of certain portable bicycles. Kingpins are mostly found in the Ackerman steering configurations of quadricycles and certain types of tricycle (see Figure 9.1).

The forces imparted to the frame by the rider (such as pedaling and weight) and by bumps, braking or other causes are transmitted by combinations of compression, tension, bending, torsion and shear along the individual frame members.

Structural principles

A detailed discussion of structural principles would be excessive but here is a basic explanation of how structures like bicycle frames support applied loads.

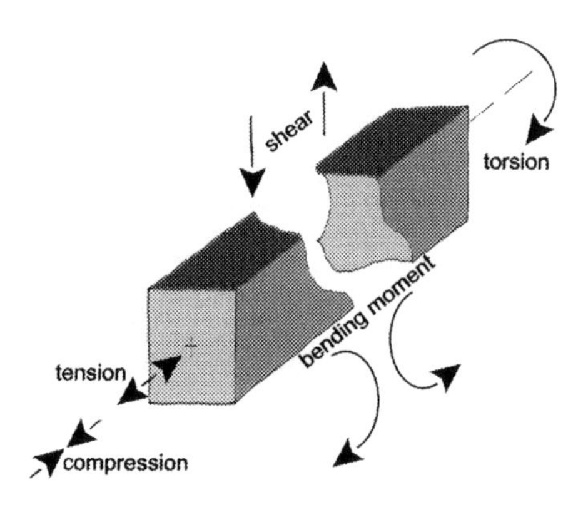

Figure 9.1 Five types of force that can act on a structural member. (copyright Tony Hadland).

DOI: 10.4324/9781003142041-13

A frame member is most efficient structurally (using the least material to support the applied load and/or provide the required stiffness) when virtually all of the material is placed where it can best resist the applied forces and when all the material is loaded almost to the material's failure stress. For simplicity, for now assume that:

- This material is as strong in compression as in tension (like many metals),
- This material has the same properties in all directions,
- That the members are symmetric cross-sections, such as cylinders, rectangular rods and the like, and that
- The members we are comparing are the same weight and length.

Resisting compression and tension

Compression and tension are simple axial (directed along the length of the member) forces distributed over the member's cross-section, resulting in a stress, e.g., pressure. The member will collapse under compression or tear apart under tension when the stress exceeds the material's strength. Greater strength increases the member's capacity, as does increasing the member's cross-section. There is more complicated behavior as forces get combined and as members and material characteristics become more complex. For example, a long, skinny member may bow or crumple locally under compression.

The centroid of an axial force is directly aligned with the centroid of the member. The centroid can be considered the average position or "center of gravity" of a cross-section or of a force applied to a cross-section. The centroid of a circular cross-section is exactly in the middle of the circle, but the centroid of a teardrop section varies in distance from its edge as you go around the circumference. Changing the material strength or size far away from the centroid will affect the centroid location more than making the same change close to the centroid.

Resisting bending

Bending a member (as books do on a heavily loaded wooden shelf) creates simultaneous compression and tension zones of equal magnitude applied to different regions of the cross-section. This results in maximum compression at one part of the cross-section, decreasing to zero halfway through the cross-section and increasing to maximum tension at the opposite part. The compression centroid will be somewhere between the maximum compression zone and zero, depending on the shape of the cross-section and similarly for the tension centroid. Increasing the distance between the compression and tension centroids by changing the cross-section increases what is known as the "lever arm" and so can increase the member's resistance to bending without any changes to material properties.

The effect of cross-sectional shape

A solid round rod does not resist bending very well, as very little of the material is highly stressed in compression or tension. A square rod of the same weight moves some lightly-stressed material from near the centerline to regions of greater compression and tension and is a little more efficient. An I-beam (with a cross-section resembling a capital letter I) moves a lot of material from near the centerline to highly stressed regions. A two-dimensional (2D) truss transfers even more material from the centerline to further improve bending capacity. A side view of a standard bicycle frame approximates a 2D truss.

Resisting torsion

Torsion is twisting a member about its long axis, like undoing a bottle's screw cap. As with bending, moving lowly-stressed material to highly-stressed regions and increasing the lever arm increases the member's efficiency. A large diameter tube is stronger and stiffer, both in torsion and bending, than a solid rod or small diameter tube of the same weight.

Resisting shear forces

Shear is applied across a member's cross-section but not directly opposed, like scissors cutting through paper. A member's shear capacity increases with increasing cross-section or material strength.

Coping with multiple loads

A member designed to efficiently resist a specific load can be weak against other types of loads. For example, an I-beam is quite weak in bending across its width and in torsion, and can buckle even when loaded in its strongest direction. A cantilever member is supported at one end and is less efficient (more flexible, with less load capacity for the same weight) at resisting applied loads than when the same member supported at both ends. Bracing the member or changing it to a tube, a truss, or a more complex shape can optimize the structure to better resist multiple loads. As noted previously, a frame and each of its members must resist multiple loads.

Most loads result in bending and shear varying along a member's length within a frame, for reasons that are beyond the scope of this chapter. Generally only tension, compression and torsion might remain the same along the entire member length. As frames need to resist a variety of loads, and often multiple loads simultaneously, different parts of the same member may need greater or lesser resistance.

Fatigue

The above discussions have been about loads and the resulting stresses that might over-stress a member and make it fail even if applied only once. Many materials can also fail slowly from fatigue, when a sub-maximal load is applied and removed many times. When the combination of stress magnitude and number of repetitions is large enough, the member can fail, usually by cracking.

Localized discontinuities in a member (such as unreinforced badly-shaped holes or abrupt thinning or thickening) can concentrate stresses at the discontinuity. This effect means that a member that can otherwise sustain the applied load easily, might over-stress or crack from fatigue unless appropriate reinforcement, tapering, or other measures are adopted at that location.

Types of frames

Traditional "diamond" bicycle frames are "triangulated" to support both ends of most of the frame members and only the fork blades are cantilevers (Chapter 8). Other frame styles employ more cantilevered members (e.g., cross frame, F-frame) to improve other frame design considerations, like low step-over height when mounting or reducing the number of welds. Some frames are more complex: space frames are a 3D truss and monocoque frames don't use individual members, rather a single large complex load-bearing shape. Examples of mono coques include Lotus track and road frames, and Trimble road and MTB frames (Figure 9.2).

Figure 9.2 Clockwise from top left – diamond frame, cross frame, F-frame, short wheelbase recumbent, space frame, monocoque. (copyright Tony Hadland).

Accommodating steering and suspension

The fork allows the cycle's front wheel (or wheels for a few multi-wheeled cycles) to be steered and almost always pivots through the headset bearings, with the handlebar above and the front wheel below. The fork may incorporate suspension (particularly for off-road use, typically telescopic but occasionally pivoted) for improved comfort, traction and shock absorption, albeit with additional complexity, cost and weight.

A cycle frame is often designed as a rigid structure, but it may incorporate pivoted suspension sub-frames for the same reasons and costs as for forks. Occasionally, controlled flex of frame members may be employed to achieve some suspension benefits. Allsop and Zipp frames cantilevered their saddles on flexible and pivoting beams respectively. Cannondale, Pinarello and many other frame designers allow the chain stays and sometimes seat stays and seat tube to flex to avoid the need for separate pivot bearings or other suspension components.

Separable and folding frames

Separable frames have connections allowing the frame to be split into sections for easy packing for travel. Folding bike frames typically employ hinges to compact the frame and handlebars into a single smaller package. To reduce the folded/packed size, separable and folding bikes are often designed around smaller wheels. Small wheels have somewhat greater rolling resistance but allow more freedom in frame design, e.g., increasing luggage capacity.

Multi-rider cycles

Multi-rider cycles generally put two or rarely three or more riders behind one another (hence 'tandem'), though a few "sociables" position riders side-by-side. Tandems are somewhat more aerodynamic than solo cycles and are often faster on flat terrain and downhill but typically climb slower. Tandem frames must sustain much greater loads than solos and their length requires additional care in their design to provide sufficient stiffness.

Frame geometry

A cycle frame's geometry must accommodate several requirements that sometimes conflict, specifically:

- Position the rider(s) appropriately for the cycle's purpose, be it for comfort, aerodynamics, visibility or power output,
- Provide sufficient strength and rigidity (Vignette B),
- Allow sufficient clearances for the wheels, other components, and accessories,
- Minimize or preferably eliminate clashes between the rider(s) and the wheels and components,
- Provide sufficient clearance to accommodate buckled wheels and/or mud on tires,
- Minimize the cycle's aerodynamic, weight and/or rolling resistance (for racing), maximize suspension's performance (off-road),
- Provide appropriate locations to carry luggage, support suspension components or mount an electric motor and battery, as appropriate,
- Handle, brake and steer in an intuitive manner that maximizes the rider's confidence while the cycle is being used.

Many standard bicycle frames position the rider broadly similarly, just varying the arrangement, number, profile, and size of the members connecting the rider's contact points (saddle, grips, and pedals/cranks). Bicycles for racing and time trialing (racing solo against the clock) tuck the rider's torso down and position their arms for improved aerodynamics, but modifications to their riding position are limited to various extents by sporting organization regulations.

Recumbents

Recumbent cycles are not limited by such restrictive regulations as they were banned from conventional cycle racing in 1934, their own racing organization being created in 1976. Recumbents position riders (to a greater or lesser extent) on their backs with the pedals in front, with the intent to improve aerodynamics and comfort. Some recumbents employ partial or full fairings to improve aerodynamics, and fairings may be structural monocoques or merely additions bolted to the cycle frame. The rider(s) may be virtually skimming the ground or perhaps 600 mm higher. Seemingly every wheel size and recumbent frame geometry combination has been tried at some point. Some recumbents have different size wheels front and back to minimize issues like foot/front wheel clashes. There are rare "supine" variations where the rider lies face down with the pedals directly behind them. Recumbents can be two-,

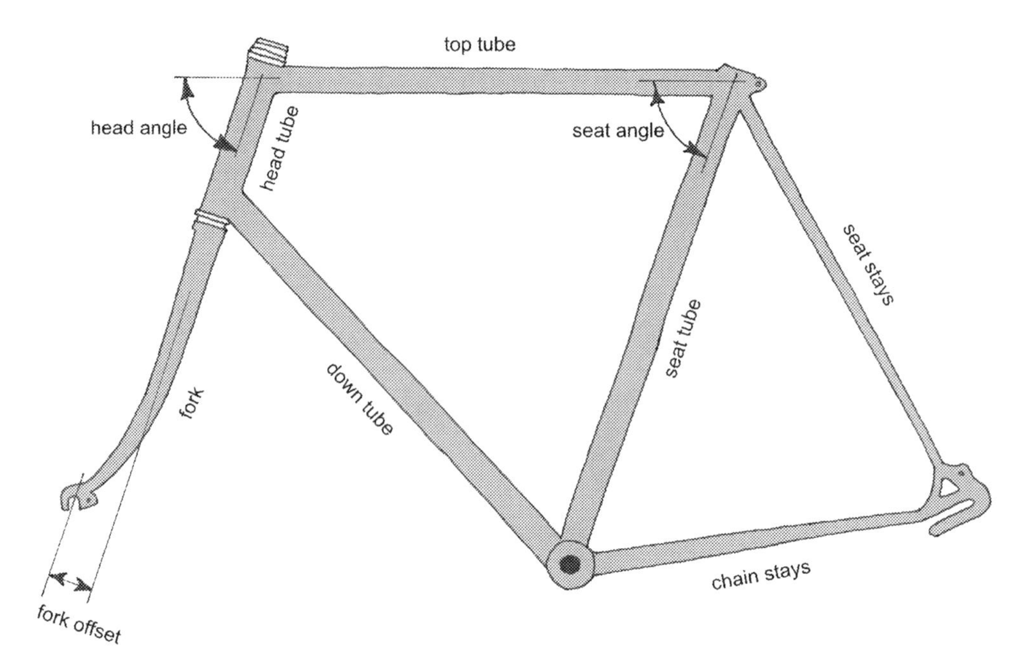

Figure 9.3 Structural members of a typical diamond frame. (copyright Tony Hadland).

three- and four-wheeled machines that may be optimized for high speed (the world record exceeds 139 km/h), load-carrying or relaxed riding. Given their immense variety, it is not possible to adequately address their complexities here.

Standard cycles

Standard cycles typically define the relationship between the saddle and bottom bracket axle by the seat tube angle, usually between 70 and 80 degrees from horizontal, most often approximating 73 degrees. Time trial and triathlon bicycles tend towards steeper seat angles for aerodynamic rider positioning, while recent mountain bikes do the same for better weight distribution while climbing. Utility and long-distance road cycles generally have shallower seat angles to reduce the body weight supported by hands and shoulders. Some frame designers define saddle position by "setback" (horizontal distance behind the bottom bracket) in combination with distance between bottom bracket axle and the seat tube/top tube junction.

Factors affecting steering behavior

Cycle steering behavior is influenced by multiple factors, including:

- Head tube angle – inclination of headset axis from horizontal,
- Rake – distance between front wheel axle and headset axis,
- Wheel size,
- Wheel mass,
- Wheelbase – distance between front and rear wheel axles,
- Total weight applied to front wheel(s),
- Tire width, inflation pressure and tread pattern,

- Stem length and handlebar shape – positioning the hands relative to the headset in both extension (occasionally known as "tiller") and width,
- Aerodynamic forces e.g., sidewinds on rider, cycle and particularly wheels
- Stiffness and profile of the surface being ridden on,

Rider position choices and component selections can significantly affect steering behavior. Aerobars (also known as 'tribars') support the forearms and hands in a narrow position to reduce the rider's aerodynamic drag but they also extend further forward from the headset compared to other handlebar shapes. Optimum rider positioning for racing against the clock can increase the weight on the front wheel compared to standard cycles. Aerodynamic front wheels are especially susceptible to lateral forces from both direct sidewinds and relative side-winds caused by steering the front wheel. The front wheel's aerodynamic center of pressure is almost always in front of the headset axis, meaning that any sidewind tends to turn the wheel downwind. Variations in the location of the centers of pressure of riders and their cycles can aid or degrade the cycle's stability in gusty crosswinds. A frame designer should take account of the likely rider position and component choices and accordingly adjust the frame's geometry to optimise its behavior.

Head angle, rake, and wheel size have the greatest effects on steering and combine to produce two measurements, trail and wheel flop. http://yojimg.net/bike/web_tools/trailcalc.php is an online calculator for trail and flop. Head angles mostly range between 65 and 75 degrees, which allows clearance between the rider's feet and front wheel and places the handlebars at an appropriate distance from the saddle.

Trail

Trail is the horizontal distance between the point where the projected headset axis contacts the ground and the center of the front tire contact area. Trail is the lever arm that helps center the front wheel, with the tire's rolling resistance being the force. Increased trail and rolling resistance increase the self-centering effect, especially as speed increases. This helps when riding in a more or less straight line, which is most of the time; but in fast corners, the cycle can feel as though it is understeering (turning less than the rider intends).

Wheel flop

Wheel flop is the vertical distance the front of the cycle drops as the handlebars are turned from straight ahead. If you imagine a bicycle with an almost horizontal headset; turning the handlebars barely affects the bicycle's direction but the front wheel naturally flops sideways, unless the fork rake matches or exceeds the wheel's radius. This effect reduces as the head angle becomes more perpendicular to the ground. At low speeds, the rider has to resist such wheel flop to ride in a straight line, until the effects of trail overpower flop as speed increases. Reducing wheel flop improves low speed handling.

Optimizing steering behavior

Except for unusual geometries that are impractical on conventional bicycles, increasing the trail measurement also increases wheel flop. Excessively high-trail bicycles tend to flop into corners at low speed while being difficult to guide round tight corners at higher speeds. Low-trail bicycles have less self-centering and can feel "nervous" at high speed if compensating factors (such as increased wheelbase and/or weight on the front wheel) are lacking. Designers try to optimize steering behavior for the cycle's intended speed range: road racing cycles and downhill MTBs are optimized for higher speeds, and children's and utility cycles for lower speeds.

Materials

Cycle frames are traditionally fabricated from various alloys of steel that vary in strength but have the same stiffness per mass. Higher strength allows less steel to be used but the resulting stiffness reduces accordingly. Alloys are metals combined with one or more other elements (usually melted into solution) that result in different properties to the original metal. Steel alloys used for cycle frames typically have good ductility (plastic elongation prior to rupture), with good resistance to cracking and fatigue. These behaviors tend to result in relatively forgiving frame failures (slowly deforming, allowing riders to stop before complete fracture), compared to the brittle failure mode (like breaking glass) that other materials can exhibit when overloaded. Most steel alloys are susceptible to corrosion (rust) and must be protected with plating (with a more corrosion-resistant metal) or paint coatings, save for stainless steel alloys with a high chromium content.

Other metals vary in density (mass per volume), stiffness (modulus of elasticity, effectively deflection per unit mass for a given force), and strength. Aluminum alloys commonly used in cycles are approximately a third as stiff as steel and about a third the density. Unprotected aluminum corrodes rapidly, and so aluminum frames are anodized (a passivation process that increases the natural oxide layer thickness) or coated.

Structural titanium alloys are about half as stiff and half as dense as steel alloys used in cycle frames. Titanium is inherently corrosion-resistant, though frames may be anodized or painted for aesthetic reasons.

It is commonplace to adjust the size, shape, and other characteristics of individual members to improve their performance. Low-density materials like aluminum can compensate for its inherently lower stiffness with larger diameter tubes and still provide a weight reduction compared to steel. Brazing and welding can adversely affect the strength and other properties of some metals and it is fairly common for metal tubes to be "butted" (tubing walls thicker near the ends than in the middle of the tube's length) to compensate for this. Butting can also accommodate the stress concentrations that may occur at and near member junctions. As such, composite frames frequently vary the amount and type of composite at those locations. Manipulating member cross-sections to ovals, teardrops, or more complex shapes (such as tapered) can allow the member to better withstand a variety of loadings or to satisfy other considerations such as aerodynamics and minimum weld lengths at junctions.

With rare exceptions, metals have the same properties in all directions, but composites may have radically different strengths and stiffnesses when measured in different directions. Composites contain at least two separate and distinct materials with different properties.

Wood is a natural composite material with cellulose fibers embedded in lignin. Wood has different strength and stiffness depending on the species and whether its properties are measured along the grain or across it. A few cycle frames are formed by wrapping the junctions of bamboo tubes with glass fiber or carbon fiber tape bonded by epoxy resin. Bamboo has also been used with steel lugs. Some frames are formed from plywood panels, solid wood, or steamed laminated timber. Biodegradable materials need to be protected with varnish or other coatings.

Carbon fiber cycle frames are commonly used for enthusiasts seeking lightweight and aerodynamic shapes. They combine thin carbon (and other material) fibers resisting tension (providing reinforcement) with resin (usually epoxy) as a binder, both gluing the fibers together and providing compression resistance. Overstressed composites can fail in a variety of ways, including fiber/binder bond failure, fiber fracture, fiber buckling and binder cracking. Some of these failure modes are brittle, meaning that little or no permanent elongation occurs, and a crack can propagate rapidly. Using different materials (such as combining aramid and carbon fibers) or other approaches can mitigate this unwanted behavior.

Composite materials, design, and manufacturing are complex subjects. Different fibers, such as glass, aramid, and carbon, have contrasting strengths and stiffnesses. So too do various binders: for example, epoxy and polyester. If all fibers are orientated parallel, the material is "unidirectional" with widely varying properties in different directions. Fibers may be woven to form fabric. Unidirectional fibers or woven fabric are often overlaid in different directions bonded with resin to adjust the structural properties. Composite production methods include hand layup, filament winding, pultrusion (which combines pulling with extrusion), and many others. Chopped composites have short fibers and are commonly resin-cast within a mold, resulting in a random fiber distribution and nondirectional structural properties.

Assembly of frames

Cycle frames can be assembled by:

- Bonding – glue which relies on high bond strength per area, sufficient bonded area, and appropriate clearances between bonded components. Applying glue, such as epoxy, too thin or thick reduces bond strength. The bond may degrade over time, such as by galvanic corrosion from joining some dissimilar materials: for example, carbon fiber and aluminum. Bonding allows most materials to be joined or combined when appropriate materials and designs are employed, and is ubiquitous for composite frames.
- Brazing – a metal glue for metal structures (commonly brass or silver alloys for steel frames). The metal being joined is not melted but nonetheless the heat involved may somewhat affect its characteristics. Brazing may be lugged (with tubes slipped into or over junction fittings) and molten braze penetrating the thin interface; or it may be a lugless fillet (pronounced "fill-it") with the molten braze applied thickly.
- Welding – the metal components (mostly tubes but occasionally forgings or pressings) are joined by being locally melted and fused, usually with an additional filler metal. The melting and solidification can significantly change the metal's characteristics and some post-weld processing may be required (particularly for certain aluminum alloys) to produce durable frames. There are many welding methods, but cycle frames are overwhelmingly TIG-welded (Tungsten Inert Gas, also known as Gas Tungsten Arc Welding), whether made from steel, aluminum, titanium, or magnesium.
- Casting – the entire frame or just the frame connections are formed from molten metal poured into molds and allowed to cool. Aluminum and magnesium have been used to cast entire frames or to cast fittings around tube junctions; but cast frames have always been rare, mostly as a result of the mold costs and material characteristics.
- Clamping – occasionally tubing was pressed or clamped into fittings and secured with bolts but this is now vanishingly rare as other technologies have surpassed it. In the past, many steel frames (typically utility cycles) that were otherwise brazed were fitted with bolted and/or riveted seat stays.
- 3D Printing – experimental frames have recently been produced by 3D printing, but the material characteristics, combined with production costs and equipment, have mostly limited this method to creating metal frame fittings (such as rear wheel dropouts) that are welded or bonded into the frame. 3D printing is particularly suited to small production runs.

Future developments

Cycle frame design and construction has advanced significantly in the recent past. Technical improvements in materials (such as carbon fiber, titanium, high-strength steels) and production methods (including hydro forming and automated welding) have been popularized with reduced costs. The introduction of mandatory standards for frame strength and fatigue, along with performance cycling's longstanding preference for light weight, have encouraged more structurally efficient designs. This trend is likely to propagate through the industry to some extent, including utility cycles.

Unusual frame designs and details aimed purely at visual appeal (occasionally masked with spurious engineering justifications) have always been a significant part of the industry. The industry requires sales, and "eye appeal" is a major incentive for much of the population. Improved materials and manufacturing methods allow thoughtful designers more freedom to combine better aesthetics with functional improvements. Plenty of opportunity exists for frame design improvements in this regard, for both utility cycles and performance machines.

The cycle industry has recently introduced a range of incompatible component/frame interface standards to optimize aesthetics and stiffness, to accommodate proprietary bottom bracket gearing systems, or for ease of assembly and construction, particularly for composite materials. This trend for new interface standards can be expected to continue, causing compatibility issues as older cycles with obsolescent standards require replacement components.

Fitting various types of motors to cycles for assistance has a long history but electric motorized cycles ("e-bikes") have recently exploded in popularity (Chapter 24). Virtually all generic types of cycle (including utility cycles, cargo bikes, road bikes and MTBs) have e-bike equivalents that mostly are legally considered to be pedal cycles. Motors may be fitted into front or rear hubs or, more recently, can be "mid-drives" mounted adjacent to or in place of the bottom bracket. Several mid-drives require proprietary interfaces with the frame. Motors and removable batteries are becoming visually integrated and sometimes literally within the cycle frame, and there is no sign of this design approach abating.

The use of 3D printing for frame fittings or sub-assemblies is likely to become more popular as material characteristics improve and prices fall.

Improvements in testing, modelling, and analysis of frame structural behavior and aerodynamics (such as use of wind tunnels and real-world power measurement) have allowed designers to optimize performance cycles for speed, within the limitations of racing organization regulations. As a result, the integration of components, luggage provision (e.g., for triathlon use), and frames to minimize aerodynamic drag has intensified, often without regard for rider comfort and adjustability, cost, and ease of maintenance: for example, cables and hydraulic hoses hidden within handlebar stems and headsets. Designers are starting to address these other considerations alongside their primary goals and this trend can be expected to continue.

Understanding of the issues of environmental sustainability and embodied energy is ever increasing throughout society. In the past, most cycle frames were metal and recycling them after their service ceased was easily carried out, in theory at least. An increasing number of frames are now made from composites and the recycling of composites is currently non-existent. Studies already exist comparing the embodied energy involved with cycle frame construction, use, and disposal for various materials. Environmental sustainability in general is starting to be considered within the cycle industry but with little significant progress with regards to frame material and construction sustainability to date, other than paint/coating systems. That is very likely to change in the future.

Additional reading

Burrows, M. (2014) *Bicycle Design: Towards the Perfect Machine* (4th edition), Haddenham: Snow Books.
Glaskin, M. (2013) *Cycling Science: How Rider and Machine Work Together*, Lewes: Francis Lincoln.
Hadland, T. and Lessing, H.E. (2019 updated reprint) *Bicycle Design: An Illustrated History*, Cambridge: MIT Press.
Wilson, D.G. and Schmidt, T. (2020) *Bicycling Science* (4th edition), Cambridge: MIT Press.

10
WHEELS AND SHOCK ABSORPTION

Tony Hadland

In its simplest form a cycle wheel comprises a simple wooden disc with a hole through its center. This hole enables the wheel to rotate on an axle fixed to the frame of the cycle. Such wheels are found today on home-made bicycles used in parts of Africa. There are also more sophisticated aerodynamic disc wheels made of carbon fiber composite. But the vast majority of cycle wheels, from the earliest times to the present day, have a hub, a rim, spokes, a tire (tyre in UK English) and often a tube.

Hubs

The hubs of all spoked wheels have an axle, wheel bearings and a rotating shell to which the spokes are attached. Hubs may also contain mechanisms involved in the drive system, braking and lighting.

Cycle wheels were originally hand-made by wheelwrights, in the same way as wagon wheels. The hub, spokes and rim were made of wood and held together by an iron tire shrunk onto the rim. Hubs were also made of wood and held together by a pair of small iron hoops. The hub was the weakest part of the wheel, heavily stressed and adversely affected by variations in humidity. Metallic hubs, which did not split and were unaffected by humidity, were introduced in the late 1860s (Hadland & Lessing 2019: 84).

The use of wire spokes in tension, rather than timber or metal spokes in compression, enabled lighter, stronger and bigger wheels. The first wire-spoke bicycle wheel was patented in France in 1868. By the mid-1880s automated grinding technology was being used to mass produce steel ball bearings accurately and cheaply (Clayton 2016: 49). Consequently, steel ball bearings with adjustable cones quickly became the norm for use in bicycle bearings of all types (Camm 1950: 113–20). They are still widely used, despite the increasing use of more reliable sealed ball bearing assemblies. Ceramic ball bearings are also sometimes used but are expensive and have questionable durability and performance advantages.

DOI: 10.4324/9781003142041-14

Hub shells

Hub shells are usually made from pressed steel or cast aluminum alloy. Flanges are provided at each end of the hub shell the for attachment of spokes via holes or slots. Most flanges are simple and of smallest practical diameter, although large diameter flanges and recessed spoke anchorages are sometimes used.

Hub gears, dynamos and brakes

Rear hubs can incorporate internal variable gears, a hub dynamo, a drum brake or a coaster brake. The hub may also have provision for the fitting of a brake disc or a screw-on roller brake.

Except where front-wheel drive is used, the rear hub must accept drive input from a chain, toothed belt or driveshaft. Most cycles have chain drive and single-speed hubs are threaded to take a sprocket, with or without an integral freewheel. If a fixed-wheel sprocket is fitted, a reverse-threaded locking ring is used to prevent the sprocket unscrewing when back force is applied to the pedals. A double-sided "flip-flop" hub has a differently sized sprocket on either side of the hub and is reversible for a choice of gear. Usually one sprocket is fixed, the other freewheel.

On cheap derailleur-geared bikes, the hub will usually have a screw-on multi-sprocket freewheel with about six sprockets, with the hub axle extending well past the bearing. Better-quality rear hubs for derailleurs have an integral splined cylinder with its own freewheel and better bearing support, onto which a cassette of up to 12 sprockets can easily be slid (Leiner 2019: 271–2).

Front hubs may include a hub dynamo, drum brake or both. Provision may be provided for a brake disc or roller brake. For electrical assistance either hub may contain a motor.

Axles

On cheaper cycles, traditional solid axle assemblies are commonplace. These are held in place by wheel nuts requiring a spanner, unless use is made of wingnuts which provide enough leverage to turned by hand. Bicycle dropouts (the slots into which the axle fits) on the fork commonly incorporate small projections, known as "lawyers' lips", to prevent a loose front wheel falling out. A cheaper wheel retention device, often used with traditional axles, is a tabbed pressed steel washer between each wheel nut and the dropout. The tab fits into a small hole in the fork end.

Wingnuts and solid axles were superseded by hollow axles containing a skewer that links the wheel nuts, one of which incorporates a lever and cam mechanism that tightens or loosens the nuts against the dropouts. In recent years, "thru axles" have become popular on many upscale bikes fitted with disc brakes and are now used across most cycle sport disciplines: they provide more secure wheel fixing and improve the stiffness of the rear end of the frame and the front fork. Thru axles are hollow but dispense with the skewer and twin wheel nut system. Instead the axle is separate from the wheel and one end screws directly into the fork or frame. This may be done using an integral lever or an Allen key (Leiner 2019: 278–9).

Axle lengths, thicknesses, threads and fixings today vary considerably, partly as a result of the need to accommodate more derailleur sprockets and front suspension. Thru axles, being a relatively new development, suffer particularly from a lack of standardization. Although most hubs are supported at both ends of the axle, front and rear hubs are available for single-sided mounting. They have the advantage of easy tire replacement without removing the wheels (Figure 10.1).

Figure 10.1 Front hub incorporating drum brake, supported by a monoblade for easy tire removal. (copyright Tony Hadland).

Rims

The earliest cycle rims were made of solid wood. By the late 19th century strong lightweight laminated wooden rims were available. They were rapidly supplanted by metal rims for most uses, but wood rims remained an option for track racing well into the mid-20th century. Today most rims for wire-spoked wheels are made of rolled steel or aluminum alloy extrusions. Rims provide braking surfaces for rim brakes and a seat for the tire. The rim also forms a compression ring to restrain the tension of the spokes and thus help maintain the structural integrity of the wheel.

For use with roller-lever (rod and stirrup) brakes, the braking surface needs to be in a near-horizontal plane so that the brake blocks can pull up against it. Westwood rims, reinforced with tubular edges, are made for this purpose, and have been in use for well over a century. Other brakes that act on the rim (such as calipers, cantilevers and direct-pull brakes) need a near-vertical face for the brake block to act on: the traditional style of rim for this purpose was the Endrick. There is also a hybrid "Westrick" rim that provides both horizontal and vertical options. Endrick-style rims originally had a simple U-shaped cross-section but most modern aluminum rims, while retaining the appearance of the Endrick rim, have a box section for greater strength. Spoke holes are often reinforced with metal eyelets.

There has been a strong move in recent decades away from steel rims and toward aluminum. This is partly because aluminum rims provide better wet weather braking. But they also wear out quicker, so they often incorporate a visual wear indicator. The braking surfaces may occasionally have a ceramic coating, which improves wet weather braking but makes it worse in the dry (Brandt 1993: 57).

Deep rims that present an aerodynamic teardrop cross-section when fitted with a tire are available in rolled or extruded aluminum or in molded carbon fiber composite, with or without a metal sub-rim (Burrows 2014: 94–5). Some composite rims used with rim brakes on long, steep descents can be heated sufficiently to soften the resin, permanently deforming the rim.

Some wheels have rims that are integral with the rest of the wheel structure. These include costly carbon fiber composite wheels (such as disc and tri-spoke), and cheap BMX wheels made from molded thermoplastic or cast aluminum.

Spokes

Traditional wire-spoked wheels are the strongest and lightest types, keeping the amount of material in compression (much of the rim) to a minimum and concentrating the forces in the tensioned spokes. Wire wheels are also generally tougher than molded or cast wheels, which tend to be relatively brittle (Burrows 2014: 93).

Standard wire spokes have a solid circular cross-section with a width in the order of 1.8 to 2 mm and are typically galvanized steel (cheaper) or stainless steel. Butted spokes have thinner cross-sections for most of their length but are thicker where strength is most needed, at each end. Bladed spokes for improved aerodynamics are available for use with deep-section aero rims. Spokes usually have a J-bend that hooks into a hole in the hub flange, but straight pull spokes are also available for hubs designed appropriately. Each spoke is secured to the rim by a threaded nipple, usually of brass or aluminum, allowing the spoke tension to be adjusted, along with the wheel's shape. Very few proprietary wheels locate the nipple at the hub.

The hub of a tension-spoke wheel can be regarded conceptually as hanging from the rim on the thin upward spoke, contrary to the compression-spoke wheel standing on a relatively thick downward spoke. A light rim would bulge under that load, however, so horizontal spokes are needed, which are under tension too. Still the rim would tend to flatten at its lower part, so additional downward spokes restrain this bulging. Thus, by using a sufficient number of spokes capable of resisting tension, the load applied at the center of the wheel can be transmitted to the ground without appreciable distortion of the rim (Sharp 1896: 338–9).

Despite this conceptual reasoning, a properly tensioned wire wheel actually behaves as a curved beam and therefore cannot be said to hang from upper spokes. This is because of the amount of pre-tension in the spokes. During assembly, when the spokes are still relatively loose, the wheel does indeed hang from the upper spokes, but the situation changes as the spoke tension is increased to the level necessary for normal use (Hadland & Lessing 2019: 210).

Wheel spokes are arranged in two arrays, starting from each of the hub flanges, and alternately meeting the rim. Thus, the spokes and the hub shell form a series of triangles, bracing the rim laterally. The earliest, simplest and lightest spoke pattern is radial, which is adequate where the hub is not used to drive or brake the wheel. But if the wheel is driven or braked, the torque causes spoke wind-up, causing repeated flexing of the spokes and possible damage to the wheel. This problem is overcome by using tangent spoking, whereby each spoke connects with the hub tangentially rather than radially. This is also called crossed spoking, because each spoke crosses over one, two or three others, depending on the required strength and duty of the wheel. The more spoke crossing, the stronger but heavier the wheel (Burrows 2014: 92–3).

Tires

The earliest cycle wheels had an iron tire that provided not only a durable running surface but also the structural compression required to hold the wooden wheel together. Such tires contributed nothing to rider comfort but were superseded by solid rubber tires which helped reduce vibration a little and improved traction and roadholding.

Pneumatic tires

By the end of the 19th century pneumatic tires had almost completely replaced solid tires. Pneumatics make a huge difference to comfort by providing "annular pneumatic suspension" at the cycle's points of contact with the road (Burrows 2014: 102). They minimize the unsprung mass of the machine and rider when rolling over bumps that are smaller than the tire cross-section – the tire soaks up the bump without significantly lifting the wheel or losing much energy. This gives a smoother ride while reducing the rolling resistance of the wheels. Compared with solid tires, pneumatics are lighter and grip better, which improves traction, roadholding and braking (Hadland & Lessing 2019: 187–98).

The problem of punctures

The biggest disadvantage of pneumatic tires is the risk of a puncture. Consequently, solid tires are still available as there is always a small demand for puncture-proof tires for certain applications. Solid tires are more correctly described as cushion tires. They often contain sealed pockets of air and various combinations of resilient materials, such as solid polymers, sponge rubber, cork and micro-cellular polyurethane foam (Camm 1950: 69; Hadland & Lessing 2019: 199–200).

Many modern pneumatic tires incorporate anti-puncture layers, using materials such as Kevlar to form a barrier or a layer of polyurethane foam (about 5mm) to trap and hold thumb tacks, flints and glass splinters. Anti-puncture barrier tapes can be bought separately and installed between the tire and its inner tube. Also, the inner tube can contain self-sealing fluid to automatically seal small punctures: the sticky fluid contains microfibers which the leaking air drags into the puncture hole, sealing it. The fluid can be installed during manufacture of the tube or retrospectively.

External devices known as flint catchers or tire protectors have a thin curved wire that rubs lightly on the running surface of the tire to flick out flints, tacks or splinters before they cause a puncture. These devices fit to fenders (mudguards) or caliper brake bolts but are rarely used today.

Single-tube tires

In the 1890s pneumatic tires for cycles settled down into several classes. These included beaded-edge tires, sew-ups and wired-ons. Another type was the now extinct single-tube tire, in which the inner tube and outer casing were inseparably vulcanized together. Being particularly difficult to repair, the single-tube tire proved unpopular in most countries, but in the USA it dominated the market from 1897 to 1933. Under the American cycle industry's business model of that era, dealers could make more money from replacing these tires than from selling cycles. Paul Rubenson has argued convincingly that this business model, dependent on the frequent replacement of the tire to the disadvantage of the cyclist, considerably retarded the popularity of cycling in the USA in the first three decades of the 20th century (Rubenson 2003: 87–97).

Sew-ups

Sew-ups, also known as tubular tires or tubs, were the traditional choice for road and track racing, being the lightest and fastest tires. Today they are still widely used, especially in the velodrome. They have a narrow cross-section and a very thin carcass which envelopes the inner tube and is sewn up on the inside, hence the name. The carcass and inner tube form a single item that can easily be folded to a small size. A time trialist can therefore carry several spares. The tire is glued to a sprint rim specifically designed for the purpose: instead of a deep well to hold the tire, a sprint rim merely has a soft curve to match the cross-section of the tire. Although sew-ups puncture easily, they are quick to change and can be ridden flat – important considerations for time trialing. But sew-ups are expensive, difficult to repair, and riding them flat can ruin both tire and rim. Also, the glue is messy, although double-sided adhesive tape is often used today (Burrows 2014: 105).

Wired-on tires

The vast majority of tires have a carcass which is open on its inside to receive the inner tube. The edges of the carcass which sit in the rim contain a reinforcing wire which holds the tire in shape. The wire is usually steel, although Kevlar is sometimes used to facilitate folding the tire.

There is a huge range of wired-on tires to suit almost any conceivable requirement. They range in width from high pressure racing tires as narrow as 19mm, to fat-bike tires in excess of 120mm. The best racing tires can match the low rolling resistance of cheap sew-ups and are much more durable and cost effective. For racing, where the lowest possible rolling resistance is required, latex inner tubes are available, but butyl rubber is much more widely used and holds air much longer.

Tires are marked with the approximate cross-section width and the all-important bead seat diameter, in the form 54–599. The outer diameter of some tires has changed considerably from the original nominal size. Hence a nominally 28-inch diameter 700C tire may have an actual diameter nearer 26.3 inches if the narrowest cross-section is used (Sutherlands 1995: 12.1–12.6). For detailed guidance on tire formats old and new, visit the Sheldon Brown website: https://www.sheldonbrown.com/tire-sizing.html.

Tubeless tires

These are similar in concept to tubeless tires used on car wheels. The carcass is open, as with other wired-on tires, but the rim, and its interface with the tire beads, are airtight. A common cause of punctures with wired-on tires is the inner tube being pinched against the rim when the wheel hits a sharp edge, such as a rock. These "snake-bite" punctures cannot happen if there is no inner tube to puncture. The tires contain a liquid sealant which is effective against most other punctures, such as those made by thorns.

Special rims are made for use with tubeless tires. Conversion kits are available to make standard rims and tires airtight, by applying a special rim tape over the spoke heads and adding the liquid sealant. The sealant makes the valve hole airtight and improves the air retaining qualities of the tire carcass (Burrows 2014: 108).

Valves

Three types of valve are used with cycle tires: Presta (also known as Sclaverand, French or Racing); Woods (also known as Alligaro, Dunlop or Easypump), formerly the standard for most non-racing cycles; and Schrader, as used on car tires, which has largely supplanted the Woods pattern.

Tread patterns

There is a huge range of tread patterns, but this has more to do with marketing than science. For road use, slick or near-slick tires roll easiest and have the best grip, wet or dry. Knobbles on the outer edges of the tire can marginally reduce the risk of slipping sideways when cornering. For off-road use in wet and muddy conditions relatively narrow open tread knobbly tires have much to commend them (Burrows 2014: 106–8).

Suspension

In common with other wheeled vehicles, cycles incorporate various means of protecting the rider from an uncomfortably rough, jarring, or bumpy ride. The benefits of such measures can be greater than merely improving the immediate comfort of the rider. As the eminent Victorian engineer Archibald Sharp pointed out (1896: 487):

> In the first bicycles made with wooden wheels and iron tires, and sometimes without even a spring to the seat, the mass [rigidly connected to the tire] included the whole of the wheel and a considerable proportion of the mass of the frame and rider; so that the energy lost in shock formed by far the greatest item in the work to be supplied by the rider.

More recently, Jan Heine compared bicycles ridden at a constant speed over a smooth road and over rumble strips. He found that over his rumble strips more than two and half times the energy was required to maintain the same speed as on the smooth road (Heine 2012).

Wheel suspension has potential for reducing the amount of energy expended by the rider by reducing the unsprung mass of rider and machine. But adding suspension is likely to add cost, complexity, weight, and the possibility of some of the rider's energy being lost in needlessly compressing the suspension, especially when riding out of the saddle. Prior to the widespread adoption of pneumatic tires some cycles incorporated wheel suspension: sprung front forks were available 150 years ago (Reynaud 2008: 124, 127). However, the "annular pneumatic suspension" and reduced rolling resistance provided by air-filled tires caused the almost complete disappearance of spring-based cycle wheel suspension for many decades (Burrows 2014: 102).

An important point to bear in mind in any consideration of bicycle suspension and anti-jarring measures is the human body's ability to absorb and dampen shocks. As Wilson and Schmidt (2020: 186–7) put it:

> By far the greatest capability for traversing or absorbing bumps is that inherent in the human body. The body's range of travel and ability to absorb energy outstrips the hardware of any ordinary bicycle suspension. In addition, human adaptability or even active compensatory motion makes a huge difference.

Much of the road shock felt by the rider is transmitted via the saddle and the handlebar. Even 200 years ago, the draisine incorporated features to reduce this. Since then many attempts have been made to improve comfort for the hands, arms and posterior: they have included sprung handlebars, sprung seatposts, hammock saddles, flexible saddle beams and pneumatic seats. The Whippet bicycle of the 1880s had the whole section of the frame carrying the handlebars, the seat and the cranks sprung as a single unit. Some of these ideas sank without trace, others periodically reappear (Hadland & Lessing 2019: 200–8).

Wheel suspension is most commonly found in four categories of bicycle: mountain bikes, hybrids (especially electrically-assisted versions), carbon fiber road bikes and upscale small-wheelers.

Mountain bikes, hybrids and e-bikes

On rough terrain, well-designed suspension can provide improved comfort, better handling (by keeping the wheels in contact with the ground) and greater efficiency. Laboratory tests have shown that front suspension can reduce horizontal forces (which slow the rider down) by 28% when passing over a 60mm bump; and that a dual suspension bike can save the rider between 30% and 60% of the power needed to maintain the same speed on the same rough track on a bike with only front suspension (Glaskin 2013: 58–9).

Mountain bikes come in many specialized variants, all with specific suspension requirements (Chapter 28). The majority of mountain bikes sold today are "hardtails," having front suspension only. By far the commonest front suspension configuration is the dual-sliding pillar, comprising a parallel pair of sprung telescopic fork blades. Such forks have been commercially available for mountain bikes since 1990: they are widely used and have become increasingly sophisticated.

The telescopic front fork comprises a crown-stanchion unit (a pair of cylindrical stanchions attached to the fork crown), on which, sliding against internal springs, is an arch-slider assembly (a pair of cylindrical legs connected at the top by an arched bracing piece and braced at the lower end by a suitably robust hub.)

The springing in each fork blade may be provided by steel or titanium coil springs, air springs or elastomers. Combinations of springs may be used: for example, a long main steel coil spring may have an elastomer core to provide damping and a short steel spring to control rebound. To limit spring oscillations, hydraulic or pneumatic damping is usually provided by forcing oil or air through a valve.

The amount of front suspension travel varies with intended usage but has increased considerably over the years: 100mm is commonplace but up to 300 mm is available. Suspension preload, which affects the amount by which the bike sags when the rider's weight is applied to it, is adjustable without the need for special tools even on inexpensive suspension forks. Some can be locked to prevent suspension movement, which can be useful under certain circumstances. In more sophisticated forks, lockout and other modifications to suspension characteristics can be adjusted from the handlebars while riding.

Many electrically-assisted hybrid or trekking bikes have simple versions of mountain bike front suspension, but it is less common on purely pedal-powered versions.

Rear suspension is less popular than front suspension on mountain bikes. It invariably involves the rear wheel hub being attached to chainstays or a trailing fork arm pivoting from a point close to the bottom bracket. The spring is usually a compact unit containing an air or coil spring and a damper. This is typically mounted in the space between the seat tube, down tube and top tube, or below the main beam, if the frame has no separate down tube and top tube.

How the chainstays or trailing fork arm connect to the spring unit varies considerably. Variants include: simple high-pivot, usually with a massive swing-arm; simple low-pivot, with a pivoted rear triangle; unified rear triangle, where the bottom bracket is integral with the pivoting rear triangle; three- or four-bar linkages, with a pivoted plate to link the spring unit to the upper stays; and parallelogram linkages, with the spring in the space between the rear wheel and the back of the seat tube. All of these configurations have advantages and disadvantages, relating to their level of complexity and the interaction of suspension movement with pedaling and chain tension (Burrows 2014: 132–3).

Road bikes

In recent years, advances in road bike frame construction, particularly in carbon fiber, have permitted the incorporation of various lightweight shock absorbing features to subtly improve comfort and control whilst reducing fatigue. These features include:

- Flexible seat stays giving a small amount of shock isolation: up to 6mm in some cases. The flexibility is obtained by narrow stay widths, thin stay walls and curving of the stays.
- Flexible or sprung seat posts giving varying amounts of shock isolation.
- High-frequency vibration damping via visco-elastic filling of parts of the carbon fiber frame.
- Decoupling of the seat tube from the top tube and seat stays via bearings, enabling the whole length of the seat tube to flex.
- Flexible seat tubes and dropped seat stays, enabling the seat tube to flex.
- Decoupling of the seat tube from the top tube via a built-in elastomer.
- Sprung handlebars using a coil spring in the headset with adjustable oil damping while riding.
- Rear wheel suspension giving 10mm of travel, using flex in the chain stays instead of a pivot, and with an elastomer spring and hydraulic damper between the seat stays and the chain stays, adjustable electronically while in use.

(Brett 2020)

Small-wheelers

Relatively high pressures are desirable in small diameter tires of narrow cross-section to reduce the rolling resistance: but unsprung small wheels give a rougher ride because they rise and fall faster over bumps and fall deeper into depressions in the road surface. On sealed but imperfect road surfaces, as commonly found in many western countries, a suspended small wheel can often give a smoother ride than a larger unsprung wheel (Burrows 2014: 97–101).

Several long-established upscale small-wheeled bicycle designs have rear wheel or dual suspension. These are mostly low-maintenance systems using elastomers, some of which have been produced for decades. For example, the Brompton folding bicycle has a rear triangle that, for folded mode, swings under the frame's main beam. On the apex of the rear triangle is a simple elastomer which, in ready-to-ride mode, bears on the back of the seat tube, providing a self-damping spring for the suspension. The pivot that enables the rear triangle to fold under the main beam doubles as a suspension pivot (Henshaw 2020: 40). A similar facility is provided on the Airnimal Chameleon folding bicycle.

Leading link front suspension has been used on some mountain bikes but is more often found on upscale small-wheelers, particularly those by Riese & Müller and by Moulton. The wheel hub is held by a pair of approximately horizontal arms (links). These are pivoted from a point behind the hub. Riese & Müller's Birdy folding small-wheeler has leading link front suspension which employs a steel coil spring with an elastomer core to provide damping. The Birdy also has rear suspension that works on a similar principle to that of the Brompton; except that the Birdy has a trailing fork arm rather than a rear triangle (Fehlau 1997: 35 et passim).

Moulton small-wheelers have been available with dual suspension since the 1960s (Hadland 1982: 10–34). Most Moultons made since the 1980s have leading link front suspension, using a steel coil spring with adjustable pre-load, and adjustable friction damping on the leading link pivots. The rear suspension has a pivoting rear triangle, the apex of which acts on a lightweight, self-damping rubber spring on the back of the seat tube (Hadland 1994: 27–9).

While all bikes incorporate some measures to mitigate jarring (if only in the saddle and handlebar grips), wheel suspension is a niche area. It is a long-established and mature technology in certain types of bicycle, particularly mountain bikes where it brings significant benefits, but almost totally absent in most others, where little or no benefit is apparent. Cycle suspension ranges from the bulky, complicated but often sophisticated systems found in mountain bikes, to the simple, lightweight, low-maintenance but effective rear-wheel suspension of the Brompton folder. It remains to be seen whether the current developments in road bike suspension become commercially successful in the long term.

Source list

Brandt, J. (1993) *The Bicycle Wheel* (3rd edition), Palo Alto: Avocet.
Brett, M. (2020) 8 Bump-taming Road Bikes that Help Stop Your Hands and Bum Getting Battered, https://road.cc/content/buyers-guide/8-bump-taming-road-bikes-257206 (retrieved June 12, 2020).
Burrows, M. (2014) *Bicycle Design: Towards the Perfect Machine* (4th edition), Haddenham: Snow Books.
Camm, F.J. (1950) *Every Cyclist's Pocket Book*, London: Newnes.
Clayton, N. (2016) *The Birth of the Bicycle*, Stroud: Amberley.
Fehlau, G. (1997) *Das Modul-Bike: Faltbare Fahrräder*, Kiel: Delius Klasing.
Glaskin, M. (2013) *Cycling Science: How Rider and Machine Work Together*, Lewes: Ivy Press.
Hadland, T. (1982) *The Moulton Bicycle* (2nd edition), Erdington: Pinkerton Press.
Hadland, T. (1994) *The Spaceframe Moultons*, Coventry: Hadland Books.
Hadland, T. and Lessing, H.E. (2019 updated reprint) *Bicycle Design: An Illustrated History*, Cambridge: MIT Press.
Heine, J. (2012) "Suspension Losses," *Bicycle Quarterly* 29 (Autumn). Seattle.
Henshaw, D. (2020) *Brompton Bicycle* (3rd edition), Wakefield: Excellent Books.
Leiner, J. (Ed.) (2019) *Fachkunde Fahrradtechnic*, Haan-Gruiten: Europa Lehrmittel.
Reynaud, C. (2008) *Le Vélocipède Illustré*, Domazan: Musée Vélo-Moto.
Rubenson, P. (2003) "Patents, Profits and Perceptions," *Cycle History 15*, San Francisco: Cycle Publishing.
Sharp, A. (1896) *Bicycles and Tricycles: An Elementary Treatise on Their Design and Construction*, Cambridge: MIT Press (1977 reprint).
Sutherland, H. (Ed.) (1995) *Handbook for Bicycle Mechanics* (6th edition), Berkeley: Sutherland.
Wilson, D.G. and Schmidt, T. (2020) *Bicycling Science* (4th edition), Cambridge: MIT Press.

11
TRANSMISSION AND BRAKES

Dan Farrell

Stop and Go, or rather Go and then Stop. The link between transmission and braking lies in motion; the former transmitting the effort of the rider into forward progress, the latter in arresting it. It is these mechanical contrivances that separate the bicycle from the scooter or draisine (hobby horse), where the rider propels the machine by the action of the feet on the ground and slows or stops in a similar fashion.

Transmission

In its simplest form, bicycle transmission has remained largely unchanged for over a century. The basic system comprises a rotating pedal and crank assembly driving the hub of the rear wheel by means of a roller chain. On early bicycles the pedals and cranks were directly attached to the (usually front) wheel and hence one turn of the crank equaled one turn of the wheel. Bicycle design up to this point had been led by the desire for speed and the limitations of direct drive; the need to travel further for each turn of the crank inevitably led to increasing the size of the drive wheel and the bicycle developed into the "Penny Farthing", "High Wheeler" or "Ordinary" with the diameter of the drive wheel only limited by the requirement for it to fit between the rider's legs. The introduction of the chain drive in the 1880s broke this direct drive layout. With a chain drive the front sprocket (known as the "chainwheel" or "chainring") could be larger than the sprocket attached to the wheel and thus each turn of the cranks could rotate the wheel several times and one could travel further and faster. An idiosyncratic hangover from the Penny Farthing is the continued use in Anglophone countries of "gear inches" to describe gear ratios; when a cyclist talks of a "90-inch" gear they are not referring to the distance travelled by a single turn of the cranks but rather the equivalent drive wheel diameter where the cranks are directly attached to the wheel hub.

Despite its apparent deficiencies and the efforts of many inventors, the simple rotary crank and pedal arrangement (i.e., with the rider's feet following an almost circular motion) has been largely unchallenged as the means of inputting human power into the bicycle. Other methods, such as systems of treadles and levers, have been tried many times but fail to match up to the efficiency offered by pedals, cranks and chains.

DOI: 10.4324/9781003142041-15

Single gear transmission

We can define the classical transmission, the single gear, with the knowledge that variable gear systems are derived from it or are additions to it. Central to the transmission is the *bottom bracket* which comprises a *bottom bracket axle* mounted in ball bearings and free to rotate within the *bottom bracket shell* (part of the bicycle frame). Affixed to each end of the bottom bracket axle is a crank arm, typically 170 millimeters long; and each crank arm has a pedal attached. The crank arms are arranged at 180 degrees to each other, so that when one of the rider's legs is pushing downwards, the other leg is being lifted.

On the inside of the right crank arm (left and right are defined from perspective of the rider) a *chainwheel (chainring)* is affixed; this has teeth profiled for a roller *chain* of half-inch pitch. The combination of the two crank arms is known as the *chainset* or *crankset*. A single *sprocket* or *cog* is attached to the rear wheel hub. The relative sizes of the chainwheel and rear hub sprocket determine the *gear ratio* – if the chainwheel is twice the size (twice the number of teeth) of the rear sprocket, then the rear wheel will turn two times for each turn of the pedals. A typical single gear would be between 60 and 70 inches; this being the equivalent wheel diameter as described earlier (to obtain the distance travelled for each pedal revolution this needs to be multiplied by *pi*, the resulting number being called the *development*). Such a gear ratio may be obtained by a chainwheel with 48 teeth and a rear sprocket with 20 teeth – with a standard 27-inch wheel this would give a gear of 64.8 inches. A variant of the single gear is the *fixed gear*, where the gear system is the same but there is no freewheel mechanism in the rear wheel hub so the cranks cannot rotate independently of the wheel and the rider must pedal all the time whilst the bicycle is moving.

Belts and shafts

A single gear chain transmission is over 95% efficient (Whitt & Wilson 1982) and this is a prime consideration as the power available is limited to that supplied by the rider. Nothing can rival the roller chain for efficiency, but a high-quality toothed belt drive can come close (friction-facts.com). Another alternative transmission is the shaft-drive, usually using bevel gears and often with the drive shaft contained with the chainstay of the frame. These are less efficient than chain drives (Berto 2009). Belt- and shaft-drive systems have other disadvantages: bicycles must be designed specifically for them, tolerances are much tighter, altering gear ratios is difficult and the use of derailleur gears is precluded. The primary advantages of belt-drive and shaft-drive are cleanliness, reduced maintenance requirements and longevity when correctly installed.

Variable gears

Whereas a single gear is often sufficient in flat terrain, where hills and mountains are encountered – and headwinds – a choice of gears is advantageous. There are two principal types of variable gear: the *derailleur* gear and the internal *hub* gear. Both are in widespread use and each has its advantages and disadvantages.

Derailleur gears

The derailleur gear is similar to the single gear, but features an array of different-sized sprockets (known by many terms – including a *rear cluster* or more simply a *block*) attached to the rear hub, and a gear-change mechanism (the *rear derailleur*) that performs two functions;

Figure 11.1 Derailleur transmission. The classical derailleur gear transmission with double chainrings. (author's photo).

to move – derail – the chain from one sprocket to another, and to take up any slack in the chain (as the chain must be long enough to fit over the largest sprocket yet remain taut when on the smallest) (Figure 11.1). In the early years of derailleurs (they were not popular outside of cycle racing until the 1950s), the sprockets were few in number (typically three or four) and usually of similar size, with only one or two teeth difference between each one ("*close-ratio*" blocks). Nowadays a much wider spread is provided with the largest sprocket having three, four or even five times the number of teeth of the smallest. Herein lies the essential conflict in all variable gear systems, the need for a large overall gear range for the steepest of climbs and the fastest of descents **and** the requirement to pedal at an efficient cadence (cadence being the speed of pedal revolutions). For the leisure cyclist or tourist the wide-ratio is ideal, but for anyone who is racing against the clock having the right gear ratios (and hence the perfect cadence) can often be the difference between victory and defeat.

To provide a greater gear range and keep the gear ratios reasonably close together, *front derailleurs* were developed. Two or three different-sized chainwheels are fitted, giving a "high range" and a "low range" of gearing ("high" gears being those that move the rider along faster, the nomenclature possibly deriving from how high the rider must be from the ground on a Penny Farthing). The front derailleur is simpler than the rear, as the latter can deal with the chain slack all the front derailleur has to do is push the chain from one chainwheel to the other – although as the chainwheels are usually significantly different in size, this is not as easy a task as rear shifting. For many years, chainwheel sizes were ten teeth apart; a *double chainset* would usually have 52 and 42 teeth chainrings, a *triple* would often be 52–42–32. In the case of the triple chainset, designed for touring cyclists, you can see that all the extra gear range is at the low end for hill climbing.

The advent of the mountain bike in the 1980s provided more impetus into the development of derailleur gears as the new sport of off-road cycling demanded much lower gears than were hitherto normally available. In more recent years, as shifting has improved, the tooth-count difference has been increased and lower gearing offered on double chainsets. The previously ubiquitous 52–42 gave way to 53–39 and the introduction of "compact" (50–34 or 52–36) and "sub-compact" (46–30 or similar) chainsets has largely rendered triples obsolete. Certainly, a triple crankset requires more thought and skill from the rider to use effectively; there is more duplication of gears and greater chain angles can result from "cross-chaining" (running the chain on the largest chainring and the largest sprocket, or from smallest to smallest), the result being reduced efficiency and more rapid wear. The proficient rider will know not to do this with a double chainset as well, of course – although the adverse effects are less significant.

The derailleur system is fundamentally brutal; the chain is pushed from one sprocket to the next. This is most apparent in the front derailleur mechanism, a metal cage that presses on the side of the chain. However, advances in technology have transformed the crude early gear shifting into slick, precise and almost silent systems. Modern derailleur systems are *indexed*, where the gear lever has a series of positive positions that correspond with each sprocket. This relatively recent innovation relies on a precision lacking in early systems where the rider had to "feel" their way into each gear, the position of the gear lever retained by friction alone. These developments include sprocket tooth profiling to aid the derailing of the chain, as opposed to single gear systems where chain retention is desirable. Initially, this was achieved using twisted sprocket teeth but further refinements, including shifting ramps and "gates" to provide an easy path for the chain were soon introduced. A refinement essential to the consistent accuracy of indexed gears is the provision of pre-stretched shifting cables – before these were introduced the primary creep in the cables would soon put the derailleur out of alignment – and longitudinally stranded cable outer housing rather than the coil-wound type previously used. Another important development was the flexible *bushingless chain*; first used in 1981, it quickly became the norm.

With modern derailleur systems becoming increasingly precise in operation, manufacturers have been able to increase the number of sprockets at the rear as well as using large sprocket tooth count differences where appropriate. Whereas the "ten-speed" road bicycle (two chainrings, usually 52 and 42 teeth, with a five-speed freewheel, typically 14–28 teeth, gear range 250%, low gear 40″, high gear 100″) was ubiquitous in the 1970s, the 1980s brought six- and seven-speed freewheel blocks (and the "cassette freehub" which replaced the traditional threaded hub and freewheel). This trend continued: eight and nine speeds before the year 2000, then ten and eleven before 2010, up to twelve and even thirteen in 2020.

Paradoxically, the increasing plethora and range of sprockets at the rear has led to a reduction of chainrings at the front. As well as the demise of the triple crankset, there is an increasing trend towards having only a single chainring and pairing this with an ultra-wide sprocket set with a largest cog of 40 or even 50 teeth. Thus a modern bicycle may have a "1×" ("One-by") transmission with a single 38-tooth chainring and thirteen sprockets from 9 to 42 teeth; it is notable to compare this 466% gear range with the 250% offered by the 1970s ten-speed. There are also benefits in reduction of weight and complication by the elimination of the second chainring, the front derailleur, its mounting and its shifting mechanism. Other advantages include increased tire and frame clearances, and more freedom in frame design to accommodate suspension, all of which are particularly useful in the mountain bike arena where the "1×" technology originated. The performance of single chainring derailleur systems is greatly improved by the use of chainrings with alternating "narrow-wide" teeth to prevent the chain wandering, and the introduction of rear derailleurs with clutch mechanisms to reduce chain movement over rough terrain. Both elements are considered essential to avoid chain drop (in the absence of any devices such as chain guides and/or chain discs to prevent the chain falling off the chainring).

Hub gears

The second type of variable gear is the *hub gear*. Hub gears are epicyclic (planetary) gears, comprising a series of cogs ("planets") rotating around a central "sun" cog. By locking different parts of the gear mechanism, a single-stage epicyclic gear provides three gears, typically

–33% and + 33% either side of the direct drive middle gear. As the mechanism is contained within the hub, the middle gear is determined by the chainwheel and sprocket sizes as per the single gear arrangement. The classic "three-speed" has been around for over a hundred years and remains in widespread use. Modern hubs provide more gears and a wider gear range by adding more epicyclic stages and, in some cases, compounding these. A hub gear with two stages provides five gears (in fact, there are six, but two are direct drive); similarly, an 11-speed hub gear comprises a six-speed drive combined with a two-speed. Each gear stage reduces the overall efficiency, and to counter this some multi-speed hubs utilize oil-bath lubrication to minimize friction and hence reduce power losses. Inevitably a modern oil-bath system requires better sealing than the semi-fluid grease lubrication used on simpler (and less expensive) hub gears (older hubs requiring regular oil lubrication worked on a "total loss" principle where the oil gradually found its way out of the hub via the wheel bearings).

Whilst a hub gear can provide as great a gear range as required by the most heavily laden touring cyclist (over 500%), inevitably these gears will be widely spread and as such one will not find a hub gear on a racing bicycle. Modern hub gears range from premium 14- and 11-speed models (400–500+% range) down to utilitarian 3-, 4- and 5-speed units, with 8-speed hubs occupying the middle ground. As hub gears are fully enclosed, they are resistant to dirt and water ingress, and less prone to accidental damage (derailleurs, by contrast, are exposed and vulnerable). These factors, combined with ease of use and minimal maintenance, mean that hub gears are the primary choice for cargo bikes and for public hire bike fleets.

Hybrid gears

Hybrid gear systems combine hub and derailleur systems (for example, a three-speed hub with a seven-speed derailleur giving 21 gears) to provide a wide gear range. These configurations often work well but introduce complexity and add weight to the bicycle.

Shifting

In most cases, shifting of gears (derailleur or hub) is achieved using Bowden cables operated by levers. Historically, these shift levers were mounted onto the frame tubes (for hub gears, the top tube, whereas derailleur shifters were mounted on the down tube), and these were later moved to the handlebars – hub gears had the option of handlebar-mounted shift levers from the early 1900s, and whilst derailleur systems had the option of drop handlebar mounted "bar end shifters" (also known as "bar cons") from the 1940s, these were of minority interest. These shift levers, and the "thumbshifters" (in effect, these were down tube shifters mounted on the handlebars) used on the early mountain bikes, are all of the direct-acting type, with a small cable drum integrated into the base of the lever. It was not until 1990 that the concept of derailleur gear levers integrated with the brake levers was introduced, and this – the "dual-control" lever (or, more colloquially, "brifter") – soon became the mainstream option on drop-handlebar bicycles. These levers required a ratchet-type mechanism with two operating levers in a push-push layout (one lever to rotate the cable drum to pull more cable in, the other to release it), with the levers returning to their original position after every shift – this ratchet-shifting mechanism having been pioneered in mountain bike shifters in 1989.

More recently, the major innovation in gear controls (and gears themselves) has been in electronic shifting. With stepper motors in the derailleurs (or hub gear shift mechanisms) and the shifters themselves reduced to microswitches, precision and reliability are improved over cable systems. Indeed, electronic control can offer technically better shifting than cables ever can (through such details as over-shifting and return, and automatic trimming of derailleur position) as well as benefits such as being able to program the sequence of gear shifts of front and rear derailleurs, lock out inefficient gears (eliminating cross-chaining), duplicate shifting switches so they are always to hand, and of course there is no cable stretch to contend with. Electronic shifting also opens up the possibilities of automatic shifting, data-logging and diagnostics, and full- or semi-wireless operation.

Continuously variable gears and non-circular chainrings

Despite the efficient nature and the fundamental maturity of cycle transmission systems, many attempts have (and continue to be) made to improve upon it. These efforts are largely misguided, but there are some worthy of note. The continuously-variable hub gear, where torque is transmitted through a ring of large ball bearings with adjustable rolling angles, has potential benefits in terms of both ease of use and durability. "Oval" chainrings (strictly speaking "non-circular") have been around for almost as long as the chain drive itself and are (usually) designed to harness more of the rider's energy during the power (downward) stroke of the pedaling cycle whilst reducing the effect of the "dead center" when the cranks are vertical. Oval chainrings are not without controversy and some implementations have been poor, but the mechanical scientific theory behind them has some merit. They are not widely used but with four Tour de France victories recorded by riders using some of the more extreme non-circular rings in recent years they are certainly established as a credible alternative to round chainrings.

Brakes

Bicycle brakes pre-date the bicycle itself. As soon as Karl Von Drais invented the *draisine* in 1817, riders needed a way to stop this from running away on inclines. Dragging their feet on the ground would work, after a fashion. Von Drais added a pivoted lever to make braking a less haphazard affair (Figure 11.2).

Spoon or plunger brakes

As bicycles developed – and particularly with the introduction of the Ordinary (Penny-Farthing), high-wheelers where the rider's feet were far from the ground – the spoon brake (or plunger brake) was widely adopted. This was a simple system of a hand-lever and a pad pressing on the solid rubber tire. The fundamental basis of bicycle brakes has not changed from these early beginnings of some form of brake pad pressing against a part of the bicycle wheel, creating friction and thus turning kinetic energy into heat. Note that early bicycles did not have a freewheel mechanism in the drivetrain, so some braking effect was possible by slowing the pedals with the rider's feet. This is still true of fixed gear bicycles today, and – in the UK at least – such machines are only required to have one other brake whereas bicycles with freewheels must be fitted with two independent braking systems. Simple mechanics informs us that the front brake performs most of the work; any significant deceleration will reduce the contact of the rear wheel on the road and make it liable to skid.

Rim brakes

The spoon brake was retained on the safety bicycles of the 1880s, but new designs were required when the pneumatic tire was adopted for cycle use as these tires wore rapidly under

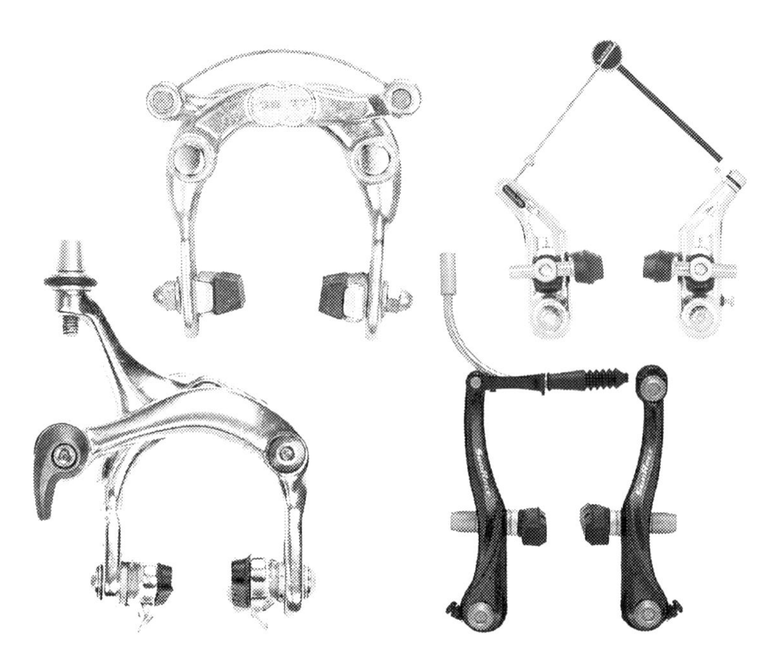

Figure 11.2 Rim brakes. Types of rim brakes, clockwise from top left: Centrepull, cantilever, direct pull 'V-brake', dual-pivot sidepull. (author's photo).

the pressure of the "spoon". The first of these was the "rod" or "roller lever" brake which, using an operating lever similar to the spoon brake, pulled a pair of brake pads onto the inside face of the shaped *Westwood* wheel rim (the pads being connected by a metal arch over the rim and tire). This was the first *rim brake* and it remains in widespread use in many areas, including Africa, India and China. It has advantages of durability and reliability; conversely, *rod brakes* are heavy and relatively ineffective in wet weather (brake pad compounds, including leather-faced rubber brake blocks, have since improved this).

Coaster or back-pedal brakes

Next came the *coaster brake* (also known as the *back-pedal brake*). This was the first *hub brake* and is still widely used (although popularity varies greatly across the world). The coaster brake is operated by the chain; when the rider moves the pedals backwards a sliding cam within the rear hub forces metal brake shoes against a braking drum (often the hub shell itself), thus causing friction and slowing the bicycle. Coaster brakes are neatly self-contained – there is no brake lever and no operating cable or rod – and require little maintenance, but they are lacking in braking power and are only suitable for use on the rear wheel. The dissipation of generated heat is a problem for coaster brakes, as the braking surfaces are small.

Rim brake problems

The fundamental difference between rim brakes and hub brakes is that in the former case the retardation effect is on the wheel rim, whereas in the latter it is on the hub in the center of the wheel. The modern rim brake, where the wheel rim and tire are embraced by a brake caliper or by brake arms operating on frame-mounted pivots and operated by Bowden cables (or, rarely, by hydraulics), uses the wheel rim as a large brake disc with the rim sidewalls being the friction surfaces. As such, the area available for heat dissipation is large, but the rim is a structural component and significant wear to the rim sides can cause it to fail. The proximity of the tire is also problematic, and in extreme cases such as alpine descents very high rim temperatures have been known to cause blow-outs.

Hub brake problems

Hub brakes have none of the above problems but have several of their own. Whilst the wheel rim is no longer part of the braking equation, the spokes and build of the wheel itself are as the retardation torque must be transmitted from the hub to the tire. Hub brakes place off-center loads on frames and forks, which have to be strengthened to compensate – further adding to the mass of the bicycle, and hub brakes themselves are inherently heavier than rim brakes. Finally, heat dissipation can be a real issue as the heat generated during heavy and prolonged braking is concentrated in a small area.

Caliper brakes

Rim brakes were transformed by the invention of the Bowden cable, which permitted a flexible link between brake lever and brake. At a stroke this removed the need for the complex and heavy linkages used by rod brakes. The classical rim brake is the side-pull caliper brake, originally in steel and later in aluminum. During the 1950s the center-pull caliper was the most effective and most desirable stopper, despite its requirement for a straddle wire, yoke and cable hanger. The inherently symmetrical action of the center-pull brake was preferred to the side-pull for many years, although some racing cyclists chose the less efficient side-pull on the grounds of reduced weight. One problem common to both center- and side-pull brakes is that larger tires require brakes with a longer reach, and these longer brake arms reduce the effectiveness of the brake.

Cantilever brakes, U-brakes and roller cam brakes

More esoteric, at least until the mountain bike era in the 1980s, was the cantilever brake. Cantilevers have no brake arch under which the tire (and possibly mudguard) had to pass, as each brake arm was separately attached to frame-mounted pivots at either side of the wheel rim (either clamped on or brazed on). Actuation is via a straddle wire and yoke, similar to a center-pull caliper brake. The cantilever brake, whilst more complex to fit, offered more powerful braking on bicycles with larger section tires (for example, touring bikes) than caliper brakes. When mountain bicycles became popular in the 1980s, cantilever brakes were very widely used on these fat-tired machines. Other rim brakes were developed for mountain bicycles, including the U-brake which was effectively a center-pull brake on brazed-on pivots, mounted underneath the rear chainstays of the bicycle (very rarely were U-brakes fitted to the front wheel). Its effectiveness as a brake was not aided by them being fitted in a vulnerable (and muddy) position on the frame. The U-brake later became popular on BMX bicycles.

Roller-cam brakes shared the same frame mounts as U-brakes and had a similar low profile – unlike cantilevers of the day that stuck out from the sides of the bicycle. A triangular cam attached to the brake cable pushed the two brake arms apart at the top, thereby acting on the rim below the pivots; the shorter brake arms were stiffer than those of a U-brake and by altering the profile of the cam the actuation ratio could be optimized to move the brake blocks quickly to meet the wheel rim and then increase the braking force for more bite. Neither Roller-cams nor U-brakes challenged the dominance of the cantilever, unlike the direct-pull brake (also known as "linear-pull" and "V-brake™") which arrived on the scene in the early 1990s.

Direct pull brakes

The direct-pull brake uses the same mounts as the cantilever and has two arms around 10 cm long rising vertically from the pivots. The brake blocks are mounted low on these brake arms, and the arms are pulled together at the top by a cable directly joining the two. As the straddle wire was eliminated, these new brakes required a greater amount of cable pull to operate correctly, new brake levers were also required. Nevertheless, the direct-pull brake soon dominated in the mountain bike world and swept the cantilever back into the relative obscurity of cyclo-cross (the direct-pull brake did not play well with drop-handlebar brake levers due to the difference in cable pull) and the traditional touring market.

Modern rim brakes, for road and mountain bicycles respectively, are of the dual-pivot side-pull caliper brake and the direct pull "V-brake" types. The dual-pivot caliper brake is a hybrid of the center-pull and side-pull brake, offering the greater power of the former and the simplified mounting of the latter. During the 1980s, significant improvements were made to road bike (drop-handlebar) brake levers – the redesign of the cable route to conceal the cables under the handlebar tape and the provision of return spring in the lever being the most notable – which, combined with the adoption of modern high-efficiency cables, greatly improved the performance of caliper brakes. The industry-wide adoption of the dual-pivot side-pull came in the early 1990s and this is still – just – the most popular road bike brake. The direct-pull brake, however, has largely been replaced by the *disc brake* on mountain bikes.

Disc brakes

The disc brake is perhaps the most versatile of the hub brake genre. In common with all hub brakes, the tight restrictions on tire size are eliminated, even to the extent that many modern bicycles are designed to accept more than one wheel size (for example, 700c and 650b). Furthermore, the wheel rims can be made lighter (and mass lost from rotating weight has more effect than elsewhere) and the wheel need not be as true as it must be for rim brakes to be used, although of course it is desirable.

Disc brakes are, in principle, miniaturized caliper brakes operating on discs much smaller than the wheel rim (typically 140 mm to 200 mm). The simplest versions are cable-actuated and are single-sided (i.e., one moving piston pushes the disc over to meet the second brake pad), whereas mid-range models have dual pistons, one on each side, which operate simultaneously to grip the brake disc. The best disc brakes are hydraulically operated, which of course requires a special brake lever (or dual-control lever for drop handlebar bikes). Hydraulic disc brakes are not cheap, but they are unmatched for modulation (control) and ultimate stopping power. Disc brakes are suitable for all types of cycles and all forms of cycling, which is perhaps why they have seen so much development in recent years. Although there is controversy over the damage that hot and (relatively) sharp disc rotors may cause in accidents, the main disadvantage of discs is high braking temperatures – a problem countered by cooling fins on the rotors and on the disc pads themselves.

Drum brakes, roller brakes and band brakes

Other hub brakes include *drum brakes*, which comprise two semi-circular brake shoes forced apart (by a cable-operated cam) to press against the inside of a steel drum built into the wheel hub. As drum brakes are fully enclosed, they are unaffected by the weather; and their large friction surfaces ensure longevity and the minimum of maintenance requirements (in normal use a drum brake may never need new brake shoes). Disadvantages include added weight and the problem of brake fade – with prolonged use the brake drum expands away from the brake shoes and reduces efficiency. An alternative to the drum brake is the *roller brake*, which is usually fitted to the side of the hub rather than being an integral part of it. Here the brake cable operates a cam which expands a set of steel rollers inside a small (and narrow) drum. The metal-on-metal friction surfaces are small and, much like a coaster brake, a high-temperature grease is required for lubrication and cooling. The *band brake*, rarely found nowadays, creates braking friction by tightening a flexible band around the outside of a drum.

The future of cycle braking

With the continuing adoption of disc brakes on all types of bicycles from road to mountain bikes, it seems probable that once-common caliper and direct-pull brakes will become outliers used on only the most specialist – or least expensive – machines. The lightest racing bikes will continue to use the dual-pivot caliper, city and public-hire bikes will opt for the low-maintenance drums or roller brakes, and the coaster brake will continue to be used in its traditional markets. For other bicycles, however, the disc brake is likely to become the standard.

References

Berto, F. (2009), *The Dancing Chain* (3rd Ed.), San Francisco CA: Van der Plas.
Whitt, F.R., & Wilson, D.G. (1982), *Bicycling Science* (2nd Ed.), Cambridge, MA: MIT Press.

Further Reading

Abbot, A.V., & Wilson, D.G. (1995), *Human Powered Vehicles*, Champaign, IL: Human Kinetics.
Ballantine R. (2000), *Richard's 21st Century Bicycle Book*, London: Pan.

12
PASSENGER CARRYING

David Henshaw

This chapter explores the settings where cycles are used to carry a passenger. Not included is the tandem, where the rider behind (often called the stoker) is a working member of a team and not strictly a passenger. The beginnings appear to be tricycles such as the Bath chair of the 18th century and sociable tricycles of the late 19th century where a couple might have an outing with the lady purely a passenger or only an intermittent contributor to forward progress. During the 20th century, bicycles and tricycles have been put to carrying passengers in three main settings. First, and the main focus here, is carrying children. Second is the cycle as taxi or rickshaw, which is still widely used in its motorized form in Asia. And third is the cycle as a means of transporting persons with an impairment (discussed more fully in Chapter 13).

Carrying children

When the bicycle was very young it was primarily an adult form of transport: with a few exceptions, the concept of carrying children as passengers seems to have come relatively late. No doubt some accommodating dandy had long before propped his nephew on his draisine's crossbar, but it was not until the safety bicycle was well established in the late 19th century that child-carrying became a practical proposition. Moreover, for many years bicycles were so expensive that ownership was largely restricted to people who could afford a nanny to look after the logistics of childcare. A bicycle was for solo use, for sociable rides with your beau, and group rides with clubs: many with need for a child-carrier could not afford the cost and complexity of owning a bicycle. People with a low income simply did not travel much, and even when they did, they were expected to get there by ducking and diving through the traffic under their own steam, an operation at least as dangerous as it is today in big cities.

The price of bicycles fell rapidly after 1895, and by the end of World War I, bicycles were easier and cheaper to buy and maintain than ever before, and a practical purchase for most people in western countries. Yet the concept of child-carrying remained somewhat obscure. The catalogues of the London department store Gamages carried every imaginable cycle accessory, with plenty of freight solutions – from trailers to giant racks and specialist cargo-bikes – but very little in the child-carrying department.

 DOI: 10.4324/9781003142041-16

Early child seats

Perhaps the most common child-carrier in the period between the wars was a small metal seat immediately behind the cyclist, with or without a folding backrest, enabling it to fold flat for load-carrying. These were uncomfortable for the child victim, noisy, and the rudimentary safety straps were pretty dangerous by modern standards. There was one big advantage, of course: they were cheap and simple to produce, and in a time of seemingly never-ending austerity this was important. But the Rolls-Royce of these early child seats were various wicker designs, either with their own frames to attach to the bike, or for bolting to a conventional rack. Wicker (usually woven from flexible "wands" of willow in temperate climates) is a light, sturdy and endurable natural material (many wicker products from the 1930s are still with us).

There does not appear to have been any research into wicker's crashworthiness, but its ability to "give" and absorb shocks without breaking or shattering would suggest that it was a superb material for child-carrying. Willow, in particular, is a renewable, more or less CO_2-neutral material from which an annual crop can be taken pretty well anywhere on the globe. Unfortunately, it was unaffordable for most people in that austerity era, because of the skilled man-hours absorbed in the weaving; it remains expensive today and has been largely replaced by plastic. However, with such impeccable "green" credentials, the wicker child seat seems even more relevant to the 2020s than the 1920s, and must surely be overdue for a comeback? Like carbon fiber, the weave and strand diameter of the willow can be fine-tuned by computer-aided design to enhance the seat's lightness, crash-safety and durability.

Trailers and side-cars

Stand-alone two-wheeled trailers were unusual in the period between the wars, but "side-cars" – with a single outrigger wheel – were very much in vogue, and in many ways, a better design solution too. Most were heavy steel machines, and nearly always aimed at the tandem market, where a little more haulage power was available, but wicker was sometimes used in this application too, although, again, these must have been quite expensive.

Other ideas were tried, but none really got anywhere. A typically innovative – but ultimately doomed – idea was the convertible perambulator designed by Warwickshire engineer Clayton Wright in 1951. This looked like a conventional four-wheeled pram, but two attachment bars could be slid out from the frame, converting it into a side-car for the journey home from the shops. It appeared to fill a niche, but was heavy, complex and too expensive for production.

In any event, times were changing. In the 1930s, the bicycle (especially the tandem) was increasingly used as a leisure machine in Britain, France and some other European countries, and more particularly as a family leisure conveyance, and future child-carrying would be explored from that angle. Cars were still luxuries and, because of the war, would remain so well into the 1950s. But as cars became more affordable, the bicycle was thrust into a secondary role (probably helped on its way by motor company PR efforts). Once a family had purchased even the humblest car (or motorcycle/side-car combination), the need for day-to-day, all-weather child transport using a human-powered machine evaporated. It was not until the concept of "green" living became popular in the 1970s that the possibilities for child transportation began to be re-examined (Hadland & Lessing 2014: 370–6). When the boom came, the manufacturers got it all wrong, and those issues have remained locked into the concept ever since. Thus, due to its *ad hoc* origins in the world of leisure cycling rather than a conventional market-led evolution, the child-carriers designed for the modern bicycle are rarely fit for serious purpose.

Child seats in recent decades

From the 1980s onwards, leisure bicycles were developed as mountain bike-style machines, with little provision for carrying anything at all. That usually meant a great big empty space over the rear wheel, and with rack-mounting lugs increasingly rare, the obvious mounting solution was a clamp around the seat pillar, with a child seat of some kind extending behind, usually suspended on sturdy spring-steel bars. This arrangement became more or less universal.

Manufacture was gradually outsourced from the West to the engineering workshops of East Asia which were price-competitive. East Asian makers built simplified cropped-back leisure bikes for a global market, most of which were made to designs outlined by western brand-owners. The largest market was the USA, where the concept of utilitarian cycling as a means of transport had never hitherto gained much traction. With bicycles designed for recreation rather than functionality, it is not surprising that a generation was growing up around the globe who had never used a bicycle as a practical means of transport. These bikes were meant for recreation and leisure. They were not intended for shopping, since they splattered mud over riders, and they were not designed to carry children.

Child seats existed, but like the bicycle itself, they were built for the market, and consumers of the 1980s and 1990s were asking for lower prices. The lifestyle message was all about "green-ness" and personal fitness but, paradoxically, bicycles and child seats were frequently driven to the park on the back of vehicles, thereby generating pollution and increasing the danger of traffic instead of accessing the healthy recreational area on the bike itself.

A child seat behind the seat-post, and suspended from it, is not a disastrous option, but nonetheless it is one of the worst. The two primary dangers for a child are a single vehicle bicycle crash (anything from falling off while stationary to colliding with an inanimate object at speed), and collision with a motor car. More by chance than design, a diamond-frame bicycle is effectively a rigid spaceframe, able to crumple and absorb some of the shock in a crash. More or less by a throw of the evolutionary dice, the rider sits partially within this safety frame, and gains some degree of protection from it. But anything carried outside the frame – be it a bag of groceries or a child – is more vulnerable, with little or no impact protection, so a child seat on the front or back of the bike is hardly the safest place.

An analysis of Swedish car/bicycle crashes found that more than three-quarters occurred at road junctions, where statistics indicate that the collision impact can be from more or less any direction (Öman et al. 2012). Some legislative work has gone into providing child seats with car-style restraint systems, which is welcome, but these are primarily aimed at keeping the child in the seat under the effect of high G-forces. This makes sense in a car, where the child is inside the crumple zone, but not on a bicycle where they are usually sitting outside that safety envelope.

Little thought seems to have been given to the actual position of the seat on the bike. By far the safest place is inside a rigid, but not solid, frame that will absorb a considerable shock: not unlike the typical diamond frame in other words. This cannot be fully achieved on a bicycle, as it is in a car, but bolting the child-seat astride the top tube is the best compromise. This is a safe place to be in a front or rear collision, and only slightly less satisfactory in a side impact, where most of the shock is taken by the rider's body and pedal/crank assembly. Rear or (more rarely) front-mounted seats leave the child completely exposed.

There are other advantages to a central mounting. The child doesn't become tangled with panniers and other (usually rear-mounted) luggage, and it is by far the most sociable option for the child and rider. The ability to engage in easy and relaxed conversation with the young passenger is a great bonus, and a safety feature in itself, as it allows the rider to keep an eye on the child's state of alertness. A sleeping child will never be as safely restrained in a harness as an alert one.

Child seats of this kind are rare, and those that do occasionally come onto the market disappear just as rapidly. This seems to be primarily a matter of consumer inertia, which tends to be a stronger force where safety issues are concerned. With no other evidence to go on, consumers observe that the majority of child seats are located behind the saddle, so this is presumed to be the safest layout, rather than the cheapest, or the easiest to design. An unproven specialist product inevitably takes some selling, and to make matters worse, it would have to become pretty mainstream to compete in terms of price with mass-produced accessories.

One legitimate argument against putting the passenger between the rider's knees is the lack of research into the possible physiological effects of riding with legs slightly apart. Generally, though, the seat need not be wide. For very small children a full seat and harness are needed, but for older children (the author has carried children aged up to ten this way), a simple yoke-shaped saddle over the top-tube and inboard grips on the handlebars work well enough.

Trailer-bikes

The trailer-bike is often chosen by inexperienced parents as a sort of balance-bike training option, but it isn't, and neither is it safe unless extremely carefully designed. A common issue is poor design and manufacture of the universal joint between the bicycle and the trailer-bike. This has to work hard in three planes: to follow the lead bike around corners in a horizontal plane and adopt the same degree of side-to-side lean as the lead bike in a vertical plane, while coping with choppy fore-aft vertical misalignments as the two machines cross depressions and bumps. This is quite a tall order for any mechanism, let alone one built down to an acceptable price for the occasional leisure ride, so it is not surprising that so many are unfit for purpose.

The most dangerous failing is slop in the vertical plane, so the trailer-bike flops one way or the other and stays there, sometimes with a child hanging on grimly, until the lead bike rounds an S-bend, causing the bike to flop suddenly the other way. With a lot of play in the mechanism this transition can be dramatic enough to throw a child off or put them into the path of oncoming traffic. Clearly, loose bearings are very dangerous, but tight bearings can be just as bad. A trailer-bike unable to follow smoothly due to seizure of a bearing will appear to function perfectly well, but it will be putting huge strain on the entire system, from the seat pillar or rack of the lead bike to the boom of the trailer-bike. Eventually, something will fail. Usually, this will be the weakest component, and the trailer-bike will then collapse onto the road.

Crash-wise, these machines are unsatisfactory. Even if the geometry of the outfit is fit for purpose, a trailer-bike effectively doubles the length of the lead bike, putting the child well outside the crumple zone of the lead bike diamond frame, and potentially confusing an approaching motorist. The poor child is also in the rather invidious position of being out on a limb some way behind and below the adult, but without the ability to get out of danger themselves should things start to go wrong.

It is an anomaly of modern transport that parents will choose a crashworthy car to transport their priceless child cargo yet invest in a cheap and mechanically dubious trailer-bike to carry the same child behind a bicycle. Most trailer-bikes are designed for gentle leisure rides in safe settings, and not for riding on busy roads.

Child trailers

A well-designed and manufactured child trailer is a good place for a child to be. The best have practical seats, and sturdy multi-point harnesses, with head restraints to protect the neck in the event of a rear impact. And nearly all place the child(ren) in the middle of a substantial metal (usually steel) outer skeleton. In a collision – provided the trailer is not crushed – it might be thrown some distance, but its skeleton should not be penetrated, with the child securely strapped in the seat. Of all the available options, this seems hard to improve upon.

Figure 12.1 shows a bicycle that is power-assisted, and the off-side (right) of the trailer is about the same distance from traffic as the off-side pedal. The trailer has rear lights, and the child-seat lies almost horizontally (small babies have weak neck muscles). Complete safety can never be guaranteed, but this is probably the best compromise for carrying a baby.

Not that long ago, American-made trailers dominated the top end of the market but, like most child-carrying options, the price, build quality and design has declined in recent years. These companies have followed consumer demand towards purely leisure solutions, so most trailers are convertible to strollers, with big heavy handlebars and clumsy stroller wheels.

Modern trailers are usually built to carry two children, with the emphasis on off-road and leisure use, but they are impractical on busy roads. Lights are mandatory in some jurisdictions, but their absence in other countries is a serious omission given their bulk.

Narrower, lighter single-child trailers are rare, and thus expensive – generally more so than a double-child trailer – and few customers are going to deliberately choose to pay more for something half the size, thus guaranteeing that they remain rare. A single-child trailer has a much

Figure 12.1 Carrying a six-week old baby in a single-seat trailer. (photo copyright of David Henshaw).

narrower track, so when lightly loaded the narrowest can – under extreme circumstances – turn over. On the positive side, they follow closer to the tow-machine, and being little wider than the envelope of the bicycle itself, they're a good deal safer on the road, particularly if set up with a bias toward the nearside kerb, leaving little or nothing protruding beyond the bicycle. They can also negotiate most cycle-path barriers or chicanes, which the bigger clumsier trailers will not; something most users only discover after purchase.

Trailers should be a serious tool in the armory of parents wishing to transport children. They are probably the safest solution for carrying very young children and even babies, with head restraints and the right multi-point harness, and they keep the child dry and tolerably warm. Older children, and even disabled adults, can enjoy a trailer ride, and keeping the passenger dry makes a lot of sense.

Cargo bicycles and tricycles

These interesting machines are particularly popular in The Netherlands and certain Nordic countries, for reasons that will become obvious. Most are based on long-established designs, but there are a few more innovative models. The concept comes with some key advantages, and quite a list of drawbacks. This is one of the most sociable child-carrier arrangements, both for the cyclist and the child/children riding inside. The human cargo usually goes in front within easy communication range of the rider, and generally low down, giving vital stability, but this arrangement usually involves a compromise to either the front wheel size and/or the length of the machine.

Some trikes get round these issues by putting the child between two conventional front wheels in the "tadpole" arrangement (putting the cart before the horse, so to speak), with a variety of steering systems, the most common being a very large pivot in the middle of the trike, effectively splitting it into two units, or more conventional – but complex – car-style Ackerman steering. The first is rarely satisfactory without some form of damping between the two parts of the trike, while the second is heavy, expensive and unsatisfactory unless carefully set up. Either option can suffer from poor or downright dangerous braking, unless the components are carefully chosen and set up to equalize brake force between the two wheels. Consequently, high-speed – meaning any speed beyond 10–12 mph – is rarely sensible.

A glaring issue common to nearly all of these machines is excessive width and cumbersome handling. This is probably the primary reason they have only really become popular in countries with a network of dedicated cycle lanes. These are not sensible designs with which to share roads busy with faster motorized traffic and even on a two-meter-wide cycle path, dealing with other cycle traffic can be quite a fraught affair, while most chicanes are impassable. The largest of these vehicles are effectively touching on rickshaw territory.

The freight-carrier with probably the most potential for carrying children is British, despite freight- and child-carrying HPVs being pretty rare in the UK. When innovative cycle designer Mike Burrows looked into cargo and child-carrying, he came up with the 8Freight. Burrows has chosen the "least bad" freight option – two 20-inch wheels (compact, but immensely tough, thanks to the pervasive influence of 20-inch BMX bikes), and a load-carrying platform slung as low as possible in the dead space behind the rider – and honed the design to eradicate most of the awkward compromises. The small(ish) wheels keep the wheelbase down to a reasonable two meters and width to just 60 cm, while an aluminum frame and carefully chosen components have resulted in unladen weight of only 20 kg. An indicator of Burrow's expertise is that the 8Freight will carry up to two and a half times its own weight, with handling and turning circle much like any other bike, with a price in the region of US$3,000, which is not bad for a hand-crafted rarity of this kind.

These figures all relate to the freight machine, as there is no child-carrier option, although several have been home-adapted. It is not a particularly sociable arrangement, but the low center of gravity, and tough space-frame around the central child-seats make it very safe – indeed arguably the safest child-carrier of all. It is a rather bitter irony that the cycle-skepticism of Burrow's countrymen has resulted in strictly limited sales, mostly in the primary cycling cities, such as London, Oxford and Cambridge. Nonetheless, worldwide the 8Freight is his biggest-selling design, despite him being most famous for his innovative high-speed bicycles and HPVs.

Lights

Lights are essential at night and are now an inexpensive option. Battery-free, fully automatic lights have been developed, which typically generate power using small permanent magnets fixed to the wheel rim, which excite a small current in fixed coils on the frame, so power is produced whenever the wheel turns. Until recently, the power produced was too small to be of practical use, but the efficiency of LEDs is now high enough for this sort of light to become practical. It is essential that lights are automatic, and low – or preferably zero – maintenance. Their positioning is already covered by legislation in most markets but needs to be improved. At least two lights are essential, and they must protect the extremities of the frame.

Tires and suspension

Free-running tires are also important, and the 20-inch nominal diameter (406 mm tire bead seat diameter) folding bike and BMX size is a good compromise. These tires are large enough to run freely, but small enough to produce a tolerable folded package, and they are cheap and very widely available.

Suspension is an interesting area. Few people realize that the jittery, jolting ride of a rigid-framed trailer can be harsh and uncomfortable for a baby or small child. The inbuilt suspension of pneumatic tires works well for adults, but hardly at all for children. It is tempting to reduce the rolling resistance of a trailer by pumping the tires up hard, but a small child might weigh a quarter of the weight of an adult. So where an adult and bike weighing 80 kg might have tire pressure of 80 psi (5.4 atm), a child and trailer with gross weight of 30 kg should not have more than 30 psi (2 atm), and probably rather less. With so little weight involved, punctures are rare on trailers, even at such low tire pressures.

Power-assistance

Power-assistance has become a rather political affair in the bicycle world. Realistically, adults may not find it easy to haul a single child and their daily needs to school, particularly where any sort of gradient is involved. In this context, it is noteworthy that the school run is more likely to be undertaken by women (UK Government 2020). A recent study by a leading European car manufacturer found that a quarter of school run trips were less than a mile in length, and the majority (53.9%) were of one to five miles (Renault Group UK 2020). These distances are not difficult on a bicycle, but more challenging when hauling a trailer. To keep these potential converts onside, the attitude within the rather sports-oriented bicycle world has to change. Power-assistance is not cheating, and the suggestion that it is can be extremely counterproductive.

Carrying adult passengers - rickshaws

Rather surprisingly, pulled rickshaws were little known or understood until they were effectively invented by the Japanese around 1869 with the lifting of a 300-year-old ban on wheeled vehicles, although they may well have appeared intermittently elsewhere in preceding centuries. Pedal rickshaws – able to haul twice the load – were an even later development, with early experimental machines following on the heels of the bicycle in the 1870s and '80s. It was well into the 20th century before the pedal rickshaw really began to take hold of urban transport in many – mostly Far Eastern – countries, with the peak probably coming as late as the 1950s. By that time they were simultaneously being withdrawn in many cities because of competition from faster public transport. By the turn of the 21st century they were becoming an endangered species, replaced through straightforward market forces by the auto-rickshaw – like a large pedal rickshaw, but powered by a cheap and environmentally disastrous two-stroke engine. Auto-rickshaws carried greater loads faster, enabling the owner to charge more per journey, and provide more revenue-earning trips. For most pedal-rickshaw operators this was a large investment, but it could yield a handsome dividend. The downside was the appalling pollution record of these machines, in terms of noise, atmospheric pollution and congestion. They appeared rapidly, and look set to disappear equally rapidly, principally through being outlawed in many cities.

A small conventional hatchback car represented an even bigger investment, but it was cleaner, quieter and safer. As legislation once again forces the hand of operators, it looks likely that small electric cars will come to dominate this private hire business worldwide within the next decade.

So, will human-powered pedal rickshaws make a comeback? They are becoming fashionable in western cities, where they provide a clean, silent zero-carbon alternative to the traditional taxi. These machines are generally pedaled by young, fit and politically-motivated cyclists, but the rather politicized concept of eliminating fossil fuels through the dignity of human labor is somewhat removed from the realities of the situation.

To understand the gentle deceit at the heart of the modern rickshaw world, it is only necessary to glance briefly at the facts and figures. A single fit human can produce 500 watts or more for short periods, but only rare individuals can maintain an output of more than 100 watts throughout a long working day. That sort of figure is more than enough to move a rider and a good-quality, lightweight bicycle at a reasonable speed, or haul a lightweight trailer at a rather lower speed, even up hill. But a substantially-built pedal rickshaw, plus the rider and passengers, possibly with luggage is another matter. Faced with a gross load of up to half a tonne, 100 watts is almost an irrelevance. Forward progress of any kind would only really be practical on a downward gradient, the machine would be slow (and thus intrinsically dangerous) on the flat, and impractical on an upward gradient. In practice, this cold-cruel reality has limited 100% human power to use over short distances on level territory, historically in city centers and more widely in theme parks and similar off-road environments, where lack of speed is not a big safety issue. Human-powered rickshaws competing with taxis – even in a very localized sense – are nearly always electrically power-assisted, and unlike basic bicycle power-assistance, where a 50/50 human/electric input can provide useful propulsion, the ratios with a heavier vehicle may need to make much greater use of electric drive.

There is nothing intrinsically wrong with the concept of a human-powered taxi, but we need to be more honest about what human-powered means in this respect. Many of today's rickshaws are really electric vehicles with pedal assistance. Modern materials and technology could help, by reducing the weight and rolling resistance, but there is no escaping the simple weight/power/speed equation. The answer, surely, is to combine cheaper Asiatic manufacture with new technologies, such as solar panels, advanced batteries and lightweight materials, guided by Western academic and scientific expertise. A global hybrid machine, lighter, smaller and cheaper than a car – electric or otherwise – would be near silent, and non-polluting at point of use, keeping the rider fit and minimizing energy consumption from other sources. The concept of a global 21st-century human/electric taxi built in very large numbers for cities east and west is an enthralling one.

This has to be a co-operative venture. Several attempts have been made to develop better machines on a small scale for western cities, but everything thus far has ended in failure, either dramatically or gradually. Plying for trade in this way is a competitive business, and not particularly lucrative, and without the necessary economies of scale, the pedal rickshaw has remained a low-tech machine reserved for events perceived by tourist passengers as whimsical and romantic.

Conclusion

The use of bicycles and tricycles to carry passengers has a long history, yet despite the potential of these machines for very local transportation, they have rarely achieved their environmental potential, particularly on the typically short-distance ride to school. As a means of neighborhood transportation for the middle class in cities in Asia, they survive in motorized form but struggle to compete with the comfort of taxicabs. And as a novel cultural experience for tourists enjoying nightlife, the rickshaw has found a new niche that is more about conspicuous consumption than saving the planet.

Source list

Hadland, T. and Lessing, H.-E. (2014) *Bicycle Design – An Illustrated History*, Cambridge, MA: MIT Press.

Öman, M., Fredriksson, R., Bylund, P-O. and Björnstig, U. (2012) Analysis of the mechanism of injury in non-fatal vehicle-to-pedestrian and vehicle-to-bicyclist frontal crashes in Sweden, *International Journal of Injury Control and Safety Promotion*, 23: 1–8. Abingdon: Taylor & Francis.

Renault Group UK (2020) Tackling air quality outside schools. https://www.press.renault.co.uk/en-gb/releases/2770 (Retrieved December 17, 2020).

UK Government (2020) *National Travel Survey: England*. https://www.gov.uk/government/collections/national-travel-survey-statistics (Retrieved December 17, 2020).

Vignette B
MICROMOBILITY IN RWANDA

Hilary Angus

Rome lays claim to seven hills, but Rwanda's claim runs into the thousands – hence its name – the Land of a Thousand Hills. There is scarcely a flat road in Rwanda. It's a country of pristine lakes, abundant agricultural land, forested mountains and a sad history. The roads wind up, down, and through towns and villages tucked in among the trees and farms. The population is largely rural – only 30% of Rwandans live in Kigali and other major towns. The remainder reside in agricultural areas that dominate the landscape.

Farming in a mountainous country is not without its challenges, especially when it comes to moving your products to market. Emerging from the shadows of its past, Rwanda still struggles with persistent poverty – many young people remain un- or underemployed and most farmers have little money to spare for transportation. Recognizing a simple solution to a complex problem, a few NGOs have been working in Rwanda over the past decades to bring bicycles to the people in order to achieve an environmentally responsible solution to their transportation needs.

The most prominent of these was *Bikes to Rwanda/ Project Rwanda*, a joint initiative between the CEO of Portland-based Stumptown Coffee Roasters and renowned mountain bike builder Tom Ritchey. During a 2006 trip to visit his coffee growers, former Stumptown CEO Duane Sorenson asked one of the farmers what he could do to improve their quality of life. The farmer responded that a bicycle would be helpful to move the beans more quickly and easily to the mills. Sorenson teamed up with Ritchey, and the Coffee Bike was born. Thus in April 2007, 400 bikes that were designed to carry heavy loads of coffee over difficult Rwandan terrain were provided by Bikes to Rwanda ahead of the harvest by providing cargo bicycles to co-operative coffee farmers, sustained by a bike workshop and maintenance program. Over a multi-year period, Project Rwanda delivered thousands of specially designed longtail cargo bikes, each with the capacity to carry up to 350 lb (Angus, 2018) (Figure B.1).

But Project Rwanda isn't the only way bicycles get to Rwanda, and not everybody is so lucky to be riding a specially-designed longtail with 18 gears. Over the years, there has been a steady influx of old bicycles to East Africa, imported on the cheap from India and China. The bikes are heavy-duty steel city bikes, single-speed, and weighing in around 20 kg. They have an elongated steel platform behind the saddle to carry cargo or people, and with their arrival, a new informal economy was born.

DOI: 10.4324/9781003142041-17

Figure B.1 A project Rwanda coffee bike (author's photo).

In any Rwandan town, young men can be seen pushing oversized bundles of firewood up mile-long hills on the backs of their bikes, sweating through their tank tops on a chilly mountain morning. On the opposite side of the street, bundled up in sweaters and beanies, men coast freely down the long inclines carrying 100 kg of sweet potatoes in a sack, or ten trays of beer bottles stacked perilously high above their heads.

If necessity is the mother of all invention, then poverty is the mother of all effort, because these men work hard. These bikes carry heavy loads but they were not equipped with electric assist, and in most cases are old, rusty and use worn-out parts. Spare parts are difficult to come by in Rwanda, so when the bikes break down they're often jimmy-rigged back together using old car parts, bits of spare wire, or whatever the owners can get their hands on.

Beyond their role in the transportation of goods, the cargo bikes have also provided employment as bicycle taxis for thousands of young men who otherwise lack marketable skills or opportunities. At any given corner in town, you will see lineups of bicycle taxis waiting for customers, ready to carry 1–3 passengers at a time for a nominal fee.

For some of the young men in this role, the bicycle presents not only an opportunity to scratch out a living, but the chance to one day ride with Team Africa Rising (TAR), Rwanda's national cycling team and the first African cycling team (possibly excepting that of Eritrea) to compete at an international level. Given the enormous effort required to ride a cargo bike daily in this hilly country, it is no small surprise that many of TAR's riders got their start as bicycle taxi drivers. In the past 20 years, Rwanda has made substantial strides in its post-genocide reconstruction effort, and this effort has resulted in, among other things, a country-wide network of well-maintained tarmac roads unlike any other in East Africa and, since 2019, a public campaign to promote the use of non-polluting e-bikes.

Walking down the street one day in the lakeside town of Kibuye, in southern Rwanda, I wondered about the separated path between the road and the sidewalk. What was this strange corridor that people use to walk their goats down? It wasn't until I saw the signpost with a bicycle emblem that I realized that this fancy goat path was in fact a designated bicycle lane. Covid-19 has given new legs to this project to improve cycling infrastructure in a number of African countries (Kuhudzai, 2020). While Rwanda may not be the next Netherlands, the simple bicycle is improving people's daily lives against difficult odds by enabling micromobility for those on the margin. The government is throwing its support behind those goals by, for instance, creating bicycle lanes in all of Rwanda's major cities. Of current interest was the founding in 2017 of GURARIDE, a Rwanda-based green e-mobility public bikeshare (PBS) transport company committed to the sustainability of micromobility in Africa.

With the possible exception of Eritrea, Rwanda is the only African country where cycling has become part of the national identity. Coffee bikes, similar longtail cargo bikes, and ancient sturdy roadsters imported long ago from Asia deliver farm products to markets at low cost, and carry back the goods needed in rural households. These same bicycles serve as bicycle taxis with the dual benefit of providing low-cost last-mile trips in towns while also serving as a training ground for Rwanda's formidable bicycle racing team. Meanwhile Rwanda's GURARIDE is pioneering public bikesharing in Africa. The resulting mobility at the microscale matches Rwanda's hilly topography, its limited energy resources, its agricultural economy, and environmental priorities.

Sources

Angus, H. (2018) "Cargo bikes abound in the land of a Thousand Hills." https://momentummag.com/cargo-bikes-abound-land-thousand-hills/

Anon. (2 August 2018) "Team Rwanda matures in Central Africa's top cycling team." *Velo News*. Retrieved 20 June 2019.

Kuhudzai, R.J. (2020) "Jumpstart Africa's bikeshare and micromobility industry." https://cleantechnica.com/2020/09/14/rwandas-guraride-looks-to-jumpstart-africas-bike-share-micromobility-industry/

13

CYCLING TECHNOLOGIES AND DISABILITY

Ron Buliung, Annika Kruse, Glen Norcliffe and John Radford

Disability and cycling have received almost no attention in transportation research and planning practice. Is this because pervasive ableism produces a lack of imagination regarding disability and cycling? As the reader will discover, the relationship between cycling and disability is actually centuries-old. Given this long-connected history, why do we find outright neglect in professional planning of the relationship between cycling and supportive infrastructure (Andrews et al. 2018; Chapter 7)? The lived experience of the first author has also shown that the production of state-of-the-art cycling infrastructure, designed for a normative "able-bodied" citizen, can produce disabling barriers for persons with disability (PWD) who are "assumed" to be transit users, passengers or pedestrians. With few exceptions, cycling and disability is largely absent from planning discourse (Andrews et al. 2018; Clayton et al. 2017; MacArthur et al. 2020).

This chapter examines cycling technologies developed for and by PWD. We look at technologies that enable disabled persons to do things they want to do using cycles, and technologies that may be used in a therapeutic context when desired or necessary (Norcliffe et al. 2021; Wong 2020). We take up the social model of disability here (Oliver 2013), and in doing so posit that adapted cycles are designed and modified to address impairment, i.e., the bio-medical situation of the user, to enable mobility in a disabling world. We follow Rocco (2011) and Oliver (2013) in viewing disability as a socially imposed category. We also endorse the social justice perspective on these matters advocated by Inckle (Chapter 7), as opposed to the special needs and medicalized/asylum approaches that overlook the universal rights of persons with disability (PWD) (Inckle 2019; Radford 2020).

While we conceptualize disability through the social model, our writing about cycles occupies a sort of liminal space between the social construction of disability, and the different ways in which every "body" engages with the environment. We also have trouble with any sort of activity–therapy binary, recognizing that cycles may be taken up for either case and are sometimes recursively based on what people want to do with their bodies, and what people want their bodies to do. The question of agency is different, of course, where children are concerned, particularly within the culture and practice of Western medicine where adults are afforded decision-making power. We also think about technology as socially constructed. Most technological research is aimed at practical issues, often by responding to the requirements of potential users (Oudshoorn and Pinch 2005). Our research shows that this is particularly true for cycling-related technologies designed to work with a range of abilities. Users, their families

DOI: 10.4324/9781003142041-18

and friends exchange ideas with makers about their needs, and have also themselves been makers as technologies are iteratively improved and copied in a process of co-construction. And makers often work in teams as ideas are bounced about.

Cycles, bodies and minds

Cycling is used by PWD for everyday mobility, for leisure and pleasure, for socializing with friends and family, and therapeutically to assist with several forms of impairment. The abilities of individuals are unique, although sometimes with overlapping requirements – what a bio-medical conceptualization would call co-morbidity. We do not claim that this taxonomy is exhaustive – we are simply making an attempt to map technological innovation and adaptation to particular user requirements since cycle technologies have been adapted or created to work with different abilities. Of course, how, why and where people engage with cycles in relation to their bodily abilities is a matter of individual choice, particularly where adults are concerned.

Mobility

Many forms of impairment, including arthritis, cardiovascular disease, spina bifida, damage to the musculoskeletal system and obesity, limit the mobility of an individual. Cycles may aid the recovery of mobility in many settings provided the needed infrastructure is installed. Increased mobility is undoubtedly the most important contribution of cycles.

Stress and anxiety

Recent research demonstrates that cycling and similar rhythmic sports have several health benefits by lowering levels of the hormone cortisol (Concordia University 2017). Regular cycling (and similar physical activities) reduces stress, depression, anxiety and tension and may forestall the need for drugs (Oja et al. 2011; Chapter 30).

Cardiovascular problems

Cardiovascular problems may limit a person's lifestyle, mobility, and the performance of routine daily tasks. Therapeutic activity programs including the use of cycles (often stationary machines) strengthen the cardiovascular system (Oja et al. 2011). Tricycles, quadricycles and stationary machines are very stable, while the seat and handlebars reduce the weight on legs and hips. E-bikes offer electronic assistance for persons advised not to exert undue effort due to cardiovascular limitations (Bourne et al. 2018).

Visually impaired

Cycling requires a rider in public spaces to see the way ahead, which poses dangers on roads shared with motorized vehicles. In the case of severe visual impairment, bikes partnered with a seeing cyclist provide a safe alternative.

Autoimmune diseases

Autoimmune diseases, including arthritis, Parkinson's disease, MS, diabetes and asthma, limit the functioning of the body, with movement limited by fatigue and muscular dysfunction. Cycling (especially with e-bikes) offers a low weight-bearing, low-impact mobility and exercise.

Balance and equilibrium

The standard bicycle obviously presents a problem for those with vertigo and related balancing issues which can be avoided using inherently stable machines such as tricycles and quadricycles.

Types of cycles for PWD

Designers of cycles used by PWD may incorporate six main attributes. The adapted machine needs to be **accessible**: this may require an open front, a low top-tube, a two-track tricycle configuration leaving the front unblocked, rear access, or a lifting pedal. **Stability** is a common priority, often requiring a wide wheel-base, inclined wheels, a low center of gravity, and/or a (semi-)recumbent riding position. **Speed** may not be a priority in some contexts, but is very important in sport, training, recreational cycling, on or off-road (Szto 2012), while **low-gearing** making for less resistance and ease of climbing provides an advantage in some contexts. Footstraps, calf braces for legs, seat belts and headrests are among the **support** devices that may be used to help riders remain in an effective (for power transfer) and stable riding position. Since the rider may be lower to the road and possibly slower than most cyclists, enhanced **visibility**, using flags, lights and brightly colored clothing, is often recommended. The ability of a cycle to **carry cargo** is frequently an advantage for PWD. **Comfort** is also important as many forms of impairment are accompanied with pain: soft materials and massage seats, a recumbent position, and arm, leg and head supports all may provide relief for the rider. Powerful shock absorbers and all-wheel suspension are often used by off-road PWD cyclists. These attributes have been adopted to varying degrees in seven main types of adapted cycles used by PWD (sites presenting images of each named machine are listed at the end of this chapter). We note that the broad labels used could apply to any type of cycle, the difference lies in how a particular element of design materializes within a particular body–machine pairing or context.

Manumotive machines (a.k.a. handcycles/hand-cranked)

Manumotive machines are driven by riders using their arms: the rider's legs are not used to propel the cycle. Typically, these vehicles were based on tricycles with a single front steering wheel. It is highly significant that Stefan Farfler, a paraplegic clockmaker based in Altdorff, Germany, designed the world's first authenticated tricycle around 1660 (Norcliffe 2012). Farfler used his knowledge of clock gears and wheels to devise a manumotive with an articulated front box housing a drive mechanism that resembled a large clock, allowing him to travel around his village; hence Farfler was both maker and user.

By the mid-19th century most manumotive tricycles were driven by levers acting on a cranked axle. An example is Charsley's *Velociman* tricycle of 1880, made for impaired cyclists by the Singer Cycle Company of Coventry from 1880 for 20 or more years. Manumotives proliferated after World War I during which numerous soldiers suffered damage to, or loss of, their legs. By this time technology had advanced and metal tube construction with chain drive was the norm. These manumotives did not require a second person to push them, although assistance in mounting and dismounting may be needed.

Today, handcycles use lightweight materials, including carbon fiber and titanium, which allow riders to move quite rapidly, with their steering mechanism incorporated into the driving levers. Seating is often custom-molded to the rider's body. For instance, STRAE Sport

Handcycles, based in Las Vegas, Nevada, are "designed by engineers and specialists seeking a safe and ergonomic cycling experience for those with spinal cord injuries, amputations, or neurological diseases." Similarly, Upright Handcycles (located close to Winnipeg, Canada) and Theraplay Mobility Tricycles (located near Glasgow, Scotland) are designed for persons (especially children) with spina bifida and/or other issues affecting the lower extremities. Off-road handcycles are now available for enthusiasts (Szto 2012). Also, for the last two decades off-road four-wheelers have been used by some riders to make mountain biking accessible. These cycles make use of hand pushing to a trail-head and then the downhill slope of the trails for movement (BikeMag 2020).

Pedomotive machines (pedal cranked/non-adapted)

For persons with a range of impairments, including those with limited hearing and those lacking a lower limb or using a prosthesis, the non-adapted pedal cycle can be a viable means of transport. In some cases, lower gears are fitted, e-assist and other minor modifications added, but broadly they resemble street cycles.

Sociables

Sociable tricycles, with riders sitting side by side, appeared in the 1880s. Of their nature, they offered possibilities for riders with disability. The first sociable, the 1878 Quadricycle, was heavy and wide and, since both sides had the same gear, its riders had to maintain the same cadence. Recently, the sociable has reappeared as a machine tailored to persons with several kinds of impairment. One user who formerly transported his daughter in a trailer behind his bicycle remarked:

> I was not able to see her reactions and was afraid she would tip over... With a recumbent sociable – I am riding with her side by side. She smiles when she sees something and I see she smiles – and she can't speak.

Sociable models can accommodate a wide range of impairments.

Recent versions of the sociable tricycle designed for people with impairment are equipped with gears on a divided axle so that two riders can maintain different cadences – each pedaling at their own rhythm allowing the work to be divided unequally; indeed, one rider can rest while the pilot does all the work.

Tandems

A tandem has two passengers, normally with a pilot up front and a stoker behind, although these positions may be reversed in the case of PWD. A skilled pilot can make tandem cycling accessible and enjoyable to a wide of disabled people. Persons who are visually impaired, have muscular dystrophy, or with various forms of dementia and memory loss, are thereby able to exercise, visit the city or countryside, or travel to a specific destination. A limitation of the classic tandem is that the stoker and pilot have to maintain the same cadence, which may require a compromise over speed between the two riders; however, newer models with clutch drives allow independent pedaling. A recent advance, technically, is Circe Cycles *Helios* tandem made in Cambridge, UK with low step-over height and easy size adjustment, allowing both young and adult riders to exercise. The Hase Bike Company of Bochum in Germany includes in its range of bikes for PWD the Pino, which places the front rider in a low semi-recumbent position, and the pilot behind and above: the Tour version of the Pino permits impaired persons to undertake long-distance tours.

Power-assisted cycles

Power-assisted cycling has a long history, pre-dating the e-bike. The Michaux–Perraux steam velocipede of 1868 was the probably the first power-assisted bicycle, followed by Lucius Copeland's prototype steam bicycles of 1881 and 1884, and a steam tricycle of 1888. These prototypes were not intended for use by PWD, but in the 20th century battery powered tricycles (using cells similar to car batteries) were adopted by persons with mobility difficulties. R.A. Harding Company of Bath (UK) launched tricycles with rear-mounted motors and batteries in 1926. More recently, e-bikes have been widely adopted, especially by older riders. About 85% of the world's e-bikes are now made in China with the remainder (more high-end) made in Taiwan and Europe. Coming full circle in many ways back to Farfler, paraplegic outdoor enthusiast, and artisanal cycle innovator, Christian Bagg, worked for more than a decade to produce the power-assisted ICON Explore trike (now the Bowhead Reach). The Explore enables access to backcountry settings for off-road cycling and hiking activities and is an assemblage of cycle technologies from many different disciplines, including Fat Bikes, BMX/trials cycling, suspended mountain bikes, and e-bikes.

Chair transporters

Chair transporters range from a variety of wheeled chairs (normally pulled or pushed by a caregiver, animal or is self-driven), to the rickshaw (where the passenger is pedaled by a rider), to bike trailers pulled by cyclists. The person being transported gains mobility largely due to someone/something else's efforts, although in some cases the rider steers the vehicle.

An early chair transporter is the Marquis of Bute's "invalid carriage" (c.1870) designed by John Ward of Leicester Square, London. Ward patented several improvements to his "invalid chairs and carriages." Bute's quadricycle box was drawn by a pony: the rear of the box drops down as a ramp so that a wheelchair can be pushed up into the box. Somewhat different is the Bath chair, invented by James Heath of Bath, England, around 1750 to replace sedan chairs. The passenger controls the steering, but relies on someone behind to push the chair. Bath chairs became a common sight following major wars as large numbers of people suffered wartime injury.

A modern variation of the box transporter, except that the motive force is a cyclist riding behind the wheelchair passenger, is a variation on the Dutch cargo cycle (short box) recently adopted by Detroit's MoGo bicycle rental company. The MoGo has three wheels for stability, with the ramp placed at the front of the box; the passenger plays a passive role. Draisin, a major maker of cycles for the impaired in Germany, has developed a chair transporter with a detachable light wheelchair fastened via a shock absorber at the front. Draisin stresses that "the design of the wheelchair-bicycle combination… has evolved and matured over the years" as feedback from users was incorporated into the design.

Stability machines

Bullmore (2018) predicts that by 2030 the biggest cause of disability will be inflammation of major organs, including the brain, caused by excess production of cytokines, a condition that he relates to obesity, diabetes and depression. Many so affected become unstable on their feet, but may benefit from non-ballistic exercise such as cycling. Two main types of stable cycles are available: stationary bicycles and tricycles. Stationary bicycles form part of the booming exercise equipment industry with apps presenting PWD an opportunity to exercise while setting their own program and pace. They play an important role in maintaining cardiovascular fitness and muscle mass improvement for PWD. Bullmore's research suggests they may soon have a wider applicability.

As early as 1897 Siegfried was using tricycles to conduct treatments on patients with ataxia to help them regain some muscle control (Siegfried 1902: 10–11). Although his work formed part of the medicalized approach to disability, his results encouraged work on therapeutic treatments. Adult tricycles typically have a wide wheelbase for stability and baskets used to carry shopping goods and assorted articles. They present moderate exercise opportunities for persons with a wide range of impairments. Persons with Down Syndrome, for instance, usually find the stability of tricycles reassuring, as do the elderly who are making increasing use of tricycles in many places.

An example of a stable tricycle is the Adventurer, made by Freedom Concepts of Winnipeg, MN. This tricycle has back support, foot straps and an extra-wide split back axle with one driving wheel and one freewheel, thus avoiding a differential. It has very low gearing, Velcro straps holding feet on the pedals, and seat belts that keep the rider upright in the high seat. Freedom stress that their designs are constructed in consultation with medical professionals, therapists, users and their families, with further customizing always possible.

Furthermore, in 1991, a three-wheeled running chair was invented in Denmark by the Danish wheelchair athletes Connie Hansen and Mansoor Siddiqi (Hornbaek Jensen 2017). Originally, it was called "The Walking Machine," whereas today it is known as "Petra bike" or as "RaceRunner." A RaceRunner is a customized 'running bike' without pedals (de Place Knudsen 2017: 9) that consists of a three-wheeled frame with handlebars, a trunk support (i.e., a chest plate), and a saddle (Donnell et al. 2010). While sitting on the saddle, users propel themselves forward by stepping their feet on the ground (Donnell et al. 2010). A RaceRunner can be used for transport purposes, leisure, or as a training device, providing an opportunity for enhanced mobility. Although originally developed for people with cerebral palsy, it can be utilized by PWD with impaired balance, reduced trunk control, restricted range of movement, and/or limited walking abilities. A RaceRunner can be fitted to the person's individual characteristics (i.e., anatomy) and even preferences (i.e., walking or running technique). Several weeks of RaceRunning training has shown positive effects related to bone health, cardiorespiratory fitness, and muscle growth in individuals with cerebral palsy (e.g., Bryant et al. 2015; Hjalmarsson et al. 2020).

Reflections on the makers

One of the specialized makers (Freedom) stresses "each bike is built to the person and their special needs," adding that "any bike can be further customized and adjusted to the exact specifications and unique requirements of the rider." Since these cycles address specific needs their users have to matter to their design. In interviews we found that makers regularly interact with purchasers (users & their families), and over time make technical modifications based on feedback. One maker (interview at RickSycle of London, Ontario) reported five rounds of modifications as users responded to their design changes, while Theraplay claims to have conducted 40,000 assessments as riders test out the suitability of their models. And most machines are offered with "options" that a user can add on if it seems appropriate. The Dutch maker of specialized cycles, Van Raam, allows purchasers to configure their bike online. Unfortunately, these adaptations often come at a steep price, and few countries subsidize their purchase, which means that class, race, age, gender and politics all come into play.

Many small makers came to the activity because of impairment of a family member or a community need; hence their interest in the project is necessarily socially constructed, with the impetus coming from individual contexts rather than from wider policy initiatives. Theraplay "was approached by a local charity working with children with Spina Bifida and was asked to design and produce a hand driven tricycle for… children with this condition," while Freedom Concepts received a request to build a cycle for a child with cerebral palsy. And information exchange between bikesharing schemes offering adaptive bicycles is growing, creating a network of organizations catering to riders with special needs.

Conclusion

For many PWD, cycles offer an opportunity to become more mobile, lower stress levels, improve health and fitness, engage safely in exercise and have fun. A range of distinct technologies have been adopted to assist PWD, most having a long history, although by incorporating present-day materials, electronics and programs and exchanging ideas and experiences these older technologies have been improved, often quite dramatically. Unfortunately, despite the possibilities these technical achievements offer, in many places planners and policy makers overlook the necessary relationship between cycling and accessible infrastructure for PWD.

Bibliography

Andrews, N., I. Clement and R. Aldred (2018) "Invisible cyclists? Disabled people and cycle planning – a case study of London", *Journal of Transport and Health*, 8, 146–156.

BikeMag (2020) "Four-wheeling in whistler bike park with Stacy Kohut", August 2014: https://www.bikemag.com/uncategorized/four-wheeling-in-whistler-bike-park-with-stacy-kohut/

Bourne, J. E., Sauchelli, S., Perry, R., Page, A., Leary, S., England, C. and Cooper, A. R. (2018) "Health benefits of electrically-assisted cycling: a systematic review", *The International Journal of Behavioral Nutrition and Physical Activity*, 15(1), 116. https://doi.org/10.1186/s12966-018-0751-8

Bryant, E., D. Cowan and K. Walker-Bone (2015) "The introduction of Petra running-bikes (race runners) to non-ambulant children with cerebral palsy: a pilot study", *Developmental Medicine and Child Neurology*, 57, 34–35.

Bullmore, E. (2018) *The Inflamed Mind: a Radical New Approach To Depression*, New York: Picador.

Clayton, W., J. Parkin and C. Billington (2017) "Cycling and disability: a call for further research", *Journal of Transport and Health*, 6, 452–462.

Concordia University (2017) "Feeling stressed? Bike to work: study shows how a pedal-powered commute can set you up for the whole day", *Science Daily* (21 June), retrieved June 26, 2020 from www.sciencedaily.com/

de Place Knudsen, S. (2017) "The RaceRunner and its Possibilities", in Siddiqi, M., Hornbaek Jensen M. (Eds) *Coaches' Manual: RaceRunning* (1st Edition) Parasport Denmark, 9–16. https://www.racerunningusa.org/uploads/1/0/0/9/100900298/racerunning_-_coaches_manual.pdf

Donnell, R. O., J. Verellen, P. van de Vliet and Y. Vanlandewijck, (2010) "Kinesiologic and metabolic responses of persons with cerebral palsy to sustained exercise on a Petra race runner", *European Journal of Adapted Physical Activity*, 3, 7–17.

Hjalmarsson, E., R. Fernandez-Gonzalo, C. Lidbeck, A. Palmcrantz, A. Jia, O. Kvist, et al. (2020) "RaceRunning training improves stamina and promotes skeletal muscle hypertrophy in young individuals with cerebral palsy", *BMC musculoskeletal disorders*, 21, 193.

Hornbaek Jensen, M. (2017) "The History of RaceRunning", in Siddiqi, M., Hornbaek Jensen M. (Eds) *Coaches' Manual: RaceRunning* (1st Edition) Parasport Denmark, 6–8.

Inckle K. (2019) "Disabled cyclists and the deficit model of disability", *Disability Studies Quarterly*, 39, 4, open source. https://doi.org/10.18061/dsq.v39i4.6513

MacArthur, J., N. McNeil, A. Cummings and J. Broach (2020) "Adaptive bike share: expanding bike share to people with disabilities and older adults", *Transportation Research Record*, 2674, 556–565.

Norcliffe, G. (2012) "Before geography? Early tricycles in the age of mecanicians", *Cycle History*, 22, 86–99.

Norcliffe, G., R. Buliung, A. Kruse and J. Radford (2021) "Disability and cycling technology: a socio-historical analysis", *Disability Studies Quarterly*, 42(1), forthcoming.

Oja, P., S. Titze, A. Bauman, B. de Geus, P. Krenn, B. Reger-Nash and T. Kohlberger (2011) "Health benefits of cycling: a systematic review", *Scandinavian Journal of Medicine and Science in Sports*, 21(4), 496–509.

Oliver, M. (2013) "The social model of disability: thirty years on", *Disability and Society*, 28(7), 1024–1026.

Oudshoorn, N. and T. Pinch (2005) *How Users Matter: The Co-construction of Users and Technologies*, Cambridge, MA: MIT Press.

Radford J. (2020) "Towards a post-asylum society", in Brown I. et al. (Eds) *Developmental Disabilities in Ontario* (4th Edition), Toronto: Delphi Graphic Communications, 25–40.

Rocco, T. S. (Ed.), (2011) *Challenging Ableism, Understanding Disability, Including Adults with Disabilities in Workplaces and Learning Spaces*, San Francisco: Jossey-Bass.

Siegfried, M. (1902) *Tricycling as an Aid in Treatment by Movement and as a Means of Carrying Out Resistance Exercise*. (Leipsic (sic); Georg Thieme) (translated by Louis Elkind: 22 pages) https://wellcomecollection.org/works/aef3z83r

Szto, C. (2012) "Mountain bikes for those with disability". http://cszto.blogspot.com/2012/03/mountain-bikes-for-those-with.html

Wong, A. (Ed.) (2020) *Disability Visibility: First-Person Stories from the Twenty-First Century*, New York: Penguin-Random House.

PART III

The cycling economy

An introduction

Glen Norcliffe

Cycling literature focusses mainly on the *use* of cycles: they are machines that people ride for pleasure, for business and for convenience. Less attention has been paid to the bicycle as a machine that is manufactured, traded, sold and re-sold, advertised, branded, repaired and used as a commercial platform. Perhaps this is because economics suffers from being dubbed – unfairly – with the sobriquet *the dismal science*. In practice, the bicycle is primarily a benign egalitarian machine promoting self-sufficiency, while economics is "a force for a more just and, crucially, less dismal world" (Thompson, 2013: 1). The ultimate goal of economics is to advance human development. It follows that, in tandem, the bicycle–economics relationship should be harmonious, and a topic of considerable importance since the bicycle advances human well-being in many situations.

For economists, bicycles are a scarce commodity, and therefore command a price determined by marginal supply and demand. Having been in demand in large quantities for about 125 years, they are mostly mass-produced in order to gain economies of scale, although very small numbers of high-end custom-made bicycles are sold for much higher prices. Historically, these clusters were found in the "old" industrial countries, in cities such as Hartford in Connecticut, Coventry in the UK, and St. Etienne in France. But the shift to globalization that began with the neoliberal turn of the late 1970s has completely re-configured the world's bicycle industry. Today, major clusters are found in Tianjin and Kunshan (China), Ludhiana (India) and Taichung in Taiwan. Since the bicycle is an assembled product, it is common to subcontract the making of specialized components, which may lead to horizontal integration, with a corporation such as Shimano making gears for many bicycle brands (Norcliffe and Gao, Chapter 14). The assembly process is organized around supply chains, as described by Cheng-Mei Tung in Chapter 15, with deliveries often just-in-time, requiring members of the supply chain to cluster geographically. Since many bicycles made in these clusters are exported, the forward linkages of these supply chains connect with importers, retailers and consumers, mainly in high-income countries in Europe and North America.

Henry Yeung (2016) has made a case that in a wide range of technology-intensive industries, a *strategic coupling* develops between key players: thus, leading computer brands have enduring strategic relations with firms that assemble their computers, while commodified sport industries such as football and hockey foster comparable links with media industries that broadcast their games. The bicycle industry, in contrast, is relatively low-tech, and brand owners and

DOI: 10.4324/9781003142041-19

bicycle makers are linked by an *opportunistic coupling* driven by price. Brand owners will visit a manufacturers' trade show to select models for the following year's line-up, and negotiate prices. Switches in suppliers are quite common, making the global production networks for bicycles unstable.

Compared with recent leaps in automobile technology, particularly following advances with electric and autonomous vehicles, bicycle technology may seem staid and slower to advance. In practice, however, bicycle makers are well aware of Schumpeter's call for creative destruction and although many of the recent innovations to cycles are cosmetic, some are technologically significant. Brand owners are adept at focussing their advertisements on such innovation in the hope of persuading consumers to upgrade to the newest model, or purchase the latest add-ons. This may amount to re-introducing old technologies and presenting them as novel ideas (for example, disc brakes were first used on bicycles in the 1890s), but sales are driven by optics, fashion and the endless quest for modernity (Slater, 1997; Norcliffe, 2001). Consumers respond to these sales stimuli, seeing a new bicycle as a way to enhance their identity as a conspicuous sporting activist, an environmentally conscious commuter, and in other diverse ways.

Cyclists are also enthusiastic purchasers of a wide range of accessories and clothing items directed at affluent cyclists. These include:

- GPS and other tracking systems that record training workouts and then share the data with other enthusiasts.
- Clothing worn not just by MAMILs (Jones, Chapter 26), and not just in winter (Vignette J); most garments use colorful modern materials and have the maker's brand conspicuously placed on them (Hilborn, Chapter 40).
- Charity and sporting events where cyclists wear "free" sponsor-labelled clothing and become roaming advertising boards.
- Hi-tech cycling equipment, such as electronic gear changing, shock-absorbing helmets, lightweight pedals and cleats, speedometers, and disc brakes.
- Footwear, gloves, sunglasses, water bottles.

According to a Mintel Press Office report of 15 July 2015, in the UK "cyclists are splashing out more on the accessories than bikes themselves." Defined to include parts, accessories and clothing, in 2014 UK bicycle sales totalled £956m, whereas sales of accessories amounted to £1.25b. https://www.mintel.com/press-centre/leisure/mad-about-the-bike-sales-of-bicycle-accessories-outstrip-sales-of-bikes (accessed 13/7/21). It seems reasonable to conclude that this relationship holds broadly true in other high-income countries.

The cycling economy is very seasonal with, in the northern hemisphere, demand cresting in spring, and a smaller peak before Christmas. The bicycle trade is geared to this annual rhythm. Three types of trade shows are identified by Michael Andreae in Chapter 16: the season begins in late spring with the manufacturers' shows which *perform* the newest models and technologies and brand owners begin their new model selections, followed by the distributors' shows in the fall, where retailers are able to pick and order models for the following spring peak of sales: retail shows take place in late winter as enthusiasts become excited at the prospect of a summer of cycling.

Bicycle retailing has undergone big changes in the post-war years, as Jay Townley explains in Chapter 17. Even as late as the 1960s, aside from children's cycles, most bicycles were functional machines used to ride to school, to work, to shop, and sold at the local bicycle shop which also did repairs, and was run by a retired racing cyclist whose days of competition were over. By the 1970s so-called "big box" stores began to compete for the low-end cycle market, and the local cycle shop was pushed out of business, except for a few successful high-end custom makers. Nowadays large sports chains and "big box" stores such as Walmart are competing with online sales, local hands-on cycle shows and curated mobile collections. Indeed, buyers now have had the option to design their own bicycle online.

In post-war years Taiwan's government made a conscious choice to develop a bicycle industry, eventually becoming the world's largest maker of bicycles with firms such as Giant and Merida becoming major Original Equipment Manufacturer (OEM) makers for western brands. As Yu-chun Lin explains in Chapter 18, post-1978 and China's Open Door policy, Taiwan's makers shifted mass production of low-end bicycles mostly to China. When Giant lost its Schwinn contract to a Chinese maker in 1987, Giant (and soon after, Merida) decided to launch their own brands. The rapid growth of the Chinese bicycle industry led to the decline of Taiwan's domestic bicycle industry, so in response Taiwan's cycle industry created a key institution, the A-Team, which re-focussed the industry on innovation and high-end production, including e-bikes. Meanwhile having launched its own brand, Giant rose to primacy among the world's bicycle makers with low-wage labour and clusters of suppliers making China the preferred location to manufacture its low-end bicycles. Technology-intensive high-end bicycles and e-bikes are made in Taiwan; and protective tariffs imposed by the European Union have led Giant also to locate production inside the EU.

The bicycle enters the economy in two quite different ways: it is a manufactured and traded product, but it also provides a platform for other forms of economic activity. As Claudio Sarmiento-Casas demonstrates in Chapter 19, it is used to sell goods, including food and beverages, tourist trinkets, farm produce and memorabilia at markets, and at sporting, musical and political events. And it provides a platform on which to sell services such as transporting people and goods, servicing other bicycles, and delivering pre-cooked meals. Such services can create opportunities for labour exploitation, with low pay, no benefits, long hours and a dangerous workplace. Claudio Sarmiento-Casas describes the trading spaces created by mobile bicycles engaged in street trade, mostly an informal activity of considerable importance for underclasses in the Global South.

Although it might seem that economics is the driving force guiding the cycling economy, in reality several other influences render pure economics unable fully to explain the geography of the bicycle trade today. Thomas Piketty (2014) argues that economics and politics are inextricably intertwined: the assumption in neoclassical economics that they can be treated separately does not provide the tools to explain the great economic inequalities of the modern world. In the case of the cycle industry, politics, particularly state policies for exporting, for protecting home industries, and tariffs, have led to the offshoring and onshoring of manufacturing, depending upon the political circumstances. The export promotion politics of Japan, China and Taiwan have all, at different points in time, led to expansion of their national bicycle industries. Although labour costs have a significant influence on choice of location to offshore bicycle production, not all low-wage countries are equally attractive. National politics and political stability also play a role: certain countries in South and Southeast Asia have attracted the lion's share of investment in the bicycle industry.

The rise of cultural economics in the new millennium challenged the supremacy of the neoclassical model (Guiso et al., 2006). For geographers, it was blindingly obvious that culture affected patterns of consumption, of work, of the treatment of capital, of political linkages, and of power relations within the household, industry, the community, and the state. For instance, the rapidly growing market for bicycles with electric-assist, particularly during the Covid-19 epidemic, meshes closely with a new culture stressing outdoor exercise. The appeal for comfort bikes/low riders and cruisers in the United States – the living room sofa in the bicycle lane – reflects a uniquely American cultural aesthetic. Tall, upright and stable Dutch bikes mirror local cycling culture, in all seasons in all weather. In each case, the purchaser is expressing his or her identity, which in turn is influenced by local and national culture.

Closely related are the concerns of the social economy which, via cooperatives, volunteering and other collectives, seeks to provide bicycles and cycle repair services at low cost, or gratis (Amin, 2009). The top-down profit motive of laissez-faire is replaced by sharing, mutual assistance, reciprocity and democratic decision-making. Sometimes referred to as the third economy, it is typically locally based and puts more emphasis on providing satisfying forms of work than on reproducing capital. Enterprises are mostly small, with substantial family inputs; many micro-enterprises seen as part of the informal sector belong to the social economy, including Tian Yun's vignette in this part on the mobile bicycle repair stall in Beijing.

In the 21st century, environmental economics has come to prominence because neoclassical economics has not historically included the environmental externalities of industry, nor public finance measures such as carbon taxes that seek to stabilize these externalities and preserve natural capital (Anderson, 2019). In the broader scheme of things, the bicycle industry may not appear to be particularly "dirty" (in an environmental sense), but manufacturing cycles requires steel, plastics, welding materials, rubber and paints, the making of which all produce noxious by-products, while battery technology for e-bikes adds another set of environmental issues. There is growing appreciation that investment in cycling infrastructure offers high rates of return when the environmental bonus derived from cycling is included in costs and benefits, and that cycle-based local mobility, whether for work or pleasure, is soundly based in environmental economics.

The main conclusion drawn from this part is that understanding the bicycle as an object of study and research requires an interdisciplinary perspective (Norcliffe, 2016). Even the making, trading and selling of bicycles cannot be satisfactorily explained in purely economic terms: politics, culture, social forces and the environment all impact where, how, when and what sort of bicycle is made, and how it is sold and used. In particular, the recent rise of political rhetoric promoting protectionism and national interests over the liberal convention of an open global economy suggest that another major reconfiguration of the global cycling economy may be in the offing.

References

Amin, A. (2009) *The Social Economy: International Perspectives on Economic Solidarity* (London: Zed Books).

Anderson, D. (2019) *Environmental Economics and Natural Resource Management* (Routledge: New York).

Guiso, L., Sapienza, P. and Zingales, L. (2006) "Does Culture Affect Economic Outcomes?" *Journal of Economic Perspectives*, 20(2), 23–48.

Norcliffe, G. (2001) *The Ride to Modernity* (Toronto: University of Toronto Press).

Norcliffe, G. (2016) *Critical Geographies of Cycling: History, Political Economy and Culture* (Abingdon, Oxon: Routledge).

Piketty, T. (2014) *Capital in the Twenty-First Century* (Cambridge, MA: Harvard University Press).

Slater, D. (1997) *Consumer Culture and Modernity* (London: Polity Press).

Thompson, D. (2013) "Why Economics Is Really Called 'the Dismal Science'." *The Atlantic*, 17 December, p. 1.

Yeung, H. W.-C. (2016) *Strategic Coupling: East Asian Industrial Transformation in the New Global Economy* (Utica, NY: Cornell University Press).

14
THE GLOBAL BICYCLE INDUSTRY

Glen Norcliffe and Boyang Gao

Background

Prior to the rise of neoliberal globalization in the late 1970s, most industrialized countries had long-established domestic bicycle industries serving their national market. Some high-quality bicycles were exported, often following earlier colonial channels – such as Dutch bikes shipped to Indonesia – but most industrialized countries had a small number of long-established bicycle makers that were revered within their national market. Import duties and restrictions on international investment protected these companies, giving them a significant advantage when selling to their home market while warding off foreign rivals. The rise of economic globalization in the first half of the 1970s coincided with the 10-speed bicycle boom. Baby-boomers in the 1960s had started riding children's muscle bikes and low-riders, and had later graduated to 10-speeds as part of their adult lifestyles. On the supply side, many long-established American and European bike makers failed to re-tool and keep up with changing demand, so they sub-contracted some of their production to rising Japanese makers such as Bridgestone, putting their own badge on the imported bicycles fitted with Japanese components such as Shimano and Sun Tour gears. Japanese makers, including Nishiki and Fuji, also started to sell their own brands in western markets, offering quality bicycles at a competitive price until the 10-speed boom subsided after 1976. By the late 1980s Japan was making 7–8 million bicycles a year but this dropped to below 1 million by 2010.

As the Japanese economy moved up market to concentrate on automobiles and electronics, the rising Asian Tigers sought to emulate Japan, with Taiwan concentrating on bicycles and electronics. Meanwhile many famous makers such as Schwinn and Huffy in the United States, Raleigh in the UK, Adler in Germany, CCM (Canada Cycle and Motor Company), Gitane in France, and Speedwell and Malvern Star in Australia faced financial difficulties. Some closed, while several others were sold to new owners who kept the brand name but sub-contracted production to South East Asia.

The rise of neoliberalism after 1980 reduced tariff and non-tariff barriers, promoted global flows of capital, and saw the creation of the World Trade Organization in 1995 to supervise international trade agreements and the ownership and transfer of intellectual property. Frobel, Heinrichs and Kreye (1980) identified the *new international division of labour* (NIDL) as a major consequence of these developments. Reduced tariffs and improved transportation (notably

DOI: 10.4324/9781003142041-20

containerization) made countries with low labour costs attractive locations for labour-intensive low-technology production such as the bicycle industry. High-income countries, in contrast, saw a major growth of services and technology-intensive industries giving rise to technopoles and knowledge development in major cities (Dicken 2015).

An organizational change advanced this global shift in production by separating production from marketing (Rosen 2002). As many bicycle manufacturers in high-income countries became insolvent their venerable brand names were sold for considerable sums to new owners specializing in marketing and (sometimes) design who delegated manufacturing to low-cost OEM (original equipment manufacturers) who undercut domestic makers' prices, thereby creating an international bicycle trade with global supply chains.

The rise of China as the world's major bicycle maker

China launched its Open Doors Policy under Deng Xiaoping in 1979 by creating 4 coastal Special Economic Zones, namely Shenzhen, Zhuhai, Shantou and Xiamen. In 1984 a further 14 coastal cities were added so that foreign companies were able to invest and source production in much of coastal China. In 1992 border cities and major inland cities were also opened to investment and trade. This opening up to foreign trade and investment complemented an important aspect of China's preceding history under Mao Zedong: Mao had declared bicycles to be one of the four household "musts" (*sì-dà-jiàn*), along with a watch, a radio and a sewing machine. In fact, China had quite a long history of manufacturing bicycles with four renowned local makers. Tianjin's Changcheng Bicycle Works was opened in 1936, and after the CPC takeover in 1949 was reconverted to make *Fēi Gē* (Flying Pigeon – but better translated as Flying Dove) cycles which were, in practice, modelled on the durable 1932 British Raleigh roadster. Also founded in 1936 was Shanghai's Tongchang Chehang Bicycle Works, making *Yǒng Jiǔ* (Forever) bicycles (see Vignette M). The other two big makers were *Hóng Qí* (Red Flag) in Tianjin, and *Fèng Huáng* (Phoenix) in Shanghai. By the end of the Maoist era, there were many small bicycle makers spread across China. With Deng's Open Door policy after 1978, bicycle production soared, from 8.5 million in 1978 to 27 million in 1983 (Zhang 1992). As Table 14.1 shows, as late as 1990 only 7 of 31 Chinese provinces had no bicycle manufacturing. This widely distributed know-how in cycle manufacturing meshed with China's new opening to foreign trade after the isolation of the Cultural Revolution, and also with the needs of western bicycle brand owners who were trying to revive profitability by sub-contracting production to China's low-cost OEM bicycle makers.

China rapidly became the world's biggest bicycle manufacturer and has accounted for around two-thirds of global production for the last two decades. Although shrinking, China's domestic bicycle market was still large, and began to grow again after 2000 as e-bikes became a popular innovation (Chapter 24).

Three elements account for this rapid growth after 2000. First, China's long-established state-owned makers expanded and modernized their product range as the industry privatized: Flying Pigeon, for instance, now makes 40 different models and around 800,000 bicycles a year (including *ofo* dockless sharing bicycles) at their re-located and automated factory northwest of Tianjin (Koeppel 2010). Second, direct investments in China were made by Taiwanese makers. And third, China invested in new bicycle factories. There are now more than 800 private Chinese bicycle enterprises, but they are spatially much more clustered with the industry concentrated in 9 provinces.

Vertical integration has created three major production regions. The Bohai Economic Rim makes around 22 million bicycles annually, with 16 major bicycle factories located in

Table 14.1 Bicycle production in China by province, 1985–2020 (excluding e-bikes) (000's)

Year	*1985*	*1995*	*2005*	*2010*	*2015*	*2020*
Tianjin	5632	5764	13067	26471	26373	22255
Hebei	993	2968	94	2118	28	609
Shanghai	6195	8388	6804	5128	3318	2650
Jiangsu	3751	7182	10987	8415	7802	9325
Zhejiang	1697	6357	6756	16373	10288	3043
Fujian	491	1018	0	0	60	193
Guangdong	2015	7449	16666	9047	6878	5709
Sichuan	712	61	498	292	249	188
All others	10791	5535	1439	351	331	3439
Total for China	**32277**	**44722**	**56311**	**68195**	**55327**	**44368**
Provinces with makers	24	22	14	13	10	9
Provinces with no makers	7	9	17	18	21	22

Source: National Bureau of Statistics of China. Thousands of e-bikes.

Wangqingtuo Town, near Tianjin, the largest being Fuji-Ta, opened in 1992, and now China's largest bicycle maker. Fuji-ta has progressively brought most steps in bicycle and component manufacture in-house and added its own house brand (Battle Bicycles), which are sold mainly in China. The second largest cluster is the Yangze Delta region producing around 15 million bicycles annually (Chapter 15). Initially, direct investments were made near Shenzhen by Taiwanese bicycle makers, including Giant, but later they re-located to the Yangze Delta where Giant now have 5 factories: 70 suppliers moved with Giant, creating a large cluster around Kunshan (Chapter 18). A key supplier, Shimano (the world's largest maker of gears), has a major factory located nearby. The third cluster found in Shenzhen, Dongguan and Foshan produces around 6 million bicycles a year.

Table 14.1 shows production in China of standard bicycles declining after 2010, but total bicycle production held steady as demand for e-bikes (Table 14.2) surged from 1 million

Table 14.2 E-bike production in China by province, 2016–2020 (000's)

Year	*2016*	*2018*	*2020*
Tianjin	3514	6409	12372
Jiangsu	4537	6058	3340
Zhejiang	6398	3513	4532
Anhui	477	522	275
Shandong	6240	1138	806
Henan	8190	3961	2118
Hubei	925	700	530
Guangdong	147	1355	3665
Guangxi	136	483	751
Sichuan	376	606	777
Shaanxi	849	906	29
All others	361	248	466
Total for China	**32150**	**25899**	**29661**
# Provinces with makers	20	19	18
# Provinces with no makers	11	12	13

Source: National Bureau of Statistics of China (Thousands of e-bikes).

(2002) to 12 million (2005), 29 million (2010) and 30 million in 2020. Table 14.2 shows a more dispersed pattern of e-bike construction. The major makers diversified into e-bikes early in the new millennium, re-enforcing the same three main bicycle clusters, but since e-bikes require more sophisticated electronics, other regions were able to enter production (Chapter 24). Consolidation of the e-bike industry is expected in the coming decade.

East Asian makers at first were contract manufacturers, taking little part in design, but they learned rapidly from the brand owner's engineers and soon developed a design capability. By 2010 major bike makers in China were shifting to ODM (original design manufacturing – also called *private labeling*) for small and low-end brands. The makers offer a number of designs from which a brand owner can pick, often negotiating minor changes plus their own badging to make the product distinctive. With the rise of the e-bike, technical and design expertise of China's bicycle industry has deepened.

Trends in the world's cycle industry today: an overview

Modern society, we are told, is besotted with the automobile, yet the data suggest otherwise. In 1965, worldwide, bicycles and autos were made in roughly equal numbers (20 million each), whereas by the mid-1980s world bicycle production was double that of automobiles (60 million vs 30 million). Today the gap remains wide with, in 2020, about 110 million bicycles made (valued at US$30 billion) compared to 78 million autos (https://www.worldometers.info/bicycles/; https://en.wikipedia.org/wiki/List_of_countries_by_motor_vehicle_production). Auto sales are tending to decline (they peaked at 97 million in 2017).

Motor vehicle sales are still rising in the global south and the BRIC countries (notably electric cars in China), but there has been a worldwide re-engagement with the bicycle, connected with three main developments. First, there is growing concern about the negative environmental impacts of internal combustion. Second, a widening clientele are persuaded there are health benefits to cycling, a trend much boosted in 2020–2021 by the Coronavirus pandemic (Chapter 30). And third, the rise of recreational cycling world-wide has become a major cultural phenomenon with aficionados of gravel bikes, hardtails, city bikes, recumbents, cyclo-cross bikes, folding bikes, MTBs, triathlon bikes, hybrids, track bikes, BMXs, tandems, cruisers and fixies all finding pleasure (and identity) in riding their chosen design.

Table 14.3 presents estimates of bicycle production in leading countries. Data on cycle production are not very reliable: there are definitional issues such as whether a country includes/excludes scooters, tricycles, e-bikes and children's bicycles; trade disputes foster secrecy; and cycles assembled in one country may use components manufactured in another. But the statistics indicate that China has made between 60 and 70 million bicycles a year for the past decade – about two-thirds of world production. India is the second-largest maker with production close to one-quarter of China's output, but India's bicycles are mostly traditional steel bicycles destined mainly for the domestic market. The United Nations Industrial Development Office (2019) reports that, at present, India lacks the skills and investment to make the type of aesthetically and technically superior bicycles currently demanded in high-income countries which would help increase exports. Taiwan is far more important than its production figures suggest, due to its two leading firms, Giant (Chapter 18) and Merida, having several factories in China and elsewhere, and owning some of the world's most famous brands. Two other regions have a significant presence: several countries in Europe produce in combination close to 10 million bicycles per year; and East Asian countries (excluding China and Taiwan) make about 6 million bicycles.

Table 14.3 Estimated bicycle production – major countries c.2020

World Total (est.)	*115 million*
China	74.0 m. [44.3 m. pedal + 29.7 m. e-bikes]
India	16.5 m
Portugal*	2.7 m
Taiwan	2.3 m. [1.54 m. pedal + 0.73 m. e-bikes]
Cambodia	2.2 m (value $473 m.)
Italy*	2.1 m
Germany*	1.5 m
Indonesia	1.0 m. (many components are imported)
Poland*	0.9 m
Japan	0.9 m
Netherlands*	0.7 m

Source: Diverse web sites. Note that production figures tend to be unreliable.

* These five countries together accounted for 70% of total EU production of bicycles in 2019.

The largest regional bicycle market is Europe, with estimated sales in 2020 of US$13 billion, although the US is the largest single market with retail sales in 2020 estimated to be US$7.9 billion (https://www.statista.com/topics/1448/bicycle-industry-in-the-us/). Fewer bicycles are manufactured in Europe (c. 10 m. p.a.) and North America (c. 60,000 p.a.) following the collapse of their domestic bicycle industries in the 1970s and 1980s (Table 14.3).

In short, since 1980 there has been a growing geographical separation of consumers from makers of bicycles which has had major economic consequences. Owners of high-end bicycles have retained design capabilities: they may send a model of their next design to OEM/ODM makers in China, Cambodia, Taiwan, Vietnam and elsewhere, increasingly as a program for 3D printing. The maker's design group usually refines the design to match their capabilities and returns it to the brand owner in America or Europe. Mass-market brand owners and big retail chains normally select low-end "off the shelf" designs to which they may make minor modifications and add their badge.

For brand owners and mass retailers, trade shows have become crucial to creating and rearranging global supply chains, and disseminating knowledge of new materials, models, and technologies by serving as points of contact between makers and sellers (Chapter 16). The resulting global separation of production and consumption has made the logistics of the cycle industry's global supply chains crucial to the success of this highly seasonal industry, with sales peaking at Christmas, and during spring (Chapter 15).

Recent trends 1: bicycle assembly

Like computers and most other durables, bicycles are assembled from numerous parts. Depending upon the model, a bicycle may have up to 1000 components, some, such as ball bearings and chain links, being very small, while the frame itself is relatively large. The resulting assembly industry operates in stages, with sub-assemblies (which may house sub-sub-assemblies) brought together on a final assembly line. The basic building block for final assembly is a frame, which is made of tubes of steel, aluminium, or titanium (or carbon fibre) that are welded (glued) together, put in a chemical bath to clean impurities, and painted. High-end makers "build" rather than assemble their bicycles with each piece carefully fitted. Until recently, bicycle making required substantial labour inputs: wheels were often spoked and trued manually; frames were built by hand; saddle making was laborious; headsets, gear clusters (blocks), brake cables and gears were physically bolted on.

The 21st century has seen automation considerably reduce labour inputs. Meanwhile, bicycles themselves have become more sophisticated with new technologies, including electronic gear changers, disc brakes, and power assist, new materials such as carbon fibre, shock absorbers, monocoque forks, and puncture-proof tires. Frames, wheels, cranks, pedals and all other parts are brought together on a moving assembly line with a detailed division of labour. Forty years ago, low labour costs were a major locational attraction, whereas today the industry requires larger capital inputs and greater technical sophistication – which is precisely the trajectory that Chinese industry in general has taken.

Recent trends 2: industrial clustering

Product architecture, as defined by Ulrich (1995), plays an important role in the geography of the bicycle industry. Each component of a bicycle has a physical function, which is simplified by standardization, and each component's performance is connected to the performance of interfaced components (such as a pedal to a crank, and gears to a chain). Proximity is therefore a big asset in improving product performance, as well as facilitating just-in-time deliveries of parts, hence bicycle industry suppliers frequently cluster around a final assembly plant. Component suppliers have to be very flexible in their output, in two ways. First, demand for cycles shows big seasonal swings, with sales peaking in spring. Makers on the other side of the world need to anticipate this rush and, if one model becomes popular and sells out, be ready to quickly assemble a second shipment. Second, consumer demand is very diversified: there are more than 14 recognized categories of bicycle, several frame sizes, differences in equipment (e.g. caliper brakes versus disc brakes; number of gears), and sometimes colour choices are offered. As a result, assembly runs for a specific sub-model tend to be quite short, followed by rapid switches to assemble another model requiring changes in configuration, size, colour and brand name. Makers have adapted Taiichi Ohno's system of lean production (Ohnoism) to a high level of flexibility, including just-in-time deliveries, *kanban* inventory control and scheduling, and *kaizen* (continuous improvements) (Chapter 18). Suppliers to an assembly plant therefore have to be equally nimble, hence the advantage of proximity and the formation of clusters practicing lean production. Large ODM firms may have several assembly lines operating simultaneously, assembling the models of rival brands in close proximity.

Bicycle clusters are found in several other countries, including Ludhiana (Punjab, India) with over a dozen makers including Hero and Raleigh bicycle factories. Merida, Giant and Fuji (formerly located in Japan) are based in Taiwan's Taichung City. Smaller clusters are found in other countries, including Bike Valley in Central Portugal.

Recent trends 3: automation

The most labour-intensive activity in the bicycle industry is spoke lacing and trueing a wheel: semi-automatic machines have been available for a while but since 2013 Holland Mechanics (and other firms) have developed robotics that lace and true a wheel to very high precision. Robotics have also been developed to weld bicycle frames (steel and aluminium). For instance, Tianjin Fuji-ta Bicycle Company Ltd. (also spelled Fushida), which is the world's largest and most sophisticated OEM/ODM bicycle manufacturer, use over 100 programmable robots for precision frame welding guided by advanced machine vision. Such technologies achieve high accuracy and a welding quality better than what can be achieved by a skilled human welder (https://www.youtube.com/watch?v=3Yn5kIpIrv0). Coupled with automated assembly lines, robotics have resulted in labour inputs to bicycle manufacture shrinking dramatically during the present century. In theory, these labour-saving technologies should make bicycle manufacture in a high-wage country cost little more than in a lower-wage economy. But firms such as Fuji-ta, with 33 bicycle and 9 e-bike assembly lines and the capacity to make 20 million bicycles a year (about one-fifth of total world production), gain economies of scale. They have also developed the logistical support to make their supply chain efficient, with input components delivered just in time, and orders shipped worldwide in a matter of weeks. The downside to such intensive capitalization is that expensive robots represent sunk capital that needs to be working most of the time to spread overhead costs.

Recent trends 4: designing bicycles

From about 1900 up to the 1970s, bicycle design progressed at a glacial pace. Some quirky ideas were patented, but because the bicycle was largely used as an everyday convenience plus a few tourers and racers, and because the Union Cycliste Internationale was a conservative institution that resisted innovation, change came slowly (Chapter 8). Traditional makers were so set in their ways that in the 1960s and 1970s they barely noticed the rising popularity of the new bikes on and off the street – the BMX and the MTB – or dismissed them as short-term fads. Quite quickly, at least for some, the bicycle was becoming not a workhorse but a source of pleasure and recreation. The dangers posed by rapidly increasing motor traffic and the failure of most countries to build good cycling infrastructure (Denmark and the Netherlands being obvious exceptions) led to many new cyclists riding off-road in parks, gyms, trails and basements for excitement, pleasure, relaxation and fitness. Since then there has been much interest in designing new bicycles, encouraged by events such as the annual design competition at the Taipei International Cycle Show. As Chapter 13 shows, in some fields (such as cycles for persons with disability) there have been some remarkable design developments.

Bicycle brand owners, such as Pacific Cycle, negotiate the design of their models with ODM makers in China and elsewhere. Typically, a brand owner's design team sends a concept to a ODM maker and asks them to quote a price; alternatively, they select from the makers' designs and ask for a price. A manufacturer may eventually turn down a potential client either because they squeezed too much on price, or want a configuration that doesn't match their production capabilities. Mass sellers such as Walmart and Decathlon select the low-to-medium-end models they plan to sell the following spring from the brand owners' offering. In turn, the importers place their orders with makers in China and elsewhere with the design and price being negotiated back and forth. Smaller brand owners select cycles offered by ODM manufacturers, usually at the major bicycle shows, TICS – Taipei International Cycle Show – held in early March, and Shanghai's China International Bicycle Fair held in early May (Chapter 16). This results in price-driven *opportunistic coupling* which is very different from the *strategic coupling* described by Yeung (2016).

The exception to this arrangement is the world's biggest bicycle maker, Giant, which has 12,000 dedicated retail shops in 80 countries (see Chapter 18). Their outlets represent a sellers' market since these franchises can only sell Giant bicycles. Giant designs its own bicycles, making low- and medium-end bicycles in the Yangze Delta region of mainland China, and high-end and e-bikes in Taiwan using designs developed at its headquarters in Taichung City, Taiwan. Giant also has factories in the Netherlands and Hungary.

Recent trends 5: consolidation of makers and brands

The relationship between makers and sellers remains in flux. Asian makers have become bigger: Giant, Merida and Fuji-ta are now firms of global importance. Makers which began as smaller assembly operations purchasing components have become vertically integrated, although the manufacture of certain specialized components, especially gears and chains, is still left to suppliers. Despite the emergence of these very large firms, the industry remains competitive with new countries such as Cambodia, Portugal and Vietnam entering the crowded field, often with state support for start-up companies. The assembly characteristics of the bicycle make barriers to entry surmountable for a new maker with just an assembly line by buying in most components.

The brand owners, who are essentially wholesalers (although some make high-end bicycles), have also consolidated. Pacific Cycles owns 17 major bicycle brands, including: Cannondale, GT Bicycles, Iron Horse Bicycles, Kid Trax, Mongoose, Roadmaster, Schwinn and Sugoi. By buying Mongoose and Roadmaster brands in 2000, Pacific gained sales access to Walmart, reputedly the second-largest US bike retailer which now retails most of Pacific Cycle's brands (Chapter 17). Statistica reports that 90% of bikes sold in the USA are currently made in China or Taiwan, with Pacific the largest distributor. The Dutch Accell Group has followed a similar trajectory by purchasing famous brands. Britain's renowned brand Raleigh, for example, having merged with Tube Investments (TI Investments) in 1960 and then undergone a management buyout in 2001 (Rosen 2002), was bought by the Accell Group in 2012. Accell owns 19 brands: it is both a manufacturer and the major distributor of bikes and e-bikes in Europe.

Recent trends 6: off-shoring vs re-shoring

Recent political, economic and technical developments show signs of initiating a new geographical restructuring of world bicycle production. Politically, the movement for neoliberal free trade has lost momentum in the face of rising economic nationalism. In 1993, the European Union imposed an anti-dumping tariff on bicycle imports from China of 48.5%, while in 2019 a higher tariff of 79.3% was imposed in imported Chinese e-bikes. Under the chaotic Trump administration there were frequent trade disputes between China and the USA, but essentially, in 2019 the US tariff rate on complete bicycles imported from China was raised from 11% to 36%, while the tariff on imported e-bikes rose from 0% to 25% (https://www.bicycleretailer.com/industry-news/2019/05/16/tariff-timeline). The current US administration seems inclined to continue protectionist policies. These developments led Bonnie Tu, chairwoman of Giant bicycles, to declare in 2019 that due to Trump's punitive tariffs the era of 'made in China' was over (at least as far as exports to the USA were concerned). Giant closed one of its 6 factories in China and brought the production of bicycles destined for the USA back to Taiwan to avoid the new tariff wall (MacMichael 2019). Meanwhile Bosch, which supplies half of the world's electric motors for e-bikes, moved its Asia-Pacific headquarters from Suzhou (China) to Taichung (Taiwan). Fuji (now owned by Ideal Bike Corporation) has branch factories in China, and since 1999 in Kutno, Poland – an OEM factory mostly serving Europe from within the tariff wall. The Taiwanese firm, Astra, moved its aluminum and carbon fibre operations from China to two frame-building factories in Vietnam. China's major bicycle makers have also moved some of their production off-shore by opening factories in Vietnam, Cambodia, Ukraine, Portugal, Russia, India, and the Philippines.

Tariffs are also creating a new geography of bicycle production in Europe and potentially in North America. A tentative re-shoring of the European bicycle industry is underway with over 60% of the c.18 million bicycles sold annually now made in the EU (https://www.reuters.com/article/us-eu-china-bicycles-idUSKCN1VJ1MI). The largest bicycle maker in Europe is Portugal producing (in 2019) some 2.9 million bicycles in 40 factories with around 8000 employees, mostly in a cluster located in *Bike Valley*, in Central Portugal. Making bicycles and e-bikes OEM and ODM for well-known European brands, Portugal also has a niche market making docking bikes for public bike shares (Chapter 23). Launched with EU funding, this cluster illustrates a *regional division of labour* within Europe, with median household incomes in Portugal in 2019 (€13,350) close to half the European average and one-third that of the richest countries. Giant Bicycles located a factory in Lelystad in the Netherlands in 1996 (to which a distribution centre was added in 2020), and a new Giant factory was opened at Gyöngyös in Hungary in 2020, which added a second assembly line making e-bikes in 2021. With median household incomes of €10,265 in 2020 (https://worldpopulationreview.com/country-rankings/median-income-by-country), Hungary offers labour cost advantages similar to Portugal.

Makers in China have not been passive in the face of such economic nationalism. Just as Taiwan (with median household income of $32,762) has retained a niche in bicycle manufacture by automating and concentrating on high-end hi-tech bicycles including carbon-fibre and e-bikes, so China is moving up market by producing more medium- to high-end bicycles in a wider range and, with rising labour costs (median household income in China was $6,180 in 2021), Chinese makers are automating wherever feasible. For example, the proposed 32-acre Guangdong Taiwan (Meizhou) Bicycle Industrial Park will feature intelligent factories and logistics; it is currently seeking investors. Meanwhile in 2010, Tianjin Municipal Quality and Technical Supervision Bureau and the People's Government of Wuqing District signed a Memorandum of Cooperation creating the China Cycle and Tianjin Institute of Product Quality Supervision to promote the adoption of new materials and new technologies, including advanced e-bikes (http://www.jjtv.cc/eng/news/n4.html#). Plans are underway for a new cycle industry park in Shenzhen's Longgang District, with an investment of 3.5 billion RMB. The plan is to build a semi-self-sufficient park with five public platforms, including R & D and testing, training, information gathering, exhibiting and trading, and logistics.

Summary and conclusions

The globalization of bicycle production launched 40 years ago resulted in China becoming the dominant manufacturer, now making about two-thirds of the world's bicycles. Indeed China's largest maker – Fuji-ta-Battle – has the capacity to produce-one fifth of the world's bicycles. Although the largest makers have become more vertically integrated, the architecture of the bicycle lends itself to suppliers clustering around an assembly plant. An assembly line making numerous models and brands OEM and ODM needs quick deliveries of components to achieve lean flexible production. Automation has reduced labour inputs, increased investment in plant, and led makers to play an increasing role in design, but the industry still remains fairly labour-intensive. As in many industries, consolidation has occurred among makers and owners of brands.

The bicycle is not normally considered a strategic good, yet it has become something of a political football, subject to protection in the major markets of Europe and North America, leading to some production being shifted from China to other countries in South-East Asia, including Vietnam and Cambodia. The outcome is a possible re-shoring of bicycle manufacture: this is already underway in Europe (Portugal and Hungary) and could happen in North

America (Mexico). China still has a very large internal bicycle market, but as China makes structural adjustments into technically more advanced industries and makers shift their factories to avoid tariff barriers, a re-organization of world bicycle production seems increasingly likely.

Acknowledgement

We thank Brad Hughes of Human Powered Solutions for sharing his extensive knowledge of the industry with us.

Bibliography

Dicken, P. (2015) *Global Shift: Mapping the Changing Contours of the Global Economy* (7th edition) (London: Sage).

Everett, H. (2019) "Export figures reveal Taiwan's strength as a cycle manufacturing nation". *Cycling Industry News*, 28 October 2019. https://cyclingindustry.news/taiwan-export-figures-analysis/

Frobel, F., Heinrichs, J. and Kreye, O. (1980) *The New International Division of Labour: Structural Unemployment in Industrialised Countries and Industrialisation in Developing Countries* (translated by Pete Burgess) (Cambridge: CUP).

Gao, B-Y., Liu, W-D. and Norcliffe, G. (2012) "Hypermobility and the governance of global production networks: the case of the Canadian cycle industry and its links with China and Taiwan". *The Canadian Geographer*, 56(4), 439–458.

Japan Bicycle Association (2016) *Japan's Bicycle Market Industry Issues and Responses*. http://ebma-brussels.eu/wp-content/uploads/2019/03/Japans-Bicycle-Market-90-PRC-imports-bad-quality-loss-of-sales.pdf

Koeppel, D. (2010) "Riding China's flying pigeon bicycle". https://www.bicycling.com/rides/a20021347/bicycling-in-china/

MacMichael, S. (2019) "*Made in China* era over says Giant, the world's biggest bike maker – and blames Donald Trump's trade war". https://road.cc/content/news/262468-made-china-era-over-says-giant-worlds-biggest-bike-maker-and-blames-donald

Rosen, P. (2002) *Framing Production: Technology, Culture, and Change in the British Bicycle Industry*. (Cambridge MA: MIT Press).

Ulrich, K. (1995) "The role of product architecture in the manufacturing firm". *Research Policy*, 24(3), 419–440.

United Nations Industrial Development Office (2019) *Technical Report – The Indian Bicycle Sector: Enhancing Productivity in the Indian Bicycle Sector*. (Vienna, Austria: UNIDO).

Yeung, H. W.-C. (2016) *Strategic Coupling: East Asian Industrial Transformation in the New Global Economy*. (Ithaca, NY: Cornell University Press).

Zhang, Z.-H. (1992) "Enterprise response to market reforms: the case of the Chinese bicycle industry". *Australian Journal of Chinese Affairs*, 28(1), 111–139.

15

THE VALUE CHAINS AND PRODUCTION CLUSTERS OF TAIWAN'S BICYCLE INDUSTRY

Cheng-Mei Tung

Taiwan bicycle industry development background

The Taiwanese bicycle industry grew rapidly from the 1950s to 2020 despite price competition from low-cost international competitors, mainly due to establishing cooperative relationships among Taiwanese manufacturers. Currently, the industry has a defined position in the global market. In addition to inter-firm collaboration, industrial policy has played an important role in the development of the bicycle industry (Wang 2011). Taiwan's bicycle industry has grown in several important stages (Chu 1997).

Industry budding and import substitution (1949–1970)

After World War II, Taiwan began to develop a bicycle assembly industry. In 1949, Taiwan traded and imported bicycles from Japan, which put pressure on the production and sales of homemade products. To protect the domestic bicycle industry, the government adopted an import substitution policy, introduced high tariffs on imported bicycles, and allowed the importation of only a limited number of key components (Wang 2011). Bicycle assembly factories and many small-scale component manufacturers were established during this period. By 1970 bicycle production was mainly supplying the domestic market. In 1962 Taiwan also commenced to assemble and produce scooters. At that time the demand for scooters as transportation vehicles was larger than that of bicycles, and it cut into the domestic demand for bicycles (Chu 1997). Faced with losing the domestic market, Taiwanese companies cooperated with the government's export-oriented industrialization policy and began exporting products overseas.

Export-oriented development (1970–1980)

As European and American brand owners moved their manufacturing networks to Asia (Gao et al. 2012), Taiwan bicycle makers took the opportunity to become OEM exporters, subcontracted to make cycles for foreign brand owners. The exported volume increased from 10,000 bicycles in 1970 to 1 million in 1972 after which demand subsided due to the impact of the 1973 oil crisis. Owing to the high sale value of export markets, low-priced and inferior

DOI: 10.4324/9781003142041-21

products that dented trust of Taiwan-made bicycles were sold in Canadian and American markets, leading to antidumping lawsuits (Chu 1997). In response, the Taiwanese government assisted manufacturers in establishing a set of standards to follow for product export compliance with international standards. The total number of Taiwanese bicycle exports in 1980 was approximately 2.98 million units (Bureau of Foreign Trade 2020).

Need for relocation, transformation, and upgrading of production (1980–1990)

Bicycle exports continued to grow during this period as Giant became a major OEM supplier to Schwinn and other American bicycle brands (Wang 2011). By 1986, Taiwan exported more than 10.2 million cycles, but the unit price of these exports decreased. Following Schwinn's abrupt decision in 1987 to switch its supplier to the China Bicycle Company in Shenzhen, Giant responded by launching its own brand. Coupled with the impact of unfavorable exchange rates, the number of cycles exported subsequently decreased. Simultaneously, because of rising wages and labor shortages, the bicycle industry was confronted with pressure to survive. Some companies decided to create new supply chains by locating their entry-level bicycle manufacturing to mainland China and elsewhere in South-East Asia (Chu 1997). Manufacturers remaining in Taiwan face extremely competitive pressures, which required transformation and upgrading. Local bicycle components and assembly manufacturers began vertical integration and adopted Original Design Manufacture (ODM) to upgrade key components and whole bicycles (Chu and Li 1999). The bicycle companies also develop horizontal specialization to increase the value-added of bicycle components.

International low-price competition (1991–2002)

In 1990, some Taiwanese bicycle manufacturers set up factories in low-wage countries such as China and Vietnam (Chu 1997) due to competition from the low-price rivals in the global market. Several of Taiwan's bicycle companies began to export high- and mid-price bicycles by adopting the original brand manufacture (OBM) strategy and adjusted their business strategies to focus on marketing and after-sales services. By 2000, the export value of the Taiwan's bicycles ranked second in the world. The sources of profit for the bicycle industry were mainly assembly factories, but it was component manufacturers that enabled the survival of bicycle assembly factories (MoneyDJ 2002). Bicycle brand owners and makers often participated in international conventions and exhibitions or sponsored international bicycle teams to grasp new technologies and user needs. Otherwise, manufacturers established industrial networks through informal connections (Galvin and Morkel 2001). From 2000 to 2003, Taiwan's bicycle exports significantly decreased (Wei 2013). The export volume, which had reached 7.53 million units in 2000, decreased to 4.21 million units in 2002 (Bureau of Foreign Trade 2020).

Research and development (R&D) to increase value-added (2003–2016)

Since 2003, Taiwan's bicycle-related manufacturers, notably Giant and Merida, have organized the Taiwan Bicycle Association known as the A-Team (Corporate Synergy Development Center 2014; Chapter 18). A collection of component suppliers formed the A-Team to compete, but also cooperate with each other. According to its rationale, R&D collaboration by the A-Team allowed Taiwan's bicycle industry to maintain its global competitiveness (Li 2013).

Taiwan's bicycle industry developed a flexible production strategy with a variety of small quantities, high value-added, inventory reduction, and customization. The A-Team also promoted the upgrading and transformation of the bicycle industry cluster located in the Taichung area, in the middle of Taiwan. A-Team members collaborated and interacted closely (Tung, Chiu, and Yu 2020) and launched a Production Management System. Production capacity and product quality were improved by the implementation of the Toyota Production System (TPS) (Corporate Synergy Development Center 2014). Since then, Taiwan's bicycle exports have gradually increased. The A-team completed its planned phase tasks at the end of 2016.

Competitive advantages of high value-added products (2017~)

Some Taiwanese manufacturers moved their high-end high unit-price bicycle production lines from low-cost countries back to Taiwan. Manufacturers actively innovated products, increased automation of production and reinforced brand marketing. Mid-to-high-end bicycle companies adopted differentiation and branding strategies to enhance their competitiveness. Furthermore, in response to increasing global environmental awareness, bicycle functions have gradually transformed from mostly sport and leisure functions back to former uses as a commuting vehicle. The rise of the electric bicycle market accelerated the growth bicycle and component exports. In 2019, the value of the Taiwan's bicycle exports was US$1.34 billion, while that of bicycle component was US$1.41 billion in 2019 (Taiwan Bicycle Association 2020).

Development of bicycle clusters

The competitive advantage of a country is often manifested in a particular industry that forms clusters which accelerate industrial development and economic growth (Porter 1990). An industrial cluster refers to a group of companies that produce similar products or have upstream and downstream relationships with each other in a geographically concentrated area (Porter 1990; Feser and Bergman 2000). Within the industrial cluster, the geographical proximity of the manufacturers provides frequent connections with each other so that knowledge circulation within the industrial cluster is easier (Pouder and St. John 1996). Companies in the cluster can more easily access information about the market and their competition, and knowledge of new technologies (Porter 1990; Pouder and St. John 1996). Porter (1990) reported that companies in an industrial cluster, including suppliers, customers, and related industry manufacturers, usually have cooperative relationships with each other, although sometimes their interests may conflict.

Most of Taiwan's bicycle industry companies are small and medium-sized enterprises, except for major assembly firms such as Giant and Merida (Chu and Li 1999), and form an industry cluster in Taichung, Taiwan. Bicycle manufacturing processes are complex, comprising a series of interconnected components, including tires, rims, wheels, frame, handlebars, cranks, pedals, saddles, chains, and gears (Galvin and Morkel 2001). Each of these components has a specific function and at least one interface with other components (Galvin and Morkel 2001). The common functions of the modules create the complete function of the product or system of the bicycle (Sanchez and Mahoney 2001). An advantage of this modularization is that, as long as the interface is the same, the components in the modular system can be improved or simply changed in a certain manner without changes to other parts of the system (Sturgeon 2002). The modular system therefore allows an independent change of each functional element of the product by changing only the corresponding components (Ulrich 1995). Component manufacturers focus on the production of niche products, which are highly compatible with modular systems and delivered to the bicycle factory for assembly (Amsden and Chu 2003).

An assembly plant obtains a stable supply from parts suppliers and concentrates on the development of core technologies. By adopting this cooperative relationship, parts suppliers do not have to develop vertical integration capability or face ferocious competition, and are able to concentrate on developing the production of components (Amsden and Chu 2003). This model created a bicycle production network.

Why did the bicycle industry form an industrial cluster in Taichung? This area already had a well-developed cluster of machine tool industries and machinery processing industries (Corporate Synergy Development Center 2014). The manufacturing of bicycles and bicycle parts requires a large amount of machining technologies, such as lathes, milling machines, stamping and pressing technologies, and plastic molding. Firms making metal products developed and accumulated hardware technologies over many years, thereby providing the foundation for the bicycle component manufacturing cluster in Taichung (Chu 1997).

In 1984, the Ministry of Economic Affairs designed a corporate synergy development system to avoid aggressive competition among bicycle component manufacturers. The bicycle industry was divided into five systems, Giant, Merida, Pacific, Xuguang, and Taihang (MoneyDJ 2002). Component and parts factories became complementary with each other in the corporate synergy development system. The system evolved over time into bicycle manufacturing clusters headed by Giant and Merida (Xiao 2019). Bicycle manufacturers in the cluster shared information and cooperated closely. Porter (1990) reported that manufacturers in industrial clusters develop complementary and cooperative relations. Regional conditions and knowledge networks create an innovation system for the industry (Fornahl et al. 2012).

One driver that promoted the improvement in industrial cluster capability was the establishment of the A-Team (Chuang and Chen 2015). In 2003, Giant and Merida invited component suppliers to form an industry alliance named the A-Team (Corporate Synergy Development Center 2014). The industry alliance accelerated cooperation in R&D, leading to the production of high-end bicycle components and bicycles sold to the global market. The achievements of the A-Team are discussed in the next section.

A-Team development and the value chain

Stakeholders in the bicycle trade include bicycle part manufacturers, importers, and domestic companies that assemble bicycles, bicycle retailers (large and small), product designers, trade agents, and trade fairs (Gao, Liu and Norcliffe 2012). These companies form the full range of activities needed to create bicycles. Taiwan's original equipment manufacturer (OEM) experience making bicycles led to the accumulation of tangible and intangible assets for companies and provided a competitive advantage in the global value chain.

The production of bicycle parts and components was originally developed mostly by the individual companies. Unfortunately, these components lacked uniform specifications, with most companies producing a wide range of different components and then having to wait for buyers to place orders. Companies were therefore obliged to carry a higher inventory of raw materials and finished products, which increased production costs. To reduce these costs, after 2002 bicycle manufacturers moved their mass production factories to low-production-cost countries. The two major bicycle companies, Giant and Merida, realized they had a cost problem and proposed that the Taiwanese bicycle industry should shift to higher-end product lines and have a more dominant role in the global bicycle value chain through intangible competitiveness, including brand, innovation, technology, and customer relations. Therefore in 2004 Giant and Merida formed an industry alliance, the A-Team, to promote the collaboration of companies in the value chain. The A-Team connected and integrated the bicycle manufacturers upstream to downstream to develop high-added-value products and product differentiation as their goal and established close partnerships with each other. The A-Team used resource-sharing and mutual learning to enhance the added value and competitiveness of member companies (Brookfield et al. 2008; Wei 2013; Yen et al. 2017).

The A-Team focused on technological innovation and regional integration to raise product value by fully using the knowledge and capabilities of its members. Simultaneously, some multinational companies use outsourcing strategies to ask Taiwan bicycle manufacturers to help with product innovation (Brookfield, Liu and MacDuffie 2008). The product value-added attracted Shimano and Colnago as foreign sponsor members of the A-Team. The government also had a role in promoting enterprises to enter global value chains (Horner 2017). The rise of foreign direct investment in bicycle production has further connected Taiwan's manufacturing network to the global network. Industrial organization innovation has been an important mechanism for the evolution of Taiwan's bicycle industry clusters and boosted regional learning and innovation capabilities.

In addition, Giant and Merida assisted A-Team members in launching a "3T production management system", including TPS, Total Quality Management, and Total Productive Management in their manufacturing process (Corporate Synergy Development Center 2014). Just-in-time (JIT) of the TPS is used in manufacturing management. Suppliers were also expected to reduce delivery times (Fullerton and McWatters 2001).

The advantage of the JIT production system is that it allows suppliers to reduce inventory so companies do not need to produce in advance, and then store a product waiting for a customer (Fullerton and McWatters 2001). After implementing the production management system, A-Team members devoted attention to improving "per person productivity" and "per level ground productivity." As a result, company revenues increased, and they became more willing to invest in new product research, development, and equipment.

The A-Team learned from the practices of Japanese car manufacturers and gradually standardized the specifications of components (Tung et al. 2020). By adopting international standards, bicycle component manufacturers were able to create a series of semi-independent industries that did not need to coordinate their activities or communicate with each other (Galvin and Morkel 2001). Modularized products led to embedded cooperation among small enterprises, reducing the need for coordination and management (Sanchez and Mahoney 2001). This promoted closer connections among manufacturers to enlarge the market, so that customers did not have to be concerned about buying the wrong parts for maintenance.

In addition, the A-Team held training sessions, excellence factory visits, and research seminars as a platform for learning and knowledge sharing among members. Most A-Team members were first-tier manufacturers of the bicycle industry. When learning 3T knowledge and capability, first-tier manufacturers shared knowledge with the second- or third-tier factories. Companies also trained their seed trainers to coach their suppliers (Corporate Synergy Development Center 2014). The A-Team acted as a knowledge platform that enabled members to obtain updated industry information. Knowledge sharing enhanced the management capability of members and promoted R&D strategies for high-end products and components (Yen et al. 2017).

The A-Team cooperated with government to launch a low-carbon-footprint project and helped member companies obtain environmental green labels. Environmental protection activities improved the A-Team members' social image. A-Team members also formed a cycling team to promote domestic cycling activities and expand the market demand for high-end bicycles.

The A-Team also formed an alliance with world-renowned bicycle manufacturers to form an international value chain that made Taiwan a global supply center for innovative R&D of advanced bicycles and components (Tung, Chiu and Yu 2020). When the goals of the A-Team alliance were met, it was wound up at the end of 2016 (Yen et al. 2017). Afterwards, a special committee of the Taiwan Bicycle Association took over all of its activities.

Consumers' role and branding

The global bicycle market is projected to grow from US$29.2 billion in 2020 to US$34.6 billion in 2027, with a compound annual growth rate of 2.4% (Mortkowitz 2020). The main bicycle market is divided into three categories: bicycles for transportation, bicycles for leisure, and bicycle for competition. The leading global bicycle vendors include Accell Group NV, Atlas Cycles (Haryana) Ltd., Derby Cycle Holding GmbH, Dorel Industries Inc., Giant Manufacturing Co. Ltd., GUANGDONG TANDEM INDUSTRIES Co. Ltd., Insera Sena. PT., Merida Industry Co. Ltd., Tandem Group Plc, and Trek Bicycle Corp. (Businesswire 2020).

Many countries have promoted the goal of sustainable and environmentally friendly transportation in recent decades (Li et al. 2013). In addition, an emphasis on health has made cycling sports and recreation a popular trend (Grand View Research 2018). Users' perceived values are closely related to behavior (Claudy and Peterson 2014; Lo et al. 2016) with, for instance, the number of people choosing bicycles to commute increasing significantly (McLeod 2018; LoBasso 2019). Policymakers and corporations have proposed policies or subsidies to encourage people to commute by bicycle (European Commission n.d.). In general, the safety and convenience of bicycling affects users' attitude toward using bicycles (Noland and Kunreuther 1995; Claudy and Peterson 2014). Demands from consumers and policymakers have driven the bicycle R&D toward lightweight, electrical, intelligent, connected, and shared services with e-bikes rapidly growing in popularity (Grand View Research 2018; Wang et al. 2018; Coherent Market Insights 2020; Marketandmarkets 2020). Electric propulsion, intelligent design, and networking are currently major trends in product development, while component manufacturers seek to reduce the weight of components. Bikesharing has become a part of the public transportation system in Taiwan and many other countries, with many cities providing shared bicycles as a means of transportation. The adoption of bicyclesharing is largely influenced by its perceived function, condition, greenness, and social value (Wang et al. 2018). In 2015, the bicycle industry was hit by the rise of the sharing economy, and exports declined. Then the rapid rise of electric bicycles replaced the market share of scooters and drove the overall value of bicycle exports to rebound. Changes in the drive system, battery and wireless connection technology of electric bicycles have brought new opportunities for Taiwanese manufacturers. Electric bicycles are now the main driving force for the growth of Taiwan's bicycle industry. The export volume was 0.76 million electronic bikes in 2020 (compared to c.1.5 million pedal bikes). The scope of electric bicycles industry covers information and communications, energy, electronics, and bicycle industries. The electric bicycle alliance is more complicated than that of the A Team. Continuously enhancing the value of electric bicycles through cross-industry cooperation will be the main challenge for the industry.

During the COVID-19 pandemic, bicycle usage has increased a lot (Goetsch and Tatiana Peralta Quiros 2020). For example, from March 2020 to mid-June 2020 the number of bicycle riders in the US urban areas increased 21%, compared to the same period in 2019 (Mortkowitz 2020). And in the post-epidemic era, cycling to work, leisure, and sports may become a part of the daily life, particularly with the adoption of e-bikes by middle-aged and older riders, and by commuters.

The bicycle market increasingly emphasizes the unique preferences of users. Makers gain a competitive advantage by offering a smaller number of diverse types of vehicles, and customized designs to attract consumers. According to market sources, it is estimated that the price of full-suspension mountain bikes increased by 92%, that of gravel bikes increased by 144%, that of sport-performance road bikes increased by 87%, and that of electric bicycles increased by 190% in June 2020 compared to the same month in the previous year in the United States (The NPD Group 2020). In terms of the bicycle market segment, the number of hybrid bicycles is expected to increase fastest in the world at a compound annual growth rate of 3.1%, reaching a value of US$13.6 billion by the end of 2027 (Mortkowitz 2020).

Market segmentation analysis is a means of increasing market share of specific transportation modes (Li et al. 2013). Precise market segmentation divides customer segments carefully so that marketing activities have larger market power. On the demand side, competition among bicycle brands for positioning, unit price, and customer segmentation is strong. The Taiwan's two largest bicycle companies also differentiate their brands based on market segmentation. For example, Giant has differentiated brands, including the well-known "Giant," woman exclusive brand "Liv," urban leisure brand "Momentum," and high-end bicycle component brand "CADEX." Merida focuses on high-end markets: their brands include "MERIDA," a joint venture with the American "SPECIALIZED" brand, and the European brand "CENTURION."

In addition to manual pedal bicycles, it is also worth noting the market opportunities for electric bicycles. The global electric bicycle market is projected to grow from US$41.1 billion in 2020 to US$70 billion in 2027, with a compound annual growth rate of 7.9% (Marketandmarkets 2020; Chapter 24). Government support and measures drive the global electric bicycle industry and increase the sales of electric bicycles. Compared to motorized vehicles, electric bicycles are cheaper, easier to charge, and do not require significant investment in supporting infrastructure.

Combined with the development of digitalization, bicycle riding can be more intelligent and related information can be connected to the cloud, including navigation, connection with social media, and theft-proofing. Related cycling data and information can be shared with others in the virtual community and customer feedback can be obtained. For example, Bosch's SmartphoneHub can connect electric bicycles to smartphones and support riders before and after riding. Bosch launched Bosch eBike systems and opened a new Asia Pacific headquarters in Taichung in 2020 (German Trade Office Taipei 2020).

Conclusion

The Taiwanese bicycle industry has been evolving since the 1950s. The industry began with the government's export promotion policy. Gradually, manufacturers have also built up component and modular production capacity and linked bicycles and component manufacturing to global value chains. In the future, the industry cannot ignore market trends and consumers changing preferences. The Coronavirus pandemic has led to a major change in transportation in people's daily lives around the world in 2020, with many governments encouraging bicycle use for commuter transportation as global demand for bicycles continues to rise. Although Taiwan has good bicycle assembly advantages, innovative R&D on key components and accessories needs to continue to maintain a competitive position.

References

Amsden, A. H. and Chu, W.-W. (2003) *Beyond Late Development: Taiwan's Upgrading Policies*. Cambridge, MA: MIT Press.

Brookfield, J., Liu, R.-J. and MacDuffie, J. (2008) Taiwan's bicycle industry A-Team battles Chinese competition with innovation and cooperation. *Strategy & Leadership*, 36, pp. 14–19. doi: 10.1108/10878570810840643.

Bureau of Foreign Trade (2020) *Trade Statistics*. Bureau of Foreign Trade, MOEA. Available at: https://www.trade.gov.tw/English/ (Accessed: 20 January 2021).
Businesswire (2020) *The Bicycle Market to Grow by $ 10.5 bn in 2020*. Available at: https://www.businesswire.com/news/home/20201117006212/en/The-Bicycle-Market-to-grow-by-10.5-bn-in-2020-Industry-Analysis-Market-Trends-Market-Growth-Opportunities-and-Forecast-2024-Technavio (Accessed: 9 February 2021).
Chu, W. (1997) 'Causes of growth: A study of Taiwan's bicycle industry', *Cambridge Journal of Economics*, 21(1), pp. 55–72. doi: 10.1093/oxfordjournals.cje.a013659.
Chu, W.-W. and Li, J.-J. (1999) 'Growth and industrial organization: A comparative study of the bicycle industry in Taiwan and South Korea', *Taiwan: A Radical Quarterly in Social Studies*, 35(1), pp. 47–73. doi: 10.29816/TARQSS.199909.0002.
Chuang, H.-M. and Chen, Y.-S. (2015) 'Identifying the value co-creation behavior of virtual customer environments using a hybrid expert-based DANP model in the bicycle industry', *Human-Centric Computing and Information Sciences*, 5(1), p. 11. doi: 10.1186/s13673-015-0028-z.
Claudy, M. C. and Peterson, M. (2014) 'Understanding the underutilization of urban bicycle commuting: A behavioral reasoning perspective', *Journal of Public Policy & Marketing*, 33(2), pp. 173–187. doi: 10.1509/jppm.13.087.
Coherent Market Insights (2020) *Bicycle & Components Market Size, Trends, Shares, Insights and Forecast – 2027*. Automotive and Transportation CMI1449. Seattle, WA: Coherent Market Insights, p. 120. Available at: https://www.coherentmarketinsights.com/market-insight/bicycle-and-components-market-1449 (Accessed: 23 February 2021).
Corporate Synergy Development Center (2014) *3T Integrate Technology*. Available at: https://www.csd.org.tw/services/preview/362.html (Accessed: 23 February 2021).
European Commission (n.d.) *5.2 Cycling Subsidies, Mobility and Transport*. Available at: https://ec.europa.eu/transport/themes/urban/cycling/guidance-cycling-projects-eu/cycling-measure/cycling-subsidies_en (Accessed: 17 February 2021).
Feser, E. J. and Bergman, E. M. (2000) 'National industry cluster templates: A framework for applied regional cluster analysis', *Regional Studies*, 34(1), pp. 1–19. doi: 10.1080/00343400050005844.
Fornahl, D. et al. (2012) 'From the old path of shipbuilding onto the new path of offshore wind energy? The case of Northern Germany', *European Planning Studies*, 20(5), pp. 835–855. doi: 10.1080/09654313.2012.667928.
Fullerton, R. R. and McWatters, C. S. (2001) 'The production performance benefits from JIT implementation', *Journal of Operations Management*, 19(1), pp. 81–96. doi: 10.1016/S0272-6963(00)00051-6.
Galvin, P. and Morkel, A. (2001) 'Modularity on industry structure: The case of the world the effect of product bicycle industry', *Industry and Innovation*, 8(1), pp. 31–47. doi: 10.1080/13662710120034392.
Gao, B., Liu, W. and Norcliffe, G. (2012) 'Hypermobility and the governance of global production networks: The case of the Canadian cycle industry and its links with China and Taiwan', *The Canadian Geographer/Le Géographe Canadien*, 56(4), pp. 439–458. doi: 10.1111/j.1541-0064.2012.00442.x.
German Trade Office Taipei (2020) *Bosch eBike Systems Opens New Asia Pacific Headquarters in Taichung, Taiwan, AHK Taiwan*. Available at: https://taiwan.ahk.de/news/newsfeed/news-detail/bosch-ebike-systems-opens-new-asia-pacific-headquarters-in-taichung-taiwan (Accessed: 9 February 2021).
Goetsch, H. and Quiros, Tatiana Peralta (2020) *COVID-19 Creates New Momentum for Cycling and Walking. We Can't let it Go to Waste!, Published on Transport for Development*. Available at: https://blogs.worldbank.org/transport/covid-19-creates-new-momentum-cycling-and-walking-we-cant-let-it-go-waste (Accessed: 17 February 2021).
Grand View Research (2018) *Bicycle Market Size & Share, Global Industry Trends Report, 2018–2025*. Electronic (PDF) GVR-2-68038-490-1. San Francisco, CA: Grand View Research, p. 120. Available at: https://www.grandviewresearch.com/industry-analysis/bicycle-market (Accessed: 9 February 2021).
Horner, R. (2017) 'Beyond facilitator? State roles in global value chains and global production networks', *Geography Compass*, 11(2), p. e12307. doi: 10.1111/gec3.12307.
Li, Y. (2013) 'Success formula for Taiwan bicycle industry alliance', *Harvard Business Review*. Available at: https://www.hbrtaiwan.com/article_content_AR0002366.html (Accessed: 31 January 2021).
Li, Z. et al. (2013) 'Bicycle commuting market analysis using attitudinal market segmentation approach', *Transportation Research Part A: Policy and Practice*, 47, pp. 56–68. doi: 10.1016/j.tra.2012.10.017.
Lo, S. H. et al. (2016) 'Commuting travel mode choice among office workers: Comparing an extended theory of planned behavior model between regions and organizational sectors', *Travel Behaviour and Society*, 4, pp. 1–10. doi: 10.1016/j.tbs.2015.11.002.

LoBasso, R. (2019) *The Death of Bike Commuting Has Been Greatly Exaggerated, Bicycling*. Available at: https://www.bicycling.com/news/a25857897/bike-commuting-decline/ (Accessed: 17 February 2021).

Marketandmarkets (2020) *E-Bike Market by Class (Class-I, II & III), Battery (Li-Ion, Li-Ion Polymer, Lead Acid, Other), Motor (Mid, Hub), Mode (Throttle, Pedal Assist), Usage (Mountain/Trekking, City/Urban, Cargo), Speed (<25 & 25–45 kmph) and Region – Global Forecast to 2027*. Available at: https://secure.livechatinc.com/ (Accessed: 23 February 2021).

McLeod, K. (2018) *New Data on Bike Commuting, League of American Bicyclists*. Available at: https://bikeleague.org/content/new-data-bike-commuting (Accessed: 17 February 2021).

MoneyDJ (2002) *Bicycle Industry Analysis*. Available at: https://www.moneydj.com/kmdj/report/reportviewer.aspx?a=2b06a77a-d60b-4a8a-a24b-881ede30c213 (Accessed: 30 January 2021).

Mortkowitz, S. (2020) *The Global Bicycle Market: What's Ahead for It, We Love Cycling Magazine*. Available at: https://www.welovecycling.com/wide/2020/11/02/whats-ahead-for-the-global-bicycle-market/ (Accessed: 9 February 2021).

Noland, R. B. and Kunreuther, H. (1995) 'Short-run and long-run policies for increasing bicycle transportation for daily commuter trips', *Transport Policy*, 2(1), pp. 67–79. doi: 10.1016/0967-070X(95)93 248-W.

Porter, M. E. (1990) 'The competitive advantage of nations', *Harvard Business Review*, 1 March. Available at: https://hbr.org/1990/03/the-competitive-advantage-of-nations (Accessed: 15 May 2020).

Pouder, R. and St. John, C. H. (1996) 'Hot spots and blind spots: Geographical clusters of firms and innovation', *The Academy of Management Review*, 21(4), pp. 1192–1225. doi: 10.2307/259168.

Sanchez, R. and Mahoney, J. T. (2001) 'Modularity and dynamic capabilities', in Voberda, H. W. and Elfring, T. (eds) *Rethinking Strategy*. London: SAGE Publications Ltd, pp. 158–171. Available at: https://experts.illinois.edu/en/publications/modularity-and-dynamic-capabilities (Accessed: 9 February 2021).

Sturgeon, T. J. (2002) 'Modular production networks: A new American model of industrial organization', *Industrial and Corporate Change*, 11(3), pp. 451–496. doi: 10.1093/icc/11.3.451.

Taiwan Bicycle Association. (2020). *Statistics*. Available at: https://www.tba-cycling.org/list/cate-206561.htm (Accessed: 31 May 2020).

The NPD Group (2020) *Plot Twist: U.S. Performance Bike Sales Rise in June*. Available at: https://www.npd.com/wps/portal/npd/us/news/press-releases/2020/plot-twist-us-performance-bike-sales-rise-in-june-reports-the-npd-group/ (Accessed: 31 January 2021).

Tung, C. M., Chiu, S.-K. and Yu, W.-H. (2020) 'Innovative activities of a logistics company in the global value chain', *Journal of Business and Management Sciences*, 8(3), pp. 77–84.

Ulrich, K. (1995) 'The role of product architecture in the manufacturing firm', *Research Policy*, 24(3), pp. 419–440. doi: 10.1016/0048-7333(94)00775-3.

Wang, Y. (2011) *Research on Bicycle Industry Policy of Taiwan*. RDEC-RES-099-021. Taipei: Research, Development and Evaluation Commission, Executive Yuan, p. 355. Available at: https://ws.ndc.gov.tw/001/administrator/10/relfile/5644/3233/0058915_1.pdf (Accessed: 9 February 2021).

Wang, Y. et al. (2018) 'Be green and clearly be seen: How consumer values and attitudes affect adoption of bicycle sharing', *Transportation Research Part F: Traffic Psychology and Behaviour*, 58, pp. 730–742. doi: 10.1016/j.trf.2018.06.043.

Wei, C. (2013) 'Strategic analysis of cluster development of Taiwan bicycle industry', *Economic Outlook Bimonthly*, (146), pp. 103–107.

Xiao, X. (2019) *Merida Leads the Way to Build an Industrial Cluster, Vision Magazine*. Available at: https://www.gvm.com.tw/event/201908_Yuanlin/bike.html (Accessed: 7 February 2021).

Yen, M.-H. et al. (2017) 'Forming a learning network across transactional networks: The case of the A-Team in Taiwan bicycle industry', *Journal of Management and Business Research*, 34(3), pp. 431–466. doi: 10.6504/JOM.2017.34.03.05.

16

BICYCLE TRADE SHOWS AS TRANSACTIONAL SPACES

Michael Andreae and Glen Norcliffe

Modern trade shows started as periodic gatherings of suppliers and consumers meeting to engage in trade in a setting where face-to-face contact created "buzz". They present an economically efficient way of selling products with buyers congregating with product sellers in one place. The final transaction may take place afterwards and elsewhere, but the object of the trade show is to *perform* a product or service and initiate a trade. This chapter examines the cycle trade shows that sell bicycles in their various forms as well as numerous related accessories and services.

The recent cessation of several bicycle trade shows has been attributed to the temporary effects of the Covid-19 pandemic, but the reality is that changes in the way bicycles are traded have been coming for some time (Chapter 17). In a post-Covid world that makes widespread use of social media, Zoom and other video conferencing techniques, can the trading of bicycles using digital communication create buzz without face-to-face contact? This chapter approaches this important question by first looking at the rise of the industrial trade fair, and then the more recent emergence of three types of cycle trade show, each with a different function, with examples of each. For some types of trade show, the general public has become less welcome as attendance has been restricted to serious players in the transactional economy. The following section elaborates on the view of a cycle trade show as a performance. The last section assesses the four features of the urban face-to-face economy identified by Storper and Venables (2004) and asks whether digital bicycle trade shows might replace the physical, in-person, show.

In the beginning

The era of the modern industrial trade fair is usually dated back to the Great Exhibition mounted in London's Crystal Palace in 1851. Since then, trade shows have been promoted as places where attendance is essential both for producers and consumers. Businesses use trade shows to display their newest products and services while paying close attention to what their competitors have on offer, and deciding whether to copy or improve upon their rival's products. Buyers see spread before them the latest technologies, fashions and services that can improve their own businesses or enhance their identity as a particular type of consumer. Attendance

DOI: 10.4324/9781003142041-22

at a trade show allows participants to observe the latest technologies and fashions and share knowledge with other participants. Consumers can embrace what Thorstein Veblen (1899) would call *conspicuous consumption* and indulge their material impulses.

Interestingly, cycles were on display at the 1851 Great Exhibition in "Class 5: Machines for Direct Use, including Carriages, Railway and Marine Mechanism" (Royal Commission 1851), and from this display we can observe several still relevant aspects of a trade show. On stand 960, William Sawyer of Dover displayed three finely crafted four-wheel velocipedes. Ritchie (1975) provides interesting details concerning Sawyer's success at the Great Exhibition. He attracted attention by achieving the highest quality – his velocipedes were light, strong and superbly crafted. He was nearly a century ahead of Alfred Sloane (of General Motors) in producing several models: First Class, Second Class, Third Class and Double Velocipede, all with a different price range. He created "buzz" (Storper & Venables 2004; Bathelt et al. 2014) which attracted a "special command" from the Emperor of Russia. Sawyer described himself as the original inventor and improver of the velocipede, and inspired others to copy his models. And it seems that they did: nearby at stall 997 was the "four-wheeled pleasure-ground Victoria chair" of J. Ward of Leicester Square, London, and at stall 995 J. Wilson's "improved velocipede constructed principally of iron adapted for exercise and amusement." Ward subsequently copied Sawyer's design, but was unable to match Sawyer's high quality.

The Great Exhibition was an expression of imperial power, the exploitation of its colonies, and of merchant capitalism, with other imperial powers, including France and the United States, later following suit with their own expositions. The character of trade shows began to change as they became more focussed on a single product, particularly if a period of innovation was resulting in lots of interesting new models or methods. For cycles "the" show in the late nineteenth century was the annual Stanley Show, held in various locations in London (including the Crystal Palace) from 1878. Birmingham's response, the Speedwell Bicycle Show, was less successful. Bicycles, tricycles, saddles, chains, gongs, lighting, pedals, oilers and much more attracted crowds: between 1878 and 1890 the number of cycles on display at the Stanley show rose from 70 to 1450 (on 230 stalls). Yet within a decade the show was fading away as the bicycle ceased to be the focus of innovation, with the last show held in 1910: it was supplanted in 1911 by the Olympia motor car and motor cycle show as consumer interest shifted to motorized transportation.

For the next half-century the bicycle attracted little interest among consumers, and if they were shown, it was at a local retail show, or as an add-on to other more fashionable consumer goods such as motor cycles. In the 1970s two sets of changes came that led to the re-birth of cycle shows. Firstly, there was rising interest in 10-speed road bicycles, MTBs and BMX bicycles and a related set of innovations: once again, bicycles became interesting. In Taiwan in 1974, TaiSPO (the Taiwan Sporting Goods Show) was launched, with bicycles included among a range of other sporting goods. Rising interest led in 1988 to the creation of the separate Taipei International Cycle Show (TICS) which was a success: by 2012, there were 400 exhibitors and 1850 booths (Taipei Cycle Show Preview 2012). Other cycle shows were launched during the same period. Secondly, a major restructuring of the global economy was underway, with much of the world's cycle production shifting from Europe and North America to Asia (Chapter 14). Often described as a new international division of labour, it also involved flows of capital and international payments as renowned bicycle brands based in the old industrial countries sub-contracted production to East Asia and, in the process, created global production networks. For brand owners, this created a need to search for makers of new models at trade shows.

Three types of cycle trade shows

Communication and trade are inexorably linked: the new shows that TaiSPO and other Taiwanese organizations fostered were specifically created to promote exports by placing Taiwanese businesses on the world stage. As international trade opened up, a new system of communication was required. When the Great Exhibition was created, London was the center of imperial commerce and Britain's regions were the main producers of the manufactured goods on show so that makers and buyers were fairly proximate. A century later, buyers and makers often found themselves geographically separated, leading the cycle trade to create three different types of trade show, each with very different objectives. At *manufacturer shows*, companies presented their newest products and accessories close to their factories so that major buyers could then make a site visit; at *distributor shows*, retailers can view the offering of importers and major brands in the regions where they were sold; local *consumer shows* are based in major cities and attract individual buyers and cycling enthusiasts. Each of these shows played a different role in the communication of information and trust which facilitates transactions and trade. Following is a synopsis of a major show in each category.

Manufacturers' shows

The original success of the Taiwanese cycle industry followed a decision made by the Taiwanese government from the first plan period (1953–56) onwards to promote export-led growth by targeting several industries with growth potential (Amsden 1979). Bicycles were among the consumer durables identified for export support during the fourth and fifth plan periods (1965–68, and 1969–1972). Taiwan's economic planners were acutely aware that the bicycle sector would have to innovate, not least because a US government tariff had exempted "Lightweight Bicycles" weighing under 35lbs with tires less than 1 5/8" wide from the protectionist import duties charged on heavier bicycles (US Congress 1962). American bicycle makers assumed that American riders' love for heavy, lumbering bikes would continue, thereby inadvertently forcing innovation abroad at exactly the time 10-speed bicycles and MTBs were becoming popular. The Taiwanese government jumped at the opportunity to develop a local industry and the push to internationalize bicycle manufacturing was underway by 1983 with the bankruptcy of CCM in Canada (Gao et al. 2012) and Schwinn in the US (Chu 1997). Supporting research into tube construction and lightweight manufacture was a part of this success. Yet another key component was Taiwan's proactive government practicing what Amsden (1979) calls étatisme, but which today might pass for state capitalism.

Taiwan's planners recognized the need for visits to Taiwan by foreign buyers if they were to purchase Taiwanese bicycles. That was the motivating force behind Taiwan establishing several trade shows in the 1970s: the aim was not to reach foreign consumers, but specifically to target brand owners and distributors by creating a setting in which to impart knowledge about their innovative cycles (Zhu et al. 2020). Manufacturers' trade shows presented makers with the opportunity for hands-on demonstrations (imparting some tacit knowledge) as well as some codified knowledge in the form of brochures (Bathelt et al. 2014). And visits to nearby factories after trade shows could extend this exchange of knowledge, and build trust between consumers and producers. Manufacturers' shows also brought the world's attention to those regions and created a cultural experience around manufacturing and exporting (Lee et al. 2018). Recently, the Taiwanese government has begun using TICS to push Taiwan as a cultural cycling destination, not just a manufacturing hub (Spinney & Lin 2019).

The Taiwan External Trade Development Council (TAITRA), created on 1 July 1970, was founded to support export industries, of which bicycles were a major component. In addition to founding the TaiSPO trade show in 1974, TAITRA spent 16 years promoting a permanent trade center in Taipei, culminating in the construction of the Taipei World Trade Center in 1986 and, two decades later, the larger Nangang Exhibition Hall.

In 1992 the Taiwanese Government created the Taiwan Bicycle Export Association, renamed Taiwan Bicycle Association (TBA) (https://en.tba-cycling.org/about-tba.htm) to work with TAITRA promoting bicycle exports. All bicycle exporters have to be a member of TBA, and in exchange the organization promotes bicycle R&D and hold a much-publicised bicycle innovation awards show as part of the annual TICS. These organizations focus on encouraging the world to see Taiwan as the preeminent manufacturer of bicycles: running the largest manufacturers show is a key part of this.

One clear indicator of a manufacturers' show is that the public is excluded. By 2008 TICS was only open to the public on the last day, and in subsequent years it was entirely closed to the public. As James Liu, president of TBEA, stated: "TICS is an image show" focused on technology, innovation, and relationship building rather than direct sales (Andreae et al. 2013). Manufacturers' shows such as TICS are scheduled in spring after manufacturers have shipped most orders and before the fall distributor shows where they are busy inking new sales contracts. TICS continues to evolve: in 2017 the organizers arranged a cycle demonstration at a large off-site space where buyers and retailers could test a range of new products. In 2021 the coronavirus pandemic forced TICS to move on-line and shifted the focus from sports and leisure to specific groups such as commuters, cargo bikes and e-bikes, with visual (video) presentations featuring prominently. Afterwards TICS reported that businesspeople from 81 countries attended, with the virtual exhibitions, business meetings and discussion groups continuing for nearly a month, collectively logging over 100,000 visits to their online platform. Innovation was foremost with moves toward autonomous bicycles and cargo tricycles, the rapid development of e-bikes and micromobility bikesharing. The apparent success of the on-line format raises the possibility (to be discussed below) of industry bike shows becoming on-line events in the future.

The main rival to TICS is the China International Bicycle Fair, held in Shanghai in May every year since 1990, with the exception of 2020, when the pandemic forced cancellation. This fair mixes bikes with automotive after-market products.

Distributor shows

While Cycle Taipei became the world's largest manufacturers' show and created a reputation for quality, not all brand owners and major retailers wanted to travel to Taiwan due to the cost, its cultural contrasts, and having to travel across many time zones. However, local distributors needed to procure their stock which was no longer being made within their own country. By the 1980s distributor-specific bike shows held closer to home markets began to form. The two largest distributor shows are Eurobike, which was started in Germany in 1991, and Interbike, founded in the USA in 1982, which moved around a little before settling for a while on Las Vegas. As manufacturing moved abroad, these shows were created to allow store staff, distributors, and larger foreign manufacturers to trade and exchange knowledge. The bicycle distributor shows were scheduled for the fall, so that orders could be placed in time for the peak sales month of April, in contrast to the manufacturers' shows, which were scheduled in the spring giving makers and distributors the time to negotiate design and price. Attendees at manufacturers' shows might negotiate possibilities before signing purchase orders at the distributor shows.

Eurobike claims to be the world's largest distributor bike show, attracting enthusiasts and retailers from across Europe since 1991. Held in the former Zeppelin manufacturing fields of Friedrichshafen, the show is gigantic, welcoming 40,000 trade visitors. Unlike the manufacturer shows, Eurobike is organized by Messe Friedrichshafen as a for-profit business.

Eurobike limits public access, only allowing non-industry attendance on the final day. Despite this, a stunning 21,000 members of the public attended in 2019 on that one day (more people than attended the largest North American and European retail shows held over multiple days). Covid-19 led to the 2020 Eurobike show being cancelled, and the 2021 show was planned for September in the hope that the epidemic would subside. Significantly, a preview show aimed exclusively at retailers called *EUROBICO – Order & Preview Show* was planned for July 2021 at Frankfurt-am-Main, a central location with space to accommodate a growing number of e-bikes.

There are two elements of Eurobike that make it very different from a manufacturer show. First unlike Taipei Cycle where the focus is on the quality of Taiwanese production, Eurobike features products from all around the world. Inside the show there are regional pavilions, including an Italian hall focused on Italian brands (many of the companies subcontract manufacturing to Taiwanese and Chinese manufacturers), and a set of booths subsidized by TAITRA to promote Taiwan as a source region. Just as firms cluster in regions to draw support from each other, these internal clusters of similar firms inside Eurobike help promote regional brands (Lee, Fu, & Tsai 2018). Secondly Eurobike featured a lot more branded experiences and giveaways; one booth featured live models being painted with the company's brand.

North America's biggest bike show, Interbike, was cancelled in 2019 after 36 years, having relocated from Las Vegas to Reno in 2018 because of declining attendance and the high cost of exhibition space in Las Vegas. The show brand was then sold by its owner, Nielsen, to Emerald Brands who placed it in Boulder City, Colorado beginning with a two-day Interbike OutDoor Demo where (like the most recent Taipei show) attendees could ride the machines on offer. An attempt to re-launch Interbike as a low-cost outdoor show was made in 2021, but its future seems uncertain. "One of the limitations of Interbike was that in the last 7–10 years, it wasn't a full gathering of the tribe. A small percentage of our key partners, retailers, suppliers, and even media were there and engaged. Interbike in its heyday had the energy and spectacle to reflect bicycling which is amazing and big and diverse, and it hasn't felt that way recently." (https://www.velonews.com/news/interbike-is-dead-now-what/).

While manufacturer shows appear healthy, three changes have put pressure on distributor shows. First, the emergence of oligopolistic distributors such as Pacific Cycles allows retailers to by-pass the shows and select from the wide range of products offered by wholesalers cum brand owners. Second, the major makers, including Trek, Specialized and Giant, have mounted their own retailer shows. And third, online shopping is growing rapidly, changing the landscape for retail bicycle operations (Chapter 17). Purchasers can now compare a multitude of products with ease and some sites allow browsers to design their own bicycle. Technology and social media have created new channels for information flowing between distributors and retailers (https://www.pinkbike.com/news/rest-in-peace-the-rise-and-fall-of-interbike-expo.html). Declining attendance made Interbike less useful as an information exchange. When key makers, retailers and cycle personalities (the "tribe") are absent, a show loses its buzz.

Consumer shows

While manufacturer and distributor shows discourage or ban the sale of goods to individuals, consumer shows are a different entity altogether. These trading shows try to sell products to every consumer they can, claiming to offer buyers a wider selection than most places, often with a "show discount." There is still a spectacle such as extreme action events to draw attendees, but the main objective is to sell goods. Consumer shows also tend to promote local activities and are usually scheduled for the very beginning of the cycling season to get consumers excited about buying a new bicycle and showing off their new clothing.

Consumer shows are hyper-localized, and named after the host city or some biking term, such as Spin London and Copenhagen Bike Show, to attract local shoppers and bike enthusiasts. The largest consumer shows include the Copenhagen Bike Show in Denmark and the Toronto Cycle Show in Canada. In 2019 these March events both drew in approximately 15,000 people to visit over 100 exhibitors. Eurobike entered into the retail show space in 2011 by founding a separate public event VeloBerlin "[to] meet the wide-range of requirements from the industry in the best possible manner and are able to offer tailored customer solutions" (https://www.eurobike.com/en/news-detail/303).

Trade shows and the information economy

Trade shows, although generally viewed as economic spaces where transactions take place, may also be viewed as performance spaces where knowledge and theater are blended. From a performance perspective, trade shows focus on the visual senses (which tend to dominate) plus sound and touch; for bicycles, the senses of smell and taste are minor considerations. The appeal to visuality is based on color (generally bright), flashing lights, lasers, and videography. Typically the visual presentation is accompanied by loud music with a strong rhythm, limited by the need not to impact performances at adjacent show booths. The most important part of a trade show performance is to create a liminal space where a client crosses a threshold to encounter a new set of ideas, while at first not fully embracing them. Once a client has entered that "in-between" space and left behind old ideas, it is up to the salesperson to "land the fish" by successfully transferring knowledge about the product.

There are eight major ways that a good performance at a trade show reduces friction in trade and therefore creates value for the participants (Andreae et al. 2013). Trade shows are periodic markets limited to short periods when face-to-face exchange can increase trust; they provide places physically to test product performance and gain hands-on tacit knowledge; consumers may gain status by testing new products and becoming first-adopters of the latest technologies; attendees themselves may be good performers, knowledgeable about the industry, and are sometimes recruited into the industry; stellar presentations attract media attention which in turn may entice personalities to attend; they create and enhance the regional brand; they allow companies to showcase the experience of their brand with distributors and interested parties; finally, by attending and performing year after year, trade show participants build trust with their trading partners outside of traditional spaces. Trust between parties facilitates transactions.

Being social creatures humans generally innovate and discover things by being together. One unexpected discovery during prohibition was that shutting bars reduced patent applications because people couldn't get together in person and discuss ideas (https://www.npr.org/2020/10/12/923123253/an-economist-walks-into-a-bar). For cyclists, the internet and related technologies may have become a source of information and even some sales, but encounters and personal experiences at trade shows and elsewhere still play an important role.

Trade shows versus digital communication

Social media and technology have dramatically changed the cycle trade. Brand owners and makers, distributors and retailers, buyers and sellers all communicate on-line using video conferencing and social media. As already noted, there are advantages to face-to-face (F2F) contact, but now there are efficient alternatives both in time and in speed. Makers and designers can exchange designs on-line and accelerate the process using 3D printers. Consumers can by-pass retailers and purchase on-line, with some vendors allowing buyers to customize their bicycles. The Covid-19 epidemic showed that trade shows could continue on-line, provided vendors took time to develop short snappy productions; since viewers view dozens of presentations, long turgid performances were to be avoided. These developments raise the question: are cycle trade shows becoming redundant given the new communications technologies? To answer this question each of the four advantages of face-to-face (F2F) communication, identified by Storper and Venables (2004) to account for the concentration of economic activity in urban areas, will be examined.

1. *F2F as a communication technology*: While codified information, for example a bicycle's schematics, can easily be transmitted digitally, F2F is the most effective way to disseminate uncodifiable knowledge using body language, emotions and hands-on demonstrations. There is also a fluidity gained from in-person communication because the speaker can respond to the listener's actions and emotions, creating a mutually enriching conversation that is lost in digital form. This makes F2F a much more efficient way to communicate tacit knowledge. Being on the test track with a potential customer where they feel the excitement of smooth e-bike acceleration helps frame and facilitate a conversation about e-bikes; it is impossible to replicate this communication experience by mail or chatting on the phone. At all three types of trade show F2F is used as an efficient communication technology.
2. *F2F creates trust*: Trust has always been a significant part of economic relationships (Lorenzen 2001). In its absence it can be very hard to validate what is real: photos are airbrushed and photo-shopped, fake videos and false news are broadcast. F2F creates trust between parties because it makes it easier to observe the full range of body-signals. And a personal factory tour after meeting at a trade show helps to assess the seller's production capabilities. Trust is also generated by repeated F2F contact, year after year, at a trade booth that has been booked annually for a lengthy period, creating an image of reliability. One example of this investment in trust at TICS was flowers and plants given to booths by their banks and suppliers to demonstrate that they were reliable clients with whom one could safely do business.
3. *Shared values and screening*: Business relationships are fraught with complexity, emotions and even distrust which shared values may help smooth out. F2F interactions, which include socializing, helps participants to identify a set of shared values. Social networks, both digital and human, help people to share experiences and create a comfort zone within which trade can be discussed. Trade shows provide public spectacles, but also private spaces where experiences are shared. Attending foreign manufacturer shows helps producers and distributors bridge cultural divides, while attending distributor shows builds strong relationships with retail stores and allows feedback about what products and brands work and which need to be changed. Consumer shows provides direct access to consumers and their values, a group which is usually blocked from the other shows.
4. *Motivation and shame*: There is a tangible excitement in attending a trade show. It presents an opportunity to try out the latest technologies and for many attendees to travel and party after a year of hard work. Conversely, attending a trade show may reveal that one has been overtaken by the latest technologies, creating pressure to update one's bicycle or its

> components. Goffman (1959) stresses the production of information that co-presence and F2F communication can create. Being told that one's bicycle is no longer the best provides strong motivation to find a better model. Of course, this information can be communicated in other ways but F2F experiences are very compelling, especially when they take place over informal social activities – a meal, a coffee, a game of golf. From drinks in private spaces on top of two story booths at industry shows, to the chance to see new bicycle designs not yet publicly available, manufacturer and distributor shows offer diverse opportunities to promote bicycle trading. Consumer shows are also bonding events where sellers are strongly motivated to develop personal rapport with customers who belong to Veblen's leisure class: they are local and will come back if encouraged.

Investigating the advantages of F2F communication, Bathelt and Schuldt (2010) undertook a comprehensive series of interviews at international trade shows, finding that "in general, we can assume that interaction in physical space is richer than interaction in Internet trade fairs, because the latter cannot transfer feelings and mediate associations in the same way as real-world trade fairs" (Bathelt & Schuldt 2010: 1969). Recent real-world experiences would seem to confirm this: Eurobike expanded to include a consumer show to intensify F2F interaction even as online communication was growing, whereas Interbike failed to generate enough F2F value for attendees and was closed. The importance of F2F was again confirmed by Lee, Fu, and Tsai (2018) when they interviewed booths at the Taipei International Travel Fair and found that F2F communication was the predominant reason given for attending. Further evidence for the value of F2F is the resumption of in-person tradeshows in 2021/2022 that had moved online during Covid-19. There is, of course, a huge space for technology in international trade, but it appears that technology will continue to complement rather than replace in person events.

Conclusion

Trade shows have always been a space for bringing people together to share ideas and exchange goods. Congregating buyers and sellers facilitates transactions, raises brand awareness, links cultures, and demonstrates new technologies. The fact that financial transactions do not always, or even often, occur at these events is of little consequence due to the benefits that F2F communication brings in the longer-run.

While the three types of shows addressed here, manufacturer, distributor and consumer, all provided F2F contact, the goals and experience of that F2F time differed significantly depending on the type of show. Manufacturer shows promote industry in host regions by inviting major brand owners. Distributor shows bring together makers, brand owners and major distributors. Consumer shows are strongly localized, focusing on moving products into the hands of final purchasers. These disparate goals are reflected in many aspects of the shows, including their location, scheduling, and the audience admitted. The distributor and manufacturer shows do not want a public audience because it takes away from face-to-face communication with commercial buyers. Manufacturer, distributor and consumer shows are all timed around the annual business cycle, manufacturers in the spring when there is no production rush, distributors in the fall so volume order contracts can be signed in person, and consumer shows in the spring when customers are excited at the prospect of a new cycling season. The shows live on because modern technologies which allow instant digital communication around the world have not replaced the F2F communication which trade shows provide.

References

Amsden, A. (1979) "Taiwan's economic history: a case of étatisme and a challenge to dependency theory." *Modern China*, 5(3), 341–380.

Andreae, M., Hsu J.-Y., & Norcliffe, G. (2013) "Performing the trade show: the case of the Taipei International Cycle Show." *Geoforum*, 49(7), 193–201.

Bathelt, H., Golfetto, F., & Rinallo, D. (2014) *Trade Shows in the Globalizing Knowledge Economy* (Oxford: OUP).

Bathelt, H., & Schuldt, N. (2010) "International trade fairs and global buzz, part I: ecology of global buzz." *European Planning Studies*, 18(12), 1957–1974.

Chu, W. (1997) "Causes of growth: a study of Taiwan's bicycle industry". *Cambridge Journal of Economics*, 21(1), 55–72. https://doi.org/10.1093/oxfordjournals.cje.a013659

Gao, B.-Y., Liu, W.-D., & Norcliffe, G. (2012) Hypermobility and the governance of global production networks: the case of the Canadian cycle industry and its links with China and Taiwan. *The Canadian Geographer*, 56(4), 439–458.

Goffman, E. (1959) *The Presentation of Self in Everyday Life* (Garden City, NY: Doubleday).

Lee, T.-H., Fu, C.-J., & Tsai, L.-F. (2018) Why does a firm participate in a travel exhibition? A case study of the Taipei International Travel Fair. *Asia Pacific Journal of Tourism Research*, 23(7), 677–690. https://doi.org/10.1080/10941665.2018.1487456

Lorenzen, M. (2001) "Ties, trust, and trade: elements of a theory of coordination in industrial clusters." *International Studies of Management & Organization*, 31(4), 14–34.

Ritchie, A. (1975) *King of the Road: An Illustrated History of Cycling* (London: Wildwood House).

Royal Commission (1851) *Official Catalogue of the Great Exhibition of the Works of the Industry of All Nations* (London: Spicer Brothers).

Spinney, J., & Lin, W.-I. (2019) "(Mobility) Fixing the Taiwanese bicycle industry: the production and economisation of cycling culture in pursuit of accumulation." *Mobilities*, 14(4), 524–544.

Storper, M., & Venables, A.J. (2004) "Buzz: face-to-face contact and the urban economy." *Journal of Economic Geography*, 4(4), 351–370. https://doi.org/10.1093/jnlecg/lbh027

Taipei Cycle Show Preview (2012) https://issuu.com/9vnel/docs/taipei_cycle_show_preview_2012_final

US Congress (1962, June 15) Hearing Before the Committee on Finance. *Stained Glass, Bicycles, Religious Articles*.

Veblen, T. (1899) *The Theory of the Leisure Class: An Economic Study of Institutions* (New York, NY: Macmillan).

Zhu, Y.-W., Bathelt, H., & Zeng, G. (2020) "Are trade fairs relevant for local innovation knowledge networks? Evidence from Shanghai equipment manufacturing." *Regional Studies*, 54(9), 1250–1261.

17
RETAILING BICYCLES

Jay Townley, Bradley Hughes and Michael Fritz

The first brand producer of bicycles, the European Sewing Machine Company (later Coventry Machinists), was documented in 1869 in Britain. The retailing and repairing of bicycles subsequently became more complex as the development of the safety bicycle in the 1880s led to the so-called "Golden Age" of the bicycle when hundreds of brands, retailers and bike shops emerged to serve consumers in Europe and North America. The retailing and servicing of bicycles continued to grow in Europe from the "Golden Age" through to 1910, but they began to decline in the US with manufacturers and retailers going out of business as the automobile became the favored mode of individual transport. Accordingly, some American bicycle manufacturers, like Arnold, Schwinn & Company, also engaged in the manufacture of motorcycles, and tried their hand at automobile manufacturing, while many bike mechanics in the US became motorcycle and automobile mechanics.

The bicycle and the retailing and servicing of bicycles also spread to the rest of the world as this human-powered invention was introduced by traders and merchants to Asia, South America and by Britain to what are now the Commonwealth countries in Africa, India, Pakistan, Australia and New Zealand. Thereafter riding a bicycle became an adult activity in Europe but largely a child's activity in the US from the 1920s through the 1940s, effecting both the bicycles themselves and the sophistication of retailing and the complexity of servicing bicycles.

As an example, the derailleur, a multi-speed shifting device, was invented and refined in Europe between 1900 and 1910 but not approved by the Union Cycliste Internationale for competitive racing until 1936. European retailers then had to become better informed about new bicycle technologies and mechanics had to learn to adjust and service derailleur shifting systems. In the US from 1910 through the 1940s selling primarily children's bicycles required minimal merchandising, and some advertising. Mechanics often expanded their skill sets to include welding, radio repair, lawn mower sharpening and small engine repair.

Logistics of bicycle supply chains

After World War II, from 1947 until the early 1980s most bicycles were manufactured in the countries where they were marketed and sold to local consumers. As an example, in 1975, at the end of the "Bicycle Boom" there were nine domestic bicycle manufacturers in the US

DOI: 10.4324/9781003142041-23

producing 5,576,000 20-inch wheel and larger bicycles representing 76% of a total market consumption of 7,294,000 bicycle units, with 24%, or 1,718,000 units imported. In 2019, 44 years later, we estimate that 9,129,000 20-inch wheel and larger bicycles were consumed by the US market with 93%, or 8,522,000 imported and 607,415, or 7% domestically produced by one primary American bicycle assembly plant and a handful of smaller manufacturers and assemblers.

In Europe after the 1980s, domestic manufacturers lost 35 to 40% of the market share to imports, depending on the country. While the US bicycle market became import-dependent after 1995, the European Community (EC) was much quicker to invoke antidumping laws and regulations and was able to retain 58 to 62% share of its bicycle markets for domestic production. The logistics for Europe have been a combination of domestic manufacturing and distribution within the EC, with importation of about one-third of the products shipped from Asia by container to ports, then distributed to large retailers and wholesalers (China's Belt and Road project is expected to switch much of this shipping trade to its trans-Asian rail system). Top-tier component suppliers like Shimano have established manufacturing and distribution inside the European tariff wall, as have larger Asian bicycle manufacturers like Giant and Merida, which has served to somewhat simplified their European supply chain and logistics.

Since 2000 the US has been import-dependent with 93 to 98% of all bicycles sold manufactured off-shore, primarily in Asia. There are no domestic or foreign component manufacturers to support large-scale bicycle production and the small number of domestic producers are dependent on the importation of the components necessary to assemble and manufacture complete products. These supply chains are dependent on offshore sourcing, primarily in China and Taiwan and elsewhere in South-East Asia and with an ocean freight component are more complex.

The two most populous countries in Asia, China and India, have more direct in-country supply chains and accordingly simpler logistics. China has both domestic and foreign component manufacturing, so its domestic and export bicycle manufacturing plants have domestic supply chains from raw material and components to wholesale distributors and retailers. Western brands like Trek and Specialized and Asian brands like Giant and Merida own extensive chains of retail stores in China that receive brand name products from their Chinese domestic production facilities. India, in contrast, remains a closed market with high import duties on components resulting in a domestic industry that depends on a domestic supply chain to produce complete bicycles for domestic consumption.

Advertising and sponsorships

Prior to World War II bicycle advertising in Europe and the British Commonwealth was aimed at the adult market. In America, advertising was aimed mainly at children and their adult parents. After World War II the consumer economies ramped up and sponsorship of bicycle racing and the events became advertising vehicles in Europe for magazines and news media. In America, print media like comic books and *Boy's Life* and television morning shows like *Captain Kangaroo* aimed at children became advertising vehicles for American bicycle brands.

During the "Bicycle Boom" from 1972 to 1974 the American bicycle market shifted from children to young adults and bicycle-specific magazines like *Bicycling*, *Bicycle Guide* and *Velo News* became the primary advertising vehicles. These print publications were joined by BMX-specific magazines for children and teens, and mountain bike-specific magazines in the 1980s. Sponsorships also evolved in Europe and North America as BMX, free-style, mountain biking, road racing and triathlons grew in popularity, along with legacy professional road races like the Tour de France.

By 2000 all of the top-tier European, American and Asian bicycle and component brands were involved in sponsorship of professional Grand Tour, triathlon, BMX, free-style and mountain bike races – and these events were admitted to the Olympic Games. Bicycle and component brand advertising became tied to professional bicycle racing sponsorships from 2000 to the present.

Mass vs specialty

In Europe, the bicycle grew in popularity as an adult activity and both amateur and professional racing became very popular. The Tour de France was founded in 1903 and, with the exception of the wartime years, has been conducted every summer since. While there was bicycle racing in North America over the 40 years between the Wright brothers closing their bike shop in 1909 and the end of World War II, the best bicycle racers went to Europe to compete, including Major Taylor, the African American who was the greatest bicycle racing champion of that era.

The retailing of bicycles flourished in Europe and both sales and service had to master multi-speed shifting systems like the derailleur, quick release hubs and rat-trap pedals and toe-clips – innovations that came out of racing. This sophistication was passed along by companies like Raleigh, the English bicycle brand. Japan and China developed their own home-grown bicycle brands with retail networks, often consisting of small bike shops servicing local areas.

Hundreds of bicycle brands are currently sold in the American market. These brands are sold through five main retail channels:

- Mass Merchants e.g. Walmart, Target
- Chain Sporting Goods (Full-Line Sporting Goods) e.g. Dick's Sporting Goods
- Specialty Bicycle Retail (7000–7400 Independent Specialty and Bike Brand Stores)
- Specialty Outdoor Retailers e.g. REI Co-op
- Other/Online e.g. Canyon

Table 17.1 shows 2019 estimated sales for these primary channels of trade in the US, by units, total value and average unit retail value (the bike shop and outdoor specialty retail (OSR) segments are shown separately in Table 17.1 but will be combined throughout most of this section).

Table 17.1 shows the mass merchant channel of trade is estimated to have captured 74% of all the new units sold in 2019. The mass merchant channel has been in transition over the last twenty years by including online sales and the introduction of higher-priced, brand name bicycles and urban, family, and electric bicycles aimed at women and adult consumers in addition to a complete selection of children's sidewalk and juvenile bicycles. The mass merchant channel continues to be the primary retail source for children's and juvenile bicycles in the US.

Clearly, mass retailers dominate the US market by units sold, although by value, specialty bike stores account for almost half of all sales: mass retailers are price-driven whereas specialty stores concentrate on high-value sales that, on average, are worth over eight times as much per unit.

Recent comparative data from France reveal a very different culture of bicycle retailing.

A comparison of Tables 17.1 and 17.2 shows the different retailing cultures in the US and France. Perhaps due to the powerful influence of the Tour de France, EU import duties, and

Table 17.1 Estimated US Bicycle Market Retail Sales by Channel of Distribution 2019

Channel	*Estimated Units**	*% Units*	*Estimated Value ($)*	*% Value*	*Estimated Average Unit Value ($)*
Mass Merchant	12,179,904	74	981,700,279	29	81
Chain Sporting	822,957	5	198,244,400	6	241
Bike Shops	1,975,120	12	1,676,876,880	49	849
Outdoor Specialty	987,560	6	420,206,695	12	425
Other / Online	493,780	3	160,478,468	5	325
Total	**16,459,330**	**100**	**3,440,000,000**	**100**	**209**

* Not including e-bikes.

Source: Human Powered Solutions.

Table 17.2 Estimated French Bicycle Market Retail Sales by Channel of Distribution 2019

Channel	*Estimated Units*	*% Units*	*Estimated Value (€m)*	*% Value*	*Average Unit Price*	
					€	$
Mass Merchant (grandes surfaces)	265,200	10	45,060	3	170	206
Chain Sporting (magasins multisports	1,776,900	67	495,660	33	279	338
Bike Shops (détaillants)	503,900	19	826,100	55	1,639	1,987
Other / Online (internet)	106,100	4	135,180	9	1,434	1,738
Total	**2,652,100***	**100**	**€1,502,000m**	**100**	**566**	**686**

* The total includes 338,100 e-bikes.

Source: https://www.unionsportcycle.com/cycle-mobilite/le-marche-du-cycle-en-france

the inclusion in France of e-bike sales, the average French bike costs more than 3 times the US bike. The average price of bikes sold in French specialty bike shops is more than double that of American ones. Specialized bike shops account for somewhat more unit sales and value of sales in France, but the biggest differences are found in mass merchants (big box) which sell almost three-quarters of all units in the US, but only 10% in France. The opposite applies to sporting chains: in France, they sell two-thirds of all units whereas in the US they sell just 5%.

Bikes vs accessories

Aftermarket sales complete the picture of the primary bicycle industry channels of trade.

Table 17.3 shows the estimated direct effect aftermarket retail dollars by channel of trade, with an estimated total of just over $2 billion based on published data. Of note is the Bike Shops+OSR channel with an estimated 53% share of US aftermarket retail dollars.

High-end custom bikes

High-end bikes have been available in the world markets for decades, but until the mid-1990s, were not popular in the US beyond the enthusiast bicycle sub-culture that included amateur and professional racing. As the data from France show, Europe has had a larger enthusiast

Table 17.3 US Bicycle Market Estimated Direct Effect Retail Dollars

Channel of Trade	*Bike Shop ($m)*	*Outdoor ($m)*	*Mass Merchant ($m)*	*Sporting goods chain ($m)*	*Other/Online ($m)*	*Total US Market ($m)*
Accessories, Helmets, Footwear, Gloves, Parts	1,062.7	140.3	381.0	200.5	220.6	**2,005.1**
Percent Share	53	7	19	10	11	**100.00**

2019 Parts, Accessories & Rubber Market Share by Channel of Trade.

(Estimates include sales at fixed locations and at websites, but do not include labor, rentals, used bikes, travel/tours or coaching)

Source: Human Powered Solutions estimates and analysis.

bicycle culture and accordingly a larger market and supply chain for high-end custom bicycles made by boutique frame makers and assemblers that until the mid-1990s were the providers of racing bicycles for Pan Am, Olympic and Professional teams and athletes.

In North America, the number of boutique high-end frame makers were limited, although Arnold, Schwinn & Company, which became Schwinn Bicycle Company in 1967, established a boutique frame manufacturing and assembly department under the Paramount brand name in 1938. Paramount became the high-end custom frame set and racing bicycle for American amateur and professional bicycle racers for the next 40 years, until around 1978 when European high-end custom brands gained market share in a growing market.

By 2005 the American high-end market had grown to the point where it supported the North American Handmade Bicycle Show, an annual trade event that attracted the 80 to 100 high-end customer frame builders that had been established in North America, along with 20 to 30 European and Asian high-end builders. The North American Handmade Bicycle Show has been held annually in American cities for 15 years, but was interrupted in 2020 by the pandemic. Paramount was bought out of the Schwinn Bicycle Company bankruptcy of 1993 and has been operating as Waterford Precision Bicycles since, being a regular exhibitor at the North American Handmade Bicycle Show. Attendees were predominantly individual consumers who had become customers for high-end custom frame sets and bicycles. The retail prices of high-end frame sets are from US$3,000 and up and complete bicycles from US$5,000 and up. Model types include road racing, track, time trial, triathlete, tandem, gravel, mountain, and cross-country. About half of the high-end frame set and bicycle boutique makers sell directly to consumers via the web, and about half sell through bike shops in North America: internet sales of high-end bikes are also important in France.

American bike shops become specialty bicycle retailers

George Garner, Sr. passed-away at age 93 in September 2016, but his son, George Garner, Jr., carries on the family tradition of owning and operating bike shops in the Northern suburbs of Chicago, Illinois. George Garner, Sr. inspired a generation of American bike shop owners and employees and was a big factor in getting bike shops out of the back alleys and side-streets and moving them to the high-shopping streets across America.

The most prolific period in American bike shop retailing started around 1950 and lasted 60 years – until the Great Recession in 2009–2010. History will show that the two greatest

Figure 17.1 Valley Cyclery, Van Nuys, California, 1963 (author's collection).

influences on this six-decade period of prosperity and growth of American bike shops were the Schwinn Bicycle Company and George Garner. In April 1947, George Garner borrowed $3,000 and purchased Valley Cyclery in Van Nuys, California. At that time, bicycles were sold in all kinds of outlets and Arnold, Schwinn & Company had over 15,000 dealers, consisting of everything from funeral homes, fixit and hobby shops to bike shops. Most were cramped, crowded, and dark. George Garner planned to turn his Van Nuys shop into the ultimate bicycle store. He started with clean stores, white walls, bright lights, and rows and rows of bicycles. Service was the key to his approach. "I wanted to make people feel *comfortable* in my store," he explained (emphasis added).

In the well-lit and organized Valley Cyclery Garner shifted the focus from price to showroom displays, product knowledge, and service. "I was always looking to adapt the displays to show off the product," he said. Garner designed and hand-built store fixtures to display bicycles and parts and accessories (Figure 17.1).

Schwinn had launched a Service School in the early 1950s and Garner displayed individual technicians' Service School Diplomas in his shops where customers could see them. Service departments had an open view of the store and customers and technicians wore white shop coats displaying Schwinn Service School patches. Garner also believed that it would be much easier to sell new bikes if the salespeople in his stores were professional and knew what they were talking about. Following Garner's example, Schwinn launched a Sales and Management School in 1964 that held its first week-long class in 1966. Building on the success of his first operation, Garner opened four more bicycle shops in the Los Angeles area within a few years and quickly became Schwinn's number one dealer.

Schwinn executives decided to create a template for other bicycle shop owners to implement called the *total concept store*, outlining how a bicycle shop should be operated, down to the finest detail, largely copying Garner's approach. Then, in 1964, Schwinn management convinced Garner to come to Illinois to open a prototype store for dealer training, but also to prove that a total concept store could be financially successful in a northern state! Located in Northbrook, Illinois, just 25 miles from Schwinn's Chicago headquarters, Garner's new store opened in 1965 and became a test bed for new displays, showroom layouts, and merchandising programs. "We changed from a price organization to one concerned about quality and service. Schwinn used my stores as an example," Garner said.

Also in 1965 Schwinn's advertising agency – George Bond and Associates – coined the name *Total Concept Cyclery*, combining "total concept store" with the term *Cyclery* that George Garner had coined. There are still quite a few stores that use *Cyclery* in their names. The *total concept store* gained acceptance as more and more dealers shared success stories in the Schwinn Reporter. George Garner was Schwinn's number one dealer for many years. The *Total Concept Cyclery* that Garner spawned influenced American bike shop store design, layout and operations for over six decades.

How many bike shops are there in America?

Ongoing research by Christopher Georger of Georger Data Services (GDS) and bicycle industry consultant, and writer Rick Vosper, agree with the industry's brand owners that there are between 7,000 and 7,400 "bike shops" in the 50 United States, defined as the number of shop locations included on bike brands dealer lists. We find this is an appropriate method for defining bike shops (GDS owns the URL www.thebikeshoplist.com).

Five types of American bike shops

There are five types of American specialty bicycle retailers:

Multi-store chain

There are at least half a dozen multi-store chains in the US, defined as retailing at more than five fixed locations as an authorized dealer of one of the top four brands (i.e. it is independent, and not owned a bike brand). The multi-store chain owners communicate and collaborate with each other including in developing and importing private label merchandise.

Brand

The four largest specialty bicycle retail brands are Trek, Specialized, Giant and Cannondale. They each have an estimated 1,400 to 1,100 authorized dealer store fronts or, in the case of Trek, a growing number of company-owned stores. It is estimated that the four largest specialty bicycle brands own or have contracted dealer agreements with just over 50% of all the specialty bicycle retailers in the US Some retailers have contracts with two or more of the "Big Four." This includes some of the independent multi-store chains. A large percentage of brand and multi-store chains have commerce-enabled websites or are Omnichannel retailers, providing 24-7 access to consumers.

Specialty Independent (IBD)

Specialty independent bike shops are the majority of the "other half" of the specialty bicycle retailers with a fixed location. These bike shops do not have contracts with the "Big Four" but

Table 17.4 US Independent Bicycle Dealers Engaged in e-commerce

US IBDs	*All IBDs*	*<$300K*	*$300–$499K*	*$500–$999K*	*$1–1.9 Million*	*$2 Million+*
Currently engaged in e-commerce (%)	58.9	44.8	32.3	53.6	77.5	82.5

Total and by Revenue Size.

Source: NBDA 2017 Specialty Bicycle Retail Study.

they tend to have authorized dealer agreements with one or more of the next ten to fifteen brands below the top four, depending on region of the country. Some of these bike shops are already offering 24-7 access through Omnichannel business models, but the National Bicycle Dealers Association (NBDA 2017) 2017 Specialty Bicycle Retail Study discloses that while 98% of all US bike shops have a website, only 59% were at that time engaged in e-commerce.

Table 17.4 uses this 2017 study to show US IBDs engaged in e-commerce by annual gross revenue size. Not surprisingly, as annual revenue increases so does the percentage of bike shops engaged in e-commerce, or with some type of Omnichannel business model.

What this table indicates is that for the traditional narrow definition of a "bike shop," those below $1 million ($999,000) in annual revenue, are in danger of being overlooked by consumers expecting 24-7 access and an experiential relationship with a bike shop when and where they want it. However, this table tells only a part of the real story because it is tied to the traditional narrow definition of a "bike shop." As a consequence, the data leave out emerging new wave bike shops, as are online bicycle retailers and direct to consumer business models. In brief, we are under-counting and under-surveying a significant and growing population of new wave "bike shops," including mobile service "shops" that we suggest need to be included to gain a more accurate picture of very important details like the real percentage of specialty bicycle retailers engaged in e-commerce!

New-wave (including mobile) bike shops

New-Wave and "Outlier" bike shops gained traction in some markets as business models that held the potential of hanging on to aging enthusiasts and attracting Gen X and millennials after the Great Recession. Over the last seven to eight years, the business model has focused on making the specialty bicycle retail business a consumer's "third-place," or as we called it "sticky," by offering beer, wine, coffee, food and a comfortable place to come, remotely work, chat with fellow cyclists and feel at home… and get a bicycle serviced or purchase a new bicycle or accessory.

New-Wave bike shops also began to offer "locally-made" bicycles to their customers. Locally-made means what it says – an urban or commuter bicycle with locally-made frame and fork and as much domestic component content as possible, painted and assembled in the local community (or at least in the US) selling for $1,800 to $3,000. The locally-made business model also frees the New-Wave specialty bicycle retailers from the traditional authorized dealer business model based on a contractual relationship between the retailer and a bike brand, which binds the retailer to the brand's retail profit structure. The ambience of a third place is calculated to deal with a growing problem for traditional bike shops – a decline of store traffic, referred to as "footfall." Providing the motivation for consumers to frequently enter a New-Wave bike shop and spend money ensures exposure to bicycle-related products and services and allows for the capture of consumer "opt-in" information and data for retailer outreach and social media contact. Mobile service businesses are included in New-Wave and Outlier bike shops because a growing number are affiliated with bicycle and accessory brands, and more New-Wave bike shops are adding and offering home and workplace pick-up, delivery and service.

Having commerce-enabled websites and being Omnichannel retailers, available to consumer and customers 24-7, is also a common component of New-Wave and Outlier specialty bicycle retail business models, along with extensive use of social media to maintain communities of interest. We estimate that there are currently hundreds of New-Wave and Outlier specialty bicycle retailers in the US, some of which are being counted because they are listed on one or more brand dealer finders – and some of which that are not being counted by any methodology because they are selling *Locally-Made* bicycles in their regional markets and via their web sites.

Online

Some online specialty bicycle retailers have been a presence for a bit over two decades, while others are more recent entrants to the marketplace, but all fall within the definition of "pure-play" since the online sales of the other four types of American bike shops are blended with their respective retail types. There is a high probability that an Online specialty bicycle retailer will be listed on a bicycle brand dealer locator and accordingly counted by GDS in the estimated 7,000 to 7,400 "bike shops" in the US.

Canyon was founded in Koblenz, Germany in 1985 as a bicycle parts maker, but in 1996 graduated to making whole bikes. In the next decade, they moved to direct to customer (D2C) sales online. The current brand name was adopted in 2001 – but an American consumer, at least officially, could not purchase a D2C Canyon brand bicycle via the Internet until August 2017. There was a lot of anticipation about Canyon officially entering the US market, but it appears that even a D2C brand as big as Canyon has created no major disruption to the online business in America.

In 2019 e-commerce, or online retail sales, accounted for 16% of total retail sales – an increase of 14.8% over 2017 – and this increase in online retail sales represented over half of the total US increase in retail sales year-on-year. Online sales are growing and they are predicted to increase to over 20% of all US retail sales by 2022–2023.

Estimated annual retail sales of specialty bicycle retailers

As shown in Table 17.1, total American Specialty Bicycle Retail is estimated to have sold 1.975 million units, or 12% of all new bicycles sold in 2019. At an average unit value of US$849, this totaled an estimated US$1.7 billion, or 49% of total annual retail dollar sales of bicycles. In addition, the Specialty Bicycle Retail channel is estimated to have sold US$1.063 billion in accessories, helmets, footwear, gloves, and parts at retail, representing 53% of total aftermarket retail sales in the US This adds up to an estimated US$2.8 billion in total retail sales.

What will the bike shop of the future look like?

In addition to the New-Wave business model that embraces the *Third-Place* strategy of Star-Bucks and Panera that we have already presented, we see more future bike shops adopting the emerging retail philosophy of *Smaller-Is-Better*! This in turn fits right into more New-Wave bike shops of the future building business models around a *Showroom* and a *Curated Collection* of both bicycles and accessories.

Figure 17.2 A New Age Bike Shop Showroom (author's collection).

The Showroom in New-Wave retailing, including bike shop retailing, adapts smaller retail square-footage, which reduces lease/rent and occupancy cost, with a Curated Collection of both bicycles and accessories that is selected by the retailer as the best selection and choices for the customers and shoppers frequenting the retail establishment (Figure 17.2).

Curated consumption is a fast-growing trend among younger consumers and goes with Omnichannel retailing. It includes a commerce-enabled website allowing consumers 24-7 access to a product selection that the retailer has carefully selected to provide the best features, function, quality and value to customers. We see a *Showroom* and a *Curated Collection* being ideal for locally-made bicycles, or a line of bicycle products that offers the highest levels of customer service and value that is exclusive to the New-Wave bike shop in their market area – and allows them to offer the brand D2C through the retailers' website or from the brand.

A *Good, Better, Best* (GBB) merchandising plan is emerging as very compatible with the *Showroom* and/or *Curated Collection* business models, with $100 to $200 price spreads between each of the three so that the up-selling features are clear, abundant and easy for sales associates to enthusiastically present and for consumers to understand. The same GBB merchandising plan is applicable to a Curated Collection of accessories, with smaller and more applicable retail price spreads.

3D printing

There is no question that bicycle service has already become more technical with the advent of electric bicycles, and the trend going forward indicates even more technical advancements such as an anti-lock braking system (ABS) and greater degrees of communication and connectivity are incorporated into both electric bicycles and human-powered-only "smart" bicycles. This will require an evolving service department and the tools and equipment that the bike shop of the future will need to invest in and the staff training that will be continually required.

One area that has already changed the service and logistics systems in the aircraft industry is 3D printing of service parts. It won't be long before 3D-printed bicycle parts will be accepted and utilized to speed up service turn-around and reduce inventory cost in more businesses and sectors, including both fixed and mobile e-bike and bicycle repair (Chapter 9).

A closing word about electric bicycles

America is not Europe or East Asia, where e-bikes have become a large bicycle category sold and serviced in many countries (Chapter 24). This big increase in electric bicycle demand and sales had, however, not reached the US in 2019, the last year for which we have complete 12-month data. The US market for e-bikes reached 263,000 units in 2017, according to the Light Electric Vehicle Association (LEVA). LEVA found that suppliers imported 215,000 e-bikes into the US in 2017 and estimates that an additional 15,000 were built from parts by various US assemblers. Human Powered Solutions (2020–21) estimates that 2018 retail sales of e-bikes increased to just over 300,000 units, resulting in a 3% market share. Unfortunately, the imposition of a punitive tariff of 25% on the FOB value of e-bikes in 2019 caused unit volume to drop to an estimated 287,000 units. In its 2019 Bicycle Market Report, HPS predicted that the break-out year for sales of electric bicycles in the US is still two to three years away.

COVID and the Pandemic have disrupted the American bicycle and e-bike markets and greatly accelerated sales and created shortages of new bicycles, e-bikes, accessories and service parts. Every category has greatly increased in sales, including e-bikes, but we still don't know by how much – although a close to a doubling is probable – it still may not push e-bikes to over a 5% market share in an American market that will probably exceed US$9 billion in sales in 2020, up from US$5.8 to US$6 billion in 2019.

References

Human Powered Solutions (2020–21) *Research: Bicycle Economy in the Pandemic* (3 reports). https://humanpoweredsolutions.com/research-report-bicycling-buying-and-riding-during-the-pandemic/ (accessed 19 January 2021).

National Bicycle Dealers Association (2017) *Specialty Bicycle Retail Study*. https://nbda.com/product/bicycle-retail-study-2017/ (accessed 19 January 2021).

18

ON THE SHOULDERS OF GIANT

Cluster innovation and entrepreneurship in the Taiwanese bicycle industry

Yu-Chun Lin

Introduction

At the beginning of Taiwan's industrialization, many industries started as original equipment manufacturers (OEM). In subsequent decades, they became the foundation of Taiwan's economic miracle. During the wave of globalization, however, these industries entered an era of meager profits due to oversupply. In response, industry recognized the need for production innovation and increased value-added, but faced bottlenecks because they lacked house brands and their own marketing channels. With reduced barriers to international trade, many companies became transnationals by outsourcing their manufacturing. This led to a huge structural transformation of industries in Taiwan with the emergence of cross-border investment and production that linked together global and local production networks. Subsequent research on industrial clusters has discounted the importance of institutional relations to their evolution. This tendency is illustrated by the bicycle industry, one of Taiwan's signature industries, which relocated production to mainland China to reduce production costs.

Taiwan's bicycle industry was among the first set of industries to move production to China after the launching of China's economic reform and "open door" policy in 1978 (Chapter 15). The success of production in China led to concern for the future of Taiwan's bicycle industry. Three factors – accumulated learning, favorable environment, and global production – were identified as conditions necessary for the survival and growth of Taiwan's bicycle industry (Chu 1997, 2001). Jian (1997) outlined the structure and position of Taiwan's bicycle industry within the global value chain. In addition, Xie (2000) examined the role of the state, arguing that government intervention was crucial to Taiwan's bicycle industry, in spite of its shift offshore. Work by Chen (2002) focused on the micro-level social processes and institutions, and proposed a network-centered perspective to analyze the development of Taiwan's bicycle industry. Unlike other traditional industries, firms in the bicycle industry survived but gradually relocated some of their production to countries other than China, while upgrading and innovating, and building their brands internationally.

The A-Team was the most important driving force behind the evolution and upgrading of Taiwan's bicycle industry. The questions now arises: how did the "new strategic alliance model" (the A-Team) work?

This chapter begins by reviewing the history of Taiwan's bicycle industry, the production crisis and the development of cross-border economic linkages. The second section argues that

DOI: 10.4324/9781003142041-24

the evolution of the Taiwanese bicycle industry was driven by two factors: firstly, China, whose lower production costs and huge market provides the base for many Taiwanese enterprises. Secondly, Giant Inc., an initiator of the A-Team, established closer interaction with other firms. Members of the A-Team shared a common vision of future development of Taiwan's bicycle industry, stressing Toyota-like efficiency in the system of production, and an up-market product shift. It institutionalized competitive and cooperative relationships among related firms, and accelerated the upgrading of Taiwan's bicycle industry. The penultimate section reviews recent developments in response to fierce global competition in the bicycle industry and difficult Cross-Strait relations. The final section presents policy implications.

A brief history of Giant

Giant was founded in 1972 near Taichung by Liu Jinbiao and his seven partners. In 1977 (when Luo Xiangan was the deputy general manager), Giant began OEM production of bicycles for the American Schwinn Bicycle Corporation, then the United States' leading brand. This led Giant to invest heavily in automatic electrostatic painting equipment. With growing bicycle sales in the United States and a wave of employee strikes at Schwinn in 1980, Giant became an important OEM supplier to Schwinn with its production increasing by two-thirds by the mid-1980s, accounting for 75% of Giant's external sales.

When Schwinn abruptly switched suppliers and signed a contract with a Chinese bicycle company in 1987 to produce bicycles in Shenzhen, Giant, under the leadership of Bill Austin (previously Deputy CEO of Schwinn), decided to launch its own brand to sell in the United States and compete in that market as well as other countries. Meanwhile, in 1984 Giant and the German bicycle manufacturer Koga-Miyata established a joint venture at Giant's European branch. Koga-Miyata's withdrawal from this venture in 1992 turned Giant's European branch into a wholly-owned subsidiary.

Having become a global player, Giant began to think about giving back to Taiwan. The goal was to place Taiwan at the commanding heights of world cycling by creating a "Cycling Island." To do this Giant has worked hard in three areas: the first is to make their premium products in Taiwan, which was achieved through the "A Team" model. The second is to turn Taiwan into a cycling paradise, with the Giant Travel Agency welcoming tourists to cycle around the island. The third is to make Taiwan's cities *cycling-friendly*. It took the lead in promoting Ubike in the Taipei Metropolitan Area, and all major cities in Taiwan have since followed up. Ubike is also recognized as an excellent shared bicycle system worldwide. Giant also became an enthusiastic promoter of the manufacturers' Taipei International Cycle Show (Andreae et al. 2013; Andreae 16 this volume).

Over the next 30 years Giant bicycles began to sell well in more than 80 countries, growing to eight manufacturing bases, 14 subsidiaries and more than 12,000 retailers worldwide. Giant Group now ranks first in Asia and among the top three in the world, with 12,000 employees worldwide, sales of 6.6 million bikes in 2017 and well over 5 million in 2020. Revenue rose from $820 million in 2007 to $1.8 billion in 2012, and $2.45 billion in 2020.

The crises and cross-national linkages of Taiwan's bicycle industry

Development, crises and the state

The bicycle industry was one of the first post-World War II manufacturing industries launched in Taiwan (Chen 2002). Beginning by producing bicycle parts in the early 1950s, the bicycle industry was under state protection, with the Taiwanese government banning the importation of complete bicycles and allowing only 12 key bicycle parts to be imported. From the 1960s to the 1970s, the volume of OEM exports increased at a dramatic pace. Meanwhile, Taiwan's government established a national standard for bicycles and bicycle parts and installed an export inspection system, prohibiting sub-standard products from being exported, which strengthened the confidence of foreign buyers in Taiwan's products (Chen 2002). By increasing the volume of production and exports, Taiwan overtook Japan's position as the number one bicycle exporting country in the world from the 1980s, as the Taiwanese government's center–satellite system gave impetus to the industry to integrate production links upstream and downstream. Hu and Wu (2011) identify 1980 as the first peak in Taiwan's bicycle innovation cycle.

In the late 1980s, domestic economic conditions including rising wages, labor shortages and escalating land prices, threatened Taiwan's low value-added, labor-intensive industries. The relaxation of domestic political control and democratization led the government to permit direct and indirect investment in China. These changes prompted the industry to upgrade its product mix, leading to the second innovation cycle around 1990 (Hu & Wu 2011). At the same time, bicycle producers began to move their production plants to China, mainly to the Pearl River Delta and then the Yangze River Delta (Chu 1997). According to the Secretary General of the Taiwan Bicycle Exporters' Association (TEBA), Chan Chao-Ching, the Chinese market offered the greatest potential for the bicycle industry. Chan stated: "The main reasons manufacturers choose to relocate to mainland China are the lower land costs, abundance of cheap labor, common culture and the potentially huge domestic market" (Trade Wind's Industry Weekly, Jul 27, 1998:1).

Following the late 1980s, a series of crises threatened Taiwan's bicycle industry and many other traditional industries. In 1986, the export of Taiwan-made bicycles peaked at around 10 million units. In 1991, the export value reached an apex of US$1.1 billion, but dropped to US$523 million in 2002, a decline of more than 50%. Between 1995 and 2002, the total value of exports fell steadily and the average unit price was between US$100 and US$120. From 2003 to 2015, the export volume of Taiwan's bicycle industry maintained about 4 million to 5 million units per year, while the value of sales continued to grow to nearly USD$1.8 billion, with a unit price of nearly US$500 (see Figure 18.1).

The role of the state in the development of Taiwan's bicycle industry was not initially apparent. Unlike some heavy industries, the central government did not intervene in the bicycle industry by direct investment. Instead, the central government provided trade protection, information and technical assistance in the early years. The Bureau of Foreign Trade and the Metal Industries Research & Development Center (MIRDC) cooperated in a project to improve the manufacturing techniques for bicycle firms and revised the outdated industrial standards in line with international standards (Chen 2002: 249). With fiscal support from the central government, the Industrial Technology Research Institute (ITRI) also developed new technology and new materials in the late 1980s.

In the early 1990s, industrial out-migration contributed to the decline of Taiwan's bicycle industry. Relaxation of domestic industry policies permitted the industry to make both direct and indirect investments in China. In contrast to traditional industries, policy in Taiwan was paying more attention to high-tech industries with the establishment of science parks, tax reductions, preferential loans, and R&D expenditure. Up to this point the state had done little to help traditional industries (including the bicycle industry) face the challenge of rising production costs: their response of outward investment to China and South-east Asia reduced their production costs, creating a cross-national production network.

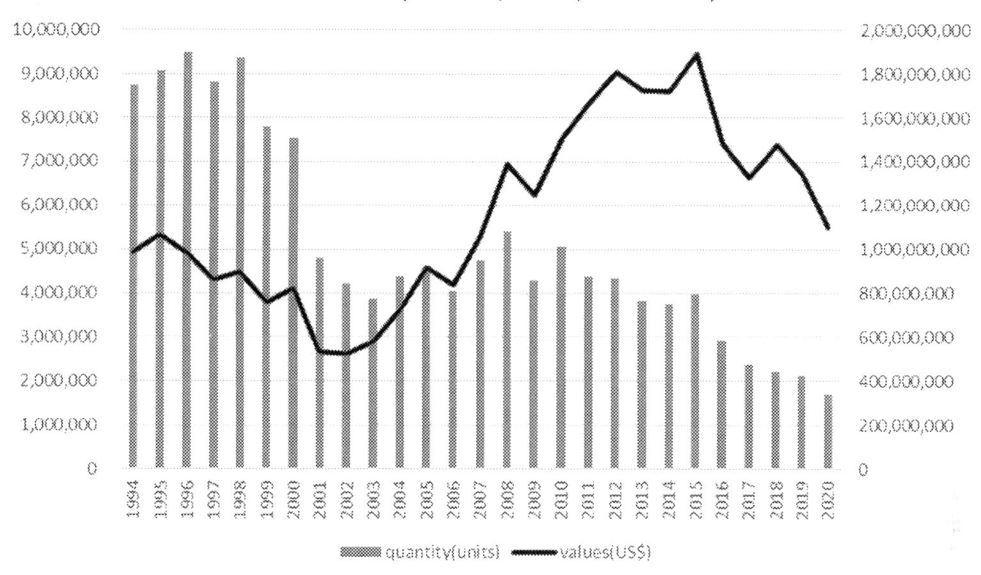

Figure 18.1 Taiwan 27-year Bicycle Export History.

Source: 2021 Taiwan Bicycle Source.

Outward investment and the formation of a cross-national production network

Cross-strait economic links have fundamentally reshaped the relationships between Taiwan and China. The cross-border production activities of transnational corporations (TNCs) have not only deepened spatial divisions of labor, but also complicated the interaction of Taiwan and China. In addition, the production system in this region forms part of a cross-national production network (CPN). Cross-national production networks are relationships among firms that organize research and development, procurement, distribution, product definition and design, manufacturing and support services in a given industry across national borders (Ernst 1995). Furthermore, Borrus and Zysman (1997) pointed out that "CPNs permit and result from an increasingly fine division of labor. The networks permit firms to weave together the constituent elements of the value-chain into competitively effective new production systems, while facilitating diverse points of innovation. But perhaps more importantly, they have turned large segments of complex manufacturing into a commodity available in the market."

Taiwan's bicycle producers are mostly concentrated in the central region of Taiwan. The structure of bicycle production and its parts involves many technologies such as applied materials, mechanics design, welding, and forging, making the industry closely related to the metal machinery-processing industry.

In 1990, Taiwan's government permitted the bicycle industry to invest in China and the first group of bicycle producers, led by Merida and KMC, started to build their plants in Guangdong, mainly in Shenzhen. According to a survey by the Shenzhen Federation of Industry and Economy in 2003, the bicycle industry chain in Shenzhen consisted of over 20 finished bicycle and 130 bicycle parts manufacturers, 90% of them funded from Taiwan (SinoCast China Business Daily News, July 2, 2003). In contrast, Giant built its plant in

Kunshan, near Shanghai, in 1992, leading to an upsurge of investment in the Yangze Delta region. At present, almost all Taiwan's bicycle producers who have built production plants in China are located in Guangdong or the Yangze Delta. These bicycle firms migrated with their entire production networks. As a result, new industrial clusters have rapidly been established in China. Meanwhile, the 'old' bicycle industry remaining in Taiwan faced serious challenges as the producers either closed down or downsized (Chen 2002).

Facing intensive competition and attracted by low labor costs, shifting production to China became the key for survival for most of the bicycle firms. The Guangdong and Shanghai regions have become the industry's second home, and more than 200 firms are now operating in China. To avoid political risks, a few firms have hedged their bets by setting up plants in Vietnam, which is becoming another production base of Taiwan's bicycle industry. The key attraction of Vietnam was that the European Union did not levy antidumping tariffs on imports from Vietnam in the early years. Led by the major assembly firms of Strongman, Asama Yuhjiun, and MT Racing, more than 20 companies have now established facilities in Vietnam, but the overwhelming majority of Taiwanese bicycle firms are located in China and South-east Asia.

Taiwan's cross-national bicycle production networks are organized for three key purposes. First, these corporations may choose to invest overseas to take advantage of lower factor prices, especially labor. Second, these corporations wish to obtain access to new local markets. And third, cross-border production takes advantage of a more intricate division of labor to create complementary production arrangements.

Taiwanese bicycle manufacturers have gradually overcome the twofold core competitiveness in recent years. They have avoided price competition with China's cheaper labor, materials, and land by focusing on raising the quality of products and services. And, after decades of development, they have accumulated enough knowledge to take over product design and development and have upgraded from OEM to original design manufacturing (ODM). They have also established their own internationally-recognized brands (above all, Giant in 1987, and Merida in 1988) and built sales networks in overseas markets. Taiwan's bicycle industry has now achieved global production and marketing (Gao et al. 2012). Besides its factories in China, in 2020 Giant opened a production plant at Gyöngyös in northern Hungary, while Merida's world design center is located in Stuttgart, Germany. In 2001, Merida bought a 49% interest in high-end US maker, Specialized Bicycles, and has since merged production.

Beginning in the late 1990s, Chinese bicycle manufacturers have been competing fiercely with the Taiwanese firms in the global OEM market and undercutting their prices. This intense price competition resulted in a loss of sales for many Taiwanese manufacturers who responded by producing higher value-added products in Taiwan (Hu and Wu (2011) identify this as Taiwan's third bicycle innovation cycle), while producing cheaper mass-market bicycles in China. In keeping with that theme and acknowledging the stiff competition in the Chinese mass market, the Taiwanese manufacturers are focused on a Three-N Policy – new materials, new functions and new uses – to highlight their R&D capabilities. "From now on, Taiwan's bicycle industry should have a clear strategy working toward differentiation. That is not sacrificing the Taiwan operations. While in China, production concentrates on low-end bicycles only," Tseng, Ting-Huang, chairman of the TBEA and president of Merida (quoted in Min Sheng Daily, Dec 8, 2004: 8). Cross-strait production has resulted in a pattern of spatial differentiation and specialization.

Taiwan has been a leading exporter of bicycles for many years, with about 90% of the bicycles produced in Taiwan exported, making this a classic export-oriented industry. But expansion of the industry in China led to a serious problem of over-supply in the global bicycle market; low-priced bicycles manufactured in China turned the market for medium and low-priced bicycles upside down. European, American and Japanese bicycle makers were put under heavy competitive pressure, while Taiwan's bicycle exports contracted steadily after 1999. In 2003 Taiwan exported only 4.6 million bicycles, compared to 7–8 million a year in the 1990s, the glory years (see Figure 18.1). The total value of bicycles and bicycle component exports in 2004 was only US$720 million, but between 2001 and 2002 it was between US$523 million and US$583 million: Taiwan's bicycle industry found itself facing the most serious crisis in its history.

This crisis also meant new opportunities. The average unit price of the bicycles that Taiwan exported rose steadily from 1997, indicating that Taiwan bicycle makers were transforming their operations to focus on high-value products such as mountain bicycles, comfort bicycles, folding bicycles, composite materials bicycles (such as carbon fiber) and electric bicycles. In particular, with environmental awareness on the rise all over the world and with the advanced nations implementing policies to reduce pollution, electric bicycles were becoming a newly important product for Taiwan's bicycle manufacturers.

At present due to the Covid-19 pandemic, there have been global stock shortages and frantic buying of bicycles, and Taiwan's bicycle suppliers have enjoyed unprecedented prosperity, working day and night to fill global orders. Following increasing product diversification and growing demand over the course of many years, e-bikes are lending new life to the cycle industry, and providing a new "blue ocean" of opportunity. According to customs statistics, after exporting fewer than 10,000 e-bikes in 2010, Taiwan's e-bike exports had risen to 83,000 units in 2015, over 130,000 units in 2016, 180,000 units in 2017, 286,000 units in 2018, and 644,000 units in 2019, before setting a new record, with 760,000 units in 2020.

Local evolution and global upgrading: role of the A-Team

Development of the A-Team

The vertically integrated cooperation system, which had kept Taiwan's bicycle industry growing, was threatened by the migration of its key bicycle assemblers to other countries. With the establishment of plants in China, the manufacture of most low- and mid-end bicycles shifted offshore. To maintain a competitive advantage, Taiwan responded by developing models featuring fine craftsmanship, creative design, and high value-added products. Increased output value more than compensated for shrinking volumes. As the industry upgraded and consolidated, companies needed to work cooperatively, but competitively (Peng et al. 2018). For example, Hsieh (2015) discovered that many breakthroughs in Taiwan's cycle industry were accomplished at the intermediate level by parts makers in a system of decentralized production rather than by large lead firms driving the system. Technological interdependence through cross-industry learning and knowledge exchange became important institutionalized mechanisms.

The A-Team was founded in 2003 by Giant and Merida, along with a consortium of other bicycle parts producers (Chapter 15). Members of the A-Team shared a common vision of the future development for the Taiwan bicycle industry, requiring competitive and cooperative relationships among related firms.

> High-volume, low-price production is not coming back to Taiwan; it is rightly produced in China. Nevertheless, some of us have banded together to create a group of manufacturers called the A-Team. We refuse to let everything go to China and we will share our resources to develop better designs, engineering and manufacturing

technology as well as just-in-time production to lift the respect and quality of our products to the highest level.

(Lo Hsian-An, general manager of Giant Manufacturing and chairman of the A-Team: Bicycle Retailer and Industry News, April 1, 2002, http://www.bicycleretailer.com/)

The A-Team was established due to both internal and external factors. Externally, China posed a threat to the Taiwanese bicycle industry. Internally, in adopting the Toyota Production System (TPS), Giant and Merida formed an alliance to increase their supply coordination with parts firms and reinforce their control power of industrial production networks. As Tseng Song-zhu, general manager of Merida, pointed out

> We cooperated with Giant because of concern with the future of Taiwan's bicycle industry. Furthermore, we found that the coordination of associate manufacturers is the key point for the implementation of TPS. If two or three assembly bicycle manufacturers get together, it would increase the proportion of order for production and the coordination incentive of associate manufacturers.
>
> *(Interviewee, May 25, 2005)*

Compared with the pre-A-Team phase, Taiwan's bicycle industry has transformed in two main ways. In respect of local evolution, the A-Team defined and positioned itself as a learning organization. Members of the A-Team shared management know-how in value-based manufacturing processes to raise production quality standards, enable faster responses to customer demands, and better brand recognition of the A-Team's member firms. The efforts led to dramatic improvement in the daily operations of team members, increasing internal production efficiency, while greatly reducing lead-time and production costs. Such results convinced other component vendors to join the A-Team.

With respect of global upgrading and implementation of the Toyota Production System, the 1-1-10 rule was put into effect. This rule sets the lead-time delivery target (*just-in-time*) at 10 days for all members, with one delivery per day. The adoption of *total-quality-control* contributed to significant improvement in cost control and product quality. Inspired by its success, other domestic firms adopted similar measures to raise their product quality. By cooperating in the development process, the A-Team introduced innovative new products that increased overall production value by shifting production dynamics from the traditional push strategy to a pull strategy based on customer orders (Liu and Brookfield 2007). Exports of finished bicycles from Giant and Merida rose 30% in 2004 and 44% in 2005. In addition, the export value of A-Team parts suppliers grew by 27% in 2005 (Chen 2005: 114). Started as an alliance for Taiwanese companies, foreign firms, including Specialized, Trek and Scott Europe, later joined as 'sponsor' members, underlining the importance of this alliance.

As a whole, the A-Team represented good quality for a team of global bicycle manufacturers. By reducing stocks, raising quality, and cutting down the lead-time, the whole production network was upgraded with TPS, and reinforced the flexibility of production. The main purpose of the A-Team, in which it succeeded, was to develop a superior product and differentiate between 'made in Taiwan' and 'made in China.'

The A-Team connected firms with a new management system stressing proximity. Two types of proximity called *organizational proximity* and *geographical proximity* were important (Rallet & Torre 2000; Boschma 2005). Organizationally, members of the A-Team shared the same set of beliefs and the same knowledge, a social relation that is mainly tacit and close, fostering trust and long-term cooperation. Geographically, the firms were all tightly embedded in Taiwan's Central Region, advancing the evolution of the local production network.

Recent developments

The A-Team alliance was wound up at the end of 2016, although the model for cooperation has continued under the supervision of the Taiwan Bicycle Association (van Schaik, 2016). It comprehensively fulfilled its goals to improve industry productivity and move production up-market. In 2016, Taiwan was ranked as the world's largest exporter of high-end bicycles and the second largest of overall bicycle products (after China). By 2020, Taiwan was manufacturing 63% of the e-bikes sold in the European market (https://www.taiwannews.com.tw/en/news/3977049 accessed 4/9/21), and "nearly 70% of the global mid-to-high-end bicycle demand" (Ou 2020). It was the major supplier of high-end mountain bikes to the US, and road bikes for competitions in Europe. In June 2020, the German firm Bosch (maker of the Bosch eBike System) moved its Asian headquarters to Taichung, to gain closed proximity to Taiwan's e-bike industry. Niddrie (2016) notes some of the hi-tech accessory manufacturers have also been attracted to Taiwan, including cycle computers, vehicle racks, electronic gear-changers, and shock-absorbers.

The wisdom of the A-Team strategy is illustrated by the unexpected impact of the Coronavirus pandemic on cycling globally. Restrictions on local and international movement have led many adults to take up the epidemically safe recreation of cycling as an outdoor activity. In North America and Europe, many citizens belonging to older cohorts have discovered that electrically-assisted cycles (both bicycles and tricycles) allow them to enjoy a ride without excessive exertion. Wishing to avoid public transport, numerous workers and students have reverted to cycling. With limits set to attendance at many mass sporting events, the bicycle presents the possibility of outdoor exercise, be it on mountain trails, road-to-rail trails or country roads. There is debate about how permanent such recent developments will be, but when the epidemic effects are compounded with a growing concern for the environmental impacts of public and private motorized transport, the renaissance of the bicycle shows every sign of being long-term in nature.

Conclusions and policy implications

In the current era of globalization, the changing nature of industrial regions due to the increased mobility of capital and labor has been a topic of extensive theoretical inquiry. Competition generated by globalization has generated a large amount of new and integrated economic activity in geographical districts to form new bicycle production spaces, particularly in the Bohai Economic Rim, the Yangze Delta, and Guangdong (Gao et al. 2012). Multinational brand-owning corporations outsource manufacturing in order to capture product innovations that then become a key strategy in their competitiveness.

The evolution of Taiwan's bicycle industry from low unit-price, OEM production to technologically-advanced ODM production producing high-quality bicycles and e-bikes for their own internationally recognized brands and for client brands, working cooperatively within supply chains, and located in efficient manufacturing clusters, has been a winning model. Taiwan's bicycle industry has morphed into a dual-track industrial system with low- and middle-end mass production activities located mainly in China, and high-end activity retained in Taiwan and, to some extent, in Europe.

In sum, both industrial policy, technological innovation and cooperative institutional structures have been important to the renaissance of Taiwan's bicycle industry. By promoting high-quality standards for the entire line of products, the A-Team placed emphasis on establishing product-development platforms, acquiring the latest technology, and strengthening R&D capabilities. In the years since the A-Team was wound up, this emphasis has been maintained and with the fortuitous widespread adopting of e-bikes has created a new product niche for Taiwan's bicycle industry.

Bibliography

Andreae, M., Hsu, J.Y., & Norcliffe, G. (2013) "Performing the trade show: the case of the Taipei International Cycle Show." *Geoforum* 9(7), 193–201.

Boschma, R. (2005) "Proximity and innovation: a critical assessment." *Regional Studies* 39(1), 61–74.

Borrus, M., & Zysman, J. (1997) "Globalization with borders: the rise of wintelism as the future of global competition." *Industry and Innovation* 4(2), 144–161. Chan (Trade Wind's Industry Weekly, Jul 27, 1998:1).

Chen, M.C. (2002) Industrial District and Social Capital in Taiwan's Economic Development: An Economic Sociological Study on Taiwan's bicycle Industry. PhD dissertation of Yale University.

Chen, M.C. (2005) "A-Team: Taiwan zixingche de chaoren tegongdui (The A-Team of Taiwan's bicycle industry)." *Cheers* 56, 108–114.

Chu, W.W. (1997) "Causes of growth: a study of Taiwan's bicycle industry." *Cambridge Journal of Economics* 21, 55–72.

Chu, W.W. (2001) "The development pattern of Taiwan's bicycle industry." *International Conference on Global Production: Global Production and Trade in East Asia*, pp. 295–304.

Ernst, D. (1995) Mobilizing the region's capabilities? The East Asia production networks of Japanese electronics firms, in E. Doherty (ed.) *Japan's Investment in Asia: international production strategies in a rapidly changing world*, pp. 25–49, Berkeley: BRIE.

Gao, B.-Y., Liu, W.-D., & Norcliffe, G. (2012) "Hypermobility and the governance of global production networks: the case of the Canadian cycle industry and its links with China and Taiwan." *The Canadian Geographer* 56(4), 439–458.

Hsieh, M.F. (2015) "Learning by manufacturing parts: explaining technological change in Taiwan's decentralized industrialization." *East Asian Science Technology and Society – An International Journal* 9(4), 331–358.

Hu, M.C., & Wu, C.Y. (2011) "Exploring technological innovation trajectories through latecomers: evidence from Taiwan's bicycle industry." *Technology Analysis & Strategic Management* 23(4), 433–452.

Jian, B.Y. (1997) Taiwan zixingche chanye quanqiu jiazhilian fenxi (Global Value Chain Analysis of Bicycle Industry in Taiwan), Master's dissertation, Taiwan University.

Liu, R.-J., & Brookfield, J. (2007) "Taiwan's A-Team: Integrated Supplier Networks and Innovation in Taiwan Bicycle Industry." http://lean.thu.edu.tw/wp-content/uploads/2018/11/Taiwans-A-Team-2007-AOM-Annual-Meeting.pdf (accessed 4/9/21).

Niddrie, D. (2016 May 24) Taiwan: The Cycling Island. Retrieved from https://momentummag.com/taiwan-cycling-island/

Ou, S. (2020) "Taiwanese Made Bicycles Steadily Grows for Global Road Bikes Demand." https://www.businesswire.com/news/home/20201123006314/en/Taiwanese-Made-Bicycles-Steadily-Grows-for-Global-Road-Bikes-Demand

Peng, T.J.A., Yen, M.H., & Bourne, M. (2018) "How rival partners compete based on cooperation?" *Long Range Planning* 51(2), 351–383.

Rallet, A., & Torre, A. (2000) "Is geographical proximity necessary in the innovation networks in the era of global economy?" *GeoJournal* 49, 373–380.

van Schaik, J.-W. (2016) "Taiwan's A-Team to end but cooperation continues." *Bike Europe*, 27 October: https://www.bike-eu.com/market/nieuws/2016/10/taiwans-a-team-to-end-but-cooperation-continues-10127959

Xie, Q.D. (2000) Cong Taiwan chanye waiyi lun guo jia jue se (The Role of State on Outward Industry Development: The Case of Bicycle in Taiwan), Master's dissertation, Donghai University.

19

STREET TRADES AND WORK CYCLES

Claudio Sarmiento-Casas

Introduction

The study and promotion of urban cycling has focused on making the bicycle a feasible mode of urban commuting, but this is changing as advocates acknowledge the need to widen their scope to include the wide variety of cycling practices in cities. Inclusive cycling must therefore consider not just the diversity of cycle users, but also how cycles are used as work tools and how they can be indispensable to the livelihoods of diverse groups of people across the globe. Those who depend on their cycles for work are often involved in marginalized or informal work, and their needs go beyond the usual calls for infrastructure provision, program implementation, and technological developments. This chapter examines these needs by characterizing the experiences of work cycles, and briefly speculating about their potential to further cycling-based economies.

By and large, cycling advocacy has considered the two-wheel bicycle as its design vehicle, obviating consideration of tricycles, cargo trailers, hand cycles, and other velomobiles. Rather than variations on commuter cycles, this technological assortment points to the multiple uses of cycling machines. Along with transporting its rider, cycles can provide loading, storage, distribution, advertisement, and power-generation capacities that facilitate work. They are *work cycles* both because of their association with working-class users (Oldenziel & de la Bruhèze 2011) and because they are intrinsically linked to the work of specific street trades (also called *working* cycles, but avoided here to avoid confusion with *functioning* cycles).

Work cycles are extensively used in cities of the so-called Global South – Africa, Asia, and Latin America – where formal policies that promote their use are limited or absent (Norcliffe 2011). Yet recent studies on work cycles have almost exclusively taken place in cities where mass motorizations led to workers shifting to automobiles, motorcycles, and trucks. Thus, work cycles have been principally portrayed as sustainable alternatives to established motor-based systems such as cycle logistics (Rudolph & Gruber 2017) or studied in contrast to conventional motorized vehicles – as with the case of cycling couriers (Kidder 2011). This chapter adds to the literature by focusing more narrowly on *informal* and *captive* work cycles, and particularly on the ways that street traders depend on their cycles to sell goods, deliver products, provide services, and transport passengers.

DOI: 10.4324/9781003142041-25

I begin this chapter with a survey of the most salient academic literature on urban cycling, particularly that which studies its economic benefits. I then couple the available academic studies on work cycles with my own research on Mexico City, particularly the street trades of its urban core. I draw on two years of participant observations to describe the technological appropriations of Mexico City's street trades and the most prevalent purposes of their work cycles and, where appropriate, I refer to practitioner and popular literature to illustrate how they are culturally acknowledged while being simultaneously marginalized from mainstream cycling advocacy.

Cycling economies

An overview of the prevailing cycling literature indicates that studies on cycling have evolved from a consideration of the bicycle as an alternative to the automobile, to a symbol of urban sustainability, to a means toward healthy lifestyles, to a global socio-cultural and distinctively urban phenomenon. Urban cycling is largely seen as an uncomplicated, affordable, healthy, zero-emission, and energy-efficient form of transportation. While most of its importance is framed around concerns for urban sustainability, the viability of cycling practices has been couched in economic terms, among them cost–benefit evaluations, street design investments, and global competitiveness (Rabl & De Nazelle 2012). They also tend to privilege quantitative and statistical analysis, as well as taking most of their cues from knowledge produced in Europe and North America (Castañeda 2021). The sustainable attributes of urban cycling are especially evident when compared to the "business-as-usual" scenarios of extensive automobile use, and only when controlling for all demographic, social, cultural and environmental factors. Indeed, sustainability is not always the main driving force behind people's decision to cycle, and this may partly explain why there has not been a more widespread adoption of cycling despite targeted efforts to promote a shift towards walking and cycling.

Cycling research and policies are generally motivated by bolstering the benefits of prospective cycling adoption. Not only are cycles marketed as "inexpensive, common, and efficient" (Weersing 2015: 2, in Kirkpatrick 2018), but they are considered less-polluting, energy-saving, quiet, and health-promoting vehicles. Most studies highlight how cycling's indirect costs are lower than those of other modes in the aggregate (Gössling 2016). For instance, one common way to encourage a shift toward more sustainable modes of transportation is to translate the environmental and health benefits of cycling into reduced societal costs (e.g. Krizek et al. 2007; Chapter 6).

Several less studied economic benefits can be derived from investment in infrastructure related to cycling and walking, such as increases in real estate values and local consumption. Cycling is also considered as a way to save on societal costs and the individual expenses of using a car, while cycling infrastructure is also seen as an attractor for outside investment, such as tourism or in business districts (Flusche 2012). Scholars have thus focused on valuing the impact of street improvements which link the increase of pedestrian and cycling flows with higher place-added values. The savings incurred by cyclists also assume that they are able to spend more money on local businesses, at a higher frequency than automobile patrons (Blue 2014). The positive economic impact of cycling can be further asserted by including the spill-over job creation of the design, construction, and materials involved in the bicycle and road infrastructure supply chain (Garrett-Peltier 2011).

While the global appeal of cycling and pedestrian facilities have indeed proven to attract further investment in other aspects of urban development, they have not necessarily led to equitable economic development (Stehlin 2019). Cities require a more widespread adoption of cycling for it to become economically attractive for cities to invest in (Vivanco 2013).

But despite evidence of small increases in cycling trips, the practice is still found to be unevenly distributed among gender, class, ethnic, and age groups. It follows that cycling investments often take place where cycling is readily adopted by elite groups, marginalizing other prospective users that have cycling-related needs and desires that go beyond sustainable mobility. Some argue that a better understanding of these marginalized cycling cultures is key to further promote cycling in cities (e.g. Grafl et al. 2018).

Indeed, cycling's role in economic progress is not limited to its role as a sustainable form of mobility. Zack Furness (2010) describes how the promotion of cycling was at the core of development aid programs aimed at both urban and rural communities of the so-called Third World. While these programs were criticized for perpetuating colonial approaches to economic and community development, they revealed the capabilities of cycling machines for more than transportation. The clearest example of this can be found in *La Bicicleta y los Triciclos: Alternativas de Transporte para America Latina* (Navarro et al. 1985), which details how cycling machines can be involved in energy production, delivery chains, local manufacturing, and pacification strategies. This guide is not a speculation on the functional possibilities of cycling machines, but an observation on how a variety of workers and cooperatives employ them in cities across Asia and Latin America. Many of these cycling practices persist today, especially among street trades that rely on them as part of their livelihoods. As marginalized groups that use cycles as an integral part of their work and continue to do so despite cycling policies that ignore them, street trades epitomize the type of work cycles that warrant further examination. Compared to conventional cycle commuting, street trades use a wider array of type and makes of cycles, including a miscellany of customizations that further expand the potentials of urban cycling.

Work cycle technologies

Cycling machines are technological artifacts that amplify human energy and work with a high degree of efficiency. Limited by the amount of effort that the human body contributes, cycles are able to perform a variety of tasks other than transporting its rider. In fact, cycles have been adapted to providing necessities related to movement from their onset. Additional load or baggage carrying, for example, was recognized as an important problem to tackle as early as the reported adaptations made to the *draisine* (Cox & Rzewnicki 2015). And as the touring bicycle was literally and figuratively handed down to service professionals and traders, this vehicle transitioned into a valuable tool for blue-collar workers, meaningfully changing its design (Oldenziel & de la Bruhèze 2011). Yet the development of the cargo tricycle runs parallel to that of the common bicycle, responding to the immediate needs to carry freight or passengers (Norcliffe 2016). In this sense, work cycles are not novel adaptations to transportation vehicles but highly flexible technological assets.

Indeed, the technological history of cycling can be understood as a set of intersecting pathways of sociotechnical adaptation. For the work cycles of the Global South, everyday adaptations were at the crux of the rapid assimilation of cycling technologies by the workforce (Arnold & DeWald 2011). Amidst the arrival of the bicycle in Mexico, William Beezley wrote: "The railway signaled the introduction of society to technology; the bicycle replicated the same phenomenon, but at the individual level. When buying a bicycle, the Mexican learned how to handle it, repair it, drive it, to change it" (1983, in García Vélez 2014). More than commuter bicycles, work cycles in the Global South exemplify how user-oriented, informal adaptations have opened up a range of possibilities for cycling.

Recent explorations into the sociotechnical and political dimensions of cycling have examined how a given cycle affords its user a set of potential faculties that they otherwise would not have access to (see Cox 2019). Thinking about cycling's *affordances* invites us to look at the fluctuant relationships between body, objects, and the settings or situations where they meet. Perceived as a system, a "cyclist" is formed by three components: the human body, the technological artifact, and the milieus they both operate in. Hence, *work cycles* are defined by the adaptations that its users make to it (and the ways that he or she uses them) and, in turn, to the geographical context in which they both are immersed.

In my observations of work cycles in Mexico City's streets, I found a combination of store-bought cycles, ready-made customizations, and self-assembled vehicles. Local brands such as Monk, Benotto, and Mercurio offer German-designed, front-loading cargo tricycles that can be personalized according to the needs of service workers, delivery companies, and street food sellers. Cargo tricycles can be seen to sport integrated sunshades, metal heating racks, wooden storage spaces, acrylic signs, glass boxes, and custom paint jobs, so that the same base vehicle can be customized to perform widely different street trades. Water jug delivery trikes, for instance, replace their front wheels with sturdier, motorcycle-standard rims to allegedly carry up to half a ton of payload (453 kg) for short distances (compared with German 400 kg limits reported in Cox & Rzewnicki 2015). In contrast, street trades that need to cover more ground, and that leave their cargo trikes unattended at various points throughout the day, resist costly customizations. They rather prefer the flexibility of use foldable variations of cargo trikes that are not only more apt to navigating narrow and winding alleyways, but are also tolerant to rust build-up, chainless gears, and broken parts, which also makes them less likely to be stolen (see Aldred & Jungnickel 2013; Vignette C).

While in-store repairs are encouraged, street traders make up for highly irregular customers due to their propensity for self-repairs and the volatility of the informal marketplaces in which they operate. Hence, it is not uncommon to see work cycles seemingly smithed from the ground up. I encountered self-built side-cars and side planks appended to standard bicycles in my observations of delivery cycles, but bicycle taxis were the most enterprising of them. Since their introduction to Mexico City's Historic Center in the early 1990s, bicycle taxi cooperatives have fashioned custom seating carts, or *calandrias*, to attach to the rear of bicycles. Cycle taxis have been continuously re-engineered to meet changing vehicle technologies, as well as the demands of new regulations that, among other things, call for the replacement of motorcycle taxis with pedal-assisted electric models (Finck Carrales 2020).

Work cycle purposes in Mexico City

Cycling machines can be involved in a myriad of economic activities. Acting as pedal-powered motors that can be activated by one or more users, cycles have been used to generate electricity, pump water, and manufacture utensils (Navarro et al. 1985). In the Mexico City context, bicycles have been known to grind coffee beans (Figure 19.1), spin cotton candy, and sharpen knives, as well as power a radio station, convey political campaigns, form part of disaster relief efforts, and become centerpieces of art installations. However, their use is more closely associated with street trades that carry out the functions of vending and the transport of goods or passengers, sometimes all at once.

Passenger transport

It is not surprising to see riders ferrying friends and family members in the front box of their cargo tricycles. But with the exception of the Historic Center's *Ciclotaxis*, bicycle taxis are still considered a largely informal operation in Mexico City. Bicycle taxis are extensively used as

Figure 19.1 A group of cargo tricycles selling food snacks in a public park in Mexico City (photograph by the author).

last-mile options to mass transit and they are more likely to be used by low-income women with children and the elderly (Pérez López & Landín Álvarez 2019). They have become an indispensable transport option in several zones of Mexico City, especially "in areas of high congestion, high marginalization, high tourist attractiveness and areas of low profitability for conventional transport services" (my translation, Carreón 2017: 4).

Especially prevalent in the periphery of the city, bicycle taxis not only increase the levels of access to work in Mexico City's center core, but also constitute job opportunities themselves. Health, safety and labor rights are pressing topics with regards to bicycle and motor taxis (Finck Carrales 2020) but, in my interviews with *Los Mosquitos* bicycle taxi cooperative, they were framed as "green" jobs and a dignified form of employment for unskilled workers. Despite their usefulness, efforts to formalize this paratransit option have always been met with resistance from bus companies, local authorities, and negative public perception (Tiwari 2014).

Goods delivery

Delivery cycles have many advantages for merchandise distribution, including a low operating cost, less driver fatigue, and the environmental benefits associated with the bicycle. Similar to bicycle taxi operators, most cycle couriers tend to be young, able-bodied men that seek a formal entryway to employment. Recently, the number of couriers has increased due to the proliferation of on-demand delivery services, both because of customer convenience and their low barriers to self-employment. While some couriers work independently with a cargo cycle of their own, I found that the majority of delivery bicycle trades are employed by store depots or established store such as bakeries that own fleets of cargo cycles.

Work cycles distribute goods between public markets, warehouses, eateries, and retail stores, especially in areas of high commercial density and vehicular traffic like Mexico City's Historic Center. In fact, up until the 1980s and before factories moved out of the urban core, it was fairly common for tricycles to distribute soda drinks and produce from door-to-door (Fernández Christlieb 1991) and, while a minuscule share of them persist, their legacy has now recently befallen to app-based food delivery services like Uber Eats, Rappi, and Didi Food. Faced with the same challenges of vehicular traffic and parking restrictions, parcel services such as DHL and Estafeta have started pilot programs with electrically-assisted cargo bicycles and tricycles. While not informal per se, these cycle couriers operate in various states of labor precarity, including being mistaken for street vendors in Mexico City.

Street vending

Street vending is a highly heterogeneous practice, and the use of cycles is assumed to be one of its *modes*. In Mexico City, street vendors tend to be classified by degree of organization, type of stand, or degree of permanence in public spaces, such that work cycles are often labeled as *itinerant vendors* or *traders* (see Chen 2016). However, the use of work cycles afford street vendors practices that facilitate their work, expand their customer base, and increase their visibility (Figure 19.2). Cargo tricycles can become a vendor's factory, warehouse, storefront,

Figure 19.2 A coffee stand built on top of a (likely unmovable) cargo tricycle, outside a Bohemian public market (photograph by the author).

and delivery truck at the same time and with a single vehicle. Street vending bicycle trades sell a variety of goods, but they also provide services on the go with the tools that they carry or with the bicycle itself. Itinerant knife sharpeners, for example, are able to use the same pedalling mechanism to power their grinding discs.

Most of the vendor cycles that I observed were food-related, highlighting the importance of peripatetic street food in Mexican culture, but also denoting their propensity to be labelled as unhygienic, dirty, and ungainly (Hayden 2014). Yet, it is a highly popular and successful endeavor. On the one hand, street vending is cheap, convenient, and expeditious for working-class populations that are on the go. Work cycles are small enough to flock to circulation paths without being a major obstruction, plus being able to meet customers where *they* are. On the other hand, street vending can be very lucrative in areas with high rates of foot traffic, such as Mexico City's Historic Center or the immediate surroundings of mass transit stations. I was surprised to know that, for some of the street trades that I interviewed, vending is their only occupation and even a full family enterprise that involved a division of labor of preparing, cooking, packaging, transporting, and selling their goods. Notwithstanding their celebration as innovative "bike businesses," work cycles selling goods and trinkets of various kinds occupy spaces of intermittent legitimacy, being stigmatized as informal while socially accepted.

Work cycle imaginaries

As socio-technical objects, cycling machines are also able to produce meaning and fix them around urban subcultures (Cox 2019; Furness 2010). While cycles gain meaning within the cultural contexts that they inhabit, they are also shaped by the spaces, technologies, and bodies involved in the practice of cycling. In his study of bike messengers, Jeffrey Kidder (2011) observed how these practices become affective rituals that help define the collective identities of these "alley-cats," simultaneously imbuing their bicycles with symbolic value. In other words, cycles provide street trades with much more than material *affordances*, helping create relational landscapes of cultural importance.

Cycle-based street trades have an important place in the cultural imaginaries of Mexican society. These itinerant merchants have been a staple of Mexican streets since pre-Hispanic times, and have always been traditionally linked to mannerisms specific to their trade. Since the advent of the bicycle, some of these popular trades have come to be fundamentally defined by their use of cycles, to the point where they are affectionately called *biciofìcios*, or bicycle trades. The image of Mexico's bicycle trades is now widely reproduced as a symbol of informal urbanity. In *Los mexicanos se pintan solos* (Cortés Tamayo & Beltrán 1986), reprinted and gradually updated since 1865, folkloric street traders are depicted through cartoons and humorful descriptions; twenty of them appear to sport some kind of bicycle or tricycle.

Work cycles hold different individual and collective meanings (Vivanco 2013) and can be simultaneously viewed as both a necessity and as a symbol of privilege (cf. Biehl et al. 2018). *Tamale* vendors, for instance, cannot be dissociated from the front-loading cargo tricycle that they use to hold their heavy pots, as much as they are immediately identified from blocks away from the sounds of the same jingle recording that they all use: "!*Ya llegaron sus ricos y deliciosos tamales oaxaqueños! !Acérquese y pida sus ricos tamales oaxaqueños!*" ("Rich and delicious Oaxacan tamales have arrived – come and order your delicious Oaxacan tamales"). Yet, there have never been public programs supporting their work; there are no publications detailing their work, their origins, and their importance to Mexican culture; and, most poignantly, there are scant efforts from the cycling advocacy community to promote their cycling-based economies (e.g., compare with the promotion of Dutch and Danish cargo cycles).

Instead of engaging with their users, the cultural capital of work cycles has been co-opted by global elites as ready-made symbols of contemporary urban life. Much like the cycle-chic hipster cultures of the single-speed bicycle, work cycles are increasingly becoming cultural signifiers that communicate a range of identities, from innovative retro-technologies to 'green' economies (Hendlin 2016). In this manner, work cycles in Mexico City also dress up the display windows and entrances of department stores. Here, street trades and their labor are removed from the work cycling equation, leaving behind a vacuous potential for an inclusive cycling-based economy.

Conclusion

This chapter has examined the everyday practices of people that rely on work cycles for their livelihoods. Using the case study of Mexico City as a departure point of discussion, I find that work cycles can both be a platform for informal street trades and a vehicle that increases access to – and the potentials for – a multitude of economic opportunities. Based on their current use, design and adaptations, work cycles have proven to be extremely versatile in fulfilling more than the needs of the conventional commuter, extending possibilities of transport, delivery, vending, servicing, and advertising, among other functions. Appreciating these possibilities reveal important gaps in the urban cycling literature that should be addressed when asserting the social, economic, and labor benefits that both practitioners and scholars presume. As a first approximation, the study of work cycles can gloss over the nuances of technological expediency, the relationship of users with this technology, and the impact that they have in both global and local economies. However, the multiple affordances of work cycles should be more than subjects of inquiry; they constitute a grounded approach to promote a more equitable urban development.

This study of work cycles in Mexico City not only provides insights into the plethora of ways that people have reshaped cycling technologies according to different needs and in diverse contexts, but answers calls for more diverse and more inclusive urban cycling policies. Marginalized groups who currently use bicycles and tricycles because of their limited access to motorized options could benefit from urban cycling policies that take into account their grievances, which go beyond mobility needs. And although there have been recent prospective forays around cycling-based micromobility, food delivery, freight logistics, and disaster relief that overlap with the affordances of work cycles (Grafl et al. 2018; Kirkpatrick 2018; Rudolph & Gruber 2017), it is necessary to locate them within issues of global social justice, rather than amidst discussions of technical challenges or shifts toward more sustainable lifestyles. The recognition of unconventional cycling not only emancipates captive users, but also ensures a wider adoption of cycling that bridges the North–South divide of knowledge, practice, and advocacy.

Researching work cycles necessitates a broadening of the scope of urban cycling research outside of the elite cycling studies that are primarily undertaken in Europe, North America, and Australia by Western scholars. And while recent research does take into consideration specific topics such as the loading capacities and latent market opportunities for work cycles, less has been studied about the knowledge of the street trades that employ them in cities of the Global South. Further research needs to be undertaken to account for the scope of the presence of work cycles in cities outside of the Global North, as well as their qualitative and quantitative connections with social inclusivity and economic development. Just as studies in informality tend to generalize the complex interplays between state and non-state actors, research on work cycles should observe how street trades straddle the line between their "informal" risks and lack of social protections, and their "formal" relationships with established businesses, local authorities, and global production chains. In this sense, we must also be cognizant of a visibility bias that acknowledges work cycles that are out in the open, but which conceal deeper social inequalities.

References

Aldred, R., & Jungnickel, K. (2013). Matter in or out of place? Bicycle parking strategies and their effects on people, practices and places. *Social & Cultural Geography, 14*(6), 604–624.

Arnold, D., & DeWald, E. (2011). Cycles of empowerment? The bicycle and everyday technology in colonial India and Vietnam. *Comparative Studies in Society and History, 53*(4), 971–996.

Biehl, A., Chen, Y., Sanabria-Véaz, K., Uttal, D., & Stathopoulos, A. (2018). Where does active travel fit within local community narratives of mobility space and place? *Transportation Research Part A: Policy and Practice*, 123, 269–287.

Blue, E. (2014). *Bikenomics: How bicycling can save the economy* (Portland, OR: Microcosm Publishing).

Carreón, A. (2017). *Ciclotaxis: La segunda revolución de la movilidad no motorizada en Ciudad de México.* Bicitekas, México. http://bicitekas.org/wp/wp-content/uploads/2017/08/BICITAXIS-BCTKS-V3.pdf

Castañeda, P. (2021). Cycling case closed? A situated response to Samuel Nello-Deakin's "Environmental determinants of cycling: Not seeing the forest for the trees?" *Journal of Transport Geography, 90*, 102947.

Chen, M. A. (2016). Technology, informal workers and cities: Insights from Ahmedabad (India), Durban (South Africa) and Lima (Peru). *Environment & Urbanization, 28*(2), 405–422.

Cortés Tamayo, R., & Beltrán, A. (1986). *Los mexicanos se pintan solos* (Segunda edición). Sociedad Cooperativa Publicaciones Mexicanas, S.C.L.

Cox, P. (2019). *Cycling: A Sociology of Vélomobility* (London: Routledge).

Cox, P., & Rzewnicki, R. (2015). Cargo bikes: Distributing consumer goods. In P. Cox (Ed.), *Cycling Cultures* (pp. 130–151). (Chester, UK: University of Chester Press).

Fernández Christlieb, F. (1991). *Las modernas ruedas de la destrucción: El automóvil en la Ciudad de México* (Mexico: Ediciones el Caballito).

Finck Carrales, J. C. (2020). *Towards Sustainable Local Transportation in the Periphery of Mexico City: The Case of Neighborhood Motorcycle Cab Service.* Roskilde University. https://forskning.ruc.dk/en/publications/towards-sustainable-local-transportation-in-the-periphery-of-mexi

Flusche, D. (2012). *Bicycling Means Business: The Economic Benefits of Bicycle Infrastructure.* (Semantic Scholar) https://www.semanticscholar.org/paper/Bicycling-Means-Business%3A-The-Economic-Benefits-of-Flusche/8ccd3bdf2a1036ad608ea0d3e9a52cbef44618d2

Furness, Z. (2010). *One Less Car: Bicycling and the Politics of Automobility* (Philadelphia, PA: Temple University Press).

García Vélez, A. (2014). Pedaleando en el siglo XIX. *BiCentenario: El Ayer y Hoy de México*, 7(26), 22–27.

Garrett-Peltier, H. (2011). *Pedestrian and Bicycle Infrastructure: A National Study of Employment Impacts* (Amherst, MA: Political Economy Research Institute).

Gössling, S. (2016). Urban transport justice. *Journal of Transport Geography, 54*, 1–9.

Grafl, K., Bunte, H., Dziekan, K., Haubold, H., & Neun, M. (Eds.). (2018). *Framing the Third Cycling Century: Bridging the Gap between Research and Practice* (Dessau-Rosslau: German Environment Agency, European Cyclists' Federation). https://www.umweltbundesamt.de/publikationen/framing-the-third-cycling-century

Hayden, T. B. (2014). The taste of precarity: Language, legitimacy, and legality among Mexican street food vendors. In R. C. V. de Cardoso, M. Companion, & S. R. Marras (Eds.), *Street Food: Culture, Economy, Health and Governance* (pp. 83–97). (London: Routledge).

Hendlin, Y. H. (2016). Bicycling and the politics of recognition. In J. M. Meyer & J. Kersten (Eds.), *The Greening of Everyday Life: Challenging Practices, Imagining Possibilities* (pp. 230–246). (Oxford: Oxford University Press).

Kidder, J. L. (2011). *Urban Flow: Bike Messengers and the City* (Ithaca, NY: ILR Press). http://go.utlib.ca/cat/8848418

Kirkpatrick, S. J. B. (2018). Pedaling disaster: Citizen bicyclists in disaster response—Innovative solution or unnecessary effort? *Natural Hazards, 90*(1), 365–389.

Krizek, K. J., Poindexter, G., Barnes, G., & Mogush, P. (2007). Analysing the benefits and costs of bicycle facilities via online guidelines. *Planning, Practice & Research, 22*(2), 197–213.

Navarro, R. A., Beck, V., & Heierli, U. (1985). *Alternativas de Transporte en America Latina: La bicicleta y los triciclos*. SKAT, Centro Suizo de Tecnología Apropiada.
Norcliffe, G. (2011). Neoliberal mobility and its discontents: Working tricycles in China's cities. *City, Culture and Society, 2*(4), 235–242. https://doi.org/10.1016/j.ccs.2011.11.006
Norcliffe, G. (2016). *Critical Geographies of Cycling: History, Political Economy and Culture* (1st edition). (London: Routledge).
Oldenziel, R., & de la Bruhèze, A. A. (2011). Contested spaces: Bicycle lanes in urban Europe, 1900–1995. *Transfers, 1*(2), 29–49.
Pérez López, R., & Landín Álvarez, J. M. (2019). Movilidad cotidiana, intermodalidad y uso de la bicicleta en dos áreas periféricas de la Zona Metropolitana del Valle de México. *Cybergeo: European Journal of Geography*. https://journals.openedition.org/cybergeo/33554
Rabl, A., & De Nazelle, A. (2012). Benefits of shift from car to active transport. *Transport Policy, 19*(1), 121–131.
Rudolph, C., & Gruber, J. (2017). Cargo cycles in commercial transport: Potentials, constraints, and recommendations. *Research in Transportation Business & Management, 24*, 26–36.
Stehlin, J. G. (2019). *Cyclescapes of the unequal city: Bicycle infrastructure and uneven development* (Minneapolis, MN: University of Minnesota Press).
Tiwari, G. (2014). The role of cycle rickshaws in urban transport: Today and tomorrow. *Transfers, 4*(1), 83–96. https://doi.org/10.3167/TRANS.2014.040107
Vivanco, L. A. (2013). *Reconsidering the Bicycle: An Anthropological Perspective on a New (Old) Thing* (London: Routledge).

Vignette C
MOBILE CYCLE REPAIRING IN BEIJING

Tian Yun

Until Mao's death in 1976, and for more than a decade afterwards, China was known to many as the *kingdom of bicycles*. Under Mao the Chinese Communist Party had identified "four musts" that every family was expected to acquire: a radio, a sewing machine, a watch and a bicycle. China mass-manufactured sturdy bicycles, and its *bicycle army* rode them multiple times per day, filling the streets with bicycles. These cycles encountered the same problems that all cycles do, with flat tires, loose saddles, punctures, worn bearings, seized chains, and flapping mudguards. To deal with these problems, neighbourhood mobile bicycle repair stands appeared in large numbers. As was the case at that time in many other developing countries (Chapter 19), these mobile bicycle stalls were located on most busy streets, and close to major destinations such as universities. Repair stands were licenced, but otherwise belonged to the informal sector run by small-scale entrepreneurs, forming part of China's economy based on working tricycles (Norcliffe 2011).

As China prospered in the post-Maoist period, automobility began to take hold, setting the stage seen in many other developing countries with growing competition for road space between motorised vehicles and pedal vehicles. Cycling declined, partly because of the growing number of cars, partly because of massive investment in good public transit, partly because of the adoption between 1995 and 2002 of bicycle-reduction policies designed to promote the car industry, and partly because Chinese cities were spreading into suburban regions with commuting distances becoming too far to cycle. The owners of many cycle repair stands were squeezed out of business, but a number survive. With the revival of cycling due to the need to reduce pollution and severe traffic congestion, and with the rapidly rising adoption of e-bikes, they continue to provide services to cyclists, although city officials may still make life difficult and the work remains precarious (Chapter 3).

Li Wenyuan's bicycle repair stall is situated in Beijing's Chaoyang District, north-east of the City centre and close to many foreign embassies. He is located on a short boulevard between a hospital and a college which generate considerable cycle traffic, although less than 20 years ago (Figure C.1). Like most Chinese bicycle stall owners, in addition to repairing bicycles Li provides several other services. He makes matching keys, is a locksmith (opening seized locks or locks with lost keys), repairs shoes and umbrellas and sharpens kitchen knives and scissors. Li has divided his stall into two parts, one part of which is his tricycle with a signboard which

DOI: 10.4324/9781003142041-26

Figure C.1 Li's bicycle repair stall in Beijing's Chaoyang District, North-East of the City centre.
Source: author's collection.

he pedals to a conspicuous place during business hours. The other part of the stall is semi-fixed in a shady corner of the alley so as to protect car batteries from being overheated by the sun (they provided electricity for his bicycle repair work).

Li is a skilled tradesman who qualified with a trade license in bicycle maintenance in 1995 and he has focused on cycle repairs ever since. Until 2015 he could earn about 100 yuan ($20) a day, but his business has been badly hit by bike sharing (the leading bike-share firms including Ofo, Mobike and Bluegogo maintain and repair their own bicycles). The rise of bike-sharing has also impacted his business because their repair requires specialized equipment. As a result, Li's income has been halved in recent years, making his living more precarious. He used to repair children's bikes a lot, but with the boom in bike sharing, parents are more inclined to let their children use shared bikes to learn cycling skills when they reach the legal age of cycling on the road (which is 14 years old). To increase his income, Li has created an e-bike recycling business. After purchasing scrapped e-bikes, he disassembles them and sells the parts for a marked-up price. Li is unmarried so does not have a family to help with his business.

Li's work pattern is geared to the hours when there is bicycle traffic. He is at his stall by 7:00 a.m. He does not rest on weekends or take any holidays, except for the Spring Festival (the Chinese Lunar New Year – *chūn jié*), because weekends and festivals mean more business. He is busy in the morning as students and hospital workers arrive and leave their bicycles with him to pick up later when they ride home. The middle of the day is usually quiet, but he is busy again in the early evening, and normally works until 8:00 pm. The daily passenger flow is not regular. On a good business day, Li may receive dozens of orders, and sometimes a few may borrow a pump to inflate their bicycles. He then locks up his stall for the night in a nearby cage to prevent theft.

Li's stall has become a small gathering place for retired friends and unemployed local residents who chat with him, complain about the City's squeeze on street trading including car repair stalls and street barbers, and share a smoke. If he needs to leave his stall for a while, they keep an eye on his business until he returns.

Li's business has fluctuated over the years depending upon city policies and technological change. In most Chinese cities, street traders selling food, fixing bicycles, repairing shoes, rickshaw taxis, and much more have been squeezed by city regulations which limit the spaces available for street trading, and favour formal outlets in shops and restaurants. But recently there has been recognition that monumental traffic congestion in China's major cities has to be addressed, and that electric cars present only a partial solution, at best, as they take up road space like any other vehicle. The coronavirus pandemic has made riders of public transit more cautious, with the result that cycling (including e-bikes and e-scooters) is experiencing a modest renaissance. Whether the benefits of that renaissance will flow to street repairers such as Li Wenyuan depends upon whether city regulators consider them to be a nuisance or a low-cost solution.

Source

Norcliffe, G. (2011) Neoliberal mobility and its discontents: working tricycles in China's cities. *City, Culture and Society*, 2(4), 235–242.

PART IV

Urban cycling

An introduction

Sheila Hanlon

Bicycle-friendly cities are incubators for thinking, action and design. This part of *The Companion to Cycling* addresses key issues and developments in urban cycling from physical, social and political perspectives and considers the future of cycling cities.

The chapters included explore cycling infrastructure, policy, space, shared micromobilities, risk and technology in relation to urban environments and experiences. What draws these chapters together and makes them innovative in their approach is that urban cycling is understood as a fusion of material, spatial, social, technical and cultural influences. How cyclists negotiate urban environments as individuals and part of society, and how design, policy and culture facilitate cycling are considered. Cyclists, cycling advocates and urban planners will find this research valuable.

Bicycles are good for cities. Cycling is low-cost, sustainable, equitable and space-efficient. It presents solutions to urban mobility issues, such as traffic congestion and insufficient public transport. From an environmental perspective, cycling cuts down on air, water and noise pollution, carbon emissions, and fossil fuel dependency (Pucher and Buehler 2010). Bicycles are arguably less dangerous than cars. The health benefits of cycling range from physical fitness to mental wellbeing to minimizing the spread of infectious diseases such as Covid-19. When urban cycling works, more people ride. Healthy cycling cities boast a broad population of riders, including women, children, older adults and people from diverse backgrounds; a variety of vehicles, such as standard pedal bikes, e-bikes and shared micromobilites; bicycle journeys for multiple purposes, such as commuting, leisure and cargo transport; and cycling community hubs. A vibrant cycling culture, pro-cycling policy and functional cycling infrastructure are key to liveable cities.

The question "What makes a city bicycle friendly?" is central to urban cycling research. Successful cycling cities, such as Copenhagen and Amsterdam, are well represented in urban cycling literature and often presented as models for pro-cycling planning and design (Larsen 2016). Not surprisingly, both capitals lead the *Copenhagenize Index* (Copenhagenize Design Company 2016), which uses 13 parameters to rank cities with a population over 600,000 and a cycle mode of 2% or more for their level of bicycle friendliness. Most of the cities in the top twenty are European, with Japan, Canada and Colombia also represented. Ruth Oldenziel et al.'s *Cycling Cities: The European Experience* asks why some capitals have become cycling

DOI: 10.4324/9781003142041-27

cities while others have not (Oldenziel et al. 2016). Five factors that encourage cycling uptake are identified: the nature of the urban environment; the availability of alternative modes of transport; cycling's place in traffic policy; the effect of social movements; and cycling's cultural status. Global cycling patterns vary significantly, as Rahul Goel et al.'s recent article analysing cycling behaviour in 35 cities confirms, ranging from over 28% cycling modal share for all trips in Amsterdam and Osaka to lows of 1.2% in New York and 0.3% in Cape Town (Goel et al. 2021).

Infrastructure has traditionally been the starting point for analysing and developing urban cycling. The "build it and they will come" school of thought theorizes that investing in physical infrastructure, such as bike lanes, increases cycling uptake (Cervero et al. 2013; Hull and O'Holleran 2014). Utrecht's well-developed bicycle lanes, bike parking at train stations and traffic calming measures, for example, can be connected to a high bicycle modal share. The European Cyclists' Federation's recent study of Covid-19 cycling measures demonstrated that bicycle traffic increased between 11% and 48% when pop-up infrastructure was added (Kraus and Koch 2021). Rachel Aldred et al.'s study of London's Mini-Holland Program, which introduced Dutch-style infrastructure to three boroughs in 2013–21, concluded that people living in areas with concentrated changes were 24% more likely to cycle (Aldred et al. 2019).

Activism, people-led movements and individual experiences also contribute to cycling uptake. The 2018 Berlin Mobility Act, for example, which commits to investment in cycling infrastructure, was in part inspired by a citizen-initiated petition with over 100,000 signatures (Volksentscheid-Fahrrad 2016). Bogota's Ciclovía open streets initiative, which is discussed in this volume, stands out as a leading social and activist movement with lasting impact and global influence. Community and collective initiatives such as clubs, bike kitchens, Critical Mass, social rides and cycle training programs play an important role in cycling uptake and advocacy. Individual motivation and deterrents are the final piece of the bicycle modal share puzzle. Even where infrastructure is poor or cycling culture non-existent, there are often a few people who choose to cycle for convenience, utility or pleasure rather than in response to interventions. Cyclists experience the city in a personal way through interactions with space, traffic, scenery, other cyclists, risk and even the weather, adding human stories to urban cycling.

Physical, social and political infrastructures work together to create sustainable urban cycling environments. In their 2021 study of what makes Amsterdam a cycling city through the experiences of newcomers, Samuel Nello-Deakin and Anna Nikolaeva identify a mix of "hard" material and "soft" social factors that encourage cycling (Nello-Deakin and Nikolaeva 2021). Amsterdam's success as a cycling city is attributed to a sociotechnical system co-constituted by activism, strong cycling traditions and pro-cycling policies. As a result, residents ride by default. In her analysis of LA's CicLAvia open streets cycling initiative, Adonia Lugo introduces "human infrastructure" as an ethnographic framework for understanding motivators for cycling. Lugo argues that mass rides such as *ciclovías*, use physical space and generate a culture around cycling in a way that encourages participation (Lugo 2013). Urban cycling is simultaneously physical, social and political.

Chapter summaries

The urban cycling part opens with an analysis of the principles of cycling infrastructure by John Parkin. Parkin identifies the key fundamentals of good cycle traffic planning and design, noting an increase in innovation and implementation since the turn of the millennium. He argues that to be successful cycle route networks must be comprehensive, attractive, comfortable, designed for bicycles as vehicles capable of speed and prioritized rather than subservient to motorized traffic. Among his recommendations are bike lanes separated from motor traffic, well-designed junctions, end of journey parking and accessibility for all riders. Parkin provides examples of mixed traffic, dedicated cycleways, junctions and crossings, signal control, roundabouts and parking that achieve these objectives.

Urban cycling policy is unpacked from a political perspective in Justin Spinney's theoretically advanced chapter. Spinney provides an overview of contemporary cycling policy with attention to the main drivers and resulting manifestations. He proposes that cycling policy is a biopolitical mobility fix focused on individual responsibilization and economic concerns. The place of capitalism is discussed as a motivator for urban cycling policy. Issues arising from current policy directions are then explored. A sustainable transformation of cycling policy can only be achieved by moving beyond logic dominated by economics and adding other measure of human success such as wellbeing, happiness, environmental protection and socio-economic equality.

Sergio Monterez and Paola Castaneda challenge traditional thinking in their innovative chapter on making space for cycling. Drawing on Lefevre's conceptualization of space as a social product and construction, a bold framework is proposed that introduces abstract thinking to cycling spaces. Latin American urban experiences and Bogata's Ciclovía are used to demonstrate how temporary experiments, policy spaces and activism contribute to a cultural shift in how cycling is perceived and experienced. They argue that producing substantive and lasting change in urban mobility needs more than physical infrastructure and encourage us to think broadly about the meaning of 'space' for cycling. It isn't, as they indicate, just about bike lanes and parking.

Shared micromobility programs are a rapidly developing aspect of modern urban mobility. Susan Shaheen, Adam Cohen and Jacquelyn Broader investigate how these programs have developed and evolved around the world, adding particular insight from the United States. They trace how shared micromobilities – short-term, on-demand access to bikes and scooters – provide flexible alternatives, first- and last-mile connections, and access to transportation for those who do not own private vehicles. These services provide personal, community, economic and environmental benefits. The growth of shared micromobilities, user demographics and impacts, policy and operations, and the future of these initiatives are discussed.

Safety and risk are concerns for cyclists, non-riders for whom danger is a deterrent and researchers or city planners who seek to understand barriers to cycling uptake. Léa Ravensbergen and Ahmed El-Geneidy discuss cycling safety as a priority for cycle planning, research and advocacy. Expanding on definitions of the dangers cycling and traffic as either objective or perceived risk to the individual or community, a mobility justice approach is proposed. Intersecting power relations related to automobility, patriarchy and racism are considered. This approach to cycling safety as mobility justice expands cycling risk beyond accidents and road hazards to include the dynamics of sexual and police harassment.

The rise of e-bikes is an important development in urban cycling. Dimitri Marincek and Patrick Rérat provide an overview of historical developments and current issues related to e-bike use, rider motivation, demographics, regulation, place as a mode of transportation, health benefits and environmental impact. Two studies carried out in Switzerland are drawn upon to demonstrate how e-bikes have extended cycling in terms of population groups and spatial contexts, especially among women, older adults, and parents, primarily in cities but also in suburban and rural areas. They note e-bikes' power to re-engage lapsed cyclists or avoid interruptions to riding over changing life-cycle stages. The future of e-bikes is considered.

Conclusion

Cycling is an increasingly important part of urban life with advantages for both individuals and society. To reap the mobility, environmental, economic and public health benefits of cycling, cities need to invest in cycling infrastructure, adopt pro-cycling policies and support social cycling networks. Individual motivations for cycling, community action, infrastructure, politics and planning work together to create the conditions necessary for a high bicycle modal share. Research in this area is integral as more and more cities embrace bicycles and rise to the challenge of planning and design to support cycling as part of commuting, leisure, utility and urban life. Future cities are cycling cities.

References

Aldred, Rachel, Joseph Croft and Anna Goodman. (2019) 'Impacts of an active travel intervention with a cycling focus in a suburban context: One-year findings from an evaluation of London's in-progress mini-Hollands programme'. *Transportation Research Part A: Policy and Practice*, 123, pp. 147–169. DOI:10.1016/j.tra.2018.05.018.

Cervero, Robert, Benjamin Caldwell and Jesus Cueller. (2013) 'Bike-and-ride: Build it and they will come'. *Journal of Public Transportation*, 16(4), pp. 83–105. DOI:10.5038/2375-0901.16.4.5.

Copenhagenize Design Company. (2016) *The Criteria for the Copenhagenize Index. Copenhagen*. Available at: www.copenhagenizeindex.eu (Accessed 1 July 2021).

Hull, Angela and Craig O'Holleran. (2014) 'Bicycle infrastructure: Can good design encourage cycling?'. *Urban, Planning and Transport Research*, 2(1), pp. 369–406. DOI:10.1080/21650020.2014.955210.

Larsen, Jonas. (2016) 'The making of a pro-cycling city: Social practices and bicycle mobilities'. *Environment and Planning A; Economy and Space*, 49, pp. 876–92. DOI:10.1177/0308518X16682732.

Lugo, Adonia. (2013) 'CicLAvia and human infrastructure in Los Angeles: Ethnographic experiments in equitable bike planning'. *Journal of Transport Geography*, 30, pp. 202–207. DOI:10.1016/j.jtrangeo.2013.04.010.

Nello-Deakin, Samuel and Anna Nikolaeva. (2021) 'The human infrastructure of a cycling city: Amsterdam through the eyes of international newcomers'. *Urban Geography*, 42(3), pp. 289–311. DOI:10.1080/02723638.2019.1709757.

Oldenziel, Ruth, Martin Emanuel, Adri Albert de la Bruhèze and Frank Veraart. (2016) *Cycling Cities: The European Experience. Hundred Years Policy and Practice*. Foundation for the History of Technology, Eindhoven.

Pucher, John and Ralph Buehler. (2010) 'Walking and cycling for healthy cities.' *Built Environment*, 36 (4), pp. 391–414. DOI:10.2148/benv.36.4.391.

Goel, Rahul, Anna Goodman, Rachel Aldred, Ryota Nakamura, Lambed Tatah, Leandro Martin Totaro Garcia, Belen Zapata-Diomedi, Thiago Herick de Sa, Geetam Tiwari, Audrey de Nazelle, Marko Tainio, Ralph Buehler, Thomas Götschi and James Woodcock. (2021) 'Cycling behaviour in 17 countries across 6 continents: Levels of cycling, who cycles, for what purpose, and how far?', *Transport Reviews*, DOI:10.1080/01441647.2021.1915898.

Kraus, Sebastian and Nicolas Koch. (2021) 'Provisional COVID-19 infrastructure induces large, rapid increases in cycling', *PNAS*, 118(15), pp. 1–16. DOI:10.1073/pnas.2024399118.

Volksentscheid-Fahrrad. (2016) 'Ahoi Volksentscheid!', Available at: https://volksentscheid-fahrrad.de/de/2016/06/14/volksentscheid-fahrrad-stellt-mit-105-425-unterschriften-rekord-auf-2281/ (Accessed 2 August 2021).

20

CYCLING INFRASTRUCTURE

Planning cycle networks

John Parkin

Introduction

Significant developments in cycle infrastructure practice and guidance have taken place since the turn of the millennium. These developments have paralleled increased policy interest and investment in cycling infrastructure. In London, for example, the cycle network increased from 435 km to 2,179 km between 2001 and 2017 (Pucher et al. 2021). When cities were being redesigned for motor traffic from the 1960s onwards, significant land area was taken from other uses. Investment and design development went into creating inner ring roads, urban dual carriageways, large-scale signal control and roundabout junctions. These changes led to a very different layout for cities, with changed routing patterns emerging as a result of roads having their status altered, and new roads having been built.

A presumption in much design guidance and practice is that infrastructure for cycle traffic is designed in relation to what already exists, and little recognition has been given to the need for space for cycle traffic. Infrastructure designed for bicycles attempts to leave other route networks as unaffected as possible. Routes for cycle traffic are often created in the liminal space between what can be taken from motor vehicle traffic movement and land uses outside the highway without unduly affecting motor vehicle movement and parking. Space is often shared with pedestrians, but with limited utility for both walkers and cycle riders as a consequence. Only recently, especially in countries and cities with low current cycle mode share, is there a growing realization that cycle routes need to be fully developed as networks in their own right, separated from routes for motor traffic and pedestrians, and with appropriately formed junctions between these different networks. These changes in the cycle network are required as a minimum necessary condition for changes in traveller behavior.

The subservience of cycle riders to other types of user in planning and design can been seen in language. Cyclists are thought of as *vulnerable*, but they would not be vulnerable if they were riding on appropriately designed infrastructure. There is little realization that the bicycle is a *vehicle*, with reference often being made to the cyclist specifically as a person. Drivers, by contrast, are de-personalized and termed 'traffic'. Cyclists are frequently defined by what they are not, i.e. *non-motorized users*. This unhelpful definition reveals the biases in thinking of some policy makers and designers.

DOI: 10.4324/9781003142041-28

Recently, design guidance has been updated in many countries, including the United States of America (National Association of City Transportation Officials 2012), the Netherlands (CROW 2017), and the United Kingdom (Department for Transport 2020). Design guidance codifies good practice, and can also codify innovation.

The rest of this chapter discusses infrastructure in the following order. First the basis for *planning cycle networks* is discussed. This is followed by a section on *design principles*. The next two sections discuss *routes with other traffic*, and then routes solely for cycle traffic, i.e. *cycleways*. This is followed by a discussion of *junctions and crossings*, and then *cycle parking*. Finally, there is a *concluding summary*.

Planning cycle networks

A number of years ago, Bracher (1989) found that countries with national cycling policies generally have higher levels of everyday cycling. National policy is especially important in providing centralized governance and guidance for local budgets. Quality networks are a necessary, but not sufficient condition for large cycle mode share. Behavior change programs help stimulate demand by assisting people to recognise the benefits of cycling. Investment must be sustained over many years and at a sufficient level to create the infrastructure necessary to encourage travel behavior change.

Transport planning methods have developed since their inception in the 1950s. These methods were often based in econometric models, with motor traffic assumed to be the sole mode of interest. In this millennium, however, these have been supplemented with a wider range of social and psychological models of travel behavior, now also including open source models for demand and infrastructure planning (Lovelace et al. 2020). Appropriate and full planning for cycle traffic has not often been undertaken. Cycling does, however, need to be accounted for using the usual stages of transport planning, which in short form may be summarized as follows: defining scope; gathering data; network planning; prioritizing; delivering (Department for Transport 2017). Whereas provision for cycle traffic has often been seen by highway professionals as one of solving a 'safety' problem, in fact the crux of planning for cycle traffic, as is the crux of any transport planning, is the development of networks for cycle traffic that are comprehensive, and also, to use the language of the Dutch design guidance (CROW 2017), attractive and comfortable to use.

Cycle traffic is intrinsically space-efficient, especially compared with motor traffic. Creation of appropriate quality networks can be achieved by eliminating motor traffic from some areas and routes using *filtered permeability* (Melia 2008). Filtered permeability can be retrofitted to existing street networks (and may be known as 'low traffic neighborhoods', or 'liveable neighborhoods'). The approach requires planning and management of motor traffic and cycle traffic at an area-wide scale. Equally, filtered permeability can be created *ab initio* in new developments. Even with satellite navigation, direction signing remains important, and particularly so where routes for cycle traffic and motor traffic are divergent.

A core feature of transport planning is the development of an understanding of demand, which will then determine necessary features of the network, such as widths and junction characteristics and sizes. Land use development planning is also an important activity to ensure the expansion and consolidation of networks for cycle traffic (Vignette D). To have the greatest beneficial effect in the shortest timeframe, investment in infrastructure should take place where the impact on cycle use will be greatest; for example, routes connecting to important destinations. This may seem obvious, but the history of the development of networks for cycle traffic suggests that the planners' thinking has been linked more with creating routes that are convenient to create, rather than those which necessarily serve core and important everyday cycling purposes. At a higher level of granularity, the sections of networks within investment projects that need to be built first should be the ones closest to important land uses that will immediately increase cycle use.

Design principles

The core fundamental reality, which seems to have frequently been overlooked by highway and traffic engineers, is that cycles are *vehicles* that are capable of *speed*. The design speed for cycles is 30 km/h (CROW 2017), and in some circumstances may be 40 km/h. This immediately indicates that cycle traffic should never normally be mixed with pedestrians, whose mean speed is approximately 4 km/h. The design speed is the principal characteristic of relevance for design of transport routes. Once it is appreciated that there is a design speed for cycle traffic, it needs to be operationalized comprehensively in all design features that are a function of design speed; for example, the geometric features of curves, visibility splays, and tapers creating lateral movements within a route. In addition, the route for cycle traffic needs to be of an adequate width for the cycle and rider's kinematic envelope (the cross-sectional outer boundary of possible lateral movements while in motion).

Cycles are heterogeneous with many different dimensions and characteristics (Chapter 8). As well as the bicycle, there are tandems, cycles with (child) trailers, cargo bikes, tricycles, hand-cranked cycles and electric assist cycles, and other generic classes. Some cycles are adapted for use by disabled people. Many people with disabilities become enabled by a cycle but disabled again by an infrastructure that is not properly designed for cycling (Chapter 7). It is important to cater for cyclists of all abilities. Frequently, this does not occur, for example where cyclists are asked to dismount. Cycle dismount signs are an admission of failure on the part of designers (signs requiring drivers to get out and push their cars over or around an obstacle are absent from the road network). Such deficiencies are discriminatory and may be legally challengeable in many countries.

Cycle users are exposed to the environment through which they are travelling. Therefore, that environment needs, where possible, to be comfortable and attractive for users (CROW 2017). These subjective design objectives can make routes feel welcoming. These requirements are less important for car drivers and train passengers, whose experience of the route is from behind a window within a speeding vehicle.

Design for cycle traffic has often been considered only in relative terms, and assuming the dominance of motor traffic. The question the designer answers is: what needs to be provided for cycle traffic *in relation* to motor traffic? The lack of understanding of the nature of cycle traffic has resulted in some truly horrendous design 'solutions'. Some have assumed that it is feasible to mix low-speed and volume motor traffic with cycle traffic, without understanding the limits of this option. It is beginning to be generally recognised that, for the highest levels of service, motor vehicle flow should be less than 200 per hour, with a majority (85%) travelling slower than 30 km/h (Transport for London 2016). A reducing proportion of the population will tolerate cycling beyond this limit.

Routes with other traffic

An initial consideration should always be the potential for reducing motor traffic dominance. The reduction of motor traffic requires area-wide traffic management, and this will create the potential for streets to be used only by cycle traffic ('cycle streets') and pedestrians. There is the potential, particularly at destination locations such as town center squares and other points where cycle traffic is slowing at the end of the journey, to have some level of sharing of space between cycle traffic and pedestrian traffic. These are the points where cyclists transition from being part of a stream of cycle traffic to being individual humans either pedalling a cycle slowly, or coming to a stop at the end of their journey.

On routes shared by motor traffic and cycle traffic, it may be appropriate to create features that calm traffic, such as lateral or vertical deviations in the route, humps or constrictions. It is critical, however, that these do not create conflict points between motorists and cyclists. Many routes on the public highway network are not attractive or comfortable for cycle riders. With no separated provision, cyclists need, for their own safety, to adopt a prominent position within the carriageway (the so-called primary position) to discourage drivers from inappropriate overtaking. There are few people willing to use a cycle for everyday travel and adopt this sort of behaviour. There are other ways of creating a slightly more benign environment for cycle riders in carriageways with lower speeds and volumes of motor traffic. These include removing centre line markings to encourage wider overtaking, removing kerbside parking, and designating streets that are two-way for cycle traffic, but only one-way for motor traffic.

At higher volumes of motor traffic, it may still be possible to provide for cycle traffic within the carriageway, but with some level of separation. A longitudinal demarcation only with paint is not very effective, and will not create attractive and comfortable riding conditions. Intermittent physical demarcation, such as low-level shaped kerbing, or wand upstands, so-called 'light segregation' is of more value. Fully separated cycle tracks provide greater safety. They would at best be one-way, but, depending on network configuration, may be two-way.

Cycleways

Cycleways are routes away from the public highway network and are important components of the overall network in countries with high cycling volumes. They may also have adjacent footways for pedestrians. Centre lines and signage help demonstrate their function as cycleways to errant pedestrians, and also help enforce passing discipline amongst cyclists. Their surfaces should be constructed to similar dimensional tolerances as carriageways to ensure comfort. The surface should not consist of discrete units, such as paving slabs or brick paviours. Cyclists may travel close to the kerb so the kerbs need to have low profiles to avoid pedal clash. Gullies used for surface drainage should be side entry at the kerb face to avoid surface irregularities in the wheel path. Access chamber covers and other irregularities should be avoided. For everyday cycle use, cycleways need to be illuminated at night, just as a highway is. Again, as with highways, there needs to be a regime of regular inspections and maintenance.

Junctions and crossings

This section discusses priority (yield) junctions, signal-controlled junctions, roundabouts, and finally crossings where a cycle route meets a highway.

Priority junctions

There is often a disconnect between the way priority junctions are laid out (with large turning radii which permit drivers to corner with speed), and the regulations in force (which suggest drivers need to give way to people crossing the side road). Better priority junction

Figure 20.1 Priority junctions on cycleways needs appropriate signing and marking (author's photo).

designs have carriageway widths and turning radii that are the minimum appropriate in order to help calm driver speed.

Where cycle routes are on cycle tracks adjacent to the carriageway, they need to be carried across the side road with design features and/or markings to prioritize cycle traffic, ensure speed reduction for turning drivers, and provide good intervisibility between all street users in the vicinity of the junction. As shown in Figure 20.1, where two cycleways meet, priority junction markings should be provided. This is more important in urban areas where the volumes of cycle traffic are higher.

Signal control

Signal-controlled junctions solve conflicts between streams of traffic by separating movements with different time periods within a sequence of stages. Such junctions can enhance safety and reduce delay in peak times for motor traffic, but they can be inefficient in periods of low flow. The only reason for the presence of a signal-controlled junction is because of higher motor traffic flows than can be accommodated at, for example, a priority junction. Of course, if motor traffic flows were lower, perhaps as a result of more people travelling by cycle, then there would be less need for signal-controlled junctions.

The focus in the design of signal-controlled junctions has been almost exclusively on the movement of motor traffic. The traffic signal cycle time (the time taken to move through each of the stages in turn before returning to the beginning of the cycle), and the times allocated to each stage in the cycle, is based on motor traffic demand. Time allocated to pedestrian and cycle movements across the carriageway (approaching on footways and cycleways) is often minimal. This creates significant delays for pedestrian and cycle movements.

There has been increased interest recently in the way that cycle traffic can be better accommodated at signal-controlled junctions, whether travelling in the carriageway, or approaching from outside the carriageway on a cycleway. Space for cyclists to wait on a red aspect more safely and in view of drivers may be provided by two stop lines, with motor traffic waiting behind the upstream stop line, and cycle traffic waiting behind the downstream stop line. This 'advanced stop line' layout is intended to allow cycle traffic to clear the junction before the motor traffic starts to flow. However, such white paint is of no value if the cycle rider arrives at the junction during the green aspect because they are just part of the flow of general traffic. The additional comfort and attractiveness of such advanced stop lines is of limited appeal to the majority of those who would not consider cycling on higher-volume routes.

One particular issue of relevance is the risk posed to cycle traffic that is proceeding straight across at a junction where there is left-turning motor traffic in countries with left-hand rule of the road (and the reverse in countries with right-hand rule of the road). The presence of a green aspect for motor traffic encourages the driver to proceed left through the junction with immediacy, despite the fact that such progress may conflict with the path of a cycle rider going straight ahead.

A method for dealing with this left-hook problem is called 'gating'. As well as the main (downstream) stop line, there is a pair of upstream stop lines: one for cycle traffic and one for motor traffic. When the downstream signals are on red, cycle traffic has a green aspect at the upstream stop line to allow progression into the reservoir between the two stop lines. When the downstream aspect turns green, the signal aspect for motor traffic at the upstream stop line also turns green, but the cycle traffic can move safely away in advance of the motor traffic. Importantly, the aspect for the cycle traffic to enter the reservoir turns red just before the downstream aspect turns green to prevent cycle traffic entering the reservoir and mixing with motor traffic when that is moving. This enhances cycle safety, but has an impact on cycle traffic progression because a cyclist will always encounter a red signal aspect either at the upstream or the downstream stop line.

A similar approach called 'hold the left' is where the left turn traffic stream is held on a red aspect for a period at the start of a stage to allow straight ahead cycle traffic to proceed. The signal aspect for the cycle traffic then turns to red during the second part of the stage during which the left turn stream is given a green aspect. So far as cycle traffic turning right is concerned (left-hand rule of the road), the Danish solution of turning in two stages is becoming more common. The cycle traffic proceeds to the far side of the junction and waits on the nearside (kerbside of the carriageway) within the junction for the following stage, during which it can then cross with the straight-ahead traffic in that stage.

Most signal controllers operate responsively to demand flows using detector loops in the carriageway surface. For signal installations to work effectively and responsively to cycle traffic, the loop detectors need to be well positioned and tuned to detect cycles. Alternatively, overhead microwave or radar detectors may be used.

Ideally, cycle traffic can approach a signal-controlled junction on a cycle track separated from the carriageway and then be able to proceed either to the left, straight ahead, or to the right in a comfortable and safe environment without any undue delay, or indeed additional delay relative to that experienced by, for example, a driver. There are now a number of much more innovative designs for junctions being developed in various countries, especially the Netherlands, but also the UK. An example is the so-called 'cycle optimized traffic signal-controlled' (Cyclops) junction (Butler et al. 2019), as shown in Figure 20.2.

Cyclists coming from the road to the bottom right-hand side of the image are considered. They turn left via a give way marking on the green cycleway. They turn right by proceeding

Figure 20.2 Cycle optimized traffic signal-controlled junction, Trinity Street, Bolton, UK. Transport for Greater Manchester (2021) Bolton opens the UK's second pioneering CYCLOPS junction with nine more set to be delivered in 2021 (News release, Transport for Greater Manchester). Available at https://news.tfgm.com/news/bolton-opens-the-uks-second-pioneering-cyclops-junction-with-nine-more-set-to-be-delivered-in-2021, accessed on 12 May 2021.

across the stop line under signal control in the centre foreground of the image, then via the marked route across the junction, and from there across the road on the far side. The road at the far side is one-way only into the junction. If it were not, straight-ahead cyclists would go the same as the right-turning cyclists but turn left. The junction shape is a little unusual, but this is of course common for urban junctions, and each installation would have different geometry, and associated signal timings.

Roundabouts

Roundabouts provide appropriate priority-based junctions where entry flows are balanced. They generally do not have capacity as high as signal-controlled junctions, but they do not create delay in off-peak periods in the way signal control does.

To accommodate cycle traffic, roundabouts need designs that constrain the speed of motor traffic. This can be achieved with single lane entry with a tight turn for traffic onto the circulating carriageway, which should also be a single lane. The single lane will allow cycle traffic to be in a prominent position, and prevent inappropriate overtaking by drivers. Some roundabout designs, notably typical British designs, apply a principle of minimal speed reduction through the roundabout consistent with being acceptably safe for motor traffic: this type of design is risky for cycle riders.

The Dutch approach to roundabouts is to use tighter geometry that encourages slower speeds through the junction, combined with an additional circulating carriageway for cycle traffic outside the circulating carriageway for motor traffic. In an urban context, the cycle circulating carriageway typically has priority, but in a rural context the circulating cycle traffic would give way at each entry arm to motor traffic entering and exiting the roundabout on

Figure 20.3 Innovative roundabout design in Zwolle, Netherlands (author's photo).

the main carriageway. There remains considerable discussion about the additional risk versus the benefits when cycle traffic is given priority.

As with signal control, creative development is taking place of new types of layout of roundabout. A good example, as shown in Figure 20.3, is a roundabout in Zwolle (junction of Wipstrikkeralle and Philosophenalle and Vondelkade) which only allows motor traffic to proceed straight ahead on the main road (achieved by a non-continuous circulating carriageway). The circulating cycleway is continuous, however, and caters for all movements. Other creative developments include roundabouts only for cycle traffic that create iconic and attractive places.

Crossings

Cycleways frequently need to cross a highway network, sometimes with an adjacent pedestrian crossing. A key issue at any crossing is to ensure that the speed of motor traffic is controlled in order to minimize risk. Crossings may have no priority for the cycle traffic, and in such cases a give way (yield) marking is needed at the crossing point to ensure that cyclists do so. Where priority is given to the crossing cycle traffic, appropriate priority needs to be shown with either give way markings for motor traffic, or, if combined with a pedestrian crossing, the appropriate country-specific markings for a zebra crossing combined with an adjacent cycle track crossing.

Signal-controlled crossings may be preferable on higher-speed and higher-volume roads, but they should not create significant delay for the crossing cycle traffic. Instead of separating in time with signal control, it may be appropriate to separate in space with grade separation (i.e. vertical separation with bridges or underpasses). This needs to be undertaken with care to limit the gradients for cycle traffic, possibly by lifting or lowering the road to be crossed as much as, or more than, the extent of lifting or lowering of the cycle track. Care should also be taken that such grade separation does not move cycle traffic too far from the desire line.

It should also be noted that crossings need to be provided for pedestrians when they meet a cycleway. These may simply be informal, but for more significant flows and conflict points, they may take the form of a zebra crossing, or even a signal controlled crossing of the cycleway.

Cycle parking

Parking frequently needs to be provided at the journey end. In countries with high volumes of cycle use, parking for cycles at significant destinations has been provided, often by constructing multi-level racking in multi-storey buildings. Some cycle parking facilities allow cycle users to undertake minor maintenance (e.g. through provision of air pumps), or they may have repair shops on site.

Where destinations are more distributed, such as along shopping streets, a larger number of smaller-scale parking areas fitted with stands that allow the cycle to be locked securely from theft may be needed. If the parking is for more than a short period, covered protection to keep the parked cycles dry may be desirable. The location of cycle parking should not obstruct pedestrian desire lines, and carriageway car parking space can often be repurposed as cycle parking space.

A key issue is to ensure that there is space and appropriate security for all types of cycle, from bicycles, through tandems, to adapted cycles and cargo bikes. Coupled with the point that disabled people may be able to cycle more easily than they can walk, it should be the case that a cycle rider can, without needing to mount kerbs or perform tight manoeuvres, access cycle parking easily.

A major planning issue is ensuring there are appropriate regulations in place to provide for cycle parking within new dwellings, and also mechanisms for retrofitting cycle parking within areas of dense housing that lack such parking space.

Concluding summary

Since the turn of the millennium, significant developments in cycle infrastructure practice and guidance have taken place. Design guidance has been updated in many countries. An undeclared approach in much design guidance and practice, however, is that the infrastructure for cycle traffic exists 'in relation' to, and subservient to, other modes, notably motor traffic. This subservience in the perception of policy makers and designers can been seen in the language used, for example, referring to cycles as 'non-motorized users', a generic and meaningless description. These behaviors on the part of decision-makers still need to be developed if cycling and cycle infrastructure is to take its rightful place.

The intrinsic space efficiency of cycle traffic (relative to motor traffic) needs to be enabled through control of motor traffic using techniques such as area-wide traffic management to create quiet routes for cycle traffic. Few people will cycle for everyday journeys in conditions with high volumes of motor traffic, especially if it travels at higher speeds. An increasing proportion will choose to cycle when there is little motor traffic or total separation, or when that traffic is travelling slowly.

In terms of design, the core fundamental which appears little understood still, is that cycles are *vehicles* capable of *speed*. Designers need to use the design speed of cycle traffic (30 km/h) to properly design the geometry of cycle routes. Additionally, because cycle riders are exposed to the environment in which they travel, cycle routes need to be comfortable and attractive.

Cycles are heterogeneous, with many different dimensions and characteristics, ranging from bicycles, through tricycles, to cargo bikes and electric assist cycles. Infrastructure needs to be designed for all, and if it is not, it may be legally challengeable.

Junctions and crossings create issues for design because they are the parts of the network where conflicts occur. These can be separated in time and space, and where they cannot be adequately separated, motor traffic speeds need to be low. This is particularly the case at all junctions where there is priority control, including side roads and roundabouts.

In order to create a comprehensive network of routes for cycle traffic, routes are needed away from the highway. Such cycleways will intersect frequently with the highways, and special consideration needs to be given to crossings that are safe. Finally, it should not be forgotten that infrastructure is needed at the journey end to safely store cycles.

References

Bracher, T. (1989) Policy and provision for cyclists in Europe. European Commission, Directorate VII.

Butler, R., Salter, J., Stevens, D., Deegan, B. (2019) CYCLOPS – Creating Protected Junctions. *Greater Manchester Combined Authority, Transport for Greater Manchester.* http://www.jctconsultancy.co.uk/Symposium/Symposium2018/PapersForDownload/CYCLOPS%20Creating%20Protected%20Junctions%20-%20Richard%20Butler%20Jonathan%20Salter%20Dave%20Stevens%20TFGM.pdf. (accessed 30/10/2020).

CROW (2017) *Design Manual for Bicycle Traffic.* CROW, Ede, the Netherlands.

Department for Transport (2020) Cycle infrastructure design. Local Transport Note 1/20. https://www.gov.uk/government/publications/cycle-infrastructure-design-ltn-120. (accessed 4/11/2020).

Department for Transport (2017) *Local Cycling and Walking Infrastructure Plans: Technical Guidance for Local Authorities.* Department for Transport, London, UK. https://www.gov.uk/government/uploads/system/uploads/attachment_data/file/607016/cycling-walking-infrastructure-technical-guidance.pdf (accessed 4/11/2020).

Lovelace, R., Parkin, J., Cohen, T. (2020) Open access transport models: a leverage point in sustainable transport planning. *Journal of Transport Policy* 97, pp. 47–54.

Melia, S. (2008) Neighbourhoods should be made permeable for walking and cycling but not for cars. *Local Transport Today*, 23 January.

National Association of City Transportation Officials (2012) *Urban Bikeway Design Guide.* National Association of City Transportation Officials, New York, NY, USA. https://nacto.org/publication/urban-bikeway-design-guide/ (accessed 4/11/2020).

Pucher, J., Parkin, J., de Lanversin, E. (2021) Cycling in New York, London, and Paris. In Pucher, J. and Buehler, R. (Eds.) *Future of city cycling.* MIT press, Cambridge, Massachusetts.

Transport for Greater Manchester (2021) Bolton opens the UK's second pioneering CYCLOPS junction with nine more set to be delivered in 2021. News release, Transport for Greater Manchester. Available at https://news.tfgm.com/news/bolton-opens-the-uks-second-pioneering-cyclops-junction-with-nine-more-set-to-be-delivered-in-2021 (accessed 12/05/2021).

Transport for London (2016) London Cycling Design Standards. Chapter 4. Transport for London, London, UK. Available at https://content.tfl.gov.uk/lcds-chapter4-cyclelanesandtracks.pdf (accessed 30/05/2021).

Vignette D
CYCLING INFRASTRUCTURE IN LUND, SWEDEN

Till Koglin

Despite being heralded as a cycle friendly city, cycling infrastructure in Lund lacks consistency, space and directness, forcing cyclists to make detours or use streets where motorised vehicles take up most of the space and travel at speeds which pose safety risks for cyclists (Kröyer 2015). This is attributable to modernistic thinking which has influenced the planning of transport infrastructure in ways that marginalise cyclists. Infrastructure is always political (Cox 2020): unfortunately, behind much transport planning, especially in Sweden, lies a rationality that often leads to the marginalisation of cycling (Koglin 2020). Notwithstanding the city's high ambitions for sustainable transport and mobility, and despite the city's good pre-existing conditions, it has not delivered the cycling infrastructure needed to develop sustainable transportation.

Lund is a university town of 91,755 in the south of Sweden and home to Lund University, Scandinavia's largest university with ca. 40,000 students. The published modal split for trips within this city in 2018 was: walking 14%; cycling 27%; train 11%; bus 12%; car 34% and other 2% (Regional Travel Survey Skåne 2018). Not counted, however, in this modal split are the 45,000 commuters driving cars daily into Lund coming from surrounding areas. Lund has been recognized as one of the top cycling municipalities in Sweden and has won several prizes for its cycling initiatives from Cykelfrämjandet, Sweden's national cycling advocacy organisation. Furthermore, cycling is considered to be an important aspect in the development of Lund's sustainable transport (Lund Municipality 2018). However, a closer look at both the modal split and cycling infrastructure, places this notion in doubt.

The modal split often presented by the municipality does not show the whole picture. Commuting trips, either by car or by public transport, are not included. This means that the share of car trips and probably also trips on public transport are substantially higher, making the share of cycling lower. Furthermore, half the population are students who rarely own cars or even have a driving licence. Overall, in the region of Scania only 27,131 cars were owned by people attending educational institutions, and only 56% of the people aged 18–25 had a driving licence in 2019 (Trafikanalys 2020). As a result, students are more likely to use public transport, walk or ride a bicycle than a car. Lund is a smallish city with short distances to all major destinations, so bicycling is easy. One might therefore hypothesise that the share of cycling in Lund is due more to external factors, such as short distances and the large student population, than to the quality of cycling infrastructure.

DOI: 10.4324/9781003142041-29

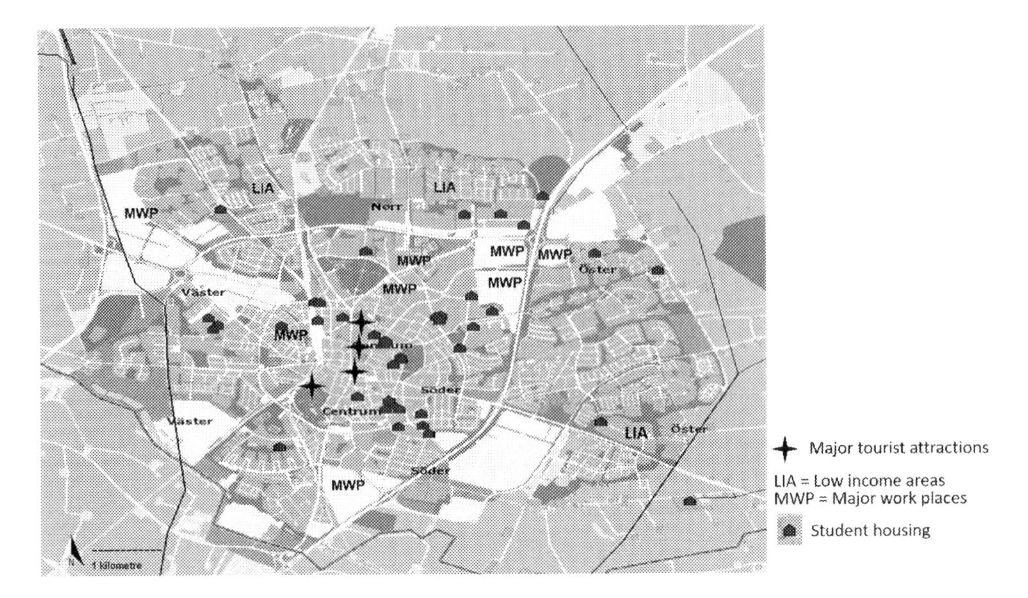

Figure D.1 Lund, Sweden.

Source: Lund Municipality, 2020.

The street layout of Lund's medieval centre contrasts with the more modern pattern in the rest of the town. Daily destinations such as food shops, schools, day care centres, etc., are located throughout the town. The university is also spread around the whole town while the major workplaces such as the city administration, hospitals and major companies such as Sony and Axis are concentrated in an axis stretching from the railway station to the north-eastern part of the town (see Figure D.1).

The town centre is more or less car-free, with access limited to buses and taxis. There is very little cycling infrastructure and cyclists share street space with buses. Elsewhere cyclists share space with pedestrians, either separated by different road materials or a painted line, or not separated at all. There is limited cycling infrastructure and sometimes, in order to follow a direct route, cyclists are forced to follow streets with motorised modes of transport that travel at higher speeds. This is not good in terms of safety since speeding vehicles pose a high risk for cyclists. Moreover, cyclists are often forced to cycle along narrow paths with two-way cycle traffic which is also shared with pedestrians. There is also little continuity in cycling routes, and limited consideration of accessibility for cyclists (Koglin & Glasare 2020). In areas adjacent to the newly built light rail system, cycling has been marginalised with space allocated to the light rail system and motorised transport instead of to cycling infrastructure (see Figure D.2).

Lund has prioritised public transport over the bicycle as the preferred mode of sustainable transport. Public transport infrastructure focuses mainly on mobility for employees working between the rail station and major workplaces. This also includes the bike sharing system, which supports the public transport system rather than creating a sustainable and just transport system for all (Koglin & Mukhtar-Landgren 2021). Cycling infrastructure is not prioritised as part of a sustainable transport system in Lund. Declared policy intentions to promote cycling and develop cycling infrastructure have exceeded their actual implementation which is seriously lacking. If Lund, or any other city/municipality for that matter, really wishes to create

Figure D.2 Cycling infrastructure beside LRT line in Lund.

Source: Photo by author.

a sustainable transport system, it will be necessary to think outside the modernistic rationality box. Urban space has to be allocated to the most environmentally-friendly modes of transport, which are walking and cycling, to develop a truly sustainable urban transport system.

References

Cox, P. (2020) Theorising infrastructure: a politics of spaces and edges. In Cox P. and Koglin T. (Eds.) *The politics of cycling infrastructure: Spaces and (in)equality*. Policy Press, Bristol, pp. 15–34.

Koglin, T. (2020) Spatial dimensions of the marginalisation of cycling – marginalisation through rationalisation? In Cox P. and Koglin T. (Eds.) *The politics of cycling infrastructure: Spaces and (in)equality*. Policy Press, Bristol, pp. 55–71.

Koglin, T. and Glasare, L. (2020) Shopping centres, cycling accessibility and planning – the case of Nova Lund in Sweden, *Urban Science*, Vol. 4, pp. 70.

Koglin, T. and Mukhtar-Landgren, D. (2021) Contested values in bike-sharing mobilities – a case study from Sweden, *Journal of Transport Geography*, Vol. 92, pp. 103026.

Kröyer, H. (2015) *Accidents between pedestrians, bicyclists and motorized vehicles: Accident risk and injury severity*. Doctoral Dissertation, Lund University, Department of Technology and society, Transport and Roads, 2015, Bulletin – 296.

Lund Municipality (2018) *För att fler ska cykla mer – Cykelstrategi 2018–2021*. Lund Municipality, Lund.

Lund Municipality (2020) https://www.lund.se/foretagare/flytta-foretag-till-lund/lunds-storsta-arbetsgivare/, https://kartportalen.lund.se/ (accessed 02/07/2020).

Regional Travel Survey Skåne (2018) http://beslutstod.skane.se/QvAJAXZfc/opendoc.htm?document=-documents%5Cresvanor.qvw&lang=en-US&host=QVS%40rspapp072&anonymous=true (accessed 17/02/2021).

Trafikanalys (2020) Vehicles 2019, https://www.trafa.se/vagtrafik/fordon/ (accessed 23/02/2021).

21

SITUATING THE MOBILITY FIX OF CONTEMPORARY URBAN CYCLING POLICY

Justin Spinney

Introduction

The last two decades have seen cycling take an increasingly central role in urban mobility policies across the globe. Cycling has made a remarkable comeback from the wilderness years of the 1970s and 80s when you couldn't pay most citizens to cycle. This comeback is in no small part due to growing dissatisfaction with the negative effects of motorized transport: congestion, climate change, poor air quality, sedentarism, and the promotion of cycling as a panacea for these ills.

This chapter gives a brief – and necessarily partial – overview of cycling policy. Whilst I do draw upon examples from around the world – Shanghai, Taipei, Amsterdam, Copenhagen (see also Chapter 37), Paris – my focus is largely on London, with which I am most familiar. Whilst there are many similarities between cycling policy and practice in cities around the world, the policies and concepts I discuss in this chapter are drawn from, and most immediately relate to, these locations.

In the first part I outline the main drivers of cycling policy and resulting manifestations. In the second part I theorize cycling policy as a biopolitical mobility fix due to its focus on individual responsibilization and its optimization toward addressing matters of economic concern. In the final part I outline some issues that arise from current policy directions. My aim in doing so is to celebrate some of the successes whilst also pointing out some of the absences and potential downsides to current directions. As a result, I give a sense of direction to future research and practice with a view to maximizing cycling's chances of contributing to a more sustainable future.

We are witnessing a transformation in the political economy of 21st-century capitalism (Harvey 1990: 121). Mobility has always been central to capitalist economies because production requires movement through time and space to overcome it (Harvey 1990: 229). As economies have become more extensive, such mobility becomes ever more central to productivity. The tendency to accelerate mobility is reflected in both private and public transport systems whose trajectories trace a developmental arc through human/animal/wind power (foot, horse and cart, bicycle); mechanical (steam train, internal combustion vehicles, jet aeroplane) and now virtual (telecommunications, internet). The rise of motorized transport in the 20th century is the most notable of these, with few societies escaping its dominance (Dennis & Urry

 DOI: 10.4324/9781003142041-30

2009: 47). There are many reasons for this dominance, but key amongst them is the ability of motor vehicles flexibly to link up and produce new sources of goods, new destinations (and thus markets), and to move people and goods quickly over large distances. The dominance of motorized vehicles has not come without cost, however, with automobility now cited as a central cause of increasing urban congestion, local and global environmental degradation, and rising sedentarism, obesity and ill health. The case for promoting cycling stems from governmental needs to address such matters.

Manifestations of cycling policy: behavior change, infrastructure and public bikes

A key goal of UK transport policy in recent years has been to intervene in social reproductive journeys – particularly commuting – by encouraging more 'active' modes of travel such as cycling. A 2007 report by SQW '*Valuing the Benefits of Cycling*' set out the economic argument to invest in cycling in the UK, stating that its overall value accrues from a unique combination of: improvement in general health and fitness (Chapter 30); reduced pollution and the emission of CO_2; and help in tackling congestion. Certainly the DfT concluded that the health benefits of cycling are the largest single reason to promote it. Whilst cycle promotion in the UK has been patchy at best, this position has been officially embodied in London's transport policy, with the GLA's current goal to increase the modal share of cycling to 5% by 2025 – a 400% increase (GLA 2014: n.p.). Sustainable Urban Mobility Plans (SUMPs) around the globe have similar goals with policies coalescing around strategies of behavior change, infrastructure and public bikesharing.

Behavior change

Behavior change has been a key pillar of UK cycling policy for some time. In its broadest sense, behavior change policy acknowledges that travel behavior is conditioned by structural, attitudinal and habitual factors (DfT 2017: 2). Advocates argue that attempts to change travel behavior are most effective when comprised of 'soft' measures such as marketing, information and coaching; and 'hard' measures such as changes to infrastructure and services (DfT 2017: 8). In practice, however, the UK government has tended to adopt an approach which is light on expensive hard measures, relying mostly on cheaper soft measures, and avoiding regulation and financial mechanisms entirely (House of Lords Science and Technology Committee 2011: 7.38/7.41). As a result, the effectiveness (and thus cost-effectiveness) of these behavior change initiatives has been limited (ibid.). Despite a potentially negligible ability to shift citizens from car use to cycle use, behavior change policy remains politically expedient in neoliberal governance because it gives citizens the choice to 'do the right thing' without forcing politicians to engage in more interventionist regulation of behavior (Pooley et al. 2013: 67).

Cycle infrastructure

Whilst the UK continues to emphasize 'soft' measures, attitudes of policy makers have shifted over the last 15 years, supported by studies finding a correlation between higher levels of bicycle commuting and the presence of cycle-specific infrastructure (Parkin et al. 2008; Chapter 20). The cycling boom in London, for example, has been characterized by substantial political and economic investment in cycling infrastructure (GLA 2013: 4 in Lam 2018: 115). On the face of it increased funding and infrastructural measures for cycling in London would appear to have been successful with a rapid increase in cycling over the past 15 years. Census data shows a 144% increase in cycling in London over the period 2001–2011 with a daily average of 131,000 cycle journeys in 2014 (GLA 2014: n.p.). More recent figures show that the number of cycle journeys in June 2018 was 187,345, a year-on-year increase of 8% (London Assembly 2017: n.p.).

Cost remains a central concern: Cycling England has noted the importance of infrastructural investment to increase cycling levels. In European cities with low levels of car use, investment in cycling infrastructure is at least £10 per head, and up to £40 per head in high-cycling cities like Copenhagen (House of Lords Science and Technology Committee 2011: 7.37). Much of London's success has been due to increased spending: in 2016 TfL announced plans to increase spending on cycling from around £8 to £18 per head (Bike Biz 2016: n.p.) compared to average spending levels outside London of around £1.38 per head) (Sutton 2016: n.p.). Compare London's expenditure on cycling to the new HS2 (rail) route which conservative estimates put at around £670 per head or on new motorways and trunk routes at £230 per head (ibid.) and the chasm between funding for cycling and other more or less sustainable modes is stark.

Public Bike Sharing Schemes

Whilst much cycle promotion has focused on encouraging private cycle use, public bike sharing schemes (PBSS) can be situated as a form of public transport (Chapter 23). PBSS is part of a broader Mobility as a Service (MaaS) revolution being driven by the increasing convergence of physical and virtual mobility (de Souza e Silva 2002: 21). Proponents argue that MaaS can reduce resource use, reduce ownership costs and barriers, and enhance community cohesion.

The blueprint for the contemporary global spread of PBSS can be found in the 2007 launch of the Velib system in Paris (Shaheen, Guzman & Zhang 2010). From 2016, the original 'docked' versions of PBSS have been joined by flexible 'dockless' and hybrid systems, generally owned and operated by private providers. Nowhere has this been more prominent than Shanghai, which saw 1.5 million shared bikes appear on its streets between April 2016 and August 2017 (Lin & Spinney 2020). According to the 'Bike Sharing World Map' in August 2021, there were 1,910 schemes in operation worldwide with 276 more planned, totaling almost 10 million shared bikes (Meddin et al. 2021). The vast majority of these schemes are located in Europe, East Asia and North America.

A number of studies have emphasized the economic, environmental and health benefits of PBSS (Shen et al. 2018). Others have been more critical. Chardon (2019), for example, argues that PBSS represents a less just and sustainable transportation option in most cases, with limited benefits accruing to city administrations. Despite such caveats, PBSS looks set to remain a fixture for some time to come. A central reason for the popularity of PBSS, particularly the kinds operated by private providers, is that with city administrations increasingly short of funding, they represent a way of providing transport services at little or no cost to the taxpayer.

Theorizing cycling policy: biopolitics and mobility fixing

Contemporary cycling policy in the UK (and also more broadly) can be understood as a biopolitical 'mobility fix' because of its tendency to responsibilize individuals to adopt new ways of moving in order to address matters of (economic) concern. These matters include speeding up the circulation time of capital and labor; enhancing human capital; boosting industrial and place image; and facilitating the production and capture of 'behavioral surplus' from user data (Zuboff 2019).

Firstly, cycling policy represents a form of 'fixing' problems of accumulation and productivity. It is notable, for example, that the subjects of cycling policy are overwhelmingly those travelling to work and, to a lesser extent, education. To give a few examples, London's cycle superhighway network and Copenhagen's cycle networks are intended to provide access to main employment locations; Taipei's U Bike PBSS is focused around Metro stations to facilitate the last mile for commuters; and Shanghai's dockless bike sharing companies have focused on specific public transport hubs and employment sites where revenues can be maximized. Such a selective geography suggests that cycling has been brought back into cities because of its potential to 'fix' the delays from congestion wrought by an overabundance of motor vehicles, and bridge the broken and missing links in existing transport systems in order to maintain faster circulation times.

However, government objectives to maximize efficiency cannot be met solely through the speeding up of labour or capital. Foucault (2010: 232) has argued that in order to understand the workings of 'efficiency', we must examine the ways in which human capital is augmented (or not) through mobility, and how this contributes to increases in productivity. Situated within a wider healthcare politics of obesity and sedentarism (Guthman 2009), cycling has become a key medium through which the individual can invest in themselves and (hopefully) improve their vitality and life expectancy. At the same time, both the state and the private sector reap the rewards of this 'improved' human capital because it is argued that such workers take less time off sick, and in the longer term place less burden on health care costs. Crucially, this is government acting through mobility (Bærenholdt 2013): through such policies cycling becomes a vehicle through which the individual recognizes and becomes equipped to deal with matters of public health concern such as obesity and sedentarism.

Cycling's role in fixing does not stop here, however: it is also mobilized at the level of inter-city image competition and place marketing. Cities increasingly compete at the level of images (which have in themselves become commodities), projecting their values, authority and power across space (Harvey 1990: 288). Images enable places to establish an identity in a crowded market of places, but as Harvey and others have been at pains to point out, as policy becomes more mobile such competition has led to a lack of distinction as particular 'best practice' initiatives are repeated and copied from place to place (Harvey 1990: 295).

As mobility is increasingly seen as a measure of how liveable a place is, one of the ways cities promote their difference is by projecting a cycling culture in a positive way. In the absence of any substantial sustainability initiatives, and an unwillingness to tackle the continued growth of automobility, the adoption of active mobilities like cycling can act as a symbol of sustainability, liveability, and health (Spinney & Lin 2019).

Research on dockless PBSS suggests a further way in which cycling policy is implicated in processes of fixing: in this case by enabling the gathering of data by surveillance capitalists (e.g. Google, Facebook, Tencent, Baidu) for new markets. Spinney and Lin (2018, 2021) have argued that data generated through usage of dockless PBSS (location, origins, destinations, socioeconomic details etc.) can be used to manufacture new predictive products. Whilst some of these products enhance provision of urban mobility (for example, through improved bicycle parking and location of cycle lanes), Spinney and Lin (2021) show that such data can also used for commercial ends. Both route data and the geo-fencing of bikes can be used to target advertising and charge retailers for 'sponsored locations' with precedents already set by location-based gaming apps like Pokémon Go (Zuboff 2019: 299). Indeed, the acquisition of user geo-location data has been a key driver of the Asian internet giants Tencent and Alibaba's investments in dockless PBSS. As Zuboff (2019: 231) has demonstrated, the reason most smartphone apps demand to access location, even when not required for functioning of the app, is because it is lucrative.

A further way in which PBSS user data can be used to drive accumulation in new markets is to utilize mobility patterns to profile users in ways similar to those of age, income, gender and ethnicity. The precedents for this come largely from insurance and financial services (e.g. credit scoring) where studies show that a small number of measurable personal characteristics can be used to predict whether people will like a brand, click on particular advertisements, or prove 'trustworthy' (Zuboff 2019: 263). The app-based credit rating function of the Chinese PBSS operator Mobike has the potential to render similar insights because data generated regarding trustworthiness and reliability are potentially valuable for businesses involved in credit scoring and personal loan provision. In its conjoining of virtual and physical mobility, PBSS 2.0 represents a new frontier in the drive to commodify aspects of cycling.

When we take all these aspects together, I argue that contemporary cycling policy is best theorized as a 'mobility fix' for two reasons: firstly because it represents a way of governing that emphasizes the production of new modes of mobile comportment rather than the production of space; and secondly, because the variants of cycling being prioritized in policy are those optimized to enable continued accumulation and address matters that potentially limit economic growth. As such, we can also view cycling policy as biopolitical because it is concerned with regulating processes – such as health – that "sustain or retard the optimization of the life of a population" (Dean 2010: 119). Accordingly, with reference to Foucault I argue that the promotion of cycling is a form of self-government and 'entrepreneurship of the self' situated within a biopolitics of obesity and sedentarism.

Questions raised by current directions in cycling policy

In this final section I want to provide a brief and necessarily selective discussion of some issues highlighted in a critical examination of contemporary cycling policy. The first – and much raised in recent years – is that of social justice and equity. Writing on transport and justice has burgeoned (Golub et al. 2016; Lucas et al. 2019; Martens 2017). Some have focused on cycling, in particular the fact that many cycling policies have disproportionately benefited able-bodied, young and middle-aged, middle-class white men at the expense of women (Chapter 2), ethnic minorities, children, older people and those with disabilities (Aldred 2015; Chapter 7). The desire to put cycling back on the agenda has led to a "reductive understanding of what facilitates a vibrant urban cycling culture" (Stehlin 2014: 36).

Cupples and Ridley (2008) refer to current approaches as a potentially dangerous 'cycling fundamentalism' that overlooks issues of difference in the scramble to encourage cycling. Whilst any increases in cycling may be viewed as good, overall numbers do not tell the full story and more needs to be done to enable a greater range of people to cycle. This requires engaging more with issues of social safety; inclusive infrastructure; inclusive cycle design; cultural barriers; spatial arrangements; child-care arrangements; age-friendliness etc.

Current manifestations of cycling also represent a narrowing of what cycling is and means because they place greatest value on qualities of directness, flexibility, speed and efficiency. As a 'dispositional' spatial rationality, cycle infrastructure enables cycling as a behavior, but it also attempts to direct and shape the qualities of cycling practice in certain directions to produce the 'right' kinds of cycling. Cycle superhighways, in particular, prioritize the performance of particular variants of cycling, most notably the fast, efficient and direct kind that contribute most to speeding up circulation whilst reducing public expenditure on health and environmental issues by enabling cycling to become a viable alternative to other commuter modes.

There are two key implications to be drawn from this regarding the range of cyclings that can be practiced. Firstly, with primarily fast commuters using narrow infrastructure, there is little accommodation for those who are – through ability or desire - slower cyclists. Such a situation is likely to disadvantage older, younger, impaired and inexperienced cyclists who may be put off using the routes even if they are conveniently located. The narrowness of cycling routes also means that more social kinds of cycling are less likely to materialize because cyclists cannot comfortably ride side by side. Whilst driving two abreast is implicitly encouraged in the design of cars and carriageway, it seems that cycling two abreast remains something that only the inconsiderate road user would entertain. Cycle superhighways as a form of material pedagogy bring into being a particular version of cycling that emphasizes a narrow range of possible qualities and marginalizes others (and the subjects who value them). Perhaps the most obvious implication of this is that those who might value cycling as a means for more playful, place-making, more social and slower forms of cycling find it much harder to perform on such a stage.

Linked to the previous points, there is also a sense that alongside the relative success of getting cycling back into mainstream transport policy, has been a narrowing of what counts as active travel, which in current policy is a euphemism for walking and cycling. Lorimer and Marshall (2015), for example, argue that any discussion of 'human locomotion' should include consideration of small-wheel modes such as inline skates, skateboards and scooters; all modes that tend to be favoured by young people (1). In their 2015 pilot study, Lorimer and Marshall found that these active modes were often faster than public transport and walking for many journeys, arguing that urban designers should therefore seek to actively include such modes as 'serious' urban 'transport' (2015: 5). As they note, the promotion of such modes could represent a "… new and innovative way of encouraging active travel among the young" (2). Whilst there are current pilot programs ongoing in the UK to look at the viability of shared electric scooters (see, for example, Sherriff et al. 2021), there remains an urgent need to research the potential contributions of forms of active travel beyond cycling.

Given that mobility is increasingly produced at the intersection of physical and virtual, there is also an urgent need to understand the implications of such hybridity on citizens. Koglin (2017: 33) has emphasized the marginalization of democratic participation in post-political physical urban planning for cycling, an issue that risks being compounded by injustices wrought in the virtual realm by a lack of transparency, as Stehlin et al. (2020) have noted. In particular, there is a need to understand dualistic processes of subjectification at work and their implications for the physical and virtual citizen. New forms of smart/shared mobility not only produce a differently mobile physical citizen to deliver the policy agenda: greener, more agile, healthier, self-responsible; but they also produce a new virtual – and much less visible – citizen largely hidden from view on the servers of governments and private corporations (Birchall 2016). Whilst we may at least have a sense of agency in relation to our physically mobile self, such agency is much less assured for our virtual self. Clearly two different versions of a citizen emerge through initiatives such as dockless PBSS, and we need to understand what purpose these subjects serve; how they are different and similar; and why have they been brought into being to ensure they are sustainable and just.

Finally, I wish to engage with a more far-reaching issue regarding the extent to which current cycling policy directions can achieve sustainability, given that they are geared toward maintaining business as usual with regard to unsustainable patterns of production and consumption. In attempting to move cycling from niche pastime to a more central role in urban transportation, activists and academics can end up contributing to the continuation of unsustainable capitalist growth, not because they are promoting cycling per se, but because their vision becomes part of a reductive agenda that foregrounds aspects and forms of cycling that enhance urban economic competitiveness. Accordingly, it is perhaps what current articulations of cycling say about how we conceptualize sustainability and value that are of most interest.

Jackson (2009: 57) has confirmed the naivety that current levels of economic growth can be maintained in the face of rising levels of affluence and population growth alongside widening levels of inequality. Indeed, the likes of Jackson (2009), Fletcher and Rammelt (2017) and Hickel and Kallis (2020) are clear that no amount of efficiencies with regard to mobility will save us if we remain focused on economic growth as our main goal. Fletcher and Rammelt (2017: 450) go on to suggest that current debates on decoupling transport emissions from economic growth – of which cycling is a key pillar - represent a delusional neoliberal fantasy because they seek to downplay tensions between economic growth, efficiency, population and inequality. There is a real danger that in its current narrow economistic framings, cycling has become part of this fantasy narrative and is being used to sustain a belief that sustainable development can be achieved despite a lack of empirical evidence supporting the possibility of green growth through decoupling (Hickel and Kallis 2020: 475). As this suggests, it is questionable whether a version of cycling that reproduces growth patterns – albeit with some reduction in environmental impact – without an underlying socio-economic transformation can be considered sustainable. In propping up current social and economic practices upon which current patterns of growth rest, contemporary manifestations of cycling represent at best a shift toward a 'weak green economy' (Ferguson 2015: 27), and at worst part of a 'green growth' agenda that fails to see any inconsistency between economic growth and sustainability.

Ferguson (2015) has argued that for a truly sustainable transformation to occur, we need to more fundamentally move away from measuring value in purely economic terms and foreground other aspects of human flourishing (Cudworth and Hobden 2011 in Ferguson 2015: 26). I argue that societal needs for wellbeing and happiness are not being met through current logics of cycle promotion which privilege wellbeing as a route toward more productive human capital, and are distributed upon a consumer logic to those of higher socioeconomic status. What is required is a new economy of worth, and current manifestations of cycling are not enabling the kinds of questioning and reorientation of social and economic life that might lead to such a new reckoning. In order to question current valuations and create new orders of worth, we need forms of mobility (including cycling) that actively enable us to question rather than reproduce current patterns of societal organization and growth. Only then might the levels of resource and energy reductions required to avoid catastrophic social and environmental breakdown be realized. Whilst current cycling policy may be a necessary precondition on the road to more transformative modes of societal organization, we cannot assume that the successes of current cycling policy will inevitably move us toward a stronger articulation of a green economy as part of a fait accompli; we must actively move it in that direction by highlighting the shortcomings of current incarnations.

References

Aldred, R. (2015) A matter of utility? Rationalising cycling, cycling rationalities, *Mobilities* 5(1), pp. 686–705.

Baerenholdt, J. O. (2013) Governmobility: The powers of mobility, *Mobilities* 8(1), pp. 20–34.

Barr, S. (2018) Personal mobility and climate change, *Advanced Review* 9(5), pp. 1–19.

Barry, J. (2012) *The Politics of Actually Existing Unsustainability: Human Flourishing in a Climate-Changed, Carbon-Constrained World* (Oxford: Oxford University Press).

Beck, U. (1992) *Risk Society: Towards a New Modernity* (London: Sage).

Bike Biz (5 December 2016) London to double spend on cycling to 5.5 percent of transport spending, Available at: https://www.bikebiz.com/london-to-double-spend-on-cycling-to-5-5-percent-of-transport-spending/ [Accessed August 2021].

Birchall, C. (2016) Shareveillance: Subjectivity between open and closed data, *Big Data and Society* 3(2), pp. 1–12.

Chardon, C. (2019) The contradictions of bike-share benefits, purposes and outcomes, *Transportation Research Part A: Policy and Practice* 121(C), pp. 401–419.

Cupples, J. and Ridley, E. (2008) Towards a heterogeneous environmental responsibility: Sustainability and cycling fundamentalism, *AREA*, 40(2), pp. 254–264.

de Souza e Silva, A. (2002) From cyber to hybrid: Mobile technologies as interfaces of hybrid spaces, *Space & Culture*, 3, pp. 261–277.

Dean, M. (2010) *Governmentality: Power and Rule in Modern Society*, 2nd ed (London: Sage).

Dennis, K. and Urry, J. (2009) *After the Car* (Cambridge, UK: Polity Press).

Department for Transport (December 2017) Enabling behaviour change – Information pack, Available at: https://www.gov.uk/government/publications/transport-business-case/enabling-behaviour-change-information-pack [Accessed August 2021].

Ferguson, P. (2015) The green economy agenda: Business as usual or transformational discourse?, *Environmental Politics*, 24(1), pp. 17–37.

Fletcher, R. and Rammelt, C. (2017) Decoupling: A key fantasy of the post-2015 Sustainable Development Agenda, *Globalizations*, 14(3), pp. 450–467, doi: 10.1080/14747731.2016.1263077.

Foucault, M. (2010) *The Birth of Bio-Politics: Lectures at the College de France 1978–79* (Basingstoke, UK: Palgrave Macmillan).

Greater London Authority (2014) *Cycle Flows on the TFL Road Network* (London: GLA). Available at: http://data.london.gov.uk/dataset/cycle-flows-tfl-road-network [Accessed 26 November 2014].

Golub, A., Hoffman, M. and Lugo, A. (Eds) (2016) *Bicycle Justice and Urban Transformation: Biking for all?* Routledge Equity, Justice and the Sustainable City Series (Routledge: New York, NY).

Goodman, A. and Cheshire, J. (2014) Inequalities in the London bicycle sharing system revisited: impacts of extending the scheme to poorer areas but then doubling prices, *Journal of Transport Geography*, 41, pp. 272–279.

Greene, D. and Wegener, M. (1997) Sustainable transport, *Journal of Transport Geography*, 5(3), pp. 177–190.

Guthman, J. (2009) Teaching the politics of obesity: Insights into neoliberal embodiment and contemporary biopolitics, *Antipode*, 41(5), pp. 1110–1133.

Harvey, D. (1990) *The Condition of Post-Modernity: An Enquiry into the Origins of Cultural Change* (Malden, MA: Wiley-Blackwell).

Hickel, J. and Kallis, G. (2020) Is green growth possible?, *New Political Economy*, 25(4), pp. 469–486.

House of Lords Science and Technology Committee (July 2011) Behaviour change, Available at: https://publications.parliament.uk/pa/ld201012/ldselect/ldsctech/179/17902.htm [Accessed August 2021].

Jackson, T. (2009) *Prosperity without Growth: The Transition to a Sustainable Economy* (London: Sustainable Development Commission).

Koglin, T. (2017) Urban mobilities and materialities – A critical reflection of "sustainable" urban development, *Applied Mobilities*, 2(1), pp. 32–49.

Lam, T.F. (2018) Hackney: A cycling borough for whom?, *Applied Mobilities*, 3(2), pp. 115–132.

Lin, W. and Spinney, J. (2020) Mobilising the dispositive: Exploring the role of dockless public bike sharing in transforming urban governance in Shanghai, *Urban Studies*, 58(10), pp. 2095–2116.

London Assembly (2017) Mayor and Commissioner set out vision for getting Londoners active, Available at: https://www.london.gov.uk/press-releases/mayoral/setting-out-a-vision-for-getting-londoners-active [Accessed August 2019].

Lorimer, S. and Marshall, S. (2015) Beyond walking and cycling: Scoping small-wheel modes, *Proceedings of the Institution of Civil Engineers – Engineering Sustainability*, 169(2), pp. 58–66.

Lucas, K., Martens, K., Di Ciommo, F. and Dupont-Kieffer, A. (2019) *Measuring Transport Equity* (Amsterdam: Elsevier).

Martens, K. (2017) *Transport Justice* (New York: Routledge).

Meddin, R., DeMaio, P., O'Brien, O., Rabello, R., Yu, C., Seamon, J., Benicchio, T., Han, D. and Mason, J. (2021) The Meddin bike-sharing world map, Available at http://bikesharingworldmap.com/ [Accessed August 2021].

Parkin, J., Wardman, M. and Page, M. (2008) Estimation of the determinants of bicycle mode share for the journey to work using census data, *Transportation*, 35(1), 93–109. doi: 10.1007/s11116-007-9137-5.

Pooley, C., Horton, D., Scheldeman, G., Mullen, C., Jones, T., Tight, M., Jopson, A. and Chisholm, A. (2013) Policies for promoting walking and cycling in England: A view from the street, *Transport Policy*, 27(5), pp. 66–72.

Shaheen, S., Guzman, S. and H. Zhang. 2010 Bikesharing in EUROPE, the Americas and ASIA: past, present, and future. *2012 Transportation Research Board Annual Meeting*. Available at: https://escholarship.org/uc/item/79v822k5 [Accessed: 31 August 2014].

Shen, Y., Zhang, X. and Zhao, J. (2018) Understanding the usage of dockless bike sharing in Singapore, *International Journal of Sustainable Transportation*, 12(9), pp. 686–700.

Sherriff, G., Blazejewski, L., Hayes, S., Larrington-Spencer, H. and Lawler, C. (May 2021) E-Scooters in Salford Interim Report, May 2021, Available at: http://usir.salford.ac.uk/id/eprint/60393 [Accessed July 2021].

Spinney, J. (2016) Fixing mobility in the neoliberal city: Cycling policy and practice in London as a mode of political-economic and biopolitical governance, *Annals of the American Association of Geographers*, 106(2), pp. 450–458.

Spinney, J. (2021) *Understanding Urban Cycling* (Oxford: Routledge).

Spinney, J. and Lin, W. (2018) Are you being shared? Mobility, data and social relations in Shanghai's Public Bike Sharing 2.0 sector, *Applied Mobilities*, 3(1), pp. 66–83.

Spinney, J. and Lin, W. (2019) (Mobility) Fixing the Taiwanese bicycle industry: The production and economisation of cycling culture in pursuit of accumulation, *Mobilities*, 14(4), pp. 524–544.

Spinney, J. and Lin, W. (2021) A vehicle for valorising the labour power of commuting: The politics of mobility fixing in Shanghai's Dockless Public Bike Sharing Sector, *Journal of Transport Geography*, 94, pp. 103–129.

SQW Consulting (2007) Valuing the benefits of cycling: Report to cycling England, http://webarchive.nationalarchives.gov.uk/20110407094607/, http://www.dft.gov.uk/cyclingengland/site/wp-content/uploads/2008/08/val-uing-the-benefits-of-cycling-full.pdf [Accessed 1 January 2014].

Stehlin, J. (2014) Regulating inclusion: Spatial form, social process, and the normalization of cycling practice in the USA, *Mobilities*, 9(1), pp. 21–41.

Stehlin, J., Hodson, M. and McMeekin, A. (2020) Platform mobilities and the production of urban space: Toward a typology of platformization trajectories, *EPA: Economy and Space*, 52(7), pp. 1250–1268.

Sutton, M. (March 2016) Analysis: How does UK's cycling spend per head compare to other transport projects? Available at: https://cyclingindustry.news/analysis-how-does-uks-cycling-spend-per-head-compare-to-other-transport-projects/ [Accessed August 2021].

Zuboff, S. (2019) *Surveillance Capitalism: The Fight for a Human Future at the New Frontier of Power* (New York: Hachette Book Group).

22
MAKING SPACE FOR CYCLING

Paola Castañeda and Sergio Montero Munoz

Introduction

We often think of "cycling space" in cities as the physical infrastructures – bike lanes, parking facilities, bike boxes, etc. – that demarcate where it is appropriate, safe or desirable to ride a bike. Conceptualized this way, "making space for cycling" seems like a straightforward process of building more bicycle infrastructure. Hence, the idea of "build it and they will come" (Cervero, Caldwell and Cuellar 2013) became a common expression in the last decade among progressive policy makers in European and North American cities to justify the need to build bike lanes and increase cycling in their cities.

In this chapter we want to think of *other* kinds of spaces for urban cycling and broaden the horizon of how bicycle space is produced. Useful here is French philosopher Henri Lefebvre's (1991 [1974]) conceptualization of space as an interrelated triad of perceived, conceived, and lived space. Perceived spaces, or spatial practices, are the physical, material spaces of daily life and social interaction. Conceived spaces, or the representations of space, are the socially constructed discourses, signs, and meanings of space – the "space of scientists, planners, urbanists, technocratic subdividers and social engineers" (Lefebvre, 1991: 38). Finally, lived space makes reference to space as experienced by people who use and occupy it, encompassing the previous two types of space. Lefebvre's triad provides a framework to understand how space is produced and how social experiences are constituted in and through space, and offers an entry point for considering what space for cycling might mean. We also understand "cycling" as more than commuting, leisure cycling, or indeed more than *pedaling* a bicycle. Multiple kinds of cycling take place in cities, often unrecognized or invisibilized: cycling economies (Sarmiento Casas 2018), urban cycling races (Gamble 2019), and community bike shops (Bradley 2018) are also ways to make space for cycling. It follows that making space for cycling need not be a top-down strategy, but can stem from citizen's initiatives and desires as they appropriate and produce cycling spaces, both enduring and temporary.

In addition to problematizing both "space" and "cycling", we want to move beyond Eurocentric interpretations of the relationship between cycling, society and space by grounding our discussion in the context of Latin America, where we both work. Over the past decade, a cycling boom has echoed in planning circles, drawing the attention of policy makers, development NGOs, citizens, and scholars (Chapter 38). However, while bicycle policies and

DOI: 10.4324/9781003142041-31

programs from the Global South have started to travel not only South–South, but also South–North (Montero 2018), cycling research and policy have largely drawn on the experiences of cities in Northern Europe. Against this, and building on scholarship primarily from and about Latin America, we heed the postcolonial call in urban studies to think about urban dynamics *from* the Global South (Robinson and Roy 2016).

Building on these ideas, we delineate three spaces for urban cycling beyond physical infrastructure. First, we discuss temporary spaces as important spaces where new imaginations and practices of bicycling can emerge. Second, we turn to cycling policies as key spaces where urban cycling becomes institutionalized, that is, the spaces where decision-makers learn about bicycle policies whilst embedded in a multi-scalar field of power and politics. Finally, we discuss civil society and activist spaces to signal the ways in which grassroots and bottom-up initiatives can bring diverse and inclusive cycling spaces and politics into being. We specifically address prefigurative cycling spaces, cycling schools and community cycling, and feminist cycling spaces, though these are certainly not the only kinds of activist cycling spaces that exist.

Bicycle space: beyond infrastructure

"Where do bicycles belong?" has been the subject of ample debate since the bicycle's appearance in the 19th century (Vignette H). In a recent article, for example, Michael Brown (2021) traces the history of cycling in South America where, owing to the multiple crashes on urban boulevards, cycling was confined to either velodromes in cities, or rural and suburban areas. More recently, a number of environmental and economic concerns have brought the bicycle into the spotlight of urban and transport planning circles, once again raising the question of where do bikes belong, or *how to make space for cycling*. Accordingly, an agenda focused on "how to get more people to cycle" has derived in policy approaches encapsulated in the adage "build [infrastructure] and they will come." Ralph Pucher and John Buehler's extensive and influential research has repeatedly drawn attention to the vast infrastructural network that makes cycling in Denmark, Germany and the Netherlands "irresistible" (Pucher and Buehler 2008). And while the authors have been careful to clarify that infrastructure is not the *only* approach for encouraging cycling and making it safer, cycle paths have become the dominant imaginary of what a space for cycling is. In this way, urban cycling has become a "mobility fix," intended to "[move] the highest number of people through a given space as fast as possible" (Spinney 2020: 10) in line with demands for continuous growth and capital accumulation, often framed as "sustainable urbanism."

The sustainability of such visions and approaches has been problematized over the past decade. Critical cycling research has scrutinized the uneven effects of cycling investments and emerging geographies (e.g. Golub et al. 2016; Stehlin 2019), demonstrating that "conflating the practice of bicycling with specific urban development projects designed to accommodate it, *limits* what can be seen as 'bike friendly' neighbourhoods and manufactures scarcity in what should be a public resource: urban streets" (Hoffmann and Lugo 2014: 45, emphasis added). This is not to say that cycling infrastructure should not be built, but that we should embrace a wider range of cycling mobilities and imaginaries to orient our efforts towards human flourishing, connection, and wellbeing (Spinney 2020). Part of this reorientation entails moving away from a persistent focus on cycle paths and toward the social, cultural, and institutional scaffolding needed to support a substantial and lasting shift in the politics of mobility (Cresswell 2010) that govern movement and the spaces where mobility is produced. In drawing on diverse experiences from Latin America, we broaden the horizon of what "space for cycling" is, and contribute to *worlding* cycling beyond the usual suspects (Montero 2017a, 2017b; Castañeda 2021).

Temporary spaces

In recent years, terms such as pop-up, tactical, or DIY urbanism have become prominent in urban planning practice to name temporary urban interventions that are often low-cost, short-term and citizen-led. These kinds of projects have been promoted by urban designers, architects and "new urbanists" as a way to quickly and cheaply transform an urban space by giving people the experience of how urban space can be used differently (Lydon and García 2015). This potential, however, has been met with skepticism by many critical urban scholars who see these projects as a rather cosmetic practice that does not challenge market-oriented urban development. Others have shown skepticism at the lack of engagement of tactical and DIY urbanism with the messy world of politics and organizing (Webb 2018). Thinking from the perspective of Global South cities, Andres et al. (2021) have highlighted how temporary urbanisms can be useful in places where formal planning is weak, but only when temporary urbanisms move from a place-shaping to a place-making approach.

Temporary bicycle programs have been at the center of this debate. Indeed, in the book *Tactical Urbanism: Short-Term Actions, Long-Term Change*, Lydon and García (2015) used Bogotá's Ciclovía as an example of tactical urbanism – a temporary program that can be easily implemented because it does not require investment in bicycle infrastructure building. Yet, although these projects can have a powerful demonstrative potential to rethink how infrastructures can have uses other than those intended by planners, in this section we argue that seeing Ciclovía as tactical, pop-up or DIY-urbanism is a poor lens to understand the complex institutionalization efforts that have shaped Ciclovía over nearly five decades (Montero 2017a). Rather, thinking these temporary programs as *experiments* in urban planning might be a better way to conceptualize not only their one-time demonstrative potential, but also the need to engage with institutions and politics to build the necessary alliances and networks to ensure the long-term impact of these programs (Sosa López 2021; Lugo 2013). Making space for experimentation is therefore key to envision alternative mobile futures in cities.

The origins of Ciclovía can be traced back to an experiment organized in Bogotá on December 15, 1974 by Pro-Cicla, a bicycle organization led by three bike enthusiasts who wanted to do something about the sprawl Bogotá was experiencing at the time (see Chapter 38). We keep here the language of "experiment" as that is the word one of Ciclovía's founders used in a 1985 article reflecting on the origins of the program (Ortiz 1985). Using their family and political connections, they were able to close to motorized traffic 80 blocks of the city's two main arteries, Carrera 7 and Carrera 11. They called the event "the great pedal demonstration" and about 5,000 people participated in it. Even though the program almost disappeared in the 1980s, the major institutional and budget changes done during the mid-1990s revived the program and expanded it significantly. Since 1995, different Bogotá mayors, planners and bicycle advocates have made important institutional changes to the program to improve and expand the length of Ciclovía (Montero 2017a). As of 2010, Ciclovía's total length was 97 kilometers, annual costs were US$1.7 million (75% coming from public funds), and attendance ranged from 600,000 and 1,400,000 users per event (Del Castillo et al. 2011). By 2017, length has expanded to 1,137 kilometers (Sarmiento et al. 2017).

While pop-up, DIY or tactical urbanism emphasizes the mindset change potential of temporarily experiencing something different, the language of experimental urbanism allows us to not underestimate the institutional changes that are needed to transform cities in the long term. Insomuch as space is lived and experienced by people in the course of everyday life, supporting and enabling a fulfilling urban environment will require a serious consideration of the multiple ways in which space is brought into being. Hence, we argue that physical cycling spaces (temporary or otherwise) ought to be coupled with an institutional framework favorable to cycling and other forms of non-motorized mobility.

Policy spaces

In the last two decades, the circulation of "sustainable urbanism" city models, policies or "best practices" has increased across the Global North and South (Chapter 20). Among those, bicycle policies have become a favorite and it is worth thinking about policy space as a critical site in which space for cycling is made. Following Lefebvre, policy can be thought of as one dimension of conceived space, in which discourses, signs, and meanings of space are coded, and mobilities governed by technocrats. How might we make space for cycling policy? Here, recent debates on policy mobilities can be helpful. This field seeks to critically examine the power-laden processes behind the movement of policies by "studying through" the everyday sites and situations of policy making to understand how and why certain policy ideas are mobilized and eventually adopted (McCann and Ward 2012). These sites and situations include "conferences, seminars, workshops, guest lectures, fact-finding field trips, site visits, walking tours, informal dinners and trips to cafés and bars, among many others" (McCann and Ward 2012: 47).

Bicycle policy models today are not only coming from Amsterdam and Copenhagen. Cities of the South have also started to become policy models for other cities, in both the South and the North. In this context, Bogotá, again, have often been the poster child as Ciclovía has been replicated in cities as diverse as Jakarta, Mexico City, Cape Town, or San Francisco (Montero 2017a). In the following paragraphs we take the case of Bogotá as our starting point to discuss three policy spaces that contribute to making space for cycling through the construction and global circulation of transport policies: conferences, study tours, and digital media platforms.

The potential of conferences to result in policy change relies on their capacity to inspire and persuade influential local actors to form broader coalitions of actors (Cook and Ward 2012). Conferences are privileged places for the mobilization of emotional elements that are hard to convey in printed or online documents. In this context, conferences and forums, particularly those with some sort of world recognition, not only help build a sense of community among participants through providing spaces of face-to-face communication and experiential learning, but also give legitimacy when these advocates are back in their city. For instance, according to San Francisco transport advocate Cheryl Brinkman, the 2008 Towards Car-Free Cities conference in Portland gave Bogotá's Ciclovía "a varnish of legitimacy" to implement it in San Francisco (Montero 2015), and policy stories, images and videos brought back from the conference helped them push San Francisco's mayor to implement the Ciclovía-based program *Sunday Streets* in the mid-2000s (Montero 2018).

A second key practice to circulate urban policies are study tours. These are short visits in which a delegation of people travels to another place to learn and experience something with potential to improve their organizations or places of origin (Montero 2017b). However, study tours are not just learning instruments; participants are often selected because of their capacity to create and expand local coalitions that would push for particular policies back home. Also known as "policy tourism" (González 2011), study tours create "a sense of being 'in tune' with what is happening elsewhere" among tour participants (González 2011: 1412). In his study of how Bogotá's transport policies were adopted in Guadalajara, Montero (2017b) showed that study tours were able to promote policy change in Guadalajara thanks to their capacity to: (1) educate the attention of influential local policy actors through hands-on "experiential learning;"

(2) expand local coalitions through the building of trust and consensus around a policy model; and (3) mobilize public opinion through references to already existing policies.

Finally, due to the increased use of digital technologies in the last decades, the circulation of policy models and knowledge increasingly happens through a virtual infrastructure in the form of media platforms, blogs and social media sites where images and videos about particular policies are mobilized by advocates and other policy actors. Gunder (2011) has shown that the media can influence how urban space should be organized through its capacity to 'engineer public beliefs.' In the context of bicycle policies, *Streetsblog*, an influential policy blog among US sustainable transportation and bicycle advocates, has been particularly important. In particular, a video of Bogotá's Ciclovía made by *Streetsblog's* sister organization *Streetfilms*, became an important resource that contributed to the global circulation of the program worldwide (Montero 2018). Sustainable transportation advocates in different cities have used this video to explain the Ciclovía concept to their communities and to persuade mayors and key urban decision-makers of its potential value. As the digital and offline world are continuously in dialog with each other, we should not underestimate digital spaces in their capacity to influence the process of making *physical* space for cycling.

Activist spaces

If we work with the premise that "cycling" is more than the physical practice of riding a bike, new imaginaries of space for cycling emerge. Fixing a bike, a mass ride, and teaching others how to cycle are some of the less obvious but thriving space-times where cycling is taking place in articulation with grassroots politics. These spaces for cycling move us to reconsider what the bicycle can do as a technology for transport *and social change*. The bicycle is being leveraged to enable projects of community integration, social and mobility justice, claims to the right to the city, and feminist resistance. Hence, some spaces for cycling are also spaces for political education and "cycling while questioning" (Casas-Cortés et al. 2008). Our intention here is not to be exhaustive – activist cycling spaces are as diverse as cycles and their users. Rather, we will focus on three paradigmatic examples: mass rides, community cycling initiatives, and feminist cycling spaces.

Mass cycle rides are a direct-action tactic and a staple of bicycle activism everywhere (Carlsson 2002). They are often referred to as "Critical Mass" rides owing to their most famous incarnation, San Francisco's Critical Mass, which is as much a celebration of cycling as it is a performative critique of automobility. Nevertheless, a more global perspective alerts us to their almost simultaneous emergence in Mexico City and Santiago de Chile in the early 1990s. While the practice has been re-appropriated by different groups and in diverse contexts in articulation with their goals (e.g. charity rides), it follows the simple formula of gathering at a specific point in the city and then riding together, taking over streets and temporarily subverting automobilized space-time. Mass rides are often accompanied by practices such as the use of music to enliven city streets where the soundscape is typically dominated by the humdrum of motor traffic. Also common is the use of costumes, and the pursuit of exhilarating experiences derived from the playful use of urban infrastructures (Castañeda 2020). What mass rides show is that space for cycling *might already exist* in the form of streets, avenues, and even highways, and in this way is closely connected to the car-free movement and its critiques of automobile-centred planning and the inequitable distribution of urban space. In other words, they draw attention to the ways in which perceived, conceived and lived space articulate to give way to space for driving, but through practice and re-signification they can also become spaces for cycling (Furness 2007).

Less visible but also significant are community cycling initiatives. This includes bicycle donation programs, cycling schools, mechanics courses, and territorial interventions that anchor cycling practice within specific socio-spatial and socio-economic arrangements (Nixon and Schwanen 2019). Here, "cycling" can be read as a constellation of practices related to the bicycle which need not include *riding* a bike, but nevertheless actualize the possibility of cycling. Leveraging these spaces is important insomuch as they provide people with the knowledge, skills, and community support that are necessary to foster cultural shifts in people's perceptions of what the bicycle is and what it can enable in terms of both transport and social inclusion. This is necessary for dismantling the culture of automobility, intervening the signs, meanings, and practices of mobility (i.e. the representations of space) associated with the modern capitalist paradigm that privileges speed, distance travelled, and individuality (Manderscheid 2014). A "Kool" Routes to school programme in Chile, for example, showed that, in addition to teaching children how to cycle, community cycling has the potential to stimulate environmental values, gender equity, and re-value sustainable transport (Sagaris and Lanfranco 2019). Cycling schools in general remain understudied, but they are ubiquitous in Latin America where they fulfill the crucial role of teaching adults (often adult women, see below) how to cycle. Like the "Kool" Routes program referred above, cycling schools are spaces for "learning and teaching of mobilities not so much as a process of disciplining as a collaborative and inventive practice that generates new ways of shaping movement" (Kullman 2015; Espíndola 2018).

Relatedly, the cycling spaces created for and by women and gender-nonconforming people ought to be a necessary referent to make more inclusive cycling spaces that consider the complex ways in which gender and mobility are co-constitutive, and the uneven outcomes of this process (Hanson 2010). Feminist, women's, and gender non-conforming people's cycling spaces draw attention to the ways in which urban cycling spaces are masculinized environments in ongoing-formation that could become otherwise (Bonham et al. 2015). On average, in Latin American cities, women represent only 25% of cycling trips (Díaz and Rojas, 2017). Issues of fear, the risks associated with automobile-centred transport planning, gendered perceptions of urban cycling, and absence of knowledge and skills necessary to cycle in cities contribute to this disparity (De la Paz Díaz Vázquez 2017). To meet the challenge of gender parity in cycling and combat patriarchal values that circumscribe women's mobilities, feminist cyclists have developed *care*ful space-times intended to create new associations between gender and mobility. For instance, mechanics workshops defy gender roles that associate bike mechanics with masculinity and foster women's autonomous mobility through the sharing of skills necessary to fix one's own bicycle. Cognizant of the gendered division of labor and its complex relationship with transport and mobility, some groups have devised space-times that take into account women's double or triple shifts of both salaried and unpaid care work to ensure a broader range of women can participate (Coyotécatl Contreras and Díaz Alba 2018). Also notable are the cycle rides that foreground girl's presence in the city and highlight their absence in most advocacy and planning discussions, typically centered on productive, adult mobilities. Other spaces are devised to challenge fear of urban night-scapes and "reclaim the night," such as the Ecuadorian feminist bike races that displace competitiveness as the driving force of participation, centering instead values of care through ludic interventions (Gamble 2019). Attention to the critiques, practices, and alternatives being enacted in these feminist cycling spaces is necessary to make spaces for cycling that cater to a diverse range of needs, skills, and mobile practices. Failure to do so will re-inscribe the inequalities that have, until now, limited women and girls' cycle mobility, even in places where cycling infrastructure is widely provided.

Conclusion

Throughout this text, we have explored a number of ways to make space for cycling that challenge the dominant approach to urban cycling – building bike lanes. Starting from a Lefebvrian spatial framework that understands "space" as produced through three interrelated processes (conceived space, perceived space, and lived space), we have analyzed here temporary cycling spaces, policy spaces, and activist spaces to underscore our central claim: that making space for cycling involves more than infrastructural interventions. In this regard, we have problematized certain trends within global urbanism, arguing that the language of "pop-up," DIY or tactical urbanism insufficiently captures the complex process of experimentation that created and expanded Ciclovía, Bogotá's famed open streets program, over the years. Rather, the concept of experimentation can be more fruitful to draw attention to the institutional frameworks that are necessary to support and sustain cycling. As lived spaces, temporary street closings depend on the resignification of automobile infrastructures, and political, technical and institutional coalitions to make them durable in time. What becomes clear is that making space for cycling often requires a vast network infrastructure of spaces for learning and persuasion to make the bicycle desirable for urban policy makers. Hence, forums, conferences and policy tours emerge as a key site in which space for cycling is made. There, particular representations of cycling space circulate amongst planners, urbanists, and activists who take back home with them experiences and stories of urban transformation in which the bicycle plays a central role. However, it is not just in these spaces of encounter that policy space for cycling is made. The internet and social media have provided pro-cycling activists and planners a rich space through which inspirational images, videos, and narratives of bicycle-driven urban change circulate. We cannot think of making space for cycling without considering the multiple actors, discourses, and sites through which cycling circulates, making space for cycling in the global policy landscape.

Finally, we discussed activist cycling spaces, seeking to elucidate how bottom-up initiatives work to enable diverse uses of the bicycle, and argue that making space for cycling need not be a top-down enterprise undertaken by mayors and policy elites. What these spaces do is broaden the scope of what we understand by "cycling" beyond commuting and leisure. Mass cycle rides, as a form of protest, work to subvert automobile space-time and make cycling a collectively enjoyed experience. Yet cycling need not involve riding a bike. Hence we also drew on the experiences of community cycling initiatives to highlight how fixing a bike, or using the bicycle as an excuse to articulate communities, can be just as important as building infrastructure. Furthermore, considering feminist cycling spaces helped us problematize the taken-for-granted assumptions about what kind of space is a space for everyone to cycle. The key lesson drawn from cycling activism here is that we should not stop at taking space to cycle, but use the bicycle to create spaces for integration and learning values of equity and sustainability. This ought to be more conducive to lasting urban transformation than simply relying on cycling infrastructure to "do the trick."

References

Andres, L., Bakare, H., Bryson, J. R., Khaemba, W., Melgaço, L., & Mwaniki, G. R. (2021). Planning, temporary urbanism and citizen-led alternative-substitute place-making in the Global South. *Regional Studies*, 55 (1), 29–39.

Bonham, J., Bacchi, C., & Wanner, T. (2015). Gender and cycling: Gendering cycling subjects and forming bikes, practices and spaces as gendered objects. In J. Bonham & M. Johnson (Eds.), *Cycling Futures*. (Adelaide: University of Adelaide Press).

Bradley, K. (2018). Bike Kitchens – Spaces for convivial tools. *Journal of Cleaner Production*, 197, 1676–1683.

Brown, M. (2021). Cycling in South America, 1880–1920. *Anuario Colombiano de Historia Social y de la Cultura*, 48(1), 287–325.
Carlsson, C. (Ed.). (2002). *Critical Mass: Bicycling's Defiant Celebration*. (Chico, CA: AK Press).
Casas-Cortés, M., Osterweil, M., & Powell, D. E. (2008). Blurring boundaries: Recognizing knowledge-practices in the study of social movements. *Anthropological Quarterly*, 81(1), 17–58.
Castañeda, P. (2021). Cycling case closed? A situated response to Samuel Nello-Deakin's "Environmental determinants of cycling: Not seeing the forest for the trees?" *Journal of Transport Geography*, 90, 102947.
Castañeda, P. (2020). From the right to mobility to the right to the mobile city: Playfulness and mobilities in Bogotá's cycling activism. *Antipode*, 52(1), 58–77.
Cervero, R., Caldwell, B., & Cuellar, J. (2013). Bike-and-ride: Build it and they will come. *Journal of Public Transportation*, 16(4), 83–105.
Cook, I. R., & Ward, K. (2012). Conferences, informational infrastructures and mobile policies: The process of getting Sweden 'BID ready'. *European Urban and Regional Studies*, 19(2), 137–152.
Coyotécatl Contreras, J. M., & Díaz Alba, C. L. (2018). Femibici: Experiencias y reflexiones feministas. *Ciudades*, 119, 47–56.
Cresswell, T. (2010). Towards a Politics of Mobility. *Environment and Planning D*, 28(1), 17–31.
De la Paz Díaz Vázquez, M. S. (2017). La bicicleta en la movilidad cotidiana: Experiencias de mujeres que habitan la Ciudad de México. *Revista Transporte y Territorio*, 16, 112–126.
Del Castillo, A., Sarmiento, O. L., Reis, R. S., & Brownson, R. C. (2011). Translating evidence to policy: Urban interventions and physical activity promotion in Bogotá, Colombia and Curitiba, Brazil. *Translational Behavioral Medicine*, 1(2), pp. 350–360.
Díaz, R., & Rojas, F. (2017). *Mujeres y ciclismo urbano: Promoviendo políticas inclusivas de movilidad en América Latina*. Banco Interamericano de Desarrollo.
Espíndola, M. (2018). ¿Quiénes quieren pedalear? La experiencia de la biciescuela de Ciclofamilia. *Revista Transporte y Territorio*, 19, 81–97.
Furness, Z. (2007). Critical mass, urban space and vélomobility. *Mobilities*, 2(2), 299–319.
Gamble, J. (2019). Playing with infrastructure like a carishina: Feminist cycling in an era of democratic politics. *Antipode*, 51(4), 1166–1184.
Golub, A., Hoffmann, M. L., Lugo, A. E., & Sandoval, G. F. (Eds.). (2016). *Bicycle justice and urban transformation: Biking for all?* (Abingdon, UK: Routledge).
González, S. (2011). Bilbao and Barcelona 'in motion'. How urban regeneration 'models' travel and mutate in the global flows of policy tourism. *Urban Studies*, 48(7), 1397–1418.
Gunder, M. (2011). A metapsychological exploration of the role of popular media in engineering public belief on planning issues. *Planning Theory*, 10(4), 325–343.
Hanson, S. (2010). Gender and mobility: New approaches for informing sustainability. *Gender, Place & Culture*, 17(1), 5–23.
Hoffmann, M., & Lugo, A. (2014). Who is 'World Class'? Transportation justice and bicycle policy. *Urbanites*, 4(1), 45–61.
Kullman, K. (2015). Pedagogical assemblages: Rearranging children's traffic education. *Social & Cultural Geography*, 16(3), 255–275.
Lefebvre, H. (1991). *The Production of Space*. (Oxford, UK: Blackwell).
Lugo, A. (2013). CicLAvia and human infrastructure in Los Angeles: Ethnographic experiments in equitable bike planning. *Journal of Transport Geography*, 30, 202–207.
Lydon, M., & García, A. (2015). *Tactical Urbanism: Short-Term Action for Long-Term Change*. (Washington, DC: Island Press).
Manderscheid, K. (2014). The movement problem, the car and future mobility regimes: Automobility as dispositif and mode of regulation. *Mobilities*, 9(4), 604–626.
McCann, E., & Ward, K. (2012). Assembling urbanism: Following policies and 'studying through' the sites and situations of policy making. *Environment and Planning A*, 44(1), 42–51.
Montero, S. (2018). San Francisco through Bogotá's eyes: Leveraging urban policy change through the circulation of media objects. *International Journal of Urban and Regional Research*, 42(5), 751–768.
Montero, S. (2017a). Worlding Bogotá's Ciclovía: from urban experiment to international "best practice". *Latin American Perspectives*, 44(2), 111–131.
Montero, S. (2017b). Study tours and inter-city policy learning: Mobilizing Bogotá's transportation policies in Guadalajara. *Environment and Planning A: Economy and Space*, 49(2), 332–350.
Montero, S. (2015). *Mobilizing Bogotá: The Local and Transnational Politics of Inter-City Policy Circulation*. Unpublished PhD Dissertation, University of California, Berkeley.

Nixon, D. V., & Schwanen, T. (2019). Bike sharing beyond the norm. *Journal of Transport Geography*, 80, 102492.

Ortiz, J. (1985). Urbanismo informal, urbanismo espontáneo, urbanismo apropiado. La Ciclovia de Bogotá. In Ricardo A. Navarro, Victor Beck & Urs Heierli (Eds.), *Alternativas de Transporte en América Latina: La Bicicleta y los Triciclos*. (Valparaiso: CESTA/SKAT).

Pucher, J., & Buehler, R. (2008). Making cycling irresistible: Lessons from The Netherlands, Denmark and Germany. *Transport Reviews*, 28(4), 495–528.

Robinson, J., & Roy, A. (2016). Debate on global urbanisms and the nature of urban theory. *International Journal of Urban and Regional Research*, 40(1), 181–186.

Sagaris, L., & Lanfranco, D. (2019). Beyond "safe": Chilean "Kool" routes to school address social determinants of health. *Journal of Transport & Health*, 15, 100665.

Sarmiento, O. L., Pedraza, C., Triana, C. A., Díaz, D. P., González, S. A., & Montero, S. (2017). Promotion of recreational walking: Case study of the Ciclovía-Recreativa of Bogotá. In Corinne Mulley, Klaus Gebel & Ding Ding (Eds.), *Walking*. (Bingley, UK: Emerald Publishing Limited). https://www.emerald.com/insight/publication/doi/10.1108/S2044-994120179

Sarmiento Casas, C. (2018). Bicioficios: Un potencial inexplorado. *Ciudades*, 119, 47–56.

Sosa López, O. (2021). Bicycle policy in Mexico City: Urban experiments and differentiated citizenship. *International Journal of Urban and Regional Research*, 45(3), 477–497.

Spinney, J. (2020). *Understanding Urban Cycling: Exploring the Relationship Between Mobility, Sustainability and Capital*. (Abingdon, UK: Routledge).

Stehlin, J. (2019). *Cyclescapes of the Unequal City: Bicycle Infrastructure and Uneven Development*. (Minneapolis, MN: University of Minnesota Press).

Webb, D. (2018). Tactical urbanism: Delineating a critical praxis. *Planning Theory & Practice*, 19(1), 58–73.

Vignette E
B2W INDONESIA AND THE RE-CYCLING OF JAKARTA

Purwanto Setiadi

It isn't an overstatement to describe Jakarta in the mid-2000s as a jungle: a cramped, cacophonous mess where might makes right. It was in such an environment that Devin Octavianus made his daily commute to his workplace on a bicycle. The distance between his (then) residence in Bintaro, Tangerang, and his workplace on Thamrin Avenue, Central Jakarta, was about 23 kilometers. Leaving home at 6:00 a.m., the ride usually took him about an hour. It was only 7 minutes faster than a trip in a private motor vehicle, but that small difference was attributable to the route he needed to take to avoid heavy traffic. "I charted my own course, through alleyways... It was not the shortest route in terms of distance, but it was time-efficient," Devin explained. Devin's routine was a rarity as the "common sense" of the time held that cycling in Jakarta could be no more than an exercise or a recreational activity on weekends, when traffic was light. Otherwise, it took nerves of steel to ride a bicycle at other times.

Devin's decision to make cycling a daily routine did not come spontaneously. He had previously been engaged in mountain biking with a community that often rides on a track along a gas pipeline near his residence. His decision to commute on a bicycle was driven by his concerns regarding Jakarta's traffic and, most importantly, air pollution in the city. As with many other major cities in the world, rapid but disorganized development and the increasing affluence of its residents burdened Jakarta with an abundance of problems. Among them are traffic congestion, fuel wastage, and air pollution, driven by increasing private ownership of motor vehicles that have been further spurred on by the lack of adequate public transportation services. By 1999 poor air quality had cost Jakarta US$220 million in healthcare. The city government attempted to introduce pollution control measures in 2001, but occasional improvements in air quality were attributable more to seasonal rainfall than to the control measures.

Devin's personal choice to commute by bicycle has inspired his friends to turn it into a movement. Sharing the same conviction that the air quality in Jakarta (along with the surrounding Greater Jakarta area, commonly known as Jabodetabek) had become intolerably hazardous to health, they committed themselves, as Devin said, to "demonstrate to urban residents that bicycles also have a place in the city; that they can be a means of transport and not just for recreation."

 DOI: 10.4324/9781003142041-32

The movement took a while to take shape, as it took time to discuss, plan its activities, and to complete other preparations. Eventually, the declaration of the movement named Komunitas Pekerja Berspeda (Community of Cycling Workers) was made on August 6, 2005, at Jakarta City Hall, in the presence of the then Vice-Governor of Jakarta, Fauzi Bowo, and over 500 cyclists. This movement would later become Bike to Work (B2W) Indonesia. The declaration was followed with a string of campaigns that culminated in the B2W Day event on August 27, 2006, at Bundaran HI in the heart of Jakarta. Attended by over 1,300 cyclists, the event commemorated the inauguration of B2W Indonesia. But of course, re-cycling a society that has abandoned bicycles as a mode of transportation is a challenging undertaking, comparable to climbing a steep slope. It involves changing habits and disrupting established conventions. Relying on self-funded operations, B2W Indonesia's strategy involves four tracks: campaigning, advocacy, education, and social networking. Taufik Hidayat, the first chairman of B2W Indonesia who recited the City Hall declaration recounted, "I deployed propaganda expressions (to effectively communicate the messages to the public)."

Experience has shown that public policy is essential to enact cultural change and to initiate new behaviors and habits. This means that government leadership has a decisive role to play. But in Jakarta, that leadership was lacking in the early days of the movement. Jakarta's successive governors tended to just continue their predecessors' programs, a trait common among regional leaders in Indonesia. Implementing new ideas is not only rare, but also restrained them from introducing actual radical changes.

When it comes to public transportation, at the time Jakarta had actually begun implementing TransJakarta, a bus rapid transport system with dedicated bus lanes – inspired by the similar TransMilenio system in Bogotá, Colombia. This system was initiated by then Governor Sutiyoso to reduce the use of private motor vehicles and the traffic congestion and air pollution that come with them. But Sutiyoso was not interested in further innovation in the transportation system by creating spaces for bicycles, although he agreed to support car-free Sundays, an event first held on May 23, 2002, and organized by a coalition of environmental civil society organizations located around Sudirman-Thamrin Avenue. Sutiyoso's successor, Fauzi Bowo, did not stray from that position. His response to the suggestion to construct bike lanes had been to subtly reject it by placing a condition: "We'll build them when there's a million (cyclists)," he said, as recalled by Taufik. Another fact that also came to the attention of B2W Indonesia's board was that there is a deep-seated belief that bike lanes do not generate revenue for regional governments.

B2W Indonesia's mission gained a stronger footing when Law No. 22/2009 on Road Traffic and Transportation was passed. As Taufik's successor as B2W Indonesia's chairman Toto Sugito said, "It gave us more confidence in running our campaigns and advocacy." B2W Indonesia had even taken part in the deliberation of the law's draft, providing inputs for articles related to bicycles as a mode of transportation and the protection of cyclists' safety.

B2W Indonesia's efforts have spurred the formation of communities focusing on bicycles as a means of mobility, such as Robek (Bekasi), Rosela (South Jakarta) and Rodex (Depok). The presence of these communities and B2W Indonesia's supporters and volunteers in them have increased public awareness of commuting by bicycle. Seeing a group of bicycle commuters from Depok, one of the satellite cities adjacent to Jakarta, has attracted Dicky Nolan to do the same. An employee at a printing plant affiliated with an influential media group, since 2008 he has commuted by bicycle from his residence in Lenteng Agung, South Jakarta, for 25 kilometers every day. He professed to have started the routine "to have some exercise while going to work," but eventually "fell in love with the bicycle." Never discouraged, before

Figure E.1 A reggae group bicycle to a performance in central Jakarta (author's photo).

Covid-19 pandemic he had participated in B2W Indonesia's campaigns and educational activities downtown before the start of office hours (Figure E1).

Dicky is just one individual among thousands of workers with varying professions who, intentionally or not, have affirmed B2W Indonesia's standing. Following his election into office in 2017, Jakarta Governor Anies Baswedan has also recognized bicycles as a part of the overhaul of Jakarta's transportation system – which foregrounds humans for sustainable urban living. B2W Indonesia has been engaged in the process of carrying it through. "We are pleased to be able to take parts in the forums as we have been working towards these positive developments for a long time," asserted the current chairman of B2W Indonesia, Poetoet Soedarjanto.

One of the strategies followed to achieve this goal has been the setting up of a cycling network on Jakarta's relatively flat roads, including establishing cycle lanes and supporting infrastructure as well as implementing a bikeshare system, and integration with public transportation services, with 63 kilometers created at the time of this writing. This is a long-term project and the pandemic has obstructed its progress, but the Jakarta administration still supports the project and its 500 kilometer target in future plans and programs for coming years. The urgency of the project has instead been boosted by the increase of cyclists during the pandemic – making this a critical moment to seize.

B2W Indonesia, with support from local bicycle manufacturers, welcomes the bike lanes as a step forward in the Jakarta administration's vision, in comparison to the period when Devin started his bike commute. But, Poetoet concluded, B2W Indonesia is committed to carry on campaigning, advocating, educating, and social networking until "cycling has become embedded in the daily lives of Jakarta's residents, and among Indonesians in general."

Further reading

Magnusson J. and Rachmita, F. (2021) "Jakarta is what resiliency looks like", *Sustainable Transport* (January) pp. 6–8. https://www.itdp.org/wp-content/uploads/2021/03/ITDP_ST32_Jakarta_Is_What_Resiliency_Looks_Like.pdf

Oktavianti, T. I. (2020) "Jakartans turn to bicycles to commute in 'new normal'", https://www.thejakartapost.com/news/2020/06/14/jakartans-turns-to-bicycles-to-commute-in-new-normal.html (Accessed 31/3/2021).

The Jakarta Post (2021) "Our bicycle friendly city", (5th March) https://www.thejakartapost.com/academia/2021/03/05/our-bicycle-friendly-city.html (Accessed 31/3/2021).

23

SHARED MICROMOBILITY

Policy, practices, and emerging futures

Susan Shaheen, Adam Cohen and Jacquelyn Broader

Shared micromobility – or short-term access to shared bikes and scooters – provides a flexible alternative for households living in urban areas, households seeking first- and last-mile connections to public transportation, and those without access to a private vehicle trying to access jobs and essential services. Up until the global pandemic, shared micromobility grew worldwide on a relatively steep growth curve, beginning in the early 2010s. Shared micromobility is a transportation strategy that enables users' short-term access to a transportation mode on an as-needed basis (Shaheen et al. 2019). Shared micromobility includes a number of operational models, including station-based micromobility (where a bicycle or scooter is picked up from and returned to any station or kiosk) and dockless (or stationless) micromobility (where a bicycle or scooter is picked up and returned to any location). Another service model, sometimes referred to as a 'hybrid model,' blends aspects of station-based and dockless systems that allows users to check out a bicycle or scooter from a station and end their trip either returning it to a station or a non-station location (or vice versa) (Shaheen and Cohen 2019).

Common shared micromobility services include bicycle and scooter sharing. Fundamentally, bikesharing and scooter sharing provide users with on-demand bikes and scooters for one-way (point-to-point) or roundtrip travel. Both bike and scooter sharing models can include a variety of motorized and non-motorized devices, including moped-style scooters. Typically, shared micromobility fleets are deployed in a network within a metropolitan region, city, neighborhood, employment center, and/or university campus. The cost of shared micromobility services usually includes maintenance, parking and, if applicable, electric charge or gasoline.

Shared micromobility has the potential to offer an array of community and individual benefits, such as enhanced mobility, greater environmental awareness, and increased use of active transportation and non-vehicular modes. This chapter discusses the growth of shared micromobility, its impacts on users and communities, and policy considerations for managing potential adverse impacts of shared micromobility, such as increased curbspace demand. This chapter is organized into four sections. The first discusses the growth and evolution of shared micromobility. The next section summarizes user demographics and shared micromobility impacts. In the third section, the authors discuss shared micromobility policies and practices for managing devices and operations in a few contrasting urban environments. The final section concludes with a discussion of the future of shared micromobility.

DOI: 10.4324/9781003142041-33

Brief history, growth, and evolution of shared micromobility

Over the last five decades, shared micromobility's evolution can be categorized into four key phases beginning with bikesharing. These include the first generation, called "White Bikes" (or Free Bikes); the second generation: "Coin-Deposit Systems;" third generation or "Information Technology (IT)-Based Systems;" and fourth generation systems: "Advanced IT-Based Systems."

Shared micromobility originated in Europe with small non-profit systems in the 1960s. First-generation bikesharing, or White Bikes, began in Amsterdam in 1965, when 50 bicycles were left unlocked throughout the city for free public use (Home 1991). This initiative failed soon after its launch, however, because bikes were often stolen, damaged, and even confiscated by the police (Schimmelpennink, L., December 2012, unpublished data). These challenges with first-generation systems led Copenhagen to launch the first large-scale, second-generation coin-deposit system in 1995. Smaller scale coin-deposit systems began launching in the early 1990s. By designating specific bicycle station locations and adding coin-deposit locks, second-generation systems were much more reliable, as users have a defined and secure space to access available bicycles. However, due the customer anonymity that is associated with coin-deposit systems, theft is still a notable concern.

These shortcomings contributed to the development of IT-based systems. With IT-based services, shared micromobility operators are able to identify and track individual users, allowing monetary and legal enforcement for damaged equipment and inappropriate user behaviors. While the technology was first associated with a bikesharing system at Portsmouth University in the United Kingdom, Vélo à la Carte, which launched in 1998 in Rennes, France, was the first IT-based system available for public access (DeMaio 2009).

In addition to these IT-based shared micromobility deployments in the developed world, IT-enabled bike and scooter sharing systems are also growing in developing economies. In Africa, the first IT-based bikesharing system, Medina Bike, started operating in 2016 in Marrakech, Morocco. In the Middle East, the United Arab Emirates (UAE) has become an epicenter of shared micromobility activity. Cyacle, an initiative of the Khalifa Fund for Enterprise Development, launched a station-based bikesharing program with 75 bicycles and 11 stations in Abu Dhabi in December 2014. This service was acquired by Careem in 2019. Additionally, a number of service providers offer bike and moped sharing in India. SmartBike is a station-based bikesharing program with more than 1,500 bicycles and 150 stations in Chennai, Chandigarh and New Delhi. Yulu is a dockless micromobility provider offering users access to shared bikes and e-bikes in India. In South-East Asia, one of the earliest bikesharing programs, Pun Bike Share, was launched in Bangkok, Thailand in October 2012. The service had an estimated 500 bikes and 50 stations in May 2021.

More advanced fourth-generation IT-based systems feature demand-responsive rebalancing (e.g., real-time information that informs the system where there are imbalances in supply and demand) and integrate – both spatially and digitally – with other transportation modes. They also can include dockless charging stations; electric bikes; public transit linkages; mobile, solar docking stations; and integration with mobility as a service (MaaS) and public transport fare payment.

In recent years, the advent of IT-based shared micromobility systems began to enable a number of new business and operational models, such as peer-to-peer (P2P) bikesharing. P2P micromobility services involve the sharing of privately owned micromobility devices where companies broker transactions among micromobility owners and guests by providing the organizational resources needed to make the exchange possible (e.g., locking mechanism, online platforms, etc.). In 2012, the smartphone app Spinlister launched a P2P bicycle rental marketplace where a bike owner could make their bicycle available to others for short time periods, enabling direct exchanges between individuals via the Internet. Spinlister eventually shut down in April 2018, but it relaunched in January 2019 with new features including remote locking and bicycle valet (where a bicycle is delivered to a user). At the same time that Spinlister was launching in 2013, another company, BitLock, created a keyless bike lock accessible via smartphone technology, enabling another P2P bikesharing option. Improvements in technology also contributed to the development of dockless shared micromobility options. In addition to P2P and station-based services provided by B-Cycle, Motivate, and others, a number of new dockless vendors entered the marketplace including: JUMP (formerly Social Bicycles), Limebike, MoBike, Ofo, Spin, and an array of smaller service providers offering dockless systems that allow users to pick up or drop off enabled bicycles anywhere within a geographic area. Although not required for a dockless system, some bikes are equipped with a locking mechanism on the device that allow the bikes to be "locked to" a bicycle rack, street furniture, or a designated bikesharing rack. Users identify bicycle availability and locations in real time through mobile or Internet applications or via bikesharing kiosk screens. The geographic proximity of bikesharing (docked and dockless systems) can be limited through "geofencing." A geofence is a virtual perimeter, which limits the range of mobility of an enabled bicycle, by comparing the GPS-satellite coordinates of the bicycle to the allowable geographic area.

As of May 2018, an estimated 1,608 bikesharing programs were operational globally (Gauequelin 2020). In 2019 in the US, 50 million bikesharing trips were taken – 40 million station-based and 10 million dockless electric trips (National Association of City Transportation Officials 2020). During the same time period, European bikesharing operators were in more than 350 cities and provided 65 million rides (Motor Insights 2020). Scooter sharing also has experienced a rapid growth. In 2019, moped-style scooter sharing was available in 88 cities across 21 countries internationally (Gauequelin 2020). Moped-style scooter sharing grew by 164% in 2019 and 44% in 2020 to reach a total of 95,000 available devices (Gauequelin 2020). In the US in 2019, 86 million trips were taken via standing electric scooter sharing services (National Association of City Transportation Officials 2020). Additionally, according to the September 2020 report from Mobility Foresights, scooter sharing was available in approximately 97 European cities. Mobility Foresights (forthcoming) also reported that in 2020 in Asia there were over 15 scooter sharing operators; however, most were piloting schemes and were not fully operational (Mobility Foresights Forthcoming, 2020).

In spite of this growth, enabled by large venture capital investment, a number of cities around the world saw a reduction in dockless shared micromobility fleets in late 2019, with some cities reporting increased use and others decreased use during the COVID-19 pandemic (Wilson 2020; Grogan and Hise 2020). There are a variety of reasons that contributed to the reduction in dockless micromobility fleets. In many cases, a large number of devices deployed in small geographic areas within a short timeframe contributed to complaints about improper parking and vandalism. This led some operators to reduce fleets, change operational strategies around parking and fleet balancing, and in some cases exit markets altogether. In response to the global pandemic, some cities have implemented "slow or healthy street" programs intended to support micromobility, outdoor dining, and other outside socially distanced activities in response to the pandemic.

User demographics and shared micromobility impacts

A number of studies have documented the demographic profiles of shared micromobility users and their impacts. Older studies of shared micromobility have found that users are typically: (1) well educated (often with a college or graduate degree); (2) younger (typically between 21 and 45 years of age); (3) childless households; (4) middle- and upper-income households; and (5) households living in more urban neighborhoods, often with limited vehicle access (e.g., zero or one car) that tend to use multiple transportation modes, such as public transportation, cycling, and walking.

These studies often reflect the demographics of early adopters, urban lifestyles, and households without children for a number of reasons. First, urban neighborhoods tend to be more walkable, bikeable, and less conducive to private vehicle use (e.g., expensive and limited parking). Moreover, the presence of children in a household is commonly associated with higher levels of household vehicle ownership. Additionally, most shared micromobility programs are not designed for families with small children. Finally, active transportation (particularly cycling) in the US can be associated with a social stigma whereas private vehicle ownership is often viewed as a status symbol. Additionally, in many lower-density North American communities limited public transport service reinforces private vehicle ownership due to its role in facilitating access to jobs, healthy food, education, and health care for individuals in communities with more limited mobility options.

In addition to user demographics, a number of studies have documented the impacts of station-based bikesharing, while studies of dockless bikesharing and scooter sharing are still emerging. Broadly, these studies of shared micromobility have documented impacts in four areas: (1) environmental impacts, (2) mode shift and substitution, (3) public health impacts, and (4) safety.

A number of studies have found that shared micromobility reduces greenhouse gas (GHG) emissions by replacing personal vehicle trips (Shaheen and Cohen 2019; Martin et al. 2020). Studies also indicate that shared micromobility has the potential to reduce congestion and fuel use, lower emissions, and increase environmental awareness. Shared micromobility also can be integral to bridging spatial and temporal gaps in the transportation network and encouraging multi-modal trips. The impacts of shared micromobility on personal vehicle and public transit use tend to vary by operational model (i.e., station-based and dockless); device (i.e., bicycle or scooter); and study location (Shaheen et al. 2012, 2014; McNeil et al. 2017; Fishman 2015). A number of studies conclude that shared micromobility could be an effective first- and last-mile strategy to connect travelers to public transportation, while others indicate that micromobility may also cause shifts away from public transit (e.g., more direct bike and scooter trips replacing long public transit routes with transfers and/or long headways between buses or trains) (Shaheen et al. 2014). Some of these studies document shifts toward public transit in situations where bikesharing tends to be more prevalent in lower-density regions on the urban periphery. In this context, station-based bikesharing may serve as a first- and last-mile connection in communities with lower densities and less mature public transportation networks. Similarly, the findings suggest that regions with higher densities and more mature public transit systems, station-based bikesharing may offer faster, cheaper, and more direct connections compared to shorter distance transit trips. For these reasons, bikesharing may be more complementary to public transit in small and medium regions and more substitutive in larger metropolitan areas. However, even competition with public transit could be a strategic goal for some transit agencies seeking to perhaps provide strategic relief to crowded public transit lines during peak periods (Shaheen and Martin 2015). In addition to public transit impacts, bikesharing and scooter sharing also have been used to replace vehicular modes – including taxis, transportation network companies (TNCs, also known as ridehailing and ridesourcing), and personal vehicles – and facilitate trips that would previously not have been taken (Martin et al. 2020). However, more research is needed to study the impacts on mode choice, particularly related to dockless micromobility services.

Shared micromobility may increase active transportation use, thereby having the potential to improve public health. A study of station-based bikesharing found an increase in physical activity among users (Martin et al. 2020). Some studies also have concluded that micromobility users report reduced stress and increased weight loss due to bikesharing. However, a key limitation associated with many of these health impact assessments is that they do not examine negative health impacts associated with ridership, such as the costs associated with increased exposure and risks to injuries and collisions from micromobility use (Alberts, Palumbo, and Pierce 2012). One study of standing electric scooter sharing found that it attracted new people to active transportation (such as walking and cycling) (Portland Bureau of Transportation 2018).

With respect to safety, studies have found that shared micromobility users tend not to wear helmets; however, more research is needed to determine the overall impacts of these behavioral differences on safety outcomes. One retrospective study of scooter sharing safety in Los Angeles, California between September 2017 and August 2018 found that scooter-related injuries are common with varying levels of severity due to high speeds, low rates of compliance with rider age requirements, and low rates of helmet use (Trivedi et al. 2019). Although studies have documented a high number of micromobility-related injuries, more research is needed to understand key risk factors such as: (1) unsafe rider behavior, (2) appropriate speeds (for vehicles and micromobility users), and (3) infrastructure design that both prevents and contributes to scooter sharing user injuries (Shaheen and Cohen 2019).

Shared micromobility policies and practices

While shared micromobility has the potential to offer individual and community benefits, the growth of shared micromobility is causing some urban centers to become increasingly congested as a variety of modes compete for space to pick-up, drop-off, and use micromobility devices (Shaheen et al. 2021). Dedicating bike lanes and curbspace for micromobility is an important policy area confronting public agencies. Key elements of micromobility policies typically include: (1) device caps, (2) service area limitations, (3) designated parking areas, (4) fees, (5) equipment and operational requirements, and (6) enforcement. Each of these are described in greater detail below.

1 **Device Caps:** Fleet caps are employed to limit the number of bicycles, scooters, or other devices that can be used for shared micromobility. Public agencies may limit the number devices in a category (e.g., dockless bikesharing, standing electric scooter sharing, etc.) or the number of devices per operator. Establishing device caps can be difficult for public agencies and operators because the number of devices needed to create an adequate network varies based on a number of factors such as: service area, built environment, density, and usage frequency. Device caps may also have unintended consequences that limit the supply of available devices, reduce the size of service areas (to ensure adequate coverage in remaining service areas), and potentially result in reduced coverage in less profitable neighborhoods (which could raise social equity concerns for low-income and minority communities).
2 **Service Area Limitations:** Some cities have geographic access zones where operators can deploy devices. Access limitations can include permissible and prohibited operational areas that may be enforced through virtual geographic boundaries (sometimes referred to as a 'geofence').
3 **Designated Parking Areas:** A number of communities have designated parking areas for micromobility devices. These parking areas can include where to park a device on the

curb, a requirement to lock or attach a device to a bicycle rack or other piece of street furniture, or a condition to return a device to a designated station or corral (e.g., a painted, barricaded, or geofenced parking location for shared micromobility devices).

4 **Fees:** A number of cities charge operator fees for allowing the placement of shared micromobility devices in the public rights-of-way. Examples of these fees include per trip taxes, application fees, and annual fees based on the number of devices deployed. In the US, for example, Portland charges a $0.25 tax per scooter ride. The funds are placed in a "New Mobility Account" that finances program administration, enforcement, infrastructure improvements, and access for underserved communities. Barcelona, Spain charges an annual fee per shared scooter (about $85 US per device) to fund micromobility infrastructure improvements (Garcia Valdivia 2020). Similarly, Stockholm, Sweden charges an annual permit fee of approximately $166 US per shared scooter (Drive Sweden 2021). In Auckland, New Zealand, scooter sharing operators pay approximately $24.75 US per device for a six-month permit in inner-city locations. Devices that operate outside the city and in the suburbs are assessed between $3.50 and $15.00 US depending on the location (Nadkarni 2019). In Singapore, bikesharing companies pay approximately $44 US per bicycle permit fee and $1,100 US to apply for the permit (Choo 2018). In the US, Chicago and St. Louis charge an application fee (typically $250 to $500) per operator. In Japan, a proposed payment system would collect fees from bikesharing operators and users to pay for micromobility infrastructure and program management (Suzuki and Nakamura 2017).

   Other fees that cities have assessed include: (1) fees per docking station, (2) performance bonds (to protect the public entity if the micromobility company goes out of business or fails to meet certain terms under a contractual agreement), or (3) escrow (bonded) payments per device (or a block of devices).

5 **Equipment and Operational Requirements:** Cities also can establish equipment requirements (such as maximum allowable operating speeds) and permissible areas of operation, such as prohibitions from operating devices on sidewalks, bicycle lanes, pedestrian malls, etc.

6 **Enforcement:** Enforcement is important to ensure that shared micromobility devices are safely parked and equitably dispersed throughout a neighborhood. Enforcement also can help ensure that devices do not impede access for people with disabilities. A variety of methods can be used to enforce micromobility use and operational guidelines. For example, Santa Monica is one of the first cities to require operators to use geofencing to improve rider compliance and safety by establishing deactivation zones around the city's beach area, effectively slowing devices to zero miles/kilometers per hour in areas of high pedestrian activity (Iglesias 2019). Some cities also have found that when fleets become stagnant (i.e., not used because they are parked in low-traffic areas) and imbalanced (i.e., too many devices located in a particular area), devices can end up congregating in areas of low activity and contributing to curbspace management challenges. Some cities have developed policies that require service providers to rebalance fleets on a particular schedule and correct parking violations within a specific time frame to address these challenges.

One North American operator, Spin, has developed a three-wheel scooter intended to help with parking compliance and reduce the need for enforcement (Figure 23.1). Because the scooter has three wheels, it is less susceptible to tipping over and blocking curbs. The three-wheel design also allows Spin operators to be able to remotely move the scooter and reposition it if a parking violation is detected. The company plans to add a feature that would allow

Figure 23.1 Spin's three-wheeled scooter.

Source: Ford Media Center, 2021.

customers to e-Hail a scooter and have it autonomously dropped off with a user. By eliminating the need for centralized parking locations, the ability to autonomously deliver a shared micromobility device has the potential to change the way curbspace is managed in the future.

Future of shared micromobility

Over the next decade, enhancements in battery range, charging times, and weight will likely contribute to the evolution and development of additional devices and/or new "form factors" (e.g., motorized quadricycles, light electric vehicles (EVs), electric auto-rickshaws, and neighborhood EVs that carry two to four passengers and operate at speeds up to 25 miles (approximately 40 kilometers) per hour). However, new form factors and operating speeds have the potential to raise a number of operational and safety challenges for shared micromobility users. Thoughtful planning and policy are needed to manage this and other emerging issues to minimize potentially adverse impacts and maximize opportunities of these new developments.

In the future, automation, safety, data privacy, and public policy could impact the evolution of shared micromobility. Automation of shared micromobility vehicles could have transformative impacts. Automating shared micromobility devices could help to simplify curbspace management and charging by allowing devices to be delivered to a person's door and returned automatically at the conclusion of a trip. Automation also could allow the parking of micromobility devices (both personally owned and shared) to be relocated to private property. While the automation of shared micromobility devices creates opportunities, vehicle automation could pose a number of risks. Shared automated vehicles could compete with micromobility for short urban trips, particularly if per trip or per mile/kilometer costs are more competitive. Early evidence from the global pandemic suggests that micromobility may be serving longer urban trips; however, more research is needed to systematically document these observations. It is important to note that vehicle automation could reinforce historic infrastructure funding and design biases that prioritize motorized vehicles over active transportation.

In addition to automation, safety also could impact community acceptance and shared micromobility growth. Several improvements could enhance safety and encourage ridership such as: (1) improved device design (e.g., larger wheels to reduce the impacts of potholes); (2) infrastructure enhancements (e.g., better pavement quality, dedicated facilities for shared micromobility use, and curbspace management); and (3) education and outreach with users (e.g., public awareness and "share the road" campaigns).

Data privacy also could impact public acceptance of shared micromobility whose operators typically track several important user data metrics, such as trip origin and destination, travel time, and trip duration. However, these data may reveal the daily routines or the residences/workplaces of users. Implementing industry-wide data protection and compliance standards could be key to protecting sensitive data; managing risk; and enhancing consumer confidence in shared micromobility (e.g., the Mobility Data Specification [or MDS], which has been adopted in the US and several other nations). MDS is a data standard that allows cities to gather, analyze, and compare real-time and historical data from shared micromobility providers. The specification also serves as a tool that can help enforce local regulations.

In the future, the growth and success of shared micromobility will be largely dependent on public policy. Prioritizing parking and high-visibility locations for bikes and scooters; enhancing infrastructure (e.g., slow lanes, multi-use trails, corrals, stations, etc.); and incorporating bike and scooter sharing into MaaS or mobility on demand (MOD) platforms could increase shared micromobility connectivity for users. By enhancing the visibility and convenience of shared micromobility and reducing rider stress, communities have an opportunity to encourage the use of bikes and scooters for first- and last-mile connections to public transit and shorter distance trip making. Since the global pandemic, micromobility has become an integral strategy for many cities across the globe to encourage safe, active transportation, while accommodating the need for social distancing. Many cities have expanded street space for active transportation to reduce traffic volumes and speeds and to expand space for pedestrians, cyclists, scooter riders, and outdoor recreation. The pandemic recovery presents an opportunity for communities to institutionalize these policies and encourage greater use of micromobility.

References

Alberts, Brian, Jamie Palumbo, and Eric Pierce. *Vehicle 4 Change: Health Implications of the Capital Bikeshare Program*. Washington, DC: The George Washington University, 2012.

Choo, Cynthia. *New licensing regime for bike-sharing operators to kick in from October: MOT*. 2018. https://www.todayonline.com/singapore/new-licensing-regime-bike-sharing-operators-kick-october-mot

DeMaio, Paul. "Bike-sharing: History, Impacts, Models of Provision, and Future." *Journal of Public Transportation*, 12, no. 4 (2009): pp. 41–56.

Drive Sweden. *Shifting Sands of e-Scooter Supervision in Sweden*. 2021. https://www.drivesweden.net/en/shifting-sands-e-scooter-supervision-sweden

Fishman, Elliot. "Bikeshare: A Review of Recent Literature." *Transport Reviews*, 36 (2015): pp. 92–113.

Ford Media Center. *Ford-Owned Spin Announces Exclusive Partnership with Tortoise to Bring Remotely Operated E-Scooters to North American and European Cities in 2021*. 2021. https://media.ford.com/content/fordmedia/fna/us/en/news/2021/01/27/ford-spin-tortoise-e-scooters.html

Garcia Valdivia, Ana. *Barcelona's Moped – Sharing New Licenses to Challenge Leading Companies*. 2020. https://www.forbes.com/sites/anagarciavaldivia/2020/02/14/barcelonas-moped-sharing-new-licenses-to-challenge-leading-companies/?sh=77270ef7592d

Gauequelin, Alexandre. "Reality Check." *Shared Micromobility*, November 3, 2020. https://shared-micromobility.com/reality-check/

Grogan, Thomas, and Phaeda Hise. *Corona Bicycle Metrics: Where Bicycling Increased and (Surprise!) Decreased*. July 21, 2020. https://www.streetlightdata.com/corona-bicycle-metrics/?type=blog/

Home, Stewart. *The Assault on Culture: Utopian Currents from Lettrisme to Class War.* Edinburgh: AK Press, 1991.

Iglesias, Miranda. *Looking Back at the Shared Mobility Pilot Program.* 2019. https://www.santamonica.gov/blog/looking-back-at-the-shared-mobility-pilot-program

Martin, Elliot, Ziad Yasine, Matthew Lin, Susan Shaheen, Adrian Witt, Belinda Judelman, and Mia Senders. *2nd Annual Shared Micromobility State of the Industry Report.* North American Bikeshare and Scootershare Association, 2020.

McNeil, Nathan, Jennifer Dill, John MacArthur, and Joseph Broach. *Breaking Barriers to Bike Share: Insights from Bike Share Users.* Portland: Portland State University, 2017.

Mobility Foresights. *Electric Scooter Sharing Market in US and Europe 2021–2026.* 2020.

Mobility Foresights. *Scooter Sharing Market in Asia 2019–2024.* Forthcoming.

Motor Insights. *Europe Bike Sharing – Market Growth, Trends, COVID-19 Impacts, and Forecasts (2021–2026).* 2020.

Nadkarni, Anuja. *Auckland Council quadruples fees for e-scooter companies.* 2019. https://www.stuff.co.nz/business/112430086/auckland-council-quadruples-fees-for-escooter-companies

National Association of City Transportation Officials. *Slow Streets for Pandemic Response & Recovery.* Last modified May 11, 2020. https://nacto.org/publication/streets-for-pandemic-response-recovery.com

Portland Bureau of Transportation. *2018 E-Scooter Findings Report.* Portland, OR: Bureau of Transportation, 2018. https://nabsa.net/2020/09/03/industryreport/

Shaheen, Susan, and Adam Cohen. *Shared Micromobility Policy Toolkit.* Palo Alto: Schmidt Family Foundation, 2019.

Shaheen, Susan, Adam Cohen, Mark Dowd, and Richard Davis. *A Framework for Integrating Transportation into Smart Cities.* San Jose: Mineta Transportation Institute, 2019.

Shaheen, Susan, and Elliot Martin. "Unraveling the Modal Impacts of Bikesharing." *ACCESS Magazine*, 1, no. 47 (2015): p. 9.

Shaheen, Susan, Elliot Martin, Adam Cohen, and Jacquelyn Broader. *Shared Micromobility and Curbspace Management.* San Jose: Mineta Transportation Institute, 2021.

Shaheen, Susan, Elliot Martin, Adam Cohen, and Rachel Finson. *Public Bikesharing in North America: Early Operator and User Understanding.* San Jose: Mineta Transportation Institute, 2012.

Shaheen, Susan, Elliot Martin, Nelson Chan, Adam Cohen, and Michael Pogodzinski. *Public Bikesharing in North America During a Period of Rapid Expansion: Understanding Business Models, Industry Trends and User Impacts.* San Jose: Mineta Transportation Institute, 2014.

Suzuki, Mio, and Hiroki Nakamura. "Bike share deployment and strategies in Japan." *Roundtable on Integrated and Sustainable Urban Transport.* 2017.

Trivedi, Tarak K., Charles Liu, Anna Liza M. Antonio, Natasha Wheaton, Vanessa Kreger, Anna Yap, David Schriger, and Joann G. Elmore. "Injuries Associated With Standing Electric Scooter Use." *JAMA Network Open*, 2 (2019): 1.

Wilson, Kea. *Why Do Micromobility Companies Keep Losing Money?* Last modified January 14, 2020. https://usa.streetsblog.org/2020/01/14/why-do-micromobility-companies-keep-losing-money/

Further Reading

Cohen, A., and Susan Shaheen. *Planning for Shared Mobility.* Chicago: American Planning Association, 2018.

Shaheen, S., and Adam Cohen. *Shared Micromoblity Policy Toolkit: Docked and Dockless Bike and Scooter Sharing.* UC Berkeley: Transportation Sustainability Research Center, 2019.

Shaheen, S., Adam Cohen, Jacquelyn Broader, Richard Davis, Les Brown, Radha Neelakantan, and Deepak Gopalakrishna. *Mobility on Demand Planning and Implementation: Current Practices, Innovations, and Emerging Mobility Futures.* Washington, DC: US Department of Transportation, 2020.

Shaheen, S., and Stacey Guzman. "Worldwide Bikesharing". *Access Magazine*, no. 39 (2012).

Wang, J., J. Huang, and M. Dunford. "Rethinking the utility of public bicycles: The development and challenges of station-less bike sharing in China." *Sustainability*, 11, no. 6 (2019): p. 1539.

Zhao, Naidong, Xihui Zhang, Shane M. Banks, and Mingke Xiong. "Bicycle sharing in China: Past, present, and future." *SAIS 2018 Proceedings*, 2018.

24
E-BIKES
Expanding the practice of cycling?

Dimitri Marincek and Patrick Rérat

Defining e-bikes

Electrically-assisted bicycles (or e-bikes) are bicycles combining muscular power with an electric assistance which activates when pedaling. E-bikes are enjoying an increasing commercial success. Whilst this success is also part of a larger trend of the rebirth of cycling, the development of e-bikes has been more rapid than the uptake in conventional cycling in many countries. E-bikes have been most successful in traditionally high-cycling countries like the Netherlands and Belgium, where they are overtaking the sales of conventional bicycles for adults (Forbes 2019).

Two main categories exist: e-bikes limited to an assistance of 25 km/h (or 20 mph/32 km/h in the United States and Canada) ("pedelecs"); and those with an assistance up to 45 km/h ("speed-pedelecs" or "s-pedelecs"). Pedelecs account for the bulk of e-bike sales and are considered legally equivalent to conventional bicycles. In most countries, speed-pedelecs are restricted to on-road use and forbidden from using bicycle paths, a legal status which has hampered their development. Indeed, the only countries to our knowledge where they are allowed on cycle paths are Switzerland and Belgium, where the highest shares of speed-pedelecs are found (15% and 4% of e-bike sales, compared to 0.9% in the Netherlands, and 0.5% in Germany). E-bikes also increasingly include recreational bicycles (for mountain biking and road cycling) and various kinds of cargo-bikes to carry goods or children.

E-bikes possess a hybrid status between strictly muscle-powered bicycles and motorized two-wheelers, combining the advantages of both. Pedaling makes for exercise and the electric assistance multiplies the rider's abilities, especially in some circumstances (going uphill, against the wind, carrying heavy loads, etc.). However, some vehicles labeled as "electric bicycles" do not require pedaling and are controlled by a throttle (Behrendt 2018; Rose 2012). They are especially popular in China (Weinert et al. 2007), where an e-bike boom at the end of the 1990s was called a "policy accident" because it resulted from a ban on gasoline-powered two-wheelers (Yang 2010). Currently, there are about 300 million electric bikes on China's roads (Bloomberg 2021), with the country being the world's foremost producer and market for electric two-wheelers. This kind of "no-pedaling" e-bike falls outside the definition adopted in this chapter.

DOI: 10.4324/9781003142041-34

In comparison to conventional bicycles, e-bikes travel faster and make it easier to cover longer distances (Cairns et al. 2017; Jones, Harms et al. 2016), reduce the barrier of topography (Lopez et al. 2017; MacArthur et al. 2014) and facilitate carrying loads or children. They also accelerate more rapidly and maintain higher speeds while requiring less physical exertion (Popovich et al. 2014). Because of their 'combination of leg and battery power' (Behrendt 2018: 64), e-bikes play an 'intermediator role' (Wolf & Seebauer 2014) or that of a 'transitional step' (Popovich et al. 2014) between conventional bikes and cars. By reducing the effort needed to cycle, e-bikes broaden the appeal of cycling to a larger spectrum of users and, due to their increased range, substitute trips made by motorized modes.

This chapter traces the history of the e-bike before giving an overview of the literature on e-bikes. It then draws on research projects the authors carried out in Switzerland to highlight how e-bikes contribute to extend the practice of cycling in terms of population groups and spatial contexts. The conclusion identifies some key questions for research and policy.

Historical perspective

Bicycles with an electric or gasoline-powered motor have existed since the start of the 20th century. Indeed, soon after the invention of the modern safety bicycle, motorized versions were developed. The Simplex, a bicycle equipped with an electric motor and battery, was for example produced in the 1920s in the Netherlands (spinningmagnets 2013). In the 1950s, as motorcycles became heavier and more powerful, cheaper and lighter vehicles which had both pedals and a motor – *Mopeds* – including the French *Velosolex* (a bicycle with a gasoline-powered motor driving the front wheel), became popular, especially among the youth. However, in contrast to today's e-bikes, they required pedaling only to start up or going uphill.

Much later, in the early 1990s, the first bicycles with an electric assistance delivered while pedaling were invented. The first mass-produced modern e-bike was the Yamaha Pedal-Assist System in 1993 which was marketed to older people. Yet it took another 10 years for sales of e-bikes to start booming, as technological improvements in battery technology, from lead-acid batteries to lithium-ion cells, enabled a reduction in weight and increased safety and performance.

Since the 2010s e-bikes have grown into a major market. Between 2014 and 2019, the number of e-bikes sold annually in Europe grew from 1.1 to 3.4 million (+150%) (CONEBI 2020). E-bikes are by far the best-selling electric vehicles. The highest market penetration rates are in the Netherlands (24 e-bikes sold annually per 1000 inhabitants), followed by Belgium (22), Germany, Switzerland, and Austria (16) (see Table 24.1). These countries also have the highest proportions of e-bikes among new bicycle sales (the Netherlands, 42%; Switzerland, 37%; Belgium, 33%; Austria, 33%; Germany, 32%).

E-cycling: main findings of the literature

Research on e-bikes addresses four main questions: Which profile do e-bike users have? What are their motivations, barriers, and experiences? How much potential do e-bikes hold for substituting other travel modes? Which impacts do e-bikes have in terms of health and environment?

Electric assistance has opened cycling to a larger base. One trait of e-bike users is their higher age compared to conventional cyclists. Users between 50 and 65 years old are over-represented (Johnson & Rose 2013; MacArthur et al. 2014; Simsekoglu & Klöckner 2019; de Kruijf et al. 2019), while some studies report a majority of retired people (Wolf & Seebauer 2014).

Table 24.1 E-bike sales per country (in thousands of units), 2014 to 2019

Sales (in 1000s of units)	*2014*	*2015*	*2016*	*2017*	*2018*	*2019*	*% change 2014–2019*
Total (Europe)	1140	1358	1667	N/A	2772	3397	150
Germany	480	535	605	N/A	980	1360	154
Netherlands	223	276	273	N/A	409	423	53
France	78	102	134	N/A	338	388	280
Belgium	130	141	168	N/A	259	251	78
Italy	51	56	124	N/A	173	195	248
Austria	50	77	87	N/A	150	143	86
Spain	18	25	40	N/A	110	143	472
UK	50	40	75	N/A	61	101	153
Switzerland	58	66	76	88	112	133	102

Source: Switzerland: Velosuisse; other countries: CONEBI, retrieved through https://bovagrai.info

The first buyers of e-bikes were older individuals, before spreading to younger cohorts, contrary to the stereotype of early adopters of an innovation being young (Peine et al. 2017). Young adults under 25 remain under-represented, due to the e-bikes' price and image (related to older users) and their stronger physical condition.

In terms of gender, e-bike users tend to follow the patterns of conventional cyclists. In cycle-friendly countries like Denmark and the Netherlands, where women are a majority among cyclists, they are over-represented (Haustein & Møller 2016), whereas a majority are men in countries such as the United States or Australia (Johnson & Rose 2013; MacArthur et al. 2014), where fewer people cycle. E-bike users live in households mostly composed of families or couples, with income and education levels above average (Johnson & Rose 2013; MacArthur et al. 2014; Wolf & Seebauer 2014).

The motivations for and barriers to e-cycling are similar to those found for conventional cycling (Haustein & Møller 2016). However, there are some differences related to electric assistance, such as the possibility to cycle despite steep gradients or long distances without sweating or feeling tired, even for people with a lower level of fitness (Dill & Rose 2012; Haustein & Møller 2016; MacArthur et al. 2014; Popovich et al. 2014). It may also be easier to complete a succession of journeys (activity chain) and to escort children with a trailer or a child seat (Jones, Harms, et al. 2016). For couples, the e-bike may present a way of working out and cycling together, as it "equalizes" differences of physical condition (Popovich et al. 2014). Improving one's health may be a motivation for e-cycling as well, because, despite electric assistance, it provides moderate-intensity physical activity and generates health benefits (Bourne et al. 2018).

However, some characteristics of e-bikes may be perceived negatively. As they are more expensive, their owners may be more concerned about theft and the need for safe storage. This may explain low e-bike ownership in cities in comparison to suburban and rural areas where housing (e.g. detached houses) may provide more adequate storage space (Ravalet et al. 2018). E-bikes are also heavier and more difficult to handle, and their weight may exacerbate 'range anxiety', which is the fear that the battery has an insufficient range to reach the destination (Popovich et al. 2014).

Another aspect is safety, with rising numbers of accidents following increases in sales (Schepers et al. 2014). The question remains open as to whether e-bikers are intrinsically more prone to accidents than conventional cyclists. Accidents may be linked to e-bikes' characteristics (higher weight and speed, more frequent use), but also to the riders (older, lower experience) or to motorists who underestimate e-bikers' speed (Petzoldt et al. 2017). Not much attention has been given to perceived safety but some accounts suggest that e-bike users feel more confident due to the ability to keep up with the flow of motor traffic and accelerate quickly from a stop (Jones, Harms, et al. 2016; Popovich et al. 2014; Rose 2012).

E-bikes have the potential for a modal shift by allowing for longer trips than conventional bicycles, though estimates of average trip distance vary from 3 to 11.5 km between studies (Bourne et al. 2020). Switching to an e-bike mostly affects either car use or conventional cycling depending on the dominant forms of mobility in the setting of the study (Sun et al. 2020). In car-centered contexts like North America or Australia, the e-bike is considered as a way to reduce car use (Johnson & Rose 2013; MacArthur et al. 2014), though this result has also been found in Sweden (Hiselius & Svensson 2017). Conversely, the e-bike mostly substitutes conventional cycling in countries where the population is already cycling at a high rate, such as in Denmark (Haustein & Møller 2016) or the Netherlands (Kroesen 2017). In China, where public transport is dominant, a modal shift has been observed from public transport to the e-bike (Cherry et al. 2016).

Although a shift from car use is more beneficial in environmental terms than one from conventional cycling (Rose 2012), both are positive as studies show an increase in the volume and duration of trips with e-bikes compared to conventional bicycles (Fyhri & Fearnley 2015). Furthermore, as we will discuss below, the e-bike allows some people to continue cycling (despite changes such as a new residential location, carrying a child, ageing, etc.) and to avoid shifting to motorized modes of transport.

Despite the electric assistance, e-bikes manage to provide a meaningful amount of physical activity (Bourne et al. 2018), especially when compared to non-active modes. Crucially, they contribute to better health for ageing users (Johnson & Rose 2015; Van Cauwenberg et al. 2019) as they improve cognitive functions and mental health through engagement with the outdoor environment, greater independence and mobility (Jones, Chatterjee, et al. 2016; Leyland et al. 2019). Health benefits concern all population groups as e-bikes are a way to incorporate physical activity in increasingly sedentary lifestyles (Gojanovic et al. 2011). Even when taking into account exposure to accidents and air pollution, the benefits of e-cycling for public health remain overwhelmingly positive (Götschi et al. 2016).

In terms of environmental impact, e-bikes consume energy through their electric assistance, but roughly half of the energy comes from the cyclist. Moreover, due to a low weight of around 20–30 kilograms, e-bikes are much lighter than most other vehicles. This translates to an overall energy consumption of around 1 kWh per 100 kilometers, which varies depending on gearing, assistance level, and hilliness. An electric car consumes 10–33 times as much energy to travel the same distance, depending on the model and its weight (Weiss et al. 2020). While the question of how lithium-ion batteries are produced and recycled is important, the amount of materials used in an e-bike is very limited in comparison to electric cars. A comparison of the lifecycle of different transport modes shows the e-bike's low ecological footprint is only beaten by the mechanical bike (or slightly by train in the case of Switzerland) (International Transport Forum 2020; OFEV 2018). This highlights the importance of the e-bike in the transition towards a low-carbon mobility.

E-cycling: expanding the practice of cycling?

In this section, we present two research projects in Switzerland to highlight how the e-bike may expand the social groups and spaces of cycling, and extend cycling over the life course. In Switzerland, 7% of all journeys are made by bike. This is higher than in most English-speaking and Latin countries and lower than in Northern Europe. As seen above, e-bikes represent more than a third of new bike sales, which makes Switzerland an interesting case study.

The first study was carried out among participants in the *Bike to Work* campaign who commit to cycling to work as much as possible in May and/or June (Rérat 2021a, 2021b). Among respondents, 10,833 (83.5%) are conventional cyclists and 2,141 e-bikers (16.5%). Folding bike and bike-sharing users (147) were removed from the sample. The e-bike widens participation in cycling across social groups: women (49% among e-bikers, 41% among conventional cyclists), people over 40 (76% vs 57%), and parents (56% vs 44%). It also shows a diversification of e-bikers in terms of gender and age. Although under-represented, younger cyclists rely on the e-bike to cover longer distances and to transport children.

The e-bike makes it possible to overcome some barriers faced by conventional cyclists, such as gradient, physical effort and distance (half of the users of a conventional bike spend 30 minutes or less on their commute both ways, compared to only one-third of e-bikers). It also expands cycling across space: People living in suburban and rural areas are much more present among e-bikers (53% and 23%) than among mechanical cyclists (43% and 14%), while the opposite is found among urban dwellers (23% vs 43%).

E-bikes reach groups that use motor transport more than average, expanding the practice of cycling as a complement or alternative to automobility. While the current cycling renaissance is mainly observed in cities, it has the potential to reach other spaces. However, both e-cycling and conventional cycling face similar challenges in terms of a lack of dedicated infrastructures and safety.

The second study (Marincek et al. 2020; Marincek & Rérat 2020) is based on 24 biographical interviews with e-bike users living in Lausanne, Switzerland. Participants had varied profiles and backgrounds in cycling, and had received a subsidy for an e-bike. The goal was to understand their cycling trajectories or "thoughts, feelings, capabilities, and actions related to cycling" over the life course (Chatterjee et al. 2012, p. 83). Adoption of the e-bike followed two main trajectories.

Restorative trajectories accounted for 14 users (58%). They had interrupted cycling by at least one year, sometimes much longer, most often in favor of the car. The e-bike enabled them to restore a regular cycling practice (Figure 24.1). Some were "returning to cycling" as they had already cycled regularly for transport before. Physical activity provided their main motivation, with e-bikes enabling an acceptable level of effort. A second group were "starting to cycle for transport". They adopted e-cycling as an efficient transport compared to car use. A third group used the e-bike as a means of "reinforcing a return to cycling" that they started with a conventional bike.

Resilient trajectories referred to 10 users (42%) who were already cycling regularly (Figure 24.2). As many chose not to own a car, cycling was their main mode of transport. The electric assistance served as a way of continuing cycling despite changes in their residential location or personal life that could have led to the reduction or interruption of utility cycling. A first group "replaced conventional cycling with the e-bike" completely. The assistance helped them to continue cycling despite the hilly topography, the need to carry children, or the advance of age. A second group alternated conventional cycling and e-cycling. The e-bike was used for utilitarian trips and in winter, whereas the conventional bicycle was preferred for summer and sport.

Both cycling trajectories show that e-bikes attract people who (re)start cycling, but, crucially, also help cyclists to sustain their practice. Moreover, they extend cycling during ageing.

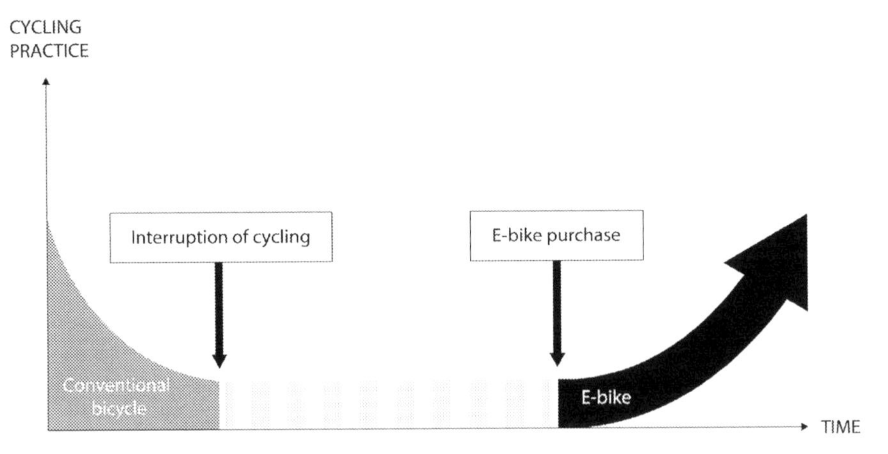

Figure 24.1 Restorative cycling trajectory. (Copyright Marincek & Rérat 2020).

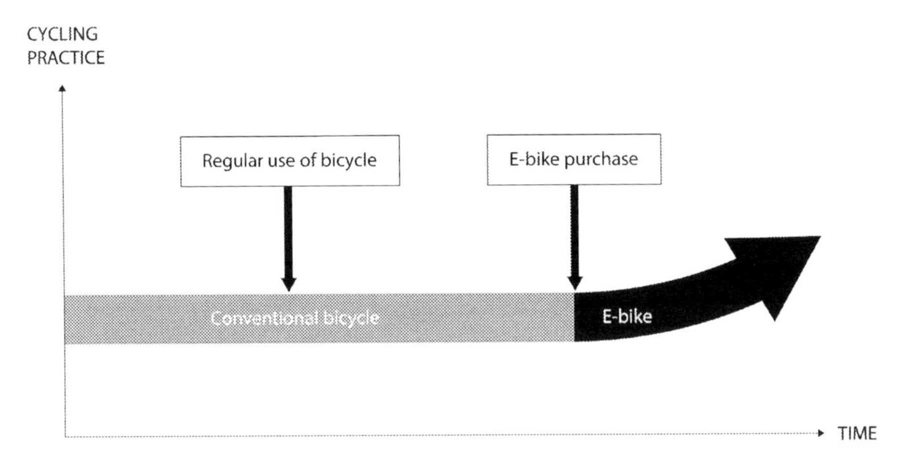

Figure 24.2 Resilient cycling trajectory. (Copyright Marincek & Rérat 2020).

Outlook: the future of (e-)cycling?

What does the future hold for cycling, and which place will e-cycling have in it? How will e-cycling evolve in years to come? Sales have been booming and have accelerated during the COVID-19 pandemic, with demand exceeding supply. What will be the effects of this trend? Will the diffusion of the e-bike within social groups and spatial contexts continue? Will e-cycling be considered not merely as a way to compensate for a lower physical condition, but also as an opportunity to cycle beyond what was previously thought possible? Will the e-cycling boom reach more rural spaces and countries outside the Global North and China? Will the current societal changes linked to the pandemic have lasting influences on travel practices, breaking routines, and inspiring users to experiment with something new, such as the e-bike? If so, how will such changes in mobility habits affect other modes of travel? The strong demand for cycling during the pandemic was linked to three phenomena: utility cycling to avoid crowding in public transport; sport cycling as traditional sport facilities are closed; and leisure cycling as international tourism has been halted. What will be the long-term effects on cycling?

A second series of questions relates to the way electric assistance changes the shapes and functions of bicycles and increases the potential of cycling. First, it increases the power, speed, and spatial range of bicycles. With advances in e-bike motors and batteries, how far can e-bikes be developed, and should legislation block or encourage these changes? Speed-pedelecs already push the boundaries of the performance of cycling, while alternative shapes such as electrically assisted velomobiles could rival cars or two-wheelers. Integrating e-bikes into bike-sharing systems may enlarge their catchment areas, foster intermodality with public transport and allow non-users to experience e-bikes. Second, electric assistance allows for larger bikes serving a wider array of uses and needs. Various forms of cargo bikes (front-loading, back-loading) strongly increase the carrying capacity of bikes used by families or businesses. Third, electric assistance is becoming more discreet and virtually invisible. Small motors and batteries can be fitted inside the frame, blurring the boundaries between existing categories of mechanical bikes and e-bikes.

Finally, the role and place of e-bikes depends on the politics of velomobility. Cycling as a practice requires infrastructures and regulations to make it safe, efficient, and enjoyable. During the pandemic, some cities have implemented pop-up bike lanes while others have expanded their networks to cope with increasing volumes of bicycle traffic. New cycling infrastructure does not go unchallenged as it implies a reallocation of space to the detriment of the dominant system of automobility. The rise of e-cycling forces planners to consider a greater diversity of bikes in terms of speed and size (as well as micromobility devices such as e-scooters), and creates new requirements for storage and maintenance services. In sum, e-cycling is set to play an increasing role in future sustainable velomobility.

References

Behrendt, F. (2018). Why cycling matters for electric mobility: Towards diverse, active and sustainable e-mobilities. *Mobilities*, *13*(1), 64–80.

Bloomberg. (2021, April 5). *E-Bikes Rule China's Urban Streets: Hyperdrive Daily*. https://www.bloomberg.com/news/newsletters/2021-04-05/hyperdrive-daily-e-bikes-rule-china-s-urban-streets

Bourne, J. E., Cooper, A. R., Kelly, P., Kinnear, F. J., England, C., Leary, S., & Page, A. (2020). The impact of e-cycling on travel behaviour: A scoping review. *Journal of Transport & Health*, *19*, 100910. https://doi.org/10.1016/j.jth.2020.100910

Bourne, J. E., Sauchelli, S., Perry, R., Page, A., Leary, S., England, C., & Cooper, A. R. (2018). Health benefits of electrically-assisted cycling: A systematic review. *International Journal of Behavioral Nutrition and Physical Activity*, *15*(1), 116. https://doi.org/10.1186/s12966-018-0751-8

Cairns, S., Behrendt, F., Raffo, D., Beaumont, C., & Kiefer, C. (2017). Electrically-assisted bikes: Potential impacts on travel behaviour. *Transportation Research Part A: Policy and Practice*, *103*, 327–342. https://doi.org/10.1016/j.tra.2017.03.007

Chatterjee, K., Sherwin, H., Jain, J., Christensen, J., & Marsh, S. (2012). Conceptual model to explain turning points in travel behavior: Application to bicycle use. *Transportation Research Record: Journal of the Transportation Research Board*, *2322*, 82–90.

Cherry, C. R., Yang, H., Jones, L. R., & He, M. (2016). Dynamics of electric bike ownership and use in Kunming, China. *Transport Policy*, *45*, 127–135. https://doi.org/10.1016/j.tranpol.2015.09.007

CONEBI. (2020). *2020 CONEBI Bicycle Industry & Market Profile*. CONEBI – Confederation of the European Bicycle Industry.

de Kruijf, J., Ettema, D., & Dijst, M. (2019). A longitudinal evaluation of satisfaction with e-cycling in daily commuting in the Netherlands. *Travel Behaviour and Society*, *16*, 192–200. https://doi.org/10.1016/j.tbs.2018.04.003

Dill, J., & Rose, G. (2012). Electric bikes and transportation policy: Insights from early adopters. *Transportation Research Record: Journal of the Transportation Research Board*, *2314*, 1–6.

Forbes. (2019). When will e-bike sales overtake sales of bicycles? For The Netherlands, that's now. *Forbes*. https://www.forbes.com/sites/carltonreid/2019/03/02/when-will-e-bike-sales-overtake-sales-of-bicycles-for-the-netherlands-thats-now/

Fyhri, A., & Fearnley, N. (2015). Effects of e-bikes on bicycle use and mode share. *Transportation Research Part D: Transport and Environment, 36*, 45–52.
Gojanovic, B., Welker, J., Iglesias, K., Daucourt, C., & Gremion, G. (2011). Electric bicycles as a new active transportation modality to promote health. *Medicine & Science in Sports & Exercise, 43*(11), 2204–2210.
Götschi, T., Garrard, J., & Giles-Corti, B. (2016). Cycling as a part of daily life: A review of health perspectives. *Transport Reviews, 36*(1), 45–71. https://doi.org/10.1080/01441647.2015.1057877
Haustein, S., & Møller, M. (2016). Age and attitude: Changes in cycling patterns of different e-bike user segments. *International Journal of Sustainable Transportation, 10*(9), 836–846.
Hiselius, L. W., & Svensson, Å. (2017). E-bike use in Sweden–CO_2 effects due to modal change and municipal promotion strategies. *Journal of Cleaner Production, 141*, 818–824.
International Transport Forum. (2020). *Good to Go? Assessing the Environmental Performance of New Mobility*. https://www.itf-oecd.org/sites/default/files/docs/environmental-performance-new-mobility.pdf
Johnson, M., & Rose, G. (2015). Extending life on the bike: Electric bike use by older Australians. *Journal of Transport & Health, 2*(2), 276–283.
Johnson, M., & Rose, G. (2013). Electric bikes–cycling in the New World City: An investigation of Australian electric bicycle owners and the decision making process for purchase. *Proceedings of the 2013 Australasian Transport Research Forum, 13*.
Jones, T., Chatterjee, K., Spinney, J., Street, E., Van Reekum, C., Spencer, B., Jones, H., Leyland, L. A., Mann, C., & Williams, S. (2016). *Cycle BOOM. Design for Lifelong Health and Wellbeing*. Oxford Brookes University, Oxford.
Jones, T., Harms, L., & Heinen, E. (2016). Motives, perceptions and experiences of electric bicycle owners and implications for health, wellbeing and mobility. *Journal of Transport Geography, 53*, 41–49.
Kroesen, M. (2017). To what extent do e-bikes substitute travel by other modes? Evidence from the Netherlands. *Transportation Research Part D: Transport and Environment, 53*, 377–387.
Leyland, L.-A., Spencer, B., Beale, N., Jones, T., & van Reekum, C. M. (2019). The effect of cycling on cognitive function and well-being in older adults. *PLOS ONE, 14*(2), e0211779. https://doi.org/10.1371/journal.pone.0211779
Lopez, A. J., Astegiano, P., Ochoa, D., Tampere, C., Beckx, C., & Gautama, S. (2017). Unveiling e-bike potential for commuting trips from GPS traces. *ISPRS International Journal of Geo-Information, 6*(7), 190.
MacArthur, J., Dill, J., & Person, M. (2014). Electric bikes in North America: Results of an online survey. *Transportation Research Record: Journal of the Transportation Research Board, 2468*, 123–130.
Marincek, D., Ravalet, E., & Rérat, P. (2020). The cycling trajectories of e-bike users: A biographical approach. In *Mobility Across the Life Course. A Dialogue between Qualitative and Quantitative Research Approaches* (pp. 221–241). Edward Elgar Publishing, Cheltenham, UK; Northampton, MA.
Marincek, D., & Rérat, P. (2020). From conventional to electrically-assisted cycling. A biographical approach to the adoption of the e-bike. *International Journal of Sustainable Transportation*. https://doi.org/10.1080/15568318.2020.1799119
OFEV [Swiss Federal Office for the Environment]. (2018). *Rapport sur l'environnement 2018*. Bern: OFEV.
Peine, A., van Cooten, V., & Neven, L. (2017). Rejuvenating design: Bikes, batteries, and older adopters in the diffusion of e-bikes. *Science, Technology, & Human Values, 42*(3), 429–459.
Petzoldt, T., Schleinitz, K., Heilmann, S., & Gehlert, T. (2017). Traffic conflicts and their contextual factors when riding conventional vs. Electric bicycles. *Transportation Research Part F: Traffic Psychology and Behaviour, 46, Part B*, 477–490. https://doi.org/10.1016/j.trf.2016.06.010
Popovich, N., Gordon, E., Shao, Z., Xing, Y., Wang, Y., & Handy, S. (2014). Experiences of electric bicycle users in the Sacramento, California area. *Travel Behaviour and Society, 1*(2), 37–44. https://doi.org/10.1016/j.tbs.2013.10.006
Ravalet, E., Marincek, D., & Rérat, P. (2018). Les vélos à assistance électrique: Entre vélos conventionnels et deux-roues motorisés? [E-bikes: Between conventional bicycles and motorized two-wheelers?]. *Géo-Regards: Revue Neuchâteloise de Géographie, 11–12*, 93–111.
Rérat, P. (2021a). *Cycling to Work: An Analysis of the Practice of Utility Cycling*. Springer Nature.
Rérat, P. (2021b). The rise of the e-bike: Towards an extension of the practice of cycling? *Mobilities, 16*(3), 423–439.
Rose, G. (2012). E-bikes and urban transportation: Emerging issues and unresolved questions. *Transportation, 39*(1), 81–96.
Schepers, P., Fishman, E., Den Hertog, P., Wolt, K. K., & Schwab, A. L. (2014). The safety of electrically assisted bicycles compared to classic bicycles. *Accident Analysis & Prevention, 73*, 174–180.

Simsekoglu, Ö., & Klöckner, C. A. (2019). The role of psychological and socio-demographical factors for electric bike use in Norway. *International Journal of Sustainable Transportation, 13*(5), 315–323.
spinningmagnets. (2013, November 9). Electric Bike History, patents from the 1800's. *ElectricBike.Com.* https://www.electricbike.com/e-bike-patents-from-the-1800s/
Sun, Q., Feng, T., Kemperman, A., & Spahn, A. (2020). Modal shift implications of e-bike use in the Netherlands: Moving towards sustainability? *Transportation Research Part D: Transport and Environment, 78*, 102202. https://doi.org/10.1016/j.trd.2019.102202
Van Cauwenberg, J., De Bourdeaudhuij, I., Clarys, P., de Geus, B., & Deforche, B. (2019). E-bikes among older adults: Benefits, disadvantages, usage and crash characteristics. *Transportation, 46*(6), 2151–2172.
Weinert, J., Ma, C., & Cherry, C. (2007). The transition to electric bikes in China: History and key reasons for rapid growth. *Transportation, 34*(3), 301–318.
Weiss, M., Cloos, K. C., & Helmers, E. (2020). Energy efficiency trade-offs in small to large electric vehicles. *Environmental Sciences Europe, 32*(1), 46. https://doi.org/10.1186/s12302-020-00307-8
Wolf, A., & Seebauer, S. (2014). Technology adoption of electric bicycles: A survey among early adopters. *Transportation Research Part A: Policy and Practice, 69*, 196–211. https://doi.org/10.1016/j.tra.2014.08.007
Yang, C.-J. (2010). Launching strategy for electric vehicles: Lessons from China and Taiwan. *Technological Forecasting and Social Change*, 77(5), 831–834.

25

CYCLING SAFETY AS MOBILITY JUSTICE

Léa Ravensbergen, Ron Buliung and Ahmed El-Geneidy

Safety concerns: a top barrier to cycling

The past two decades have witnessed an increased interest in cycle planning in many cities from a vast and diverse array of stakeholders with vested interests in urban transport. In some cases, the hard work of advocacy groups has been realized through the planning and building of cycling-supportive facilities; the political will necessary to accomplish such things has also appeared in some cities, and the recent pandemic has, in some places, accelerated the installation of both temporary and more permanent cycling facilities. Concurrently, participation in urban cycling has grown considerably. The most dramatic increases have occurred in cities where cycling was not previously a regular means of daily travel, such as North American cities where cycling's mode share has tripled or even quadrupled since 1990. Cycling is also re-emerging in some Chinese cities (Chapter 36), and is on the rise in South America, Europe, Australia, and even in some cities that have historically had high cycling rates such as Amsterdam and Copenhagen (Pucher and Buehler 2021; Chapter 37).

When it comes to encouraging city cycling, a key policy priority has been to foster a safe cycling city. Indeed, concerns about traffic safety are consistently identified as a barrier to uptake (Elvik 2021). Improving cycling safety is also important for reducing cyclist injuries, a public health priority. Given this elevated status of "safety" in cycling discourse of all kinds, we explore cycling safety in multiple contexts in this chapter. We begin with a brief literature review, highlighting key research findings and debates in studies on cycling safety. Much of this work focuses on traffic safety and considers either 'objective' or 'perceived' safety at the scale of the individual or the neighborhood. Then, a different approach for thinking about cycling safety is introduced: reimagining cycling safety as mobility justice. This approach involves a more comprehensive and holistic view of safety that considers cyclists' safety not just from cars, but from other forms of danger such as crime or sexual or police harassment. In so doing we explore the intersectional power relations that can shape one's safety while on the move.

 DOI: 10.4324/9781003142041-35

Table 25.1 Approaches to studying cycling safety

	Objective	*Perceived*
Individual	– Helmet design, regulation, and use – Cyclist or driver behavior – Safety or health risk due to cycling – Variations across demographic groups	– Self-reported safety – Effects of 'near-misses' on cyclists' – Variations across demographic groups
Neighborhood	– Urban planning – Infrastructure presence & design – Effect of built environment features on cyclist safety or health	– Self-reported ratings of safety, risk, or danger of infrastructure of street design

Approaches to studying cycling safety

Diverse disciplines have contributed to cycling safety research, including geography, epidemiology, health sciences, planning and engineering. Much of this literature focuses on bicycle–motor vehicle interactions, and key topics include cyclist behavior (helmet usage, distracted riding), crash causation, evaluation of injury-prevention solutions, urban planning, and infrastructure design (Dozza et al. 2017). Most scholars working on cycling safety take up positivist or post-positivist onto-epistemological approaches, where knowledge about safety is produced through measurement, survey, hypothesis testing, application of the scientific method. This work often focuses on measuring injuries, crashes, and conflicts, or 'objective' safety studies, or analyzing self-reported or subjective safety, or 'perceived' safety research, at the scale of the individual or the neighborhood (Table 25.1).

Objective safety

Work on 'objective' bicycling safety typically focuses on measuring and modelling injuries, crashes, or conflicts (Reynolds et al. 2009). Three sources of data are commonly used: police records, hospital (or other medical facility) admission records, and cyclist surveys where incidents are self-reported. Official crash statistics notoriously under-represent cycling incidents, as many go unreported to the police and/or do not require formal medical care (Elvik 2021).

Much of the literature on objective safety focuses on the individual-level protection of cyclists (Winters et al. 2012). Many of these studies examine the role of cyclist and driver behavior factors (e.g. violations, substance use, training, etc.) on collisions. In the North American context in particular, cycling safety research has focused on helmets, for instance their design, regulation, and/or use (Reynolds et al. 2009). A key finding is that helmet use can reduce the severity of injury (Attewell et al. 2001). However, previous work has found that mandatory helmet use can discourage cycling (Reynolds et al. 2009). Many studies have also found that injury rates decrease as cycling increases, a phenomenon referred to as the "safety in numbers hypothesis" (Prati et al. 2019). Some argue that helmet laws may work against safety in numbers by deterring cycling uptake.

A sub-set of studies on cycling safety specifically examine probabilistic risk. Much of this research identifies the probability of a safety or health-related metric (e.g., injury, collision, all-cause mortality, life-years gained, etc.) resulting from a measure of cycling exposure (e.g., number of days of cycling, total number of trips, time spent cycling). A recent literature review found that cycling is consistently associated with reduced risk of all-cause mortality (Zhao et al. 2021). Another main finding in this body of work is that cyclists experience greater risk of injury in places where cycling for transport is not the norm (Elvik 2021). However, when studies compare the safety risks and benefits of cycling; many report that the health benefits of cycling far outweigh the health risks (Pucher et al. 2010).

Objective safety can also be examined at the scale of the neighborhood. Many papers explore intersection safety, or the impact of bicycle lane design on safety. Longitudinal studies have explored how safety changes before and after infrastructure is built (Thomas & DeRobertis 2013). Overall, the presence of bicycle infrastructure is consistently associated with greater safety and lower risk, including reduced collisions, decreased injuries, and less severe injury (Prati et al. 2019). Taken together, key topics in objective cycling safety research include the role of helmets, the safety in numbers hypothesis, the health and safety risks of cycling, and the protective effects of infrastructure. By focusing on injuries, crashes, and conflicts, this body of work does not necessarily reflect whether cyclists feel safe. Therefore, the following section reviews another approach commonly used in cycling research: studies examining perceived safety.

Perceived safety

Research on cycling safety has also examined perceived safety, or self-reported or subjective safety. Perceived safety from traffic can be measured at the individual level in travel surveys and has been found to impact individuals' decision to cycle, as well as their route choice (Winters et al. 2011). Low perceived traffic safety has also been identified as one of the most important barriers to cycling uptake by potential cyclists (Manaugh et al. 2017). More recently, perceived safety has been studied at the individual level by examining the impact of 'near-misses', i.e., non-injury incidents between cyclists and other road users, which have been found to negatively impact cyclists' perceptions of safety (Sanders 2015). Finally, research on individual-level perceived safety has also focused on concerns over personal safety from crime (Willis et al. 2015).

Many studies examine perceived safety with a focus on neighborhood features. For example, Delmelle and Delmelle (2012) found that bicycle infrastructure is associated with lower perceived concern over safety. Others have compared perceived safety across different types of streetscapes (Winters et al. 2012). For instance, Winters et al. (2012) found that major streets with shared lanes and no parked cars have the greatest perceived risk while multiuse paths have the lowest. Overall, these studies indicate that cyclists perceive greater safety when cycling within infrastructure separated from traffic.

Though objective and perceived safety are measured differently, there is evidence that they are closely related. For example, Winters et al. (2012) compared the objective and perceived risk of different types of cycling infrastructures in Toronto and Vancouver and found that there was little discrepancy between perceived and objective measures. Further, regardless of how safety is measured, cycling infrastructure has consistently been associated with higher rates of cycling participation (Manaugh et al. 2017), which, in turn, can increase the effect of the "safety in numbers" trend discussed earlier. There is also evidence that both objective and perceived safety are not experienced evenly across the population, which will be explored in the next section.

Who is safe? Who feels safe?

Research on both objective and perceived safety has also examined how these factors vary across social and demographic factors. For instance, risk of fatal injury per kilometer cycled has been found to be higher for people aged 65 years or older (Elvik 2021). Past experience cycling, measured as number of bike trips per week, number of years' experience cycling, or distance cycled per week, is also associated with reduced risk of injury (Hollingworth et al. 2015). The type of bicycle you ride may also influence safety. Notably, recent work has examined safety considerations for electric bike (or e-bike) riders, an increasingly popular emergent technology that may be associated with greater injury risk and severity than conventional bikes (Elvik 2021).

The literature on cycling safety has focused considerable attention on "sex" and/or "gender," largely depending on the disciplinary home of the scholar(s). When controlling for past experience and/or distance cycled, some studies have found that women experience lower injury risk than men (Hollingworth et al. 2015). Much of the work on gender and safety has focused on perceived safety and many studies have found that women express greater concern over safety than men in places with low cycling rates (Ravensbergen et al. 2019). Indeed, women's great concern over safety has been put forth as an explanation for the gender gap in cycling, the trend whereby women represent approximately one third of cyclists in low-cycling areas (Ravensbergen et al. 2019). In other work, we have taken issue with such normative, almost biological, explanations for this gender gap that often rely on the gender binary and fail to capture social constructions of gender and how gender interacts with other axes of social difference (Ravensbergen et al. 2019). There is some evidence that safety concerns are larger barriers to other social groups as well, such as children (Clayton and Musselwhite 2013). Research has also begun considering the relationship between race, ethnicity and/or income and cycling safety in recent years. Cycling and walking are often examined together here, and evidence indicates that income and ethnicity-based disparities exist in safe walkable and bikeable environments (Yu 2014). Taken together, research considering equity and cycling safety has found that cycling experience, bicycle type, age, gender, race, and income, can influence one's risk of injury and perceptions of safety.

Cycling as mobility justice

Research on cycling safety has had important impacts on the ways in which we plan and design cycling infrastructure. Little of the work on cycling safety, however, has yet to engage fully with more critical approaches to mobilities, such as recent developments of mobility justice. Though there is no consensus on the definition of mobility justice, Sheller (2018) theorizes it as "an overarching concept for thinking about how power and inequality inform the governance and control of movement, shaping the patterns of unequal mobility and immobility in the circulation of people, resources and information" (p. 23). The remainder of this chapter advances a conceptualization of cycling safety as mobility justice, whereby the ability to feel safe cycling is constructed as a product of intersectional and multi-scalar mobile power relations.

Centering power in cycling safety

Viewing cycling safety as mobility justice requires thinking through intersecting power relations at multiple scales (e.g., among bodies, streets, cities, nations, and global systems). While cycling safety has been examined at both the scale of the individual and the neighborhood, much of the current cycling safety literature has yet to consider how power relations permeate one's ability to feel safe while cycling at these scales. For instance, the domination of motorized vehicles on public streets greatly impacts safety. Bicycle activism has a long history of questioning the logics of unequal road space allocation and governance, and some researchers have examined this dominance on road safety as well. For instance, some have argued that campaigns promoting helmet use place the responsibility for being safe on the cyclist (rather than the driver), all the while constructing the cyclist as the road user in need of protection (i.e., the unsafe road user). Cycling is not intrinsically a very dangerous activity (crashes that do not involve motor vehicles tend not to cause serious injury); instead, the danger is imposed on cyclists in dangerous environments (Jacobsen and Rutter 2012).

The role of the cyclists in policing their own safety has been found empirically as well. In a qualitative study of commuter cyclists in the UK, interviewees discussed how they fought for space and respect on the streets and mentioned having strategies to keep safe from cars, the dominant form of travel. These cyclists also expressed frustration about the lack of emphasis on driver behavior in cycling promotion (McKenna 2007). Similarly, Aldred (2013) has shown that some cyclists expressed how they got out of the way of traffic in order to perform as the 'good cyclist', a practice that "indicate[s] the power of the discourse that it is cars that belong and take precedent on the roads" (p. 265).

Perhaps unsurprisingly, this work on cyclists' internalized responsibility for their safety tends to originate from places characterized by car-oriented design where the dominance of motorized vehicles is particularly strong. In these places, car-oriented design results in negative externalities at multiple scales. At the local scale, cities have undergone incredible restructuring to make room for the automobile (e.g., nearly one-half of Los Angeles is devoted to car-only environments). Furthermore, approximately one million people are killed each year, and many millions more are injured, in road crashes. Cars also produce emissions responsible for localized air pollution and global ozone destruction, and greenhouse gas emissions contributing to climate change (Urry 2004). While the car's countermovement has grown in recent years, some have argued that it tends to place blame on the car itself, rather than challenge the source of the problem: our car dependence, or what Urry (2004) coined the 'System' of Automobility.

This System of Automobility is made up of numerous interdependent relationships between the automobile and social, political, cultural, geographical, historical, and technical systems. For instance, the System of Automobility is deeply embedded with other industries such as road building, oil production, car repair and sales, and highway gas and rest stations. As such, dismantling automobility has social and economic effects beyond the automobile industry. The System of Automobility is also social, for example the car comprises dominant discourses around what constitutes a 'good life'. This system produces and reproduces the domination that the car holds on many cities (Urry 2004). In many contexts, cyclists (as well as pedestrians and transit riders) exist within this System of Automobility where motorized vehicles are a powerful driver of complex and diverse processes. Cycling safety as mobility justice views cyclists as embedded in automobility. Therefore, cyclist's safety is not solely a question of being safe from individual cars but being marginalized in a social system in which motorized vehicles are dominant. This System of Automobility frequently places the responsibility for cyclist safety in the hands of the cyclist – the 'vulnerable road user'– resulting in victim-blaming when cyclists are injured. The solution lies in proving safe cycling environments, in other words in dismantling the very system where cars put cyclists at risk.

Considering intersectionality

When considering cycling safety as mobility justice, one must not solely consider how power relations due to the System of Automobility influence cyclist safety, but also how these relations intersect with other axes of privilege, domination, and oppression. Doing so broadens discussions of safety to include other important safety considerations on city streets such as sexual and police harassment. Coined in 1989 by Kimberlé Crenshaw, intersectionality offers an analytical framework for understanding how one's identities combine to create different experiences of discrimination. This work originated from Crenshaw's critique that feminist and antiracist activism could omit the experiences of black women (Crenshaw 1989). In the realm of transport, for instance, proposed policies meant to help women feel safe such as increased police presence can overlook the fact these policies can have the opposite effect on many women, particularly BIPOC (Black, Indigenous, and people of color) women.

The Untokening (untokening.org), a multiracial mobility justice collective, has argued that framing safety as Mobility Justice requires not solely focusing on protection from cars (and the people driving them), but to consider how safety is experienced differently by people embodying intersecting marginalized identities. Indeed, placing the responsibility for safety on the vulnerable road user does not solely exist for cyclists, as discussed above. Women and gender minorities, BIPOC, and members of LGBTQ+ groups are also forced to take disproportionally more responsibility for their personal safety on city streets. For instance, women have been socialized to feel fear in public space, an experience compounded by experiences of sexual harassment (Valentine 1989). Many women use coping strategies to deal with safety concerns, such as avoiding 'dangerous places', traveling with an escort, and avoiding confrontation (Dunckel Graglia 2016). Black youth have also been found to employ strategies to navigate and avoid police contact (Fox-Williams 2019). These strategies exemplify the internalization of responsibility for personal safety. When vulnerable road users, be it cyclists, women, or BIPOC, take on responsibility for their own safety, not only does the responsibility for safety wrongly fall on the vulnerable road user, but the strategies employed by vulnerable road users to stay safe affect how people use public space and can constrain their mobility.

There is nothing inherent or biological about the internalization of safety – rather this is a result of unequal power relations, lived experience, and generational aspects of both. In her seminal work "The Geography of Women's Fear", Valentine (1989) calls women's inhibited use and occupation of public space a "spatial expression of patriarchy" (p. 389). Similarly, the ability for BIPOC or members of the LGBTQ+ community to feel safe in public space is shaped by spatial and relational expressions of racism and/or homophobia. Quite simply, why wouldn't it be the case that what we might often think of as spatially fixed experiences of racism and misogyny would translate into mobile settings? Indeed, a study of cyclist experiences in Portland, Oregon, found that racial profiling and gender-based violence can discourage bicycling amongst women and people of color (Lubitow et al. 2019). Mobility justice opens avenues to consider that cyclists' ability to feel safe is a combined spatial expression of automobility, patriarchy, racism, classism, and other systems/forms of discrimination and oppression.

Figure 25.1 outlines these dynamics. Framing cycling safety as mobility justice, automobility can be posited as one of the many societal intersecting axes of power, domination, and oppression. The ability to feel safe while cycling is a product of where one sits in this complex intersection of mobility, gender, race, and other axes of power. The power relations included herein are not comprehensive – they merely present three of the power relations discussed in this chapter.

Empirical evidence supports this framing as well. A study examining the impact of a bicycle course on women found that participants had to not only develop competencies to share the road with motor vehicles (automobility), but also to navigate sexual harassment and negotiate masculinized public bicycling space which requires assertiveness or tolerate

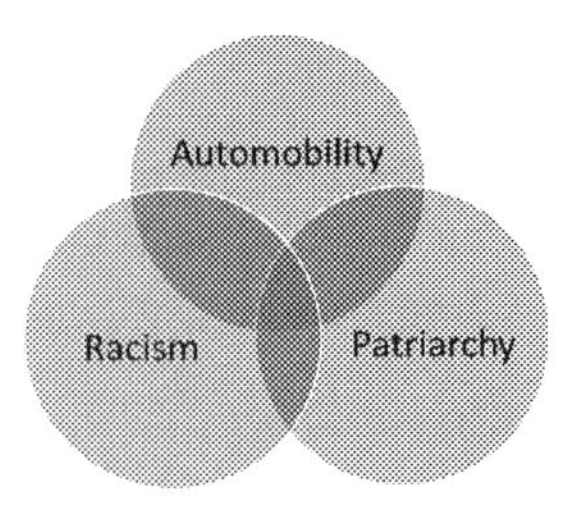

Figure 25.1 Cycling safety as the outcome of intersectional power relations.

Source: author.

aggressiveness (patriarchy) (Sersli et al. 2021). In another study focused on newcomer and refugee cyclists, safety concerns included negotiating road space, sexual harassment, and worries about interactions with law enforcement – highlighting how safety was produced at the intersection of automobility, patriarchy, and racism (Ravensbergen et al. 2020a). These results are also supported by Charles Brown's recent work on #Arrestedmobility which focuses on the constraint to Black people's mobility, including over-policing. The death of Dijon Kizzee following an alleged cycling violation at the hands of Los Angeles police demonstrates the importance of considering policing in cycling safety (Olzer, September 29 2020; Ravensbergen et al. 2021).

Context matters greatly, as the experience of gender, race, and riding a bicycle is shaped by a place's geography and history. For instance, depending on where you are, riding a bicycle in public can be perceived as a loss in status or as a trendy way to commute. These associations intersect with identity as well. In a study in Toronto, Canada, participants associated cycling with masculinity and as a means of travel for people from lower-income households (Ravensbergen 2020). In another study, this time in London, UK, gendered, ethnic, and class identities shaped participants' decision to cycle (Steinbach et al. 2011). Further, the dominance of the automobile varies greatly geographically. For example, cars are less dominant in cities famous for cycling such as Amsterdam and Copenhagen than more car-dominated cities, including London and Toronto. Scale is important here too, as within these cities automobility is experienced unevenly. For example, parts of Copenhagen are car-centric while select districts of Toronto have of over 30% cycling mode share (Ravensbergen et al. 2020b).

Cycling safety as mobility justice requires thinking through how automobility, patriarchy, racism, and other axes of power influence cycling safety, as well as how these dynamics are shaped by, and shape, place across these scales. For instance, take women's concern over cycling safety. As discussed earlier, women consistently report greater concerns over traffic safety than men in places with low cycling rates, a trend which has been put forward to explain the gender gap in cycling. In places with high cycling rates, such as Amsterdam, the gender gap appears to disappear (Ravensbergen et al. 2019). As Aldred et al. (2016) note, cycling in car-centric places can be perceived as a high-risk activity associated with bravery and confidence – characteristics that comply with hegemonic masculinity. Therefore, both automobility and patriarchy shape, and are shaped by, gendered concerns over safety. Using Toronto as an example again, similar rates of male and female cyclists can also be found within high-cycling neighborhoods (Ravensbergen et al. 2020b). Within car-dominant Toronto, there are sections of the city that have begun to dismantle automobility. This does not mean, of course, that patriarchy has been dismantled in these communities. For example, many women may experience sexual harassment in these high-cycle areas. And within high-cycle areas, not all women have the same experiences of cycling. For instance, a recent study in Grenoble, France, found that racism and Islamophobia resulted in challenges creating bike programming that disproportionately targets Muslim women (Vietinghoff 2021). Cycling safety as mobility justice requires thinking deeply about how, in this example, women's experiences of safety lie at the intersection of safety from traffic, from harassment, and from exclusion and are the product of complex context-specific relationships between automobility, patriarchy, and racism.

Conclusion

Framing cycling safety as mobility justice presents opportunities for holistic and comprehensive studies of cycling safety that are historically and geographically rooted. This approach also dampens the tendency to essentialize cycling safety, i.e., to view differences between subjects as natural, or biological and to overlook power relations between these social categories. Just as different people do not have to experience mobility differently, there is no biological, natural reason for them do so, cycling is not innately dangerous – and cities are not inherently destined to be car-dependent. Indeed, recent shifts toward people-focused planning demonstrate that cities can take back their streets from automobile dominance. By tackling the socially constructed and multi-scalar power relations shaping cyclists' safety, possibilities emerge not only for the cycling city, but for a truly safe and just cycling city which takes steps to dismantle systems of automobility, patriarchy, racism, and classism. We invite not really a departure from the ongoing and important work focused on other types of cycling safety – but rather we encourage greater reflection on the possibilities lying at the intersection of work on safety, mobility justice, anti-racist praxis and so on. This would involve broadening discussions of safety to include crime and sexual and police harassment, amongst other concerns, alongside safety from motor vehicles. Further, this holistic approach would require understanding the power struggles at the root of cycling safety, power struggles which currently result in the internalization of safety and victim blaming of vulnerable road users. Mobile bodies are also raced, and gendered bodies, and indeed experiences of cultural and other forms of violence perpetrated against immobile subjects may translate into the mobile realm – producing safety concerns that reach well beyond the physics of vehicular collision.

References

Aldred, R. 2013. Incompetent or too competent? Negotiating everyday cycling identities in a motor dominated society. *Mobilities*, 8, 252–271.

Aldred, R., Woodcock, J., & Goodman, A. 2016. Does more cycling mean more diversity in cycling? *Transport Reviews*, 36, 28–44.

Attewell, R. G., Glase, K., & McFadden, M. 2001. Bicycle helmet efficacy: A meta-analysis. *Accident Analysis and Prevention*, 33, 345–352.

Clayton, W., & Musselwhite, C. 2013. Exploring changes to cycle infrastructure to improve the experience of cycling for families. *Journal of Transport Geography*, 33, 54–61.

Crenshaw, K. 1989. Demarginalizing the intersection of race and sex: A black feminist critique of antidiscrimination doctrine, feminist theory and antiracist politics. *u. Chi. Legal f.*, 139.

Delmelle, E. M., & Delmelle, E. C. 2012. Exploring spatio-temporal commuting patterns in a university environment. *Transport Policy*, 21, 1–9.

Dozza, M., Schwab, A., & Wegman, F. 2017. Safety science special issue on cycling safety. *Safety Science*, 92, 262–263.

Dunckel Graglia, A. 2016. Finding mobility: Women negotiating fear and violence in mexico City's public transit system. *Gender, Place & Culture*, 23, 624–640.

Elvik, R. 2021. Cycling safety. *In:* Pucher, R. B. (ed.) *Cycling for Sustainable Cities*. Cambridge, MA: MIT Press.
Fox-Williams, B. N. 2019. The rules of (dis)engagement: Black youth and their strategies for navigating police contact. *Sociological forum (Randolph, N.J.)*, 34, 115–137.
Hollingworth, M. A., Harper, A., & Hamer, M. 2015. Risk factors for cycling accident related injury: The UK cycling for health survey. *Journal of Transport and Health*, 2, 189–194.
Jacobsen, P. L., & Rutter, H. 2012. Cycling safety *In:* John Pucher, R. B. (ed.) *City Cycling*. Cambridge, MA: MIT Press.
Lubitow, A., Tompkins, K., & Feldman, M. 2019. Sustainable cycling for all? Race and gender-based bicycling inequalities in Portland, Oregon. *City & Community*, 18, 1181–1202.
Manaugh, K., Boisjoly, G., & El-Geneidy, A. 2017. Overcoming barriers to cycling: Understanding frequency of cycling in a University setting and the factors preventing commuters from cycling on a regular basis. *Transportation (Dordrecht)*, 44, 871–884.
McKenna, J. W. M. 2007. Qualitative accounts of urban commuter cycling. *Health Education*, 107, 448–462.
Olzer, R. 2020, September 29. When you're black, "just ride" isn't an option: Why representation matters, part 3. *BIKE Magazine*.
Prati, G., Puchades, V.M., DeAngelis, M., Fraboni, F., & Petrantoni, L. 2019. Factors contributing to bicycle–motorised vehicle collisions: A systematic literature review. *Transport Reviews*, 38, 184–208.
Pucher, J., Dill, J., & Handy, S. 2010. Infrastructure, programs, and policies to increase bicycling: An international review. *Preventative Medicine*, 50, S106–S125.
Pucher, J. R., & Buehler, R. 2021. Introduction: Cycling to sustainability *In:* Buehler, R. (ed.) *Cycling for Sustainable Cities*. Cambridge, MA: MIT Press.
Ravensbergen, L. 2020. 'I wouldn't take the risk of the attention, you know? Just a lone girl biking': examining the gendered and classed embodied experiences of cycling. *Social & Cultural Geography*, 1–19. doi: 10.1080/14649365.2020.1806344.
Ravensbergen, L., Buliung, R., & Laliberté, N. 2020a. Fear of cycling: Social, spatial, and temporal dimensions. *Journal of Transport Geography*, 87, 102813.
Ravensbergen, L., Buliung, R., & Sersli, S. 2020b. Vélomobilities of care in a low-cycling city. *Transportation Research: Part A, Policy and Practice*, 134, 336–347.
Ravensbergen, L., Buliung, R., Sersli, S., & Winters, M. 2021. Guest editorial: Critical Vélomobilities. *Journal of Transport Geography*, 92, 103003.
Ravensbergen, L., Buliung, R., & Laliberté, N. 2019. Toward feminist geographies of cycling. *Geeography Compass*. doi: 10.1111/gec3.12461.
Reynolds, C., Harris, M.A., Teschke, K., Cripton, P.A., & Winters, M. 2009. The impact of transportation infrastructure on bicycling injuries and crashes: A review of the literature. *Environmental Health*, 8(47). doi: 10.1186/1476-069X-8-47.
Sanders, R. L. 2015. Perceived traffic risk for cyclists: The impact of near miss and collision experiences. *Accident Analysis and Prevention*, 75, 26–34.
Sersli, S., Gislason, M., Scott, N., & Winters, M. 2021. Easy as riding a bike? Bicycling competence as (re)learning to negotiate space. *Qualitative Research in Sport, Exercise and Health*, 10–21. doi: 10.1080/2159676X.2021.1888153.
Sheller, M. 2018. Theorizing mobility justice. *In:* Cook, N., & David, B. (eds.) *Mobilities, Mobility Justice and Social Justice*. London, UK: Routledge.
Steinbach, R., Green, J., Datta, J., & Edwards, P. 2011. Cycling and the city: A case study of how gendered, ethnic and class identities can shape healthy transport choices. *Social Science & Medicine (1982)*, 72, 1123–1130.
Thomas, B., & DeRobertis, M. 2013. The safety of urban cycle tracks: A review of the literature. *Accident Analysis and Prevention*, 52, 219–227.
Urry, J. 2004. The 'System' of automobility. *Theory, Culture & Society*, 21, 25–39.
Valentine, G. 1989. The geography of women's fear. *Area*, 21, 385–390.
Vietinghoff, C. 2021. An intersectional analysis of barriers to cycling for marginalized communities in a cycling-friendly French City. *Journal of Transport Geography*, 91, 102967.
Willis, D. P., El-Geneidy, A., & Manaugh, K. 2015. Cycling under influence: Summarizing the influence of perceptions, attitudes, habits, and social environments on cycling for transportation. *International Journal of Sustainable Transportation*, 9, 565–579.
Winters, M., et al. 2012. Safe cycling: How do risk perceptions compare with observed risk? *Canadian Journal of Public Health-Revue Canadienne De Sante Publique*, 103, S42–S47.

Winters, M., Davidson, G., et al. 2011. Motivators and deterrents of bicycling: Comparing influences on decisions to ride. *Transportation*, 38(1), 153–168.

Yu, C.-Y. 2014. Environmental supports for walking/biking and traffic safety: Income and ethnicity disparities. *Preventive Medicine*, 67, 12–16.

Zhao, Y., et al. 2021. Association of cycling with risk of all-cause and cardiovascular disease mortality: A systematic review and dose-response meta-analysis of prospective cohort studies. *Sports Medicine*, 51(1), 1439–1448.

PART V

Sport, health and lifestyle

An introduction

Tim Jones

Cycling relies on a combination of mind, body and machine. It provides the opportunity to test athletic ability at different levels, from amateur road riders riding socially in groups on a weekend, through to professional elite riders competing on the world stage at the Tour de France. With historical roots in Europe and North America, cycle sport is popular across the world. Different cycling disciplines have emerged over time with their own unique rituals, clothing, machines, technology, codes and governance. The following six chapters and two vignettes provide a glimpse into the world of amateur and professional cycle sport, including road, off-road and track cycling. There is widespread agreement on the benefits to health for those engaged in its practice and these, as well as the relative risks, are covered in this part. Cycle sport is now a complex commercial enterprise where the stakes are high. It is also important to acknowledge the darker side of the sport and the risks taken in the search for success.

The part begins with my own contribution on the rise of the so-called Middle-Aged Man in Lycra (MAMIL) who emerged in the Anglosphere in recent decades. Although road cycle racing has its roots in working-class communities, amateur sports cycling is now regarded as a high-end sport and has been described as the 'new golf'. The sport has seen the growth of amateur cycling clubs, often seen riding in groups on rural roads on weekends, and a profusion of events and activities, including mass participation 'challenge'-style charity rides, timed 'sportives' on iconic routes traversed by professional elite sports cyclists, and package-style, supported 'sporting tours', at key cycling destinations. The chapter investigates what is behind the growth of high-end bicycle sales and increasing participation among middle-aged men whose new-found engagement in the activity has been pejoratively described as "the noughties version of a mid-life crisis". Charting events and cultural trends that have influenced this upsurge, I also delve inside the lifeworld of the MAMIL, drawing on my own investigations with riders in southern England to reveal cycling trajectories and the motivations and meanings associated with practice. I reflect on how new markets and representations of amateur sports cycling reinforce and legitimize new forms of masculine identity and the possible impacts on broader issues, such as health and cycling for transport.

Attention then turns from the amateur to the professional rider. Daam Van Reeth maps out the terrain of professional cycle sport, highlighting the peculiarity, often misunderstood by the passive observer, that cycling is a team sport won by individuals. Competitive road racing is also heterogeneous, with a variety of different events all taking place on public roads

DOI: 10.4324/9781003142041-36

and in different formats. Reeth discusses the negative impact this has on the business model of the sport and the complex relationship between agents involved in its governance. All of this presents challenges for professional road racing that threaten the long-term viability of the sport. Reeth highlights what is regarded as the most significant threats including the doping legacy, a theme that is expanded upon by Charlotte Smith in the final chapter of this part. The vignette that follows by Michael Barry gives the reader an experiential account of being in the peloton of one of the professional races.

Moving from tarmac roads to more varied terrain, Karen McCormack and Ben Osborn's chapter provides an overview of the growth in off-road riding and the different categories from cyclocross, to mountain biking and the recent craze of gravel riding. They look at how off-road riding has unique and multiple identities associated with different equipment and practices vis-à-vis road riding. This also presents different risks and impacts on the environment, particularly public parks and wilderness areas, where this type of cycling sometimes takes place. McCormack draws attention to the battles over trail building and land use which have parallels in relation to planning urban space for cycling for transport particular with the advent of electrically assisted (mountain) bikes (see Part IV: Urban Cycling).

Whereas on-road and off-road cycling inevitably take place outdoors, with riders on free-wheel geared bicycles of different configurations, that have the ability to stop by using different braking systems, track cycling predominantly takes place indoors on a single (fixed) gear bicycle with no brakes! Michael Jordan's chapter explains the peculiarities of track cycling starting with the configuration of the track bicycle and the evolution and nature of the site of track cycling, namely, the velodrome. The vagaries of the different types of professional track cycling events and competitions are discussed including one venue that has contributed to the rich history of track cycling, Madison Square Garden, New York. Cycling shapes physiology and the range of track cycling disciplines means that no physiological type of athlete will succeed across all disciplines. Jordan illustrates the trained rider capacities that are required to support endurance as well as explosive efforts. The Hour Record is the perhaps one of the most prestigious records in cycling and the ultimate test of the endurance track cyclist. Jordan takes us through the history of record holders, including the challenges posed by developments in bicycle technology to the main organization governing the events and how these have been accommodated. Keizo Kobayashi's vignette, which follows, describes the unique culture of Japan's keirin races.

Cycling is associated with many health benefits, both for the individual, but also at a societal level for the reasons set out in Norcliffe's introduction to this volume. Bauman, Titze and Oja take a broad view of the benefits of cycling to include gains to physical health, mental and psychological health, and community health. They provide a comprehensive overview of the outcome of recent systematic reviews of studies of cycling and health (including among children and adolescents), highlighting how these consistently demonstrate the substantial benefits of all modes of cycling to physical health. Whilst they highlight that research into cycling and mental health is less well developed, the studies that do exist indicate a positive association between riding and positive mental health – a finding that perhaps won't be lost on readers who already find pleasure in cycling! Bauman, Titze and Oja note that mental health benefits may be mediated through physical activity, but other mechanisms may be in play, such as the aesthetic pleasure of riding though different environs. There is also the sociality of cycling that allows people to connect and feel part of the world as highlighted in my own chapter on middle-aged amateur sport cycling at the beginning of this part. Of course, despite health benefits, cycling does not come without risk. The authors review the evidence, weighing up the risks of cycling against the benefits, and come to a clear positive conclusion leading them to call for effective strategies that increase population levels of cycling among all age groups.

The final chapter in the part turns to the elite end of the sports cycling spectrum and the topic of performance enhancement through illicit means. Charlotte Smith's chapter addresses the often-discussed subject among cycling and non-cycling fanatics alike: do (all) sports cyclists 'dope' and how and why do they do it? This is one of the most discussed topics in relation to cycle sport and therefore it is only right that a compendium of this nature addresses it. The chapter begins by outlining different methods of doping and shows how they enhance performance. It discusses the complex triggers that cause cyclists to dope and the difficulties encountered in establishing the prevalence of doping within the sport. There are reports that the growth in amateur road cycling has also fuelled curiosity about the benefits of doping and suspicion due to the rise of the 'super-MAMIL' within this community (see Usborne, 2016). Smith delves into the alternative motivations of these riders relative to their elite professional counterparts for whom cycling is essentially an occupation. The chapter also covers the Festina Affair and the Lance Armstrong saga, two of the best-known scandals in professional road racing, and the anti-doping detection methods and the processes that governing bodies have subsequently put in place to try to act as a deterrent. To conclude the chapter, Smith reflects on the future of 'clean cycling', particularly in light of mechanical doping and gene therapy, and whether professional cycling can ever be 'clean', given the demands of the sport and what is at stake, including the large monetary incentives, the importance of soft power to international competition, and the collective pressures in a team sport.

Reference

Usborne, S. (2016). Dope and glory: The rise of cheating in amateur sport. *The Guardian*, 1 June 2016. Available at: https://www.theguardian.com/lifeandstyle/2016/jun/01/dope-and-glory-the-rise-of-cheating-in-amateur-sport (Accessed: 10 September 2021).

26

AMATEUR SPORT CYCLING

The rise of the MAMIL

Tim Jones

Introduction

In 2014, lexicographers added the noun, and acronym, 'MAMIL' ('Middle-Aged-Men-in-Lycra') to the Oxford English Dictionary on the grounds of its common usage in representing a particular community: "A middle-aged man who is a very keen road cyclist, typically one who rides an expensive bike and wears the type of clothing associated with professional cyclists."

In this chapter I explore the rise of the MAMIL and its idiosyncrasies, and, in particular, the reasons for the upsurge of interest and engagement in amateur sports road cycling across the Anglosphere (and beyond) and the meanings associated with cycling practice. I conclude with some thoughts on what this might mean for the future of amateur cycling, particularly in relation to social relations and cycling more generally.

The rise of the MAMIL

In the UK, there was a significant upsurge in cycle sales around 2010, with quarterly expenditure on cycles increasing from £350 million in 2011 to £375 million in 2012 (Statista 2020). Retail analysis, by the British market-research company Mintel, reported that cycle sales growth was being driven by 35 to 45-year-old middle-aged men, new to cycling, purchasing high-end bicycles for pursuit of recreation (Mintel 2010). This group were likely to have a household income of at least £50,000 per year, read broadsheet newspapers, and shop at premium brand supermarkets. The report author, Michael Oliver, coined the term MAMIL, 'Middle Aged Men in Lycra', somewhat pejoratively, to represent this phenomenon stating, "Thirty or 40 years ago, people would ride a bike for economic reasons, but our research suggests that nowadays a bicycle is more a lifestyle addition, a way of demonstrating how affluent you are" (Pidd 2010).

That the report stated the trend was 'the Noughties version of a mid-life crisis' was soon picked up by the British press. The BBC's Dominic Casciani quipped that the market for flash road bikes had 'expanded faster than a 45-year-old's waistline' and that 'no mid-life crisis is complete without a souped-up road bike' (Casciani 2010). Meanwhile, other commentators leapt to the defense of the 'much-mocked' MAMIL. Writer, journalist and cyclist, Matt Seaton,

DOI: 10.4324/9781003142041-37

argued how being a MAMIL was far more benign than the indulgence in the old mid-life crisis cliche of the sports car, exclaiming, "I own my MAMILhood: I'm a middle-aged man in Lycra – and proud" (Seaton 2012).

The emergence of the MAMIL phenomenon largely manifested itself on weekend morning rides, either in standalone groups, or integrated into new or existing cycling clubs. There was also growth in mass participation rides, or noncompetitive sportives, effectively timed rides with the objective of posting a personal time and to go on the record as a 'finisher'. The most iconic of these, Ride London, was first held in 2013, designed as an annual legacy from the London 2012 Olympic Games and cycling's equivalent to the London Marathon. The event involves including a mass participation ride for the public and professional racing for elite riders.

Originating in Italy in the 1970s, and still relatively new in the UK, the Gran Fondo provides the opportunity for amateur sports cyclists to 'follow in the footsteps of giants' and ride many of the routes and iconic climbs traversed by the Grand Tours. Every year, keen UK-based cyclists travel to Italy for the experience, which involves the relative novelty of the freedom of closed roads because UK sportives, with the exception of Ride London and a few others, rarely exclude vehicle traffic. Given the lack of 'serious' mountains and guaranteed sunshine in the UK, many also pack up their bikes in protective boxes with friends and head off for a week or more to ride the Pyrenees or the Alps. There are also the islands of Mallorca and Lanzarote, where the lure of smooth tarmac and mountain climbs can be combined with sun, sea, sand and Sangria.

British cycling success and the 'Wiggo Effect'

The focus of attention on the MAMIL phenomenon is linked to trends in the UK, North America and Australia, but of course, the concept of middle-aged men emulating elite riders in the three Grand Tours (i.e. the Tour of Italy, Tour de France and the Tour of Spain) has been embedded in Europe for decades. So what led to the rise of the British MAMIL towards the end of the noughties? There is speculation that this was due to the confluence of British success at the Beijing Olympics, the development of road bike technology, and smart advertising (Casciani 2010). Indeed, the pattern in the UK was similar to that in the USA and Australia: 'success' for US cycling included seven successive Tour de France wins, Lance Armstrong from 1999 to 2005 and Floyd Landis in 2006 (both subsequently stripped of titles due to doping); success for Australia came with Cadel Evans who won the Tour de France in 2011.

The seed of cycling fervour in the UK can perhaps be traced to British Cycling success at the Summer Olympic Games in Beijing in 2008 when the GB Cycling Team dominated track cycling winning 7 gold, 4 silver and 2 bronze medals, while on the road, Nicole Cook became the first British woman to win an Olympic Gold medal in any cycling discipline. This captured the popular imagination. That same year, sprinter Mark Cavendish had four stage wins in the Tour de France and in the women's UCI Road World Championships, Nicole Cooke won Gold – Cooke had already won two years earlier and was rated number one female rider in the world. In December 2008, Chris Hoy (Gold in the Sprint and Keirin and Team Pursuit) was voted BBC Sports Personality of the Year and the GB Olympic Cycling Team was voted Team of the Year. By 2009, Cavendish had won a further six stages of the Tour de France and the Milan–San Remo Classic followed by five stages of the Tour de France in 2010 while Emma Pooley won the World Women's Time Trial Championship. The following year led to further success for British Cycling with Team GB winning six events at the 2011 World Road Race Championship in Denmark, including Men's Elite Road Race (Mark Cavendish) and Lucy Garner winning Women's Junior Road Race.

British cycling off the track, as well as on the track, was in the ascendancy and was receiving more widespread coverage in the British media and on terrestrial television. This was the prelude to a historic moment in the making. In 2012, following his success at Paris–Nice, Bradley Wiggins became the first British rider to win the Tour de France. Wiggins crowned this glory by taking Gold in the Men's Time Trial at the London Summer Olympics in 2012, with hundreds of thousands turning out to line the route through Surrey.

The growth in enthusiasm among the British public for cycling, in the wake of Wiggins' win at the Tour de France and success at the 2012 Olympics, was indisputable. UK media channels reported how Wiggins was a positive role model for the sport (Gupta 2012) and how his newfound celebrity status had helped spark a growth in UK bike sales and the popularity of cycle sport (Davies 2012). The popular journal among UK cycling circles, *Cycling Weekly*, began to run a series of articles under the banner 'The Wiggo Effect'. This centered on interviews with people who had been inspired to ride following the summer of success for Bradley Wiggins and Team GB's elite cyclists. In the meantime, British Cycling, the governing body for cycle sport in the UK, reported reaching a new record level of membership. The crescendo was Wiggins being voted 2012 BBC Sports Personality of the Year, beating 11 other contenders, including tennis star Andy Murray, and Olympic athletes Jessica Ennis and Mo Farah, to the prestigious title (BBC 2012).

The changing face of amateur road cycling

From its roots among working-class communities, road cycling in the UK, and indeed across much of the Anglosphere, is now regarded as a high-end sport. In 2013, *The Economist* magazine declared cycling 'the new golf', proclaiming that road cycling had become the preferred way of networking for the modern professional (G.D. 2013). In an interview in *Business Insider* in 2015, Max Levchin, co-founder of PayPal, explained how, while golf was traditionally the sport of choice where ideas are developed and relationships forged, younger executives had lost interest in golf. They had become were more interested in cycling as something you can quantify and show off fitness and equipment, or what he termed 'road jewelry' (McMahon 2015). Meanwhile, the *Financial Times* reported how corporate biking events were a massive industry providing opportunities for corporate sponsorship and engagement (Wallop 2016). This trend was not dissimilar to what had happened in the world of off-road mountain biking in the 1990s. Urban renewal academic, Richard Florida, commented on this phenomenon in his book *The Creative Class* (2002: 174).

> For the creative people and high-tech professionals in my field studies, riding a mountain bike has become almost a *de rigueur* social skill – much as horseback riding was for the members of the old elite. And this is not just among the young. I have come across countless forty and fifty somethings who are avid mountain bike riders. Quite a cultural transformation: forty years ago the bicycle was a childish symbol of small-town squareness; today it is cool.

The fashion industry also capitalized on renewed interest in amateur road cycling. Wiggins was a self-confessed *Mod* (short for 'Modernist'), a subculture originating in London in the 1950s focusing on music and fashion, that was embedded in the 1960s 'swinging London' aesthetic. Wiggins linked up with Fred Perry in a new collaboration to launch a range of clothing mixing classic Fred Perry and old school cycling style (Renwick 2016). Meanwhile, up-market London based global cycling clothing firm, Rapha, aimed to redefine cycling's 'image problem' by fostering a new aesthetic. In a feature in *The Observer Magazine* (2013) on the emergence of style-conscious British cyclists, writer Rob Penn reflected on the 'new breed' of cyclists who were increasingly adopting a new cycle chic aesthetic.

> What began as 'heroin-chic' cycle messenger subculture in the late 80s has been crowned by the MAMIL (Middle-Aged Man in Lycra), often see wearing full Rapha. It is a wonder how these men – and I rank among them – have collectively overcome the traditional British fear of shopping and turned cycling into a slavishly stylish pursuit.
>
> *(Penn 2013)*

Rapha has ridden the wave of cycling success and capitalized on consumer culture, generating panache and ire in equal measure. This is not surprising given its over-the-top mastery of selling luxury goods, including an espresso tamper priced at nearly £100 with the sales pitch, "Designed in collaboration with the America Barista & Coffee School, [it] offers precise dimensions, an exceptional feel in the hand and the iconic styling of a Chris King headset" (Acquire Mag 2015). Some have taken to creating spoof websites parodying the apparent pretentious and exclusive nature of the brand and its loyal followers, with *Secrets of the Peloton* website being just one example. Under the banner, 'Crapha', it became celebrated for touting fake products like 'bottled air from Mt Ventoux' and 'endangered goat scrotum leather gloves' (Bike Biz 2011). For cycle sport Brand Manager, James Fairbank this is unfair,

> Strong brands with a clear identity are an easy thing to parody. Given that our overarching objective is to help make road cycling the most popular sport in the world, being exclusive for any reason certainly isn't in our interest.

Rapha company CEO, Simon Mottram, argues that the rise of the brand simply parallels the emergence of the style-conscious British cyclist who fits with a broader cultural trend towards that of the "metrosexual man". For Mottram men were, "... more body conscious, fitter, better toned, able to talk about skincare and more thoughtful about what we wear and how we present ourselves than ever before" (Penn 2013).

That road cycling had become a way of male self-expression and a site of conspicuous consumption is nowhere more evident than in the growing online availability of 'gifts for him' that now include, not just 'beer and golf' branded trinkets, but also MAMIL pampering kits – see, for example, Men's Society Middle Aged Man in Lycra gift set (Men's Society 2021). The pervasiveness of cycling style magazines, including *Rouleur* (2021), which uses fine photography and essaying 'to convey the essence, passion and beauty of road racing', is just one conduit for the commodification of road cycling by association with, and through the seduction of, high-end luxury cycling goods.

MAMILs, of course, come in all shapes and sizes, both in terms of body, and wallet. Many have a long history of road cycling experience, including racing, or they may have switched from other disciplines, including the 1990s craze for off-road mountain biking (Chapter 28). They take issue with being lumped with a *nouveau cycling elite*, and can sometimes secretly mock their 'inauthentic' counterparts with phrases such as 'All the gear and no idea!' Not all buy into, or indeed can afford, high-end carbon – or, more recent, 'graphene' – technology or are found accoutered in Rapha. Some can be seen astride classic old racing bicycles squeezed into outmoded spandex, not so bodily conscious and less obsessed with their waist size and body stats. The impression that 'cycling has become the new golf' and the enclave of a new business elite, therefore, may be acknowledged, but it is also contested as constitutive of all MAMILs.

Delving inside the lifeworld of the MAMIL

In the remainder of this chapter I draw on my own research from interviews with MAMILs conducted in 2013, recruited through an Oxford UK cycling club and through the website *Wheelsuckers* (2021). The aim was to understand the trajectory into the world of amateur road cycling, the motivation, as well as the meanings associated with its practice. I then provide some closing remarks on what this might mean in terms of gender and women's role in the proliferation of the MAMIL phenomenon and the chances of this translating into practical everyday cycling as part of a broader move toward environmental sustainability. All participants were male and aged in their forties and fifties and were company directors or working in higher managerial and professional occupations. All names used are pseudonyms.

Motivation

I started by asking my interviewees, "Please tell me the story of your path to taking up cycling and about those events and experiences that were important to you personally." It was apparent that the fervor for sports cycling in the UK following the Tour de France and Olympic Games success had acted as a catalyst to try cycling.

> There was nothing conscious... There probably was some renewed interest in cycling... because of Brad[ley Wiggins]... I remember, sort of, following bits of the Olympic cycling with some interest because I was, sort of, interested about the whole concept of marginal gain, the sort of sports psychology of the whole thing, and quite impressed with... you know, here's this sport where they're actually doing a brilliant job of pushing the boundaries...
>
> *Brad*

The people I interviewed were conscious that cycling was becoming more popular and there was a sense that cyclists were an increasing presence on the landscape. Peter's growing awareness of the increased visibility of cycling prompted him to give it a go while Ben was prompted by an impromptu challenge by a friend to partake in an organized long-distance cycle event.

> ... you know we'd see all these Lycra-clad cyclists, and, you know, I was very intrigued by all of that, and you know, eventually decided to give it a go.
>
> *Peter*

> I was wondering back from the pub one night and I bumped into [a friend] and he said, 'Do you fancy cycling to Paris?' and I said, 'Yeah I'd love to!' That was actually the first time that I'd ever thought seriously about something exciting to do with cycling in all of those years...
>
> *Ben*

Some participants reflected on how their body had changed shape as they had reached middle age. Family responsibilities and the forfeiture of former field sports such as rugby and football left a sense of having 'let go' of their fitness and of the desire to re-engage with physical activity. Cycling appeared to provide the opportunity to reconfigure the body and mind.

> I suppose in the background there was [my friend] Matt who... two or three years ago he was racing and he was really into it and he was in great shape. I suppose I had a residual memory of what it was doing for him... his body shape, his fitness, his self-esteem.
>
> *Brad*

Becoming

I was interested to understand how participants felt about their new-found MAMIL status. As a prompt, I asked them to respond to a quote by BBC's Dominic Casciani from the article he wrote in 2013 on the rise of the MAMIL, "... is this a 21st-century mid-life crisis? Has the silence of skinny tires and carbon fibre frame-sets replaced the thunderous noise of motorbikes?" There was general acknowledgement that this was probably a valid statement (often accompanied with a wry smile) before my participants explained their own personal trajectories.

Chris, for example, explained how he had reached a point where he had started to reflect on his health and had come to the conclusion that cycling was a hobby that could fulfill his desire to follow a path towards a much healthier lifestyle.

> ... you kind of get to a certain age where you make a decision – am I going to carry on with any of the stuff that's detrimental to my health and my longevity, or am I going to do something to try to eke out an extra 10 or 20 years at the end of my life.
>
> *Chris*

Ben was philosophical about life in general and the apparent paradox between personal health and personal economy across the life course. For him, becoming a MAMIL was a *phenomenon* rather than a crisis as he tried to explain:

> I mean life's the wrong way round in that sense isn't it? When you are seventeen and you've got the body to die for you need to have the machinery to die for with it. Inevitably you cannot afford that 'til you're fifty, so you know it's the wrong way round, but you shouldn't knock it as a result of that... with cycling, it's not a mid-life crisis... it's a mid-life phenomenon...
>
> *Ben*

O'Connor and Brown (2007) have highlighted that amateur sports cyclists in Australia, 'distance themselves from serious competition and its associated layers of formality'. This also seemed to be the case among my interviewees as emphasis was placed on *not* partaking in competitive cycle races or obsessing over bodily stats. The concept of man-as-machine was eschewed and riding was, instead, focused on personal challenge and not competition. Chris articulated this, juxtaposing the role of the successful elite British rider Mark Cavendish in terms of the *job* of winning races and his *personal (re)connection* with the true nature of cycling:

> I read an interview with Mark Cavendish where he still goes back to his blokes that he started cycling with on the Isle of Man... 'Cos he hasn't got 30 gels up his shorts and the tactics, and the guy who's got to lead him out and he's got 1.5km from the

end... you know all the chess of it... And to just be able to whizz round all his old haunts, that must be lovely... and that's what I'll always figure I'll get out of it. Yeah, I don't want to be the bloke in a shed you know looking at my heart rate or my power wattage or something like that – that just doesn't appeal.

Chris

Materials and technology

Discussions of the materials associated with cycling centered around bicycle technology and the seduction of purchasing high-end equipment and clothing. Speccing out with carbon or titanium cycle frames and quality equipment was justified on the grounds of evolution. Many reflected on the improvements in cycle technology compared to when they last rode a racing bike in their youth. This included pedal technology and how the feeling of having achieved status as a 'legitimate cyclist' was reached when having purchased and mastered clipping into 'clipless pedals' – essentially a bind of the sole of the foot to the pedal similar to ski boot to ski.

There's something about having a light carbon fibre bike. You know when you get on it it goes, and you know that's a great feeling. Whereas the other two that I tried when I was buying the bike, you know I mean they'd obviously... obviously good bikes, but they just felt like the sort of thing I used to ride when I was a kid, but you know obviously evolved. So you know it's getting on a bike like that... and I still feel that every time I get on and sort of clip into the cleats, you know that's a good feeling.

Peter

In terms of clothing and performing cycling, there was initial reticence about 'squeezing into spandex' and it feeling 'alien'. Participants espoused the benefits of this 'performance apparel' and initially adorned themselves in such gear with a sense of humor. Indeed, the ability to laugh at oneself and not take oneself too seriously, particular in the face of non-cycling friends or companions, was important, as Ben reflected:

INTERVIEWER: What about your wife when you sort of said... you turned up in your bib shorts whatever and your tight lycra what was her response?

BEN: I think she was fairly cynical, used to take the piss actually. I mean she just accepts it now. I mean... I have become so much fitter and it is a very engaging thing, you know I'm really enthusiastic about it. So um... it's not a chore, it's hugely pleasurable, so she indulges me completely now.

The luxury cycling brand, Rapha, often came up in discussion. There was a sense that the company had hit the zeitgeist and had provided 'something bigger' for non-elite cyclists to belong to adding to their authenticity as 'real cyclists':

They've got the whole style haven't they? They've got something that's really hit the zeitgeist [and it makes you think that you're not just a cyclist] yeah, something broader, something wider, that's right. It's heritage and there's a little bit of luxury in there as well. They're making you feel like you've been doing this for a long time even if you haven't. Which makes you feel like you have permission to do it, rather than you're the new boy.

Chris

Experience

The beneficial experience associated with taking up amateur road cycling in middle age centered on three themes: engaging with self, engaging with others, and engaging with the outdoor environment. Cycling provided a sense of purpose. Some enjoyed the solitary experience of riding alone, cognizant of the rhythmic nature and sense of flow:

> I love the solitary space you can get into in your head when you're on a bike. You have a sort of… it's not robotic, but you alternate much of the eye on the road, watching for the potholes, dealing with the traffic kind of stuff on a bike and just close off and you're in your head dealing with things. And the rhythm of it means that you do [what] you like, you can just phase it to the point where your brain's just kind of asleep effectively mulling stuff over, or you can be spinning through at a million miles an hour – I find it really good for that.
>
> *Ben*

Some participants had a heightened perception of competition and demand for success at their company and how cycling fitted within the demands of their professional life. For high flyer, Brad, cycling allowed him to pursue a shared passion in a semi-competitive environment while experiencing comradeship:

> I'm competitive and ambitious at work as well, but… but actually you know as you get more senior the working world gets a bit more political. And I think I see sport as a more honest way of competing in a way, you know. So you know particularly at [company], which is quite a toxic and political environment, it's not quite enough to be good at your job, you also have to be a bit of an arsehole to really succeed. So in parallel to sort of… while subsisting in that environment, the honesty of cycling which is you know it genuinely rewards effort… you know you put the hours in, you put the effort in, you get the results… there is a straightforwardness to it I've found refreshing.
>
> *Brad*

The crux of being a MAMIL and enjoying the company of other MAMILs was the combination of shared interest in technology and enhancement of the mind and body through shared enjoyment of cycling through the landscape. In summary, cycling offered the chance to escape everyday stresses of work and family life:

> A big part of it for me is getting outside, you know. Spend your life in an office, spend my life in a car, on a train, on a plane… everything else in my life is preordained – you've got to do the school run, you've got to go to work and I've got to have this meeting. [With cycling] You know you're outside, it might be a beautiful day, it might be a shitty day, but you're still out there, you know, breathing decent air.
>
> *Chris*

Gender roles

One of the tensions of being a MAMIL is negotiating time, usually a three-hour-plus ride on a weekend morning, away from family. This has led to stories in the press of 'cycling widows' and the tolerance employed by abandoned partners (Cook 2009; Price 2014).

> Cycling isn't inherently macho but it does have an element of suffering and purgatorial challenge that appeals to a certain male psyche. It is about escape – freedom from family; seizing time to be solitary, beyond responsibility. Not an easy concept to embrace when you're the cycling widow left to look after small children.
>
> *Helen Cook, 'Help, I'm a Cycling Widow!', The Telegraph 17 June 2009*

Women are traditionally under-represented in amateur road cycling, but there are signs that that is changing with shifting composition of cycling clubs and the rise of the middle-aged woman in Lycra (Falcous 2017). The cycling industry is attuned to this trend with the growth in availability of 'women-specific design' bikes and technology as well as garments. Whether this is reshaping gender roles and how MAMILs mediate joint desires to cycle with their partners warrants further study.

Impact on everyday practical cycling

What of the potential of the popularity of sports cycling to transfer to everyday utility cycling? High-profile cyclists acting as role models and the elevation of cycle sport in general has not translated into cycling's status as an ordinary way of moving around (Bauman et al. 2018). Elite cycling events such as the Grand Tours and the Olympics may have captured the public imagination, but as a spectator sport rather than a participatory practice. Sports cycling's general approach to roads is to minimize risk rather than challenge the car system (Falcous 2017). And the so-called 'war between cyclists and drivers' is a misnomer as people who cycle are most likely to own at least two cars (Pidd 2010). As Horton and Parkin (2012: 309) have pointed out, 'faith in events such as the Olympics to boost cycling need to be decoupled from the much broader and more radical task of restructuring cities to support a culture of everyday cycling'.

Summary

Does the rise of MAMIL represent a gentrification of cycling? Certainly over the past decade there have been changes in the social patterns of cycling and increasing commercialization, at least in the UK, on the back of British success in the Tour de France and the London Summer Olympic Games of 2012. This has followed trends seen in other parts of the Anglosphere, notably North America and Australia. The profile of cycle sport has shifted to one which legitimizes the activity for certain privileged groups who may otherwise not have engaged. It provides the opportunity for active aging, social connection and opportunities to indulge in high-end technology and display certain cultural capital (Sirna, 2016). It is not, however, the only image of cycling and it would be unfair to state that it has wholly eclipsed traditional images of road cycling.

Perhaps the sport, at least in the UK, has departed from its image of working-class 'hard man' masculinity and has reconnected with its early roots and represents more the escapades of early bourgeois cycling clubs. This 're-embourgeoisement' is reinforced by a 'technology of class privilege' in the form of a growing market for over-specced and overpriced carbon bikes that signified new cultural status and wealth (Cox 2008). The representation of road cycling that has evolved has allowed men to diversify the available options of masculinity. They are able to construct their identity through consumption within a global consumerist society as part of a wider phenomenon of what has been called a 'flexible masculinity' (Gee 2014). When all is said and done, MAMILs are relatively benign. Perhaps the investment of time and money on technology that provides a gateway to a healthy mind and body, as well as camaraderie and shared enjoyment of the outdoors, and not the old midlife crisis cliche of the sports car, should be celebrated as much as it is mocked (Seaton 2012).

Bibliography

Acquire Mag (2015). Rapha x Chris King Espresso Tamper. [Online] Available at: https://www.acquire-mag.com/lifestyle/rapha-x-chris-king-espresso-ta [Accessed: 30 April 2021].

Bauman, A. E., Blazek, K., Reece, L., & Bellew, W. (2018). The emergence and characteristics of the Australian Mamil. *The Medical Journal of Australia, 209*(11), 490–494. https://doi.org/10.5694/mja18.00841

BBC (2012). Cyclist Wiggins wins BBC trophy. [Online] Available at: https://www.bbc.co.uk/sport/sportspersonality/20748902 [Accessed: 30 April 2021].

Bike Biz (2011). Rapha responds to spoof website. [Online] Available at: https://www.bikebiz.com/rapha-responds-to-spoof-website/ [Accessed: 30 April 2021].

Casciani, D. (2010). Rise of the Mamils (Middle-Aged Men in Lycra). *BBC News*. Available at: https://www.bbc.com/news/magazine-10965608 [Accessed: 30 April 2021].

Cook, E. (2009). Help! I'm a cycling widow. *The Telegraph*. Available at: https://www.telegraph.co.uk/sport/othersports/cycling/5550817/Help-Im-a-cyclingwidow.html [Accessed: 30 April 2021].

Cox, P. (2008). Class and competition: The gentrification of sport cycling. Presentation at *5th Cycling and Society Symposium*, University of the West of England, Bristol, UK. 8–9 September 2008.

Davies, L. (2012). Bradley Wiggins London 2012 triumph brings boost to British cycling. *The Guardian*. Available at: http://www.theguardian.com/sport/2012/aug/02/bradleywiggins-london-2012-triumph-boosts-cycling [Accessed: 30 April 2021].

Falcous, M. (2017). Why we ride: Road cyclists, meaning, and lifestyles. *Journal of Sport and Social Issues, 41*(3), 239–255. https://doi.org/10.1177/0193723517696968

Florida, R. (2002). *The Rise of the Creative Class and How It's Transforming Work, Life, Community and Everyday Life*. (New York: Basic Books).

G.D. (2013). Cycling is the new golf. *The Economist*. Available at: https://www.economist.com/prospero/2013/04/26/cycling-is-the-new-golf [Accessed: 30 April 2021].

Gee, S. (2014). Bending the codes of masculinity: David Beckham and flexible masculinity in the new millennium. *Sport in Society, 17*(7), 917–936. https://doi.org/10.1080/17430437.2013.806034

Gupta, T. (2012). London 2012: 'Wiggins effect' sees cycling enthusiasm grow. *BBC News*. Available at: https://www.bbc.com/news/uk-england-surrey-19093616 [Accessed: 30 April 2021].

Horton, D. & Parkin, J. (2012).Towards a revolution in cycling. In Parkin, J. (Ed) *Cycling and Sustainability*. Bingley: Emerald Publishing Ltd.

McMahon, D. (2015). Millionaire entrepreneur explains why cycling—And not golf—Is the new sport of choice for young professionals. *Business Insider Australia*. Available at: https://www.businessinsider.com.au/cycling-is-the-new-golf-2015-2 [Accessed: 30 April 2021].

Men's Society Middle Aged Man in Lycra pampering set. [Online] https://www.menssociety.com/products/middle-aged-man-in-lycra?variant=13675569873001 [Accessed: 30 April 2021].

Mintel (2010). *Bicycles UK*. London: Mintel Group.

O'Connor, J. P., & Brown, T. D. (2007). Real cyclists don't race: Informal affiliations of the weekend warrior. *International Review for the Sociology of Sport, 42*(1), 83–97. https://doi.org/10.1177/1012690207081831

Penn, R. (2013). Pedalling fashion: The rise of cycle style. *The Observer*. Available at: http://www.theguardian.com/fashion/2013/jun/29/bike-fashion-rise-of-cycle-style [Accessed: 30 April 2021].

Pidd, H. (2010). Cyclists v drivers? They're often the same people. *The Guardian*. Available at: http://www.theguardian.com/environment/blog/2010/aug/10/cycling-boom-survey [Accessed: 30 April 2021].

Price, J. (2014). The shame of being married to a MAMIL, a Middle Aged Man in Lycra. *Daily Mail Online*. Available at: https://www.dailymail.co.uk/femail/article-2869069/Oh-shame-married-MAMIL-s-Middle-Aged-Man-Lycra.html [Accessed: 30 April 2021].

Renwick, F. (2016). Fred Perry And Bradley Wiggins are back with a new collection. *Esquire*. Available at: http://www.esquire.co.uk/style/fashion/news/a9806/two-britishicons-collaborate-on-a-new/ [Accessed: 30 April 2021].

Rouleur. [Online] https://www.rouleur.cc/ [Accessed: 30 April 2021].

Seaton, M. (2012). The humble Mamil: Why we need 'middle-aged men in Lycra'. *The Guardian.* Available at: http://www.theguardian.com/commentisfree/2012/sep/11/mamil-middle-aged-men-in-lycra [Accessed: 30 April 2021].
Statista (2020). Quarterly expenditure on bicycles in the United Kingdom (UK) from 1st quarter 2011 to 1st quarter 2020, based on volume. [Online] https://www.statista.com/statistics/492738/bicycle-purchase-trend-united-kingdom-uk/ [Accessed: 30 April 2021].
Wallop, H. (2016). Cycling beats golf in the new world of networking. Available at: https://www.ft.com/content/98eb4072-7a98-11e6-ae24-f193b105145e [Accessed: 30 April 2021].
Wheelsuckers [Online] https://wheelsuckers.co.uk/ [Accessed: 30 April 2021].

27
PROFESSIONAL ROAD CYCLING

Daam Van Reeth

Introduction

The first road cycling races date back to the late 19th century and some of cycling's best-known events are now well over 100 years old. Paris–Roubaix was first organized in 1896, the Tour de France was created in 1903 and the first edition of the Tour of Flanders took place in 1913. This earns professional road cycling the title of one of the oldest professional sports. The popularity of road cycling has steadily grown throughout the 20th century and the sport gathered a stable core of fans in France, Belgium, Italy and Spain, the heartlands of cycling. Cycling's most prestigious events, the "Grand Tours" (the Tour de France, the Giro d'Italia and the Vuelta a España), and the major classics called "the monuments of cycling" (Milan–Sanremo, Tour of Flanders, Paris–Roubaix, Liège–Bastogne–Liège and Tour of Lombardy), now attract millions of fans who watch the spectacle from the roadside or in front of their television sets (Van Reeth and Larson 2015). But in spite of its popularity in some European countries and a growing global fan base, from a worldwide point of view road cycling is still a relatively small and commercially underdeveloped sport. A lackluster business model creating financial distress, internal conflicts of interest and the doping legacy are among the problems professional road cycling currently faces.

In this chapter we first explain the peculiar context of road cycling. Next, we introduce the current institutional setting and we take a closer look at the economics and finances of the sport. We conclude with an overview of professional road cycling's main challenges for the future.

The peculiar context of road cycling

Road cycling differs critically from many other sports: it is a team sport that is won by individuals, its competitions are heterogeneous and the sport is practiced on public roads by teams without a home base. These differences have a significant impact on the business model of the sport.

DOI: 10.4324/9781003142041-38

A team sport won by individuals

Professional road cycling shows multiple characteristics of a team sport. At the organizational level, the participants in cycling competitions are the cycling teams. Individual riders can only compete if selected as members of the team to whom they belong. At the financial level, the salary a rider gets paid from the team is his main source of income because prize money is relatively small. This has strong repercussions on his behavior given the fact that most contracts in professional cycling are relatively short term, seldom exceeding two years. At the competition level, it is virtually impossible for a rider to win a race without a strong team support. Team strategies protect the leader, reducing his workload and making sure that an opposing team's leader must do as much work as possible. Even the best riders thus need the team's strength and tactics to win races. Few other sports see athletes completely sacrificing themselves to help a team member win.

But in contrast to classical team sports, cycling races are won by individual riders. It is the only Olympic sport where the team contributes to the victory of an individual, with a medal only for the winning athlete (Jutel 2002). Therefore, professional road cycling is probably best compared with Formula One or MotoGP racing, where the results of individual pilots are also highly dependent on the technical support of the team's engineers and the team strategy. However, unlike the case in motorized sports, cycling teams are not given much credit for their contribution to the result. In cycling, the individual winners receive all the glory and the prize money, although the latter is usually shared between all team members, including staff. Consequently, professional road cycling is neither a typical individual sport, like golf or tennis, nor a pure team sport, like football or basketball (Benijts, Lagae and Vanclooster 2011). Professional road cycling could, in fact, be described as the most individual of all team sports.

A sport with heterogeneous competitions

Many sports have homogeneous events. All football games are played on similar-sized pitches in games of equal duration and with universal rules and a 100-meter sprinter performs the same act throughout his entire career (Rebeggiani and Tondani 2008). In some sports, however, competitions are heterogeneous to a certain extent. Tennis games are played on grass, gravel or hardcourt and alpine skiing courses or Formula One racing tracks are different at every event. Such differences between individual events are even more prominent in road cycling competitions. Not only is there a difference between stage races – which can run from a couple of days up to three weeks – and one-day races, but all competitions also have their own identity and race trajectory. Since climbing a steep mountain pass requires totally different skills than riding on cobblestones, certain top riders (like Chris Froome) will never participate in one-day classics (such as Paris–Roubaix) that do not very well match their strengths. The heterogeneity in races also makes it difficult to determine the best overall rider. The rider who wins the most races (usually a sprinter type of rider) is unlikely to be the winner of the most important cycling competition (the Tour de France) (Van Reeth and Lagae 2015).

Additionally, there is a heterogeneity in prizes. In stage races, multiple prizes are at stake at the same time. Thus, competitors in the same competition may have very different objectives which do not necessarily correspond with trying to win the race. For instance, in the Tour de France, next to the competition for the overall win in the general time classification (yellow jersey), there are four other competitions for secondary prizes: the general points classification (green jersey), the king of the mountain classification (polka dot jersey), the best young rider classification (white jersey) and the team classification. Since there is an individual winner as well in each of the 21 stages, there are at least 26 major prizes to be won in a single Tour de France.

A sport practiced on public roads by teams without a home base

Most sports are played in stadiums or take place on closed tracks. This allows organizers to charge admission fees, creating a revenue stream that supports the sport. Since road cycling takes place on public roads, this source of revenue is unavailable to race organizers. Consequently, although the Tour de France attracts millions of spectators along the streets every year, this does not lead to any revenues for the organizers (Rebeggiani and Tondani 2008). The road nature of the sport also prevents cycling teams from directly collecting revenue from their fans. The absence of a home base makes it harder to create long-term fan loyalty, especially given the fact that cycling teams regularly change sponsors and thus change names. Instead of following a team, cycling fans rather identify with riders. While a fan of Barcelona will usually dislike a former Barcelona player the moment he is transferred to a rival team, a fan of Mathieu Van der Poel will remain a fan no matter what team he is in. In general, most professional cycling teams lack the regional fandom that is typical in many other sports. Consequently, cycling has always wrestled with massive financial challenges. Today, cycling teams still almost exclusively depend upon sponsorship for their financial viability (Van Reeth and Lagae 2018).

The use of public roads also has other important consequences. First, it implies that cycling competitions cause negative externalities to society. Main roads, and sometimes entire city centers, are blocked for hours to the discomfort of the local inhabitants and causing traffic congestion in the wider surroundings of the race trajectory. Second, the use of public infrastructure also imposes direct costs to society. To warrant safety within the race as well as along the race trajectory, a significant amount of public money must be spent on, for example, municipal staff and policing. Traditionally borne voluntarily by local and regional entities, these costs are now increasingly passed on to the race organizers. Third, safety investments in road infrastructure pose a significant problem to the riders. It is a paradox that the introduction of measures to increase road safety, like speed humps, roundabouts, or separate bike paths, have adversely increased the risk of crashes within the peloton. Finally, as a sport run on the road, TV coverage of cycling is complex and expensive. It requires a huge array of specialized equipment and personnel, such as helicopters, motorcycles, mobile high-definition cameras, satellite uplink trucks and a TV relay aircraft. Especially when compared to 'on site' sports, the production costs of a cycling race quickly become relatively high. This is one of the reasons why, except for a small number of top races like the Tour de France, TV revenues are virtually non-existent in professional road cycling and race organizers often pay broadcasters to cover their events, rather than the other way around (Van Reeth 2015).

Professional road cycling today

We first describe the current institutional setting. This is followed by a discussion of the league structure and we conclude with some observations on the economic value of the sport.

Institutional setting

Professional road cycling is organized around four main types of agents: the governing bodies, the race organizers, the cycling teams and the riders. Already in 1900, the International Cycling Union (UCI) became the governing body of the sport (Mignot 2015) and is recognized as such by the International Olympic Committee. It now groups close to 200 member federations and operates as a traditional regulatory body, licensing races and riders, providing referees and enforcing rules. The UCI not only overlooks male professional road cycling, it is also responsible for junior and women's road cycling, as well as for seven other cycling disciplines, such as track racing and BMX (Rebeggiani 2015). The UCI has often been described as a weak institution (Long 2012; Chapter 8), especially in comparison to powerful organizations as the FIFA in football or the FIA in Formula One.

The main reason for UCI's assumed weakness is the missing market power since the most prestigious and commercially most interesting races are controlled by private race organizers. Race organizers have a prominent role in road cycling because of their ability to control access to their events unregulated by ex-ante defined rules (Rebeggiani 2015). Most race organizers are jointly organized in the Association Internationale des Organisateurs de Courses Cyclistes (AIOCC) representing over 100 competitions. The most influential member is without a doubt Tour de France organizer Amaury Sports Organisation (ASO), with Tour de France director Christian Prudhomme also being the president of the AIOCC.

The basic organizational unit of professional road cycling is the cycling team. These teams are supported by one or two principal commercial sponsors and are referred to by these sponsor names (Brewer 2002). The interests of the cycling teams are, at least theoretically, protected by the Association Internationale des Groupes Cyclistes Professionnels (AIGCP). But in contrast to the AIOCC, the AIGCP is a poorly organized and internally divided association with many feuds, resulting in teams constantly leaving and rejoining the organization. Therefore, some cycling teams started to create their own interest groups. In 2007 a number of primarily French teams founded the Mouvement Pour un Cyclisme Crédible (MPCC) to defend the idea of clean and healthy cyling (Rebeggiani 2015). In 2014 about a dozen teams created the Velon Group, a joint-venture company that aims for a business model that ensures a sustainable future for the teams (Van Reeth 2015). The fact that only a couple of teams joined both projects clearly exposes the continuing schism between the cycling teams.

The most important organization representing the riders is the Cyclistes Professionnels Associés (CPA). The general aim of the CPA is to defend the riders' interests before the UCI, the race organizers and the teams (Rebeggiani 2015). A major problem with the CPA is that it only represents riders from about a dozen countries with, for example, no Australian, Colombian or Russian cyclists, for various reasons. Furthermore, it is led by former cyclists and there is almost no engagement from current professional riders. Therefore, although the CPA occupies seats in several UCI commissions, overall the riders are still a rather powerless group in cycling, especially when compared to athletes in other sports. The situation in men's cycling is in stark contrast to the case in women's cycling, where most of the riders joined the very active and highly profiled Cyclists' Alliance rider union.

League structure

The UCI WorldTour is the top league of the pyramidal three-tier competition structure of professional road cycling. In 2021, the UCI WorldTour calendar listed 35 cycling events, although some of these races were eventually cancelled because of the Covid-19 pandemic. In total, 553 riders from 19 cycling teams were part of cycling's top competition (Table 27.1). Below the WorldTour level are the ProContinental level (19 teams, 417 professional riders) and the Continental level (168 teams, about 2,000 semi-professional or amateur riders). The barriers between the three levels are not 100% strict though. ProContinental teams can be invited to participate in WorldTour races and all three types of teams can start in Continental

Table 27.1 The competition structure of professional road cycling in 2021

	World Tour level			*Pro-Continental level*	*Continental level*
	Total	*Europe*	*Rest of the world*		
Teams	19	11 (58%)	8 (42%)	19	168
Riders	553	430 (78%)	123 (22%)	417	± 2000
Races	35	29 (83%)	6 (17%)	± 350	

Sources: www.cqranking.com and www.uci.ch

races. The approximately 350 non WorldTour races in 2021, are divided over five continental circuits: the America Tour, the Asia Tour, the Africa Tour, the Oceania Tour and the Europe Tour, with the vast majority of races belonging to the latter circuit.

One of the main goals of the UCI WorldTour is to globalize professional road cycling. Yet Table 27.1 illustrates that professional road cycling at the top level is still very much a European sport. Of the 35 races at the WordTour level, only six (17%) are not organized in Europe: two each in Oceania, Asia and North America. Furthermore, still close to 80% of the riders in the WorldTour are European. The bulk of the non-European WorldTour cyclists originate from Australia, North America and South America (mostly Colombia). The fact that eight out of the 19 WorldTour teams (42%) are non-European hides the fact that many of these teams are still dominated by a European staff and rider group. In fact, just two teams (EF Education-Nippo and Team BikeExchange) have a majority of non-European riders.

The current league structure, the UCI ProTour as it was initially called, was introduced in 2005. At the time it was promoted as the Champions League of cycling, but it has failed to live up to its expectations. The overall opinion is that the WorldTour is a poorly managed and badly promoted product, illustrated by the fact that never during its now already 15 years of existence a sponsor for the competition was found (Van Reeth and Lagae 2015).

Economic value

Little information on the finances of cycling teams, cycling races or rider contracts is publicly available. It is therefore difficult to assess in detail the economic value of professional road cycling. Nevertheless, in Table 27.2 some key financial data about professional road cycling are presented for the 1990–2020 period. The data on the team budgets and the broadcasting rights are estimates based on a variety of sources while the information on the prize money is obtained from the official Tour de France website.

Team budgets kept increasing continuously in the past 30 years. In nominal terms, the average budget of the best ten cycling teams almost grew almost sixfold, from about €4 million in 1990 to €23 million in 2020. This represents an annual increase of about 6%, which is significantly higher than the long-term inflation rate of 2–3% in western countries in recent decades (Van Reeth 2015). The average is to some extent distorted, however, by the richest team in the peloton, INEOS Grenadiers, the successor to the famous and highly successful Sky team. Its rumoured budget of €46 million is 50% more than that of UAE Team Emirates (€30 million), the second-richest team. Without INEOS Grenadiers, the average budget of cycling's ten richest teams in 2020 would be around €20 million.

In 2021 ten out of 19 WorldTour teams are financed exclusively by traditional commercial sponsors: Ag2R-Citroën, Bora-Hansgrohe, Cofidis, Deceuninck-Quick Step, EF

Table 27.2 Key financial data for professional road cycling (1990–2020, in nominal euros)

	1990	*2000*	*2010*	*2020*
Average budget top 10 teams	4,000,000	7,000,000	10,000,000	23,000,000
Broadcasting rights Tour de France	2,500,000	18,000,000	50,000,000	65,000,000
Prize money Tour de France winner	305,000	335,000	450,000	500,000
Total Tour de France prize money	1,535,000	2,363,000	3,377,000	4,000,000

Sources: www.letour.fr, Sportune.fr, Van Reeth (2015) and Mignot (2014).

Education-Nippo, INEOS Grenadiers, Intermarché-Wanty-Gobert Matériaux, Jumbo-Visma, Movistar Team and Team DSM. The other teams are sponsored entirely or partially by either bike (equipment) manufacturers (Team BikeExchange, Team Qhubeka-Assos, Trek-Segafredo) or national lotteries (Groupama-FDJ, Lotto-Soudal). A more recent phenomenon on the sponsor side is the rise of teams that are funded by states or state conglomerates. In 2006, the Kazahk Astana team was the first such team and it remained the only one for over a decade. In the past five years, however, three similar teams joined the WorldTour: the Bahrain and the United Arab Emirates teams in 2017 and the Israel Start-Up Nation team in 2020.

French television began to pay for the right to cover the Tour de France in the 1960s. The real explosion in TV broadcasting rights started from the 1980s on. In the early 1980s, Tour de France TV broadcasting rights were valued at €250,000 only (Mignot 2014). Forty years later, they are estimated at about €65 million, of which €26 million is paid by France Télévisions. This growth is, of course, not that much different from what happened to the value of the broadcasting rights in many other sports in the recent past. Yet the situation for the Tour de France is exceptional in professional road cycling. Although, as in other sports, television is how 99% of fans follow the competitions, in cycling, outside a handful of prominent events, TV revenues are virtually non-existent.

Table 27.2 also illustrates how in cycling the commercial benefits generated by a race like the Tour de France are hardly redistributed to the main actors in the sport, i.e. the riders or the cycling teams. While broadcasting rights (in nominal terms) more or less increased by a factor of 25 between 1990 and 2020, total Tour de France prize money, which also includes the participation fees to the teams, only increased by a factor of 2.5 from just over €1.5 million in 1990 to €4 million in 2020. Since total Tour de France revenue (broadcasting rights, sponsorship, subventions and fees by hosting cities) is estimated to be over €150 million (Van Reeth 2015), less than 3% of the revenue generated by the event is currently redistributed to riders and teams. Moreover, the prize awarded to the overall Tour de France winner only increased from €305,000 to €500,000 over a 30-year period, which does not even compensate for inflation. In fact, the winning prize was set at €450,000 in 2006 and remained unchanged for almost ten years until 2016, when it was raised to its current level of €500,000. It is remarkable, as well, that over the years the share of the winner in the total amount of prize money decreased from 20% in 1990 to only 12.5% in 2020.

The challenges for professional road cycling

Professional road cycling faces numerous challenges that threaten the long-term viability of the sport. Four of the most significant threats are: the fragile business model, ASO's dominance; the relatively small and aged TV audience; and the doping legacy.

The fragile business model

The Tour de France is not only by far the best-watched cycling event (Van Reeth 2013); it is probably also the biggest free professional sporting event in the world (Wille 2003). From a fan's perspective, professional road cycling is indeed among the cheapest spectator sports. Apart from travel costs, watching a race from the roadside is completely free and in many European countries most major cycling races are broadcast on subscription-free free-to-air TV channels (Van Reeth and Lagae 2015). But there is no such thing as a free lunch. The free nature of the sport comes at a cost, namely the high dependence of teams and races on sponsor revenue. Since cycling teams are funded for over 90% by sponsorship money (Van Reeth 2015) and few companies are willing to commit themselves for more than a couple of years, the future of a cycling team is always insecure. But also race organizers face insecurity. In most sports, television revenues are a major source of revenue but in cycling, as explained before, they are limited. The fact that there is no gate revenue either means that also the race organizers mainly depend on sponsorship money to cover the organizational costs of the races. One of the main challenges for professional road cycling is how to adjust this sponsorship-driven business model, i.e. how to create better financial stability through a diversification of revenue.

ASO's dominance

Cycling's most prestigious event, the Tour de France, is absorbing most of the media attention for professional road cycling and in many countries it is actually the only cycling race that is widely covered. The commercial importance of the race, with a market share of two-thirds of all television revenue and sponsorship income generated by other WorldTour organizers, threatens to reduce the cycling season to basically just a single event. As explained, the Tour de France is organized by ASO, a for-profit company that specializes in the organization and promotion of, among others, motor sports, athletics, equestrian sports, golf events and cycling. ASO also owns other well-known cycling races in France and Belgium, such as Paris–Nice, Paris–Roubaix, la Flèche Wallonne and Liège–Bastogne–Liège, it holds a majority stake in the Vuelta a España and it is involved in the organization of many more smaller cycling races around the world. The company is without a doubt road cycling's most powerful actor and its dominant position is a mixed blessing for professional road cycling (Van Reeth and Lagae 2018). The commercial success of the Tour de France allows ASO to invest in less profitable races, to support new races logistically and to award a significantly higher amount of prize money than other races do. This way, ASO indirectly subsidizes professional cycling. But its dominant position is also a threat to the further development of the sport. As a private company, above all it defends the commercial value of the Tour de France and it uses its market power to block any reform plan that is likely to threaten its prime position. Consequently, while theoretically the UCI is responsible for managing world cycling, in practice it is ASO that holds the key to the future of the sport.

The relatively small and aged TV audience

TV audiences for road cycling are much smaller than most of the stakeholders in cycling like to believe or even communicate. For example, while ASO claims a multibillion TV audience for the Tour de France, the actual per-stage average audience is only about 20 to

25 million (Van Reeth 2019). Furthermore, cycling is confronted with an ageing fan base. In 2020, the average Tour de France TV viewer in France was 62 years old. In Flanders, 74% of the audience was over 55 while only 15% was under 45 (Van Reeth 2021). Although cycling fans in Anglo-Saxon countries are, on average, younger than their counterparts in continental Europe, professional road cycling seems to lose touch with new generations, largely because of the lengthy broadcasts in which regularly little of interest happens until the last couple of minutes. Race innovations, such as shorter stages or the use of dirt roads and providing live content through other media platforms that are much more adapted to the habits and needs of younger sports fans, are just a few of the changes cycling needs to consider more seriously.

The doping legacy

Professional road cycling has a long association with doping, from the sport's inception in the late 19th century, throughout the 20th century and into the 21st century (Vandeweghe 2015). Although the introduction of, respectively, the whereabouts system, a blood passport and a no-needle policy did improve the situation significantly in the past 15 years, the sport continues to find it difficult to deal with this historic legacy. Consequently, doping remains a constant threat to the development of road cycling because, much more than the forgiving European cycling fans, international sponsors and broadcasters are very vigilant of any doping violation (Lagae and Van Reeth 2015; Chapter 31).

Why is doping such a persistent problem? One explanation is often given by the exceptional physical requirements that are needed to compete in professional road cycling races (Rebeggiani and Tondani 2008). Especially the Grand Tours require almost super-human efforts from the riders. A three-week stage race like the Tour de France consists of 21 sequential stages with usually only two days of rest. About a third of the stages are held in the mountains. Athletes also face a high number of competition days, unlike what is the case in other endurance sports such as a triathlon or marathon running. During an average year, a professional cyclist covers a distance of up to 40,000 kilometers in training and competition.

However, it has also been argued that the doping problem partially has its roots in cycling's organizational structure and dysfunctions. Three aspects seem to be of particular importance (Rebeggiani 2015). First, riders face financial instability because, as mentioned earlier, unlike most other professional team sports, contracts in cycling are usually short-term in nature (no longer than two years) and insecurity about their team's future is much higher, given the almost complete dependence on the main sponsor. This adds pressure to deliver strong results, especially for younger and weaker riders, which might in turn foster doping practices. Second, the strong dependence on team support for producing individual top results, rather unique among endurance sports, has shaped a very particular social environment with huge peer pressure. The cycling peloton has often been described as a very close community, where each rider has to observe tacit rules and where non-cooperative behaviour is undesirable (Vignette F). It seems probable that such a system of personal ties and informal rules has considerably contributed to the growth and the persistence of the doping phenomenon by making it difficult for team members to express any possible dissent. Third, another singularity of cycling is that the winner of a race or a classification usually shares his prize money with all other team members, and sometimes even with the staff. This phenomenon of prize sharing generates another peer effect in doping. There is not only an incentive for team leaders and potential winners to dope, but also for the support riders.

References

Benijts, T., Lagae, W. & Vanclooster, B. 2011, The influence of sport leagues on business-to-business marketing of teams: the case of professional road cycling, *Journal of Business & Industrial Marketing*, 26(8), 602–613.

Brewer, B. 2002, Commercialization in professional cycling 1950–2001: institutional transformations and the rationalization of doping, *Sociology of Sport Journal*, 19(3), 276–301.

Jutel, A. 2002, Olympic road cycling and national identity: where is Germany, *Journal of Sport and Social Issues*, 26(2), 195–208.

Lagae, W. & Van Reeth, D. 2015, Paradoxes in professional road cycling: a plea for a new cycling industry. In: Vanden Auweele, Y., Cook, E. & Parry, J. (Eds.), *Ethics and Governance in Sport: The Future of Sport Imagined* (pp. 121–128). Abingdon, UK: Routledge.

Long, J.G. 2012, Tour de France: a taxpayer bargain among mega sporting events? In: Maennig, W. & Zimbalist, A. (Eds.), *International Handbook on the Economics of Mega Sporting Events* (pp. 357–385). Cheltenham, UK: Edward Elgar.

Mignot, J.F. 2014, *Histoire du Tour de France*. Paris: La Découverte.

Mignot, J.F. 2015, The history of professional road cycling. In: Van Reeth, D. & Larson, D. (Eds.), *The Economics of Professional Road Cycling* (pp. 7–31). London, UK: Springer.

Rebeggiani, L. 2015, The organizational structure of professional road cycling. In: Van Reeth, D. & Larson, D. (Eds.), *The Economics of Professional Road Cycling* (pp. 33–54). London, UK: Springer.

Rebeggiani, L. & Tondani, D. 2008, Organizational forms in professional road cycling – efficiency issues of the UCI ProTour, *International Journal of Sport Finance*, 3(1), 19–41.

Vandeweghe, H. 2015, Doping in cycling: past and present. In: Van Reeth, D. & Larson, D. (Eds.), *The Economics of Professional Road Cycling* (pp. 285–311). London, UK: Springer.

Van Reeth, D. 2013, TV demand for the Tour de France: the importance of stage characteristics versus outcome uncertainty, patriotism and doping, *International Journal of Sport Finance*, 8(1), 39–60.

Van Reeth, D. 2015, The finances of professional cycling teams. In: Van Reeth, D. & Larson D. (Eds.), *The Economics of Professional Road Cycling* (pp. 55–82). London, UK: Springer.

Van Reeth, D. 2019, Forecasting Tour de France TV audiences: A multi-country analysis, *International Journal of Forecasting*, 35, 810–821.

Van Reeth, D. 2021, *Road Cycling TV Viewing Report 2020*. KU Leuven.

Van Reeth, D. & Lagae, W. 2015, The future of professional road cycling. In: Van Reeth, D. & Larson D. (Eds.), *The Economics of Professional Road Cycling* (pp. 313–341). London, UK: Springer.

Van Reeth, D. & Lagae, W. 2018, A blueprint for the future of professional cycling, *Sport, Business and Management: An International Journal*, 8(2), 195–210.

Van Reeth, D. & Larson, D. 2015, *The Economics of Professional Road Cycling*. London, UK: Springer.

Wille, F. 2003, The Tour de France as an agent of change in media production, *The International Journal of the History of Sport*, 20(2), 128–146.

Vignette F
IN THE PELOTON

Michael Barry

On the start line the peloton waits impatiently. With the fire of the starter's pistol the group of a hundred or so riders clip into their pedals and roll away from the crowd of spectators, the emcee whose voice echoes through the town over the loudspeakers and the rest of the circus surrounding the bike race. Over the next hours, the peloton will build in tension, in speed, and ferocity as it tears through towns, careens down mountainsides, hammers over torn up farm roads and surges towards the line. The nerves the riders try to digest are borne in the dozens of unknown variables ahead of them, and in their competitive drive to win. Crashes will send some to hospital. Without the fitness or motivation to follow, others will struggle. Some will surprise themselves with their stamina as they surge to the front, and for the fortunate few, the day will go as planned.

As they wait for the start, the riders sit on their top tubes, arms resting on the brake lever hoods, stoically appearing calm and relaxed masking their nerves. The aroma of liniment, rubbed into their smoothly shaven tanned legs, wafts through the peloton. Their bodies are lithe and tuned over tens of thousands of kilometers to race. Rivals and teammates chat about the race course, the lousy hotel they stayed in the night before, the muscle striations and veins in the legs of another rider, their families, the weather. Most of it is inane banter which overlies the nerves each feels. The spectators lean over the barriers which line the road, watch, point, encourage, or even, jeer. Few of them know or understand what the riders are contemplating.

When the starter pistol fires, the anxious energy the riders feel dissipates within the first few pedal strokes. The peloton immediately finds its formation. The riders move together as one, only splitting and regrouping as they negotiate a round-about. It thins into a long line as it accelerates out of a corner, or as the pace builds or the wind blows from the side. Within the group, the riders are within centimetres of each other. Their elbows touch as they roll through the countryside, sometimes at over 70 km/h. On the descents they'll break 100 km/h. It is a proximity they become accustomed to with years of experience. They place faith in their own skills and those of their rivals. They trust the riders in front of them will point out a curb, a pothole, or any other impediment, and they'll have the innate skills to avoid it.

The peloton bulges as the pace slows, or as the road narrows, or in a moment when every rider knows they need to be near the front. The riders at the front are doing roughly thirty

DOI: 10.4324/9781003142041-39

percent more of the work, as they cut through the wind, the greatest of hundreds of variables they will face in every race. Even on a calm day, air resistance influences the outcome. Almost all of their team tactics and the race dynamics are planned in accordance with the wind. The team leaders will sit in the slipstream of their teammates, their *domestiques*, for as long as possible to conserve energy until their final burst to the finish line. The domestiques share the work, blocking the wind for their leaders. Each rider has a distinct role and goal within the race to place the leader in a position to win.

Every half hour the riders will reach into their pockets for a bite of an energy bar, or gel, or small sandwich to fuel their effort. Behind the peloton the team cars follow in a tight caravan with spare wheels, bikes, parts, drinks, food, and everything the riders might need. From the driver's seat, the team's sports director barks commands into a radio that transmits to the riders' earpieces. A mechanic sits in the back with wheels and tools ready to fix the bikes, while a team doctor sits up front with his medical kit ready to patch up the riders or treat an illness.

In a calmer moment, when the pace is steady, riders will return to their team cars, when they require more food and drink, or to discard or pick up more clothing. A domestique will carry the load of bottles and food for the entire team, so the others don't have to expend the energy returning to the peloton after the trip back to the team car.

On a long hard climb, the group will quickly thin into a line and then split into smaller groups. The riders' chatter is killed by the pace and gradient. Their breath accelerates and deepens as they need more oxygen to feed the effort. Some riders will ascend alone while others will find rivals and teammates to follow or pace. Each will try to find a rhythm that can be sustained for the duration of the climb. Those with the legs to accelerate and distance themselves from their rivals will hold back until the opportune moment where the attack will be its most potent and damaging.

As the summit is reached, they'll take a moment to ease off on their pedals, take a few deep breaths, eat, take a sip of water, and prepare for the descent. If it's cold, they'll look to the roadside to grab a newspaper from a spectator to push down the front of their jerseys or, if they can get one from their team staff, pull on a race jacket. Focussed on every meter of road they'll corner like race car drivers, braking at the last second, and accelerating hard to regain speed. A moment of inattention and they'll careen off the road. At 90 km/h, the fall could be fatal. In the valley, they can take a moment to rest and regroup, and prepare for the final push to the finish line.

In the peloton, the flow of movement is incessant, as riders jostle for position. They fight for their place in the group out of the wind, near their teammates and key rivals and close to the front. Teams of riders group together within the peloton so they can react as one. They quickly coordinate to follow if another team accelerates, if a rider crashes or if a leader has a mechanical problem. They're ready if the sports director orders them to take charge of the race, to up the speed or to attack their rivals.

Although only one rider will cross the line in first, there is a team of riders who helped orchestrate the victory. As the leaders race towards the line, with their job for the day done, the domestiques conserve their energy for the coming races.

The race over, the riders will embrace a leader who has won or console one who has lost. Their mouths parched from the effort, they'll gulp down a drink handed to them by the team staff, and pedal slowly to the team vehicles, now feeling the ache of the hours in the saddle in their legs. For a few minutes they'll be able to relax, elated when the job is done, before they have to refocus on the next race where they'll once again push their bodies to the limit.

References

Benijts, T, Lagae, W & Vanclooster, B 2011, The influence of sport leagues on business-to-business marketing of teams: the case of professional road cycling, *Journal of Business & Industrial Marketing*, 26(8), 602–613.

Brewer, B 2002, Commercialization in professional cycling 1950–2001: institutional transformations and the rationalization of doping, *Sociology of Sport Journal*, 19(3), 276–301.

Jutel, A 2002, Olympic road cycling and national identity: where is Germany, *Journal of Sport and Social Issues*, 26(2), 195–208.

Lagae, W & Van Reeth, D 2015, Paradoxes in professional road cycling: a plea for a new cycling industry. In: Vanden Auweele, Y, Cook, E & Parry, J (Eds.), *Ethics and Governance in Sport: The Future of Sport Imagined* (pp. 121–128). Abingdon, UK: Routledge.

Long, JG 2012, Tour de France: a taxpayer bargain among mega sporting events? In: Maennig, W & Zimbalist, A (Eds.), *International Handbook on the Economics of Mega Sporting Events* (pp. 357–385). Cheltenham, UK: Edward Elgar.

Mignot, JF 2014, *Histoire du Tour de France*. Paris: La Découverte.

Mignot, JF 2015, The history of professional road cycling. In: Van Reeth, D & Larson, D (Eds.), *The Economics of Professional Road Cycling* (pp. 7–31). London, UK: Springer.

Rebeggiani, L 2015, The organizational structure of professional road cycling. In: Van Reeth, D. & Larson, D (Eds.), *The Economics of Professional Road Cycling* (pp. 33–54). London, UK: Springer.

Rebeggiani, L & Tondani, D 2008, Organizational forms in professional road cycling – efficiency issues of the UCI ProTour, *International Journal of Sport Finance*, 3(1), 19–41.

Vandeweghe, H 2015, Doping in cycling: past and present. In: Van Reeth, D & Larson, D (Eds.), *The Economics of Professional Road Cycling* (pp. 285–311). London, UK: Springer.

Van Reeth, D 2013, TV demand for the Tour de France: the importance of stage characteristics versus outcome uncertainty, patriotism and doping, *International Journal of Sport Finance*, 8(1), 39–60.

Van Reeth, D 2015, The finances of professional cycling teams. In: Van Reeth, D & Larson D (Eds.), *The Economics of Professional Road Cycling* (pp. 55–82). London, UK: Springer.

Van Reeth, D 2019, Forecasting Tour de France TV audiences: a multi-country analysis, *International Journal of Forecasting*, 35, 810–821.

Van Reeth, D 2021, *Road Cycling TV Viewing Report 2020*, KU Leuven.

Van Reeth, D & Lagae, W 2015, The future of professional road cycling. In: Van Reeth, D & Larson D (Eds.), *The Economics of Professional Road Cycling* (pp. 313–341). London, UK: Springer.

Van Reeth, D & Lagae, W. 2018, A blueprint for the future of professional cycling, *Sport, Business and Management: An International Journal*, 8(2), 195–210.

Van Reeth, D & Larson, D 2015, *The Economics of Professional Road Cycling*. London, UK: Springer.

Wille, F 2003, The Tour de France as an agent of change in media production, *The International Journal of the History of Sport*, 20(2), 128–146.

28
OFF-ROAD CYCLING

Karen McCormack and Ben Osborn

Off-road cycling takes many forms, from gravel riding to cyclocross to riding on single-track trails to jumping over purpose-built jumps. A wide array of activities ranging from professional bikers racing down steep mountain cliffs to families and recreational cyclists riding on fire roads and road-to-rail trails (a.k.a. trail riding) are often grouped together as mountain biking. We would argue that the two things that separate mountain biking from other forms of cycling are the terrain, riding on dirt or gravel most often on trails, and the bikes themselves (Chapters 9 and 10), typically with wider knobby tires, flat handlebars, and frequently a suspension system.

Most forms of mountain biking resemble other action or lifestyle sports more than competitive or team sports, despite the fact that it can be both competitive and done in teams. Wheaton (2004) identifies the characteristics of lifestyle sports as:

a relatively recent in origin with an emphasis on grassroots participation;
b often involving the consumption of new objects and requiring a commitment of time, money, and lifestyle;
c having a participatory ideology promoting fun and emphasizing creativity and aesthetics over speed and strength;
d made up of a predominantly white, middle-class, and Western participant group;
e predominantly but not exclusively individualistic and nonaggressive but often fetishizing risk and danger;
f performed in outdoor spaces with an appreciation of nature.

These characteristics accurately describe many versions of mountain biking in different parts of the world.

Mountain biking differs significantly from road cycling because the desire among many mountain bikers to ride on narrow, single-track trails in wilderness areas requires both active advocacy work to secure access to land and volunteer labor to build trails. While many do simply ride, there is a widespread acknowledgment that this activity would be impossible without the work of a large volunteer base, creating advocacy groups with large memberships. These groups often feature group rides and other social activities, supporting and sustaining the community alongside local bike shops.

DOI: 10.4324/9781003142041-40

In this chapter, we will briefly look at the emergence of the sport and its advocacy groups, then explore the most common varieties/types of mountain biking and trails, the competitions and festivals that often bring together large numbers of riders, the demographics of the community, and finally the environmental impact of riding in wilderness areas.

History of the sport

The history of off-road/mountain biking begins in 1896 with the 25th Infantry Bicycle Corps in the United States. The Corps, a unit made up of 20 Black soldiers with a white commanding officer, was testing the use of bicycles as an effective mode of transportation in warfare. These soldiers rode through challenging terrain and weather conditions, from Montana to Missouri, a route that today would be about 1600 miles by road. The military soon dismissed the idea of using bicycles in warfare in part because of the difficulty (Anacker 2018).

Riders continued to ride off-road, but Marin County, California in the 1970s is often credited with the origins of mountain biking precisely because of the group of riders that began building better adapted off-road bikes. What began as the collection of old bike frames to retrofit for dirt riding led over the decade to the building of unique frames and then to the first company, MountainBikes, building purpose-built bikes for trails. The name "mountain bike" may have first been used in the application for a business license for MountainBikes (Berto 2014). Sales of mountain bikes grew quickly in the late 1970s and early 1980s, identifying this period as a critical moment in the development of the sport.

While Marin County is the home of the development of the mountain bike, the North Shore of British Columbia, near Vancouver, is central to the history of trail building. According to the Marin Museum of Cycling (2021), early builders began building bridges to cross over swampy areas in 1984, followed by more extensive and challenging wooden features such as "skinnies" – thin bridges or logs with shaved tops, log rides, wall rides, jumps, drops, and high bridges that now feature in many trail systems across the world. Beginning in the late 1990s, these builders shared video of trails, inspiring bikers through the US and beyond to mimic these features on their own trails.

Mountain biking spread widely throughout the 1980s and onward across the globe. The International Mountain Bicycling Association (2021) was formed in 1988 to advocate for trail access and promote mountain biking through educating riders and trail builders to limit erosion and environmental impacts. More than other types of cycling, grassroots organizations are necessary to advocate for land access and to organize and recruit volunteers to build and maintain trails. Trail use disputes between bikers, hikers, and equestrians have been common, especially in countries like the United States with little sustained public investment in trail building. This need for volunteers and coordinated effort have made mountain biking a more inclusive sport with a strong community and shared culture (McCormack 2017).

Many local advocacy groups organize trail days that invite all riders in an area to participate in trail maintenance and building. This work is done with at least two objectives: to develop trails and ensure that they are rideable, but also to educate riders about trail building techniques, including building to ensure water drainage and minimal erosion. One unintended but important consequence of this work is forging stronger ties among riders and creating a sense of belonging to a local biking culture.

Types of off-road biking

There exist several subcategories of the sport of off-road biking, each with its own unique culture, history, equipment, and philosophy. For years, these subgroups have been evolving and splitting, making it nearly impossible to identify every one, but several main genres of the sport can be identified. We would argue that rather than try to identify every subcategory, it is helpful to think of these types as existing on a spectrum. We have chosen to use terrain as the organizing feature, beginning with the varieties most similar to on-road cycling.

Before delving into the different types of mountain biking, it is important to clarify a few facts about the continuum of terrain used to group these riding styles. First, several types of riding blur the line between road and mountain biking. One example is gravel riding, in which riders typically cover long distances on backcountry gravel or unpaved roads. This type of riding requires wider forks than road bikes to accommodate wider tires, stiffer spoking, frames with good damping as hybrid (road and gravel) bikes are designed to allow for long distance riding on rougher surfaces.

Classifying riding types on a continuum is also challenging because the differences between some forms of terrain – a manicured dirt jump and a rugged downhill trail for example – are considerable, but cannot easily be placed on this spectrum. As such, the order of this section does not recognize every type of mountain biking but we contend that this is the clearest way to organize the various types. With this in mind, the major types of riding can be explored with some semblance of order.

Cyclocross: The first variety of riding, and the most similar to on-road riding, is cyclo-cross. Riding on tracks or trails interspersed with a variety of obstacles, cyclocross riders focus on covering terrain as quickly as possible. To this end, it is standard for a cyclocross rider to dismount their bike when approaching an obstacle and continue on foot, toting their bike over their shoulder. With less focus on technical biking ability and more on physical endurance and speed, the discipline of cyclocross presents as some combination of on-road cycling with an obstacle course race. While the riding portions of cyclocross tracks vary greatly in difficulty, they are generally fairly smooth and technically straightforward.

Cyclocross bikes are designed to maximize efficiency not only in the riding portion of these courses, but also on the sections through which the bike is carried. As such, these bikes are lightweight and stiff, resembling a slightly more robust road bike with thicker tires. These bikes do not typically have any suspension, and often employ the drop handlebar design otherwise only seen on road bikes. The purpose of this type of handlebar is to allow the rider to get into an aerodynamic crouching position, reducing drag. Due to mud often sticking to the tires, the space between the tire and fork crown is larger than normal. Those who participate in cyclocross riding (it is popular in Western Europe) can often be seen wearing skintight Lycra suits, similar to those worn by on-road cyclists. Typically, a cyclocross rider will wear a helmet but no other protective equipment.

Due to the requirement of purpose-built trails for this type of riding, most cyclocross riders participate in races. It is fairly uncommon for a cyclocross track to exist absent a race or event requiring its presence, and most are disassembled after an event. For this reason, many cyclocross riders also participate in another form of riding, including mountain biking and on-road cycling. This fact makes it a costly pursuit, as these riders often need to own multiple bikes.

Cross-country mountain biking was pushed into the mainstream by the Olympic event and is now the most widely recognized form of the sport. Much like an on-road cyclist, a cross-country mountain biker is primarily focused on physical endurance, efficiency, and grit.

Typically, cross-country riders will opt for a less technically challenging route than enduro and downhill bikers in order to cover more distance and test their endurance.

Cross-country bikes are lightweight and minimalist in construction, with limited suspension. Typically they have a front "fork," or front suspension that cushions the impact of riding on rough surfaces. It is not uncommon for a cross-country bike to have no rear suspension at all, a style known in the community as a "hardtail." These bikes are fast and efficient, but have limitations on harsher terrain. Due to the generally undulating terrain encountered by cross-country riders, helmets are typically the only type of protective equipment worn. Some cross-country riders, and most racers, can be seen sporting Lycra clothing, while many riders opt for the more relaxed look offered by athletic clothing.

The cross-country mountain biking community is fairly diverse. The health benefits of this cardio-intensive variety of biking attract many middle-aged and older riders, who may be hesitant to delve into the other, higher-risk, varieties of mountain biking (Chapter 30). These same benefits attract younger riders looking for a fun workout, as well as those interested in getting into the sport of mountain biking. The low technical difficulty of this style allows beginner riders to get hours of experience on a bike without the frustration of technical trails with a steeper learning curve. Entry-level cross-country mountain bikes are also among the cheapest mountain bikes on the market, making it possible for riders from a wide range of socioeconomic backgrounds to enjoy the sport. The low risk, relatively low initial investment, and health benefits offered by cross-country mountain biking have made it the most popular form of mountain biking.

Enduro or trail riding is more technical than cross-country riding. Enduro is a rapidly growing segment of the sport covering a broad range of riding styles. Riders look to combine the technical difficulty of downhill riding with the endurance aspects of cross-country. With bikes capable of efficiently climbing as well as descending over difficult terrain, enduro mountain bikes are a favorite for those who want to enjoy days at the downhill bike park as well as cross-country rides. These bikes have both front and rear suspension, and in order to make them climb efficiently, frame construction often uses lightweight materials. Carbon fiber bikes are popular for their lightweight construction, yet these materials – along with high-quality brakes, suspension, and other components – make enduro bikes quite expensive. Because of their versatility, enduro bikes occupy a fairly large segment of the bike market today.

Due to the loose definition of enduro riding, it is hard to pinpoint any one demographic to which it appeals most. As such, it is probably best understood as some amalgamation of downhill and cross-country riding, with riders falling anywhere along the spectrum between. Enduro bikes tend to be expensive, but riders in this category will likely need less protective equipment and fewer repairs than downhill riders, and can ride on trails that do not have an entry fee, making it a somewhat less expensive option than downhill riding for many. The versatility of enduro bikes also helps more serious riders to enjoy different kinds of riding without buying several bikes.

Fat biking is another increasingly popular style of riding in cold climates (Vignette J). It involves retrofitting enduro bikes with wide forks, and very wide, fat tires to enable riding on snow and ice. Dedicated fat-tire bikes are also now available for purchase, and increasingly these bikes are found on trails in all kinds of weather. Fat tires enable riders to clear obstacles more easily but do not offer the maneuverability of cross-country or enduro bikes. Technical trails, once packed with snow, become smoother and more rideable though winter riders must contend with slippery conditions as an additional challenge.

Downhill mountain biking is a subset of the sport which has seen rapid growth in recent years. As the name suggests, downhill riding typically contains very little or no climbing at all.

To achieve this, many downhill riders travel to bike parks designed specifically for downhill riding. These parks are often out-of-season ski mountains, with chairlifts retrofitted to carry bikes. Despite the large investment required to update lifts and build trails down the mountain, increasingly unpredictable winters have driven many ski resorts to invest in the sport as a means of diversification, helping to push the sport into the mainstream.

Much like downhill skiing, downhill mountain biking requires a substantial financial investment in equipment, travel, and lift tickets. While entry-level mountain bikes may survive lower-intensity cross-country riding, a relatively high-end bike is all but required for downhill riding. This cost is partly due to the need for downhill mountain bikes to have robust full-suspension designs with a large amount of travel (distance from the bottom to the top of the suspension). Downhill riding takes a toll on even the most high end of mountain bikes, and riders accept that they will be constantly repairing and replacing components of their bikes, something which can prove extremely costly. Additionally, many downhill riders opt for the additional protection afforded by body armor, ranging from the nearly ubiquitous ensemble of knee pads and a full-face helmet to full suits of protective padding. The cost of this equipment, along with ever-increasing lift ticket prices and the limited number of downhill parks, makes downhill riding a very expensive sport.

One might think that the high price tag associated with downhill mountain biking would lead to an older group of participants with more disposable income, but this is not the case. Downhill riders tend to be quite young, as the danger associated with this discipline of riding deters many older riders. The core group of dedicated downhill riders has been composed primarily of white men in their twenties, but as the discipline has grown and more mountains have embraced the model for summertime business, the demographics of this group have begun to expand. Many downhill riders opt for purpose-built mountain biking attire, including pants similar to those worn by motocross riders, and shirts made of a thick but breathable fabric, called jerseys.

Dirt jumping is a unique segment of the sport of mountain biking, more similar to some forms of BMX than to the aforementioned disciplines. The participants in this subgroup, called dirt jumpers, focus entirely on riding jumps. A dirt jumping trail, often referred to as a set or a line, consists of smooth dirt jumps and rollers. Dedicated to learning new tricks and riding as smoothly as possible through the features, dirt jumpers develop excellent control of their bikes and comfort in the air. These skills help many dirt jumpers to improve at other types of mountain biking as well, and many riders are not exclusively focused on dirt jumping.

Dirt jumping bikes are typically light and small, and most only have front suspension. Less aggressive tires, in conjunction with these lighter and stiffer bikes, allow riders to carry speed more easily. This means that a line of dirt jumps does not necessarily have to travel downhill, as riders can manipulate the terrain to maintain speed even when there is little or no vertical drop from start to finish. These techniques of speed management resemble those seen in the more widely recognized sport of BMX racing; riders push their bikes into downhill sections of trail, lifting over the uphill sections. Those who participate in this form of riding often embrace a style derived from skateboarding and BMX riding. Rather than wearing purpose-built biking clothes, many dirt jumpers can be seen sporting streetwear. Some producers of mountain bike clothing have embraced this style, and now create lines of mountain bike clothing made to look like street clothes.

Dirt jumps require a great deal of work to create and maintain, and since most dirt jump parks are not operated for-profit, most of this work is done by riders. This means that many dirt jumpers become quite attached to the jumps which they helped to build. Many builders of dirt jumps see their work as a form of art, taking pride in the shape of each feature and the way they flow together. Constant wear on the jumps from riding and weather mean that many hours have to be spent to keep a line functional. A common mantra in the dirt jumping community is "no dig no ride," calling on those who enjoy the trails to do their part in the creation and maintenance of the jumps.

Similar to dirt jumping in many ways, the sport of **slopestyle** mountain biking includes large jumps, smooth trails, and in-air tricks. What sets slopestyle riding apart from dirt jumping is the tendency for slopestyle courses to descend vertically from start to finish. This descent allows riders to attempt bigger tricks, as they don't need to land perfectly to gain enough speed for the following feature. Slopestyle courses are fairly rare due to the liability associated with such large features.

Mountain biking competitions and events

While much mountain biking is done for recreation, mountain bike races and competitions bring attention to the sport and are increasingly recognized as important sporting events, evidenced by the inclusion of mountain biking in the Olympic Games beginning in 1996 (Atlanta). Bikers race in nearly every category of the sport, with cross-country, enduro, and downhill racing perhaps most widely practiced. The Union Cycliste Internationale (UCI 2021), an international organization that hosts events for road and mountain cycling, hosts world championship races in 5 disciplines: cross country, downhill, 4X (in which 4 riders begin together and work to overtake each other in a short track with numerous obstacles), XCC/Short Track (with roughly 20 minute sprint races), and a category added in 2019 for E-MTBs, pedal-assisted electric bikes.

Along with UCI and Olympic events, bike races are hosted at the national and state/regional levels by local bike parks and mountain bike groups. These can take the form of the races at the UCI level but also run the gamut from amateur and kids races to multi-day, grueling competitions like the British Columbia (BC) Bike Race, 24-hour races, and more. Races can serve as gathering spots for mountain bike enthusiasts.

Other types of competitions highlight additional disciplines. Crankworx, which *International Mountain Bike Magazine* calls the biggest mountain bike festival in the world, creates festivals in multiple locations each year, bringing together some of the best riders in downhill and slopestyle biking, featuring big features and risky tricks. Along with the elite-level competition, Crankworx features kids' races, pump track competitions, and amateurs, along with booths and activities to celebrate mountain bike culture. Crankworx hosts events at some of the hot-spots for mountain biking, including Whistler-Blackcomb Mountain in British Columbia and Rotorua, New Zealand.

Other festivals, like Rampage sponsored by energy drink company Red Bull, make for compelling video footage of riders doing extremely risky riding on the mesas (and cliffs) of Utah, US. These events are frequently the most viewed televised imagery that links mountain biking to a particular version of risk-taking masculinity in the public eye. While this coverage advertises mountain biking to the public, it projects an adrenaline-fueled, risk-taking version of the sport that discourages some potential riders.

Who mountain bikes?

It's nearly impossible to know how many people participate in mountain biking or who those people are given that many riders have no group affiliations and simply ride trails for pleasure on their own. The Outdoor Foundation, which surveys Americans to learn about their recreation in the outdoors, reported that 17.3% of Americans participated in some form of cycling, including road, mountain, or BMX biking at least once in 2021. This is not a good measure of the number of regular riders even if we combine all types of road and mountain biking, since someone riding around their neighborhood once a year was counted.

One measure of the popularity of the sport comes from bike sales and market analysis. Global Industry Analytics (2021) estimates that 44.2 million bikes were sold in 2020 across the world. *The New York Times* (Sloss 2021) reported that mountain biking, and the sales of bikes, skyrocketed in the United States during the COVID-19 pandemic, with bike sales increasing at every price point and style.

Gender is perhaps the most studied variable in understanding the population of mountain bikers. While the Outdoor Participation survey in the US suggests that 1 in 3 mountain bikers are women, other studies, including the research done by the first author, suggest that it is closer to 1 in 5 participants (Outdoor Industry Association 2021). Based on survey data from 2,363 mountain bikers recruited through mountain biking groups on Facebook, mostly from the United States, Canada, Australia, New Zealand, and Europe, 19% identified as women.

In this survey, about 85% of respondents participated in cross-country riding, nearly 30% in Enduro, just over 20% rode downhill, about 12% rode cyclocross, and then less than 10% rode dirt jumps and BMX. 78% of the sample were between the ages of 26 and 55 and nearly two-thirds had an education level of a bachelor's degree or higher. Attempts to measure racial and ethnic diversity in mountain biking reinforce the contention that action sports are disproportionately white and male.

Over the past decade or so, made possible by the explosion in social media, many women's riding groups and forums have emerged to support female riders. Along with these grassroots organizations, many bike companies now have women-specific lines of bikes (the Juliana bikes from Santa Cruz, Liv from Giant, for example). The number of skill clinics focused on women and girls have likewise grown significantly in the past decade, encouraging more girls and women to take up the sport.

Programs have also emerged to support Indigenous riders, like the Indigenous Youth Mountain Biking Program in British Columbia, though there are significantly fewer programs targeting Black or Indigenous riders or other under-represented communities than those targeting women. While mountain biking organizations try to recruit a diverse rider group and biking companies attempt to expand their customer base, the cost of mountain bikes and proximity to trails can be significant barriers to entry.

Environmental impacts

While some debate around trail use comes from competing user groups (hikers, equestrians, bikers), much debate around opening public land or allowing for trail development concerns the environmental impact of mountain biking. The effects of all trail-based activities include reduction in vegetation, decline in types of living species, soil compaction, and erosion (Evju et al. 2021). Most trail activities have a curvilinear impact on the environment: initial use has increasing impact only to a certain point at which time increased use has less effect. Initially biking packs down a trail and compacts soil, but once the trail is packed, increasing use has less impact.

Much of the research that has been done suggests that mountain biking has a similar effect as hiking on the environment with similar ecological risks. Both of these activities have less impact than horseback riding. In a recent study that relies on both experimental and observational methods, Evju et al. (2021) did find some differences between user groups, particularly in muddy and moist terrain; their findings suggest that a higher percentage of mountain bikers on wet, muddy trails is correlated with widening of the trail (as bikers attempt to avoid the mud). They recommend strategies like trail hardening and rerouting to avoid additional erosion or impact. Organizations like the International Mountain Bicycling Association and local mountain biking clubs promote sustainable trail building techniques including drainage systems and raising rider awareness of riding etiquette.

The introduction of electric or pedal-assisted bikes raises additional concerns about potential environmental impact since battery power could enable many more riders to go further into the wilderness in a shorter time and to ride faster than pedaling alone. There is currently no consensus among trail managers about whether to allow e-bikes onto trails.

Conclusion

Off-road biking, especially mountain biking, is a growing sport that encompasses a very wide range of riding styles, terrain, bike types, and abilities. Many riders are drawn to the sport through their love of nature, enjoying the solitude of the woods and forests, and others are drawn to the challenge of rocky and rough terrain, jumps, and drops, and still others for the cardiovascular exercise and conditioning provided by mountain biking. While downhill riding and dirt jumping may be disproportionately done by the young, other forms of mountain biking provide a source of fun, exercise, and community across the lifecycle. Advocacy, trail building, and rider groups (often individuals fall into all of these categories) create strong local off-road bike communities which provide important social connections for many participants.

References

Anacker, Caelen. 2018. 25th Infantry Bicycle Corps. *BlackPast.org*. https://www.blackpast.org/african-american-history/25th-infantry-bicycle-corp-1896-97/, accessed 8/5/2021.

Berto, Frank. 2014. *Birth of Dirt: Origins of Mountain Biking*, 3rd edition (San Francisco: Van der Plas/Cycle Publishing).

Evju, Marianne, Dagmar Hagen, Mari Jokerud, Siri Lie Olsen, Sofie Kjendlie Selvaag, Odd Inge Vistad. 2021. Effects of mountain biking versus hiking on trails under different environmental conditions, *Journal of Environmental Management*, 278(2). https://doi.org/10.1016/j.jenvman.2020.111554.

Global Industry Analytics. 2021. Mountain Bike – Global Market Trajectory and Analytics.

International Mountain Bicycling Association. 2021. https://www.imba.com, accessed 8/3/2021.

McCormack, Karen. 2017. Inclusion and Identity in the Mountain Biking Community: Can Subcultural Identity and Inclusivity Coexist? *Sociology of Sport Journal*, 34(4), 344–353.

Marin Museum of Cycling. 2021. https://mmbhof.org, accessed 8/3/2021.

Outdoor Industry Association. 2021. *2021 Outdoor Participation Trends Report*. https://outdoorindustry.org/resource/2021-outdoor-participation-trends-report/, accessed on 8/5/2021.

Sloss, Lauren. 2021. The Mountain Bike Cure: Exercise, Fresh Air, and Fellowship. *The New York Times*, May 20. https://www.nytimes.com/2021/05/20/travel/mountain-biking-vacation.html, accessed on 8/5/2021.

Union Cycliste Internationale. 2021. https://www.uci.org, accessed on 8/6/2021.

Wheaton, Belinda. 2004. Mapping the Lifestyle Sport-scape, in Wheaton, B. (ed.), *Understanding Lifestyle Sport: Consumption, Identity, and Difference*, pp. 1–28. London: Routledge.

29
TRACK CYCLING

Michael Jordan

Track bicycles

A track bicycle has one, fixed, gear. A chainring is attached to the cranks and a sprocket, or cog, is attached to the rear wheel. The ratio of their sizes dictates the gear that is ridden. There is no freewheel on a track bike, so once selected, screwed and bolted on, there is a directly proportional relationship between the pedaling speed, called cadence, and the velocity of the bicycle and its rider.

A track bike also has no brakes. Acceleration is achieved by pedaling faster, deceleration by resisting the inertia of the pedals. This permits the machine a great simplicity in its design allowing manufacturers to have fewer constraints in the pursuit of superior aerodynamics: frames can be designed specifically for one single Olympic performance.

As with other types of bicycle, there is a range of different materials used in track frame manufacture (Chapter 9). Steel was the mainstay for the 1900s and remains so in Japanese and Korean keirin racing (Vignette G), as well as the single-speed, or "fixie" market. For the 2000s carbon is the material of choice for high-level racing, due to its light weight and comparative stiffness, improving the transfer of power applied to the pedals, through to the rear wheel. Aluminum became popular in the 1980s due to being lighter than steel yet maintaining the strength required, and it remains a popular choice at the entry level and for fleets of track bikes for public use at open Velodromes.

The wheels selected have fewer constraints than those selected for other cycling activities. With no brakes there need be no brake track for rim brakes, nor location to house a brake disc. Wheels are commonly attached with threaded nuts. These are used as opposed to quick release systems as they have greater resistance to the forces the rider applies which could otherwise pull a rear wheel out of line. The range of wheels, whilst not as diverse in shape as the frames, remains wide in terms of monetary cost, material selection and visual appearance. In the Tokyo 2020 Olympic Games (held in 2021 because of the global pandemic) Team Pursuit riders rode with carbon disc wheels on the front and back of their bike. Even within the scope of disc wheels there are choices between a flat disc and lenticular, which have the shape of a double convex lens.

At the entry level of track cycling, and assumed lower speeds, the gains made from enhanced aerodynamics can be of less importance than the affordability of light, yet strong, aluminum wheels. The upgrade from these can be to carbon wheels, equally likely to have a deeper rim, the likes of which are used at international-level competition in bunch sprint and endurance track racing.

DOI: 10.4324/9781003142041-41

Traction on the smooth wooden boards or, still smooth, outdoor concrete tracks, comes from the main two types of tire used. Tubular tires have the external tire casing sewn closed around the inner tube and are glued to the rim. Clincher tires have the same separate inner tube and outer tire system most cyclists will use in everyday riding and are usually ridden at approximately 100 to 140 psi, whereas a tubular tire may be inflated up to 220 psi!

The handlebars represent the "cockpit" and are often referred to as such. The basic shape of rounded, road-bike-styled, bars are commonplace in bunch endurance events. For timed events, for individuals or teams the normal is time-trial bars which may be an integrated unit formed as a single item emanating from the headset, or, more simply, additional bars which clip on to the standard road handlebars. With no gear shifters or cables, and no brake levers, the engineering of these handlebars can be dedicated to the pursuit of ever higher speeds, encapsulating the very essence of riding a bicycle on a track.

Velodromes

In Brighton, on the south coast of England, lies an unassuming paved track around parkland, containing a field mowed for cricket and adjoining the modern clay and hard-court tennis courts of Preston Lawn Tennis Club. The 633-yard-(579-m)-long, 6-yard-wide track is neither a perfect circle nor has it two perfect curves joined by the parallel straights of the modern normality. It is, however, a working velodrome and the oldest working velodrome in the world. Opened in 1877 (during the era of highwheel bicycles), Preston Park Velodrome is the home of the Sussex Cycle Racing League and has modestly banked curves helping the cyclists withstand the forces pushing them to the outside of these bends. The banking there, however, bears almost no similarity to the banking of a modern track designed for Olympic competition, measuring close to 45 degrees at its steepest point. These curves have numbers. After the start line a rider enters "turn 1;" "turn 2" is the second half of that same curve. This numbering convention appears more rational in the context of tracks like Preston Park with four distinct bends after four distinct straights, but such a convention remains even for the most modern of track designs.

The current standard for Olympic and World Championship competition is an indoor 250 m-length track and a constant width (often 7 m) for its entirety, regardless of the angles involved; however, different track lengths are used in other Union Cycliste Internationale (UCI)-sanctioned track events, ranging from as little as 133 m, up to 500 m. Keirin racing, a stand-alone professional sport in Japan and Korea, is raced on outdoor 333.33 m- and 400 m-length tracks within large stadia (Vignette G).

The rationale for the design of the tracks is often determined by the building in which the track is to be housed or by the need to accommodate other sports, with longer outdoor circuits able to contain a full 400 m running track, and a football field within that. On competition days, the infield of an indoor 250 m track is full of athletes warming up for their races and recovering for the next, whereas the infield of a Japanese keirin track, for example, may contain lakes and ornamental gardens with athlete preparation happening out of public view.

Different length tracks will, almost by definition, dictate different banking angles. The Forest City Velodrome, in London, Ontario, is the shortest permanent velodrome in the world, measuring 138 m in length with 50 degree banking and 17 degree straights (Figure 29.1).

Velodrome surfaces, whether longitudinally planked timber, concrete or a high traction asphalt which allows racing to continue in the rain, are typically smoother than normal roads. Lines are marked on the track surface denoting distances or the start and finish of events. The

Figure 29.1 Track racing at the velodrome in Cali, Columbia, 2014.

Source: https://en.wikipedia.org/wiki/Track_cycling

côte d'azur is the name of a light blue strip, measuring at least 10% of the width of the track and represents the inside edge of the track. The next colour used is the black line on a light-coloured track, or white on a dark-coloured track, which is drawn 20 cm from the inside of the track and is known as the "measuring line" in the UCI rules as the track's length is dictated by the inside of this line. The red line, the "sprinters' line," is 85 cm from the inside of the track and creates the "sprinters' lane." In the rules of the sprint a rider may not challenge or pass on the left an opponent riding in the sprinters' lane, so it is there for safety in this high-speed sport. A blue or "stayers' line" is drawn at a third of the width of the track or 2.45 m (whichever is greater) from the inside edge of the track.

There are also lines marked perpendicular to the track. Pursuit races start and finish at the midpoint of the straight; other races nearer the end of a straight. The 200 m line is drawn 200 m from the finish and so the variability in the track dimensions will alter a timed result, which would be inconceivable in sports like track and field athletics. The Krylatskoye Sports Complex Velodrome in Moscow was built for the 1980 Summer Olympics. The Flying 200 m event sees athletes attempt their fastest time for 200 m with the aid of a moving or "flying" entry. The theoretical optimum racing line for a Flying 200 m on the 250 m track which will be used at the next Olympics is to follow the black "measuring" line. The Moscow Olympic track is 333 m in length and 10 m wide and descending from the top of the banking down to the sprinters' lane has been compared to descending a giant wave. On this track, the 200 m line is near the end of turn 2, so athletes riding on the optimum racing line are closer to the track's external perimeter fence than the inside, affording them a downhill sprint.

It is of no surprise that the list of record progression for the Flying 200 m has many entries from the Krylatskoye Sports Complex!

Olympic track events

It is as recent as 2012 that, for the first time in Olympic history, the women's program of track cycling events was the same as that of the men, with the introduction of women's keirin, team sprint and team pursuit. For the Tokyo 2020 Olympic program the Madison was introduced for women and reintroduced for men for the first time since 2008, increasing the number of events to six for both men and women. Of these six, three were sprint events and three were endurance.

Sprint events

Two sprint events in the Olympic Games are individual: the sprint and the keirin. The team event is simply called the Team Sprint.

Sprint – The qualification phase is a timed Flying 200 m. Athletes ride for three and a half laps of the 250 m track where the last 200 m of this is timed. This ranks the riders and a predetermined number of these riders progress to sprint against each other in match sprints of two riders, or occasionally three, depending upon the stage of the competition. These matches are raced over three laps and have minimum speed at the outset of "walking speed" before the high-speed action occurs. As rounds progress through to knock-out stages, four riders are left in the competition to create semi-finals. The two athletes beaten in each of the two, best-of-three-race semi-finals compete for the bronze medal and those who win race for the gold. The rider coming second in the final secures the silver medal.

Keirin – The keirin is raced over six laps at the Olympic Games and the World Championships. The first three laps are behind a motorized pacer, known as a 'Derny', which begins at 30 km/h. Riders draw lots to see the order in which they begin, with the first rider moving directly behind the pacer as the race begins, and they must remain in these positions for at least the first lap, then may not pass the pacer before the pacer leaves the track having increased speed to 50 km/h! Athletes may well begin a long sprint for the finish at that time, or may save their energy for the final stages of the race deploying their tactics at high speed as late in the race as they can (see Vignette G).

Team Sprint – Teams of three riders start side by side competing against another team on the opposite side of the track and aiming to record the faster time. Each rider starts at the same time with the first of the three riders of the team moving to the right after one lap to leave two riders for lap two, after which the second rider also moves to the right to leave one last rider. The third rider in the team rides the third and final lap to complete the race.

Endurance events

In the Olympic Games, the three endurance events are the Team Pursuit, the Madison and the Omnium.

Team Pursuit – In this 4 km event of four riders, it is the third rider whose front wheel stops the clock for the team. Starting together, side by side, they quickly form one line and take turns at the front of the line before the lead rider swings up the track, as their team mates then pass on the inside, to then rejoin the line as the last rider, giving them the opportunity to try to recover after their effort ready for the next. Average race speeds are now in the region of 60 km/h for men and women.

Omnium – Three of the four races comprising the Omnium - the Scratch race, Points and Elimination race - are World Championship events in themselves, though this can constantly adapt and evolve. The other Omnium event is a Tempo race, a bunch event in which a single point is awarded to the first rider across the finish line on every lap after the first four laps.

Madison – Named after cycling events held at Madison Square Garden, this event is raced in pairs. With no physical baton to pass, the handover is completed by hand contact between the riders seeing one rider have the opportunity to recover by riding around, towards the top of the track, at a reduced speed. Once it is that rider's turn again, the rider commonly races around at the bottom of the track to follow the shortest distance possible. The race result is determined by points won, just as the Points race itself.

There are events at the World Championships which are not, or are no longer, in the Olympic program. These are: the Time Trial, an individual timed event over 1 km (men) or 500 m (women); the Individual Pursuit, a longer time trial styled event of 4 km (men) and 3 km (women); and the Scratch, Points and Elimination races.

The Scratch race is a "first past the post bicycle race" of a predetermined length. In the Points race, there are points awarded every ten laps on a 250 m track for the first four riders across the finish line. If a rider is able to ride away from the field to such an extent that they catch the back of the remaining bunch they gain 20 points which may be enough for an otherwise low-scoring rider to claim a medal or win the race! The elimination is also known as "Devil take the hindmost." On a 250 m track on every second lap the rider who is last in the bunch, and any who have significantly dropped off prior, is eliminated leaving the final two riders to duel the last 500 m to decide the winner.

Physiology and training of track cyclists

In track cycling there is no single physiological type of athlete who will succeed more than another athlete across all of the wide range of events. A successful sprinter on the track will have the capacity to produce very high, short-term power, and a peak power in excess of 2500 watts has been recorded. Track endurance cyclists are more similar to road cyclists and will often transition quickly and easily between the two disciplines. Though there may be no possibility of their matching the track sprinter's score for very short-term power, their success is born out of their capability to maintain high power for several minutes and a repeatability of a multitude of sprints within one track endurance race.

There are three principal systems of energy supply in the human body. There are two which supply short-term power – the dominant fueling mechanism for work of less than two to three minutes in duration – and a third dominant in longer duration, termed endurance. None of these systems is used independently, as they are all contributing at any given point in time, but their relative contribution is based upon the duration and intensity of the physical task.

Adenosine triphosphate and creatine phosphate are grouped together, and termed ATP-CP. This is the immediate energy system used to power many rapid actions, whether that be a golf swing, an ice dancer's jump or the track cycling sprinter's launch from the starting gates. For maximal effort of up to six to ten seconds, this energy system will dominate the supply of energy to the muscles. The body can store just one or two second's worth of ATP and six to ten of CP, which, once used, are replenished in recovery.

Anaerobic glycolysis is the energy system dominating the power supply of maximal efforts between ten seconds and the two- to three-minute mark. Anaerobic, whilst it literally means without oxygen, does not mean that there is insufficient oxygen required to break down glucose, the glycolysis part of its name, but that the body's demand for energy to supply the working muscles is beyond the rate of supply which the aerobic system can meet. Whilst there are many different factors which affect fatigue, this high-intensity work will see an accumulation of hydrogen ions and a decrease in the blood pH, making it impossible for the body to work at such high intensities for long.

Once a singular "effort," as track cycling training repetitions are known, lasts beyond two to three minutes the contribution from the anaerobic glycolysis system markedly decreases and the dominating energy system is the aerobic system. With training the human body can store enough glycogen for about 180 minutes of aerobic activity, and enough fat for 1–3 months! The intensity at which the cyclist can ride must decrease once these glycogen stores are depleted as the rate of ATP production is slower when fat is the main source of fuel, but technically there is then enough energy to ride for a very long time!

Sprinters using a standing start, where the track bicycle is held in a gate to begin the race, will train their ATP-CP system by lifting heavy weights and riding their bikes, if only for a few pedal strokes, against a very high resistance. The anticipated physiological adaptation is an increase in the muscle's resting level of ATP and CP to fuel their efforts. Force alone will not translate into track sprinting success as highly coordinated application is required for efficient pedaling to provide the speed of motion to turn this force in to power. Track sprinters will commonly compete at cadences above 140 rpm.

Each of the sprint events, discussed in more detail earlier in this chapter, will take less than two minutes at high speed, so training the anaerobic energy systems to produce and maintain as much power, watts, as possible is of vital importance to the athlete. Multiple World and Olympic champion Sir Chris Hoy describes training sessions of four efforts of 500 m in a track training session. These could be with as much as 30 minutes between! Riding a bicycle for 2 km over the course of a little over an hour and a half may not sound an arduous task, but the recovery duration is required for each to be at maximal speed. Hoy is the current holder of the men's Flying 500 m world record, with a time of 24.758 seconds, and an average speed of nearly 73 km/h!

The men's points race can be as long as 40 km and the women's as long as 25 km, so these endurance events are well into the domain of the aerobic athlete. The adaptations to the training that the track endurance rider undertakes include cardiovascular, pulmonary and metabolic changes, as per all endurance sports, but must be complemented by the skill acquisition required so that the athlete can pedal their fixed gear bike at more than 120rpm. These high cadences can be for an extended period of many laps, or in shorter bursts, as there can be great variations in speed as a race unfolds.

A sprint track cyclist will always require a certain degree of aerobic capacity and an endurance rider will need a capability to accelerate to a sprint, so whichever type of athlete a rider may be, an appropriate balance of training of each energy system, and the skills to apply them, is a necessity for any athlete to succeed.

Madison square garden, six days of new york

Madison Square Garden, now the home of New York Knicks (basketball) and New York Rangers (ice hockey), was once the home of some of the most extreme feats of endurance track cycling the sport has known, beginning in the late 1880s. In these early years there were many different forms of track racing, but that which became the most famous, if not infamous, was the Six-Day race where competitors would ride as far as possible in that working week. Racing beginning at midnight on a Monday morning would conclude at midnight the following Saturday night, a full 6 days later, allowing predominantly Christian audiences to respect the Sabbath. The sight of riders able to distance a hound, a horse or a locomotive, as described in a *New York Times* article entitled "A Brutal Exhibition," as fine a thing as it was, caused outcry due to the risks to the riders racing solo, without enforced rests. In 1899 Collins Law was passed legislating that a competitor may not ride for more than 12 hours a day.

The result of the law was that these Six-Day races became events contested by teams of two (or occasionally three). The building hosting these events, and indeed its actual location, has changed since the first Madison Square Garden was built; however, the name remains in each incarnation. This event required at least one of the pair to be on the track at all times as the rider completing their turn would relay their partner into the race, originally with hip slings via shorts made to offer a simple handle, and then later the hand sling, which remains in use today.

The principle of the event was that the winners were those who had ridden the furthest in this six-day week, though organizers in 1899 paired long-distance riders with sprinters and introduced points accumulated in two-mile sprints and other short races during peak viewing hours to add to the spectacle for spectators. Early motorbikes were introduced, called a Derny, and racers would draft behind them, increasing speeds to over 50 mph (80 km/h) on the tightly banked track.

Pairs in a New York Six would complete distances of up to 2,800 miles (4,500 km!) further than a modern Tour de France, made possible, in part, by the rest areas set up in the middle of the track. This was not a peaceful place of recovery as this area was also full, with minimal segregation, of spectators and even jazz bands creating matinee and evening performances.

Whilst in ever-changing formats, the New York Six ran, albeit not each year, to a total of 73 editions until the final race in 1961. The name of Madison, however, remains very topical, as the eponymous event is in the current Olympic and World Championship program, though not over a six-day duration. Women elite Madison finals are upto 30 km and men race upto 50 km, in pairs, relaying each other into the race, in not too dissimilar a manner to that of 1899.

Six-Day racing has seen a resurgence of interest in recent years with the creation of the Six Day Series, though the concept of multi-day carnivals of cycling has never been absent from the racing calendar. Now it is no longer a singular event of endurance, but a series of individual endurance and sprint races from which competitors accumulate points through each evening of the event.

The hour record

Perhaps the simplest of track cycling events to understand, and least complex to replicate if not to achieve, is the Hour Record. For the unpaced, solo, Union Cycliste Internationale (UCI) record it is simply the furthest distance a cyclist can ride, though it must be on a UCI-approved track bicycle.

The Human Powered Vehicle (HPV) one-hour record is now over 90 km for men and 84 km for women. These were set with recumbent bicycles which place the rider in a reclined position. A recumbent introduced by Charles Mochet of France in the 1930s won races against the best professional athletes of the era who were riding traditional bikes of the time. The radical reduction in aerodynamic resistance resulted in far higher speeds than riders on a conventional bicycle could achieve and so in 1934 the UCI announced that recumbent bicycles were no longer permitted to be used in racing, as they did not look, to the eye of the UCI, as a bicycle should. That ruling, setting the tone for subsequent regulations determining what represented an approved bicycle, meant that alterations in rider position in later years could be quickly banned once implemented. Distances achieved with the benefit of motor-pacing are also ineligible for this described hour record. As early as 1900, motor-paced cyclists were riding beyond 65 km in 60 minutes, which was 24 km/h further than the best un-paced, single, cyclists of the day.

Whilst hour records have been recorded faithfully since 1876, when F L Dodds of England rode his highwheel bike 25.508 km, surpassing J T Johnson's 22.785 km ridden in 1870, the first hour record regarded as such by the UCI was by Henri Desgrange of France, creator of the Tour de France, riding 35.325 km in May of 1893. The record was increased by a succession of elite cyclists, including many different Tour de France winners, including Eddie Merckx, who rode 49.431 km in 1972, a record which lasted for over 11 years until it was broken by Francesco Moser in January 1984 on an aerodynamically improved bicycle with a sloping top-tube enabling a low crouched position and which saw the first modern use of disc wheels.

In the years following Moser's record – and he was to break it five times in all – extreme aerodynamic riding positions were adopted by British athletes Chris Boardman and Greame Obree especially. Obree placed his arms by his sides and hands under his shoulders and rode in this tucked position to claim the record in 1993 (51.596 km). In 2000, the UCI introduced a ruling so that traditionally styled "drop handlebar" track bikes, therefore very much in the visual appearance of that used by Eddie Merckx in 1972, were to be used for future attempts at setting the hour record. The last of the records set prior to the change of rules, called the Lugano Charter, was Boardman's 56.375 km in 1996 for men and 48.159 km for women, set by Jeannie Longo-Ciprelli in that same year.

Boardman's 56.375 km distance received the title "Best Human Effort" when the rule changes meant that it no longer could be considered an Hour Record. In achieving this mark Boardman rode in the "superman" position, with his arms extended out in front of him, just as Obree had done in riding a then record of 52.713 km two years prior. Despite ever-increasing distances being achieved currently this is yet to be beaten, though Longo-Ciprelli's record has since been eclipsed.

In 2014 rulings were altered allowing a track bicycle used in Individual or Team Pursuit events (which is significantly more aerodynamic than a standard, drop bar, track bike) to be used in UCI Hour Record attempts, resulting in the record changing hands many times since 2014 in both the men's and women's disciplines. It remains to be seen how ever-evolving training techniques and modernization of track cycling clothing will help set new records and what rules are implemented to protect the integrity of the sport.

Technological advancements in the creation of velodromes, optimizing its angles, curvature and atmospheric environment, are part of a constantly evolving science in the pursuit of higher speeds, as are those of the track bike itself, simple in its concept yet increasingly as complex as the environment in which it is ridden.

Bibliography

Anon (1897) A Brutal Exhibition. *New York Times*, Dec. 11, p. 8.

Bassett DR Jr, Kyle CR, Passfield L, Broker JP, Burke ER (1999) Comparing cycling world hour records, 1967–1996: modeling with empirical data. *Med Sci Sports Exerc* 31(11):1665–1676.

Hoy C (2018) *How to ride a bike: From starting out to peak performance*. London: Hamlyn.

Kyle C & Bassett D (2003) The cycling world hour record. In: Burke, E, ed., *High-tech cycling*. 2nd ed. Champaign, Illinois: Human Kinetics.

Hadland T & Lessing H-E (2014) *Bicycle design: an illustrated history*. Cambridge, Massachusetts: The MIT Press.
Heijmans J & Mallon B (2011) *Historical Dictionary of Cycling*. Lanham, Maryland: Scarecrow Press.
McArdle WD, Katch FI & Katch VL (2015) *Exercise physiology: Nutrition, energy, and human performance*. 8th ed. Philadelphia, Pennsylvania; Lippincott Williams & Wilkins.
UCI Cycling Regulations (2021) *Part 3 Track Races*. [online] Available at https://www.uci.org/docs/default-source/rules-and-regulations/3-pis-e_english.pdf
de Wilde A (2012). Six-day racing entrepreneurs and the emergence of the twentieth century arena sportscape, 1891–1912. *Journal of Historical Research in Marketing*, *4*(4), 532–553.

Vignette G
KEIRIN CULTURE

Keizo Kobayashi

In Japan, the keirin means bicycle racing on the track with betting. In Europe, North America and Australia horse racing is the most popular betting sport, while in Great Britain betting is also found at dog races. But in Europe, it seems that there is no tradition of betting on human races except maybe in Denmark where betting is permitted at certain track cycle races, but only in a small scale and only in some localities.

In Japan, there are four authorized betting races: horses, speedboats, motorbikes and bicycles. Keirin is managed by the Ministry of Economy, Trade and Industry through Japan's Keirin Autorace Foundation (JKA). Keirin are held every day and at any time (day and night) at 43 velodromes in total. Velodromes, whether open air or covered by a roof, are located in all the principal towns of Japan. Races are transmitted by TV and by internet and may be run on outdoor velodromes in dry or wet weather.

Origins of keirin

The first bicycle race in Japan was held in 1895 at Yokohama City. By 1897 there is a report of 20 Japanese riders participating in a race in Tokyo. By the time of the Second World War bicycle racing had become very popular in Japan. After the War, the Government invested heavily in reconstruction and in the development of a bicycle industry. The Government was inspired by the success of the pari-mutual system used in horse racing in generating public revenues, and decided to apply the same system to bicycle races which were very popular at that time. To that end, a law of keirin was promulgated in 1948 and the same year the first keirin was started on 20th November at Kokura on the island of Kyushu in the south of Japan.

Keirin racing

All keirin races are assigned one of six grades: GP, GI, GII, GIII, FI or FII. Keirin Grand Prix (GP) races, which are the most important, are held 30th December each year with a winning prize of about one million US Dollars: the winner becomes the year's keirin champion. There are six keirin racer rankings: SS rank is the highest ranking for which all keirin racers battle fiercely, followed by S1, S2, A1, A2 and A3. At present there are 2330 racers who all belong to Japan's Professional Cyclist Union.

DOI: 10.4324/9781003142041-42

Keirin is a very physical competition. In addition to having well developed thighs, Keirin racers need to have powerful upper bodies. While leg power is needed to move the bicycle, racers also need considerable upper body strength to control the handlebars. Upper bodies need to be as well developed as legs. Keirin racers' thighs are extremely important since thigh diameter is equated with a cyclist's ability. The racer Kojima has the biggest thigh circumference of 2.43 feet (74 cm)!

Those aged more than 17 who want to become professional keirin racers (there are no amateur keirin racers) must go to the Japan Institute of Keirin for about one year to acquire knowledge and skills related to keirin races, and make it through the demanding training course. When I was in high-school, I rode my bicycle a lot. My neighbors (Niwa brothers) were keirin racers and their father advised me to become a keirin racer. I was strong enough on the bicycle but unfortunately I was myopic, so I was barred from becoming a keirin racer! Keirin is one the largest professional sports in Japan, and successful racers earn salaries comparable to those of professional sportsmen in baseball, sumo and football.

Koichi Nakano

Koichi Nakano was born at 1955 in Kurume, not far from Kokura where the first keirin was held. He is the most famous keirin racer and is also among the best track sprinters of modern times. From 1977 to 1986, he won an unprecedented 10 consecutive gold medals in professional sprint events at the UCI Track World Championships. Nakano beat Canadian Gordon Singleton in 1982 in the UCI World Pro Sprint final. Having retired from keirin, he is currently the director general of Japan's National Track Cycling team for the Japan Cycling Federation.

Women's Keirin

Women's keirin races, which had been held for 15 years from 1949 to 1964, were not very popular because of the monotony of the races: the strongest racer always won so there was little interest in betting. But this kind of race was reborn in 2012 as Young Women's Keirin designed to make keirin more attractive. This type of keirin race takes place with rules and equipment different from a men's keirin race. There is also the betting for women's keirin. Women aged more than 17 who want to become professional keirin racers must, like men, go to the Japan Institute of Keirin for about one year.

Uniform

Uniforms and helmets are color coded to help identify each racer using clothing codes that were adopted in 2002. The color of the shorts worn by each keirin racer indicates rank, with racers in keirin's highest SS rank wearing special uniforms. A keirin racer's jersey is basically the same as a road-racer's one but without back pockets, only long sleeves; a helmet cover is the same color as the jersey.

Keirin bicycles

The bicycles used in keirin are all custom-made machines with each part fine-tuned to the unique characteristics of the racer. Keirin bicycles weigh 7 to 8 kg (15.4 to 18.6 pounds) and are designed purely for speed. All parts not needed for attaining high speeds have been stripped off. They have no brakes so a racer's speed is adjusted by pedal power alone. The racer and each part of the bicycle work as a single unit for optimum handling on a track's banks.

The frames of bicycles for keirin are made only of chrome-moly steel designed to combine suppleness and strength, and are handmade by custom frame builders. Unlike street bicycles, the rear gear on keirin bicycles is fixed and constantly engaged with the wheel: the absence of a freewheel makes backpedaling impossible, but provides the only available brake. The tires of keirin bicycles have an outer diameter of 26.57 inches (675 mm) and a thickness of 0.87 inches (22 mm). They are designed to reduce friction by minimizing the surface area in contact with the track, resulting in higher speeds.

A Keirin race

A keirin racer's life is sometimes compared to the spiritual life of a monk in a Buddhist monastery. They live an austere life while training. For one day before a race and during the race they are isolated as a group, without mobile phones and any communication with the outside world to prevent race fixing.

A keirin race is about to start! Racers are introduced before the start and immediately after the end of the previous race. This start introduction is also called "ground riding" or "leg show". The racers then return to the waiting room where many athletes purify their bodies and bicycles with salt before racing. Keirin begin with the racers on the starting blocks. Every racer is assigned a number and a color for identification and betting purposes. Most often there are 7 or 9 racers. At the sound of the gun, the racers leave their starting blocks and settle into a position behind the pacer, who is another keirin racer wearing purple with orange stripes. The pacer drops out between the second and third corner at one and a half laps to the finishing line (Figure G.1).

Figure G.1 Keirin race: Goal sprint. (Copyright JKA with permission).

Keirin races have three common lengths: 365 yards (333.3 meters), 437 yards (400 meters) and 547 yards (500 meters). The track is wide with 4 numbered corners.

The distance of each race depends on gender and rank. All races for women are currently run at 0.99 miles (1,600 meters). For the Keirin Grand Prix, it is 1.74 miles (2,800 meters). The number of laps depends of the type of track, and varies between 3 and 6 laps.

During the race, the speed of the bicycle is around 37.29 miles/h (60 km/h), reaching about 43.5 miles/h (70 km/h) before the finish. The track is made of concrete so that races can be held when it is raining. The race is very physical; pushing and elbowing is common so there are crashes. That's why riders wear protection on their upper body.

Betting: line and tactics

Keirin is a very special type of bicycle racing. Each racer must announce their race strategy in advance! Keirin racers need more elements than just speed to win. The strategies they use against each other are an important part of the sport, increasing its subtlety and appeal. Racers use various strategies to win. The battle of wills among the racers can be complex, often creating dramatic ups and downs during the race. Once you understand the nuances of keirin, trying to predict race outcomes becomes a fascinating exercise.

A tight pack of racers teaming together in single file is called a 'Line'. Reviewing the lines is key to understanding keirin. The strongest lines are formed by racers who often ride together, know each other's habits and performance characteristics, and can give each other support. Lines therefore usually consist of closely acquainted racers registered in areas near each other.

There are four winning techniques: Breaking away; Chasing; Insert (drive in); and Mark. Breaking away indicates 1st or 2nd place finishes by front-running. Chasing indicates 1st or 2nd place finishes by responding to a breakaway. The Insert (drive in) indicates to pass the leading rider in the final straight after passing the 4th corner. The Mark indicates to stay in 2nd place when this racer fails to drive in.

Wager and odds

All tickets are sold by the pari-mutuel method, and you can purchase a voting ticket that predicts the winning race target and receive a dividend if the prediction is correct. One can buy tickets at the race meet, or by phone or by internet. Keirin has five types of wager in all. Perfecta: selecting the first two finishers in correct order. Quinella: first two finishers in either order. Trifecta: first three finishers in correct order. Trio: first three finishers in any order. Quinella place ("wide"): selecting two to finish in the top three, in any order.

Punters check the odds, and decide how much to bet. Odds (payout multiples) are very important information since they indicate the payout amount for a winning wager. The lower the payout multiple, the more popular (closer to favorite) a given wager is, and the higher the multiple, the less popular (more of a long shot) it is. Comparing the payout multiples of different wager types and deciding how much to bet is part of the fun of keirin.

Keirin in the world

International Keirin has been held since 1981 by inviting generally eight foreign (essentially European) pro-track racers to improve international friendship and competitiveness through races with Japanese keirin racers.

In South Korea, after the Seoul Olympics held in 1988, the Seoul Olympic Committee built the Seoul Velodrome. In 1991 the Bicycle Racing Law for keirin racing was enacted. At present there are three velodromes for keirin in South Korea: Seoul, opened in 1995; Changwon, opened in 2000; and Busan, opened in 2003. There are about 500 keirin racers. South Korea's keirin races are held with 7 racers, in contrast to the 9 racers in Japan.

Keirin races for men (of course without betting) were included in the 2000 Sydney Olympic Games and became the second Olympic sport after judo to originate from Japan. Keirin pacers use a power-assisted bicycle and at the Sydney Olympic Games, Nakano was selected as the first pacer! Keirin for women was also included at the 2012 London Olympic Games where Victoria Pendleton (GBR) won the gold medal. In the delayed 2020 Tokyo Olympics, the Keirin race will be held at Izu Velodrome which is indoor having wooden 250 meters (273 yards) track with a maximum inclination of 45 degrees. It is located beside the Japan Institute of Keirin.

Since 1975, Japanese leisure consciousness has shifted away from "indoor sport" to "outdoor sport" and Keirin has been called an "outdated leisure". The total sales of Central Horse Racing continued to increase year-on-year, but sales of the other four groups (regional horse racing, speedboats, motorbikes and keirin) have continued to fall below the previous year's level since 1981. In 1991, the total revenue of keirin races was US$18.80b but in 2019 it declined to US$6.35b which is only one third of the 1991 total.

Government revenues raised by keirin contribute to the public good in a wide range of social welfare activities such as amateur and schools athletic programs, health care, education and disaster relief. At weekends there may be a crowd in the velodrome, but during the week many punters follow races on television or on-line and the velodrome is quite quiet. Keirin has become a settled element of Japanese culture and society combining an ascetic lifestyle for racers, commercialization with elite racers earning substantial salaries, and an associated gambling economy which is taxed to support a range of social programs.

Further reading

McCurry, J. (2020) *The War on Wheels: Inside the Keirin and Japan's Cycling Subculture* (New York, NY: Pegasus).

Keirin race navigator: http://Keirin.jp/pc/dfw/portal/guest/campaign/navi/2017_english/HTML5/pc.html#/page/1

A Beginner's guide to keirin racing: http://Keirin.jp/pc/static/beginner/en/abcs/

30
HEALTH BENEFITS OF CYCLING

Adrian Bauman, Sylvia Titze and Pekka Oja

Introduction and framework

This chapter focuses on the health benefits of cycling, taking a broad definition of health to include physical, mental and psychosocial health, and community health, and assesses cycling benefits across the lifespan. The approach uses a conceptual model developed as Figure 30.1 to illustrate the benefits of cycling. Benefits are influenced by the purpose or mode of cycling as well as in which environments (and settings) cycling occurs. The purposes of cycling trips often overlap, but include getting to and from work (commuting), cycling for recreation or sport, and cycling for purposes other than recreation. The health benefits of cycling will also be influenced by the intensity, duration and frequency of cycling (total dose), which may contribute to reaching the World Health Organization (WHO) recommended physical activity threshold for health. This is typically expressed as at least 150 minutes per week of moderate intensity physical activity, or equivalent combinations of moderate and vigorous activities (WHO 2020a). Benefits will vary by the amount cycled, and by demographic factors such as age, shown in the left-hand part of Figure 30.1. Many benefits of cycling are physical health benefits, but in addition, mental health benefits may result. Other, non-health benefits may accrue through social connections (community benefits), and through active transport (air quality benefits, reduced car emissions, reduced overall climate change greenhouse gases). Overall, benefits substantially outweigh the risks of cycling in terms of population health (Figure 30.1).

Physical health benefits of cycling in adults

During the past 30 years cycling has become recognized as a form of physical activity that can contribute to public health (Chapter 7). Research-based evidence on the health effects of cycling has grown steadily from a few physiological intervention studies to multiple large-scale population-based cohort studies. The purpose of this section is to describe briefly how the cycling-specific research evidence for adults has developed during this period. In this section we only considered intervention and cohort studies as they provide the best evidence (cross-sectional studies were excluded).

Recent research interest on the health effects of cycling began in the 1990s. Finnish and Dutch research groups conducted randomized controlled trials on the physiological effects

DOI: 10.4324/9781003142041-43

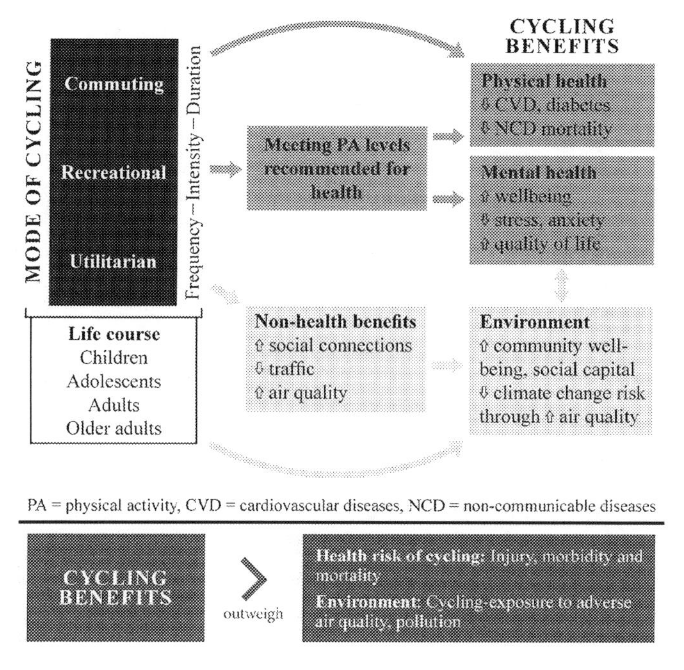

Figure 30.1 Health benefits of cycling.

Source: authors.

of commuter cycling (e.g. Oja et al. 1991; Hendriksen et al. 2000). These controlled studies found distinct beneficial effects in improving cardiorespiratory fitness and cardiovascular disease (CVD) risk factor profiles. Prompted by the emerging evidence of the health benefits of cycling, WHO soon adopted cycling in their transport policies (WHO 1999). Subsequently this led to the development of the Health Economic Assessment Tool (HEAT) to assess the health impact of walking and cycling as a public health measure (WHO 2011).

Growing research interest on the health effects of cycling continued in the 2010s. An initial systematic review identified four cross-sectional, four intervention and eight cohort studies (Oja et al. 2011). Based on this limited evidence, the authors concluded that there was strong evidence for benefits in aerobic fitness, moderate evidence for benefits in CVD risk factors but inconclusive evidence for benefits in all-cause mortality, coronary heart disease (CHD) mortality and morbidity, cancer mortality and overweight and obesity.

Recent systematic reviews

Growing research interest continued in the 2010s, with a 2014 systematic review of studies on cycling (and walking) and all-cause mortality (Kelly et al. 2014) as part of the HEAT development. They found seven cohort studies with 157,000 individuals and 2.1 million person years of exposure, with an overall 10% decrease in the risk of all-cause mortality attributable to cycling participation. Another systematic review (Mueller et al. 2015) reported on 30 studies examining the health impact of mode shift from motorized transport to active transport, which consisted of walking and cycling complemented by public transportation or any other physically active transport mode. Among these studies the health benefit–risk or benefit–cost ratio of transport mode shift from motorized to physically active ranged widely between –2 and 360 with median ratio of 9. This study did not differentiate between the benefits of cycling and walking, leaving the cycling-specific benefits unspecified.

A recent systematic review by Dinu (2019) analyzed 23 prospective cohort studies on active commuting. They found 7 cycling studies, 8 walking studies, and 11 mixed-mode studies. Among the cycling commuting studies, the risk of all-cause mortality of cyclists was reduced by 24% and cancer mortality by 25% compared to non-cyclists.

Finally, a systematic review has shown metabolic and training benefits of indoor cycling similar to those found in outdoor cycling (Chavarrias et al. 2019), but this is not explored further as this setting for cycling is beyond the scope of this chapter. Furthermore, this area potentially includes decades of cycle ergometer training studies, as well as gym users and cycling 'spin classes' as this form of cycling is often carried out concomitantly within other indoor exercise and fitness programs.

Summary of most recent intervention and cohort studies

Many studies have focused on active commuting (combining walking and cycling together), leaving a gap in health benefits specially of cycling. In addition to the overall cycling effects, it can be hypothesized that the health effects may be different due to the differences in the physiological characteristics of commuting and recreational cycling.

An ongoing data search has located 18 cycling studies published after 2014, all of which, with the exception of one randomized controlled trial, were cohort studies. Of these studies, 13 focused on total cycling, five on commuter cycling, two on recreational/leisure-time cycling, two on utilitarian cycling, and six on changes in cycling over time.

Total cycling has been reported by two or more cohort studies to be inversely associated with all-cause mortality (Andersen et al. 2015; Oja et al. 2016; Porter et al. 2020; Schnohr et al. 2018), incident CHD (Blond et al. 2016; Koolhaas et al. 2016) and type 2 diabetes (Mielke et al. 2020; Rasmussen et al. 2016). Single cohort studies show inverse associations between cycling and cardiovascular, respiratory and diabetes mortality (Andersen et al. 2015), incident and recurrent myocardial infarction (Kubesch et al. 2018), asthma and chronic obstructive pulmonary disease (Fisher et al. 2016), and hypertension and obesity (Mielke et al. 2020). Positive associations have been reported between cycling and life expectancy (Dhana et al. 2016), and cycling and upper limb fractures (Armstrong et al. 2020).

The randomized controlled trial of a six-month cycling program demonstrated improvements in glucose tolerance and VO_2 max, and reductions in body weight, fat mass, intra-abdominal adipose tissue and clamp insulin (Blond et al. 2019). These cardiometabolic benefits are thought to be intermediary factors contributing to long-term mortality and morbidity benefits. Recent cohort studies on **commuter cycling** have shown inverse associations with all-cause mortality (Östergaard et al. 2018), incident CHD (Blond et al. 2016), type 2 diabetes (Rasmussen et al. 2016), and obesity, hypertriglyceridemia, hypertension and impaired glucose tolerance (Gröntved et al. 2016). We have found two studies specific to **recreational/leisure-time cycling**. Östergaard (2018) reported reduced all-cause mortality, and Blond et al. (2016) reduced incident CHD. **Utilitarian cycling** other than for commuting has been less studied. Reduction in BMI was reported by Dons et al. (2018).

Several cohort studies have reported associations between **changes in cycling** behavior and health outcomes. Comparison between stable cycling and stable non-cycling has shown the former to reduce all-cause mortality (Östergaard et al. 2018), incident CHD (Blond et al. 2016), BMI (Dons et al. 2018; Turrell et al. 2018), and overweight and obesity (Rasmussen et al. 2018). Comparison between new cyclists and passive controls indicated cycling-related reductions in all-cause mortality (Östergaard et al. 2018), obesity, triglyceridemia and impaired glucose tolerance (Gröntved et al. 2016), and type 2 diabetes (Rasmussen et al. 2016).

One recent addition in the cycling field is the growth in e-bikes, and their health benefits are now recognized, particularly for cycling among older adults (see text box) (Chapter 24).

Benefits of e-bikes

Increases in e-bikes and their health benefits are a recent phenomenon. Sales of e-bikes have increased in the past decade, with adoption particularly among older adults (Van Cauwenberg et al. 2018). One reason for the adoption of e-bikes is to maintain physical activity into older ages, and allow maintenance of fitness, functional capacity and activities of daily living and social interactions (Höchsmann et al. 2018). E-bikes can achieve energy expenditure reaching the moderate intensity threshold (Bourne et al. 2018), with more than half of 17 studies showing that e-bike users achieved higher MET values (higher intensity) than walking, and achieved improvements in cardiorespiratory fitness (Bourne et al. 2018). Users achieved moderate physical activity goals also because they cycled for longer or rode up hills, compared to regular bike users. E-bikes appeal to older people interested in staying active, but are also useful for low active or overweight older adults (Castro et al. 2019, Sundfør and Fyhri 2017; Van Cauwenberg et al. 2019). Finally, e-bike users are not more likely to suffer injuries (Schepers et al. 2020) compared to regular bike riders of the same age. The adoption of e-bikes by older adults or those with injuries or co-morbidity may maintain some health-enhancing physical activity in these population groups.

This brief overview of research shows multiple, consistent and substantial benefits on cardiometabolic mortality and morbidity and related risk factors. The evidence suggests that all modes of cycling benefit health.

Cycling and physical health benefits among children and adolescents

Among children and adolescents, the effects of commuter cycling on health is more often investigated than leisure-time cycling. Furthermore, in many studies walking and cycling are combined, as is the case with some adult studies. In only a few countries (Denmark, Norway, Sweden, the Netherlands) is cycling behavior prevalent enough to stratify active commuting into walking and cycling. As the intensity of commuter cycling is typically higher than the intensity of commuter walking (e.g. Bourne et al. 2018), we excluded studies or systematic reviews in which the outcome variable "active commuting" also included walking.

Health benefits associated with regular physical activity among children and adolescents include cardiovascular and muscular fitness, cardiometabolic health, weight status, bone health, cognition, and reduced depression. Other health conditions, such as chronic diseases (type 2 diabetes, hypertension, cardiovascular diseases, cancer), are infrequent among children and adolescents and are rarely reported. Interest in health outcomes related to cycling among children and adolescents started roughly two decades ago. It became apparent that active commuting is a health-enhancing physical activity behavior separate from vigorous-intensity leisure-time physical activity.

One of the first studies on the **fitness effects** of commuter cycling among children and adolescents was published in 2006 (Cooper et al. 2006). In Denmark, data were available from children (around 9 years of age) and adolescents (about 15 years of age). Among children, some 38% and among adolescents 66% cycled to school. Cycling to school was associated with higher levels of cardiorespiratory fitness compared with those traveling to school by passive (motorised) transport. Follow-up of the nine-year-old children for six years (Cooper et al. 2008) showed those who cycled to school at one or both time points compared with those who did not cycle at either time point had 9% higher cardiorespiratory fitness. Similar results were found in Sweden and Estonia (Chillón et al. 2010). Only 1% of the Estonian and 20% of the Swedish 9- and 15-year old children/adolescents cycled to school. Swedish boys and girls who cycled to school had significantly better cardiorespiratory fitness compared to those who commuted passively or walked to school. At six years of follow-up, those reporting cycling to school at baseline or those adopting cycling showed greater fitness compared to the passive commuters and walkers (Chillón et al. 2012). In a randomized controlled trial in Norway, adolescents aged 10–13 years were either in a 12-week cycling to school group or in a non-cycling group. Cyclists improved their fitness significantly more than non-cyclists (Børrestad et al. 2012).

Among adolescents (around 13 years old) in Rotterdam in the Netherlands and Kristiansand in Norway about one-third cycled to school on at least three days a week, for an average of 29 and 12 minutes respectively. In both cities, cyclists were less often **overweight** than non-cyclists, even after adjustment for sport participation. This indicates a possible independent effect of cycling to school on weight status (Bere et al. 2011a). These cross-sectional analyses were followed by a two-year longitudinal analysis, with the prevalence of being overweight increasing for those who stopped cycling to school (Bere et al. 2011b).

Adolescents' physical activity is important for bone health as periosteal surfaces (bone deposition) is maximal during this period of life (Gordon et al. 2017). In general, it is not expected that the **bone health** (i.e. areal bone mineral density and bone mineral content) improves with cycling because cycling is not a bone-loading physical activity. In two cohort studies, the relationship between sport cycling and bone health has been investigated in young males aged 12–14 years (Ubago-Guisado et al. 2018; Vlachopoulos et al. 2018). DXA measures of bone density and bone mineral content at 12 months was lower in the cyclists (and also in swimmers) compared to footballers.

In commuter cycling studies among children and adolescents, the most frequently assessed health benefits were fitness and BMI status. Other health benefits, such as academic performance, depression/mood or cardiometabolic risk factors, were rarely investigated and only in cross-sectional cycling studies. Overall, commuter cyclists have a better fitness compared to non-cyclists.

Cycling and mental health

There is much interest in cycling and mental health in the general media, in online discussions, in the popular print media and in materials and resources distributed by non-government organizations and cycling groups and societies. There is a paucity of studies to corroborate popular perceptions, other than numerous cross-sectional surveys, where cyclists report increased wellbeing that they attribute to cycling. For that reason, this section is divided into two parts; firstly, a review of physical activity and mental health generally, and then a review of recent published evidence of cycling and mental health specifically. As cycling is an efficient form of moderate to vigorously intense physical activity, one can assume that any physical and mental health benefits of cycling are mediated through achieving greater amounts of physical activity (Figure 30.1).

There are numerous systematic reviews and meta-analyses demonstrating that all forms of leisure-time physical activity and active commuting contribute to positive mental health (White et al. 2017; Ahn and Fedewa 2011). Given the heterogeneity in measurements of exposure and outcome, these studies and reviews cannot be directly compared, but the overarching evidence consistently shows a positive association. This evidence is reviewed for children and adults.

Physical activity and mental health in children and adults

Systematic reviews of physical activity interventions in children showed some beneficial effects on preventing depression (Dale et al. 2019), increasing resilience and improving mental health and wellbeing (Andermo et al. 2020). A recent review of more than 40 reviews identified moderate intervention benefits on depression, stronger intervention effects on cognitive performance, and a lack of evidence on self-esteem (Biddle et al. 2019). Overall, there is moderate evidence for the benefits of physical activity, and therefore cycling, in children and adolescents, particularly for measures of positive wellbeing.

There have been numerous epidemiological reviews of physical activity and mental health outcomes in adults. Recent studies have pointed to a clear effect of moderate or vigorous intensity physical activity on reducing anxiety (McDowell et al. 2019), and other research has identified that physical activity reduces the incidence of depression (Mammen and Faulkner 2013). Physical activity interventions are also considered effective in reducing depression and other symptoms in those with diagnosed mental health conditions. For older adults, there is some evidence that physical activity in older adults helps maintain cognitive function, especially executive memory (Engeroff et al. 2018), with mixed evidence on reducing dementia risk (Brasure et al. 2018). In summary, more physical activity is associated with less anxiety, depression, improvements in mental health symptoms and improved cognitive function among older adults.

Cycling-specific studies of mental health

The net sum of evidence suggests that through becoming physically active, cycling has mental health benefits. There are few cycling-specific studies that have examined this question. The weakest evidence comes from cross-sectional surveys which only indicate associations, not causal relationships. Studies in 3567 adults across Europe noted cross-sectional beneficial associations between bicycle usage and all mental health measures, including increased vitality and reduced loneliness (Avila-Palencia et al. 2018). Another study carried out among 788 adults in Barcelona suggested that regular bicycle commuters reported less stress (Avila-Palencia et al. 2017). An Australian study showed that 864 regular cyclists reported improved mental health and quality of life, especially amongst men (Crane et al. 2014). A Japanese study of 670 adults reported a contrary observation, that the General Health Questionnaire Mental Health scale was lower among active commuters, especially men, compared to those who did not report active commuting (Ohta et al. 2007). Qualitative research has been conducted to assess the "joy of cycling" (Zander et al. 2013), where older adult cyclists reported joy and increased confidence attributed to cycling. In addition, there are some intervention studies, such as a controlled trial of light cycling and mood among 60 adults (Lindheimer et al. 2017). This psychological study controlled for the expectation of mood benefits among cyclists, and compared to controls, found that mood and cognitive performance were not improved by active or passive cycling.

Finally, a summary meta-analysis of domain specific physical activity and mental health measures was reported by White et al. (2017). A total of 98 studies were used published between 1988 and 2015, of which 15 examined transport-related physical activity. Three-quarters of included studies used a cross-sectional design. The transport domain did not distinguish between walking or cycling for transport. Multi-level meta-analyses showed that leisure-time physical activity (r = 0.13) and transport physical activity (r = 0.13) both had small, but positive and significant associations with measures of mental health.

Aesthetic (pleasurable) experiences may also be a form of cycling benefit (Whitaker 2005), as is may reflect an interaction between the natural environment, cycling and mental health. Some studies have noted that cycling in a natural environment was associated with positive mental health more than cycling in other (urban) environments (Mitchell 2013; Glackin and Beale 2018; Zijlema et al. 2018). This raises the hypothesis of the environment as a moderator that may partly explain the mental health benefits of cycling.

Finally, there may be multiple factors involved in the relationship between cycling and mental health. One study has proposed these factors for commuting cyclists include convenience, journey reliability, sensory experiences in nature, the direct pleasurable effects on mood and the benefits of social interaction (Wild and Woodward 2019). This complex situation could apply in different ways to recreational cycling, cycle tourism, and group club rides, suggesting the need for more integrated research into the causes and mechanisms of mental health benefits of cycling.

In summary, cycling is likely to contribute to mental health benefits, probably mediated through physical activity. Other mechanisms, such as aesthetic pleasure and natural environment engagement, may contribute to direct cycling benefits to mental health and wellbeing.

Health risks of cycling

There are some risks posed by cycling. The greatest risk is the high numbers of cycling injuries, leading to hospital attendance, admission and, rarely, death. Epidemiologically, the injury field is complicated by different populations in which injury rates are reported; some studies provide self-report survey estimates from cycling groups; others provide hospital-level information; and a few provide population-wide surveillance data on cycling-related injuries. This makes trends and cross-country comparisons difficult, but WHO estimates that 3% of road traffic deaths annually are attributable to cycling (WHO 2020b), with a total number of global deaths between 40 and 50,000/year.

Cycling injury rates have been reported for decades, with early data from the USA (Thompson et al. 1990; Thompson and Rivara 2001). Recent numbers have been reported by the US Centers for Disease Control, with around half a million injuries each year due to cycling, and around 800 deaths/year nationally (~2 deaths/million people), with rates highest among adults over 65 years (2.8 deaths/million, CDC 2020). Cycling injuries resulted in around 7000 admissions/year to hospital in the USA, among the leading injury-related causes of hospital attendance for children and adults (Hamann et al. 2013). Canadian data suggest around 70–80 deaths/year, or 2.6/million population, and around 4000 hospital admissions year for cycling injuries (Statistics Canada 2017). Australian population data are similar, with around 38 deaths/year (2 deaths/million, mostly in older adults) and around 9000 hospital admissions/year (AIHW 2019).

UK data indicate that around 100 people died from cycling injuries each year (4 deaths/million people), with around 4000 being seriously injured (Cycling UK's Cycling Statistics 2018). European data indicate a **decline in cycling fatality** numbers from 3000 in 2007 to around 1800 by 2016 (ERSO 2018). The highest rates of injury and mortality occurred in Eastern European countries. Although high absolute numbers of cyclists were injured in high-cycling countries such as the Netherlands and Denmark, these rates expressed as injuries per million kilometers cycled indicated a lower individual risk in these countries. Similarly, countries with good infrastructure report lower overall cycling deaths, with cycling fatalities per 100 million kilometers cycled being 4.7 in the USA, and 1.3 in Germany, with a similar differential ratio for cycling injuries (Buehler and Pucher 2012). For all European and North American data, there was a several-fold greater risk of injuries in men than women, and risk increased with age, especially for adults over 65 years (ERSO 2018; Ekman et al. 2001).

Cycling injuries in children and adults have **declined** since 2010 in high-income countries (Buehler and Pucher 2012). For children, the decline may be related to declines in the number of children cycling to and from school (Buehler et al. 2020). For children, cycling still contributes globally to 2–8% of all childhood traffic accident deaths, and 3–15% of all childhood traffic-related injuries (Embree et al. 2016). Injury rates are generally higher among early adolescents, compared to earlier childhood (Embree et al. 2016). Rates vary by country and across differently-designed studies, but the issue of preventable cycling morbidity remains an unresolved concern in the disbenefits of cycling.

Other health risks of cycling are rare. Frequent, but usually minor issues for cyclists include sunburn, blisters, dehydration and low blood sugar levels, all of which are preventable by actions before and during cycling. One established cycling-related clinical syndrome is due to perineal pressure on the saddle for males, and may result in perineal pain, numbness, and even erectile dysfunction (Thompson and Rivara 2001; Sommar et al. 2020) but does not contribute substantially to population levels of cycling-related morbidity.

Overall, the major health risks are cycling-related accidents. An additional health risk is the **air quality** to which urban cyclists may be exposed. The World Health Organization reports that the transport sector contributes 23% of the greenhouse gas emissions globally, which, in turn, result in 3.7 million deaths (WHO 2020b). Urban cyclists have direct exposure to poor air quality when cycling through traffic, increasing exposure through an increased respiratory rate when engaging in cycling activity (Bevan et al. 1991; Ramos et al. 2016). Specifically, poor air quality exposes urban cyclists to more particulate matter (PM2.5), carbon monoxide and several volatile organic compounds (Ramos et al. 2016).

A key public health question is whether the risks of accidents and air quality are outweighed by the benefits of increased physical activity among regular cyclists? To answer this "benefit: risk question," several studies have modelled the physical activity health gains compared to injuries and air quality exposure; all studies have noted **"benefit: risk ratios"** from around 9:1 (de Hartog et al. 2010) to 77:1 (Rojas-Rueda et al. 2011) with any ratio greater than 1:1 considered "good." Clear evidence of both health and economic benefits could be accrued through switching sufficient car users to cycle for transport, even in high-pollution cities (Rojas-Rueda et al. 2011; Tainio et al. 2016). These studies suggest that only on the most extreme pollution days should cycling be discouraged (Tainio et al. 2016).

Conclusion

The benefits of cycling are well recognized, as cycling is an excellent and accessible form of physical activity, which leads to physical and mental health benefits across the lifespan, and if cycling increased across societies, it would contribute to improved air quality and climate change mitigation (Chapter 6). All types of cycling provide these benefits. The biggest disbenefits of cycling are the deaths and injuries in traffic, but there is a clear positive "benefit: risk" ratio in all studies that have considered this. The benefits of cycling are not for improved cardiometabolic health, weight loss, and non-communicable disease prevention, but also to mental health and wellbeing. The benefits of cycling are present at all age groups, show gender equity, have the potential to narrow socioeconomic differences in physical activity, and can be integrated into everyday life in many countries. For older adults, the benefits of maintaining cycling will contribute to maintaining physical health, and also to social and community connections. We have sufficient evidence for public health action. Now we need to implement effective strategies that increase population levels of cycling, especially in "cycling-resistant" countries with poor cycling culture, excessive traffic and limited cycling infrastructure.

References

Ahn, S. and A. L. Fedewa (2011) A meta-analysis of the relationship between children's physical activity and mental health, *Journal of Pediatric Psychology* 36: 4, 385–97.

AIHW: R. Kreisfeld and J. E. Harrison (2019) Pedal cyclist deaths and hospitalisations, 1999–00 to 2015–16. Injury Research and Statistics Series no. 123. Cat. no. INJCAT 203. Canberra: AIHW.

Andermo, S., M. Hallgren, T. T. D. Nguyen, S. Jonsson, S. Petersen, M. Friberg, A. Romqvist, B. Stubbs and L. S. Elinder (2020) School-related physical activity interventions and mental health among children: a systematic review and meta-analysis, *Sports Medicine* 6: 1, 25. https://doi.org/10.1186/s40798-020-00254-x

Andersen, Z. J., A. de Nazelle, M. A. Mendez, J. Garcia-Aymerich, O. Hertel, A. Tjønneland, K. Overvad, O. Raaschou-Nielsen and M. J. Nieuwenhuijsen (2015) A study of the combined effects of physical activity and air pollution on mortality in elderly urban residents: the Danish Diet, Cancer, and Health Cohort, *Environmental Health Perspectives* 123, 557–563. https://doi.org/10.1289/ehp.1408698

Armstrong, M. E., J. Lacombe, C. J. Wotton, B. J. Cairns, J. Green, S. Floud, V. Beral, and G. K. Reeves (2020) The associations between seven different types of physical activity and the incidence of fracture at seven sites in healthy postmenopausal UK women, *Journal of Bone and Mineral Research* 35, 277–90.

Avila-Palencia, I., A. de Nazelle, T. Cole-Hunter, D. Donaire-Gonzalez, M. Jerrett, D. A. Rodriguez and M. J. Nieuwenhuijsen (2017) The relationship between bicycle commuting and perceived stress: a cross-sectional study, *British Medical Journal* open 7: 6, e013542.

Avila-Palencia, I., L. I. Panis, E. Dons, M. Gaupp-Berghausen, E. Raser, T. Götschi, R. Gerike, C. Brand, A. De Nazelle and J. P. Orjuela (2018) The effects of transport mode use on self-perceived health, mental health, and social contact measures: a cross-sectional and longitudinal study, *Environment International* 120, 199–206.

Bevan, M. A. J., C. J. Proctor, J. Baker-Rogers and N. D. Warren (1991) Exposure to carbon monoxide, respirabie suspended particulates, and volatile organic compounds while commuting by bicycle, *Environmental Science and Technology* 25: 4, 788–791.

Bere, E., S. Seiler, T. A. Eikemo, A. Oenema and J. Brug (2011a) The association between cycling to school and being overweight in Rotterdam (The Netherlands) and Kristiansand (Norway), *Scandinavian Journal of Medicine & Science in Sports* 21: 1, 48–53. https://doi.org/10.1111/j.1600-0838.2009.01004.x

Bere, E., A. Oenema, R. G. Prins, S. Seiler and J. Brug (2011b) Longitudinal associations between cycling to school and weight status, *International Journal of Pediatric Obesity* 6: 3–4, 182–187. https://doi.org/10.3109/17477166.2011.583656

Biddle, S. J. H., S. Ciaccioni, G. Thomas and I. Vergeer (2019) Physical activity and mental health in children and adolescents: An updated review of reviews and an analysis of causality, *Psychology of Sport and Exercise* 42, 146–155.

Blond, M. B., M. Rosenkilde, A. S. Gram, M. Tindborg, A. N. Christensen, J. S. Quist and B. M. Stallknecht (2019) How does 6 months of active bike commuting or leisure-time exercise affect insulin sensitivity, cardiorespiratory fitness and intra-abdominal fat? A randomised controlled trial in individuals with overweight and obesity, *British Journal of Sports Medicine* 53: 18, 1183–1192. https://doi.org/10.1136/bjsports-2018-100036

Blond, K., M. K. Jensen, M. G. Rasmussen, K. Overvad, A. Tjönneland, L. Östergaard and A. Gröntved (2016) Prospective study of bicycling and risk of coronary heart disease in Danish men and women, *Circulation* 134, 1409–1411.

Børrestad, L. A. B., L. Ostergaard, L. B. Andersen and E. Bere (2012) Experiences from a randomised, controlled trial on cycling to school: does cycling increase cardiorespiratory fitness? *Scandinavian Journal of Public Health* 40: 3, 245–252. https://doi.org/10.1177/1403494812443606

Bourne, J. E., S. Sauchelli, R. Perry, A. Page, S. Leary, C. England and A. R. Cooper (2018) Health benefits of electrically-assisted cycling: a systematic review, *International Journal of Behavioral Nutrition and Physical Activity* 15: 1, 1–5.

Brasure, M., P. Desai, H. Davila, V. A. Nelson, C. Calvert, E. Jutkowitz, M. Butler, H. A. Fink, E. Ratner, L. S. Hemmy, J. R. McCarten, T. R. Barclay and R. L. Kane (2018) Physical activity interventions in preventing cognitive decline and alzheimer-type dementia a systematic review, *Annals of Internal Medicine* 168: 1, 30–38.

Buehler, R. and J. Pucher (2012) Cycling to work in 90 large American cities: new evidence on the role of bike paths and lanes, *Transportation* 39: 2, 409–432.

Buehler, R., J. Pucher and A. Bauman (2020) Physical activity from walking and cycling for daily travel in the United States, 2001–2017: Demographic, socioeconomic, and geographic variation, *Journal of Transport and Health* 16. https://doi.org/10.1016/j.jth.2019.100811

Castro, A., M. Gaupp-Berghausen, E. Dons, and the Pasta consortium (2019) Physical activity of electric bicycle users compared to conventional bicycle users and non-cyclists: insights based on health and transport data from an online survey in seven European cities, *Transportation Research Interdisciplinary Perspectives* 1: 100017.

CDC, USA. (2020) Fatal Injury Reports, National, Regional and State, 1981–2019. https://webappa.cdc.gov/sasweb/ncipc/mortrate.html

Chavarrias, M., J. Carlos-Vivas, D. Collado-Mateo and J. Pérez-Gómez (2019) Health benefits of indoor cycling: a systematic review, *Medicina* 55: 8, 452. https://doi.org/10.3390/medicina55080452

Chillón, P., F. B. Ortega, J. R. Ruiz, K. R. Evenson, I. Labayen, V. Martínez-Vizcaino, et al. (2012) Bicycling to school is associated with improvements in physical fitness over a 6-year follow-up period in Swedish children, *Preventive Medicine* 55: 2, 108–112. https://doi.org/10.1016/j.ypmed.2012.05.019

Chillón, P., F. B. Ortega, J. R. Ruiz, T. Veidebaum, L. Oja, J. Mäestu and M. Sjöström (2010) Active commuting to school in children and adolescents: an opportunity to increase physical activity and fitness, *Scandinavian Journal of Public Health* 38: 8, 873–879. https://doi.org/10.1177/1403494810384427

Cooper, A. R., N. Wedderkopp, R. Jago, P. L. Kristensen, N. C. Moller, K. Froberg et al. (2008) Longitudinal associations of cycling to school with adolescent fitness, *Preventive Medicine* 47: 3, 324–328. https://doi.org/10.1016/j.ypmed.2008.06.009

Cooper, A. R., N. Wedderkopp, H. Wang, L. B. Andersen, K. Froberg and A. S. Page (2006) Active travel to school and cardiovascular fitness in Danish children and adolescents, *Medicine and Science in Sports and Exercise* 38: 10, 1724–1731. https://doi.org/10.1249/01.mss.0000229570.02037.1d

Crane, M., C. Rissel, C. Standen and S. Greaves (2014) Associations between the frequency of cycling and domains of quality of life, *Health Promotion Journal of Australia* 25: 3, 182–185.

Cycling UK's Cycling Statistics (2018) https://www.cyclinguk.org/statistics / https://www.gov.uk/government/statistics/walking-and-cycling-statistics-england-2019

Dale, L. P., L. Vanderloo, S. Moore and G. Faulkner (2019) Physical activity and depression, anxiety, and self-esteem in children and youth: an umbrella systematic review, *Mental Health and Physical Activity* 16, 66–79.

de Hartog, J. J., H. Boogaard, H. Nijland and G. Hoek (2010) Do the health benefits of cycling outweigh the risks? *Environmental Health Perspectives* 118: 8, 1109–1116.

Dhana, K., C. M. Koolhaas, M. A. Berghout, A. Peeters, M. A. Ikram, H. Tiemeier, A. Hofman, W. Nusselder and O. H. Franco (2016) Physical activity types and life expectancy with and without cardiovascular disease: the Rotterdam Study, *Journal of Public Health* 39, E209–E218. https://doi.org/10.1093/pubmed/fdw110

Dinu, M., G. Pagliai, C. Macchi and F. Sofi (2019) Active commuting and multiple health outcomes: a systematic review and meta-analysis, *Sports Medicine* 49, 437–452.

Dons, E., D. Rojas-Rueda, E. Anaya-Boing, I. Avila-Palenciaet, C. Brand, T. Cole-Hunter et al. (2018) Transport mode choice and body mass index: cross-sectional and longitudinal evidence from a European-wide study, *Environment International* 119, 109–116.

Ekman, R., G. Welander, L. Svanström, L. Schelp and P. Santesson (2001) Bicycle-related injuries among the elderly – a new epidemic? *Public Health* 115: 1, 38–43.

Embree, T. E., N. T. Romanow, M. S. Djerboua, N. J. Morgunov, J. J. Bourdeaux and B. E. Hagel (2016) Risk factors for bicycling injuries in children and adolescents: a systematic review, *Pediatrics* 138, 5. https://doi.org/10.1542/peds.2016-0282

Engeroff, T., T. Ingmann and W. Banzer (2018) Physical activity throughout the adult life span and domain-specific cognitive function in old age: a systematic review of cross-sectional and longitudinal data, *Sports Medicine* 48: 6, 1405–1436.

ERSO (2018) Traffic Safety Basic Facts 2018 European Road Safety Observatory www.erso.eu

Fisher, J. E., S. Loft, C. S. Ulrik, O. Raaschou-Nielsen, O. Hertel, A. Tjönneland, K. Vervad, M. J. Nieuwenhuijsen and Z. J. Andersen (2016) Physical activity, air pollution, and the risk of asthma and chronic obstructive pulmonary disease, *American Journal of Respiratory Critical Care Medicine* 194: 7, 855–865. https://doi.org/10.1164/rccm.201510-2036OC

Glackin, O. F. and J. T. Beale (2018) 'The world is best experienced at 18 mph'. The psychological well-being effects of cycling in the countryside: an Interpretative Phenomenological Analysis, *Qualitative Research in Sport, Exercise and Health* 10: 1, 32–46.

Gordon, C.M., B. S. Zemel, T. A. L. Wren, M. B. Leonard, L. K. Bachrach, F. Rauch, V. Gilsanz, C. J. Rosen and K. K. Winer (2017) The determinants of peak bone mass, *Journal of Pediatrics* 180, 261–269. https://doi.org/10.1016/j.jpeds.2016.09.056

Gröntved, A., R. W. Koivula, I. Johansson, P. Wennberg, L. Östergaard, G. Hallmans, F. Renström and P. W. Franks (2016) Bicycling to work and primordial prevention of cardiovascular risk: a cohort study among Swedish men and women, *Journal of the American Heart Association* 5, e004413.

Hamann, C., C. Peek-Asa, C. F. Lynch, M. Ramirez and J. Torner (2013) Burden of hospitalizations for bicycling injuries by motor vehicle involvement: United States, 2002 to 2009, *Journal of Trauma and Acute Care Surgery* 75: 5, 870–876.

Hendriksen, I. J. M., B. Zuiderveld, H. C. G. Kemper and P. D. Bezemer (2000) Effect of commuter cycling on physical performance of male and female employees, *Medicine and Science in Sports and Exercise* 32: 2, 504–510.

Höchsmann, C., S. Meister, D. Gehrig, E. Gordon, Y. Li, M. Nussbaumer, A. Rossmeissl, J. Schäfer, H. Hanssen and A. Schmidt-Trucksäss (2018). Effect of e-bike versus bike commuting on cardiorespiratory fitness in overweight adults: a 4-week randomized pilot study, *Clinical Journal of Sport Medicine* 28: 3, 255–261.

Kelly, P., S. Kahlmeier, T. Götschi, N. Orsini, J. Richards, N. Roberts, P. Scarborough and C. Foster (2014) Systematic review and meta-analysis of reduction in all-cause mortality from walking and cycling and shape of dose-response relationship, *International Journal of Behavioral Nutrition and Physical Activity* 11, 132.

Koolhaas, C. M., K. Dhana, R. Golubic, J. D. Schoufour, A. Hofman, F. J. A. Van Rooij and O. H. Franco (2016) Physical activity types and coronary heart disease risk in middle-aged and elderly persons: The Rotterdam Study, *American Journal of Epidemiology* 183, 729–738. https://doi.org/10.1093/aje/kwv244

Kubesch, N. J., J. T. Jørgensen, B. Hoffmann, S. Loft, M. J. Nieuwenhuijsen, O. Raaschou-Nielsen, M. Pedersen, O. Hertel, K. Overvad, A. Tjønneland, E. Prescot and Z. J. Andersen (2018) Effects of leisure-time and transport-related physical activities on the risk of incident and recurrent myocardial infarction and interaction with traffic-related air pollution: a cohort study, *Journal of the American Heart Association* 7: 15, e009554.

Lindheimer, J. B., P. J. O'Connor, K. K. McCully and R. K. Dishman (2017) The effect of light-intensity cycling on mood and working memory in response to a randomized, Placebo-controlled design, *Psychosomatic Medicine* 79: 2, 243–253.

Mammen, G. and G. Faulkner (2013) Physical activity and the prevention of depression: a systematic review of prospective studies, *American Journal of Preventive Medicine* 45: 5, 649–657.

McDowell, C. P., R. K. Dishman, B. R. Gordon and M. P. Herring (2019) Physical activity and anxiety: a systematic review and meta-analysis of prospective cohort studies, *American Journal of Preventive Medicine* 57: 4, 545–556.

Mielke, G. I., T. G. Bailey, N. W. Burton and W. J. Brown (2020) Participation in sports/recreational activities and incidence of hypertension, diabetes and obesity in adults, *Scandinavian Journal of Medicine and Science in Sports* 30: 12, 2390–2398. https://doi.org/10.1111/sms.13795

Mitchell, R. (2013) Is physical activity in natural environments better for mental health than physical activity in other environments? *Social Science & Medicine* 91, 130–134.

Mueller, N., D. Rojas-Ruede, T. Cole-Hynter, A. de Nazalle, E. Dons, R. Gerike, T. Götschi, L. I. Panis, S. Kahlmeier and M. Nieuwenhuijsen (2015) Health impact assessment of active transportation: a systematic review, *Preventive Medicine* 76, 103–114.

Ohta, M., T. Mizoue, N. Mishima and M. Ikeda (2007) Effect of the physical activities in leisure time and commuting to work on mental health, *Journal of Occupational Health* 49: 1, 46–52.

Oja, P., P. Kelly, Z. Pedisic, S. Titze, A. Bauman, C. Foster, M. Hamer, M. Hillsdon, M. Stamatakis (2016) Associations of specific types of sports and exercise with all-cause and cardiovascular-disease mortality: a cohort study of 80.306 British adults, *British Journal of Sports Medicine* 51: 10, 812–817. https://doi.org/10.1136/bjsports-2016-096822

Oja, P., A. Mänttäri, A. Heinonen, K. Kukkonen-Harjula, R. Laukkanen, M. Pasanen and I. Vuori (1991) Physiological effects of walking and cycling to work, *Scandinavian Journal of Medicine and Science in Sports* 1: 3, 151–157.

Oja, P., S. Titze, A. Bauman, B. de Geus, P. Krenn, B. Reger-Nash and T. Kohlberger (2011) Health benefits of cycling: systematic review, *Scandinavian Journal of Medicine and Science in Sports* 21: 4, 496–509. https://doi.org/10.1111/j.1600-0838.2011.01299.x

Östergaard, L., M. K. Jensen, K. Overvad, A. Tjönneland and A. Gröntved (2018) Associations between changes in cycling and all-cause mortality risk, *Preventive Medicine* 55: 5, 615–623.

Porter, A. K., C. C. Cuthbertson and K. R. Evenson (2020) Participation in specific leisure-time activities and mortality risk among U.S. adults, *Annals of Epidemiology* 50, 27–34.el. https://doi.org/10.1016/j.annepidem.2020.06.006

Ramos, C. A., H. T. Wolterbeek and S. M. Almeida (2016) Air pollutant exposure and inhaled dose during urban commuting: a comparison between cycling and motorized modes. *Air Quality, Atmosphere and Health* 9: 8, 867–879.

Rasmussen, M. G., A. Gröntved, K. Blond, K. Overvad, A. Tjönneland, M. K. Jensen and L. Östergaard (2016) Associations between recreational and commuter cycling, changes in cycling, and type 2 Diabetes risk: a cohort study of Danish men and women, *PLOS Medicine* 13: 7, e1002076.

Rasmussen, M. G., K. Overvad, A. Tjönneland, M. K. Jensen, L. Östergaard and A. Gröntved (2018) Changes in cycling and incidence of overweight and obesity among Danish men and women, *Medicine and Science in Sports* 50: 7, 1413–1421.

Rojas-Rueda, D., A. de Nazelle, M. Tainio and M. J. Nieuwenhuijsen (2011) The health risks and benefits of cycling in urban environments compared with car use: health impact assessment study, *British Medical Journal* 343, d4521. https://doi.org/10.1136/bmj.d4521

Schepers, P., K. Klein Wolt, M. Helbich and E. Fishman (2020). Safety of e-bikes compared to conventional bicycles: What role does cyclists' health condition play? *Journal of Transport and Health* 19; Article 100961.

Schnohr, P., J. H. O'Keefe, A. Holtermann, C. J. Lavie, P. Lange, G. B. Jensen and J. L. Marott (2018) Various leisure-time physical activities associated with widely divergent life expectancies: The Copenhagen City Heart Study, *Mayo Clinic Proceedings* 93: 12, 1775–1785. https://doi.org/10.1016/j.mayocp.2018.06.025

Sommar, J. N., C. Johansson, B. Lövenheim, A. Markstedt, M. Strömgren and B. Forsberg (2020) Potential effects on travelers' air pollution exposure and associated mortality estimated for a mode shift from car to bicycle commuting, *International Journal of Environmental Research and Public Health* 17: 20, 1–16.

Statistics Canada (2017) https://www150.statcan.gc.ca/n1/pub/82-003-x/2017004/article/14788-eng.htm. Health Reports: Cycling in Canada. Ramage-Morin P, April 19, 2017.

Sundfør, H. B. and A. Fyhri (2017) A push for public health: the effect of e-bikes on physical activity levels, *BMC Public Health* 17, 1–12.

Tainio, M., A. J. de Nazelle, T. Götschi, S. Kahlmeier, D. Rojas-Rueda, M. J. Nieuwenhuijsen, T. H. de Sá, P. Kelly and J. Woodcock (2016) Can air pollution negate the health benefits of cycling and walking? *Preventive Medicine* 87, 233–236.

Thompson, D. C., R. S. Thompson and F. P. Rivara (1990) Incidence of bicycle-related injuries in a defined population, *American Journal of Public Health* 80: 11, 1388–1389.

Thompson, M. J. and F. P. Rivara (2001) Bicycle-related injuries, *American Family Physician*, 63: 10, 2007-2014+2017-2018.

Turrell, G., B. A. Hewitt, J. N. Rachele, B. Giles-Corti, L. Busija, and W. J. Brown (2018) Do active modes of transport cause lower body mass index? Findings from the HABITAT longitudinal study, *Journal of Epidemiology and Community Health* 72: 4, 294–301.

Ubago-Guisado, E., D. Vlachopoulos, I. G. Fatouros, C. K. Deli, D. Leontsini, L. A. Moreno et al. (2018) Longitudinal determinants of 12-month changes on bone health in adolescent male athletes, *Archives of Osteoporosis* 13: 1, 106. https://doi.org/10.1007/s11657-018-0519-4

Van Cauwenberg, J., I. De Bourdeaudhuij, P. Clarys, B. De Geus and B. Deforche (2018) Older e-bike users: demographic, health, mobility characteristics, and cycling levels, *Medicine and Science in Sports and Exercise* 50: 9, 1780–1789.

Van Cauwenberg, J., I. De Bourdeaudhuij, P. Clarys, B. de Geus and B. Deforche (2019) E-bikes among older adults: benefits, disadvantages, usage and crash characteristics, *Transportation*, 46: 6, 2151–72.

Vlachopoulos, D., A. R. Barker, E. Ubago-Guisado, F. B. Ortega, P. Krustrup, B. Metcalf, et al. (2018) The effect of 12-month participation in osteogenic and non-osteogenic sports on bone development in adolescent male athletes. The PRO-BONE study, *Journal of Science and Medicine in Sport* 21: 4, 404–409. https://doi.org/10.1016/j.jsams.2017.08.018

Whitaker, E. D. (2005) The bicycle makes the eyes smile: exercise, aging, and psychophysical well-being in older Italian cyclists, *Medical Anthropology* 24: 1, 1–43.
White, R. L., M. J. Babic, P. D. Parker, D. R. Lubans, T. Astell-Burt and C. Lonsdale (2017) Domain-specific physical activity and mental health: a meta-analysis, *American Journal of Preventive Medicine*, 52: 55, 653–666.
Wild, K. and A. Woodward (2019) Why are cyclists the happiest commuters? Health, pleasure and the e-bike, *Journal of Transport and Health* 14. https://doi.org/10.1016/j.jth.2019.05.008
WHO (1999) Regional Office for Europe. *Transport environment and health*. WHO Regional Publications, European Series No. 89.
WHO (2011) *Health economic assessment tools (HEAT) for walking and cycling. Methodology and user guide*. WHO Regional Office for Europe Publications.
WHO (2020b) *Cyclist safety: an information resource for decision-makers and practitioners*. Geneva: World Health Organization; Licence: CC BY-NC-SA 3.0 IGO.
WHO (2020a) *Guidelines on physical activity and sedentary behaviour*. Geneva: World Health Organization. https://www.who.int/publications/i/item/9789240015128
Zander, A., E. Passmore, C. Mason and C. Rissel (2013) Joy, exercise, enjoyment, getting out: a qualitative study of older people's experience of cycling in Sydney, Australia, *Journal of Environmental Public Health* 2013, 547453. https://doi.org/10.1155/2013/547453
Zijlema, W. L., I. Avila-Palencia, M. Triguero-Mas, C. Gidlow, J. Maas, H. Kruize, S. Andrusaityte, R. Grazuleviciene and M. J. Nieuwenhuijsen (2018) Active commuting through natural environments is associated with better mental health: results from the PHENOTYPE project, *Environment International*, 121, 721–727.

31

DOPING IN CYCLING

Past, present and future trends

Charlotte Smith

Introduction and overview

This chapter explores doping in competitive cycling, perhaps the most heavily discussed topic throughout its history and the sport most often associated with prohibited performance enhancement. In the public's eye, and both popular and academic discourses (Lopez 2015; Schneider 2006), the default perception is that the majority of riders dope (Sefiha 2017). Yet despite many doping confessions, the public remains loyal, on the roads and TV screens (Mignon 2003). Thus far, there has been a particular narrative throughout cycling's history of mixed support for and against doping with many circumventing the anti-doping regulations (Schneider 2006). Doping can occur in all types and levels of competitive cycling; however, the focus herein is primarily road cycling as this is combined with the highest financial rewards and fame.

The chapter starts by outlining the different means of doping and how cyclists enact this in practice. The middle section moves on from the pragmatic aspects of doping toward the detection of cheating, and of cyclists' behavior circumventing this. It is argued that doping detection policies and education have had some success in cycling and there have been some changes regarding riders' attitudes and usage of doping since the turn of the century (Waddington and Møller 2019). Despite this, overall cycling still struggles with demonstrating a new approach to its varied stakeholders, given its turbulent backstory (Marty et al. 2015). Therefore, the last part of the chapter reviews "clean" cycling and considers some of the future predicted aspects of doping that may threaten this position.

Doping for strength and endurance

According to Dr. Dumas at a European Colloquium held in 1963 on the initiative of doping, it is defined as "the use of substances or of all means designed to artificially enhance performance, in preparation for or on the occasion of competition..." (Mignon 2003: 235). Doping primarily enables cyclists to enhance their performance by increasing how much oxygen their blood can carry to muscles. Increasing oxygen-carrying capacity gives a higher VO2 max and a bigger associated maximal aerobic output, measured in watts (Lodewijkx 2014). Doping is also

DOI: 10.4324/9781003142041-44

Table 31.1 Popular doping methods in cycling compared

	Physiological effects	*Method*	*Timing*
Blood doping	Increases the amount of oxygen-carrying red blood cells in the body which increases oxygen levels and boosts aerobic power	Withdraw blood, store it for a period and then replace with that blood or someone else's	Out of competition as it initially weakens a cyclist
Epoetin Alpha (EPO)	More EPO results in more red blood cells giving more oxygen to muscles	Injection	Taken chronically, relatively long-lasting anabolic effects
Stimulants	Enhances alertness, increases blood pressure, heart rate, and depth of respiration. Reduces tiredness	Tablets	Immediate effects, not taken chronically
Steroids	Increases muscle bulk and power output, supports changes in fat mass and distribution	Tablets or injection	Taken chronically over long period as no immediate pharmacological effects

used to stimulate muscle growth and aid muscle recovery in this demanding sport (Marty et al. 2015). Doping is not something that normally happens by accident. It requires cyclists' sustained deliberation, self-regulation, planning, and commitment, encased by significant protective and risk factors for their careers (Petroczi and Aidman 2008). Some of the key techniques can be seen in Table 31.1.

The need for doping in cycling

The triggers causing cyclists' doping are vast, but there is an often-underlying narrative of the impossibly hard nature of cycling, being labelled as one of the most demanding sports (Vandeweghe 2016). There are many cultural aspects influencing doping. For example, if a cyclist believes that another cyclist is doping then they think they are justified in cheating as well, though riders often overestimate others' doping (Marty et al. 2015). The majority of substances taken suppress fatigue and relieve pain (Vandeweghe 2016), whilst restoring the cyclist's health (Mignon 2003) which they see as necessary to attain peak performance (Brissonneau 2015). Overbye et al. (2013) add that a top incentive for doping is that it carries a low risk of being caught and has little chance of threatening an elite cyclist's career. The big prize money for the few who win grand tours is also an incentive to dope. Today, just like other professional sports, cycling is an occupation complete with its own norms, values, and codes of practice (Sefiha and Reichman 2016). There is a clear culture whereby cyclists see doping as legitimate because of their job (Mignon 2003), strongly connected to the cycling environment and team members and staff (Lentillon-Kaestner 2013). There are key socio-economic factors at play. As Brewer (2002) argues, *professional cycling is commercial cycling*. The implications are that the necessity for success comes with substantial pressure for many who come from lower social class groupings, having quit school early to become professionals and consequently have few other options to make a living (Schneider 2006; Smith 2017). These socio-economic factors highlight the pressure for cyclists to dope; to try and encourage their bodies to repair quicker, for career survival, to sustain an elite life world (Smith 2017). There has in the past been tremendous pressure to cheat from team managers who set the rules for competing in their team. These pressures are reinforced with "The Omerta", an unspoken rule whereby cyclists can only talk about their own doping and importantly never implicate another rider if they have not already been implicated (Hardie 2015). Some research has focused on hypothetical effective deterrents for cyclists not doping and these primarily concern legal and social sanctions, the side effects of consumption, and riders' moral positions upheld (Overbye et al. 2013).

The prevalence of doping

Despite the perception that doping is systematic across cycling, it is inherently difficult for researchers to determine how many athletes dope and who do not get caught. The upshot is a distinct underestimation of its prevalence (Healey 2013). Accurate statistics on doping are also difficult to attain because of the consequences for the athlete of admission in a research setting (Hackney 2017). De Hon et al. (2015) estimated that between 14 and 39% of elite athletes intentionally used some means of doping and that on average 1–2% of athletes overall are caught with doping substances or metabolites in their system. In May 2021 the Union Cycliste Internationale (USADA 2021) listed 93 professional cyclists (across all disciplines) who are currently serving a period of ineligibility due to an anti-doping rule violation (ADRV). As Yesalis and Bahrke (2002) have noted, it is reasonably assumed that the use of anabolic steroids, stimulants, blood doping, or EPO use by female athletes occurs at rates similar to males' usage, though evidence suggests males are more willing to dope (Overbye et al. 2013). Marty et al.'s (2015) report also confirmed that doping occurs in women's cycling, though it is not as widespread or systemic compared to male cycling as it does not carry the same status or financial backing. As female cycling develops and given the history of the effects of cycling's commercialization in the past, it is anticipated that their doping may end up in the same situation as men's cycling, unless preventative measures are put in place (Marty et al. 2015).

Following difficulties determining the amount of doping, there are also difficulties in establishing how much faster doping may make cyclists. For example, Lodewijkx's (2014) study was unable to determine whether riders in the height of the EPO era were able to ride much faster than those in subsequent years.

Doping amateurs

Over the past decade, there has been a rise in serious amateurism, coupled with the continuing buoyancy of the bicycle industry (Møller 2008). Subsequently, there has also been an increase in Middle-Aged-Men-in-Lycra (MAMILs), whom often have spare disposable income and a competitive spirit ripe for doping. Knowledge of amateur's doping practices is equally difficult to attain, with research thus far primarily focusing on bodybuilders, gym goers, and non-elite sports competitors (Henning and Dimeo 2018). A key issue when monitoring doping amateurs is that the majority of participants do not compete to win, but instead participate for alternative motivations and a wide variety of behaviors and situations (Henning and Dimeo 2015), like benefitting their appearance, offsetting the aging process (Henning and Dimeo 2018), and raising charity money. Amateurs are also a heterogeneous group of individuals, varying in age, experience, motivation, and lifestyle (Henning and Dimeo 2018). The same rules that are applied to preventing doping in professional-level athletes and which will be discussed below are transferred similarly to amateurs (Henning and Dimeo 2018). However, amateurs tend to have less detailed knowledge of where to get drugs, doping policies and regulations. For example, they may assume that legal medications are permitted, that they have less chance of being tested and less chance of a positive test result (Henning and Dimeo 2018). As Henning and Dimeo (2015) found from their study of amateur US cyclists, many infractions by non-professional cyclists derive from recreational drug use, supplements containing banned substances, or medicines prescribed by a medical practitioner.

Doping in the Tour de France

The Tour De France (TDF) remains the most gruelling professional cyclists' race; typically covering over 3000 km across 20 stages averaging 4–5 hours daily. Given the centrality of the TDF in riders' careers and its spectacular nature, it is not surprising that many of the biggest doping scandals have occurred within it.

The Festina Affair and Operation Puerto

Perhaps the biggest scandal occurred in the 1998 TDF when the Festina team's soigneur (Willy Voet) was found by French customs in possession of 500 doses of EPO and growth hormone, highlighting how routinely and normalized doping was within professional cycling teams (Soule and Lestrelin 2011). Following this discovery, Festina's sporting director, Bruno Roussel, admitted he had personally and rationally organized for his riders to take drugs under medical supervision (Mignon 2003). This scandal was a key factor in the decision to set up the World Anti-Doping Agency (WADA), which will be discussed below. Another major doping scandal to shake professional cycling in May 2006 was "Operation Puerto" which identified a network of cyclist's blood sample use. The home of Dr. Fuente (a Spanish sports doctor) was raided, which led to the identification of anabolic steroids, blood transfusion equipment, and some 200 refrigerated bags of blood of around 60 professional cyclists (Marty et al. 2015).

Anti-doping policies and detection methods

Multiple texts document the history of the "doping problem" and how it began: space limitations prevent detailing these. However, there are some notable points that can be traced to decisions made during the 1960s by the International Olympic Committee (IOC), which primarily portrayed all performance enhancement as an existential threat to the ideal and spirit of sport (Hunt et al. 2012). The IOC's attention and action towards Performance Enhancing Drugs (PEDs) was notably increased when the Danish cyclist Knud Enemark Jensen died in the 1960 Rome Olympic Games as a result of amphetamine use (Hunt et al. 2012).

The key means of doping deterrence in sport broadly can be summarized as a two-pronged approach consisting of repressing athletes from taking banned substances (through detecting usage) and education. The purpose of anti-doping policies is to preserve what is intrinsically valuable about sport (health and fairness), often called the "spirit of sport" emanating from the essence of Olympism (WADA 2021). In 1999 WADA was established and tasked with harmonizing anti-doping regulations worldwide (Fincoeur et al. 2020), which it does mainly through *The WADA Code*. This code is a set of rules that binds governments and their national sporting organizations globally to ensure a harmonized approach to the fight against doping (WADA 2015). A substance is on the WADA prohibited list if it potentially enhances sport performance, using the substance or methods represents a health risk, and using the substance or method violates the spirit of sport. Though previous doping affairs have shown the wide range of stakeholders influencing a rider's doping, riders themselves hold the responsibility for knowing what substances are prohibited under the *Strict Liability Principle*. This principle means that every rider is responsible for all products in their body (including non-specific products), regardless of how they got there. An example of a breach was when Alberto Contador won the 2010 TDF, but later tested positive for clenbuterol. The court ruled that the positive test was due to a contaminated food supplement (Marty et al. 2015).

The biological passport and the athlete whereabouts reporting system

In the last ten years, anti-doping regulations have shifted away from the earlier means of education and repression, toward a refined focus of detection and punishment (Houlihan 2015). This newer approach can be seen with WADA's introduction of *The Biological Passport* and *The Athlete Whereabouts Reporting System*. The former is a rider's individual electronic record that contains the results of all urine and blood tests and their hematological and steroid levels profile. The purpose of this process is that their "normal" levels can then be used to identify any possible variations resulting from doping. This technique has faced criticism as dopers may "microdose" instead, which involves taking a much smaller dose of a banned substance that is barely detectable. Doping in this way keeps their blood parameters constant and thus they most likely avoid detection (Marty et al. 2015). The whereabouts system enables in-and-out-of-competition testing as athletes have to provide a one-hour time slot every day when they are available for testers.

Legal definitions of doping and grey zones

"Grey zones", understood as acts of doping/cheating that are not formally defined as such, have become a prevalent issue circumventing anti-doping policy (Fincoeur et al. 2020). The most frequently occurring grey zone is when an athlete uses a prohibited substance (according to the WADA code) for medical reasons. In connection with the WADA code, there is often reference to Therapeutic Use Exemptions (TUE) which are when an athlete applies for and demonstrates that they need to use a prohibited substance to avoid significant health problems. For example, Bradley Wiggins was granted a TUE to take corticosteroid triamcinolone (for respiratory issues) before the 2011 and 2012 TDF and 2013 Giro d'Italia (BBC 2018). However, in reality, some cyclists have been accused of using the TUE system as a vehicle for legitimate doping and there is suspicion around which athletes get what exemptions and why. Other acknowledged grey zones are the use of legal enhancers such as painkillers, hypoxic chambers, and supplement use (Fincoeur et al. 2020).

Implementing doping rules is not problem-free, and loopholes also frequently arise. For example, loopholes can often be found when regulation is badly implemented or if insufficient scientific knowledge is known about specific doping substances or techniques which develop frequently (Soule and Lestrelin 2011). WADA, therefore, retains urine and blood samples for up to eight years so that they can be re-tested for substances that are banned but currently have no scientific test (Møller and Dimeo 2014). Lastly, there has always been ambiguity about what constitutes a "clean" cyclist. For instance, some riders consider themselves clean if they take performance-enhancing drugs but have not been caught; others believe they are clean if they take substances that are not yet detectable or listed in the WADA code. Some riders take substances and stop before a big event and consider that not cheating (Marty et al. 2015).

Sanctions for doping

If a rider intentionally cheats, regardless of the substance, then they face four years' ineligibility from competition (WADA 2021), missing an entire Olympic Games schedule. If their doping was unintentional then the sanction is two years' ineligibility, and this can be reduced to a minimum of one year's ineligibility. If the doping violation involved a specified substance or contaminated product and the athlete can demonstrate that they had no significant fault, then ineligibility may range from two years to a reprimand.

Hiding and evading doping

Much of the story so far shows how cyclists go to great lengths to hide doping. There are a variety of ways of doing so, including: urinating before the sample; injecting saline to decrease the hematocrit level within just 20 minutes notice of an imminent test; adding a few grains of washing powder to urine (makes EPO undetectable); using substances such as EPO in the evening (meaning not positive by morning) when testers rarely visit (USADA 2021); dosing outside of competition; providing vague whereabouts information; not answering the door to testers; and colluding with the team doctors abusing the TUE system (Dimeo 2014). As Marty et al. (2015) note, at the elite level, doctors are a key part of supporting the rider in performance enhancement whilst also minimizing the risk of them getting caught. If an athlete evades sample collection, refuses, or fails to provide a sample after being notified that they have been selected for doping control then this counts as an ADRV, carrying the same sanction as a positive test for a prohibited substance does (WADA 2015). Three missed tests over 18 months also constitute a doping offense (WADA 2015).

Confessing doping

There have been many confessions of doping scandals (e.g. "Festina Affair" in 1998, "Operation Puerto" in 2006, David Millar in 2004, Floyd Landis in 2006, and Bjarne Riis in 2007, to name but a few). These narratives exemplify the omnipresence of doping, particularly EPO use, and occur in riders' books. In these, as Sefiha (2017) has interpreted, riders use many techniques such as distancing, education, and normalization of doping to repair their tarnished identity in the media and the public eye. A key position that emerges from such autobiographies is that confessing to doping is the morally right thing to do after years of cheating (Sandvik 2019). For more information and examples, please see the list of further reading.

The Lance Armstrong saga: "Pursuing the American dream"

This chapter would be incomplete without discussing Lance Armstrong, who won the TDF seven times between 1998 and 2005 but who was found later to have systematically doped throughout his career and stripped of his winning medals. Whilst Armstrong and his team manipulated doping tests on every level, he was nevertheless tested around 275 times in his career (USADA 2021). In 2013, he confessed and outlined much of the details in an Oprah Winfrey documentary. Apart from the extent of doping, one of the distinguishing features of Armstrong's case was that it was primarily dealt with via a police investigation and not the usual scientific testing regime that had dominated anti-doping narratives until that point (Hardie

2015). As Hardie (2015: 197) states: "his [Armstrong's] was a business model that changed the face of professional cycling, perfecting techniques of racing, doping, media management and being the vehicle, even the pawn by which cycling administrators globalized the sport". It was also found that when Armstrong tested positive, he was protected via the complicity of the International Cycling Union (Hardie 2015).

"Clean cycling" and doping in the future

Evidence suggests the widespread tolerance toward doping in elite cycling prior to the millennium is eroding (Fincoeur et al. 2018). However, whilst anti-doping efforts have been reasonably successful in the past two decades, history indicates that the horizon should be scanned for future alternative means of cheating, some of which are outlined below.

There is no agreed definition of what "clean" cycling means and entails – the general consensus though is that it is a rider who does not take products or engage in any of the techniques of cheating on the WADA list (Marty et al. 2015). There are now teams that publicly advocate a strong anti-doping culture, and which concentrate on fostering an environment where their riders can ride clean (Marty et al. 2015). A prominent example is the former Team Sky (now Ineos Grenadiers), led by Dave Brailsford, that had an established rule that any rider with any ADRV would not be invited to join the team (Dimeo 2014). The spirit of this portrays an important message that cheating is not necessary for success in cycling, though many argue the caveat that Team Sky had access to some of the best available legal enhancements (Dimeo 2014). Despite this, and as Englar-Carlson et al. (2016) have rightfully noted, anti-doping efforts will never catch all cheaters and anti-doping cannot test its way to clean cycling. Englar-Carlson et al. (2016) have also argued that a revitalized approach is necessary whereby prevention and education measures raise the profile of clean sport to the extent that doping is no longer viewed as either the norm or an acceptable part of cycling.

Mechanical doping

Innovation has always been integral to cycling history. Speculation that cyclists may now be using machines concealed in their bikes (termed "mechanical doping") are being contended as a new form of cheating. Minuscule motors may be hidden in the bottom bracket area, or magnets hidden in wheels which help propel the crank arms and add around 50 watts to a rider's pedal stroke, enough to produce a race-winning result (Cycling News 2018). The first instance of mechanical doping was confirmed at the World Cyclo-Cross Championships by European Champion Femke Van Den Driessche, where a motor was found in her spare pit bike (Fotheringham 2016). One of the reasons speculated for these developments toward a more "technical cheating" has been the result of successful improvements in traditional anti-doping (Marty et al. 2015).

Genetic therapy/engineering

Recently, the principles and practices of gene therapy have been noted for their potential abuse as a doping technique. There is now the possibility of manipulating human genetic materials and regulating gene expression to increase or decrease the production of certain enzymes and proteins (Cantelmo et al. 2020). In cycling, it could be used to increase a rider's production of proteins such as EPO (Bird et al. 2015). As it stands, anti-doping agencies have not yet approved a test for this type of doping (Cantelmo et al. 2020), and whilst there are no recorded cases of gene doping thus far in cycling, it is an important addition to the WADA list of banned substances (Bird et al. 2015).

Conclusion

This chapter has shown that professional cycling is a difficult occupation to work in, particularly owing to the sport's commercialization (Chapter 27). Much of the narrative of the chapter has highlighted, however, that regardless of cycling authorities stepping up and advancing doping controls, the effects of this can be questionable (Dimeo 2014). There is a pattern in play whereby riders who dope are usually one step ahead of the detection techniques (Vlad et al. 2018) and some confess at some later stage in their career. The chapter also reviewed some of the newer techniques of doping that could come into play. Having outlined these, it is difficult to argue that cycling will not follow a similar pattern of doping followed by confessions in the future. Vlad et al. (2018) argue that, since doping is the result of the human, social and economic pressures on professional sports, the end of doping, especially in cycling, is sadly an unrealistic proposition.

References

BBC. (2018) Sir Bradley Wiggins says he '100% did not cheat' after damning MP's report. Last accessed 19th May 2021, https://www.bbc.co.uk/sport/cycling/43293645

Bird, S., Goebel, C., Burke, L., and Greaves, R. (2015) Doping in sport and exercise: Anabolic, ergogenic, health and clinical issues, *Annals of Clinical Biochemistry*, 53(2): 196–221.

Brewer, B. (2002) Commercialization in professional cycling, *Sociology of Sport Journal*, 19(3): 276–301.

Brissonneau, C. (2015) 'The 1998 Tour de France; Festina, from scandal to an affair in cycling'. In *Routledge Handbook of Drugs and Sport*, edited by Møller, V., Waddington, I., and Hoberman, J., 181–192. Abingdon: Routledge.

Cantelmo, R. A., Da Silva, A. P., Mendes-Junior, C. T., and Dorta, D. J. (2020) Gene doping: Present and future, *European Journal of Sport Science*, 20(8): 1093–1101.

Cycling News. (2018) Mechanical doping: A brief history. Last accessed 19th May 2021, https://www.cyclingnews.com/features/mechanical-doping-a-brief-history

De Hon, O., Kuipers, H., and Van Bottenburg, M. (2015) Prevalence of doping use in elite sports: A review of numbers and methods, *Sports Medicine*, 45(1): 57–69.

Dimeo, P. (2014) Why Lance Armstrong? Historical context and key turning points in the 'cleaning up' of professional cycling, *The International Journal of the History of Sport*, 31(8): 951–968.

Englar-Carlson, E., Gleaves, J., Macedo, E., and Lee, H. (2016) What about the clean athletes? The need for positive psychology in anti-doping research, *Performance Enhancement & Health*, 4(3–4): 116–122.

Fincoeur, B., Henning, A., and Ohl, F. (2020) Fifty shades of grey? On the concept of grey zones in elite cycling, *Performance Enhancement & Health*, 8(2–3): 1–9.

Fincoeur, B., Cunningham, R., and Ohl, F. (2018) I'm a poor lonesome rider. Help! I could dope, *Performance Enhancement & Health*, 6(2): 69–74.

Fotheringham, W. (2016) Rider implicated after motor found on bike at world cyclo-cross championships. Last accessed 19th May 2021, https://www.theguardian.com/sport/2016/jan/30/hidden-motor-bike-world-cyclo-cross-championships

Hackney, A. (2017) *Doping, Performance-Enhancing Drugs and Hormones in Sport*. Chapel Hill: Elsevier.

Hardie, M. (2015) 'Lance Armstrong'. In *Routledge Handbook of Drugs and Sport*, edited by Møller, V., Waddington, I., and Hoberman, J., 193–205. Abingdon: Routledge.

Healey, J. (2013) *Doping and drugs in sport*. Thirroul: The Spinney Press.

Henning, A., and Dimeo, P. (2015) Questions of fairness and anti-doping in US cycling: The contrasting experiences of professionals and amateurs, *Drugs, Education, Prevention and Policy*, 22(5): 400–409.

Henning, A., and Dimeo, P. (2018) The new front in the war on doping: Amateur athletes, *International Journal of Drug Policy*, 51(1): 128–136.

Houlihan, B. (2015) 'The future of anti-doping policy'. In *Routledge Handbook of Drugs and Sport*, edited by Møller, V., Waddington, I., and Hoberman, J., 249–260. Abingdon: Routledge.

Hunt, T., Dimeo, P., and Jedlicka, S. (2012) The historical roots of today's problems: A critical appraisal of the international anti-doping movement, *Performance Enhancement & Health*, 1: 55–60.

Lentillon-Kaestner, V. (2013) The development of doping use in high-level cycling: From team-organized doping to advances in the fight against doping, *Scandinavian Journal of Medicine & Science in Sports*, 23(2): 189–197.

Lodewijkx, H. (2014) The EPO fable in professional cycling: Facts, fallacies and fabrications, *Journal of Sports Medicine and Doping Studies*, 4(3): 141–154.

Lopez, B. (2015) 'Drug use in cycling'. In *Routledge Handbook of Drugs and Sport*, edited by Møller, V., Waddington, I., and Hoberman, J., 89–103. Abingdon: Routledge.

Marty, D., Nicholson, P., and Haas, U. (2015) Cycling independent reform commission: Report to the president of the Union Cycliste Internationale. Last accessed 19 May 2021, https://www.sportsintegrityinitiative.com/wp-content/uploads/2016/03/CIRCReport2015_Neutral.pdf

Mignon, P. (2003) The Tour de France and the doping issue, *The International Journal of the History of Sport*, 20(2): 227–245.

Møller, V. (2008) *The Doping Devil*. Copenhagen: Books on Demand.

Møller, V., and Dimeo, P. (2014) Anti-doping – The end of sport, *International Journal of Sport Policy and Politics*, 6(2): 259–272.

Overbye, M., Knudsen, M., and Pfister, G. (2013) To dope or not to dope: Elite athletes' perceptions of doping deterrents and incentives, *Performance Enhancement & Health*, 2(3): 119–134.

Petroczi, A., and Aidman, E. (2008) Psychological drivers in doping: The life-cycle model of performance enhancement, *Substance Abuse Treatment, Prevention and Policy*, 3(1): 1–12.

Sandvik, M. (2019) The confession dilemma: Doping, lying and narrative identity, *Sports, Ethics and Philosophy*, 13(2): 213–226.

Schneider, A. (2006) Cultural nuances: Doping, cycling and the Tour de France, *Sport in Society*, 9(2): 212–226.

Sefiha, O. (2017) Riding around stigma: Professional cycling and stigma management in the "clean cycling" era, *Communication & Sport*, 5(5): 622–644.

Sefiha, O., and Reichman, N. (2016) When every test is a winner: Clean cycling, surveillance, and the new preemptive governance, *Journal of Sport and Social Issues*, 40(3): 197–217.

Smith, C. (2017) Tour du dopage: Confessions of doping professional cyclists in a modern work environment, *International Review for the Sociology of Sport*, 52(1): 97–111.

Soule, B., and Lestrelin, L. (2011) The Puerto affair: Revealing the difficulties of the fight against doping, *Journal of Sport and Social Issues*, 35(2): 186–208.

USADA. (2021) Consequences imposed on license-holders as result of anti-doping rule violations (ADRV) as per the UCI Anti-Doping Rules (ADR). Last accessed 18 April 2022, https://www.usada.org/wp-content/uploads/ReasonedDecision.pdf

Vandeweghe, H. (2016) 'Doping in cycling: Past and present'. In *The Economics of Professional Road Cycling*, edited by Van Reeth, D., and Larson, D., 285–311. New York: Springer.

Vlad, R., Hancu, G., Popescu, G., and Lungu, I. (2018) Doping in sports, a never-ending story? *Advanced Pharmaceutical Bulletin*, 8(4): 529–534.

WADA. (2015) Anti-doping textbook. Last accessed 19th May 2021, http://antidopinglearninghub.org/sites/default/files/supporting-material/Anti-Doping%20Textbook%20-%202015%20Code.pdf

WADA. (2021) World Anti-Doping Code 2021. Last accessed 19th May, https://www.wada-ama.org/sites/default/files/resources/files/2021_code.pdf

Waddington, I., and Møller, V. (2019) WADA at twenty: Old problems and old thinking? *International Journal of Sport Policy and Politics*, 11(2): 219–231.

Yesalis, C., and Bahrke, S. (2002) History of doping in sport, *International Sports Studies*, 24(1): 42–76.

Further resources

"Dopeology": Online resource containing all doping cases in European professional road cycling since 1980, listing the riders, substances, and techniques based on reliable press sources. https://www.dopeology.org

There are many books and resources that share confessions and tales of doping use amongst professional riders. The two below are particularly worth reading:

Hamilton T and Coyle D (2012) *The secret race. Inside the hidden world of the Tour de France: Doping, cover-ups and winning at all costs*. Bantam: New York.

Tyler Hamilton was a teammate and confidant of Lance Armstrong. His memoir details the doping practices (e.g., EPO and blood transfusions) that he used with Armstrong on the U.S. Postal Service Team. He was also given an eight-year ban for doping.

Millar D (2011) *Racing through the dark*. Touchstone: New York.

Written by David Millar, this book illustrates his development as a professional cyclist prior to his doping. Unlike similar texts, Millar carries on the journey and discusses his development as a clean rider and key advocate in the fight against doping.

United States Anti-Doping Agency – Lance Armstrong and US Postal Service Pro Cycling Team, Reasoned Decision

Provides the overwhelming evidence of over 1000 pages and 26 testimonies (including 15 riders) in the Lance Armstrong formal police investigation – https://www.usada.org/wp-content/uploads/ReasonedDecision.pdf

PART VI

Places of cycling

An introduction

Luis Vivanco

As many observers have noted, one of the interesting and important things about bicycles is how they elicit and enable the expression of powerful human universals. Throughout their history and across the world, bicycles have been experienced as a physical manifestation of individual freedom and independence; enabled the embodied pleasures of effortless speed; expanded human potential through athletic achievement; symbolized ideals of industrial modernity and technological progress; and channeled aesthetic appreciation for beautiful objects. But "the rubber meets the road" somewhere, as they also say, which is a convenient bicycle-related metaphor that recognizes that the meanings of bicycles and the practices involved in riding them are also always grounded in particularities of place, infrastructure, community, and time.

The purpose of this part is to explore bicycling as a place-based and socially-embedded activity, drawing on case studies from North America, Latin America, Europe, Africa, South Asia, and East Asia. How and why have certain places become closely associated with bicycling? What is the "sense of place" that bicycles enable, and how have bicycles connected distantly related places, through relationships of colonialism, travel and tourism, or patterns of economic globalization? How do local dynamics of landscape, urbanization, social hierarchy, politics, and community affect not just who cycles, but also cycling's spatial practices, cultural meanings, and everyday experiences? In addressing these questions, we highlight the relativity of the bicycle, the place-making and spatial dimensions of bicycling, and the historical, geographic, political-economic, and cultural patterns that contribute to how and why different groups around the world have appropriated and adapted bicycles in distinctive ways.

This part opens with two chapters exploring the bicycle's role in the construction of new geographic imaginaries during the late 19th and 20th centuries. As historian Robert McCullough observes in his synoptic overview chapter, "Cycling's Symphony of Place," cyclists have always been "geographically-minded," cultivating a distinctive awareness of surroundings, temporality, and distance. Early cyclists were acutely aware of how their rambles through the countryside were conducive to the creation of novel ways of seeing landscapes and cultural imprints on the land. They drew on new languages of sensibility, techniques of illustration, and photographic technologies to record and share their observations. According to McCullough, contemporary cyclists inherit the keen observational skills of cycling's pioneers, allowing them to confront the "dreary placelessness" of contemporary post-industrial landscapes and creatively reconnect with neglected and marginalized urban and suburban spaces.

DOI: 10.4324/9781003142041-45

In the chapter that follows, historian Duncan Jamieson focuses on the powerful allure of the bicycle journey as an end in itself. Throughout cycling history, bicycle journeys have been associated with sensations of freedom, excitement, exploration of the unknown, camaraderie, and self-realization. At the same time, the precise meanings of those sensations are unstable and dynamic. For example, though bicycle journeys were once a mirror for Anglophone exploration and imperialism and associated with attitudes of civilizational superiority, Jamieson asserts that contemporary meanings of bicycle journeys emphasize their ability to open oneself to cross-cultural encounters and understanding.

The remaining five chapters substantiate the importance of cross-cultural encounters and contexts for making sense of cycling's contemporary spatial and social specificities. The first of these chapters, one focused on Africa and the other on India, consider the complex intersections of local bicycle cultures with postcolonial social conditions, poverty, and national development priorities. In his chapter on the meanings and uses of bicycles throughout Africa, anthropologist Hans Peter Hahn emphasizes that the bicycle should not be understood as an icon of colonial oppression (the automobile has served that role), but rather as a practical and malleable technology appropriated and ingeniously transformed by local communities to meet specific individual and collective transportation tasks, especially in rural communities. The "Africanized" bicycle designed and modified for local use demonstrates the enduring creativity of local cultures to redefine what a bicycle will look like, be used for, and whose needs it will fulfill. It also confirms two of the central points we hope to make in this part. One of these is the truism that distinct times and places value the same object differently. The other is that while bicycles have been and continue to be disruptive technologies, they also do not enter into social or historical vacuums and passive communities. People are just as likely to disrupt the technology itself.

In their chapter on cycling in Indian cities, urban and transportation planning scholars Rutul Joshi and Jacob Baby assess the emphasis in current Indian planning and policy on themes of transportation sustainability and the promotion of non-motorized transport and bicycle use to meet national carbon reduction goals. But these policies confront patterns of urban development that are based on a view of the bicycle as an illegitimate form of everyday transportation. The result is that urban bicycle users – many of them India's impoverished rural-to-urban migrants – have been pushed to the dangerous edges of roads where risk of death and injury are high. At the same time, interest in recreational cycling among India's urban upper-middle class has recently exploded, not just revealing stark differences between these different users in terms of their relationship with bicycles, but also raising concerns over whether future transportation development plans will be equitable, producing safe and accessible accommodations for all and not just those with greater political and economic access.

Each of the three remaining chapters examines places recognized for their iconic associations with bicycles – China, Denmark, and Colombia – giving us a set of perspectives on how and why certain societies have gained that symbolic distinction as well as the nuanced cultural politics that persist around bicycles. In his chapter, historian Xu Tao explains that popular interest in bicycles in China was slow to develop at the turn of the century; as he humorously observes, the story of the bicycle in China is like a suspense movie with many wonderful ups and downs as the European technology, at least initially, collided with the country's cultural and political histories. But by mid-century, skepticism had melted toward the bicycle among the middle class, and the government of Mao Zedong realized that the significant problem of mass transportation could be solved through widespread everyday bicycle use. It ramped up an industrial system that eventually, through various twists and turns, came to dominate global bicycle production today. Although access to automobiles has grown dramatically in recent decades and created new challenges for cycling, Xu emphasizes that the popularity of new bicycle technologies – electric bikes and bike sharing systems, especially – are generating new energy and interest, ensuring that the country's status as the "Kingdom of Bicycles" persists.

The story of Denmark and its capital Copenhagen that follows offers intriguing echoes of the China story. As urban planning scholar Malene Freudendal-Pedersen writes, formal governmental commitment and investment in making cycling an everyday practice have contributed heavily to making it a normal and unremarkable activity, even as the increasing automobilization of the Danish transportation system creates new pressures on cycling. She shows how carving out a formal place on the road is transformative for growing cycling and enabling cyclists to feel like they are, as some have reported to her, part of the "organism of the city." But protecting what's been achieved and the possibility for cycling's expansion are highly politicized and by no means inevitable, even in a country with a declared commitment to bicycle transportation.

In my own chapter on Bogotá, Colombia – the self-declared "World Bike Capital" – I draw on ethnographic fieldwork among everyday bicycle riders and bicycle transportation activists to consider what it means to ride a bicycle in a city full of contradictions. Bogotá is famous for its weekly Ciclovía (open-streets) program and is home of the largest cycle track system in the Americas, which have facilitated the expansion of recreational and everyday riding in the city. But getting around the city by bike is full of challenges, dangers, and inconveniences. The result is that cycling's pleasures and opportunities intersect in dynamic and complicated ways with embodied experience of the city's uneven and fragmented spatial environment, the political-economic and social inequalities that play out on city streets, and cultural perceptions and ideologies of urban public space as fearful.

Readers will find that juxtaposing any two or three chapters here will yield numerous themes that are ripe for connection and comparison: The importance of accessing cycling's complexities through stories and narratives. The emotional and embodied relationships cyclists create with specific places. The heterogeneity, not just of who uses bicycles and how they use them, but also of the diversity of bicycle forms that they create and mount. The local ambivalences and contradictions associated with the bicycle. The non-inevitability of the bicycle's adoption in certain places, in spite of its exciting potential; but also of its demise, in spite of predictions that it will disappear in the face of mass automobilization. I can go on, but it's probably best to just let the rubber meet the road.

Vignette H
EARLY CYCLING IN THE BOIS DE BOULOGNE, PARIS

David V. Herlihy

Paris is arguably the birthplace of the bicycle (a pedal-powered two-wheeler), and no part of the city is more closely connected to that vehicle's early history than the leafy Bois de Boulogne on its western edge. This vast playground was created in 1852 by the ruling monarch, Emperor Napoleon III. Fifteen years later, when his regime was at its apogee, hosting the Universal Exhibition, *Boulogne* became the training ground of the "vélocipédistes de la première heure" ("The first-hour bicyclists"). And it would remain the epicenter of the Parisian velocipede scene throughout the so-called "boneshaker" era.

To be sure, its sister park, the *Bois de Vincennes*, on the opposite end of the city, would also be invaded by bicycles in the same timeframe. It, too, offered broad, smooth alleys that accommodated these cumbersome vehicles with iron frames and tires, weighing around seventy pounds. Indeed, in September 1868, *Vincennes* was scheduled to host a novel women's race.[1]

But it was *Boulogne* that would become the sport's true hotbed. The park, after all, was renowned as the place where fashionable Parisians went to show off their latest acquisitions. Moreover, it was within convenient striking distance of the pioneer bicycle company, Michaux, on the Avenue Montaigne off the Champs-Elysées, about 1.5 miles to the east.

In August 1867, *Le Sport* reported that a small group of prominent men – including its own contributor 28-year-old Georges d'Orgéval and the 56-year-old Comte de Nieuwerkerke, the Emperor's art minister – were gathering near the park's artificial cascade every morning to teach themselves the challenging new art of riding a bicycle.[2] Large crowds gathered to watch their friendly competitions and to guffaw at their occasional spills. The spectacle was so novel and curious that numerous foreign newspapers and magazines picked up *Le Sport*'s paragraph, giving the bicycle its first extensive international exposure.[3]

Henry Michaux, a son of Pierre Michaux, the blacksmith who was overseeing the bicycle operation, would credibly claim in 1893 that he had regularly led the Cascade group from the Michaux shop to *Boulogne* (he would have been about 13 at the time). Indeed, most, if not all, of these early bicyclists were Michaux clients and the route from his family's workshop to the park was relatively short and smooth. Henry recalled an incident when the entourage propped up their bicycles along the park fence in order to patronize a nearby café, eliciting the strenuous objections of a guard.[4]

These pioneer riders promptly formed the *Véloce Club de* Paris, joining the skating and pigeon shooting clubs that were also based in the park. Its headquarters were at the Pré-Catelan,

DOI: 10.4324/9781003142041-46

an enclosed garden near the center of the park that was reserved for special seasonal events, complete with an administrative building and a grandstand. Its director, Théobald Saint-Felix, always eager to draw a crowd, looked favorably upon the new sport.[5] The club elected a president, the noted photographer Edouard Delessert, whose opulent residence in Passy was also conveniently located near *Boulogne*.[6]

The club announced plans for an inaugural race to be held that autumn from the Round Point at the Champs-Elysées to Saint Cloud, a western suburb of Paris, which presumably would have passed through *Boulogne*.[7] Although that event did not come off, on 8 December 1867 at least a hundred velocipedists rode from the Rond Point to Versailles. This remarkable "caravan" likely followed the originally planned route to Saint Cloud, via *Boulogne*.[8]

The following spring, rather than fade from view, the velocipede reappeared in *Boulogne* in even greater numbers. A correspondent with the British review *The Field*, in a letter dated 15 April 1868, reported on a *concours hippique* (horse show) held in the *Palais d'industrie* near the Michaux shop: "The carriage manufacturers had got up on one side of the *palais* a small show. The most interesting vehicle I saw there was Michaux's velocipede. There is quite a mania for it here. At the Bois de Boulogne, in the morning one may see quantities of them shooting along the well-macadamized roads."[9]

For all its success within a few short months, the *Véloce Club* – said to have grown to sixty members – had yet to stage an "official" velocipede race. The club announced that competitions would be held at the Pré-Catelan on Sunday 24 May, in conjunction with the "Rose festival."[10] The planned event, however, apparently fell through. The races held the following Sunday in the park of Saint-Cloud would thus go down as the first official velocipede races, sanctioned by the Emperor himself.

One possible factor for the annulation of the planned velocipede races at Pré-Catelan was a pronounced backlash against velocipedists from the park's administration. On 28 May, *Le Figaro* announced that velocipedes had been banned from the park after noon, due to complaints from terrified carriage drivers. "What will they do when bicycles become as common as umbrellas?" the paper scoffed, adding. "One of these days they will ban pedestrians."[11]

A week later the paper published a defensive response from park officials explaining that the ban was in fact due to "the fear that the carriages will knock down the velocipedists, or that the latter will injure pedestrians." The spokesperson cited the bicycles' "speed and the difficulty of stopping instantaneously," adding that the ban applied only to the most congested alleyways.[12]

The velocipede nonetheless maintained a strong presence in and around the park. That July one paper commented on the presence of female velocipedists "dressed as jockeys, undertaking steeple chases without horses, on the great avenues along the Bois de Boulogne."[13] By fall, the hippodrome abutting the park was holding regular velocipede races (Figure H.1). Meanwhile, a growing presence of the vehicle on theatrical stages helped to keep the novelty fashionable.[14] By early 1869, "velocipede mania" had spread to the United States and was beginning to make inroads in Great Britain and elsewhere. With the movement showing no sign of slackening, Saint-Felix announced velocipede races, "innovation" contests, and exhibitions to start that spring and run throughout the entire year.[15]

The opening races on 11 April drew a large crowd, but the velocipedists still had to contend with antagonistic park officials. The competitors had to transport their vehicles in cabs from the park entrances to the Pré-Catelan, and they were not allowed to exit the park on their wheels until 7 p.m.[16] One would-be entrant that day clashed with a park guardian on his way to the Pré-Catelan and was reportedly arrested, fined 300 francs, and sent to prison for three days.[17]

After several more velocipede events at the Pré-Catelan that spring and summer, the 1869 season closed in late October with an "international races" won by J. T. Johnson, an Englishman,[18]

Figure H.1 A velocipede race at the Pré Catelan, May 1868.
Source: *L'Univers Illustré*, 6 June 1868, p.348.

and a "retrospective" exhibition organized by the *Compagnie Parisienne des Vélocipèdes*, the successor to the Michaux company. The display included a kick-propelled "hobby horse" from 1818, and an early wooden bicycle from 1865 that had purportedly accomplished "long voyages."[19] The company also sponsored, a few days later, the first city-to-city road race from Paris to Rouen.[20]

In the spring and summer of 1870, the Pré-Catelan continued to host velocipede races until the outbreak of the Franco-Prussian War effectively ended the phenomenon known as *vélocipédomanie*, not to mention the Empire itself.[21] Nevertheless, despite the war's disastrous outcome from the French perspective, the sport would remerge in Paris in the early 1870s.

In July 1872, an organization called *Vélo-Sport* held races at the Pré-Catelan every Sunday. The winners, including women, were awarded "medals of gold, silver, and bronze."[22]

By this time, however, the second generation bicycle, the fleet but precarious "high wheeler," was gradually supplanting the original bicycle and changing the nature of the sport. The bicycle, in effect, had outgrown the park.

Notes

1 *L'Opinion nationale* of 6 September 1868 announced that the suburb of Charenton would sponsor races that day in the *Bois de Vincennes*. The program was to include two speed races, one slow race, and a "course en amazone," but it is uncertain whether the event came off.

2 *Le Sport*, 4 August 1867. *The London Evening Standard* of 1 August 1867 reported that the group numbered around twenty.

3 For example the (London) Morning Post of 5 August 1867 and *The Evening Star* (Washington, D.C.) of 22 August 1867.

4 *Le Vélocipède Illustré*, 4 July 1893.
5 *Le Vélocipède Illustré*, 1 April 1869.
6 *Le Figaro* of 7 October 1867 reported that Delessert rode every morning from Passy to Paris.
7 *Le Sport*, 4 August 1867.
8 *Le Petit Journal*, 10 December 1867.
9 *The Field*, 18 April 1868.
10 *Le Siècle*, 18 May 1868.
11 *Le Figaro*, 28 May 1868.
12 *Le Figaro*, 3 June 1868.
13 *L'Eclair*, 18 July 1868.
14 *L' Annuaire de la Société des auteurs et compositeurs* of 1868 notes that *Dagobert et son vélocipède* debuted on September 1, and *Paris-Velocipede* on December 28. A number of other plays opening also presented the velocipede on stage.
15 *Le Manuel du vélocipède*, le Grand Jacques [Richard Lesclide], 1869, pp. 48–50.
16 *L'Opinion Nationale*, 17 April 1869.
17 *Le Soir*, 5 May 1869.
18 *La Presse*, 2 November 1869.
19 *Le Vélocipède Illustré*, 18 November 1869.
20 *Le Vélocipède Illustré*, 11 November 1869. The winner, James Moore, covered the 70 miles in ten hours and forty minutes.
21 *Le Petit Journal* of 6 June 1870 announced a "Fête des Apprentis" ("apprentices' festival") on that day that was to include velocipede races.
22 *Le Journal de la ville de Saint-Quentin*, 21 July 1872.

32
CYCLING'S SYMPHONY OF PLACE

Robert McCullough

It were useless to attempt to tell the delight of a tricycle ride through a pleasant country, where Nature invites the eye to dwell upon her charms, when the whole body tingles with exhilaration born of quickened circulation and speedy movement through the air. To experience is to know. The half cannot be told.

Helen Drew Bassett (pseud. Daisie), "The Ladies' Eastern Tricycle Tour from the Merrimac to Naumkeag," Outing (December 1888)

… wending my way alone to West Medford, slowly riding up and over Winter Hill, I made a stop directly at the right and at the top of the hill. It is a favorite spot of mine, when not in a hurry, to look over the Mystic Valley and recall what scenes have taken place there since it had been ground out by the great ice-floe that scratched out its course over the entire New England states during the glacial period. Land-marks have an awe-inspiring history, which is gradually more and more unfolded… whether they have been left by nature, or the agency of man.

William S. Beekman, "A Geological Ride," in *Cycle Gleanings, or Wheels and Wheeling for Business and Pleasure and the Study of Nature* (1894)

Awheel: our senses of place, time, distance, memory, and now

Adventurously athletic travelers *awheel* during the 1880s, perched atop elegant bicycles with grand front wheels, pedaled boldly beyond commercial avenues or suburban streets and into open countryside, hoping to observe new places from a distinctive perspective and pace. These cyclists imagined the experience to be similar to flying – a quiet glide past startled onlookers – and the impression was vivid enough for members of the League of American Wheelmen, or LAW, to adopt a winged wheel as their insignia.

When riders on safety bicycles with rear-wheel chain-drive and pneumatic tires on wheels of equal size overtook high-wheelers, cycling's tourists multiplied. Wheelmen and wheelwomen traced out new vicinages among the shifting margins between urban and rural, the two often juxtaposed in cycling's narratives during a dynamic era when expanding industrial or commercial interests collided with pastoral ideals. Always seeking paths to scenic or unusual locale but usually tethered in some way to the city, these cyclists surveyed corridors that can

DOI: 10.4324/9781003142041-47

tell us much about the murky, sequential growth occurring at urbanity's edges. Unaware, these cyclists also recorded impressions of landscapes soon altered irreversibly by the automobile.

Having mastered a newly engineered means of unbounded independence, whether astride high-wheel or safety bicycles, cyclists from that era became discerning landseeërs. A few notable writers have altered the meaning and spelling of the word *seer* to *seeër* – a person whose sense of sight is penetrating rather than mystic, and who achieves a clarity of awareness, whether through contemplation or imagination (Stevenson 1882). The word *landseeër* is particularly apt for 19th-century cyclists such as Beekman, who deciphered not only the types of literal landmarks described in his geological journeys, but also cultural imprints on the land. In a series of articles for *Bicycling World*, Beekman offers architectural critique, opining that the costly new houses he encounters lack the dignity of older landmarks. He also urges cyclists to "use the wheel to see what there is around" and to scan and compare "the peculiarities of each town" on the road to becoming practiced observers (Beekman 1895).

Passages such as those by Beekman or Bassett abound in cycling's great catalog of journals. Bassett, who used the pseudonym "Daisie" to pen a column titled "From a Feminine Point of View," describes a segment of the ladies' North Shore tricycle tour through Essex Woods in Massachusetts, as a "poem in cycling," with the senses of travel and view and place accentuated by sun-dappled shade beneath a bower of arching trees, all garnished by autumnal foliage. She also encouraged wheelwomen to "strike out into new territory, explore new fields of observation, try unfamiliar paths" (Bassett 1886, 1888). Many other cyclists expressed the sentiments of unconstrained mobility in simple landscape prose – *the paradise of open country*.

Surprisingly, most of the geographers, historians, and writers who have shaped human awareness of our landscapes have overlooked cycling's contributions. Among such figures, several are renowned. Carl Sauer, who forged cultural geography into an important branch of science, acknowledged the value of exploratory travel to the study of historical geography. In a 1925 essay, "The Morphology of Landscape," Sauer points to the earliest forms of geographical literature – the sagas and myths of humankind's contests with nature – and proffers the phrase *sense of place* to characterize chorology, or areal knowledge. He also stresses that inquiries fundamental to historical and cultural geography are quests for origins and knowledge – a rigorous aspect of science and not a form of antiquarianism. Yet he confesses, too, that certain aspects of landscapes, their symphonic quality, might lead to a level of understanding beyond the reach of science – the types of exquisite settings that captivated Daisie and her companions (Sauer 1925).

Although in that same essay Sauer laments the absence of exploratory zeal from his own generation, he neglects to consider cyclists' straightaway journeys and travel narratives compiled during the 1880s and 1890s. The global trek by Thomas Stevens, beginning in 1884 astride a Columbia high-wheel bicycle, became a venturous quest dominated by Stevens' struggle to overcome nature's forces, especially in America's western territories – precisely the type of adventure saga that Sauer assigns to much earlier periods. Equally important, Stevens' narratives provide vivid descriptions of the American scene and belong in any catalog of nature-culture writing or geographical exploration (Stevens 1887).

A generation younger than Sauer, landscape historian J. B. Jackson also studied cultural texts evident on the land, but he chose to read the chapters of ordinary as well as extraordinary experience. The inaugural issue of *Landscape*, the journal he founded in 1951, proclaims: "A rich and beautiful book is always open before us. We have but to learn to read it." As well, Jackson imparted added meaning to landscapes by accentuating the ability of place to harbor memory or shared experience, *a sense of time*, in turn prompting the impulse to protect traces of that shared experience and, in the process, helping to broaden the environmental movement to include cultural as well as natural resources (Jackson 1951). As did Sauer, Jackson pushed the boundaries of landscape-related studies far beyond emphasis on purely aesthetic qualities; he also studied roads, the avenues through which change finds its way onto the land. Although he astutely recognized the significance of such corridors in a multitude of varied contexts, the lexicon for roads that he offered in an early issue of *Landscape* would have benefitted greatly from the road books circulated by the LAW after 1885 (Jackson 1952). Few people were as well informed about roads as cyclists during that period, and the editors of these pocket-sized travel companions recorded information that exists nowhere else in compiled form: road types, grades, surfaces, long climbs, steep descents, sidepaths, and adjoining landmarks.

In addition to providing a body of descriptive writing, imagery and other geographic evidence that informs understanding of our surroundings, cyclists' contributions to landscape-related studies often address the temporal aspects that Jackson and other writers have attached to land places, but without attribution. For instance, in his study, *What Time is this Place*, urban planner Kevin Lynch seeks a means to dispose of the obsolete past in our environments, substituting instead a temporal collage that combines both past and present to create what he describes as a *sense of now* (Lynch 1972). Similarly, in *The Past is a Foreign Country*, historical geographer David Lowenthal points to the implacable forces of change. Yet cyclists have long understood that perceptions of time and distance when considered from a unique point of view – above the handlebars – are influenced greatly by the pace of travel. Thus, the temporal collage Lynch commends is likely to be very different from that envisioned by cyclists, whose awareness of surroundings, sense of time, sense of obsolescence, and sense of place is easily lengthened in both duration and distance. As we confront change and contemplate choices that may alter our perceptions of time, distance, space, and landscape – or the invisible but redolent weave of memory evoked by such places – the dimensions of our choices loom large.

Other theorists have also explored the relationships between the pace of travel and perceptions of time, distance and change without regard to the bicycle. For example, in *The Railway Journey*, Wolfgang Schivelbusch explains how the development of railroads compressed our concepts of spatial expanse and time, and he describes the experiential quality of travel as an emotion created by a continuous sequence of impressions formed during the traverse of space (Schivelbusch 1977). Yet he simply restates what one of cycling's quixotic itinerants, Karl Kron (born Lyman Hotchkiss Bagg), explained almost a century before in his encyclopedic diary of travel, *Ten Thousand Miles on a Bicycle*, observing:

> A tourist on foot moves too slowly to see the country on a grand scale; a tourist by train moves to swiftly to see the individual significance of any particular features of it; and a tourist on horseback or in a carriage would probably find more physical pain than intellectual pleasure.
>
> *(Kron 1887)*

Kron continued to ride a high-wheel bicycle long after safety bicycles had rendered the former redundant, explaining to a Canadian reporter that his high-wheel provided a better perch from which to view the countryside, especially "on a level above the hedges" ("Why Kron Rides the Ordinary" 1904).

Similarly, in Warren Belasco's work, *Americans on the Road*, the author explains how automobiles altered American mobility and geographic consciousness, offering new experiences of motion, time, and distance; releasing long-suppressed desires of vagabondage; and fostering nostalgia for picturesque villages (Belasco 1979). Yet most of the cultural shifts that Belasco attributes to the automobile had begun three decades prior and on two wheels rather than four. Writer and photographer Wallace Nutting's views from the road, first by bicycle, then by carriage, and eventually by automobile, also chronicle that important transition and the changing attitudes about visual quality in American landscapes (Nutting 1936). His peregrinations began in 1897, but cycling's geographic explorers already were adept with cameras, fostering a market for photographic kits for high-wheel bicycles during the early 1880s (Figure 32.1). Cyclists also transformed nostalgia into heritage tourism at least forty years before America's historic preservationists recognized the economic value concealed behind aging architectural facades (Carrington 1896).

Landscape historian John Stilgoe understands the extraordinary relationship between pace of travel by bicycle and the ability to observe surroundings that remain invisible to most people (Chapter 48). In *Outside Lies Magic*, a little book that brims with observation, Stilgoe detects that pedestrians moving past a picket fence see only narrow glimpses of features located behind the fence, but a bicycle rider moving at a careful pace mentally collects those images rapidly enough to assemble a complete picture (Stilgoe 1998). Coupled with the capacity to lengthen a sense of place, time and memory over linear distances, cyclists' ability to notice, recall, and connect land features that are unseen by others holds considerable present value, especially when cyclists become fluent at reading the cultural imprints that shape our landscapes. As those cyclists explore the forgotten and often-neglected places in our urban and suburban environments – the nearby unknown – creative opportunities to connect and reclaim such

Figure 32.1 Photographic Outfits. *Outing* 7 (October 1885): 129. (Author's collection).

places via the narrow traces best traversed by bicycle are bound to occur. In doing so, cyclists can thus confront placelessness, the dreary voids that pervade our environments and reveal the often-destructive relationships between humanity and its environments.

Urban exploration, or in today's vernacular *roof-and-tunnel hacking*, has enjoyed surging popularity, but the lure of discovery in the nearby unknown is also far from new. Cycling's 19th-century wanderers often traveled close to home, and Charles Pratt, one of the founding members of the LAW, tutored cyclists that although distance lends enchantment, true enchantments are not all distant, wagering: "Ten to one you, Reader, unless you be a wheelman, do not know your own county" (Pratt, February 1883a). Another of cycling's urban roamers, Jay Howe Adams, ventured into Philadelphia's mysterious nether regions, a lowland area called "The Neck" formed by the confluence of the Delaware and Schuylkill rivers. There, poor market farmers and squatters eked out a meager existence in old stone houses or shanties built close to the road, many within view of coal docks, grain elevators, oil tanks, refineries, and gas works. Rather than expressing disdain for these *Neckers*, Adams envies their seclusion and accepts the trappings of industry with objectivity (Adams 1885).

Cycling's illustrators

Cycling's journalistic narratives also benefitted from imagery of exceptional quality, which transported readers to imagined places. Cycling's photographs and wood-engraved illustrations helped to change attitudes about landscapes worthy of study by American artists, furthering the trend toward drawing or painting from nature during the closing decades of the 19th century, and turning attention toward ordinary as well as transcendent settings. Among cycling's many artists, F. Childe Hassam illustrated numerous issues of *The Wheelmen* during the early 1880s, a journal funded by bicycle manufacturer Albert Pope and edited by young S. S. McClure (Figure 32.2). Hassam, who became one of the country's celebrated Impressionist painters and one of the founders of an exhibition society, The Ten, worked with artist Edmund Garrett and engraver H.E. Sylvester for many of those issues during a period when American wood-engraved illustration excelled (Pratt, March 1883b).

In Philadelphia, artist Frank Hamilton Taylor and illustrator Joseph Pennell also gained renown. Taylor formed his own printing and publishing business and eventually worked as a special artist for publishers such as *Harpers Monthly*. His valuable work for cycling includes an 1896 publication, *Cyclers' and Drivers Best Routes in and Around Philadelphia*, for which he served as author, illustrator, cartographer, and photographer. Pennell, one of the founders of the Germantown Cycling Club in 1879, gained prominence as an illustrator after he and his wife, Elizabeth Robins Pennell, toured Europe by tandem tricycle and bicycle, publishing a series of successful tour books – she the author and architectural critic and he the illustrator of the cathedrals, abbeys and monasteries she studied (Pennell 1887; Chapter 48). Although he spent much of his life abroad, Pennell contributed to cycling closer to home for journals such as *Scribner's Monthly*, exploring Philadelphia's backwater regions and sketching oil refineries and other industrial sites – distant in both place and thought from that era's academic standards for art (Egan 1881).

After photography superseded wood-engraved illustration during the late 1880s and early 1890s, cycling's photographers exploited that medium enthusiastically. The increasing facility of cameras permitted compositional studies, some of which are thematically geographic, including a portfolio of sixty images by Cline Rogers, a professional photographer from Rochester, New York commissioned by the Monroe County Sidepath Commission to document the county's extraordinary network of bicycle paths (Figure 32.3). Unique in its focus and scope,

Figure 32.2 "By the Roadside," a wood engraving by H. E. Sylvester after a painting or drawing by F. Childe Hassam for *The Wheelman* (September 1883). (Author's collection).

the collection reveals two separate transportation networks leading outward from a major city at a time of rapid growth – a tableau of urban, suburban and rural in the midst of change (Rogers 1899).

Today's cyclists have inherited the skills of 19th-century landseeërs and are among our most discerning observers of natural and cultural imprints on the land – possessing an innate awareness gathered from vantage points that are remote to those who move about by other means: the quiet, carefree and car-free passages; the sheltered routes on windy days; the few summits where a breeze might be circulating on a sultry, summer evening; the precise distances to country stores with adequate food; the best vistas to reward a long climb; and the knowledge that the grade of a steep hill always seems more vertical from the hill opposite than it does at the base of the climb – a useful reminder when cyclists are ticking off the number of climbs left at the end of a long afternoon's ride.

Cycling's geographers

The distinctive melding of sense of place, time, memory and distance that shapes cyclists' perceptions of our built, cultural, and natural environments represents only one aspect of cycling's contributions to landscape-related studies. Equally important, the substantial corpus

Figure 32.3 Scottsville Sidepath, Monroe County, New York. Photograph by Cline Rogers, 1899. Courtesy Rochester (N.Y.) Public Library, Local History and Genealogy Division.

of books, periodical literature, road guides, maps, photographs, illustrations, travel narratives, and perceptive reflections amassed by cyclists belongs among the many historical disciplines that touch landscape studies. In many ways, that body of work establishes cycling as its own subset of geography, much in the same way that railroads and automobiles created parallel bodies of land study.

Cycling's road books and maps are important but woefully neglected parts of this geographical archive. Carl Sauer allowed that the most precise expression of geographic knowledge is found in the map, "an immemorial symbol" (Sauer 1925), and some of cycling's compilations are original in conception – specifically designed for cyclist's unique exploratory outlook. Unlike earlier gazetteer-like guidebooks with narratives describing well-travelled circuits, cyclists' treks required the ability to find one's way along often unnamed and unmarked roads, making cyclometers essential. Not surprisingly, the travel experience differed as well, changing the relationship between cyclist and his or her surroundings and forcing riders to discern land features that other travelers confined to route-bound transport overlooked.

Accurate depiction of landmarks thus became essential, and the LAW developed its own genre of road books, beginning with a cycler's companion by Charles Pratt, titled *American Bicycler*, that delineated specific routes with careful directions (Pratt 1879). The league soon asked local consuls to plot specific routes, verify mileage, and send the data to the league's national headquarters for circulation to members. To improve access to routes among the league's rapidly growing constituents, state divisions appointed road book committees and began publishing guides in bound, pocket-sized form (barely larger than today's phones) after a league consul from Philadelphia, Henry Wood, devised a method for organizing route data on a single page in horizontal format, a style called *tabulated routes* (Pennsylvania 1886).

Hotel or Restaurant.	POINTS ON ROUTE 60	Total Distance from Start. Miles.	Distance Between Points.	Material of Road.	Grade of Road.	Condition of Road at its best.	Turns, Forks, General Instructions. T. L.=turn to left. L. F.=left fork. T. R.=turn to right. R. F.=right fork. T. P.=telegraph poles. X R.=cross roads. See also "System of Abbreviation" facing Route (1).
Wyoming Valley H.	WILKESBARRE	0.00	2.00	clay path or pike	1	A2	About 2½ m. to walk in going to the Lake.
	KINGSTON	2.00	2.00	clay path or pike	3	A2	
	MILL HOLLOW	4.00	2.00	clay path or pike	3	A2	Stop at Ice Cave.
Ice Cave H.	ICE CAVE	6.00	3.00	clay path or pike	3	A2	
	TRUCKSVILLE	9.00	5.00	clay path or pike	3	A2	Chicken and waffles at Raub's.
Raub's	DALLAS	14.00	4.00	clay path or pike	3	A2	Good bass fishing at Lake.
Rhodes'	HARVEY'S LAKE	18.00					Fine coasting on the return.
	ALSO SHORT TRIP						Wilkesbarre to White Haven 18 m. direct, via Laurel Run 3½ m and Bear Creek 9¼ m.
Wyoming Valley H.	WILKESBARRE	0.00	2.00	clay	1	A2	Cross Susquehanna.
	KINGSTON	2.00	5.00	clay	1	A2	North along river.
Leacock's	WYOMING	7.00	5.00	sidepath	2	A3	From Wyoming House northwest to Drake's Mill P. O. T. R. to Camp.
Boarding H	M. E. CONFERENCE CAMP-MEETING	12.00		74			Fine view of Wyoming and Lackawanna Valleys from Lookout Point near Camp.

Figure 32.4 LAW Pennsylvania Division, L.A.W., Road Book of 1893, Tabulated Route 60. (Author's collection).

The *Pennsylvania–New Jersey Road Book*, published in 1885, became the first to use Wood's system, and other state divisions adopted the same style, exchanging stereotype printing plates (Pennsylvania, August 1886) (Figure 32.4). Massachusetts and New York adopted Wood's method, and by 1887 the *New York Hand-Book and Road-Book*, compiled by Albert Barkman, president of the Brooklyn Bicycle Club, tabulated routes in New England, New Jersey, Delaware, Pennsylvania, Maryland, Virginia, and Ohio. Later printings for various states included pockets for larger, folded maps. Notably, both the style of tabulated routes and the title, *road-book*, distinguished those unique circulars from other LAW guides described as *hand-books*, which included gazetteer-like narratives and occasionally maps, but without the all-important route tabulations.

Although Wood's design proved to be ingeniously suited for field use with easily turned pages, the value of maps could not be ignored, particularly for cyclists whose exploratory impulses veered away from common routes. In Connecticut, league consul Charles Huntington rejected Wood's tabulated system after the number of routes and spur-routes in that state became too confusing, and he retained the J. B. Beers Company to prepare county maps showing all public roads as well as hills, railroads, town boundaries and a few other features. Huntington, who criticized Wood's system as little more than a *route book*, delineated the best bicycle corridors in red and coded them by number according to quality and grade, and he proclaimed Connecticut's 1888 *Cyclist's Road-Book* to be the country's first true road book. Although the maps are advantageous, they nevertheless require several folds, making them unwieldy.

Road books with serviceable maps became indispensable as bicycle touring increased during the 1890s, and most of the league's state divisions adopted elements of all three styles: tabulated routes; maps with routes accentuated by color or heavy line; and narrative description.

All the while, the league continued to improve formats, and New York, New Jersey and Pennsylvania developed two successful styles: *skeleton* and *sectional*. New York's Road Book Committee introduced a facile type of skeleton map, each page showing a single route with maps aligned vertically to correspond to cardinal compass points, and each map usually coupled with a second map on a facing page, thus making the entire route visible. Easy to read, and thus making color unnecessary, the maps show match lines very clearly, and mileage and cross-streets tally progress along each route.

In 1897, New Jersey's Road Book Committee hired Frank Hamilton Taylor to prepare individual maps for seventy-nine different routes together with four, coded index maps and an index of towns. By that time, United States Geological Survey maps were available, and Taylor relied on that work to draw maps in horizontal format only slightly larger than the pages of Wood's tabulated routes and requiring no folding. Taylor and the committee improved that format the following year by devising a fifth style of road book using section maps, which divided the state into numbered sections to scale, aligned the maps vertically in coupled format to show a complete section, depicted all the roads in each section rather than just single routes, and referenced adjoining sections in the margins. In Philadelphia, engineer and cartographer Carl Hering adopted a similar method for Pennsylvania's road book of 1898, devising a unique system for distinguishing road conditions without resort to color, and then coding varying terrain with symbols in the margins.

Landseeërs in a digital age

Cyclists have always been geographically minded, and as the popularity of touring bikes increased dramatically after 1970, route books of various types and formats began circulating. True, geographical study focused less on the surface of roads and general terrain than on the extent of car traffic and quality of scenery, but the experience of exploratory adventure that Carl Sauer found missing in American society has never waned among cyclists. Today, aided by sophisticated technology, cyclists using Strava and GPS data not only can map their own routes, upload and track fitness records, and monitor performance along the way, but also share that information instantly on social media, all from their phones. And, to encourage self-propelled travel on two wheels in the digital age, the United Kingdom's Global Cycling Network (GCN) *YouTube* channel sponsors weekly contests for inspirational photographs. Most winning images, likely taken with phones, are chosen from views of natural scenery, especially sunrises or sunsets, rather than significant built or cultural environments. Yet some cyclists who feel overwhelmed by digital data have wondered whether the weight of technology hinders the often fleeting, momentary awakening to our surroundings so common to cycling – or dulls the individual creativity that can guide exploratory choices. The advice of Abbot Bassett, offered more than a century ago, may help. Bassett, long-time secretary of the LAW, editor of its bulletin, husband to Daisie, and one of cycling's "Old Guard," urged wheelmen and wheelwomen to explore new territory, investigate nature's wonders, visit places of historic interest, and to tour with a worthy purpose – a goal as laudable today as then.

References

Adams, J.H. (September 1885) "Through 'the Neck' on a Bicycle," *Outing* 6, 682.

Bassett, A. (October, 1921) "Wheel About the Hub," *Bassett's Scrap Book* 18, 116.

Bassett, H.D. (June 18, 1886) "From a Feminine Point of View," *Cycle* 1, 213–24.

——— (June 18, 1888) "The Ladies' Eastern Tricycle Tour from the Merrimac to Naumkeag," *Outing* 13, 260–65, at 262.

Beekman, W.S. (1894) *Cycle Gleanings, or Wheels and Wheeling for Business and Pleasure and the Study of Nature*, Boston: Skinner, Bartlett and Co., 41.

——— (June 7, 1895) "Nature from the Standpoint of the Observant Cyclist," *Bicycling World* 31, 231–235.

Belasco, W.J. (1979) *Americans on the Road: From Autocamp to Motel, 1910–1945*, Cambridge: MIT Press, 15–17, 20–23, 26–29, 48–63.
Carrington, J. (June 1896) "Through Virginia Awheel," *Outing* 28, 204.
Connecticut State Division, League of American Wheelmen (1888), *The Cyclist's Road Book of Connecticut*, C. G. Huntington, ed., Hartford: Case, Lockwood & Brainard Co.
Egan, M. (June 1881) "A Day in the Ma'sh," *Scribner's Monthly* 22, 343–352.
Jackson, J.B. (Spring 1951) "The Need of Being Versed in Country Things," *Landscape* 1, 1.
——— (Autumn 1952) "A Brief Lexicon of Road Words: Together with a Selection of Topics Suitable for Meditation by the Earnest Taxpayer-Motorist," *Landscape* 2, 32–33.
Kron, K. (1887) *Ten Thousand Miles on a Bicycle*, New York: The author, 303.
Lynch, K. (1972) *What Time is this Place?* Cambridge: MIT Press, 65.
New Jersey State Division, League of American Wheelmen (1897), *Road Book of New Jersey*, F.H. Taylor, Mappist, (no pl., no pub.).
——— (1898) *Road-Book of New Jersey*, F.H. Taylor, Mappist, Philadelphia: Alfred M. Slocum, Co.
New York State Division, League of American Wheelmen (1887), *Hand-Book and Road-Book of New York: Containing Also the Principal Routes of Maine, New Hampshire, Vermont, Massachusetts, Rhode Island, Maryland, Virginia and Ohio*, A. B. Barkman, Ed., Philadelphia: Stanley Hart and Co.
Nutting, W.K. (1936) *Biography*, Framingham, MA: Old America Company, 70.
Pennell, J.E. (1887) *Two Pilgrims' Progress*, Boston: Roberts Brothers.
"Pennsylvania, New Jersey and Maryland Road Book" (August 30, 1886) *The Wheel* 10, 10.
Pennsylvania State Division, League of American Wheelmen (1898) *Road Book of Pennsylvania: Eastern Section*, C. Hering and L. Fay, Eds., (no pl., no pub.).
Pratt, C. (1879) *The American Bicycler: A Manual for the Observer, the Learner, and the Expert*, Boston: Houghton, Osgood and Co., 130–149.
——— (February 1880) "A Wheel Around the Hub," *Scribner's Monthly* 19, 481.
——— (February 1883a) "Echoes and Shadows," *The Wheelman* 1, 321.
——— (March 1883b) "Our First Bicycle Club," *The Wheelman* 1, 409.
Rogers, C. (1899) *Photograph Album of Sidepaths in Monroe County*, Rochester, NY: Monroe County Sidepath Commission.
Sauer, C. (1925) "Morphology of Landscape," *University of California Publications in Geography 2*, C. Sauer, Ed., 19–54, at 21, 48.
Schivelbusch, W. (1977) *The Railway Journey: The Industrialization of Time and Space in the 19th Century*, Berkeley: University of California Press, 52–69.
Stevens, T. (1887) *Around the World on a Bicycle 1*, New York: Charles Scribner's Sons, 46–69.
Stevenson, R.L. (1882) "A Gossip on Romance," *Longman's Magazine* 1, 79.
Stilgoe, J. (1998) *Outside Lies Magic. Regaining History and Awareness in Everyday Places*, New York: Walker and Co., 110.
Taylor, F.H. (1896) *Cyclers' and Drivers' Best Routes in and Around Philadelphia*, Philadelphia: Author.
"Why Kron Rides the Ordinary" (July 9, 1904) *Bicycling World and Motorcycle Review* 49, 468.

Further Reading

McCullough, R.L. (2015) *Old Wheelways. Traces of Bicycle History on the Land*, Cambridge: MIT Press. Portions of the above essay are drawn in thought and substance from this earlier work.

Vignette I
CONSTRUCTING PEACEFUL PLACES THROUGH BICYCLES

Jeanette Steinmann, Mitchell McSweeney, Lyndsay Hayhurst and Brian Wilson

For many cyclists, the bicycle is a source of inner peace, found when pedalling leisurely along a favorite trail. But is giving someone a bicycle and helping that person learn to ride safely also an act of peace? Is the world a safer and more peaceful place because of the transfer of bicycles from affluent societies to marginalized communities? These are questions that scholars in the burgeoning 'bicycles for development' (hereafter BFD) field are pondering. The United Nations (UN) hailed the bicycle as an effective tool for catalyzing social change. Non-governmental organizations (NGOs), corporations, international organizations, and local communities use bicycles to achieve various aims related to development and peace. 'Place' is central to BFD because the bicycle has different connotations in different places. In this vignette, we explore key features of the BFD movement, outline how it creates peaceful places, discuss some critiques, and offer examples drawn from Uganda, Canada, and Nicaragua.

Issues pertaining to development are inextricably related to peace. Helpfully, Johan Galtung (1996) distinguishes between 'negative peace', which is merely the absence of direct violence, and 'positive peace', found when the state, institutions and society actively engage in the removal of structural violence in order to integrate society. *Violence* in this case relates not just to physical violence but also to various forms of social, psychological and emotional violence. Positive peace is promoted through access to education, poverty reduction, and social, economic, racial and gender equity initiatives, assisted, in some cases, by bicycle mobility.

In recent years, a number of BFD studies have shown how bicycles may be vehicles for social change. For example, we have been part of a five-year study that focuses on the 'BFD movement' in Uganda, Canada, Nicaragua, India, and South Africa. In Uganda, Ardizzi et al. (2020) found that the meaning of the bicycle changed, depending on the context in which it was situated. In rural areas, the bicycle was commonly associated with wealth and access to resources whereas in cities cycling was sometimes associated with poverty. In some regions, the health implications of having a bicycle were major, because a bike made it easier for people to access medical services and medication, and allowed health care workers to travel greater distances to communities in need. These meanings positively and negatively influenced the willingness of some to use bicycles. In this case, the bicycle was integral to efforts to promote positive peace.

DOI: 10.4324/9781003142041-48

In Canada, researchers explored participants' experiences in an inner-city bicycle program where youth learned how to build a bicycle and ride safely on the road (Steinmann et al., 2022). The bicycle program created a safe space where cycling-related knowledge was accessible to young, racialized, low-income people who had experienced barriers to cycling. The interviewed youth initially felt that cycling was something white urbanites did. Through the program, however, many realized that a cyclist's identity is one they, too, could claim – fostering a more inclusive and peace-promoting understanding of urban life.

McSweeney et al. (2020) use actor-network theory (ANT) to explore non-human as well as human factors that facilitate and/or hinder BFD work. Drawing on information provided by practitioners working within bicycle-related development organizations, they found that factors like government regulations, the sturdiness of the bicycle, and environmental conditions affected the bicycle's potential as a developmental tool. The places where bicycles are used are particularly important: rugged terrain may cause costly breakdowns making bicycles more a burden than a pro-development 'object', although where bike mechanic training is offered (e.g. Uganda) low-cost repairs are possible.

Like sport-for-development programs, BFD programs are limited when they focus on individuals without linking to advocacy for broader responses to the structural issues that need fixing, such as accessible education, healthcare provision in rural areas, and cycling infrastructure. Some pro-development programs *appear* to address important societal issues when the root causes are not addressed. This is where the pursuit of positive peace among more advocacy-oriented BFD organizations – especially those concerned with promoting women's and girl's mobility and rights – is notable (Ardizzi et al. 2020; Hayhurst et al. under review).

There are important barriers to overcome for any BFD organization. Do bicycle-related interventions have the sustainable impact that is often claimed? For example, in Uganda and Nicaragua, although the bicycle may be effective in enhancing the mobility of women and girls, it may increase their domestic and economic work (Hayhurst et al. under review). Thus, there is a need to address more comprehensively the developmental role of the bicycle. The mobility of women and girls may benefit on an individual level but it may not deconstruct patriarchal relations or address gender inequities. Put another way, BFD in many cases does not address the root causes of the problems that it is meant to address. It is important to consider socio-political contexts and power relations relating to class, race, ethnicity, gender, sexuality and ability, while studying how bicycles shape daily lives.

BFD presents an interdisciplinary approach to enhancing positive peace in many places. Some BFD organizations, such as 'Bike for Peace', organize bicycle rides promoting a more peaceful, non-violent world. More generally, by promoting peoples' interactions with bicycles, we can learn more about how the bicycle can become a tool for peaceful and convivial social change and economic development.

References

Ardizzi, M., Wilson, B., Hayhurst, L., & Otte, J. (2020). "People still believe a bicycle is for a poor person": Features of "bicycles for development" organizations in Uganda and perspectives of practitioners. *Sociology of Sport Journal*, *38*(1), pp. 36–49.

Galtung, J. (1996). *Peace by Peaceful Means: Peace and Conflict, Development and Civilization*. (Oslo, Norway: PRIO).

Hayhurst, L.M.C., McSweeney, M.J., Otte, J., Bandoles, E., del Soccoro Cruz Centeno, L., & Wilson, B. (Under review). "Women shouldn't ride the bicycle": Investigating gender, technology, mobility and 'Bicycles for Development' in Nicaragua and Uganda.

McSweeney, M., Millington, B., Hayhurst, L., Wilson, B., Ardizzi, M., & Otte, J. (2020). "The bike breaks down. What are they going to do?" Actor-networks and the bicycles for development movement. *International Review for the Sociology of Sport*, 56(2), pp.194–211.

Steinmann, J., Wilson, B., McSweeney, M., Bandoles, E., & Hayhurst, L.M.C. (2022). An exploration of safe space: From a youth bicycle program to the road. *Sociology of Sport Journal*. DOI: 10.1123/ssj.2020-0155.

33
IN QUEST OF ADVENTURES

Duncan Jamieson

Every morning I set off on a journey…
celebrating the freedoms and simple joy
of a bicycle ride and the marvels of travelling.

Roff Smith, *The Art of the Ride*

The adventure of cycling

For 150 years men, women and children have cycled, in quest of adventures. The cycle is the integral part, making the journey possible, encouraging the rider to savor all the sights, sounds and smells along the way (Chapters 32 and 45). More than a mere collection of tubes, wheels and cables, the cycle becomes one with the rider. Setting out in 2013 on a 14,000-mile journey from Oregon to Patagonia, Jedediah Jenkins realized "my bike became a part of me," a sentiment repeated by many other cyclists, myself included (Jenkins 2018). Jane Bennett theorizes that objects are also actants, having interrelationships with sentient beings (Bennett 2010). Throughout its existence cyclists have come to know the wheel as a faithful companion that carries them forth; if necessary, they would reverse the role. Together cycle and rider enjoy all that is experienced on the journey. While cycling in the early 1900s, J.W. Allen, who believed the bicycle the only good invention of the 19th century, encountered a fellow cyclist who eight years earlier had left everything behind, becoming a cycling nomad, traveling without maps or guidebooks as they eliminated the sense of wonder and adventure (Allen 1909). As it is used here, journeying describes the entire experience in the following sections: The culture of cycle venturing; Urban journeys; Imperialism to acceptance; Introspective journeys; and The past is present.

The culture of cycle venturing

Once I had learned to ride a two-wheeler, adventuring began with the school year's end when my friends and I got our bicycles and enjoyed our first taste of freedom. Traveling through New York's Queens County we climbed the last ice age's terminal moraine to reach the East River and the Whitestone Bridge. When traveling the same way by car I never noticed the change in elevation and certainly not the energy displaced to reach the summit of the glacial

DOI: 10.4324/9781003142041-49

fill, though each time I pedaled it I knew what I had earned. Unlike the automobile, a cocoon to transport its occupants from point to point, cyclists absorb the journey as an end coequal with the destination. I cycle to work, to run errands, to take a relaxing ride after a stressful day, to complete hundreds of centuries (one hundred miles in a day) and a continental crossing from Los Angeles to Boston. Without the cycle, the journey and the destination would not exist; together they form a triad.

Alfred Chandler and John Sharp of the Suffolk (Massachusetts) Bicycle Club spent a month in the 1870s riding their high wheels through England and Wales (Chandler 1881). Jim and Elisabeth Young set off from the Pacific on their tandem, "The Spirit of Fun," on a round-trip ride to commemorate the 75 anniversary of the Battle of Gettysburg (Young 1940). In the 1950s Louise Sutherland, a New Zealand nurse, ventured around the world on a drop frame bicycle she bought at a jumble sale (Sutherland 1960). Towing a small trailer she completed the circumnavigation in six years, working along the way to replenish her funds. Willie Weir believes his "love affair with [bicycle] travel began" in the 1970s when he and a buddy "left Seaside, Oregon on our... touring bikes," reaching New York City two months later (Weir 1997). Tim Travis and his wife left their traditional life in Arizona in 2002, going to Mexico to become cycling journeyers. On their multi-geared, loaded touring bicycles, they met farm workers going to the fields on one-speed cruisers. Despite differences in equipment, nationality, language and purpose, Travis understood they "were all cyclists, a common bond that transcends cultures" (Travis 2004). Luis Vivanco has spent "countless hours" commuting, competing and traveling by bicycle, prioritizing "the bicycle in [his] lifestyle" (Vivanco 2013). Hundreds of others who wrote of their travels and countless more cyclists unknown seek adventures by the roadside, perhaps on a short commute home (Vignettes E and J), or thousands of miles away riding around the world.

The words freedom, adventure, excitement, exhilaration, exploration, knowledge, joy, fun and camaraderie are associated with cycling. People bicycle to historic sites and sights of natural beauty, while creating or renewing an appreciation with nature and landscape, using cameras to provide a visual record. In the 1880s, A.J. Wilson suggested that "it is only upon taking to cycling that the adult discovers a locomotive recreation of limitless extension for novelty, restricted to no season, and which can be pursued in any locality" (Wilson 1887). A decade later, Dr. Arthur Conan Doyle offered this prescription: "When the spirits are low, when the day appears dark, when work becomes monotonous, when hope hardly seems worth having, just mount a bicycle and go out for a spin down the road without thought of anything but the ride you are taking" (Doyle 1896). As a 1930s American exchange student, Fred Birchmore bought a bicycle to travel when not in class. Over the winter break he rode from Germany to Egypt where he lost his passport. Unable to return for classes he continued bicycling east around the world on what he believed the ideal means of transportation. "When [his bicycle] found it impossible to plow through jungles or surmount rugged peaks, I merely turned carrier instead of rider – that to me was a feeling of real power, freedom, and a mastery of one's fate" (Birchmore 1939). In the mid-1990s Erika Warmbrunn wanted to run away. On her way to buy a train ticket in Munich, Germany, "on a whim [she bought] a bicycle instead. When I started pedaling, I thought I knew where I was going. I thought I knew France and Germany pretty well, but over the next five weeks, my department-store bicycle's two wheels suddenly opened up the lands' remote corners – the farmhouses, the mountains, the villages where the trains didn't stop" (Warmbrunn 2001). She had a goal of exploring Mongolia, China and Vietnam, but how to travel? She wanted the freedom to roam where and when she pleased, unencumbered by the limits and vagaries of commercial transportation. Even though not a "hardcore cyclist," she knew the bicycle was her only option. Around the same time a "hardcore cyclist" waited at a traffic light in Idaho as an old, blind rancher, stopped crossing the street to tap Joe Kurmaskie's bicycle with his cane. "'Ah, metal cowboy,'" he said to a "long-distance cyclist pedaling the open road in search of adventure" (Kurmaskie 1999).

The 19th-century industrial revolution increased urban populations while reducing working hours, thus creating more free time. The 1870s high wheel bicycle, followed by the diamond frame a decade later, offered the opportunity for leisurely, independent, personal travel to explore wherever the cyclist chose, whenever time permitted, alone or in groups. The bicycle gives the rider an escape to think, to create, to revel in the simple pleasures of life, to relax and recharge. Riders enjoy a spin around the neighborhood or an extended trek far from the madding crowd. Either provides a departure from the everyday stress and problems, allowing riders to engage all their senses. They see their surroundings, hear the bird songs, smell the flowers, touch the wind or the rain, and if it is a supported club ride there are bound to be homemade cookies to taste. Savoring the relaxed pace under their own exertion, they cycle through neighborhoods or past woods and meadows, perhaps stopping for breakfast across town or visiting a quaint village or an historic home. In 1893 Elizabeth and Joseph Pennell wrote of just such an adventure, careful not to disclose its location to protect its anonymity (Pennell 1893). A century later, Tom Vernon used "the most personal and democratic means of transport" to travel through France, offering detailed directions on six separate routes (Vernon 1994).

Technological advances – smartphones with cameras, energy snacks and drinks, high-tech sleeping bags and tents, bright lights fore and aft, and modern, synthetic fabrics for reflective, breathable clothing – make today's cycling journey safer and more comfortable, but they have not changed its fundamental allure. Riders remain close to the environment and the elements, focused on their surroundings as they move sedately in the slow lane. When Thomas Stevens left Oakland, California, in 1884 on his odyssey around the world, he became the first of thousands of cyclists, singly or in groups, to circumnavigate the globe. He rejoiced "in the sense of boundless freedom... that comes of speeding across open country where nature still holds its primitive sway" (Stevens 1888). He reveled in the freedom of movement without concern or restraint. Hugh Callan followed Stevens' wheel tracks across Europe, privileged to see "new scenes and scenery" (Callan 1887). He delighted in riding in the morning, resting in the middle of the day and continuing the trek as the stars came out. Purely for pleasure, George Burston and Harry Stokes cycled their native Australia before heading to Asia and Europe, freed from timetables, able to go where tourists could not (Burston 1890: Figure 33.1). A century later, Anne Mustoe bested Stevens by circling the globe in both directions (Mustoe 1998).

For the first hundred years when cyclists ventured forth to study distant climes, they had little opportunity for contact with those left behind. It could take weeks between the posting of a dispatch and its arrival if it arrived at all. Thomas Stevens "disappeared" in China for weeks before resurfacing. A few years later, Frank Lenz left Tabriz, Persia, headed west, only to vanish. Rumors circulated as to his whereabouts; not until another around-the-world cyclist went in search did the truth of Lenz's murder surface. In 1981, to celebrate her sixtieth birthday, Dervla Murphy intended to disappear on her bicycle, being not only physically separated from the day-to-day world but mentally and emotionally free as well by creating "an oasis in time" (Murphy 1993). Fifteen years later, Mark Beaumont set a Guinness Book world record for round-the-world cycling; throughout the journey he remained in constant contact via cellphone and internet with his mother who ran his "base camp" in Scotland (Beaumont 2009).

A solo adventure of any length is a joy, as are sponsored rides where the numbers can reach the thousands; both are equally appealing. The League of American Bicyclists, Adventure Cycling, The Cyclists' Touring Club, Cycling UK, or Cycling Canada Cyclisme all sponsor rides. Frank Elwell offered bicycle tours beginning in 1881. Though he preferred solitary riding

Figure 33.1 George Burston and Harry Stokes. (Courtesy of John Weiss).

Karl Kron joined some three dozen others on one of Elwell's Maine tours in 1883 (Kron 1887). A two-day, roundtrip ride with a few friends from Columbus to Portsmouth, Ohio in 1962, morphed into the Tour of the Scioto River Valley (TOSRV), one of the largest annual bicycle rides in the United States. Today's *Register's* Annual Great Bike Ride Across Iowa (RAGBRAI) is a multi-day festival for approximately 8,500 through cyclists and an additional 1,500 day riders. With a new route each year, in 2019 riders completed 427 miles, with 14,735 feet of climbing from Council Bluffs to Keokuk. California's Davis Bike Club has been offering the Hilly Double Century for over fifty years. Numerous for-profit organizations offer bicycle tours almost anywhere in the world, varying in length, price and amenities.

Wanderlust is often reflected in the titles riders choose for their cycling books: *A Journey to the Heart of America*; *Roll Around Heaven All Day*; or *Discovering America*, each focusing on the riders' experiences as they travel through the United States. For those more adventuresome who wander further afield: *Where the Pavement Ends*; *Into the Remote Places*; *Miles from Nowhere*; *Pedaling the Ends of the Earth*; *Off the Map*; *Travels Along the Edge*; or *Riding Outside the Lines* add to the sense of exploration awaiting those who leave the beaten path far behind.

Urban cycle journeys

Experiences rather than distance or time define a journey, which means city bicyclists can find adventure right outside their doors. Thomas Stevens learned to ride his high wheel in San Francisco's Golden Gate Park. Frederick Law Olmstead, who designed Boston's Emerald Necklace and New York's Central Park, kept cyclists out, until Colonel A.A. Pope, manufacturer of Columbia Bicycles, forced New York City to open Central Park. Pope addressed the question still being debated today – who has rights to the road? Beginning in the 1890s, cyclists started lobbying for safe, improved riding surfaces. New Yorkers have miles of safeguarded spaces for cycling. While cities across the continent are expanding protected bike lanes to improve safety and hopefully reduce congestion and pollution caused by automobile traffic, they represent a conundrum – some cyclists fear separate rights of way could lead to denying them access to the city streets.

The first organized urban tour in the United States was the 1879 two-day "Wheel Around the Hub," where cyclists paraded through Boston and its environs. Repeated until the 1930s, it was reenacted in 1965 and revived in 2005 as the "Hub on Wheels." Beginning in May 1977, the Great Five Boro Bike Tour grew from a few hundred riders to over 32,000 in 2019. Cyclists wheel through Gotham, from the urban canyons of Lower Manhattan to Central Park before crossing the Harlem River to the gritty South Bronx. Returning to Manhattan, riders cruise down the FDR Drive with the placid East River on the left and bustling Manhattan on the right. "Feeling groovy," riders cross the 59th Street Bridge to Queens, the borough of homes before entering Brooklyn, the borough of churches. After climbing onto the Verrazano Narrows Bridge and cycling through Staten Island riders return to Manhattan on the Staten Island Ferry, disembarking at Battery Park to complete the 42-mile tour. Throughout the United States, cyclists have similar opportunities to gather and explore their city. Residents of Milwaukee's Riverwest initiated an annual 24-hour bikefest, inviting all comers to enjoy the local sites and businesses as they cycle loops through the neighborhood to further community spirit. The camaraderie exposes tourists to this area while allowing the locals to advertise their home to all participants (Hoffman 2016).

New York City's Department of Transportation sponsors "Biketober," offering maps for self-guided rides throughout the city. "Our rides are perfect for spending time with family, practicing physical fitness, or exploring your neighborhood." In the mid-twentieth century, Daniel Behrman enjoyed cycling through Paris and New York, believing the bicycle could save them from the automobile. Whereas the car "inhibits human contacts, the bicycle generates them, insinuating itself unseen into the innermost tissue of a large city where there is so, so much life that cannot be sensed through a windshield." Instead of separating drivers from the environment, the bicycle draws its rider into contact with the surroundings to explore the unknown (Behrman 1973).

Some interesting perspectives on the bicycle's place in the urban environment are: who is riding; why are they cycling rather than driving; and where are they going? Today's urban officials are realizing, as Behrman did fifty years ago, that automobiles clog the roads, choking cities. More people using bicycles for light shopping, commuting to work or school, or traveling around the city, would reduce vehicular congestion. That might result in more bicycle lanes, the downside being gentrification, changing the dynamic of the neighborhood. Many cities witness bicycle use growing exponentially among young, upwardly mobile whites who bicycle by choice, not necessity. Portland, Oregon's Albina district had bicycle lanes added to the streets, encouraging cyclists to move in, raising rents and displacing long-time residents. On a more positive note, when Minneapolis built greenways and bicycle paths along an abandoned railroad corridor, it displaced the homeless and drug dealers who had occupied the area. This environmental gentrification created amenities that attracted retail businesses, including bike share. But since this requires a credit card it does not aid the poor.

Bike share began in Amsterdam in the 1960s as a free service, but ended when the bicycles were stolen (Chapter 23). Still, the idea to provide an alternative transportation opportunity caught on in the 1990s when smart technology allowed people to check out and return bicycles.

Today there are bike share companies in most major American cities. They offer visitors the opportunity to engage their tourist gaze while residents rent them to move about the city for work or for running errands or to explore new neighborhoods without the necessity of owning a wheel. Because of the need for a credit card or a membership, these programs further marginalize the poorer urbanites. While the bicycle enhances the quality of urban life, the advantages may not be distributed evenly to all.

From imperialistic chauvinism to accepting cultural differences

Until the 1930s cycling mirrored the late 19th-century Anglophonic imperialism. It furthered the belief in the West of superior societies with both the technology and the fortitude to undertake journeys to those less civilized regions. Thomas Stevens and other English-speaking cyclists vaunted their superiority to the uncivilized, benighted natives in the wilds of Eastern Europe and Asia. Stevens prided himself on being monolingual and forcing others to understand him through gestures and sign language. At night it became his custom to demand entry for a place to sleep. When threatened by a mob in China he vowed to kill as many as possible while saving the last bullet for himself. John Foster Fraser, F.H. Lowe and Edward Lunn belittled the native laborers who carried themselves and their bicycles across large swaths of China (Fraser 1899). They delighted in leaving Chinese villages in the fusillade of rocks launched by the angry residents. H. Darwin McIlrath viewed the Chinese as an uncivilized, barbaric race (McIlrath, 1898). These chauvinistic attitudes reinforced the belief in the "white man's burden" and did little to further amicable relations between the privileged cyclist and the locals. As much as these hegemonic cyclists credited India's cultural achievements on the British presence and loathed the Chinese, all the riders found the Japanese much more enlightened and advanced, simply because they had adopted Western ways. In a similar vein, Joseph Pennell echoed these sentiments with his decidedly anti-Semitic outlook (Pennell 1925).

Softening attitudes toward other cultures began to appear in the 1930s when Bernard Newman wrote glowingly of the people and the cultures he encountered while bicycling throughout Europe (Newman 1960). He did regret his inability to speak Yiddish, which would have served him well when engaging the local populations. In the 1970s, David Duncan and two companions distanced themselves from the earlier imperialists (Duncan 1985). When Barbara Savage saw Egyptian street children, shoeless and begging in rags, the thought of American parents buying designer shoes for their children overwhelmed her (Savage 1983). Though the conditions in the Far East continued to be wretched and primitive by Western standards, the journeyers had softer reactions, deploring the poverty but accepting the different cultures and lifestyles. American tourists who haggled with poverty-stricken Vietnamese mortified Erica Warmbrunn. Unlike the 19th century imperialists, Irishwoman Dervla Murphy, who had family ties to the IRA, contrasted issues of social justice, environmental destruction and the horrors of inhumanity against the joys and enlightenment of mingling with locals. She found the Tibetan scenery stunningly overpowering, but the dreadful treatment of refugees sickened her. She bicycled to Northern Ireland during the height of "The Troubles," interviewing Catholics (the minority) and Protestants, along with the British military caught in the middle. Her words and photographs documented the urban carnage caused by abandoned houses, burned-out pubs and no-man's lands separating the warring groups. She studied racism and poverty in two industrial cities in the English Midlands, and later wrote of life in two post-Soviet Union countries. Using the firsthand experiences of her journeys through Africa, she documented the devastation of AIDS and the elections that brought an end to apartheid in South Africa. She next wrote of the impact of war and the West in Laos.

Introspective journeying

Many cyclists write of their own being as they reflect on their experiences and the lives of those met along the way. Freed from the normal routine, they turn inward to explore personal growth and development while at the same time enjoying the cyclist gaze. In the 1890s, at the encouragement of her physician, middle-aged Frances Willard, president of the Women's Christian Temperance Union, learned to cycle for her own enjoyment and to further women's rights (Willard 1991). A century later, Daryl Farmer made two journeys through the Rockies, first in his 20s and later as an out-of-shape 40-something. On the second adventure his cycle was stolen. Knowing it had brought him peace of mind as it improved both his mental and physical health, he felt the loss of a part of his identity (Farmer 2008). In his autobiography, William Saroyan credits his youthful association with the bicycle as the defining aspect of his life and career. In addition to riding for pleasure he spent three years delivering telegrams by bicycle (Saroyan 1952). As David Lamb entered middle age his contemplations on the stress of aging in modern America took him on a transcontinental journey where he tested himself through risk-taking (Lamb 1996). While cycling on a Fairbanks, Alaska bike path, Peggy Shumaker was hit and seriously injured by the teenager illegally riding a four-wheeler. Her memoir explores the emotions of loss and anger that morphed in time to discovery, healing and forgiveness (Shumaker 2007).

Two journeys past and present

For the first two decades of its existence, the bicycle was the best available option for independent travel. Once the automobile appeared it pushed self-propelled venturing to the sidelines, though it did not stop people from riding for the same various and sundry reasons. Perhaps more than anything else cycling allowed people to reset. In a 1903 English sporting novel, the feminine protagonist often took long bicycle rides in the countryside to clear her mind and focus her priorities (Kennard 1903). More than a century later Vivanco interviewed a 30-something woman in Burlington, Vermont, about her reasons for riding: "I do it because it feels good… I feel refreshed. It clears my head." One fictional, one real, and generations apart, their thoughts about wheeling and its importance are virtually identical.

In 1896 Elizabeth Robins Pennell left London with her husband, Joseph, to bicycle to and over the Alps (Chapter 48). Though she wore traditional dress she was a quiet feminist, challenging the Cult of True Womanhood. When she rode over nine passes, six in one week, she became the first woman to conquer the Alps on a bicycle, but she never intended to be the last. She described in detail the demanding climbs and descents, the weather and scenery, as she encouraged other women to follow in her wheel tracks, sure that they could accomplish the journey (Pennell, 1898a). In 2014, Alison Stone, accompanied by her husband, bicycle historian John Weiss, did just that, riding the mountain passes that Elizabeth had ridden. Both women cycled for the joy it brought to challenge the popular stereotypes of women's physical abilities. Alison marveled at Elizabeth's strength to complete the journey on a single-speed bicycle while she rode a modern, multi-speed touring wheel. Both couples rode through the Simplon Pass and the Tete-Noire, which Alison described as Elizabeth's "Ride from Hell" (Figure 33.2). Like cycle travelers everywhere, she felt a sadness at the end of the ride. Sometimes smoother pavement, sometimes not, though separated by over a century, the experiences and journeys remain remarkably similar (Stone 2021). It took Thomas Stevens 109 days to reach Boston from California; I crossed the continent in 47. And like Alison and

Figure 33.2 Alison Stone reaches the summit of the Simplon Pass, 2014. (Courtesy of John Weiss).

John, I felt the same sadness creep in. Thrilled with the freedom and camaraderie, I wanted to turn right at the Atlantic and continue wheeling to Florida rather than rejoin the world. Such is the allure of the bicycle.

Bibliography

Allen, T. (1897) *Across Asia on a Bicycle*. (New York: The Century Company).
Allen, J. (1909) *Wheel Magic*. (London: John Lane).
Beaumont M. (2009) *The Man Who Cycled the World*. (London: Bantom Press).
Behrman, D. (1973) *The Man Who Loved Bicycles*. (New York: Harper's Magazine Press).
Bennett, J. (2010) *Vibrant Matter*. (Durham, NC: Duke University Press).
Birchmore, F. (1939) *Around the World on a Bicycle*. (Athens: University of Georgia Press).
Burston, G. (1890) *Round the World on Bicycles*. (Melbourne: George Robinson and Company).
Callan, H. (1887) *From the Clyde to the Jordan*. (New York: Charles Scribner's Sons).
Chandler, A. (1881) *A Bicycle Tour in England and* Wales. (London: A. Williams & Co).
Doyle, A. (1896) *Scientific American*. (New York: Scientific American).

Duncan, D. (1985) *Pedaling the Ends of the Earth.* (New York: Simon & Schuster).
Farmer, D. (2008) *Bicycling Beyond the Divide.* (Lincoln: University of Nebraska Press).
Fraser, J. (1899) *Round the World on a Wheel.* (London: Methuen & Company).
Herlihy, D. (2010) *The Lost Cyclist.* (Boston: Houghton Mifflin).
Hoffman, M. (2016) *Bike Lanes are White Lanes.* (Lincoln: University of Nebraska Press).
Jenkins, J. (2018) *To Shake the Sleeping Self.* (New York: Convergent).
Kennard, E. (1903) *The Golf Lunatic and His Cycling Wife.* (New York: Brentano's).
Kron, K. (1887) *Ten Thousand Miles on a Bicycle.* (New York: Karl Kron).
Kurmaskie, J. (1999) *Metal Cowboy.* (New York: Breakaway Books).
Lamb, D. (1996) *Over the Hills.* (New York: Random House).
McCullough, R. (2015) *Old Wheelways.* (Cambridge: MIT Press).
McIlrath, H. (1898) *Around the world on wheels.* (Chicago: Inter Ocean Publishing Company).
Murphy, D. (1967) *The Waiting Land.* (London: John Murray).
Murphy, D. (1978) *A Place Apart.* (London: John Murray).
Murphy, D. (1987) *Tales from Two Cities.* (London: John Murray).
Murphy, D. (1992) *Transylvania and Beyond.* (London: John Murray).
Murphy, D. (1993) *The Ukimwi Road.* (Woodstock, NY: The Overlook Press).
Murphy, D. (1997) *South from the Limpopo* (London: John Murray).
Murphy, D. (2002) *Through the Embers of Chaos.* (London: John Murray).
Mustoe, A. (1998) *Lone Traveler.* (Swan Hill: Swan Hill).
Mustoe, A. (2001) *A Bike Ride.* (London: Virgin Books).
Newman, B. (1960) *Speaking from Memory.* (London: Herbert Jenkins).
Pennell, J. (1893) "The Most Picturesque Place in the World. (*The Century Illustrated Monthly Magazine*: New York).
Pennell, E. (1898a) *Over the Alps on a Bicycle.* (T. Fisher Unwin: London).
Pennell, J. (1898b) "In Andalusia with a Bicycle." (*The Contemporary Review:* New York).
Pennell, J. (1925) *The Adventures of an Illustrator.* (Little, Brown and Company: Boston).
Pratt, C. (1880) "A Wheel Around the Hub." (*Scribner's Monthly*: New York).
Saroyan, W. (1952) *The Bicycle Rider in Beverly Hills.* (New York: Charles Scribner's Sons).
Savage, B. (1983) *Miles from Nowhere.* (Seattle: The Mountaineers).
Shumaker, P. (2007) *Just Breathe Normally.* (Lincoln: University of Nebraska Press).
Stevens, T. (1888) *Around the World on a Bicycle.* (London: Century).
Stone, A. (Retrieved July 7, 2021). https://www.crazyguyonabike.com/doc/page/?o=1mr&page_id=409188&v=LS&src=page_first
Sutherland, L. (1960) *I Follow the Wind.* (London: Southern Cross Press).
Travis, T. (2004) *The Road that Has No End.* (Indianapolis: Down the Road Publications).
Vernon, T. (1994) *Fat Man in France.* (London: BBC Books).
Vivanco, L. (2013). *Reconsidering the Bicycle.* (New York: Taylor and Francis).
Warmbrunn, E. (2001). *Where the Pavement Ends.* (Seattle: The Mountaineer Books).
Weir, W. (1997) *Spokesongs.* (Halcottsville, NY: Breakaway Books).
Willard, F. (1991) *How I Learned to Ride the Bicycle.* (Sunnyvale, CA: Fair Oaks Publishing).
Wilson, A. (1887) *The Pleasures, Objects, and Advantages of Cycling.* (London: Liffe & Sons).
Young, J. (1940) *Bicycle Built for Two.* (Portland, OR: Binford).

Vignette J
WINTER CYCLING
Montreal's four-season bicycle network

Bartek Komorowski and Kevin Manaugh

Year-round bicycle use

The recent expansion of cycling has taken several forms, both in space and time: enlarged bike networks, off-road bicycles, and cycling at night are some of the notable developments. Perhaps less obvious to residents of tropical and temperate regions is the expansion of winter cycling in regions which experience seasonal snow and ice such as northern parts of Europe, Asia, and North America where previously cycling was mostly restricted to three seasons. Two developments in regards to equipment have assisted this. Bicycles designed for winter riding have become more robust, with fenders reducing splash, and wide tires with knobbles or studs that grip a slippery surface better. Meanwhile winter clothing and footwear – often adapted from designs used in snowshoeing and skiing – have become lighter, warmer and more wind-resistant. The design of accoutrements, including visors, helmets and gloves, has also improved. Meanwhile, winter riding has been promoted by changes in infrastructure and policy, including the design and maintenance of bike paths. The latter development is illustrated by the subject of this vignette, Montreal, a large conurbation of over four million people who experience a cold and snowy winter from December to March.

Journey to work data from the Canadian long form census and the quinquennial regional origin-destination (household travel) survey show that bicycle use in Montreal has been steadily increasing in both absolute and relative terms for at least two decades (Statistics Canada 2016; Autorité Régionale de Transport Métropolitain 2018). Data from automated counters scattered across the bicycle network confirm this trend too, but also show that seasonal variations in bicycle use have been evolving. Namely, there is evidence that more Montrealers are using bicycles later into the autumn than before and getting back in the saddle earlier in the spring, and an ever-growing number never stop riding (Vélo Québec 2021), not unlike residents of other Canadian cities (Nahal & Mitra 2018).

Pre-pandemic data from a set of bicycle counters that have been in continuous operation for several years show that wintertime bicycle use is growing much faster than fairweather use. Between 2015 and 2020, average bicycle counts during the peak season, from May to October, increased 10%. Average counts for January and February, the two coldest and snowiest months, increased 415% over the same period. Another way to look at counter data is to compare average fairweather use to average winter use. In 2009, the average rate of bicycle use from

DOI: 10.4324/9781003142041-50

November to April compared to May to October, what can be called the *retention rate*, was about 6%. A decade later, in 2019, the retention rate was approaching 15%.

Towards a four-season cycling network

The idea that the cycling network or parts thereof should be maintained for wintertime use, the so-called *réseau blanc* (or 'white network'), had been pushed by local cycling advocates since the 1990s but took a while to gain traction with the City of Montreal.

By the mid-2010s, the *réseau blanc* concept was abandoned and replaced by the somewhat more aptly named *réseau quatre saisons* (four-season cycling). Some 400 km of shared lanes and paint-only bicycle lanes instantly became "four season" facilities simply by being on roads that are maintained throughout the winter. The change in nomenclature provided little added value for cyclists, as conditions for cycling on these types of facilities are generally less than ideal. In the case of shared lanes, the presence of snow banks and ruts limit the usable width of the road. Where they run along parked cars, they end up serving as storage spaces for snow plowed from the roadway, forcing cyclists into the adjacent travel lane. Even when snow is removed from the bicycle lanes, they tend to get coated with packed snow and ice, ironically because cars do not use them. Standard winter road maintenance techniques depend on heat and friction from cars to activate salt, which dissipates residual snow left behind by plows. Bicycles provide neither, and the packed layer of snow left behind by plows tends to linger and turn into ice.

The situation has significantly improved in recent years, mainly thanks to the increasing availability of well-maintained, physically separated cycling facilities. An ever-growing number of delineator-post-separated facilities are no longer being shut down for the winter season and many kilometers of new separated facilities have been added. Most of the latter feature a wide, lateral buffer space between them and the adjacent parking or travel lane that is used to store plowed snow and prevents snowbanks from encroaching into the bicycle lane.

Some paint-only bike lanes, squeezed between parked cars and a travel lane, have been "flipped" – i.e. placed between parked cars and the curb – and now undergo dedicated snow clearing and de-icing operations (Figure J.1). Some of Montreal's boroughs have been experimenting with new snow removal techniques on cycling facilities, such as using rotating sweepers, which leave less residual snow than plows, and de-icing with liquid melters rather than rock salt.

With more physically separated bicycle lanes and better maintenance, Montrealers using bicycles during the winter are likely to encounter better surface conditions and find themselves riding in a less stressful environment. As more well-maintained, separated facilities are connecting, a genuine "four-season network" is beginning to take shape, at least in Montreal's densely-populated central boroughs.

Evolving perceptions and discourse

Perhaps another sign of the coming of age of winter cycling in Montreal is the evolution of public discourse on the subject. Less than a decade ago, it was still not uncommon to hear and read disdainful commentary in the early winter about crazy and irresponsible yahoos out on their bikes in the snow, getting in the way of cars. In the latter half of the 2010s, the tone changed and coverage shifted to topics such as of which parts of the city were most accessible by bicycle during the winter, how well different boroughs were maintaining their cycling facilities, and whether BIXI, the heavily-used bicycle sharing system, should remain open during the winter. In short, winter cycling is increasingly treated as a legitimate practice.

Figure J.1 Conventional (left) and parking-protected (right) bicycle lane configurations. (authors' collection).

All this being said, in terms of the absolute number of users, cycling in the Montreal winter remains a fairly marginal practice. Cities such as Oulu in northern Finland, which manages to keep more than half of its (very large) number of fairweather bicycle users riding during a much colder and darker winter, suggest that the ceiling for winter cycling is much higher than the current retention rate of 15%. The available research and local experience suggests that the key is to densify and expand the network of well-maintained, low-stress cycling facilities if we are to capture and retain more fairweather bicycle users during the winter months. This is easier said than done: a significant portion of the population and some local politicians still see winter maintenance of cycling facilities as an expensive frivolity for the benefit of a handful of car-hating extremists. If a true four-season cycling network is to develop, public perceptions will need to further evolve.

References

Autorité Régionale de Transport Métropolitain, Enquête Origin-Destination, (2018).

Nahal, T. & Mitra, R. (2018). Facilitators and barriers to winter cycling: Case study of a downtown university in Toronto, Canada. *Journal of Transport & Health, 10*, 262–271. https://doi.org/10.1016/j.jth.2018.05.012

Statistics Canada, (2016). Census of Population, Journey to Work.

Vélo Québec, (2021). *Cycling in Quebec in 2020.* https://www.velo.qc.ca/en/media/cycling-in-quebec/cycling-in-quebec-in-2020-2/

34
THE AFRICANIZED BICYCLE

Hans Peter Hahn

> It matters little who it was made for; what we should learn from is the imagination that produced it. [The bicycle] is not there to be Other or the Yoruba Self; it is there because someone cared for its solidity; it is there because it will take us further than our feet will take us;
>
> *(Appiah 1992: 139)*

Introduction

The role of the bicycle in Africa is inextricably intertwined with the paradoxes of modernity, with the violence of colonialism in the nineteenth and twentieth centuries and with the enduring creativity of local societies, in Africa and elsewhere. The history of bicycle usage in Africa reveals the obstacles that existed on this continent to develop culture and society and to find its own place in the dynamics of modernity and progress. Nevertheless, in many regions of Africa, users and owners of bicycles have found ways to appropriate the technology, modify bicycles for their own purposes and thus establish new local traditions (Vignette B). Drawing on the bicycle as an example, the great African philosopher Anthony K. Appiah used the motif of a bicycle to illustrate the paradoxes of the 'own' and the 'foreign'. In his novel "*In my Father's House*", he refers to a wooden sculpture of a cyclist described by the museum curator as 'neo-traditional'. Why, Appiah asks, does one not simply see this figure as an expression of African contemporaneity in the twentieth century with regard to the use of an almost ubiquitous technology? Quite rightly, he argues for a consideration of the cyclist and his vehicle for what it is: an expression of contemporary everyday life in Africa.

The bicycle – unlike the automobile – was not an icon of colonization. While the car was used as a status symbol by colonial administrators as well as wealthy individuals and local political officials (Alber 2002), from an administrative perspective the bicycle was hardly ever seen as a component of important infrastructure. Paradoxically, even in recent accounts of the history of mobility in Africa (Callebert 2016: 119), automobiles are celebrated as a breakthrough to freedom and economic empowerment of the predominantly agrarian population, although automobiles initially lead to new dependencies, on petrol or diesel for instance. The one-sided emphasis on a form of mobility dependent on fossil fuels is all the more surprising when one takes into account the significant economic benefits that can be observed with the use of the bicycle in

DOI: 10.4324/9781003142041-51

local economic practices. The bicycle as a sustainable instrument of mobility has the potential to play a special role in Africa (Parkin 2012).

The bicycle in Africa has also been neglected in historiography (Hopkins, 1973). While automobiles and railways are described as significant economic factors, bicycles are not even mentioned (McCracken 2011). This is partly because a disproportionate amount of attention has been paid to the macro perspective. This, in turn, is due to the lack of sources on bicycle history. A valuable exception is John McCracken's well-informed article on the bicycle in Malawi's colonial history, which is partly based on the analysis of local daily newspapers (McCracken 2011: 5–6).

It is important to note that, in contrast to other technologies of mobility including car, railway, telegraphy, the introduction of the bicycle was almost contemporaneous; spread of the bicycle in Africa occurred around the same time as the adoption of this means of transport in many regions of Europe. Therefore, it is significant to look also at the similarities of the historical, technical and social developments of the bicycle in Africa and in Europe. In Europe, too, the status of the bicycle had often been contested. Contexts of the bicycle ranged from its marginal position as a particular kind of sports equipment, to its status as a symbol of the working class, to its consideration as an obstacle for motor vehicles in the period of growing automobility in cities. This broader perspective invites us to see the history of the bicycle not only in its technical-historical and economic dimensions, but also as a history of lifestyle and gender (Lavenir 1998). This also includes the question of the right transport infrastructure, whose one-sided focus on railways and motorized transport in urban regions needs to be questioned urgently (Foster & Briceo-Garmendia 2010).

Wherever smallholder households dominate in Africa, bicycle use has been an important advancement in economic development (Hanlon & Smart 2008; Olvera, Plat & Pochet 2008; Vignette B). Even though the acquisition of bicycles is a challenging act of consumption and cycling is always linked to the necessity to purchase spare parts, the ongoing need for financial resources when using a bicycle remains significantly lower than for any other means of transport. In other words, the specificity of cycling is its relative autonomy in day-to-day-use.

Early traces and evaluations

There are no reports about the arrival of the first bicycles in Africa. But early photographic documents prove that around 1900 bicycles were quite common in many urban contexts, for example in Lomé, Togo. Moreover, there is clear evidence of the great popularity of the bicycle in the period of the first decade of the twentieth century. Three years after Harry Green completed his record-breaking ride from Land's End to John o'Groats on a Raleigh bicycle (Bowden 1975: 30), there was a bicycle race for the indigenous population in colonial Lomé in 1911, documented by the photographer Alex Acoladse (Hahn 2004: 269).

Nancy Hunt (2005: 134) reports how in the then Belgian colony of Congo, Prince Albert himself used the bicycle to cover part of his travel route in 1909. Obviously, bicycles quickly gained a certain popularity. Moreover, they were not exclusively used by Europeans, but were quickly adopted by the local population as well. Urban scenes from Lubumbashi (Congo) describe a sometimes conflictual coexistence of pedestrians and local cyclists in 1911 (Esgain 2001: 62). We know that in this period (1910) both "Nigeria" and also "West Africa" are mentioned on Raleigh's list of export destinations (Bowden 1975: 132). This date is again confirmed by a report about a village in southern Nigeria where the first bicycles came into use in 1912 (Ugorji and Achinivu 1977: 242).

What is less clear, however, is the overarching assessment of this phenomenon by the colonial administration that governed many countries on the continent at the time. It is perhaps significant that while Europeans used bicycles, the European cyclists – as far as the documents reveal – were predominantly not formally members of the administration of these territories. Rather, they were marginal figures, such as a missionary who undertook a journey stretching 1000 km over several months by bicycle through what is now Ghana and Togo. Or a wildlife trapper and filmmaker who discovered the advantages of the bicycle while searching for certain animals.

The two European travelers mentioned above did reflect the advantages of the bicycle as a form of mobility in their travelogues. Rudolf Fisch, a missionary, was travelling in the north of present-day Ghana in 1909. As he highlights, he was impressed by how much the bicycle interested the population of various villages he visited (Fisch 1911: 113). Hans Schomburgk travelled through Liberia by bicycle in 1911 and he underlines that it was precisely this means of transport that enabled him to ask questions and ultimately to enter in conversations with the inhabitants of villages he visited (Schomburgk 1922). In both cases, according to their reports, it was of great importance not to be considered part of the colonial administration. For both, direct access and communication with the rural population was of particular importance, and the bicycle proved to be a suitable tool for this.

Even though these incidental research findings cannot claim to be a representative testimony of early bicycle use in Africa, they do give clear indications that bicycles were by no means held in low esteem. The appreciation probably applied to European travelers as well as to local bicycle users, even if the latter have not provided direct textual evidence for this.

Economic relevance of cycling during colonial times

Another scene from Nigeria dates about twenty years later and describes a completely different context. In this case, as Anthony Nwabughuogu (1984: 96) reports, it is about the transport of palm oil on a bicycle rack. As reported by the colonial administration, in the years after 1930, an estimated 20,000 cyclists each year brought barrels of oil from remote villages in various parts of southern Nigeria to Lagos. In this way, the legions of cyclists made transport by lorry redundant. The colonial authorities then considered a bicycle tax to generate revenue from this activity. In fact, there was such a tax on the ownership of bicycles in Mali in the 1970s (Löfwander 1983).

Slightly earlier than this, in the 1920s, fish traders in Malawi were transporting their perishable goods to market using their own bicycles (McCracken 2011). Similarly, Luther P. Gerlach (1963) reports how in the 1950s in the British colony of Kenya traders with a wide variety of goods built their economic existence on the use of bicycles. Dried fish also plays a role here, but long-distance trade with imported goods is also significant: bicycles are loaded onto trucks for the longer distances and can thus be used in different places for "last mile" deliveries.

Only at first glance does another example of everyday bicycle use show greater proximity to the state. According to Nancy Hunt's (1999, 2005) accounts, bicycles were standard equipment for nurses, who often used them to visit their clients. But these bicycles were more than *official transport devices*: they were seen as signs of moderate wealth. Owning a bicycle was an important goal in the consumption pattern of the emerging colonial middle class. Soldiers in the Congo had been equipped with bicycles as early as 1912, while nurses were also given these means of mobility from the 1920s onwards (Hunt 2005: 135). First of all, bicycles were outstandingly valuable objects that could not easily be put down or stored anywhere.

They were intended, as Hunt quotes verbatim, to be used (Hunt 2005: 136). They remained the transport of choice for nurses, even though from the 1930s at the latest, local chiefs and other dignitaries preferred automobiles as status symbols. It is as if, after the layer of meaning of 'status and modernity', a second layer of meaning came to the fore that continued to make the bicycle desirable, but for a different reason. Now it was about the practical advantages: the ability to reach patients quickly and to facilitate the transport of medical equipment on the carrier. The transition to a widespread, everyday vehicle in the Congo in the 1950s was also supported by massive advertising campaigns that promoted not only the bicycle but also other consumer goods. Hunt (2005: 139) describes in detail the downsizing of prestige and social recognition for ownership and use. Hunt's study is particularly significant because it shows different perspectives on the bicycle. It becomes clear that these different valuations may well relate to the same object at different times.

Terence Ranger (2003: 80) reports from Bulawayo in western Zimbabwe that around 1950, 15 bicycles were registered and taxable in this city. But already twenty years earlier, in 1929, there was a lively traffic of bicycles in the same city. The owners of these vehicles were white-collar workers who used their bicycles to commute between their homes on the outskirts of the city to their places of work (Ranger 2003: 78). Similar to Hunt's account of the Congo, the bicycle was undergoing a transformation: it shifted from a prestige object to an everyday utilitarian object that asserted its economic importance.

Current uses and cultural appropriation

To the present day, the bicycle is economically significant. It is a means of survival for many who have found ingenious uses for the bicycle. As Ignasio Jimu (2008) points out, bicycle business is by no means a preferred activity in Malawi. However, many have no alternative because there are too few 'regular' jobs, so they find uses for the bicycle in the informal sector. But when these people create a cycling business, it does not seem to be without profit. They become 'bicycle taxi entrepreneurs', using their bikes themselves and owning other bicycles that are rented to other taxi drivers. The average daily earnings reported in interviews are about the minimum set by the government but could actually be higher. Still, according to Jimu (2008), many bicycle taxi entrepreneurs are 'semi-literate', having completed primary school only. Some of these entrepreneurs are homeowners and pay for their house with the income from the bicycle taxi business. It is therefore an unjustly despised source of income that does not receive any support from the government.

The term *bodaboda*, mentioned repeatedly in the literature, refers to bicycle (and, more recently, motorcycle) taxis in Kenya and Uganda, where bicycles have been commonplace for a long time due to the extremely high prices of fossil fuels (Bos, Koster and Mulder 2003; Bloemink and Smith 2007: 87). According to Zägel (2012: 73–76), the term *bodaboda* emerged in the 1980s, but – since the 1990s – also includes motorbike taxis. Initially known only in Uganda, where it was used as a means of transport to export locally harvested coffee beans to Kenya, this innovation soon spread further west and as far as Nairobi. While motorbikes subsequently dominated the market in the big cities, the bicycle has remained the key element of the *bodaboda* phenomenon in smaller towns to the present day. At the end of his study, Zägel presents a mixed assessment especially with regard to the motorized taxis: despite the obvious demand for the *bodaboda* and its economic appeal, he considers the health hazards of the drivers and the environmental pollution caused by motorbike taxis to be so serious that he does not see any future prospects for this mobility practice. This puts him at odds with other authors (Heyen-Perschon and Kisamadu 2000), who highlight the development opportunities of bicycle transport in Uganda 20 years ago. In a similar vein, Ken Banks (2008) argues that the use of mobile phones could be an important step towards more efficient use of bicycle taxis as a form of Uber ride-sharing.

The history of the *bodaboda* is also a history of the adaptation and transformation of the bicycle. The industrially produced bicycles that have been delivered to Africa for decades, mainly from China, are not always suitable for the sometimes heavy loads and often-difficult roads or pathways. More recently, bicycles have been promoted as life-saving devices, when used as ambulances (Wallrapp and Faust 2008). The appreciation of the bicycle is therefore based on the possibility of adapting this device to a multitude of purposes.

As the author explored in more detail while conducting ethnographic fieldwork in Burkina Faso, West Africa (Hahn 2012, 2016), bicycles undergo various stages of transformation from the moment of acquisition, adapting them for everyday use. Undoubtedly, new bicycles are highly valued, they are cared for, people avoid lending them out, and as the most expensive consumer item in many rural households, they are given special attention. But after a while, practical issues come to the fore: for example, when the harvesting season reaches its peak, and numerous journeys between the fields and the households become necessary, it may make sense to remove certain parts of the bicycle, so as not to be held up on the road by a detaching mudguard, a jammed rim brake or a broken bicycle stand, which is anyway unable to cope with the heavy load. This first step, the removal of the parts that are not immediately necessary, is followed by a second step, which involves the addition of certain individual parts. This second step is also received as a strengthening for use with heavy loads and on bad roads. The second step also includes less obvious aspects, such as inserting a second tire to better protect the inner tube or attaching a torch to the handlebars with wide rubber straps. *In toto*, this creates a local form of bicycle that is more trusted for durability and ease of repair under conditions of intensive use. Without mudguards and chain guards, a chain is easier to put back on and a bicycle tube is easier to pull out of the tyre and repair. The bicycle has become an 'Africanized bike'.

There is no bicycle that does not need repair from time to time. However, it is important to note that the robust and simplified 'Africanized bicycle' can be easily and quickly repaired by both the users themselves and the repairers. Bicycles are not only used to transport the harvest or to visit the market. An essential role also concerns the transport of goods to the market, and last but not least the daily trips of children to school, as well as the occasional trips to the health center.

As highlighted in the other case studies, the valuation of the bicycle in rural Burkina Faso oscillates between the high esteem of an expensive consumer good and the pragmatic recognition of its many everyday uses. One of the widespread practices of cultural appropriation is the decoration of bicycles, using brightly colored plastic bands around the handlebars and tubes in bright colors to embellish the brake cables. This is current among young people in Burkina as well as in Nigeria (Renne and Usman 1999). Other elements of decoration are brightly patterned saddle covers and small pennants attached to the frame. These changes take place under the eyes of the village public. Sometimes the owners are criticized. Those who invest too much in beautifying their bicycles instead of gradually giving them over to more utilitarian purposes sometimes receive heavy reproaches. They are accused of marrying the bicycle – instead of spending the money on their fiancée or wife – (Hahn 2004). Obviously, the proper place of the bicycle in society is not only a matter of individual preference. Precisely because of its multiple economic contexts, the family and community make a certain claim that bicycles serve for the needs of the family as a whole at a certain point in time, i.e. that they are transformed from luxury objects into everyday tools.

Conclusion: bicycles between appreciation and pragmatism

Although bicycles are used by both men and women in many West African countries, Dorte Thorsen (2007) accurately describes how young men in Burkina Faso, in particular, pursue the acquisition of a bicycle as their first and most important consumption goal for years. Even today, the desire to own a bicycle is a strong and widely recognized motive for participating in labor migration to the plantations in southern Ghana and Côte d'Ivoire. So-called 'targeted migration' describes the situation in which people are engaged in labor migration until they have saved up the necessary amount of money to buy a bicycle as a consumption target.

This ethnographically documented and subjectively relevant desire of many persons to acquire a bicycle is contrasted with the economic evaluation that has already been carried out several times in development contexts. It is undoubtedly correct to differentiate transport tasks and transport costs according to distance, time, gender of the person responsible and the goods regularly transported. Such studies produce a differentiated picture of all transport tasks that practically occur in a given place (Kipke 1991; Dawson and Barwell 1993; Barwell 1998; Calvo 1994). No less important, however, appears to be the perspective of small farmers, whose production often yields little more than the necessary subsistence for the family. From this perspective, trips of short distance on mostly poorly maintained roads are dominant. The peasant perspective combines very different transport purposes (harvest, market, school, ambulances) with a principle of minimizing pecuniary expenditure. Wherever possible, an attempt is made to obtain transport services from neighbors or members of the extended family, thus dispensing with the expense of using a commercial transport. All these criteria increase the relevance of the bicycle in many farming households, even if on the macro level, e.g. in relation to the transport performance of a nation, this one means of transport does not play a prominent role. For this reason, Paul Smethurst (2015: 131–132) must be contradicted: the bicycle is neither predominantly a symbolic object, nor is its economic impact marginal. As the examples gathered for this chapter show, the bicycle has a significant economic impact, especially on peasant households, even if this has only led to a professionalization of bicycle use in a few cases. The value of the bicycle is not always easy to quantify economically, especially when used by the informal sector where accounts are rarely kept. The practices of transforming this vehicle, embedding it in local contexts and cultural appropriation can hardly be represented statistically.

Cultural appropriation and redesign of the bicycle by the users themselves is superior to all attempts to develop specific bicycle models for Africa from the development context (Bloemink and Smith 2007; Bernard 2009). The future of the bicycle in Africa depends on those users who manage to intelligently mediate between the attractiveness of a not entirely cheap but still affordable consumer good and its multiple usages. The bicycle in Africa is neither marginal, nor is it a symbol of colonial oppression. This vehicle has acquired its special intermediary role because there have always been ingenious ways of turning the globally widespread means of transport, standardized by industrial norms, into an Africanized bicycle.

It is a strong sign of the vitality of these locally appropriated technologies that, as we have recently learned, the global company Uber is now moving to include bicycle taxis in its network in Kampala.[1] The innovative nature of African practices of cycling is also evident in the recent introduction of cargo bikes in Europe with the model name *bodaboda*.[2] Engineering has recently developed a new interest in the local production of bicycles in Africa, based on bamboo frames, but also with the intense study of enhanced carriage models (Oberholzer et al. 2017; Oyesiku 2018).

It is no exaggeration to state that the bicycle in Africa is an ambivalent or even contradictory object. While this means of transport is highly valued, especially in many rural areas, and is used intensively, it can hardly claim the role of a prestige object in the more urbanized areas, where the automobile and motorbike play more prominent roles (Pochet 2002). The ambivalence of the bicycle is a characteristic that also describes the situation in Europe and the USA very well. Contradictions are evident in the case of debates in the industrialized countries with regard to the expectations of a prominent contribution by the bicycle to the de-carbonization of mobility and the still moderate intensity of its use. Presumably, neither Africa nor Europe will succeed in overcoming the ambivalences of bicycle use and its discourses. If more attention is paid to practices, this would be the best starting point for a proper understanding in both contexts and both continents.

Notes

1 https://qz.com/africa/983567/in-kampala-uber-and-safeboda-are-trying-to-convince-ugandans-they-dont-need-a-car/
2 https://www.yubaeurope.com/bikes-add-ons/boda-boda/electric-boda-boda-ausverkauft

References

Alber, Erdmute (2002) "Motorization and colonial rule. Two scandals in Dahomey, 1916." *Journal of African Cultural Studies*, 15(1): 79–92.

Appiah, K. A. (1992) *In my Father's House. Africa in the Philosophy of Culture*. Oxford: Oxford University Press.

Barwell, Ian (1998) *Le transport et le village. Conclusions d'une série d'enquêtes-villages et d'études de cas réalisées en Afrique*. Washington: Banque mondiale.

Bernard, Philippe (2009) Des vélos en bambou pour l'Afrique. *Le Monde*, (21.1.2009).

Bloemink, Barbara, and Cynthia Smith (2007) *Design for the Other 90%*. Washington: Cooper-Hewitt, National Design Museum, Smithsonian Institution.

Bos, Erwin, Freek Koster, and Frank Mulder (2003) *The Bicycle Sector in Uganda: An Overview*. (http://www.jugendhilfe-ostafrika.de/wp-content/uploads/2011/12/Overview-bicycles-Uganda.pdf)

Bowden, Gregory H. (1975) *The Story of the Raleigh Cycle*. London: Allen.

Callebert, Ralph (2016) "African mobility and labor in global history". *History Compass*, 14 (3): 116–127.

Calvo, Christina M. (1994) *Case Studies on Intermediate Means of Transport Bicycles and Rural Women in Uganda*. (SSATP Working Paper 12). Washington, DC: World Bank.

Dawson, Jonathan and Ian Barwell (1993) *Roads Are Not Enough. New Perspectives on Rural Transport Planning in Developing Countries*. London: Intermediate Technology Publications.

Esgain, Nicolas (2001) Scènes de la vie quotidienne à Elisabethville dans les années vingt. In: Vellut, J.-L. (Ed.), *Itinéraires croisés de la modernité. Congo belge (1920–1950)*. Paris: L'Harmattan, S. 57–70.

Fisch, Rudolf (1911) *Nord-Togo und seine westliche Nachbarschaft*. Basel: Basler Missionsbuchhandlung.

Foster, Vivien and Cecilia Briceo-Garmendia (Ed.) (2010) *Africa's Infrastructure. A Time for Transformation*. Washington, DC: The World Bank.

Gerlach, Luther P. (1963) "Traders on Bicycle. A Study of Entrepreneurship and Culture Change among the Digo and Duruma of Kenya." *Sociologus*, 13(1): 32–49.

Hahn, Hans P. (2004) "Die Aneignung des Fahrrads". In: Beck, K., T. Förster, and H.P. Hahn (Eds.), *Blick nach vorn. Festgabe f. G. Spittler zum 65. Geburtstag*. Köln: Köppe, S. 264–280.

Hahn, Hans P. (2012) "The appropriation of bicycles in West Africa: Pragmatic approaches to sustainability". *Transfers. Interdisciplinary Journal Mobility Studies*, 2(2):31–48.

Hahn, Hans P. (2016) Use and cycling in West Africa. In: Oldenziel, R. and H. Trischler (Eds.), *Cycling and Recycling. Histories of Sustainable Practices*. New York: Berghahn, S. 15–32.

Hanlon, Joseph and Teresa Smart (Eds.) (2008) *Do Bicycles Equal Development in Mozambique?* Oxford: Currey.

Heyen-Perschon, Jürgen and Richard Kisamadu (2000) *How Can the Bicycle Assist in Poverty Eradication and Social Development in Africa?* Amsterdam: Vélo Mondial.

Hopkins, Anthony G. (1973) *An Economic History of West Africa*. New York: Columbia University.
Hunt, Nancy R. (1999) "Nurses and bicycles". In: Hunt, N.R. (Ed.), *A Colonial Lexicon of Birth Ritual, Medicalization and Mobility in the Congo*. Durham: Duke University, S. 159–195.
Hunt, Nancy R. (2005) Bicycles, birth certificates, and clysters: Colonial objects as reproductive debris in Mobutu's Zaire. In: Van Binsbergen, W. (Ed.), *Commodification. Things, Agency, and Identities. (The Social Life of Things Revisited)*. Münster: Lit, S. 123–141.
Jimu, Ignasio M. (2008) *Urban Appropriation and Transformation: Bicycle Taxi and Handcart Operators in Mzuzu, Malawi*. Mankon: Langaa.
Ken Banks (2008) "Africa's Grassroots Mobile Revolution: A Traveller's Perspective." *Receiver (Vodafone)*, 20: 1–19.
Kipke, Barbara G. (1991) Bicycle usage in two cities of Africa. In: Kipke, B.G. (Ed.), *Bicycle Reference Manual for Developing Countries. African Experiences*. Eschborn: GATE (GTZ), S. 155–173. (http://www.mobility.de/brm/brm_pre.htm)
Lavenir, Cathérine B. (Ed.) (1998) *La bicyclette* (= Cahiers de médiologie, 5). Paris: Gallimard.
Löfwander, Torild (1983) *Die sozialökonomischen Verhältnisse der bäuerlichen Bevölkerung in der Republik Mali*. Berlin Akademie.
McCracken, John (2011) "Bicycles in colonial Malawi: A short history". *Society of Malawi Journal*, 64(1): 1–12.
Nwabughuogu, Anthony I. (1984) "The Role of bicycle transport in the economic development of Eastern Nigeria, 1930–45". *Journal of Transport History*, 5: 91–98.
Oberholzer, J. F. et al. (2017) "Evaluation of resource efficient process chains for primary manufacturing processes of bamboo bicycles". *Procedia Manufacturing*, 8: 36–43.
Olvera, Lourdes D., Didier Plat, and Pascal Pochet (2008) "Household expenditure in Sub-Saharan African cities: Measurement and analysis". *Journal of Transport Geography*, 16: 1–13.
Oyesiku, Olukayode O. (2018) "Development of bicycle and motorcycle carriage for goods mobility in rural areas of Nigeria". *African Journal of Science and Nature*, 6: 114–130.
Parkin, John (Ed.) (2012) *Cycling and Sustainability*. (= Transport and Sustainability, Vol. 1). Bingley: Emerald Group Publishing.
Pochet, Pascal (2002) Le vélo ou le grand absent das capitales africaines. In: Godard, X. (Ed.), *Les transports et la ville en Afrique au sud du Sahara. Le temps de la débrouille et du désordre inventif*. Paris: Karthala, S. 343–355.
Ranger, Terence (2003) Bicycles and the social history of Bulawayo. In: Morris, J. (Ed.), *Short Writings from Bulawayo*. Bulawayo: amabooks, S. 76–81.
Renne, Elisha P., and Dakyes S. Usman (1999) "Bicycle decoration and everyday aesthetics in Northern Nigeria". *African Arts*, 32(2): 46–51.
Schomburgk, Hans (1922) *Bwakukama. Fahrten und Forschungen mit Büchse und Film im unbekannten Afrika*. Berlin: Deutsches Literarisches Institut.
Smethurst, Paul (2015) *The Bicycle. Towards a Global History*. Basingstoke: Palgrave.
Thorsen, Dorte (2007) *'If Only I Get Enough Money for a Bicycle!' A Study of Childhoods, Migration and Adolescent Aspirations Against a Backdrop of Exploitation and Trafficking in Burkina Faso*. (= DRC Working Paper T 21). Brighton: DRC.
Ugorji, Rex U., and Nnennaya Achinivu (1977) "The significance of bicycles in a Nigerian village". *Journal of Social Psychology*, 102(2): 241–246.
Wallrapp, Corinna and Heiko Faust (2008) "Bicycle ambulances in rural Uganda: Analysis of factors influencing its usage". *World Transport Policy & Practice*, 37(3): 10–11.
Zägel, Alexander B. (2012) *Boda Boda. Bicycle and Motorcycle Taxis in Uganda and Kenya. Innovation, Diffusion, Livelihood, Risks*. (= Diploma thesis). Heidelberg: Institut für Geografie.

Vignette K
WHEELS OF FIRE
Women cycling in the Middle East

Alon Raab

On a cool autumn morning Fatma Aliye Topuz rides her tricycle past Constantinople's houses, markets, government buildings and mosques. Men jeer and sometimes try to pull her off her vehicle and prevent her from continuing on her journey but she ignores them.

Born in 1862, Fatma, like most women of her time and place, knew from birth familial and societal restrictions. Daughter of a high government official and a homemaker, she was given away in an arranged marriage to an army officer who, fearing her exposure to dangerous ideas, forbade her to read books in foreign languages, and demanded that she fulfill her destiny of motherhood and housekeeping. Fatma was not destined however to follow a script written by others, carving out for herself a life of independence as an activist for women's rights and a successful journalist and novelist. From her first novel, 1892's *Muhadarat* (*Useful Information*), to her death in 1936, Topuz, a devout Muslim, depicted women creating lives of their own, free of male domination.

Fatma Aliye Topuz's struggle for liberation was aided by the bicycle, a vehicle that became for her, and for many other women across the Ottoman Middle East, a path of personal and social transformation. Bicycles were an important part of the emerging feminist movement in the region.

The first cyclists in the region were western travellers, starting in the 1880s, including several women, notably globe-trotting American Annie Londonderry in 1894–95. Bicycles elicited curiosity and a desire by many to participate. European bicycle manufacturers quickly noted a potential market.

Due to the vehicles' high cost, Constantinople's first riders were predominantly European merchants and the local bourgeoisie. As cost went down, the vehicle was embraced by students and intellectuals. Cyclists soon appeared in other large regional urban centers and in eastern Mediterranean port cities with diverse populations, vibrant economic and cultural exchanges, with many citizens adopting European practices, including mixed gender sociability, conditions conducive to the introduction of bicycles.

As in other lands, opposition to cycling was quick to appear, mostly from conservative elements and religious fundamentalists who often labeled it *the Devil's Chariot*. Several Muslim religious authorities designated it as *bid'ah* (any technological innovation deemed heretical) with bans on cycling in Yemen and Saudi Arabia. In other places most of the attacks – expressed in the press but also in laws and physical assaults on riders – were directed at women cyclists.

 DOI: 10.4324/9781003142041-52

Claims that cycling harms reproductive organs, encourages sexual permissiveness and the destruction of the family were common. Unstated was the desire to confine women to their homes and to prevent unsupervised meetings between men and women.

Still, women cyclists persisted. Across the empire bicycles became prevalent in studio photography, a symbol of middle-class respectability and openness to modern ideas. By travelling unchaperoned to school, work, exercise and sometimes assignations with their beloved, the bicycle offered Ottoman women cyclists a new sense of freedom of mobility that extended to other areas of life.

The collapse of the Ottoman Empire and the birth of the Republic of Turkey were a boon to women's greater participation in public life and greater visibility in public space. The latter included a dramatic increase in the number of women cycling, encouraged by the inclusion of physical education classes in schools and more positive portrayals of female riders in the press, literature and films.

In Palestine and Lebanon, British and French rule also led to more women cyclists among the urban and educated elites. The Jewish women's cycling community in Palestine was bolstered by the arrival in the 1930s of German and Central European refugees who had previously cycled and continued in their new environment, often in kibbutzim where cycling was safe and popular.

The decades from the 1950s saw some slow increases in the number of women cyclists, but also some setbacks. Periods of greater growth coincided with openness to western values, greater security and improvement in the status of women. These included Lebanon in the 1950s, Iraq in the early 1960s, and Iran before the Shah's fall.

In the new Millennium, the number of women cyclists in Jordan, Lebanon, Israel and Turkey have grown. *Bisikletli Kadın İnisiyatifi*, Turkey's first women's cycling group, was established in 2017: through talks and rides, with chapters in several cities, it has increased public awareness of accomplishments and created a supportive community. Women from across the region (and from other lands) have embarked on *Follow the Women* rides for peace. Cycling through Lebanon, Syria, Jordan and Palestine, women highlight suffering and the need for justice. Rides carrying food to refugees and material aid and support to Palestinians have also taken place.

Iran and Saudi Arabia remain lands where being a woman cyclist remains dangerous. A 2013 ban on women cycling, enacted by Saudi's *Committee for the Promotion of Virtue and The Prevention of Vice*, was finally overturned, but social mores and local laws banning riding are still powerful. Women must wear Abayas (long modesty robes), ride only in parks or other confined areas for recreational purposes and must be accompanied by a male guardian.

In Iran, a 2016 Fatwa by Ayatollah Khamenei, the supreme leader, banned women cycling as "Riding a bicycle often attracts the attention of man and exposes the society to corruption, and thus contravenes women's chastity." Local governments have threatened violators with "Islamic punishment". Yet across the Islamic Republic women resist these edicts, despite arrests and public censure.

Cycling found powerful filmic expression in the 2000 Iranian film by Marzieh Meshkini, *The Day I Became a Woman*, and the 2012 Saudi *Wadjda* by Haifaa Al-Mansour in which a young girl enamored with cycling wins a Quran recital competition but is denied the money for a bike by her father. Her father's new wife rebels and gives Wadjda the money and in the film's final scene she races and beats her friend Abdullah.

Middle Eastern women cyclists will be inspired by early trailblazers and by contemporaries such as: the Baghdad artist Marina Jaber, whose ride, part of an art project, blossomed into rides of dozens of women and men; the Iranian Poupeh Mahdavi Nader, who has been cycling around the globe, raising funds for refugees "in the name of love, friendship and global peace."; the Saudi sisters Fatima and Yasa Al-Bloushi, Dina Al-Nasser and Anoud Aljuraid, who founded the HerRide group in Jeddah; and Sana'a photographer Bushra Al-Fusail who started the *Yemeni Women Bike Group*.

The latter organization drew to its initial 2014 ride only 14 women, of which 10 had never cycled before, but, in the words of its founder, "*Biking was our way of showing that nothing can stop us – not bombing, not cultural taboos, this is our right; we have a right to live and the right to movement.*" These cycling projects challenge social norms and legal restrictions and open discussions about women's lives, planting seeds of independence and rebellion. The wheels of fire keep turning.

Bibliography

Al-Saltan, Taj (2003) *Crowning Anguish: Memoirs of a Persian Princess: From the Harem to Modernity*. Washington, DC: Mage Publishing.

Arzu, Öztürkmen (2013) "The women's movement under Ottoman and Republican rule: A historical reappraisal". *Journal of Women's History*, 25(4), 255–264.

Ross, Hannah (2020) *Revolutions: How Women Changed the World on Two Wheels*. New York: Plume Books.

35

CYCLING IN INDIAN CITIES

Between everyday cyclists and affluent cyclists

Rutul Joshi and Jacob Baby

Introduction

A news story broke in the Indian media in May 2020 that a 15-year-old girl, Jyoti Kumari, cycled with her injured father (an auto rickshaw driver) as a pillion rider close to 1200 km between Gurgaon near Delhi to her village in the Darbhanga district, Bihar (Ray 2020). It was a remarkable journey, but at the time not an especially unusual one. In March–April 2020, severe lockdowns were imposed all over India due to the COVID-19 pandemic bringing all kinds of transport services to a standstill. The pandemic wrecked livelihoods, especially those of urban informal sector and low-income workers – many of them among the country's 455 million migrants – who found themselves out of work and surviving on meager savings (Pandey 2020). A mass exodus began and while some who could afford it travelled in buses or trains (if they were available), many simply walked or cycled to their villages (Rukmini 2020). Media reported migrant workers' efforts to purchase or rent bicycles (one of them stole a bicycle and left a letter of apology), travelling up to 2000 km to their home villages (Hindustan Times 2020). Some bundled up their belongings on a bicycle and travelled at night to avoid the summer heat. They cycled close to 100 km a day, taking shifts in the afternoon, surviving on the limited supplies of food they carried or on the charity of strangers on the way (Al Jazeera 2020). For many poor workers like Jyoti Kumari, a cycle is their most reliable means of mobility.

A year later, a different kind of pandemic-related news story broke from Ahmedabad (a city of about eight million people), where the city police initiated a special drive to catch late-night cyclists violating the pandemic evening curfews (Times of India 2021). This is another side of cycling in Indian cities, as wealthier individuals cycle for health or recreational reasons (Chapter 26). During the pandemic, bicycle sales doubled as people switched to cycling as an exciting alternative, raising important questions around sustainability, resilience, and mobility in Indian cities.

The news stories of Jyoti Kumari and affluent night cyclists in Ahmedabad show contrasting practices of cycling in the Indian city by different income groups. While one story touts cycling as a healthier and sustainable alternative to motorized transport and encourages a modal shift, the other sheds light on the limited choices of the urban poor and their dependence on a non-motorized cycle. In both cases, cycling is a subversive practice that defies the norms of the lockdowns or curfews. In the case of the affluent cyclist, subversion takes place with a sense of entitlement and in the case of Jyoti, with a sense of desperation. The dichotomy

DOI: 10.4324/9781003142041-53

between cycling as a special activity for better health or recreation versus cycling as a practice of everyday life is an apt starting point to think about the state of cycling in Indian cities.

Census data for 2011 show around 40% of all trips in urban India involve walking and cycling, while a small percentage occurs on private motorized transport. However, conventional transport planning and governance prioritize building more space for these private vehicles to facilitate their movement (Tiwari 2007; Abhishek 2020). Pedestrians and cyclists are reduced to the margins, often finding limited or no space/infrastructure to support their movement (Badami 2009). In the process, walking or cycling on Indian roads becomes precarious, making pedestrians and cyclists vulnerable to fast-moving traffic (Tiwari 2011). Given these realities, the renewed interest in cycling in the city is welcome, but it is important to unpack what this entails.

In this chapter, we explore the histories, practices and socio-economic groups engaged in cycling in Indian cities. We also examine the various policy-level focuses and the newer initiatives that support cycling. Finally, we argue that there needs to be concerted efforts toward making cycling safe, accessible, and affordable for all kinds of user groups. These efforts will sustain the current momentum for cycling and in the long run leads towards sustainable and equitable urban transportation.

Who cycles in the Indian city?

Cyclists in Indian cities can be broadly categorized into two groups: 'no-choice everyday' cyclists and 'affluent recreational' users. Neither of these categories are homogeneous, and their social identities are an important element in defining their mobility practices. In large cities of India, around 7–15% of trips are made on bicycles, while in smaller cities this is around 13–21% of trips, most of which are by lower-income households, who cannot afford other modes of transportation. (Arora 2013; Tiwari & Jain 2008; TRIPP, IIT Delhi, & Shakti Foundation 2012). Most low-income households in urban India spend between 50 and 70% of their income on food and housing. The urban poor tend to locate close to their workplaces so as to spend less time and money on commuting. Public transport fares are expensive, so the choice is to walk or cycle or rely on intermediate public transport such as shared autos.

Faiz (2009) captures the characteristics of low-income everyday cyclists in Ahmedabad. They earn around Rs. 2800–4500 rupees ($38–$61) a month and are employed in informal daily wage occupations such as delivery and door-to-door salesman (Das Gupta & Puntambekar 2016; Chapter 3). Their average trip lengths are between 3.5 and 5 km, cycling around 20–25 minutes to reach their destination. Joshi (2014) captures the qualitative experiences of these cyclists in the city of Ahmedabad. Cyclists navigate heavy traffic and they avoid certain busy routes to save time and effort. Everyday cyclists find cycling becoming increasingly difficult due to longer distances, increasing traffic and reduced cycling space on the street. It is noteworthy that most women from low-income households prefer walking to cycling given the restrictions on using a cycle forced upon them by their male family members (Chauhan 2016). If there is any mode of transport in a low-income household, men would claim them as their first right. Arora (2008) also point out how street designs and paths discourage women from cycling on these streets. Such narratives indicate that women do not enjoy the same levels of freedom as men in determining their mobility choices.

Lack of safe cycling paths, and incomplete or absent pedestrian walkways, means cyclists and pedestrians are part of the mixed traffic and exposed to potential fatalities caused by fast-moving vehicles. The no-choice cyclist is often a male, the income-earner of a poor household (Faiz 2009), and carries enormous risk when he moves in this mixed fast traffic. A single fatality or accident is a massive blow to these households, hence everyday cycling correlates with the vulnerability of its users.

The second category of cycle users are the recreational cyclists who generally belong to upper-middle-class groups whose primary mode of transport is a private automobile (taxis included). Suh (2015) notes the direct economic benefits of cycling, including physical fitness and a recreational value (Chapter 30). The indirect benefits include cleaner air due to zero emissions and a low carbon footprint. Anantharaman (2017) points out how middle-income groups in Bangalore saw both the direct and indirect benefits of cycling and adopted the practice. At the same time, there are concerns about everyday cycling related to safety and urban pollution that dissuade these groups from adopting the cycle for commuting.

There is a stark difference between 'everyday cycling' practiced by 'captive' cyclists and 'recreational' cycling practiced by affluent groups. The latter ride high-end cycles, buy trendy equipment and clothing, and form cycle groups to promote health by engaging in cycling activities such as long-distance night rides. The 'captive cyclist' depends on his bicycle as all other modes of transport, with the exception of walking, are unaffordable. The 'recreational cyclists' see the cycle as a recreational activity rooted in new ethical and cultural norms. As Joshi and Joseph (2015) mention, both these groups are heterogeneous, and have their own class and gender distinctions, and face varying degrees of inclusion and exclusion. However, it is crucial to ask whether everyday cyclists and affluent cyclists can be allies while lobbying for better cycling infrastructures.

Evolution of cycling in Indian cities

In pre-independent India, census and linguistic reports state that bicycles were a major mode of transportation and found in increasing numbers in towns and villages (Census of India 1931; T.C. Hodson 1937; cf. Chapter 36). The first set of 35,000 cycles in India were imported from Europe in 1910 (Arnold 2015). Some Indian states even issued bicycle licenses and the traffic police regulated bicycles as mainstream traffic. Traffic police required cyclists to affix dynamo lights on their cycles while traveling at night.

India's recent urban history can be understood across three timelines. The first of these begins in the early 1950s, when an independent India grappled with poverty and civil strife while trying to modernize rapidly. Cities like Mumbai, Kolkata and Ahmedabad, along with industrial townships such as Jamshedpur and, Bhilai, emerged as new industrial centers. For the working class cycling was a popular choice to move around in the city. Students, office workers and even professionals cycled in Indian cities during the 1940s and 50s. Nottingham-based Raleigh collaborated with an Indian entrepreneur to set up the Sen-Raleigh bicycle manufacturing in Asansol, West Bengal in 1952 (TTW 2021). In the same year, Atlas Cycle Industries produced the first fully Indian-made bicycle in Sonepat, Haryana. Atlas became the largest manufacturer of cycles in India by the mid-1960s (Karelia 2021). The cycles were marketed as a 'good choice for hard working men'. It was rare to find women cycling in cities. Cycling has always been gendered in India, with their use dominated by men.

The notion of a "modern Indian city" was conceived around this period and was realized in the planning and development of Chandigarh, designed by Le Corbusier as a futuristic city with the car as the dominant mode and a street network conducive to fast, free-flowing traffic: cycling and walking streets were in the network's 'lowest hierarchy' (Joshi & Joseph 2015). Chandigarh was designed with inward-looking neighborhood units, separated by wide roads with dead public edges, setting the precedent for a modern Indian city. This "automobile-centrism" symbolized speed, progress, and modernity. Nevertheless, outside of a handful of large cities, India's major towns still had a considerable percentage of bicycle users.

The second timeline is situated between the 1970s and the late 1980s. Escaping poverty, migrants from the countryside flocked to cities with new factories opening and informal service industries scaled up. Many settled near to their workplaces and walked or cycled to work (Joshi 2014). At the same time, the number of private cars and motorbikes increased from 0.3 million in 1951 to 1.9 million in 1971 (MoRTH 2019). The bicycle was an affordable alternative and was used by people across various income categories. Five-Year Plans at the national level and master plans at the city focused on improving shelter for the urban poor and provision of infrastructure in urban areas. Urban transportation was focused on building or widening roads that connected cities, towns, and villages, or improving suburban commuter rail or public bus systems (Vaidyanathan et al. 2013), although Pune city's master plan developed in the early 1980s imagined Pune as a bicycle city with dedicated cycle lanes. Unfortunately, it never materialized due to lack of funds and political will.

The third timeline runs from the late 1980s to the 2000s, with a liberalizing economy and a surge of urbanization and motorization. Private motorcycles and cars became affordable to a growing middle class. These private modes offered comfort and convenience over public transport and became a symbol of prosperity. Cars and motorcycles were marketed as an aspirational good, so most users who used to walk or cycle and who could afford them shifted to motorcycles and scooters. This meant that walking, cycling, using public transport or other shared modes was now being used only by those who could not afford the private modes. Streets in most cities were being developed as 'conduits' for these private vehicles, in stark contrast to the Indian street that was once 'vibrant, diverse & accommodating' (Edensor 2021; Anjaria 2012). The left-over marginal spaces shrank or were appropriated for 'free parking', as all other users, including pedestrians, cyclists, vendors and hawkers, had to contest for space.

Cycling moved from being a legitimate and preferred mode of movement in the 1950s to an unimportant, neglected and some cases, illegitimate, mode of transport in the Indian city by the early 2000s. Since then, there have been many efforts to change the direction of the urban transport policies and infrastructure building towards sustainable modes of transport.

Cycling in transport policies and programs

In recent years (2005–present), policies such as the National Urban Transport Policy (NUTP) and the Jawaharlal Nehru Urban Renewal Mission (JNNURM) emphasized building cities that support non-motorized transport modes that "move people and not vehicles." In 2005, the National Urban Transport Policy (NUTP) had emphasized the need to improve urban transportation networks by integrating land use and transportation, investing in high quality public transport, and '*enabling safe, affordable, quick, comfortable, reliable and sustainable access to city residents*' (MoUD 2006). The policy identified the key problems plaguing the urban transportation sector, including increasing numbers of private vehicles, congestion and decreasing road space for non-motorized modes. It also highlighted the increasing cost of travel for the poor as they find it riskier to walk or cycle. NUTP urged cities to promote walking and cycling as low-cost and low-emission modes of transport, much suited to short trips and emblematic of sustainable mobility. The NUTP was supported by public finance available under JNNURM, an urban renewal program introduced in 2005, designed to develop '*cities as engines of economic growth*' by improving urban infrastructure, with financial support from the central and state governments. All JNNURM funded projects had to meet the objectives of the NUTP, opening up a space for India's first-generation sustainable mobility projects.

Several cities had proposed adding Bus Rapid Transit Systems (BRTS) to improve their public transportation system. It was also proposed that the BRTS corridors be developed on the basis of 'complete streets' approach where the proposed street design included the needs of all road users. Ahmedabad was one of the first cities to develop a 90-kilometer-long BRTS network. 65 km of footpaths and 20 km of cycle paths were built beside the BRTS tracks to facilitate a tight integration between NMT modes and BRTS. The authors found that these footpaths or cycle paths have not created safe infrastructure: the paths are broken, absent, poorly maintained, not universally accessible and frequently blocked by debris, garbage, and vehicular parking.

Despite the statistics supporting the relevance of NMT (non-motorized transport), and even though India has a very low per capita car ownership rate (22 per 1000 people) (UITP 2015), motorized traffic was prioritized rather than radically changing the infrastructure to accommodate all road users. Cyclists, pedestrians and others are 'invisible' or 'disadvantaged citizens' on the street (Joshi 2014; Chapter 38). In an interesting turn of events, Kolkata, a city often dominated by the left-wing politics and termed 'the bicycle capital', moved a resolution to ban cyclists from entering 176 major roads 'for their own safety' (Tandon 2013; Sur 2017). It was argued that cyclists create traffic jams, congest city roads and slow down traffic (Ghosh & Sharmeen 2021). In a city where close to 2.5 million cycle trips happen on a daily basis (Bera & Down to Earth 2013), such legislation is indicative of the deep-rooted bias of urban transport policy toward recognizing automobiles as the 'only' legitimate users of road space. Ghosh & Sharmeen (2021) point out how, despite the repeal of the cycle ban in Kolkata, police exploited the poor cyclists' ignorance of the law.

A review of the JNNURM programs show that the majority of funds (50–55%) allocated for transportation infrastructure were utilized in road-widening and flyover construction projects and around 35–37% devoted toward building Bus Rapid Transport System (BRTS). Little was spent in improving pedestrian pathways (around 1.3% of allocated funds) or for cycling infrastructure (IIHS 2012). However, the National Urban Transport Policy combined with the urban renewal mission supported India's first-generation sustainable mobility projects, including a few attempts at building cycling infrastructure or designing 'complete streets'. Some of these attempts paved a way to the second-generation cycling projects – the bicycle master plan of Pune, the emergence of public bicyclesharing systems and, more recently, the Cycle4Change program at the national level.

Planning for cycling at city level – Pune bicycle plan 2017

Pune was one of the first cities in India to have a dedicated cycling master plan. The Pune Development Plan of 1980 had proposed pathways to facilitate cycling. The Pune Bicycle Plan (PBP) of 2017 notes that although Pune cyclists had a 27% modal share during the 1970s–80s, growing use of private vehicles brought the cycling share to a meager 3% by 2017. The Pune Bicycle Plan of 2017 was the culmination of several efforts by various stakeholders to promote cycling as an efficient alternative to motorized transport. The plan identifies issues such as a lack of dedicated cycling infrastructure, friction with motorists and a lack of institutional support to promote NMT. Many stakeholders stressed the lack of safe bicycle tracks dissuaded them from cycling. The PBP 2017 proposed 800 km of cycling paths along major city streets and dedicated infrastructure to promote a modal shift to cycling. The plan proposes a 25% modal share increase for cycling from the current 3% by 2031.The most important shifts would be motorized two-wheelers, long-distance walkers and a 1% shift of trips by people in cars, autos, buses of trip lengths 0.5 to 5 km. A key takeaway from this process is the need for institutional support at the municipal level with various civil society groups promoting cycling as an affordable safe alternative for moving around the city of Pune (Chapter 6). Alas, despite continuous lobbying, four years after making the plan, its implementation has not started apart from some pilot projects on a couple of streets.

Making cycling mainstream – Cycle4Change program

In June 2020, the Smart Cities Mission of the Ministry of Housing and Urban Affairs (MoHUA), along with ITDP India, announced the Cycle4Change program. The program focused on promoting the cycle as a major mode of transportation in post-pandemic cities. The program brief reads as "*An opportunity for Indian cities to create safe, attractive, inclusive and quick cycling solutions*" (Ministry of Housing & Urban Affairs et al. 2020). The program, which was open for cities with populations greater than half a million under the Smart City Mission, was designed as a participatory pilot project where citizens and city officials identified preliminary cycle tracks, and later developed a cycle plan that could be scaled up across the city. The first phase of the program carries out a pilot intervention and the second phase focuses on scaling up and working towards institutional support by establishing NMT cells at the municipal level. Cities are scored on their performance, with the top scorers given adequate funding from the Ministry and support from ITDP India to further develop their cycling plan.

The program rolled out with a series of technical assistance and capacity-building initiatives from ITDP India and the Ministry for participating cities. 107 cities registered for the program. Cities were encouraged to hold citizen consultation campaigns and conduct outreach programs outlining the benefits of cycling in the city. More than 40 cities conducted handlebar surveys and identified major 'pain points' and used this information to conduct their pilot interventions. As of February 2021, 41 cities had submitted their proposals and 25 were selected as finalists for a detailed review process that took place in March 2021. Of these 25 cities, 11 cities were selected to be the first-stage awardees. Awareness about personal health and urban recreation in the post-Covid world has contributed to many cities being excited about promoting cycling. Cities like Kohima, Nagpur, New Town (Kolkata) and Warangal were making their first-ever attempt to build cycling infrastructure. Critical examination of the Cycle4Change challenge reveals that some of the crucial institutional mechanisms, sustained financing or developing quality infrastructure for cycling are still not addressed. But these are second-generation sustainable mobility projects where the idea of taking cycling seriously is scaled up to more than 100 cities.

Apart from the Cycle4Change program, several programs and policies have attempted to tap into these choices to persuade users to shift from motorized modes and consider cycling and walking. Car-free days, safe streets, and cycle to work are amongst the many diverse initiatives in Indian cities. Several state governments have introduced welfare schemes where cycles are given free of cost to a large segment of the low-income population to help them reach various opportunities. While the idea of cycling is gaining popularity in Indian cities, it is far from becoming a sustained movement. Urban cycling still remains a marginal urban activity or a subculture. If cycling is to be a mainstream commuting choice, it must offer safety and convenience when moving from one place to another with necessary dedicated infrastructure and continuing policy support.

Making cycling available & affordable – public bicycle sharing system

Mass transit systems operate on fixed routes with first and last-mile connectivity posing significant challenges in many Indian cities. Cycling can easily overcome many of the challenges provided there is adequate infrastructure and easy access to a supply of bicycles.

Everyday cyclists obtain bicycles by purchasing second-hand ones, often via informal vendors present in many low-income settlements as mentioned by Faiz (2009) and Joshi (2014). They operate on networks of trust and charge for renting and maintenance. Affluent cyclists purchase sophisticated cycles, built to withstand difficult terrains and long distances. Owning a cycle is advantageous but they need to be maintained and parked in a secure place given the risk of theft. These burdens of ownership dissuade some users from participating in everyday cycling. In cities where mass transit plays an important role, users need to be able to take their bicycles on the transit so that they can cycle to their first and last mile connections.

To cater to the growing demand for cycling and to promote a cycling culture across all groups, Public Bike Sharing (PBS) schemes present a viable option (Chapter 23). PBS is a flexible transportation system where cycles are rented for shorter periods at low cost at convenient places such as public transport nodes (MoUD & ITDP 2012). A short-distance trip of 5 km in Mumbai costs Rs. 120 on Uber Go ($1.64), Rs. 60 on an auto, Rs. 5 on a bus or a second-class local train, and Rs. 2 on a cycle (Diwan & George 2021). This shows that short-distance trips on cycles are really affordable across all income segments. Potential PBS trips include trips to transit points, trips for education (to and fro to college or hostels), trips for recreation, trips for retail shopping and trips to work. Many Indian cities are installing PBS systems. Some PBS schemes, such as MyByk, have partnered with transit agencies such as the Kochi Metro to provide cycles in the station area, allowing metro riders to complete the last leg of their journey on a cycle.

PBS is a capital-intensive model and needs a critical mass of users to sustain its revenues. The costs for a typical PBS system with a strength of 2500 cycles can go up to Rs. 0.1 million per cycle (MoUD and ITDP 2012). They need major subsidies and risk sharing with city governments to stay afloat. PBS costs a fraction of public transport, but creative risk-sharing models and policy support is required for it to flourish. While the extremely low user fees are attractive, most PBS requires upfront security deposits from users to use the system and works entirely on a digital-cashless system. This could exclude a large number of everyday cyclists who cannot afford to pay the deposit and have limited access to the digital-cashless system. The reliance on the "internet of things," operating in service areas that are largely city centers and limited entry points to the system can impede PBS from achieving it's intended outcomes.

Secondly, A successful PBS system involves considerable risk-sharing between public and private agencies. While the public agencies take care of policy and financial support, the private agencies take care of the day-to-day operation of the system. Several PBS operators have wound up their operations due to dwindling revenues from user charges and the inability to sustain the initial momentum driven by discounts and rewards. PBS systems cannot be financially self-sufficient: they need major subsidies to stay afloat. The key takeaway is investing in PBS and cycling infrastructure costs a fraction of the cost when compared with usual transportation projects. It is also important to link cycling infrastructure upgrades while expanding the PBS system.

Conclusion

Joshi & Joseph (2015) found cycling in Indian cities was steadily declining in spite of the first-generation cycling projects in India. Given the recent winds of change supported by the second-generation cycling initiatives, everyday cycling has seen a resurgence amongst affluent recreational cyclists. It is also noteworthy that urban local bodies and civil society organizations have recognized the importance of cycling in many cities. Programs such as Cycle4Change and Public Bike Sharing are working toward mainstreaming cycling. With the shift from saying, "Who cycles these days anyway?" to a senior government official cycling to work, supporting cycling initiatives in various cities, cycling is achieving some legitimacy in the eyes of the state and the citizens.

It is important, however, to continue to examine some of these initiatives from the point of inclusion/exclusion, accessibility, and the larger paradigm of sustainable mobility in cities. The urban poor are the 'everyday cyclists' riding in mixed traffic, incurring the wrath of motorists, and faces the risk of an accident. While data reaffirm the struggles of everyday cyclists, they are still missing from major policy initiatives. They are at the mercy of law enforcement who consider them illegal and not part of the traffic or the street. Their continued invisibility from the state apparatus deprive the poor of accessibility to destinations, opportunities and their right to the city.

In contrast, affluent cyclists have more political influence to 'get things done'. They are mobilized for planned rides and cycling clubs, but they are yet to campaign for better cycling infrastructure in a sustained manner. It is not enough to publicize the benefits of cycling in Indian cities, there needs to be tangible efforts to make cycling safe and accessible by infrastructure building and policy support. Could the recreational cyclists form an alliance with low-income everyday cyclists in demanding better infrastructure and safe spaces for cycling? When the identities of the cyclists merge so that the leisure rider becomes an everyday cyclist big change may be possible. An important pathway to better conditions for urban cycling in Indian cities requires an alliance between affluent cyclists and the everyday cyclists to demand a better cycling environment.

Bibliography

Abhishek, Vineet. (2020). Inadequate urban transportation facilities leave the poor in India high and dry. *EPW Engage, 55*(28–29). https://www.epw.in/engage/article/inadequate-urban-transportation-facilities-leave

Al Jazeera. (2020, October 19). India's 'cycle girl': How pandemic changed a Dalit family's life. https://www.aljazeera.com/features/2020/10/19/indias-cycle-girl-how-pandemic-turned-fortunes-for-dalit-girl

Anantharaman, M. (2017). Elite and ethical: The defensive distinctions of middle-class bicycling in Bangalore, India. *Journal of Consumer Culture, 17*(3), 864–886. https://doi.org/10.1177/1469540516634412

Anjaria, Jonathan Shapiro (2012, August). Is there a culture of the Indian street? *Seminar, 636*, 21–27.

Arnold, D. (2015). *Everyday Technology: Machines and the Making of India's Modernity* (Paperback edition). The University of Chicago Press. https://doi.org/10.7208/chicago/9780226922034.001.0001

Arora, A. (2008). A gendered perspective on cycling, in: Tiwari, G., Jain, H., Arora, A. (Eds.), *Bicycling in Asia. Interface for Cycling Expertise (i-CE)*, The Netherlands, p. 160.

Arora, A. (2013). Non-Motorized Transport in Peri-urban Areas of Delhi, India (Global Report on Human Settlements 2013, p. 23) [Case Study]. UN Habitat. http://www.unhabitat.org/grhs/2013

Badami, M. G. (2009). Urban transport policy as if people and the environment mattered: Pedestrian accessibility the first step. *Economic and Political Weekly, xliv*(33), 43–51.

Bera, S., & Down to Earth. (2013, July 31). Is cycling a crime? https://www.downtoearth.org.in/coverage/is-cycling-a-crime-41625

Census of India. (1931). Census 1931 volume 1. Census of India. http://piketty.pse.ens.fr/files/ideologie/data/CensusIndia/CensusIndia1931/CensusIndia1931IndiaReport.pdf

Chauhan, R. (2016). *Gender, mobility and everyday cycling: The case of Ahmedabad.* [Master's Thesis, CEPT University]. http://hdl.handle.net/20.500.12725/14210

Das Gupta, P., & Puntambekar, K. (2016). Bicycle Use in Indian Cities: Understanding the Opportunities and Threats. *12th Transportation Planning and Implementation Methodologies for Developing Countries*, Mumbai.

Diwan, P., & George, A. (2021, January). How can public bike-sharing initiatives thrive in Indian cities? *WRI INDIA*. https://wri-india.org/blog/how-can-public-bike-sharing-initiatives-thrive-indian-cities

Edensor, T. (2021). The culture of the Indian street. In *Public Space Reader*. Routledge. https://www.taylorfrancis.com/chapters/edit/10.4324/9781351202558-66/culture-indian-street-tim-edensor

Faiz, S. (2009). *Bicycle as a mode in low-income households of Ahmedabad: Practices and prospects.* [Master's Thesis, CEPT University]. http://hdl.handle.net/20.500.12725/15317

Ghosh, B., & Sharmeen, F. (2021). Understanding cycling regime transition and inequality in the global south: Case study of an Indian megacity. In Zuev, D., Psarikidou, K., Popan, C. (Eds.), *Cycling Societies Innovations, Inequalities and Governance* (1st ed., p. 18). Routledge: London. https://www.routledge.com/Cycling-Societies-Innovations-Inequalities-and-Governance/Zuev-Psarikidou-Popan/p/book/9780367336615

Hindustan Times. (2020, May 16). Migrant worker steals a cycle to reach UP, leaves a moving apology note. *Hindustan Times*. https://www.hindustantimes.com/india-news/migrant-worker-steals-a-cycle-to-reach-up-leaves

Hodson, T.C. (1937). *India Census Ethnography 1901—1931. Government of India.* The Manager of Publications: Delhi. https://archive.org/details/in.ernet.dli.2015.226791/mode/2up

Hull, A., & O'Holleran, C. (2014). Bicycle infrastructure: Can good design encourage cycling? *Urban, Planning and Transport Research*, *2*(1), 369–406. https://doi.org/10.1080/21650020.2014.955210

IIHS. (2012). JnNURM: An opportunity for sustainable urbanisation. https://iihs.co.in/knowledge-gateway/wp-content/uploads/2017/11/JNNURMAn-Opportunity-for-Sustainable-Urbanisation-Final-Report-_final-1.pdf

Jain, H. (2012). *Development of a bicycle demand estimation model incorporating land use sensitive parameters: A case of Pune city* [PhD Thesis]. Indian Institute of Technology, Delhi.

Joshi, R. (2014). *Mobility practices of the urban poor in Ahmedabad (India)* [PhD Thesis, UWE Bristol, University of the West of England]. https://uwe-repository.worktribe.com/output/810291/mobility-practices-of-the-urban-poor-in-ahmedabad-india

Joshi, R., & Joseph, Y. (2015). Invisible cyclists and disappearing cycles. *Transfers*, *5*(3), 23–40. https://doi.org/10.3167/TRANS.2015.050303

Joshi, R., & Mahadevia, D. (2012). *Promoting low carbon transport in India and the challenges of social inclusion: The bus rapid transit (BRT) case studies in India.*

Karelia, Gopi. (2021, January 25). Born in a Tin Shed, Swadeshi 'Atlas' became India's largest cycle manufacturer by 1965. *The Better India*. https://www.thebetterindia.com/247725/atlas-cycles-made-in-india-atma-nirbhar-brand-childhood-nostalgia-gop94/

Ministry of Housing & Urban Affairs, ITDP, Smart Cities Mission, & National Institute of Urban Affairs. (2020). *India cycle 4 change challenge*. https://smartnet.niua.org/indiacyclechallenge/wp-content/uploads/2020/11/04112020-C4C-Brief.pdf

Ministry of Road Transport & Highways. (2019). *Road transport year book 2016–17*. Ministry of Road Transport & Highways, Government of India.

Ministry of Urban Development, & ITDP India. (2012). *Toolkit for public cycle sharing systems (DRAFT).*

Ministry of Urban Development, Government of India. (2006). National urban transport policy. http://www.indiaenvironmentportal.org.in/files/TransportPolicy.pdf

Pandey, Vikas. (2020, May 20). Coronavirus lockdown: The Indian migrants dying to get home. *BBC News*. https://www.bbc.com/news/world-asia-india-52672764

Pune Municipal Corporation. (2017). *Comprehensive bicycle plan for Pune.*

Ray, Umesh Kumar (2020, May). From Gurgaon to Bihar, 15-year-old girl cycles 1,200 km with injured father. *The Wire*. https://thewire.in/rights/jyoti-kumari-bihar-gurgaon-cycle-covid-19-lockdown

Rukmini S. (2020, May 28). Why India's "migrants" walked back home. https://www.livemint.com/news/india/why-india-migrants-walked-back-home-11590564390171.html

Suh, J. (2015). Economics of everyday cycling and cycling facilities. In *Cycling Futures* (pp. 107–130). https://digital.library.adelaide.edu.au/dspace/bitstream/2440/107696/2/hdl_107696.pdf

Sur, Malini. (2017, February). In Kolkata, citizens defy police attempts to squeeze bicycles off the road. *The Scroll*. https://scroll.in/article/828176/in-kolkata-citizens-resist-police-attempts-to-squeeze-bicycles-off-the-road

Tandon, Rahul. (2013, September). Why has India's Calcutta city banned cycling? *BBC News*. https://www.bbc.com/news/world-asia-india-24237390

Times of India. (2021, June 21). Ahmedabad: Cyclists booked in Bodakdev, Vastrapur. *Times of India*. https://timesofindia.indiatimes.com/city/ahmedabad/cyclists-booked-in-bodakdev-vastrapur/articleshow/83698883.cms

Tiwari, G. (2011). Key mobility challenges in Indian cities. *International Transport Forum Discussion Paper Series*, OECD.

Tiwari, G., Arora, A., & Jain, H. (2008). *Bicycling in Asia.*

Tiwari, G., & Jain, H. (2008). Bicycles in urban India. *Institute of Urban Transport (IUT) Journal*, 7, no. 2, 59–68.

Tiwari, Geetam. (2007). Urban transport in Indian cities. *Urban Age, LSE Cities*. https://LSECiti.es/u33081366

TRIPP, IIT Delhi, & Shakti Foundation. (2012). *Planning and Design Guideline for Cycling Infrastructure*. https://shaktifoundation.in/wp-content/uploads/2017/06/NMT-Guidelines.pdf

TTW (2021, February). 19th century Bengali hero who introduced bicycles to Indians. *Get Bengal*. https://www.getbengal.com/details/19th-century-bengali-hero-who-introduced-bicycles-to-indians

UN Environment Program. (2019, June 11). Cycling, the better mode of transport. https://www.unep.org/news-and-stories/story/cycling-better-mode-transport

Union Internationale des Transports Publics (UITP). (2015). Mobility in cities database. *UITP*. https://cms.uitp.org/wp/wp-content/uploads/2020/06/MCD_2015_synthesis_web_0.pdf

Vaidyanathan, V., King, R. A., & de Jong, M. (2013). Understanding urban transportation in India as polycentric system. *Policy and Society*, *32*(2), 175–185. https://doi.org/10.1016/j.polsoc.2013.05.005

36

THE RISE OF THE *KINGDOM OF BICYCLES*

Xu Tao

The story of the bicycle in China is not one of linear development: it is more like a suspense movie, with wonderful ups and downs, and the plot is always unexpected, yet reasonable. For a century and a half, this two-wheeled vehicle has continuously encountered conditions specific to China. It has been modified by this impact and continues to evolve. Eventually, under Mao, this western innovation became fully integrated into the daily life of the people. More recently, with globalization, the East wind now affects cyclists in almost every corner of the world.

The early years

Who the original inventor of the bicycle was remains an inconclusive issue. Both the East and the West have several candidates. There are suggestions that Chinese pioneering engineer, Lu Ban, invented a proto-tricycle 2500 years ago. Subsequently, according to some documents, around 1790 during the Qing Dynasty, Huang Lu Zhuang 黄履庄envisioned a two-wheeled machine (http://www.famimobi.com/etagid34744b0/). Many people in China think Huang was the true originator of the bicycle. In any case, the importation of bicycles into China from the 1860s to the end of the 19th century was discouraged, and mostly resisted by the Qing elites who subscribed to a conservative anti-scientific branch of Confucianism. What can be confirmed is that the technology of the bicycles that we are familiar with today really started in Europe in the early 19th century and became more sophisticated in the early 20th century. Their development had taken over 100 years, and is the result of the joint efforts of technicians from many countries.

The first encounter between a Chinese person and a bicycle followed the development of the *cranked velocipede* in the mid-1860s, and a brief velocipede craze took place globally (Vignette H). It is not yet possible to verify the exact time when the first bicycle entered China, but it is certain that the bicycle had appeared on the streets of Shanghai no later than 1868. The *Shanghai Xinbao* published on November 24, 1868 wrote:

> I see that there are a few bicycles in Shanghai. With a draisine, a person sits on the bicycle that has one wheel in the front and the other behind, and the person uses his toes to touch the ground. The wheels turn as he walks along. With another kind

DOI: 10.4324/9781003142041-54

> (a velocipede), it is like stepping on a balance, with cranks attached to the front wheel, turning like flying, people can travel without walking. More than one person has seen these bicycles, lots of us saw them.

This article ends: "China's long-distance travelers can buy them and use them. It's no inconvenience." The authors of this article were promoting the sale of bicycles in China.

Nevertheless, during the late 19th century, bicycles were not popular in China; most of the cyclists were Westerners, among them traders, missionaries and diplomatic/consular staff. Three reasons can be offered to explain why the bicycle did not fit with "Chinese ownership" at this early stage: first, the bicycle itself was not technically perfect, and it was difficult to learn to ride; second, road conditions were not good, making bicycle riding dangerous; third, the price was expensive. But behind this, Chinese mandarins and businessmen were not happy. As Huai observed, "Because of the habits and customs of the Chinese people, those who do not serve others are served by others, and they are rarely willing to use their own strength to serve themselves. Therefore, they only use a carriage or a sedan chair... It [the bicycle] will be laughed at" (Huai 1942). For this reason, although bicycles had been introduced to China, they encountered great resistance from diverse quarters at the beginning, and the road to popularization suffered significant setbacks.

From the Imperial entourage to the general public (late 19th century–1949)

Following the Sino-Japanese War of Jiawu (1894–5), and especially after the *gengzi* year of 1900 that saw the Boxer Uprising and foreign military occupation and looting of Beijing, traditional Chinese structures and beliefs were challenged (*gengzi* years occur every 60 years in Chinese cosmology and presage disaster). The combined effect of the invasion of the Orient by the West and the eastward expansion of Western learning (technology, in particular) shattered the self-reliant conservative mentality of Chinese scholars. Chinese resistance to things foreign changed at this time from the initial "a strange surprise" about all things from the West, to "it will deeply add joy". Thereafter, news reports about Chinese cycling began to appear in newspapers, although the technical disadvantages of cycling were still there. The road conditions in China were still bad, but the Chinese people's perspective had changed. Bicycles were no longer viewed as "extremely laborious and rarely seen" useless tools (Ge 1989: 17); instead, they were seen capable of "enhancing the mind", "strengthening the body", and even as an "artifact of civilization".

Still, at the beginning of the 20th century, bicycles were rare and expensive, and the people who owned them were upper-class Chinese people living in big cities. A typical representative of people who loved bicycles was the last emperor Aixinjueluo Puyi. In 1911, the Revolution replaced the Qing Dynasty with the Republic of China, but the Manchu Emperor of the Qing Dynasty was allowed to stay in the Forbidden City to continue his daily life thanks to an agreement "The Edict of Abdication". At that time, many members of the Eight Banners Army (Banners were Manchu military-social units) were fond of bicycles. When Puyi (1906–1967) got married in 1922, several went to the palace to accompany him to study at a young age. Pu Jia (1908–1949), a cousin who had the closest relationship with Puyi, gave him the gift of a bicycle. (Chinese People's Political Consultative Conference... 1982) Puyi was immediately attracted to bicycles, and his love for bicycles is a well-known story in China (Puyi 2007: 99). Puyi became emperor at the age of 3 and abdicated at the age of 6. He was used by various political forces throughout his life and had no commendable achievements. When he revisited the Forbidden City in his later years, as he passed the gate without a threshold, he pointed to people and said with a smile: "This is my achievement. In order to ride a bicycle, I dared to saw off the threshold placed so that my ancestors spirits could not leave!" (Wenda 1979).

Mechanical power replaced human and animal power, trams and cars eliminated horse-drawn carriages and sedan chairs, and modern technology changed the face of public transportation. But it was not until the 1920s that bicycles were commonly seen on the streets of major cities in China. Post offices, telegraph bureaus, telephone companies, public bureaus, police stations and other agencies throughout China equipped their staff with many bicycles to perform official duties and improve work efficiency. Also, a large number of ordinary citizens purchased bicycles through various distributors. Bicycles were used as a means of transportation and delivery. Among users were the staff of various foreign firms, but also teachers and students of various universities, and reporters of various newspapers. Regardless of public or private use, China really had its own bicycle class at this time. They are roughly the middle class living in large and medium-sized cities, and they were distinguished from the arrogant car class on the streets and the rickshaw class who sold their labor.

During World War II, especially after the outbreak of the Pacific War, because of the extreme scarcity of strategic resources such as gasoline and electricity, public transport including cars and trams were almost paralyzed, and the advantage of bicycles moving without using fuel was revealed. Demand increased day by day. As one writer of the time put it, "bicycles save money and are very convenient… as long your feet are on one, you can go everywhere!" (Anonymous 1942a). Instead of walking, bicycles suddenly became the most popular form of transportation during wartime, and the most accepted means of transportation in China. During the Japanese occupation, bicycles transported much-needed rice from the countryside to starving Chinese in the cities. As the "sedan chair for civilian traffic", bicycles became popular in cities and provided many conveniences for citizens' lives. Some people even shouted the slogan "Long live the freedom-vehicle (the bicycle)!" (Anonymous 1942b).

Almost all of the many bicycles used by Chinese people were imported from foreign countries. In other words, foreign bicycles imported into China were the source of the popularization of bicycles in Chinese urban society (Xu 2015, 67–103). Although there were many bicycles called "domestic products", before 1949 they were actually bicycles assembled in China using foreign parts. Most of the key parts and the materials used were still imported from abroad, and China had not yet completely got rid of its dependence on the foreign bicycle industry (Xu 2007).

Extreme scarcity to full saturation (1950–1990s)

Like all countries in the world, the popularization of bicycles in China also encountered fierce competition from other means of transportation. From the middle of the 19th century to the middle of the 20th century, the most powerful competitor of bicycles in China was the rickshaw. Compared with bicycles, rickshaws had many insurmountable drawbacks, especially the "inhumanity" of people pulling people, which obviously could not survive after the establishment of the Socialist Republic (Chapter 12). In addition, during the Cold War with capitalism dominated by the United States, the Western world imposed comprehensive economic sanctions on China with its nascent Communist regime, closing the channels for foreign-made bicycles imported to China from Europe, America, Japan and other countries. China had never planned to establish a national bicycle industry, but it did in the 1950s.

The bicycle industry in New China did not grow from zero. It benefited from factories established by Japan when it invaded China. In 1936, Kojima Kazusaburo, a Japanese veteran, recruited workers in Shenyang, Tianjin, and Shanghai to set up three Changhe Works factories. After victory in the Anti-Japanese War, the Kuomintang government took over the Changhe Works. The three factories became the Shenyang Automobile Factory, the Resources Committee Tianjin Machinery Factory No. 1 Branch, and the Resources Committee Central Machinery Co., Ltd. Shanghai Machinery Factory No. 2. In 1949, they were taken over by the Communist regime and became the state-owned Shenyang Bicycle Factory, Tianjin Bicycle Factory, and Shanghai Bicycle Factory. Based on these three large factories, by the end of 1952 when the national economic recovery period ended, the annual output of bicycles in China had reached 80,000, which was eight times as much as in 1949. In the 30 years since then, the national bicycle industry has maintained this momentum of rapid growth (Zhang, 1992, p. 8). Yang (1985) reports that by 1979 about 100 million bicycles were in operation, compared to only two million motor vehicles. Details are shown in Figure 36.1.

Even though more and more bicycles were produced, domestic demand for bicycles seemed never to be met. "Demand exceeded supply" has always been the main theme of the Chinese bicycle story in the Mao Zedong era. There was much talk by that generation about the "tight ticket economy" (a government ticket was required to purchase a major consumer good – a form of rationing). With constraints on supply, young people in Chinese cities did everything possible to secure a home for marriage, and mark the high-quality life with the "three turns and one ring", the so-called "four major pieces" (watch, bicycle, sewing machine, radio) (Gerth 2020).

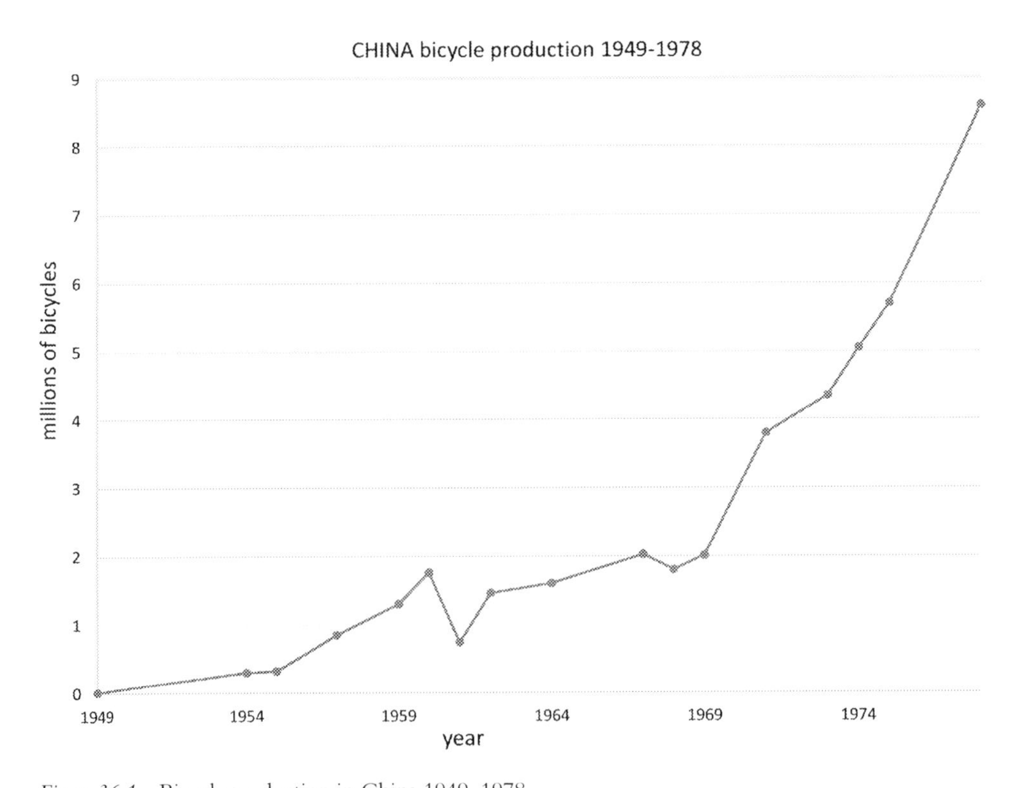

Figure 36.1 Bicycle production in China 1949–1978.

Source: Data provided by the Chinese Bicycle Industry Technology and Information Office.

What I want to emphasize are two more far-reaching changes that have been neglected by previous writers. One is the unified national bicycle production standard and the generally improved manufacturing level. The bicycle industry, under the unified management of the Ministry of Light Industry, formulated the first "Ministry-Promulgated Standard" in 1958. For the first time, the *standard bicycle* using uniform component names, dimensions and quality indicators nationwide was produced, which made it possible for specialized production (with horizontal integration) and national collaboration. China's bicycle industry expanded from Shanghai, Tianjin, and Shenyang, and assembly sites and parts manufacturers were established in Guangdong, Jiangsu, Zhejiang, Shandong and other provinces. Second, the popularity of bicycles rapidly spread from cities to rural areas. In 1955, the Xinhua News Agency said:

> According to a survey conducted by the Ministry of Commerce in 1954, in 17 provinces including Jiangsu, Shandong, Yunnan, Sichuan, Guangdong, and Xizang... the growth rate of rural purchasing power is faster than that of cities, and the proportion is increasing year by year; also that the purchasing power of the countryside is larger than the cities.
>
> (*Anonymous 1955*)

Rural areas in China increasingly became a broad market for industrial products such as bicycles. But in 1961 during the famine, national bicycle output plummeted from 1.76 million in 1960 to 740,000 in 1961. At the same time, the selling price of some agricultural products rose rapidly, and the income of farmers increased. In order to accelerate consumption and discourage farmers from saving their newly rising incomes, bicycles were placed on open supply (implying anyone could buy one, without a "ticket") from 1962 to 1964, but only at high prices. A total of 279,358 high-priced bicycles were sold nationwide, 85% of which were purchased by farmers and 15% by urban workers (Shanghai Bicycle (Group) Company 1990: 238). In order to adapt to the rural market, the "Forever", "Phoenix" and "Flying Pigeon" brand bicycles produced in Shanghai, Tianjin and other places were re-designed as heavy-duty bicycles more suitable for rural settings and use by farmers. In rural areas, farmers affectionately called this kind of heavy-duty bicycle "little donkeys that don't eat grass" and praised the "little donkey" for its adaptation to the needs of the countryside (Anonymous 1972).

During the 1960s and 1970s, the period when China became known as the *Kingdom of Bicycles* (zìxíngchē wángguó – 自行车王国), the People's Republic became the leading bicycle riding country in the world. Road infrastructure was adapted to match this modal mix. Major thoroughfares were divided by a low moveable fence about 2 to 3 feet high. The central lanes were for motor vehicle trucks and buses, and the side lanes for slower-moving traffic such as bicycles, ox carts, donkey carts, rickshaws, delivery tricycles, and for street trading. At major intersections, traffic police on raised podiums often controlled the flow of traffic. Large bicycle sheds were built at apartment buildings and places of work.

Cyclists moved at a uniformly steady rate – it was difficult to move at a faster speed than the mass of cyclists. They rode to work, to shop, to school, to visit relatives, often carrying purchases and things to sell. Cyclists wore everyday clothes, bundling up in warmer clothes in winter in northern cities. Carrying children as passengers was a common practice (Figure 36.2).

After the reform and opening up beginning in December 1979, China paid more attention to light industrial production, and bicycle manufacturers began explosive growth, from 46 firms in 1979 to 140 three years later. In 1985, China's bicycle industry had its first joint venture linking the British Company making Emmelle bicycles with the Shenzhen China Bicycle Co., Ltd. After that, a large amount of overseas capital flowed into the industry (Chapters 15 and 18).

Figure 36.2 People riding bicycles across a railway bridge in Beijing, 1991, as the Kingdom of Bicycles approaches its end. (Photo by Wang Wenlan 王文澜 : see https://www.pinlue.com/article/2017/05/2221/381926780814.html).

Stimulated by the policy shift and the market economy, by 1986 national bicycle production reached 32.3 million in the first 11 months alone. Four years later, the number of bicycles in China exceeded 300 million, ranking first in the world. In 1993, the number of bicycles in China reached its peak, with an average of 197 bicycles per 100 households.

In the 1980s and 1990s, the share of all trips on bicycles in urban traffic was about 45% (http://opinion.haiwainet.cn/n/2016/0411/c345439-29820226.html, accessed 24/6/2021). During commuting peaks, bicycles converged into a torrent of cyclists, which shocked the world and helped China to win the reputation as the *kingdom of bicycles* (Anonymous 2021). At that time, whether it was a remote rural area or an increasingly prosperous metropolis like Beijing and Shanghai, bicycles could be seen everywhere, and bicycle bells could be heard all day. Adults rode bicycles to go to work, children use bicycles to play, lovers went out on bicycles. Bicycles became deeply rooted in people's daily life and were the most common means of transportation. Every Chinese person has a story about his or her beloved bicycle, which accumulated in the collective memory of that generation.

E-vehicles, the internet, and globalization: 2000 to the present

In 1994, the State Council promulgated the first Auto Industry Policy, which became an important turning point in the history of China's transportation. During the following 20 years, China experienced an economic miracle rarely seen in the world. Various types of motorized vehicles such as private cars, subways, and magnetic levitation trains were introduced into the country, and they grew like mushrooms after the rain. From policy makers to ordinary people, bicycles were regarded as the antithesis of motorized transportation, and a hangover from the previous generation of chaotic road grabbers and accident triggers. Interest in the bicycle faded, and levels of use dropped steadily. But with the help of the latest technologies, many new bicycle variants were developed that promised to reverse the decline. Meanwhile, concern over atmospheric pollution in major cities led to a reassessment of the non-polluting bicycle.

The emergence of electric bicycles

Since the 1990s, when electric bicycles first appeared, they have quickly found a market in China. In 1985, the first real electric bicycle was developed in China, but it was not popular. In 1998, the national output of electric bicycles (excluding Taiwan Province) was about 60,000. In 1999, the output doubled to more than 130,000 (Zhu 2000). The emergence of electric bicycles has partly changed the way the Chinese people travel, and caused a series of chain reactions, including the internal restructuring of the bicycle industry and the reformulation of traffic laws in various cities.

Since the invention of the electric bicycle, there has been a problem of how to define it. Is it a member of the bicycle family, or does it have the attributes of a "light-riding" motor vehicle? Many discussions followed. On May 28, 1999, the State Bureau of Quality and Technical Supervision formulated the first national standard – *The General Technical Requirements for Electric Bicycles* (GB17761-1999). The degree of acceptance of electric bicycles varies widely. In 2001, Fuzhou City, Fujian Province, banned the use of electric bicycles on the streets (Anonymous 2001). Subsequently, Beijing, Guangdong Province, and Hainan Province all issued local regulations prohibiting electric bicycles from being licensed or on the road. The blocking order in a few areas has not affected the popularization of electric bicycles nationwide. In 2004, more than 1,000 enterprises in China were allowed to produce electric bicycles, with an output of 6.76 million units (Guo 2005).

With the widespread adoption of lithium batteries, the range and service life of electric bicycles has greatly improved, and the upgraded version has become a better alternative to pedal bicycles and fuel-burning motorcycles. Electric bicycles are increasingly becoming the first choice for Chinese people to travel short and medium distances. From 2008 to 2019, electric bicycle sales as a proportion of all two-wheel vehicles sales rose from 40.7% to 73.4%, an increase of more than 30% in 12 years. More than 90% of domestic electric bicycles are made and sold within China (Li and Meng 2020). Its popularity is also closely related to the rise of e-commerce, express delivery, and food delivery services that have been spawned by Internet technology.

Bicycle sharing

Since the turn of the 21st century, the problem of urban congestion caused by the popularity of motor vehicles has troubled many citizens. Out of the demand for green transportation and low-carbon travel, in early March 2008, Hangzhou City, Zhejiang Province took the lead in proposing the creation of a public bicycle-sharing system (Chapter 23). On September 16, it officially opened. In the following years, public bicycle sharing spread to many Chinese cities. By the end of 2014, the number of shared bicycles (docking and dockless) exceeded the total of any other country in the world, with a total of more than 400,000 shared bicycles. By 2015, 215 cities and counties had started the construction of these systems. Under the government-led and enterprise-operated model, the disadvantages of public bicycles were gradually exposed. Even Hangzhou, with the earliest and best-developed system, has had problems, such as the difficulty in applying for a membership card and difficulty in returning a bicycle to a docking station. There are still too few public bicycles and docking stations in many cities, with the result that the system fails to facilitate the daily travel of most citizens and runs at a loss (Anonymous 2015, 2016).

2014 saw the launching of a new system of dockless shared bicycles based on innovative technologies including mobile phone bike location and payment, and satellite positioning to locate the nearest bicycle (Wang et al. 2019). The *ofo* small yellow bicycle (the two o's of the name are intended to resemble a bicycle), first appeared on university campuses and quickly gained favor with venture capitalists. In April 2016, Mobike dockless bikes were launched in Shanghai. In contrast to public bicycles in docking stations, the shared dockless bicycles can be registered and used by downloading the mobile phone APP. The initial deposit ranges from 99 yuan (US$18) to 299 yuan (US$46). Subsequently, the 'no deposit' system became available: you simply scan the frame's QR code with your mobile phone to unlock the dockless bicycle. The cost of renting a bicycle ranges from 0.5 yuan to 2 yuan per half-hour, and the entry threshold is lower; in addition, the shared bicycle has a stylish appearance and integrates mobile payment, GPS positioning, explosion-proof tires, axle transmission and other high-tech features, and is available anytime, anywhere. According to the 2017 White Paper on *Shared Bicycles and Urban Development*, one year after the emergence of shared bicycles, the travel structure of China's urban traffic was changed. The proportion of bicycle trips in urban traffic trips increased from 5.5% to 11.6%, realizing the claim to be the *kingdom of bicycles* (China Social Science Net 2017).

Since the end of 2016, shared bicycles have triggered a carnival of capital, with at least 25 new brands joining the game. As if overnight, shared bicycles reached the point of overwhelming the city, with colorful shared bicycles seen everywhere in the streets, subway stations, bus stations, and community gates in Chinese cities. Oversupply became a problem. Bike sharing was once considered a way to promote the complete transformation of government-led public bicycle services to market-led shared bicycle services, helping to fundamentally solve the "last mile" problem worldwide as Mobike, Ofo and other dockless bikes were introduced in many world cities (Guo et al. 2017). However, bike sharing is not a panacea. After the capital bubble burst at the end of 2018, abandoned bicycles piled up, long queues formed for refunds of rents, collective price hikes were made to cycling fees, and random parking crowded public spaces. The popularity of shared bicycles in China is rapidly cooling down.

Made in China and sold all over the world

The export of bicycles from China started in 1953, after which about 200 bicycles were exported each year. In the 1960s and 1970s, when domestic supply was still in short supply, few completed cycles were exported overseas. In January 1986, the State Council established the *Mechanical and Electrical Products Export Office* to expand the autonomy of bicycle companies in foreign trade, and introduced favorable policies such as "using exports as a breakthrough point to fully revitalize light industry". Driven by this, China's bicycle export numbers rose steadily, reaching 500,000 in 1986 and breaking through 10 million in 1992 (Li 1993; Chapter 14). In 1998, China became the world's largest bicycle exporter for the first time. By the end of the twentieth century China was leading the world in total output, but it still had not caught up developed countries in terms of product quality, variety and technical content (Figure 36.3).

In the 21st century, China's bicycle industry has entered a new period of consolidation, with e-bikes and bike sharing broadening the domestic market, while exports were helped by China's accession to the World Trade Organization (WTO) in 2001. At the same time, more and more attention was paid to the role of bicycles in sustainable mobility worldwide. The bicycle

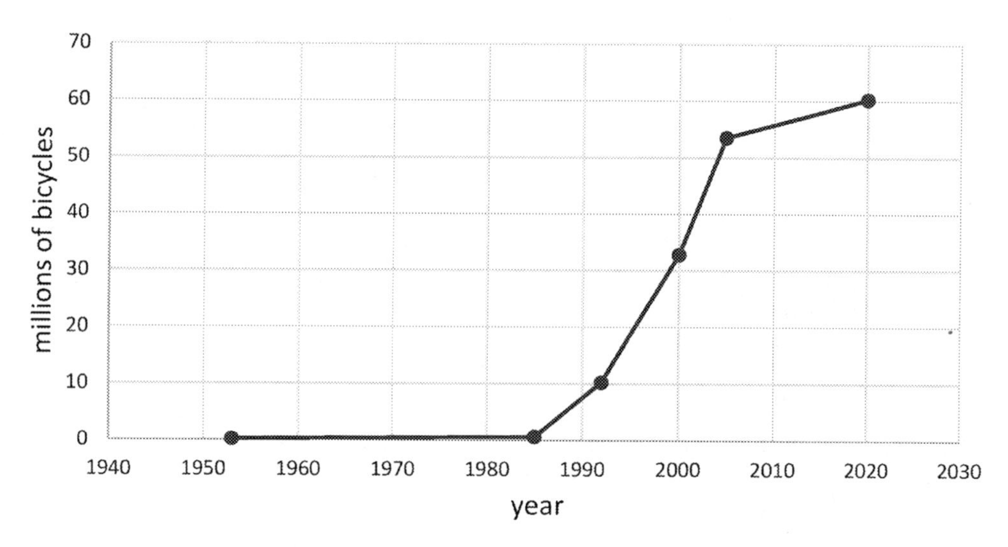

Figure 36.3 China – bicycle exports 1953–2020.

Source: Data provided by China Bicycle Association.

market in most countries has expanded (especially during the Covid-19 epidemic), and China's domestic bicycle industry is rapidly integrating with world markets, in commerce, technology, standards and culture. Mutual contacts and exchanges have been further enhanced, and outward-looking characteristics have become more and more obvious. Despite the European Union market imposing 30.6% antidumping tariffs on bicycles and parts produced in China since 1993, *Made in China* bicycles have occupied a dominant position in the United States, Japan, and the rest of Asia (Zhou 2010). China is now the world's main producer and exporter of bicycles (Zhang 2003). Since 2007, more than 60% of the world's international bicycle trade volume has come from China. China has almost completely occupied the international market for low- and mid-range bicycles and parts, and is showing a trend toward high-value-added products. Since the beginning of the 21st century, the quality of bicycles in China has also improved significantly. According to the statistics of the China Bicycle Association, the domestic proportion of the technology sources of products in the Chinese bicycle industry has increased from 60% in the 1980s to 90% in 2020 (Anonymous 2021).

In 2020, the total output of China's bicycle industry was 116 million units. Since 2018, China's bicycle exports have accounted for more than 80% of the world's annual output for two consecutive years. China has once again become "the *Kingdom of Bicycles*".

References

Anonymous (1942a) "Hu Feng also learns to ride a bike." *Mass Media*, 2(36), March 28, p. 1.

Anonymous (1942b) "The Rise of Free Cars, the Beloved of the Times (Part 2)." *Shen Shen*, April 26.

Anonymous (1955) "The purchasing power of farmers across the country continues to grow, and rural areas have become a broader market for industrial products." *People's Daily*, January 7.

Anonymous (1972) "Serving the workers, peasants and soldiers is our direction – The story of Shanghai Light Industry Enterprises increasing product quality and varieties based on the views of the masses." *People's Daily*, March 20.

Anonymous (2001) "Let the dignified and imposing electric bicycle on the road." *China Bicycle*, No. 7, p. 1.

Anonymous (2015) "Growth worries, behind the world's large number of public bicycles in Chinese cities part 1." *China Bicycle*, No. 12, 78–81.

Anonymous (2016) "Growth worries, behind the world's largest number of public bicycles in Chinese cities – part 2." *China Bicycle*, No. 1, 58–61.

Anonymous (2021) "Years are like songs, the hearts are like rocks, and the new journey is forging ahead – China's bicycle industry celebrates the 100th anniversary of the founding of the party." *Consumer Daily*, June 3, p. A1.

China Social Science Net (2017) *The 2017 White Paper on Bike Sharing and Urban Development*. April 12. https://www.sohu.com/a/133672773_468661 (accessed 31 July 2021).

Chinese People's Political Consultative Conference, National Committee Literature and History Materials Research Committee (1982) *Information on Palace Life in the Late Qing Dynasty* (Beijing: Literature and History Materials Publishing House) p. 129 (in Chinese).

Ge, Yuanxu (1989) *Miscellaneous Notes on Shanghai Travel, Volume 1*, (Zheng Zuan punctuation). (Shanghai: Shanghai Ancient Books Publishing House) (in Chinese).

Gerth, Karl (2020) *Unending Capitalism: How Consumerism Negated China's Communist Revolution* (Cambridge: Cambridge University Press).

Guo, Haiyan (2005) "Development and current status of China's electric bicycles." *China Bicycle*, No. 12.

Guo, Peng, et al. (2017) "Shared bicycles: Cooperative governance in internet technology and public services." *Journal of Public Management*, 14(3), 1–10.

Huai, Lang (1942) "Bicycle gossip", *Vientiane*, No. 2 (in Chinese), p. 7.

Li, Tong and Meng, Lingxi (2020) "Sell the traditional bicycles abroad, leave the electric ones to China." *Jiemian News*, November 25. https://www.jiemian.com/article/5313488.html

Li, Yuankai (1993) "Worries, troubles and countermeasures in the *Bicycle Kingdom*." *China Bicycle*, No. 5, 31–36.

Puyi, Aixinjueluo (2007) *The First Half of My Life* (Beijing: Mass Publishing House).

Shanghai Bicycle (Group) Company (1990) *Shanghai Bicycle Industry History* (Shanghai: Publisher unknown). p. 238.

Wang, Jiaoe, Huang, Jie and Dunford, Michael (2019) "Rethinking the utility of public bicycles: the development and challenges of station-less bike sharing in China." *Sustainability*, 11(6), 1–20.

Wenda (1979) "Puyi touring the Forbidden City." *Travel*, No. 1, 18–19.

Xu, Tao (2007) "Research on the National Bicycle Industry in Shanghai, 1897–1949." *Social Science*, No. 11, 164–173.

Xu, Tao (2015) *The Bicycle and Modern China* (Shanghai: Shanghai People's Publishing House) pp. 67–103 (in Chinese).

Yang, J-M (1985) "Bicycle traffic in China." *Transportation Quarterly* 39(1), 93–107.

Zhang, Peisheng (2003) "Strong growth of my country's bicycle exports." *China Bicycle*, No. 6, pp. 8–9.

Zhang, Xun-Hai (1992) *Enterprise Reforms in a Centrally Planned Economy: The Case of the Chinese Bicycle Industry* (London: Palgrave Macmillan).

Zhou, Zhou (2010) "The road to exporting bicycles to Europe: Seventeen years of tragedy." *International Business Daily*, July 28.

Zhu, Xinggen (2000) "Thoughts and suggestions on the development of electric bicycles." *China Bicycle*, No. 3, 12–17.

37

COPENHAGEN IS A GOOD PLACE TO BIKE

But it could be better...

Malene Freudendal-Pedersen

Having cycled in Copenhagen all of my adult life, this chapter is inevitably based both on research and on the embodied experience of this way of moving around the city. I cycled, as a young person, as an expecting mother, as a mother (often with children), and as an older adult (without accompanying children) (Chapter 2). This is not unusual; it is rather common among Copenhageners. Many Copenhageners ride all year long in all kinds of weather. In Denmark, there is a phrase stemming from a satirical show that is at least 30 years old: "There is no such thing as bad weather – there are only wrong clothes." It has a melody so you can sing it (with an accent on the wrong syllables – which makes it fun) and is commonly used as a phrase when people complain about the weather. It provides a good picture of the local approach to cycling: when it rains you put on rain clothes, and in winter you just need a warm jacket, gloves, a big scarf and a hat (Vignette J). For many people, cycling is taken for granted in Copenhagen and everyday cyclists are not visually identifiable (except by their bikes) or associated with stereotyped images or visual signifiers. Cycling is not segregated by class, age and gender and those cycling the streets of Copenhagen can be seen in all kinds of outfits, including high heels and suits; it is not at all uncommon to see ministers or mayors cycle to and from work. Of course, this has not come out of the blue, so before discussing the culture of cycling in Copenhagen, background into how Copenhagen has become a cycling city will be provided.

Infrastructure and planning matters

Copenhagen has the infrastructure that supports everyday cycling due to a long history of not excluding bicycles in favor of cars, which happened in many cities at the end of the 1950s onwards. The first cycle path in Copenhagen was developed in 1892 and further developed when former bridle paths were turned into cycle paths. Since then, there has been a constant construction of cycle paths even as the car entered into everyday life during the 1950s and 1960s, when the infrastructure for cars became the main priority and thus sidelined the cyclists. With the oil crisis in the 1970s the picture changed, especially when Copenhagen introduced car-free Sundays in 1973 and 1974 to conserve oil reserves. This marked the beginning of the turn toward more cycling as we can see it in Copenhagen today; in the 1980s cycle paths became the norm in Copenhagen traffic planning, including the curbed path, providing the

DOI: 10.4324/9781003142041-55

Figure 37.1 A curbed bike lane in Copenhagen (Photo – Jesper Pagh with permission).

cyclist with their own separated infrastructure (the curb has the form of a raised line of blocks separating cyclists from motorized traffic) (Chapter 20; Vignette D). The curbed cycle path is of great significance for cyclists – not least concerning safety, since it presents a physical barrier blocking cars from driving or parking on bike paths (Figure 37.1).

As the capital of Denmark, Copenhagen today has half a million residents in the city and 1.2 million in Greater Copenhagen. The *Bicycle Account* that comes out every second year (https://cyclingsolutions.info/wp-content/uploads//2020/12/CPH-Bicycle-Account-2018.pdf) is a biennial travel survey addressed to over 1000 residents of the city. It shows that in 2019, 44% of Copenhageners commuted on bikes to work or school/university. The city has 385 kilometers of cycle tracks alongside roads and many green routes and super cycle highways crisscrossing the city, separated from the car infrastructure. In many places, especially on major roads, there are separate traffic lights for bikes where cyclists get to start 4 seconds (in some cases 12 seconds) before the cars start. Also, many "simple" improvements have been made to the infrastructure which help cyclists; for instance, allowing cycling against one-way motorized traffic. On busy streets there is a curbed bike path, and on streets with a 30 km/h speed limit, it is based on a shared space concept that permits visual contact between cars and bikes, thus increasing the safety.

The first *Bicycle Account* was presented in the mid-1990s. This was part of an increased focus on cycling in the municipality. This was cemented when the municipality of Copenhagen created a separate Cycling Department in the municipality at the end of the 1990s with the main goal to get more Copenhageners to cycle. Before this, the project of increasing cycling had been undertaken by a few planners located in different departments with no common strategy. The Cycling Department no longer exists as an independent unit; today, cycling is instead implemented in all areas of the environmental and traffic departments. The Cycling Department was committed to making cycling an everyday practice for everybody. The main goal was to remove typologies about who cycles, and instead promote cycling as the easiest and fastest way for everybody to get around in Copenhagen. As part of this strategy, the municipality created a promotional video and pamphlet *City of Cyclists – Copenhagen Bicycle Life*, where one of the main slogans was "We don't have cyclists in Denmark – We merely have people transporting themselves by bike." This communicative approach was aimed at Copenhageners, but it was also supposed to be an international strategy to put Copenhagen on the map. One of the famous Copenhagen tourist attractions is the little mermaid (of the Hans Christian Andersen fairytale) and the Cycling Department used this actively in the communication strategy. The little mermaid is not a huge monument but 'life sized' and often tourists are quite surprised when they see the statue. With this 'life size' metaphor as a reference, they write in the *City of* Cyclists pamphlet: "If it's monuments you're after in Copenhagen, don't look up. Look all around you, right there at street level. Our greatest monument is motion. It is a massive, constant, rhythmic and life-sized legacy." Apart from this, the Cycling Department created slogans and social media platforms like "I bike Copenhagen" and made a large festive event celebrating the new city bikes of Copenhagen (the first ones were introduced in 1989). The Cycling Department played a large role in reinforcing the cycling culture in Copenhagen and reinstalled it with a right to the city in relations to the still dominating cars. Also, in these years (2009), Michael Colville-Andersen founded the firm Copenhagenize, that has played a large role in making Copenhagen cycling known globally. Even if his influence on the direct planning for cycling in Copenhagen is limited, Colville-Andersen's advocacy for cycling has been agenda setting. The chic cycle blog featuring people (many women) wearing everyday clothes (high heels, dresses, suits, jeans, shirts etc.) has been part of emphasizing that cycling in Copenhagen does not make you a cyclist – but merely a person transporting yourself by bike.

In recent decades, Copenhagen has become a wealthy city. The city has undergone gentrification and the housing market has become increasingly expensive. Still Copenhagen has a large share of non-profit housing that makes it possible for less affluent people to live in the city (if they are able to get one of these apartments). The focus on promoting cycling has been a part of the strategy for Copenhagen to attract more wealthy residents and the large population increase during the last 20 years has resulted in growing car ownership. Growing car ownership does not mean that people necessarily use the car for commuting, in fact only 22% thereof is done by car. More than 25% of all families with two kids in Copenhagen own a cargo bike or a bicycle trailer and it is a common sight around schools and daycare facilities to see children transported this way. Another important aspect in Copenhagen is the opportunity to combine biking with other modes of transport. On the S-train (the urban–suburban rail connection) there is no charge for a bike, and the trains have flex wagons with bike racks. In all other trains and the metro, the bike can be brought along for a small fee. Also, taxis in Copenhagen have racks for carrying bikes.

Even though car ownership in Copenhagen is increasing, the number of trips done by car is declining. This creates issues around parking when 12% of road space in Copenhagen is used for car parking and only 1% is used for cycle parking. Also 54% of the road space is used for car lanes and only 7% for bike paths. Only 40% of car trips across the municipal border are made by Copenhageners and 55% of Copenhageners do not have access to a car, so the space allocated for cars is an issue. The majority of Copenhageners (87%) agree that better infrastructure for cyclists is needed and 76% agree that some of the road space for cars should be used for improving cycling (Copenhagen Municipality, 2019). These numbers highlight one of the problematic issues in Copenhagen, because, even if many people are cycling and it might seem to be a success story, Copenhagen has also become a victim of its own success. I have previously written articles focusing on the fight for space and the unintended safety issues for children and the elderly (Freudendal-Pedersen, 2015a, 2015b, 2018, 2021).

Teaching children to cycle safely is a very important part of creating sustainable mobility; the way childhood mobility is learned determines adult life mobility. This is perhaps one of the biggest challenges. Many parents feel it is too unsafe to let their children bike to school because of the numbers of cyclists in the bike lanes. During rush hour you need to handle three lines of people cycling at different paces, consisting of cargo bikes, children, and commuters travelling at both high speed and moderate-to-slow pace. Despite the focus on cycling, there are still issues when it comes to limiting accessibility for cars on a political level, and a significant part of strategic planning still focusses on the car. Also, new property developments in Copenhagen are built with parking spaces due to the firm belief that you cannot sell an expensive apartment without a parking space and there is still status embedded in driving (or at least owning) a car and the car remains a symbol of power and wealth (Freudendal-Pedersen, 2014a, 2014b).

In spite of the dominance of the car, it is still cycling that is used strategically as part of the urban development in Copenhagen (Jensen, 2013), and even if cycling is declining in most Danish cities (with the exception of Copenhagen, Odense, Aarhus and Aalborg), the health and safety of the cyclist is ingrained in Danish traffic laws, and the government takes responsibility for providing resources for cycling. But there is definitely antagonism at play between motorists and cyclists when Copenhagen sets major goals for future cycling, while simultaneously expanding the number of car parking places in the inner city.

One of the big developments for easy mobility for Copenhageners is the large system of inner-harbor pedestrian and cycle bridges. The bridges entail significant shortcuts for cyclists as well as the opportunity for beautiful routes away from heavy traffic. These bridges are a big part of the promotion of the best cycling city in the world, which Copenhagen has as an official goal. Here an interviewee explains the difference between cyclists and car drivers in relations to the construction of the bridges in Copenhagen:

> It is sort of as if the car is the grown up transport mode. Investment in roads and expansions in roads for cars, I guess is crazily more expensive than anything made for cyclists. And it is also much more a subject discussed, much more than what we should do for cyclists. It is interesting with Bryggebroen (Copenhagen's first cycling bridge over the harbor). The first time I crossed it, it was dark and I couldn't find my way. In the beginning something was missing that took you to Dybbelsbro. And making that lane thing (Cykelslangen). I think it cost the same as the bridge. And then people discuss if it is OK to spend that much money on this. But before, when you crossed the bridge, you ended up at a staircase and needed to get off the bike and carry it up the stairs. One wouldn't do this with cars – then there would be chaos.

Bryggebroen, which opened in 2006, has around 12,000 people crossing it on an average weekday. Cykelslangen ('The Cycle Snake'), which opened in 2014, connects Bryggebroen with a train station (Dybbelsbro), a main road, and the neighborhood on the other side of the road. From 2006 to 2014 cyclists needed either to go on an 800-meter detour or to get off their bikes and push them up the stairs on a ramp. So even with the importance attached to cycling in Copenhagen, cyclists still had to wait eight years before there was a direct connection to the road and neighborhood that the bridge was intended to connect to. This was due to fights over which specific project to prioritize, and then securing the funding for the connecting bridge. It is hard to imagine a situation like this happening if new connections were built for cars.

In the same period when the bridges in Copenhagen were being constructed, Copenhagen municipality implemented a new planning concept where mobilities are planned based on a principle that public transport, cycling and walking has the main priority. The concept emerged in 2008 as part of a traffic experiment in a neighborhood street with problems due to congestion. During the test period, red patches were painted on the roadway to indicate that the street's function had changed. Drop-off and load zones were marked, and new lines for the bike paths were drawn. This approach falls under the idea of tactical urbanism coined by Mike Lydon and Anthony Garcia (2015) in their book *Tactical Urbanism: Short Term Actions for Long-term Change*. By providing test scenarios (traffic experiments) it is easier to get political acceptance and, when people realize they can still continue their everyday life as they are used to, the road is subsequently paved, leading to long-term implementation of the plan. This planning strategy also became possible due to the demands of a large number of cyclists and an increasing number of inhabitants in Copenhagen belonging to the millennial demographic (born in the early 1980s to the early 2000s) who either moved back to (or stayed in) cities that promote walking, cycling and public transportation, while at the same time offering cultural, commercial and recreational opportunities (Lydon and Garcia, 2015).

Before the traffic trial started, the street had twice as many cyclists as cars and 50% more bus passengers than passengers in cars. The way the infrastructure was distributed was according to these transport demand numbers (though in the first part of the trial, cars were completely forbidden). During the trial period three concept principles were developed:

1 The urban space must be beautified, and urban life strengthened;
2 Cyclists' conditions must be improved on stressed routes; and
3 Public transport needs to be improved, allowing for shorter travel time and increased regularity for the buses.

Not mentioned was the anticipated reduction in car traffic, even though it is clear that if these conditions were met, a downward prioritization of car traffic would be the result. Instead, the focus was on livability and promoting the cycling city. The neighborhood which has this street as its major artery was one of the first neighborhoods in Copenhagen going through an urban renewal process. Today the area (Nørrebro) is the most mixed ethnic area in Copenhagen, also housing a lot of students with expensive and less expensive accommodation. One of the unexpected things that happened with the redesign of the neighborhood street was that a bridge across the Copenhagen lakes connecting the inner city with the neighborhood streets has become a new urban area. With the majority of cars removed, the bridge has become the second most used space for dwelling, second only to Nyhavn (an old harbor district in the inner city) (City of Copenhagen, 2013). The storytelling about this transformation of road space has turned into a positive story about creating better city space and making improvements for cyclists which led to its adoption as a planning concept for neighborhood streets in Copenhagen.

Afterwards the planning concept has been completed in another neighborhood street in Copenhagen and it has even been decided to build another one. The lessons learned from the first neighborhood street have increased the focus on the dialog with residents and users before refurbishing car space. When opening up a dialog about what features residents want to keep and what they want more of, it becomes clear that not everything is possible. If there is a wish for more life on the street, more public places, playgrounds and safety for cyclists, it is not possible to also have cars with a 40 km per hour speed limit, car parking and less noise. Through this dialogue it became clear that keeping the same opportunities for cars meant giving up most of the other wishes for the use of the urban space (Freudendal-Pedersen, 2020).

Focusing on the stories that have an impact

As another strategy to prioritize the bike in relation to overall traffic investments nationally and locally, the Copenhagen municipality was the first one to develop a (simple) socio-economic model that showed how cycling is good for the economy. The model demonstrated that cycle investments had similar or better returns than projects such as the enlargement of a highway. Also, due to costs for the health system, it showed that when a person chooses to cycle, the society has a net gain of 0.16 euro per kilometer cycled. In contrast, the society has a net loss of 0.1 euro per kilometer travelled by car and the socio-economic benefits on health and life from cycling are seven times higher than the accident costs. And most importantly, in comparison to driving children around for their activities, it showed that children who cycle to school are nearly 10% more fit than their classmates who walk or are transported by cars (Copenhagen Municipality, 2009). This focus on health led to increasing attention being given to the significance for overall health of getting daily exercise by using cycling as a means of mobility. This has been calculated in relation to mortality which is significantly lower for bike commuters and on top of this, the numbers show how companies save money from cycling commuters, as they have fewer sick days.

In my research on cycling in Copenhagen the emotional and embodied practice of cycling comes out strongly and favorably, not least in relation to health: interviewees comment that "I have a bodily restlessness which comes out through biking, when I go by car to work I can feel that it doesn't work for me" or that "My body feels heavy all day if I do not go to work by bike". The health issue is today picked up by various organizations that promote cycling; they plan to develop a sophisticated socio-economic model showing the opposite outcome to the usual ones where more car infrastructure is claimed to be the most beneficial.

A related concept of great potential impact is the idea of freedom which is normally associated with the car, but cyclists present a different understanding. In my research, car drivers and cyclists revealed high convergence in the way they describe the feeling and significance of freedom. What is different is the embodied relation to the feeling that cyclists express: "The best thing about cycling is the sense of freedom; it is the sky above my head which is very, very big when I bike." There is a tremendous potential in combining cycling health and freedom as a strategy to promote more everyday cycling. Often claims are made that you cannot organize everyday life without the car – or when you have kids you need the car – and the freedom feeling attached to the car is often used as the extra argument. But with approximately 30% of European car trips being less than 3 km and half less than 5 km there is no doubt that Copenhagen serves as an example that these trips can be made by bike without our everyday life falling apart. Through focusing on the experiences and everyday practice of cycling, stories also emerge on cycling as a cultural symbol stemming from childhood memories. In Denmark most children know how to bike – not only in parks but also in traffic. It is part of the cultural upbringing that you learn it and in primary schools it is part of the education to learn about traffic rules. That also means that moving around in Copenhagen, you see children as young as 3–4 years riding their own bikes, accompanied by an adult. In this sense Copenhagen works as an example of families functioning perfectly well on a daily basis without using a car.

Cyclists in Copenhagen describe themselves "as part of the city's organism" (Chapter 4). When cycling, there is the opportunity to "hop off the bike to enjoy a pause outside a sundrenched café". Flexibility is the point here, as well as being able to smell, feel and see the environment that becomes quite different when experienced from a bike. Also, you are part of a cycling movement that makes Copenhagen what it is:

> The best thing about Copenhagen is that there are so many people who are used to cycle. They create a special flow; sometimes it is almost poetic when everybody knows what to do and how to behave. When the flow is working it gives an atmosphere of a very careless life, for instance when you can see millions of bikers waiting for the green light and then they start moving and it is the kind of movements where everybody knows what to do – that's beautiful.

Copenhagen cyclists emphasize a cycling culture which contributes to the quality of the city. The feelings and emotions they attach to this mobility are about many different aspects of life, from waking up, having your own private time when moving, moving the body and so forth. When cycling, you are not trapped in a metal box, instead you are part of city life. Of course, this is nicer when the sun is shining, but these emotional embodied attachments might also be a big reason why the majority cycles all year round – even when it snows.

Concluding remarks

In my research I have tried to avoid painting a glossy picture of cycling, but to point out the problems and difficulties for cyclists in Copenhagen. Many of these problems are related to the continuous dominance of the car and the resulting lack of space for cyclists. But when all is said and done, the city is still a good place to be a cyclist. I spend a big part of my everyday life in Munich, where biking is also part of my mobility. And even if this city is improving its facilities for cyclists, it is also a reminder of the difference to the infrastructure in Copenhagen. The curbed bike lanes in Copenhagen are quite important: cycling in Munich without curbed lanes means constantly having cars in the bike lane. Also, when cycling in Munich, it is very common that a bike path suddenly ends and cyclists needs to figure out where to place themselves in traffic. The worst thing about cycling in Munich is being squeezed in between parked cars and moving cars and being doored is a constant threat. This difference between the two cities is also related to the cycling culture in Copenhagen where cycling is not just a secondary mode of mobility but respected by cars and pedestrians. In Copenhagen, ghost riding (riding against traffic) rarely happens and (most) pedestrians look up before they step onto the bike path to cross the street.

More optimistically, biking in Copenhagen has a large potential to show how cycling can be an integrated part of everyday life in a sustainable mobility system. Focusing on improving infrastructures for cycling makes a lot of sense from a socioeconomic perspective and not only in dense cities like Copenhagen. Making cycling part of intermodal mobility planning also adds to the possible range of trips that can be made partly by bicycle. People often comment that the reason why so many people cycle in Copenhagen is that it is so flat there, and that it might be different if it was hillier. To that I always respond that Copenhagen is on the coast – the harbor runs through the city – and it can be very very windy. And yes – sometimes this is very annoying, especially when it rains as well. This means that all places have potential, but Copenhagen was worked on by several fronts to make its cycling culture as strong as it is today. It has not happened overnight, but one cannot help wondering what would happen if we were able to replace all car trips of less than 5 km with bike trips – that would make quite a difference for our cities and their future.

References

City of Copenhagen, 2013. *Evaluation of the Nørrebrogade Project – Stage One*. Copenhagen.

Copenhagen Municipality, 2019. *Copenhagen City of Cyclists*. Copenhagen.

Copenhagen Municipality, 2009. *Bicycle Account 2008*. Copenhagen.

Freudendal-Pedersen, M., 2021. Vélomobility in Copenhagen – a Perfect World?, in: *The Politics of Cycling Infrastructure*. https://doi.org/10.46692/9781447345169.010

Freudendal-Pedersen, M., 2020. Sustainable urban futures from transportation and planning to networked urban mobilities. *Transportation Research Part D: Transport and Environment*. https://doi.org/10.1016/j.trd.2020.102310

Freudendal-Pedersen, M., 2018. Engaging with sustainable urban mobilities in Western Europe: Urban utopias seen through cycling in Copenhagen, in: Low, S. (Ed.), *The Routledge Handbook of Anthropology and the City*. Routledge, New York and London, pp. 216–229.

Freudendal-Pedersen, M., 2015a. Cyclists as part of the city's organism – Structural stories on cycling in Copenhagen. *City & Society* 27, 30–50.

Freudendal-Pedersen, M., 2015b. Whose commons are mobilities spaces? – The case of Copenhagen's Cyclists. *ACME* 14, 598–621.

Freudendal-Pedersen, M., 2014a. Tracing the super rich and their mobilities in a Scandinavian welfare state, in: Birtchnell, T., Caletío, J. (Eds.), *Elite Mobilities*. Routledge, London and New York, pp. 209–225.

Freudendal-Pedersen, M., 2014b. Searching for ethics and responsibilities of everyday life mobilities: The example of cycling in Copenhagen. *Sociologica* 8, 1–23. https://doi.org/10.2383/77045

Jensen, A., 2013. The power of urban mobility: Shaping experiences, emotions, and selves on a bike, in: Witzgall, S., Vogl, G., Kesselring, S. (Eds.), *New Mobilities Regimes in Art and Social Sciences*. Ashgate, Farnham, pp. 273–286.

Lydon, M., Garcia, A., 2015. *Tactical Urbanism: Short-term Action for Long-term Change*. Island Press, Washington, DC.

Vignette L
BEACH ROAD, MELBOURNE

Charlie Farren

Every Saturday and Sunday morning Melbourne's Beach Road is transformed into a river of cyclists of all types – fast and slow, young and old, small or massive groups, beginners to A grade competitors. This is a place to see and be seen, test your fitness, socialise or just take in the view (Figure L.1). Numerous cycling clubs use the road to train for road races and triathlons, while thousands of others are out for fitness and fresh air as they form an almost endless snake along the eastern shore of Port Philip Bay. Organised events sometimes close sections of the

Figure L.1 Beach Road Melbourne, Australia, Sunday morning (author's collection).

DOI: 10.4324/9781003142041-56

road, much to the annoyance of local residents. In 2006 the Commonwealth Games 40 km time trial used this route. After eight years of lobbying by cyclists, in 2010 on-road car parking was restricted along Beach Road on weekends from 6 to 10 a.m. Meanwhile cafés along the route have exchanged car parking for bike parking. There is also a separate bike path adjacent to Beach Road for recreational riders.

This short 30 km stretch of road in Melbourne connects with many aspects of Australian cycling culture, one with a long cycling history encompassing such immortals as Mulga Bill, Anna Meares and Cadel Evans. Australians took to the bicycle at an early date, with Victoria becoming the fulcrum of Australian cycling after the first velocipedes were imported from Europe in 1868. Australia's first bike race was held at the Melbourne Cricket Ground (MCG) in 1869. The race was contested by just three riders, and spectators were sorely disappointed at their sluggish performances, which they considered a far cry from the excitement of horse racing. But thirty years later huge crowds were attending exciting track races, including 32,000 for the 1901 Austral Wheel Race at the Exhibition Building track.

The first long-distance ride took place in October 1869 when William Kernot rode from East Melbourne to Geelong (75 km) in around 10 hours. Thereafter the mania for endurance riding saw many intrepid riders traversing the continent in search of glory. Track racing also started early. The Melbourne Bicycle Club (MBC) was the first Australian club, formed in 1878. MBC's first ride took place along Beach Road on 3 October 1878. The launch of its Austral Wheel Race in 1887 (the world's oldest still-operating track cycling handicap race) placed Australia on the world cycling map. Generous prize money attracted many international riders.

By the late 1890s, scarcely three decades after the first velocipedes arrived, cycling had permeated most areas of life in Victoria – sport, recreation, business, everyday transport and work – in other words, it had become a cultural phenomenon. The first professional cycling body, the League of Victorian Wheelmen, was formed in 1893, while in 1894 the Victorian Amateur Bicycle Club was founded to advance amateur interests.

By the 1960s, Australia had solidified its international identity as a sport-fixated country. Indeed, in a 1962 edition of *Sports Illustrated*, Australia was named the most sports-obsessed country in the world (Adair and Vamplew 1997). Two geographical circumstances contributed to this: a young and active immigrant population, and a land with a climate favouring bicycling, barbecues and swimming.

Beach Road soon became hugely popular with cyclists, and by the end of the 19th century its numerous riders were in conflict with other road users. The mid-20th century saw many conflicts between cyclists, motorists and residents, leading to citizen activism. A key player was the Bicycle Institute of Victoria, established in 1975 in Melbourne. In 1976, it presented the Melbourne Bikeway Plan, and in 1978 the Geelong Bike Plan, forerunners of today's bicycle infrastructure.

For many years, every Saturday at 7 a.m. an informal training ride called the Hell Ride sets off from Black Rock clock tower 32 km along Beach Road for Mount Eliza. Fallout from a 2006 accident leading to the death of a pedestrian during this ride led to further legislation. The Victorian State Coroner's investigation into this death resulted in a review of the literature on group riding on public roads, and a police presence monitoring both motorists and cyclists during the weekly Hell Ride. Resulting research released by Monash University's Accident Research Centre in January 2009 led to the introduction of "culpable cycling" laws by the Victorian State Government. The *Road Legislation Amendment Bill 2009* introduced new penalties for cyclists who cause death or injury in a collision, for failing to stop and render assistance, and for "dangerous and careless riding".

An investigation by the Amy Gillett Foundation (founded to promote cycle safety) found that cyclists are among the most physically vulnerable road users. Legally, they are recognised as legitimate road users, but the space for riders on Victorian roads ranges from fully separated to symbolically separated (with bike lanes or wide kerbside lanes) to complete intermingling with motorised vehicles, ideally in slow-speed streets (Levasseur 2014). Australia's cycling environment contrasts with the extensive networks of physically separated facilities for cycling that exist in many European countries. The interaction with vehicles on Australian roads contributes to a higher rate of bike rider fatality and serious injury crashes than in Europe (Garrard et al. 2010). In response, mandatory helmet laws were introduced in Australia in the 1990s and are actively enforced, with a fine for not wearing a helmet of A$207 in Victoria. Recently, legislation enacted throughout Australia (Victoria was the last conforming state in 2021) sets a 1.5 m (5 ft.) minimum passing distance for vehicles. The ban on weekend street parking on Beach Road was another response to this problem.

In brief, the Beach Road cycling phenomenon embodies recognised Australian national characteristics of mateship, endurance, obsession with sport, teamwork and resilience, while exemplifying the enduring problems resulting from mixing cyclists with motorized vehicles.

Bibliography

Adair, D. and W. Vamplew (1997) *Sport in Australian History*. Melbourne: Oxford University Press.

Garrard, J., S. Greaves and A. Ellison (2010) "Cycling injuries in Australia: Road safety's blind spot?" *Journal of the Australasian College of Road Safety* 21(3), 37–43.

Levasseur, M. (2014) *Cycling Aspects of Austroads Guides: Quick Reference*. https://trid.trb.org/view/1309692, accessed 27/10/2020.

38

BOGOTÁ

Perspectives on the "World Bike Capital"

Luis Vivanco

Residents of Colombia's capital of 8 million people, Bogotá, tend to be very spare with praise for their city. Most are quick to emphasize its dangers, disorder and dirt, the pervasive sense of fear and mistrust. But solitary flowers do sometimes erupt in sidewalk cracks. It once happened while I was riding a bicycle during the city's recreational open streets program, Ciclovía, *that closes 126 kilometers of city streets between 7 a.m. and 2 p.m. on Sundays and holidays. I was with Juancho, a thirty-something working-class Bogotano of Afro-Colombian descent who cobbles together a modest living working various jobs in the formal and informal economies. The bicycle plays an important role in his life. He uses it to get around the city and he sometimes earns money carrying packages or guiding visitors. His riding technique involves a lot of standing on his pedals (to be more alert to his surroundings, he once told me), and he sings when he sees something that gives him pleasure.*

We had come deep into the southern part of the city, and we stopped for salpicón, *the ubiquitous fruit salad sold by ambulant vendors who line the avenues during* Ciclovía. *As we ate and chatted with the vendor about the cargo bicycle from which she sells snacks, Juancho suddenly broke into a popular song with a chorus of "*Hola, ¿qué tal?*" ("Hi, what's up?"). The vendor and a bystander joined in. As the song finished, smiles all around, Juancho commented, "The nice thing about* Ciclovía *is that I can interact with the neighbor here. I didn't know her until now, but next time we'll recognize each other and we create relationship." Smiling broadly the woman responded, "Other cities have the sea. We have* Ciclovía*!"*

As we rode away, Juancho waxed poetically about Ciclovía *as "a thread in the tapestry that connects people here." The weekly closure of several major avenues to motorized vehicles, he believes, creates through-lines in this deeply divided city, connecting rich and poor who otherwise conduct their daily lives in geographically and socially separated worlds. As he said: "it allows us to intermingle in each other's spaces and encounter each other as equals… it humanizes our city."*

But then, just as suddenly as he had earlier broken into song, Juancho's tone grew dark. "Just don't come here tomorrow," he said. "It's placid now, but during the week, the perros callejeros[1] *are out and it's dangerous." He described the dramatic difference just a few hours, or crossing a few blocks, can make in this impoverished and notoriously fearful part of the city. "But it's not just here," he said. "Riding a bicycle anywhere in Bogotá is an odyssey." Wherever you go, he said, cyclists confront a harsh urban milieu of attacks, robberies, and other street threats; deaths and injuries due to collisions with pedestrians, buses, and automobiles; and stolen and broken bicycles due to physical obstacles, terrible parking options, and crumbling, inconsistent roads.*

***

DOI: 10.4324/9781003142041-57

In recent years, Bogotá has gained an international profile as a beacon of bicycle friendliness, once even dubbed by the BBC a "bicycle paradise" because of its investments to support bicycle transportation (Wallace 2011). *Ciclovía*, which can draw a weekly participation of as many as 1.5 million residents, occupies a special symbolic niche as a popular program and is credited with inspiring several hundred other cities to organize open streets initiatives of their own (Montero 2017). Other elements include the construction of the largest *cicloruta* (protected bicycle lane) system in the Americas, currently at 550 kilometers; the establishment of an annual car-free day in February; and the creation of bicycle policies and promotional programs. Official delegations from cities across the Americas make treks there to understand how these things were achieved. Building on this reputation, city leaders a few years ago confidently declared Bogotá to be the *Capital Mundial de la Bici*, or the "World Bike Capital." Such figures and imagery are quite remarkable. They can even seem magical for a city that just a few decades ago was considered one of the world's most violent and desperate places.

But the closer one gets to the actual streets, the more unsettled and less magical – and a lot more interesting for the cycling ethnographer – that beacon, "world capital," or whatever the label assigned to it, becomes. Bogotá is not a bike paradise, and whether or not it is as dystopic as Juancho expressed depends on who you talk to. Bicycles have been positioned in Bogotá's public sphere as an innovative tool to transform urban social conditions and the built environment. But between poles of salvation and damnation lie certain nuanced realities. One of these is that most residents are largely indifferent to activities like *Ciclovía* and everyday bicycle use. They have myriad reasons, among them the impractical distance of these things from where they actually work or live, the quotidian dangers and fears of being in public space with an object that is easy to steal, and the challenges of scraping together a living which just doesn't permit space and time for bicycles. While more than half of Bogotano households own at least one bicycle, the vast majority of people walk or ride the bus to get places. About 15%, a wealthy minority, drive private automobiles.

For those who do actually get around by bicycle – some 800,000 trips a day, which constitutes approximately 6.6% of overall trips – it is not uncommon for them to express positive associations of bicycling with freedom, social relationship, and intimate knowledge of the city. But these sensations intersect in complicated ways with the embodied experience of the city's uneven and fragmented spatial environment, the political-economic and social inequalities that play out on city streets, and cultural perceptions and ideologies of urban public space as fearful. As a result, riding a bicycle can concentrate dissonant feelings of joy *and* fear (Vignette I); possibilities of social innovation and connection *and* confrontation with intractable problems of poverty and violence; and the hopefulness of intentional planning *and* the inevitability of entropy in a crowded and mistrustful city.

***

Tomás was born in a small town, but several decades ago his family moved to Bogotá. The family is originally from the province of Boyacá, where bicycle racing is taken seriously and the poor move around the countryside by bicycle. At their Bogotá home, his father keeps several extra bicycles around the house to lend to a visitor, like me, who drops by. Tomás offered to show me various routes around the city that he uses on a daily basis. He is college-educated and he likes to think analytically about the role of bicycles in the city and in his life. He was the first to share with me the popular notion, among bicycle riders at least, that they are hares and motorized vehicles are tortoises; the former zipping around the city freely, the latter moving ever so slowly through trancones*, traffic jams (c.f. Salazar 2013). He also told me about what he called the "diverse imaginaries" of Bogotá cyclists. One in particular appealed to him* *personally, connecting his own everyday use of the bicycle to Colombian accomplishments in international cycling sport. He told me, "Although riding as an urban cyclist isn't sport, there's a tie many of us feel to this history and its glories. It gives you passion and can help you in certain moments, like an extra push up a hill, or through a tricky intersection. It's the same heart, the same feeling."*

***

Bicycles carry important cultural and political resonance in Colombia. For example, there are few other countries in the world where a presidential candidate would run a campaign centered on bicycle symbolism, presenting himself, like the bicycle, as a great unifier of a fractious nation. Yet this is exactly what the former mayor of Bogotá, Enrique Peñalosa, did in 2014. He lost the election, but not because of the bicycle. Colombia hosts one of the world's most grueling bicycle races, the *Vuelta a Colombia*, and takes pride in producing some of the world's great cycling athletes. The international bicycle racing sensation with humble countryside origins is a celebrated national type, among them Nairo Quintana, who grew up as a potato farmer in Boyacá, and Lucho "El Jardinerito de Fusugasugá" Herrera, who transported himself by bicycle to and from jobs as a gardener in the mountains outside of Bogotá. As one prominent commentator notes, the bicycle can be viewed as a hopeful yet vengeful symbol: in a country of savage socio-economic inequalities, the universal accessibility of the bicycle and the international victories of cyclists from humble backgrounds symbolizes revenge against the country's oligarchs who are indifferent to the extensive suffering they have perpetuated (Abad Faciolince 2014).

In Bogotá, bicycles are largely vehicles of the poor. About three-quarters of everyday bicycle users are low-income men (with an additional smaller percentage being low-income women), their imaginary of the bicycle being largely one of convenience, saving money or making it, and saving time, of being free of getting stuck in buses or traffic jams that affect getting to and from menial jobs. For others, many of them young, educated, and new to non-recreational use of the bicycle, it is connected to efforts to reimagine Bogotá as a positive city of progressive values like quality of life, sustainability, and mobility justice. For an even smaller number, it is about associating with a "Copenhagenized" vision of transnational urban trendiness and chic fashion. For still others, it is imagining themselves as active and healthy through their participation in recreational spaces like *Ciclovía*. The diversity of bicycles one sees on the streets, many of them Colombian-made – mountain bikes, fixies, racing bikes, BMX, cruisers, gentleman's bicycles, cargo bicycles, etc. – mirror and express these imaginaries. But those imaginaries are also not mutually exclusive. More important is that they are all filtered and framed through a dominant narrative of life in Bogotá, that it is a "city of fear" (Salcedo 1996), which shapes bicycle riding in important ways.

***

For several months, my daily experience riding on the city's famed ciclorutas *mostly happened in the center of the city, to and from my fellowship at the Universidad Nacional. These lanes were the first ones built during the early 2000s, and, consequentially, they were placed on sidewalks because of political resistance to taking street space. It is striking how inconvenient they can be: you have to duck, dodge, and weave between pedestrians and vendors who have strayed onto, or simply have no alternative to walking or working in, the bike lanes; stop or slow down at the end-of-block intersections where priority tends to go to motorized traffic; navigate steep ramps or drop-offs between sidewalk and street that require the skills and strength of mountain biking; and manage uncertainty where* ciclorutas *suddenly end without any indication of where to go next. Sometimes I would just hop on the road with cars and buses, as I saw countless other cyclists do, and I would get places faster.*

When I mentioned this to Pedro, a school bus driver I know who also rode his bicycle to and from work, he confirmed the selective use and avoidance of ciclorutas*. But he also told me I needed to experience a different kind of* cicloruta*, the 80th Street* cicloruta *that passes through the western side of the city. It is like a bicycle superhighway and is one of the busiest in the city. Pedro lives in Suba and uses it himself during weekdays to get to his job, so he offered to show it to me one Saturday morning.*

He told me that, like many others, he was motivated to try out bicycle transportation when they first built this cicloruta *that connected with his lower-income neighborhood. It felt safer than riding on streets with autos and buses, and offered the promise of uninterrupted flow, of not sitting on a bus in traffic jams. During much of our ride, instead of talking about flow and traffic jams, however, Pedro wanted to share with me the insecurity of the* cicloruta *itself. He had been robbed of his bicycle by knifepoint three times on the 80th, and witness to numerous attacks. The bridges, he said, are especially dangerous chokepoints, especially after dark. As we paused at one, he explained that he times his daily commute home to coincide with the 4 to 6p.m. rush hour, before the darkness settles in and when there is still safety in numbers. Stragglers who are just a bit late getting out of work make for an easy target for predators who can block both ends of the bridge to cut-off escape, which is how he lost two of his bikes.*

As we rode on, I realized Pedro did not take for granted that a bridge is simply a bridge, a cycle track is not simply a cycle track. They don't just ease the flow; they also can render him vulnerable and disrupt the flow. I began to appreciate his acute sense of spatial awareness and his ability to read subtle cues among others on and near the cicloruta*. I told him as much. He agreed, "it gets quite crowded here at times, with people on bicycles. There is no talking; we are all in our heads thinking about our day. But we all have that alertness. Sometimes, that's not even enough, though. Danger is lurking and you also have to rely on your premonitions. That's Bogotá."*

★★★

Since the middle of the 20th century, Bogotá experienced massive population growth as millions of impoverished and traumatized refugees moved there to escape civil war and armed conflict, drug cartels, and death squads in the countryside (Vivanco 2013). The city was weakly prepared to handle that growth. As the country's capital, it did not have strong municipal institutions, not even a mayor elected by the residents, and was under the influence of modernist urban planning traditions that emphasized national economic and social priorities over local concerns and histories (Montero 2017). After the 1948 riots known as the "Bogotazo," the wealthy and middle classes abandoned the historic downtown to create secure residential subdivisions and private social and commercial spaces to the north of the city, contributing to the physical fragmentation of the city into distinct sectors. The city that developed was characterized by pervasive social and spatial exclusion, inadequate housing for the poor, the privatization of public spaces, indifference to civic life, and conspicuous insecurity and violence (Berney 2010). By the 1980s and 1990s, the combined intensity of street crime, the violent activities of urban guerillas, and bombings related to Pablo Escobar's reign of terror to prevent his extradition to the U.S. earned the city a reputation as one of the most violent and troubled in the world (Rivas Gamboa 2007).

The city's historical patterns of urbanization had important effects on mobility. During the early 20th century, Bogotá developed a small tram system for mass transit, which was eventually displaced by the development of a private bus system (Montezuma 2008). The inefficient and unregulated buses pushed unplanned development at the city's margins, yet also provided an accessible means of transit for the growing city. Beginning in the 1960s and 1970s, the explosion of private automobile use among the wealthy exacerbated the city's problems. Minimal investment was made to maintain street infrastructure or even enforce traffic rules, and entitled drivers took over sidewalks and parks, already few in number and neglected, as parking areas for their cars (Ardila and Menckhoff 2002). Ever-increasing quantities of buses, taxis, and cars competed for space, creating a daily chaos of traffic jams, crashes, noise, and pollution.

A turnaround began to take shape during the mid-1990s, a few years after constitutional change allowed Bogotá to elect its own mayor (Martín and Ceballos 2004). In 1995, the philosopher, educator, and political outsider Antanas Mockus was elected mayor. He framed the city's severe violence and mistrust as a lack of "citizen culture" and he sought to implement policies and pedagogical programs to cultivate civic values, especially those encouraging direct encounters in public space (Buendía 2010). Dysfunctions of urban mobility were a paramount concern, and his administration implemented a number of creative (and effective) policies to address it, one of the most famous being the short-term replacement of the traffic police with mimes. Mockus was himself an everyday cyclist who emphasized the dignity of getting around under one's own power, and one of his most important initiatives in relation to bicycling was the expansion of *Ciclovía*. *Ciclovía* was a modest experiment begun in 1974 by politically-connected middle-class activist group Pro-Cicla to close several kilometers of city streets for a "pedal demonstration." It attracted several thousand participants, and by 1976 it was institutionalized in the city's sport bureaucracy and promoted as a recreational program – even though its original intention was to problematize car-centric patterns of urbanization typical of U.S. cities that were in vogue in Bogotá at the time, and to demonstrate the bicycle's legitimacy as a form of getting around the city (Montero 2017).

With its wide avenues; flat topography; cool, mostly pleasant, climate; dense concentration of work and residential opportunities; and tight territorial extension of about 35 kilometers from north to south, Bogotá does, at least in the abstract, have agreeable everyday cycling conditions. Although initial plans for a bicycle network were drawn up and began construction during the Mockus administration, the Dutch-designed *ciclorutas* were mostly financed and constructed during Enrique Peñalosa's first administration (1998–2001). Emphasizing the quality of life benefits of reorganizing the transportation system, Peñalosa's administration also raised gasoline taxes and implemented "*pico y placa*" (use of automobiles on alternate days); expanded the number of car-free activities and further expanded *Ciclovía*; developed the public bus rapid transit system *Transmilenio*; built public libraries and parks in poor neighborhoods; and recuperated and expanded hundreds of public squares, parks, and sidewalks (Ardila and Menckhoff 2002; Montezuma 2005; Cervero et al. 2009).

As Sergio Montero (2017) demonstrates vividly, these images of Bogotá city leaders taking bold and visionary action began to circulate internationally during the early 2000s. Promoted by some of those same city leaders – especially Peñalosa, who has become a recognizable figure in international transportation policy circles – and enabled by the transnational flow of universalizing discourses about what constitutes "a good city," these experiences have symbolically transformed what were, in essence, experiments, highly contingent and contested processes, and ad hoc decisions into "best practices" intended to be imitated in other cities.

But just as Bogotá's reputation for urban mobility innovations and "best practice" crested internationally, the sense in the city itself after Mockus left office from a second term in 2003 was the end of "a golden age" (Pardo 2010). Succeeding mayors ignored or deprioritized the bicycle. The city's legendary traffic jams roared back, multiplied horrifically because the wealthy simply purchased more cars to get around *pico y placa* rules. And the city's new transportation infrastructure – especially its bus rapid transit system, *Transmilenio*, and the *cicloruta* network itself – couldn't handle the demand (the former) or showed wear and tear as maintenance faltered (the latter). During those difficult times, new forms of politicizing the bicycle and citizen activism began to catalyze.

***

Several dozen members of the Bogotá Bicycle Users' Assembly gathered outside the Consejo de Bogotá (city council), waiting to be let in for an evening meeting. They were a mixture of men and women in their twenties and thirties, counting among their ranks lawyers, university students, artists, and other middle-class professionals. Security was extremely tight, as it is at all public buildings that have been the target of violent attacks. We were subjected to various identity checks and each of us had to go through the common and laborious ritual of registering our bicycles in a ledger by brand, color, and serial number so they could be parked in a guarded area. Later we would have to show I.D. and recite those details back to the guard to release our bicycles. We were eventually let into the building and led to the chambers where two city councilors, who were developing a legislative agenda for bicycle transportation in the city, were awaiting us.

The meeting's goal was to coordinate efforts and reach agreement about strategies to push the Mayor's office to recognize the severity of two problems: bicycle theft due to lack of secure parking facilities, and pervasive insecurity on the ciclorutas. *Having just gone through the rigamarole of simply parking a bicycle to attend a public meeting that really wasn't open to "the public," it was clear why issues of security and parking might be at the forefront of activists' priorities.*

Before the meeting started, a lawyer and leader of the Bicycle Users' Assembly gave me the most concise version I've heard on new forms of bike activism taking shape in Bogotá: "We had the gift of two strong mayors who expanded Ciclovía *and gave us the* cicloruta. *But after them there was only stagnation. Maybe some rhetoric about how beautiful bicycles are. But no execution. One administration even took the position that bicycles are not a viable form of transportation in this city. That's when my friends and I started organizing rides, to get people out there and build a force. But now we are also organizing a political agenda, to promote a public policy for the bicycle because one doesn't exist. We feel like it can go either way here for the bicycle."*

***

In spite of international recognition of Bogotá's efforts to position bicycles as key protagonists in its urban transformation, there is a widespread sense of fragility about it amongst those who would like to expand it further or just not see it all fall apart. A lot of wariness lies in the simple fact of wear and tear on infrastructure and in patterns of bureaucratic neglect in a crowded and underfinanced city with abundant crises. But it also lies in knowledge of the powerful resistance that pro-bike administrations have faced, including the dogged commitment of the city's elites to ensure that their use of private automobiles will not be diminished by efforts to promote bicycles. It also comes from the knowledge that in Colombia there are always dark and violent forces of political manipulation at work that can undo efforts at progressive change.

Nevertheless, during the past 15 years, citizen-based *bici-activismo* and *colectivos* (collectives) began to form with the intention to get bicycles back on a public agenda. Not all bicycle *colectivos* are explicitly political; some are simply hobbyists, like one that restores old bicycles, and others that are education- or arts-focused. But through their creation and involvement in *colectivos*, new categories of bicycle riders in Bogotá – the middle-class urban cyclist, feminist cyclists, artists on bicycles, etc. – have been gaining new presence and influence on city streets and in city mobility politics. Pro-Cicla was a predecessor in this vein, but the rise of *bici-activismo* is still viewed as a novel development, mainly because, as one prominent alternative transportation advocate has described, "Colombia has never been characterized for its bursting citizen participation. Like the good conservative country it is, people are fine wearing a pro-something t-shirt or donating $10 a month to UNICEF and saying 'I participate.' It took a fifty-year war of the FARC to produce a massive mobilization, but that only happened once. For other problems, not much has happened" (Pardo 2011).

Bogotá's *bici-activistas* work in spaces like the wood-paneled chambers of the city council, and some activists have found their way into official city positions in recent years to work on bicycle issues. But their spaces are mostly on the streets and in social media, where numerous *colectivos* can organize and participate in mass rides to make themselves and the bicycle more visible (and more audible; one sign of the Bogotano bicycle activist is a whistle hanging by string on their helmet, whose purpose is to signal to others to stay out of their way as they speed through the city). A high-profile *colectivo* that strives for such visibility is *Ciclopaseos de los Miércoles* (Wednesday Bicycle Rides), which is run by some of the same and young professionals who were at the meeting described above. Every Wednesday evening, it conducts a Critical-Mass-style ride on routes that often go into areas of the city avoided by the young middle-class constituency that gravitates toward CPM. It uses social media to great effect, projecting imagery of urban cycling cool, good times, and adventure, and can attract more than four or five hundred participants on any given Wednesday. But leading several hundred cyclists, many of them inexperienced in the demanding spaces of Bogotá, is a risky and complicated endeavor, requiring tight organization, logistical savvy, and a willingness to manage inevitable conflicts between cyclists, automobile owners, and bus drivers. It also requires discretion, and organizers exercise great secrecy about their weekly routes to prevent tipping off street criminals. Their overall goal is partly to help participants gain confidence riding in urban traffic, to cultivate new (and influential) adherents to everyday bicycle riding, and also to help participants experience parts of the city and landmarks they might not otherwise be aware of because of fear, or distance from their home. In that sense, it is Urban Bicycling 101, and civic tourism for a class of people unfamiliar with many areas of the city in which they live. But the other part is explicitly political, an assertion of the right to free travel on city streets.

This style of bicycle activism is very familiar to North American eyes, and, in fact, is inspired by it. Most *bici-activistas* are cosmopolitan elites, fluent in and responsive to government policy debates, attentive to the actions and limitations of official institutions, and engaged in transnational conversations themselves about developments in urban bicycle activism. They have also had, since 2017, friendlier access and more entry points in the mayor's office as Peñalosa returned to office (2017–2019), and efforts by the current mayor to promote bicycling as an effective response to the challenges of socially-distanced transportation during the Coronavirus pandemic. But as much as imagery of Bogotá's autonomous successes constructing itself as a bicycle-friendly city circulate internationally, it is important to recognize that the elites who have sought to make it that way have always drawn inspiration, ideas, plans, and reference points from elsewhere.

In taking stock of Bogotá and its complex relationships with bicycles, it is striking to note that, in striving for visibility and staking their claim for status and rights on city streets, today's *bici-activistas* stand in sharp contrast from the majority of the city's everyday bicycle riders. These latter would just as soon not draw attention to themselves as they quietly travel through the city, hoping to avoid entanglement and threat from criminals and rumbling busses on pothole-ridden streets. Indeed, many of those individuals are literally invisible, as they travel to and from their menial jobs in distant parts of the city early in the morning or late at night. As expressed eloquently in a recent book about bicycling in Bogotá (Alcaldía Mayor de Bogotá 2013),

"For those people whose lives are profoundly tied to this technology – the security guards, the plumbers and masons, bakers, and of course, the maids – the motivation to get on a bike is not health, or to save the planet of hydrocarbons, nor is it to improve mobility and claim spatial equity from the rich in their cars, or for the poor smashed into buses. In this case, riding a bicycle is the everyday struggle for existence, the struggle for the love of one's children, and of parents who wait anxiously for their loved ones to arrive home safe and healthy every night." *Bici-activistas* and those riders may live in the same city, but they live, quite literally, in different worlds. And this raises a big and persistent question: can the bicycle overcome Bogotá's dogged problems and social divisions?

Note

1 "Street dogs," referring to petty criminals and thieves.

References

Abad Faciolince, Héctor (2014) "El Desquite de la Bicicletas" *El Espectador* 31 May 2014. World wide website accessed on May 31, 2021. https://www.elespectador.com/opinion/columnistas/hector-abad-faciolince/el-desquite-de-las-bicicletas-column-495802/

Alcaldía Mayor de Bogotá (2013) *El Libro de la Bici Bogotá.* With Secretaría de Cultura, Recreación, y Deportes. Bogotá, D.C., Colombia.

Ardila, Arturo and Gerhard Menckhoff (2002) "Transportation Policies in Bogota, Colombia: Building a Transportation System for the People." *Transportation Research Record* 181: 130–136.

Berney, Rachel (2010) "Learning from Bogotá: How Municipal Experts Transformed Public Space." *Journal of Urban Design* 15(4): 539–558.

Buendía, Felipe Cala (2010) "More Carrots than Sticks: Antanas Mockus's Civic Culture Policy in Bogotá." *New Directions for Youth Development* 125: 19–32.

Cervero, Robert, et al. (2009) "Influences of Built Environments on Walking and Cycling: Lessons from Bogota." *International Journal of Sustainable Transportation* 3: 203–226.

Martín, G. and Ceballos, M. (2004) *Bogotá: Anatomía de una Transformación, Políticas de Seguridad Ciudadana 1995–2003.* Bogotá: Editorial Pontificia Universidad Javeriana.

Montero, Sergio (2017) "Worlding Bogotá's Ciclovía: From Urban Experiment to International 'Best Practice.'" *Latin American Perspectives* 44(2): 111–131.

Montezuma, Ricardo (2005) "The Transformation of Bogotá, Colombia, 1995–2000. Investing in Citizenship and Urban Mobility." *Global Urban Development* 1(1): 1–10.

Montezuma, Ricardo (2008) *La Ciudad del Tranvía: 1880-1920. Bogotá: Transformaciones Urbanas y Movilidad.* Bogotá: Editorial Universidad del Rosario.

Pardo, Carlos Felipe (2010) "Transport Policy in Bogotá Ten Years after the 'Golden Age:' The Challenge of Being an Example." Presentation to GTZ (German Organization for Technical Cooperation), Stuttgart.

Pardo, Carlos Felipe (2011) "Activismo pro-bicicletas en Bogotá (y Colombia): Necesitamos Más." *Distintas Latitudes*, 14 April 2011. World wide website accessed on 19 May 2014: http://www.distintaslatitudes.net/activismo-pro-bicicletas-en-bogota-y-colombia-necesitamos-mas

Rivas Gamboa, Angela (2007) *Gorgeous Monster: The Arts of Governing and Managing Violence in Contemporary Bogotá.* Saarbrücken: AV Akademikerverlag GmbH & Co.

Salazar, Oscar Iván (2013) "De Liebres, Tortugas, y Otros Engendros: Movilidades Urbanas y Experiencias del Espacio Público en la Bogotá Contemporánea." *Revista Colombiana de Antropología* 49(2): 15–40.

Salcedo, Andrés (1996) "La Cultura del Miedo: La Violencia en la Ciudad." *Controversia*, No. 169. Centro de Investigación y Educación Popular.

Vivanco, Luis (2013) *Reconsidering the Bicycle: An Anthropological Perspective on a New (Old) Thing.* New York: Routledge.

Wallace, Arturo (2011) "Bogotá – Latin America's Biking Paradise." 20 July 2011. World Wide Website accessed on July 2, 2012: http://www.bbc.co.uk/news/world-latin-america-14227373

PART VII

The visual culture of cycling

Nicholas Oddy

It is a difficult job to put together a part on such a huge topic, a process disrupted by the Covid pandemic. The result contains six essays. They are varied in approach, but it is interesting to see how they link. Their binding feature is surely that all represent the bicycle as an emblem of modernity. This is no surprise as, although it vies with the sewing machine in being the first modern 'consumer durable', its place in the public realm and its complicated relationship with gender made it an object of far more public and critical attention during the second half of the nineteenth century, the period in which Baudelaire's concept of modernity was defined.

We start with a study of the machines themselves, their overtly technological aesthetic and their quite uncompromising black finish that lasted from the 1870s until the 1930s as the default for European machines, regardless of intended market. This was the period when ownership of a bicycle was a statement of conspicuous consumption available to an ever-enlarging market. It could be said that the bicycle displayed a 'machine aesthetic' that made it not only a symbol of modernity, but also of modernism; yet, ironically, during the inter-war period, when modernism was truly established, bicycle makers adopted more decorative strategies of presenting machines to their markets.

On the other hand, almost the opposite seems to have happened when cycling was represented on objects within the home, such as ornaments and table-wares. Here the machines and the activity were represented using tried-and-tested historicist styles. Of course, while these might not look 'modern' in stylistic terms, they were certainly so in terms of a class of good. Factory-made in large numbers and freely available to the new middle classes with disposable income to spend on non-essentials representing what was usually little more than a pastime. Their successors continue to be sold in huge quantities to those who want to declare their personal interests in outdoor activities through the things they display at home or, possibly more importantly, to those who feel that such items would be fitting gifts for those they know.

It seems appropriate to move from a study of the machines themselves, to the visual culture of their riders. Emma Hilborn provides us with a consideration of the way in which cycling clothing developed from its inception to the present, but largely focusing on the 'dress problem' faced by female cyclists in the late 19th century. Again, the very issue was one of modernity: how to retain personal and expected cultural values in a time of rapid change and widening opportunities? The topic is a common one in cycling history, but Hilborn comes at it from a refreshingly pragmatic angle based on the actual evidence, rather than the romantic one of

DOI: 10.4324/9781003142041-58

'new womanhood' that is repeated to this day – that somehow cycling was the catalyst for dress reform in which the end product was that it became acceptable for women to wear 'bifurcated garments'. Rather, the greatest issue for many women cyclists was how to conduct the activity *without* having to change out of a long skirt and thus not seriously challenge their socially constructed femininity. The technological issues that presented the skirt wearer with a 'dress problem' were addressed by the first years of the 20th century and female cyclists almost entirely reverted to long skirts for another twenty years until hemlines rose, rather than anything else.

We then move on to Nadine Besse's analysis of 'art posters' issued by French cycle agencies in the 1890s. It is here that some of the myth that Hilborn addresses originates. Poster artists usually tried to capture the positive experience of cycling, rather than the cycle itself as a product, particularly in those posters that depicted female cyclists, often dressed in the most fashionable bifurcated garments (or sometimes no garments at all). The modernity of the cycling art poster is overt. The art poster's very being relied on the structures of modernity, many advertised UK- and US-made products, relying upon the latest manufacturing, communication and distribution methods to reach a remote, anonymous market. The product itself was one of conspicuous consumption for little more than its owner's pleasure. Firmly lodged in addressing the new urban middle classes, the art poster blurred the divisions between high and popular culture. Had he lived to see them, Baudelaire would probably have approved. However, art posters remain an excellent example of mythology being created round a product that would be more the stuff of Roland Barthes a century later, a mythology that remains alive and well and will probably continue to do so for as long as art posters remain a desirable collectible.

If Besse explores a fruitful collaboration between art and cycling, Scotford Lawrence looks at the somewhat more problematic relationship cycling seems to have had with fine art. Here, it seems the cycle was at odds with 19th-century fine art practice, often mired by the norms of the 'academy' and preconceptions of what was suitable subject matter for painting and sculpture. It was the overt rejection of the academy initiated by Courbet's 'realism' that allowed artists like Lautrec to produce the art posters of the 1890s; but Lawrence points out that it is really not until the 'avant garde' of the early 20th century that cycling and fine art come together, often in quite conceptual ways, beginning with Duchamp's 'ready-mades'. However, by this time it was too late for cycling to develop a 'genre' that one would associate with, for instance, horses as a subject of formal depiction. Again, the way in which cycling seems almost naturally to ally itself with modernity and distance itself from those values that preceded it is notable. Had cycling become a subject for 'academic' artwork of the nineteenth century, rather than that of Impressionism, one would almost be at a loss to explain why.

The relationship between cycling and art is brought right up to date with Hilary Norcliffe's analysis of Ai Weiwei's *Forever Bicycles*. This huge sculptural installation is posed at a moment when China is in the midst of mass motorization in the context of the climate crisis and a general, if unfulfilled agreement amongst leading world nations to try to promote low energy and preferably human powered transport.

Finally, Bruce Bennett provides an overview of cycling and films. As Lawrence points out earlier, cycling quickly allied itself with photography in the velocipede era of the 1860s and the two remained closely interlinked. Indeed, by the 1890s magazines such as *Cycle and Camera* graced the shelves of railway bookstalls. Bennett points out cycling's move to cinema was immediate on its public introduction by the Lumière brothers and, since then, it has remained a part of both 'popular' and 'art' cinema, adapting to much cultural repositioning throughout the twentieth century and into the present.

Bennet's essay points out just why cycling and film are so suited to one another. Photography tended to depict cyclists when stationary: posing with their machines in a studio; waiting for the starting gun on the track; with their trophies after a race. While the art poster usually attempted to capture movement, it too was in freeze frame; but the movie camera allowed that movement to be depicted in all its excitement. Thereafter the film built narratives around the changing status of the machine that would be familiar to many, from the workhorse of the working man, to the freedom machine of children and, indeed, extra-terrestrials.

39

THE MACHINE AESTHETIC

The visual identity of the bicycle and its representation in advertising and artefacts

Nicholas Oddy

In Nadine Besse's contribution to this volume (Chapter 41) we will see how bicycles were presented to the French market through chromo-lithographed posters of great visual power. On their first printings they were acclaimed by collectors (many being acquired by them, both by request and theft) and ever since had a place at the high table of graphic arts. But what exactly were they advertising?

In this essay I will explore the visual culture of the bicycle and cycling in general over the time of the machine's history from 1817 to the present, largely focusing on the century from 1870 to 1970, in which three distinct 'machine aesthetics' emerged, a period when design and manufacture was still largely focused in Europe, particularly the United Kingdom and North America (mainly the eastern USA), and the machines largely responded to the markets in these places. I consider the machines not only in their own right, in terms of appearance and finish, but also the way in which they were presented in advertising and domestic ornaments. To a great extent this is the bringing together of a number of papers I have presented to the International Cycling History Conference, each considering different ways in which the bicycle was understood as a visual object, often symbolizing modernity, but at the same time hedged in by the tropes of the domestic sphere (Oddy 1992, 1994, 1997, 2001, 2006, 2007a, 2007b, 2013). The machine itself awkwardly posed between inside and outside. It is this this visual and metaphorical ambiguity that I intend to focus upon here.

Let us start with the French posters so ably outlined by Besse. When they began to appear in the mid-1890s they were an extension of an established genre that had its origin in performance arts, particularly cabaret, using lithographic techniques developed by artist Jules Chéret. Chéret's own approach was to present the viewer with a picture not necessarily of the product, but of the atmosphere surrounding the product. There was no attempt at photographic realism in the depiction; what made Chéret and his followers so significant was the very loose, often impressionistic drawing style of many of their posters. In real terms many of his posters could advertise any number of *fin-de-siècle* venues or events, the determining factor being what the lettering says, rather than the image or the product. When called upon to create a poster for a specific performer, the style lent itself to capture the individual. But for hard products, this was very different. For bicycle makers, this was particularly significant because, as a product, bicycles are remarkably anonymous and therefore the visual generality that creates an atmosphere and links a name to it was advantageous

DOI: 10.4324/9781003142041-59

Figure 39.1 Hampden/Barber poster by Burch (1895). Single colour stone litho, 22½″ × 35″ printed by Macbride & Macintyre, Glasgow.

Source: Author's collection.

To make the point, here is an 'art poster' for the Hampden Cycle Co of Glasgow from 1895, by 'Burch', a local artist and cartoonist whose work appeared in the *Scottish Cyclist* (Figure 39.1).

There is no part of this image that actually represents either the products of Hampden or its sister tyre manufacturer, Barber; the names of both could be replaced by those of any other maker and the poster would be equally effective. However, the artwork not only lodges Hampden and Barber at the cutting edge of poster design and therefore being up-to-date to the point of being *avant garde*, but also implies a particular style of riding and attitude to authority that places them clearly in the world of the young, daring and fashionable. Much the same, with different nuance in terms of style of riding and attitude, could be said of almost all the French 'art' posters. How many would be able to differentiate a 'Gladiator' bicycle from any other on the evidence of the famous depiction of it by PAL (Jean de Paleologue), no matter what the accuracy of depiction (Dodge 2006: 117; Rennert 1973: 42)? What makes it memorable is the naked woman with a mane of red hair who clutches the machine as it flies through the night sky. There is no doubt that she, rather than the bicycle, is the subject of the poster, given that she is brought forward of the product. An exception to this approach is the poster designed by Toulouse Lautrec for Simpson Chains (Dodge 2006: 172; Rennert 1973: 28). The Simpson Lever Chain was so idiosyncratic that it stands out in the artwork.

Whether or not it would do so quite so obviously had Lautrec not been commissioned by L. B. Spoke (Simpson's Paris agent) is unimportant. It was possible for Lautrec to emphasize the nature of the product over its context because of its overt visual difference from any other. On the other hand, the bicycles themselves could be any bicycles, not necessarily Simpsons, even though the company did offer their own machines.

I raise this as an opening because it illustrates so well the 'invisibility' of the bicycle in relation to its rider and its context. To sell a bicycle it was best to sell its context and hope that the customer would remember to relate it to the product's name.

As an object, a bicycle's individual identity was not helped by the tendency for the design of the product to quickly stabilize, even when it went through major upheavals in pattern. The Velocipede bicycle of the 1860s was soon defined by a single format, the so-called 'diagonal'; the high bicycle of the 1870s was so simple in form it was hardly possible to adjust it; while it took less than a decade for the diamond framed safety to develop to a form that would define the bicycle for a century to come, something that arguably it still does. In such situations, makers had to either follow the archetype, or take a serious dare by adjusting the design, by radically changing the wheel sizes and/or frame. As this carried a high risk of ridicule by the established market, few dared. The products of those makers who did are highly collectible today, but at the time were often considered eccentric.

If the bicycle posed problems for brand identity because of its anonymity, it also inhabited an unusual position in terms of its relation to the domestic sphere in the 19th century and, remarkably, still does. No other form of transport was likely to enter the home, its place was outside, or in a separate building. But the bicycle, often owned by middle- and (later) working-class urban dwellers without the benefit of land and outbuildings, was often brought into the house for safe storage. From the outset it was awkwardly posed between being seen as a domestic appliance and/or a form of transport and/or a piece of sporting equipment and continues to be so today.

Finally, there was the disconnect between the bicycle as an icon of modernity, which the 'art posters' so often try to capture, and the expectations of appropriate design within the domestic sphere, which it and its riders occupied when not out cycling.

When the predecessor of the bicycle was introduced, it fitted firmly into the established traditions of carriage building and blacksmithing. To an extent, this was defined by its wheels, the work of traditional wheelwrights. The move from the all-wooden frame of the Draisienne to Johnson's 'Pedestrian Hobbyhorse' with wooden 'backbone' with metal axle guards to the rear and forks to the front made no real impact on the way the machine was perceived, or built. Johnson himself was in the carriage building trade. They were painted in bright colours with broad lines, yellow with black being particularly popular because of its association with express post coaches, amongst the fastest vehicles on the road.

The carriage aesthetic was repeated in the velocipede era of the late 1860s. Whether it carried into this time is debatable, given the seemingly inconspicuous nature of human-powered transport in the intervening forty years. However, like Johnson, the significant maker of this time was Michaux, a carriage builder. It is Michaux machines that set the pattern for others to follow. Pattern books survive to tell us that manufacturers were geared up to provide machines in a bewildering range of colours and qualities of lining, as they would carriages. Today most surviving machines are finished black with unpainted wooden wheels, but this is anachronistic, the consequence of it not only being the easiest to achieve in the 'restoration' of an abandoned and deteriorated machine, but also reflecting the huge and long-lasting change of aesthetic that took place in the early 1870s. In 1869 a gathering of velocipede bicycles would have been colourful indeed.

A combination of factors is probably at play that turned the bicycle from carriage to machine. The invention and rapid adoption of the wire spoked, all metal, tension wheel may be the key one. It had no relationship to any previous wheel type and demanded an entirely different technology from traditional wheelwrighting and the related trade of blacksmithing. Although the tension wheel developed in France, the 'centre' of bicycle production shifted to the UK with the eclipse of the Parisian makers and market during the Franco-Prussian War and their wide-scale loss in the chaos that followed. Moreover, the faddish and fashionable uptake of the velocipede bicycle saw other quickly developing markets, particularly in the USA, wither in the early 1870s as the fashion passed. These factors that led to the new bicycle type being adopted only in the UK are beyond the scope of this essay, but they could be argued to have allowed the bicycle industry to restart from scratch using entirely new technology for a fairly strictly defined market that set a precedent for the future.

The machine that was to develop in the UK was the 'high bicycle' (later termed 'ordinary', later still 'penny farthing'). It was all-metal and its nature was defined by its large driving wheel that was fitted to the inside leg of the rider, commonly about 50 to 54 inches in diameter. The lightness of the wheel gave it a huge advantage over the heavy, compression spoked, wooden wheel, where weight outstripped efficiency at large diameter. A rider could easily pedal a 50-inch tension wheel at a pace that would tax the same rider on a wooden wheel at 38 inches, thus achieving far greater speeds, while the tension spokes absorbed a lot vibration, giving rise to the term 'bone-shaker' to describe wooden-wheeled machines thereafter.

However, it was the appearance and nature of the machine that could be said to define both its market and therefore the finishes that were deemed acceptable for it. Both the hobbyhorse and velocipede had spawned fashionable crazes that lasted briefly. Although concerns regarding their safety were always expressed by their critics, they did not *look* unsafe and thus invited a fairly general (if almost entirely male) market. The high bicycle was another matter, it *looked* unsafe, almost fantastic, and still does. To ride it required a leap of confidence that many young men, even, did not have. Moreover, it had developed round one determinant, to increase speed in competition. On its introduction to the UK its place was to be found in sport and competition. While this did not bode well for its reception on the public road, it did mean that it had a stable and slowly increasing market of young, confident, usually affluent, athletic men who found exhilaration in the speeds it could achieve. This swiftness sustained the competition the high bicycle offered, the athletic prowess it demonstrated and, not least, the fact that riding one made one the focus of attention on the public road whether ridden competitively or not.

The aesthetic that developed was one that was to define the look of cycles for two generations or more: black enamel with more or less bright metal on working surfaces. This finish had first become common on sewing machines (if with a lot more 'relief' in applied gold decoration) and it is notable that a number of early bicycle makers (including the Coventry Machinists' Company, the first large-scale UK maker of velocipede bicycles) were originally sewing machine makers. The factory-based, mass-made, precision metalworking demanded by sewing machines was well aligned to the wire spoked bicycle. However, the black-and-brightwork look was probably as much determined by its market as its makers. Black was seen as practical and robust, appropriate for the demands made of it and the gendered nature of its market. Truly the bicycle had become a 'machine', the word favoured by cyclists to describe their mounts ever since. Even by the mid-1870s any machine not finished in black was exceptional (Oddy 2007b: 107–108). This is important in understanding the consistency of the finish of machines through the huge changes both in the machines and their market that were to take place over the next fifty years.

Even by the time that James Starley began to develop light, high-performance tricycles that could be ridden by women wearing full skirts, he had been making high bicycles for long enough to not think that a tricycle aimed at the female market, with non-competitive touring in mind, should differ in aesthetic from the high bicycle aimed at competitive sport. When his seminal 'Coventry Lever' tricycle was introduced in 1877 it was enamelled plain black, like the bicycles, rather than more firmly addressing the female market by having the black elaborately relieved with gold ornament, like a sewing machine. Almost all others followed in its wake.

I would argue that the 'machine aesthetic' of bicycles and tricycles was stable *because* of the seismic shifts in design and market, particularly the latter, which reflected almost entirely the social hierarchies of the period. The bicycle had become lodged in a particularly powerful one, affluent young men. From then its finish trickled down, first to upper- and middle-class women and then to progressively lower social classes. No change in the design or format of the machines themselves would disrupt the machine's aesthetic as each looked up to the last. Only once was it disrupted, and this was during the 'bicycle boom' of the mid-1890s.

The boom was percolated by a number of factors, most obviously the development of the low-wheeled, rear-driven 'safety' bicycle of a format that remains familiar today. Its format offered possibilities of 'dropping' or 'opening' its frame to allow the rider to mount from in front of the saddle, while its height made dismounting forward of the saddle with legs straddling the frame equally easy. This allowed the machine to be ridden in full skirts, while the lower riding position was far less challenging to men who previously would not contemplate a high bicycle. The second was the application of the pneumatic tyre, which made the smaller diameter wheel more shock absorbent than a high wheel, while significantly adding to its energy efficiency. Indeed, a pneumatic tyre applied to a high wheel had no significant effect on speed or ride quality, while expensive to make and store, whereas a pneumatic-tired safety bicycle could be geared up to achieve speeds impossible on any other cycle type. This was to make the high bicycle obsolete, even in competition. It took only about eight years for the new safety bicycles to move from novelty in 1885 to penetrate their potential market and even achieve 'closure' in 1892 (Pinch and Bijker 1984; Bijker 1995). The last high bicycles seem to have been made in 1891 and manufacturers' catalogues thereafter were devoid of them. Rapidly increasing sales of machines to mainly middle- and upper-class purchasers, eager to be seen riding what was a publicly presented status symbol encouraged rampant speculation in and expansion of the industry from 1893 until 1897 when the fashion began to falter. Cycle makers then began to look at exploiting those markets with less disposable income by reducing production costs, often by employing stamping and pressing technologies instead of machining, in addition to economizing on material quality. At the same time, purchase by instalment began to be more widely offered both by makers' agents and independent retailers.

During the 1890s bicycle boom a particularly notable, if short-lived development was that of 'society cycling' in metropolitan areas, where privileged riders promenaded in parks and boulevards. Here color appeared, with female riders in particular having machines finished to complement their dress, or merely to catch attention. The importance of this phenomenon is exactly that; it was a phenomenon, entirely at odds with cycling in general, and only possible in the context of the very wealthy and powerful being able to flaunt the unwritten rules. However, though it gained a lot of attention, it was in fact very limited to a few cities and was quite at odds with the less conspicuous bulk of the cycle market in Europe who remained faithful to black with bright-work.

However, it might be noted that the 1890s boom did something quite remarkable considering the bicycle's relationship with competition. It established its main market as middle-class adults who intended to cycle for pleasure and, possibly, utility and who would never dream of being confused with competitive riders. The emphasis for the next thirty years would be on 'roadsters', machines designed for a consistent, comfortable pace of about 10–12 mph over long distances without their riders looking like they were 'getting into a sweat'. On these respectable members of the bourgeoisie could enjoy the suburbs and countryside, calling in on wayside hostelries and arrive at hotels without looking like they were underdressed, a bicycle equivalent of pleasure-driving in a motor car. While the market remained aspirational there was a clear disconnect between how it perceived cycling as an activity differentiated from sport and competition; rather, the same as today's market generally differentiates a family car from a racing car. In the UK, at least, this differentiation has become increasingly blurred in more recent times where many riders consider it necessary to dress in Lycra merely to ride a few miles to the workplace (Chapter 26).

While the cycle itself could be seen as an exercise in inconspicuous practicality of form following function that would please modernist commentators a generation later, the opposite could be said of those things that took cycling into the home, or represented it off the road. A fruitful case study here is cycling trophies (Oddy 2013). Those that survive (many were melted for their bullion value) are usually ornate and follow in the wake of the traditional decorative arts using historicist styles, sometimes incorporating figurines of cyclists with their machines. Whatever the modernity of the machines and activity, even newer decorative styles such as Art Nouveau are rarely to be found in the repertoire of cycling trophies of the 19th and early 20th centuries, nor indeed anything else, very much at odds with the way the bicycle was presented in art-posters.

The range of cycling-inspired domestic goods is bewildering, reflecting the social status of the machines. Many were designed as appropriate for gifts and indeed were often used as presentation items. They range from mantel clocks, usually adorned with heroic-looking cyclists where otherwise might be a more traditional figure grouping, to novelties often in the form of bicycles and tricycles. Many were smokers' companions where, say, the lamp is a cigar cutter and milestone a vesta stand. A whole range of items had things incorporated into the wheels of miniature bicycles, such as timepieces and barometers. In general, what can be said of most is that they were gendered male and intended for the exclusively masculine spaces of studies, offices and personal libraries. Those items gendered female were less common, although German porcelain makers, such as Heubach, made extensive ranges of flower-encrusted and romantic figurines, clearly intended for the more feminine spaces of the drawing room, parlour or boudoir. Most frequently seen today are matched him-and-her fireside figurines waving at one another over the expanse of the mantelpiece or hearth while clutching their machines in their other hands.

In this context, the bicycle in the house was problematic. Just as today, many middle-class urban dwellers had no convenient outbuildings and could only stable their machines in the porch, or even the body of the house. While more or less ornate stands were made for the purpose (again, often using established historicist forms over something more modern), it was possible to buy bicycle cabinets, disguised as dressers and sideboards, in which to store machines that were not meant to have a presence in the domestic sphere. Again, the pattern of these was usually traditional and historicist (Wheelwoman 1896: 11).

This makes the art posters particularly interesting for their embracing of the latest modern styles in the graphic and even fine arts. However, while they were quickly acquired by the art-aware, resulting in so many surviving today, their *intended* place was outside on a billboard, and in the commercial context of the cycle agency. Modernity, as expressed by the machines themselves and the daring ways they were represented in the public realm, was something the public seems to have warmed to when it was outside of the home; when inside, the security of tradition seems to have won out. The same challenge was to face a generation of early modernists. To the larger public, modernist architecture and design looked exciting, so long as one did not have to live in it.

When did the look of the bicycle begin to change from the black-with-brightwork of the roadster to something different? And, why? In previous papers I have proposed that in Europe, at least, it can be symbolically placed with the products of the Hercules company of Birmingham (UK). It was this maker which systematically bought the price of a new machine well within the buying power of the non-skilled working class of the inter-war years and placed the final nail in the coffin of aspirational cycle purchasing in the mass market (Millward 1995). The black-and-brightwork 'roadster' was now something cheap and ubiquitous, available to all, while money and class followed motor vehicles.

During the 1930s, and particularly the early post-war years, 'serious' cycling moved to a different style of machine, the 'lightweight' closely aligned to competitive sport, ridden by enthusiasts, not unlike the market for the high bicycle. Interest in democratizing health and fitness across all social groups increased substantially in the inter-war period and this was reflected in a rising interest in cycle touring and personal performance. Touring was no longer the bourgeois model of going from respectable hotel to hotel, but a more economic version of camping and youth hostels (first introduced in the UK in 1931). At the same time, the uniquely UK competition of time-trialling (which had replaced mass-start road racing in the 1890s) gave a model of the individual rider competing with him- or herself. Often both touring and personal time trialling were combined. All this introduced large numbers of people to the possibility of serious cycling for pleasure, something that increased substantially in the early post-war era with petrol rationing keeping large number of motor vehicles off the road.

Lightweight machines were costly, but in a period of greater disposable income, more affordable to the working- and lower-middle-class social groups, particularly as those who wanted them were generally committed to the activity as enthusiasts, rather than just users, and were often willing to go to great lengths to save for one. The machines themselves were finished in bright colors, very different from the ubiquitous black of the roadster. While the roadster was designed to be ridden in everyday clothes, the lightweight machine tended to demand an element of dressing up (or perhaps down) to ride it efficiently. At the very least this was likely to resemble hiking garb, often with both male and female riders donning loose shorts and open neck short-sleeved shirts, reflecting inter-war ideals of a healthy outdoors lifestyle and, whenever possible, removing oneself from the city, but many favored the tight fitting woolens (Chapter 40) favored by track riders of the time (from which the famous 'yellow jersey' of the Tour de France derives). This type of self-imposed sumptuary requirement determined the rider as a 'cyclist', somehow different from other road users. This remains a problematic in road politics and culture to this day. Why the change? There is no certain reason, but one can propose a number of possible ones. The first is that the makers and users of lightweights wanted to distance themselves from the established form. While their machines were different to an expert, both roadsters and lightweights were similar in format to the casual viewer. The second is that lightweight manufacture was generally smaller scale, a lot being turned out by 'craft makers' more or less to order. This allowed for personalization, not only to the demand of the purchaser, but also to provide identity to the maker. Finally, strict rules about publicity often barred the use of trade names on machines being ridden in competition; again, bright colours made machines different from one another, while a few makers, such as Hechins' 'curly', Baines' 'flying gate' and Rench's 'Paris Galibier' tried visually adjusting the design of the frame.

The roadster, in the meantime, was often seen as 'poor-man's transport', for those too mean even to pay for a bus or tram to take them to work. This attitude became more entrenched as access to motor cars increased in the post-war years, reaching its height in the 1960s and 1970s. In the UK the stereotype roadster cyclists of this time were either hardened manual laborers going to or from 'the works' (for instance, as depicted in the contemporary, now nostalgic popular artworks of Alexander Millar) or eccentric elderly ladies in picturesque villages (as depicted by Margaret Rutherford in the 1945 film *Blithe Spirit*). Both stereotypes, at the bottom end of the aspirational ladder, tell us just how marginalized cycling as a form of day-to-day transport had become.

In the USA and Canada, cycling had gone down a somewhat different trajectory. The extraordinary collapse of the US industry in the period 1897–1903 has often concealed the fact that cycle making and use continued on a wide scale, but it did not have the same position as middle-class leisure transport it had in Europe (Herlihy 2004: 317–324). Moreover, the adoption of motor vehicles was faster than in Europe, transferring the allegiances of the comfortably off somewhat earlier, too. North American makers increasingly looked to a particular segment of the market, that of adolescents and teenagers excluded by age from motoring and here the machine adopted an aesthetic entirely different from elsewhere, largely mimicking that of motorcycles. These machines were not designed for distance, although they were certainly used for utility by those who had jobs such as newspaper deliveries, probably the longest distance most of them covered was suburban home to town center and/or school. Their aim was to visually impress by their scale, incorporating wide 'balloon tires' (sometimes made even less efficient by suspension systems), equally massive fenders, broad handlebars, wide cushion saddles, even would-be petrol tanks (that were often really tool-boxes or electric horns). The finish of these 'balloon tire bicycles' reflected this. Colourful enamels, usually three-colour with dynamic triangular end flashes and broad lining. While European cyclists perceived them as some sort of aberration where form entirely displaced function (Herlihy 2004: 354–363; Ritchie 1973: 174–175 referring to the Raleigh 'Chopper' in the same way), from the post-modern position of social construction of technology (Bijker 1995) they are far from that. The function of the machine was to visually impress friends and onlookers while taking their users short distances; this they did very effectively, continuing to do so today. In real terms, it might be worth considering how much, if any more use most bicycles have today? Rather like those who insist on driving high-performance motor vehicles for utility, it seems odd that cyclists often seem to think it necessary to have similarly high-specification bicycles when the most use they get is a few miles into work or along a local cycle path.

Effectively, then, through much of the 20th century, there were only three machine aesthetics applied to bicycles. The first, an inheritance of the 19th century, was the modernist dream in which form and finish truly followed function, but the machines themselves were slowly being reduced in status and in build quality. The other two begin to appear when the first reaches saturation in the inter-war period. The lightweight aesthetic was in reaction to it, and in the modernist world the finish was at odds with the machine. Highly colorful and often quite decorative (some machines even had rococo curlicues fretted into their lugs, ostensibly for heat displacement when being brazed), it could be seen to contradict the idea of the 'purity' of the precision and highly crafted engineering of the lightweight machine itself. Meanwhile, the North American balloon tire aesthetic was completely about the look of the machine where the form was following a different function from what many commentators considered to be the machines' function and the finish followed.

This was all to end with the rise of postmodernism; which, in bicycles, can be placed firmly with the development of the mountain bike (Chapter 28). It is interesting to observe that this type of machine developed from North American balloon tire machines that were robust enough to be plunged down mountain tracks at high speed (Berto 2014: 34–54). The mountain bike was significant because it appealed to young adults, particularly young male adults, rather than the more adolescent market of the balloon tire machines in the 1930s–1960s. Moreover, it was designed round an activity that was entirely alien to 'normal cycling'. Both these factors made it difficult to compare to established norms and broke the assumptions as to what cycling was about. The robust nature of the machines was quick to be exploited for normal road use. So-called 'All Terrain Bicycles' (ATBs) brought the mountain bike on to the urban street. Their market was generally perceived to be 'yuppie' (Young Upwardly-mobile Professional), the sort of people who already owned cars and bought the bicycles for enjoyment (Piesman et al. 1984; Ash and Piesman 1984). However, though often extremely high-specification and consequently expensive, the machines were in no way suitable for established competition. Effectively, all previous stereotypes were broken.

Is there a machine aesthetic today? I would argue no. In truly postmodern style there are multiple machines and multiple aesthetics, rather than the three easily defined ones of the past. Those three are still in place, of course. One can buy new high-grade black roasters and fun-coloured balloon tire machines, traditional lightweights and far more advanced versions of the same, in addition to myriad new classes of machine that range into a wider grouping of Human Powered Vehicles (HPVs). In an essay for a book called *The Invisible Bicycle* (Oddy 2019), I argued that the machine aesthetic for cycles is now largely irrelevant, attention being placed on the aesthetic of the rider, rather than the machine, the opposite of motoring where the driver is invisible behind tinted glass, something of serious importance to road politics, but another story from the one outlined here.

References

Ash, R. and M. Piesman (1984) *The Official British Yuppie Handbook: The State-Of-The Art Manual for Young Urban Professionals* (London: Ravette).

Berto, F. (2014) *The Birth of Dirt – The Origins of Mountain Biking*, 3rd Edition (San Fransico: Cycle Publishing).

Bijker, W. (1995) *Of Bicycles, Bakelites and Bulbs – Towards a Theory of Technological Change* (Cambridge, MA: MIT Press).

Dodge, P. (2006) *The Bicycle* (New York: Flammarion).

Herlihy, D. V. (2004) *Bicycle: The History* (New Haven, CT: Yale University Press).

Millward, A. (1995) "The Founding of the Hercules Cycle and Motor Co" in Van der Plas, R. (ed), *Cycle History – Proceedings of the 5th International Cycle History Conference 1994* (San Francisco: Bicycle Books) 99–106.

Oddy, N. (1992) "The Machine Aesthetic – Marketing the Bicycle in the Late 19th and the Early 20th Centuries" in *Proceedings – 2nd International Conference of Cycling History 1991*, 66–75.

Oddy, N. (1994) "Non-Technological Factors in Early Cycle Design" in Herlihy, D. (ed), *Cycle History – Proceedings – 4th International Cycle History Conference 1993* (San Francisco: Bicycle Books) 63–67.

Oddy, N. (1997) "An Invaluable Refinement: The Aesthetic of the British Cycle Accessory in the Late 19th and Early 20th Centuries" in Van der Plas, R. (ed), *Cycle History – Proceedings of the 7th International Cycle History Conference 1996* (San Francisco: Bicycle Books) 63–67.

Oddy, N. (2001) "Cycling in the Drawing Room" in Ritchie, A. (ed), *Cycle History 11 – Proceedings of the 11th International Cycling History Conference 2000* (San Francisco: Van der Plas) 169–176.

Oddy, N. (2006) "The Cycle on Display" in Ritchie, A. (ed), *Cycle History 16 – Proceedings of the 16th International Cycling History Conference 2005* (San Francisco: Van der Plas) 169–176.

Oddy, N. (2007a) "Cycling – A Game for All Players" in Norcliffe, G. (ed), *Cycle History 17 – Proceedings of the 17th International Cycling History Conference 2006* (San Francisco: Van der Plas) 143–146.

Oddy, N. (2007b) "The Flaneur of Wheels?" in Horton, T., Rosen, P and Cox, P. (eds), *Cycling and Society* (Aldershot: Ashgate) 97–112.

Oddy, N. (2013) "Cycling Trophies – An exercise in mediocracy" in Ritchie, A. (ed), *Cycle History 17 – Proceedings of the 23rd International Cycling History Conference 2012* (Birmingham: Cycle History Publishing) 130–138.

Oddy, N. (2019) "History, Tweed and the Invisible Bicycle" in Männistö-Funk, T. and Myllyntaus, T. (eds), *Invisible Bicycle – Parallel Histories and Different Times* (Leiden: Brill) 215–232.
Piesman, M., M. Hartley and R. Ash (1984) *The Yuppie Handbook: The State-Of-The Art Manual for Young Urban Professionals* (New York: Pocket Books).
Pinch, T. and W. Bijker (1984) 'The social construction of facts and artefacts: Or how the sociology of science and the sociology of science might benefit each other' in *Social Studies of Science* 14, 319–441.
Rennert, J. (1973) *100 Years of Bicycle Posters* (London: Hart-Davis McGibbon).
Ritchie, A. (1973) *King of the Road* (London: Wildwood).
Wheelwoman (1896, September 5) *The Wheelwoman and Society Cycling News* 'A Boon for Wheelwomen' 11.

40
DRESSED TO RIDE

Emma Hilborn

To ride a bicycle is a particular way to show oneself off in public. Due to the inevitable visibility of the cyclist, the ways in which cyclists choose to present themselves come into focus. The appearance of a fleet of cyclists navigating busy roads in cities like Copenhagen or Amsterdam might not gather much attention from passers-by, but the recent pejorative term "MAMIL" – middle-aged men in lycra – indicates that a cyclist's dress is not entirely without consequence (Chapter 26). As Elizabeth Wilson writes in her pioneering work on fashion as a social phenomenon, clothes are "unspeakably meaningful" and inevitably communicate certain ideas about its wearer (Wilson 2003: 3). The main focus of this chapter will be on some of these silent but meaningful garments that have attracted the most attention in both contemporary debate and scholarly works: women's first bicycle dresses. The chapter draws on my previous research published in Swedish: this article will be referenced primarily in my analyses of the source material.

Sporting men

During the late 1860s, cycling became a popular sporting activity for wealthy young men in France and Great Britain. The fashionable exercise of cycle racing spread first in those countries and from there across Europe and to North America. Riders mainly wore equestrian clothing, the closest sport to cycling. This enthusiasm for cycling waned somewhat in the early 1870s, but soon gained new momentum as the high bicycle, where the rider was poised over the large front wheel, replaced older velocipedes. The high-wheeler was more dangerous than its predecessors, but it was also significantly faster and more comfortable. After that, cycling was firmly established as a sport for well-to-do men, and bicycle races became a popular spectator sport (Herlihy 2004; Hadland & Lessing 2014; Strange 2002; Norcliffe 2001; Vignette A).

According to Kat Jungnickel (2018: 20), these racing men were often shown in advertisements wearing very tight-fitting, even revealing, matching shorts and tops, though members of cycling clubs were recommended to wear a "jacket, breeches or knickerbockers, stockings, and a cap or a helmet" (Pratt 1880: 110). Bicycle fashion usually aligned with sports clothing from other, related sports: riding clothes and boots were standard when young men straddled the first striding machines in 1818 and 1819. However, during the 1870s cyclists began to adopt specially designed wool clothing when riding the high-wheeler and, later, the safety bicycle.

DOI: 10.4324/9781003142041-60

The material was thought to possess unheralded properties of breathability and capacity to facilitate the elimination of the poisonous and unhealthy substances that were emitted from a person's skin (Meinert 1998). When used in sports clothes, wool could keep the body warm in winter and cool in summer and the creator of these special wool garments, the German naturalist and physician Dr. Gustav Jäeger, became immensely influential. The commercial success of the Jäeger shop opened by Lewis Tomalin in the center of London in 1884, guaranteed that a great many cyclists wore wool for the next 60 years, until the adoption of new synthetic fibers during and after World War II (Meinert 1998).

Outside of the circle of dedicated athletes, however, men found that their everyday clothes could be adapted to cycling with relative ease. People complained about their recklessness in traffic and the dangerously high speeds of bicycle races, but criticism directed at male cyclists seems not to have been particularly preoccupied with their dress. Men's cycling clothes were perhaps not considered appropriate in all social settings, but they were not a frequent topic of discussion either. If men's bicycle attire made them recognizable as cyclists all the better – middle- and upper-class men were usually keen to be associated with the masculine sport (Jungnickel 2018: 19–20).

Sporting women

With a few notable exceptions among female adventurers and circus artists, the high-wheeler had never really been an option for women (Macy 2011; Jönsson 2014). In addition to the female cyclist's legs being scandalously exposed in the elevated position above the front wheel, the machine was almost impossible to ride in long skirts. The introduction of the safety bicycle – in most respects identical to today's bicycles – changed this situation and made cycling available to women by the end of the 1880s. A few years later, when the safety bicycle was equipped with pneumatic tires and then with an open frame, it gained even more followers. The low-step frame, chain cover and skirt guard allowed for a more comfortable ride for women, ankle-length skirts intact, as did the freewheel (Chapter 39). Falling prices also broadened the market to include the middle class. The following influx of middle-class women cyclists made people speak of a female "bicycle craze" starting in the 1890s (Herlihy 2004; Hadland & Lessing 2014; Strange 2002; Norcliffe 2001).

In fact, it was through fashionable cycling that middle-class women first gained access to a kind of sport in their spare time, while aristocratic women had been able to take up suitable hobbies like riding, golf, croquet and tennis (Cahn 1995: 15; Hargreaves 1993: 74). Naturally, this required a socially acceptable dress. Although variants of trousers or short skirts were considered acceptable sporting attire during gymnastic exercises in the home, the sharp line between public and private required a completely different outfit for cycling outdoors than for exercising indoors with only women present (Riegel 1963). The same was true for all sports: as tennis began to attract women in the 1870s, for example, the game had to be played in bustles, long skirts and corsets (Wilson 2003). Traditionally, such conventional and restricting dress ensured that upper-class women moved with a sense of decorum even when engaging in sports.

From the very beginning, the quest to find a similarly suitable dress for women cyclists was accompanied by a set of problems that arose from the perceived unfeminine nature of the activity itself. Since the bicycle was closely associated with the masculine domain of sport, women who rode the new safety bicycle ran the risk of being accused of emulating men and thus endangering their own femininity. This was in keeping with the general view of sport: between 1890 and World War I, moderation was the guiding principle for all types of exercise for women (Cahn 1995; Hargreaves 1993). Even though this notion of the female body as delicate was really only applied to upper- and middle-class women, the ultimate fear was that strenuous activity would make the graceful female body gross and masculine. Of course, high-speed races and bicycle competitions were out of the question for women. However, removing the element of competition was not enough to wholly free the bicycle from all lingering masculine connotations. Unlike gymnastic exercises that could claim to enhance female beauty via a series of movements perfectly calibrated for the female body, there was nothing definite that separated men's bicycle rides from women's (Lykke Poulsen 2005; Cahn 1995; Hargreaves 1993, 1994).

One might think that cycling should have been relatively easy to modify in order to for it to be more suitable for women, since elements such as speed and distances could be adapted to the individual. Scholars who have studied women's increased participation in organized sports during the 20th century stress that these sports without exception had to be modified in a way that underlined fundamental differences between the sexes. In her study of media representations of women athletes, Helena Tolvhed sums this up as the constant need to off-set the inherent masculinity of sports by repeated displays of conventional femininity. Most importantly, women were not allowed to compete with men. In sports where results could be compared – for example, running and swimming – distances and rules of the game were slightly different for women (Tolvhed 2008, 2015). Cycling for pleasure, on the other hand, had no established rules that had to be changed in order for women to be able to participate, which might seem like a situation that would encourage women to take up cycling. On closer inspection, however, the absence of clear rules and recommendations meant that there was no established way of "feminizing" the activity. When introduced in ladies' magazines in the 1890s, advice on how to cycle "the right way" was often conflicting or vague. This could become a source of worry: it was not possible to definitively say at what speeds or distances the bicycle ride would exceed the limits of femininity and instead become dangerously masculine (Hilborn 2018).

The dress problem

Since cycling women had to distance themselves from the reckless young men using the bicycle to test their strength, the quest to find suitable cycle wear could not end with a bicycle dress designed for great physical exertion, even if it had to allow for the light exercise of a bicycle ride at a more leisurely pace. Kat Jungnickel defines the "dress problem" as trying to figure out "how to cycle safely and comfortably yet also evade looking too much like a cyclist, so to minimize social hostility from onlookers who disproved of newly mobile women" (Jungnickel 2018: 4; Sims 1991). One way of trying to solve this problem was to design new clothes that combined functionality with elegance (Jungnickel 2018). Most women, however, had to rely on more conventional sources for their fashion advice.

In fact, it is notable how concerns over women wearing 'bifurcated garments' cease in the press after the late 1890s, when the free-wheel clutch became standard, allowing riders to cycle in skirts with far less danger of being wound in. This largely eliminated the major practical reason why many women riders adopted bifurcated garments during the mid-1890s. This should also be seen in the context of fashion. The 'bloomer girl' was a Parisian phenomenon of the mid-1890s and she set a fashion that had passed by the late 1890s. Taking account of both these factors, there is very little evidence of bifurcated garments being worn by women cyclists in the first three decades of the 20th century.

Fashion scholars Julia Christie-Robin, Belina T. Orzada and Dilia López-Gydosh have identified two main strands in the view researchers have taken of the importance of women's cycling attire at the turn of the century. According to the dominant perspective, the bicycle gave women a new freedom of mobility, and this freedom manifested in the looser lacing and shorter skirts of the bicycle dress (Christie-Robin, Orzada & López-Gydosh 2012). The major changes in women's dress necessitated by the bicycle proved long-lasting and paved the way for the active women of today (Riegel 1963; Harmond 1971/1972). This interpretation of bicycle fashion as a form of emancipation has been further developed by later researchers, such as Lisa S. Strange, who argue that the late 19th-century bicycle craze led to a breakthrough for functional and practical costumes with loose bodices, less voluminous skirts, or even trousers: a development that was analogous to fundamental societal changes in gender roles. These studies describe how women, often with ties to the women's movement, used their bicycles and their newly designed bicycle dresses as a means to undermine the bourgeois ideal of the complacent housewife (Strange 2002; Grossbard & Merkel 1990; Hurst 2009; Christie-Robin, Orzada & López-Gydosh 2012).

The second perspective, which Christie-Robin, Orzada and López-Gydosh join, argues that the conclusion that the bicycle brought about great changes in women's dress is based more on a general conviction that the bicycle represented something completely new, than on actual sources. Changes in both dress and gender roles in the wake of the bicycle craze have either been greatly exaggerated by researchers, or are in reality the result of many different factors (Christie-Robin, Orzada & López-Gydosh 2012; Rush 1983; Park 1989). According to Patricia Campbell Warner, turn-of-the-century media created a simplistic image of the female cyclists who preferred to dress in so-called "bloomers", a notion that has been passed on as fact by later historians. The fact that bloomers – a kind of oriental-inspired trousers, sometimes under a calf-length skirt – got its name after being worn by the American feminist Amelia Bloomer in 1851, probably contributed to the tendency to portray the allegedly mannish cyclist as wearing them. During the 1890s, the eccentric cyclist wearing bloomers or trousers became a staple in satirical cartoons and humorous magazine illustrations showing the absurdity of modern life (Figure 40.1) (Warner 2006; Shapiro 1991; Marks 1990).

Even if cartoons tended to pay the greatest attention to cyclists who could choose to wear loose-fitting garments, bloomers or trousers, Warner suggests that this outfit was in fact extremely unusual. Women's everyday cycling dresses first and foremost sought to adhere as closely as possible to prevailing fashion trends (Warner 2006). Cultural historian Ellen Gruber Garvey, too, finds that the mannish women of satirical cartoons have very little in common with women pictured in illustrations printed in fashion editorials and advertisement in women's magazines. On the contrary, both the advertisements and the American magazines in which they were published projected an image of cycling that consistently emphasized traits that corresponded with traditional femininity. Cycling women were usually depicted wearing skirts that almost covered the cyclist's foot and would have nearly dragged in the dust if she had been standing (Garvey 1995).

Garvey argues that the manufacturers of bicycles and clothes for women strove to make cycling appear as attractive as possible to female consumers, and therefore had no interest in presenting a bicycle ride as something that could ever call the cyclist's respectability into question. The references to the familiar concept of consumption might well be what convinced many women to give cycling a chance. In a study of cycling in late 19th-century Canada, Phillip Gordon Mackintosh and Glen Norcliffe point out that cycling women could consider cycling to be an opportunity to demonstrate their tasteful consumption habits. For middle-class women, cycling came with its own set of rules regarding clothing, behavior and decoration; a phenomenon that the authors refer to as "the bicycle as parlor". In this context, there was no real contradiction between cycling and the archetypal housewife (Mackintosh & Norcliffe 2007).

258
PUNCH, OR THE LONDON CHARIVARI.
[June 4, 1898.
NOSCE TEIPSUM.
Lady Cyclist (touring in North Holland). "What a Ridiculous Costume!"

Figure 40.1 A cartoon satirizing women's cycling clothing.

Source: Punch 4 June 1898, p. 258.

The problem with the dress problem

Even though the cultural presence of early cycle dress for women has been very differently assessed by scholars, the studies share one important observation: that the clothes were central to how society as well as women themselves perceived cycling. In different ways, the bicycle dress poses a problem regardless of which perspective one choses. Studies that stress the importance of unusual outfits like trousers, loose garments or short dresses, leave the reader with the impression that the choice to go for a bicycle ride would probably come at such a great social cost that it would be hard to explain the sudden popularity of the bicycle among middle-class women. Additionally, they seldom account for the discrepancy between the most notoriously avant-garde costumes, and the conventional dresses recommended by advertisers and ladies' magazines. Studies that describe both the bicycle dress and cycling as rather uncontroversial parts of bourgeois consumer culture, on the other hand, give few hints as to why the bicycle dress was discussed with such fervent in contemporary media. If it was "business as usual" and regular clothes only needed a few adjustments – as was the case with men's dress – why would there be any need to repeatedly explain how women should dress when riding the bicycle?

I would argue that interest in fashionable clothing could be a way to endow cycling with femininity and thus disconnect it from the masculine world of sport. In this chapter, the examples of how this was done are mainly from Swedish and Danish ladies' magazines, but in light of previous studies, it is clear that the Scandinavian debate paralleled that of other western countries. An inventory of the cycling dresses mentioned in Swedish and Danish ladies' magazines shows that all magazines published detailed descriptions and sewing patterns for the fashion-conscious female cyclists. Judging by the inquiries and reprints, these patterns were very popular from the 1890s and at least until the 1910s.

The dominant fashion trends in the years leading up to the First World War generally prescribed ankle-length skirts in various models and a waist that was clearly corseted. During the 1890s, wide puff "*leg of mutton*" sleeves were combined with a narrow waist and wide, draped skirts. At the turn of the century, the sportier *Gibson Girl* became popular. This trend included flowy blouses, skirts that had less volume and were relatively snug around the hips, as well as large, wide-brimmed hats. This ideal was replaced during the first decade of the 20th century by a narrow, slightly S-shaped silhouette that accentuated the bust (Cunningham 2003). In general, the silhouette of the magazines' bicycle dresses stayed close to mainstream fashion, with the possible exception of the hat, which would never be large or extravagantly decorated. A short corset was recommended, for example, because the magazines reasoned that ladies probably did not want to cycle completely without a corset. The fact that it was inconceivable to cycle without a fashionable, although modestly decorated, hat highlights that this form of cycling was not comparable to competition cycling but could be placed comfortably within the boundaries of traditional femininity (Hilborn 2018).

Most important, however, was the skirt, which on no conditions whatsoever was allowed be too short. The preferred skirt was one that was only slightly shorter than usual. The strategy of portraying cycling women with reassuringly long skirts that Garvey (1995) observes in the American press, has clear parallels in the Danish and Swedish ladies' magazines, where the skirts at most revealed the ankles of the cycling women. Most of the patterns, illustrations and advertisements show women either standing coquettishly next to the bicycle or riding with impeccably straight posture in an almost stationary pose, wearing skirts that ended at their ankles or just above. Although sport, mobility and freedom were mentioned in connection with cycling, the illustrations of bicycle fashion showcased the ability to maintain stability and correctness even on a bicycle. The cycling woman in these pictures was not a rebel who carelessly pedaled her bike wearing outrageously modern clothing (Figure 40.2). On the contrary, she was as elegant, well-dressed and restrained as she was in her own parlor. The thesis that the exercise and adventure that came with the modern safety bicycle resulted in a reformed fashion for women is thus not supported by either the Swedish or the Danish source material. Experiments with bloomers, divided skirts or trousers in combination with a skirt, are practically nonexistent (Hilborn 2018).

Still, the cycling skirts were described as more close-fitting and slightly lighter than ordinary skirts, and it was recommended that they be made of a relatively robust fabric that could withstand the wear and tear on the bicycle. Notably, the danger of tangled skirts was hardly touched upon at all, in spite of being a serious danger in the 1890s. While skirt guards and chain covers were considered effective enough in preventing skirts getting caught, the problem of the pedals 'winding in' a skirt when coasting were very real before the wide-scale adoption of free-wheel clutches after c.1900. But the magazines seem to have refrained from referencing this problem altogether, possibly because in the 1890s it was considered unseemly for a woman to coast and feet should remain firmly on pedals.

Worries about the dress were limited to reminders to choose a skirt that did not cover the foot and to secure it so that it did not flap in the wind. Both the magazines' articles and adverts recommended a dress that was especially made for cycling, but the emphasis was on the presumed pleasure in owning a fashionable dress, rather than on functionality. Ladies' *could* choose to alter their old skirts, but that meant that they would miss out on a crucial part of the

Figure 40.2 A well-dressed cycling woman.

Source: Svensk Damtidnings modebilaga 1901.

joy of cycling. For example, one Swedish magazine (*Svensk Damtidning*) published a short story in 1898, in which a young woman managed to overcome her initial distrust of the bicycle by focusing on how pretty she would look in a particularly attractive bicycle dress she had seen in the shop (Hilborn 2018).

Evidently, the magazines believed that women cyclists were as interested in fashion as in cycling – probably even more – and fashion was a recurring theme not only when publishing sewing patterns, but also when reporting on cycling. One role model featured in the magazines was the Danish princess Maud, who was described as gracefully whizzing along a country road wearing a coffee-colored skirt, an Eton jacket and a sailor hat decorated with a single ribbon. In one article on upper-class women cycling in Hyde Park, cycling was described as a kind of fashion show, complete with a large audience who hoped to catch a glimpse of the latest, chic bicycle fashion. The ladies were wearing close-fitting skirts weighed down by lead weights, and short jackets. Silk or cotton blouses were common, although deemed less elegant than the stylish dresses. However, there were some limits to which costumes aroused admiration: "Only one lady is dressed in trousers," the writer added, "and it is an older, ugly, scrawny 'miss', so you cannot expect her costume to find much appeal" (Hilborn 2018: 20).

References to the nobility and royalty not only assured readers of the respectability of cycling – it also placed the bicycle in a completely different context than that of sports. In comparison to the variety and vagueness that marked suggestions for speeds and distances suitable for the female physique, there seemed to be a striking consensus regarding the design of cycling suits. Cycling and dressing for cycling thus appeared to be without major social risks. As noted above, cycling's origin as a sport for men resulted in some confusion over how to modify it in a way that clearly separated it from a masculine (and potentially masculinizing) hobby. In fashion, on the other hand, there were established role models and identifiable boundaries, which could also be passed on in great detail through sewing patterns. Within these boundaries, one could allow for a degree of personal taste and creativity without risking neither femininity nor respectability.

The commitment to finding the perfect, most tasteful cycling dress was not necessarily a sign of concern over an apparently unsolvable problem. The fact that recommendations were formulated with such a sincere interest in details and different styles, indicates that fashion offered a pleasurable and safe way of approaching an activity that might otherwise appear risky. The cycling woman, characterized by a femininity drawn from the elegant world of fashion, conveyed a view of female cycling as completely different from the competitive and energetic male cycling. While the implied masculinity of other sports could be compensated by feminine practices such as the sportswomen showing a feminine demeanor outside the arena or only participating in modified versions of different sports, cycling could integrate the elegant symbols of femininity into the activity itself. Who could possibly mistake a cycling woman for a man if she was clad in the very symbols of femininity?

The new dandies

The unspeakably meaningful messages conveyed by the clothes of the men and women described in this chapter helped define the cyclist as belonging to a certain gender, but naturally, this was not the only identity that could be expressed using the bicycle dress. The elegant clothes of the ladies and gentlemen who strived to combine the bicycle with the latest fashion, instantly set them apart from the lower classes. Occasionally, there were reports on workers using the bicycle, but their class seems to have excluded them from the discussion on bicycle fashion altogether. As bicycles became more affordable during the 1920s, the activity morphed from a leisurely hobby into an unremarkable necessity associated with those who could not afford a car. This, and the fact that everyday clothes became more casual and easy to cycle in, meant that the interest in bicycle fashion faded away. In later years, however, we might have seen a return of the bicycle fashion enthusiasts, as certain bicycles (and the opportunity to ride them in perfect conditions) have become new status symbols in urban areas. Interestingly enough, this trend seems to be centered around men: one study of the Australian cycling culture appoints MAMIL the new dandies: preoccupied with their appearance and their extravagant expression of masculinity (Ferrero-Regis 2018; Chapter 26). Another trend, *tweed runs*, require riders to wear retro cycle clothing like woolen tweed plus-four suits, Norfolk jackets and cloth caps. Perhaps it is only a matter of time before we will see these modern-day dandies experiment with bloomers?

References

Cahn, Susan K. (1995) *Coming on Strong: Gender and Sexuality in Twentieth-century Women's Sport* (Cambridge, MA: Harvard University Press).

Christie-Robin, J., Orzada B. T. & López-Gydosh, D. (2012) "From bustles to bloomers: exploring the bicycle's influence on women's fashion 1880–1914." *Journal of American Culture* 35(4), 315–331.

Cunningham, P. A. (2003) *Reforming Women's Fashion, 1850–1920: Politics, Health, and Art* (Kent, OH: Kent State University Press).

Ferrero-Regis, T. (2018) "Twenty-first century dandyism: fancy Lycra® on two wheels." *Annals of Leisure Research* 21(1), 95–112.
Garvey, E. G. (1995) "Reframing the bicycle: advertising-supported magazines and scorching women." *American Quarterly* 47(1), 66–101.
Grossbard, J. & Merkel, R. S. (1990) "'Modern wheels liberated' The Ladies' 100 years ago." *Dress: The Journal of the Costume Society of America* 16(1), 70–80.
Hadland, T. & Lessing, H.-E. (2014) *Bicycle Design: An Illustrated History* (Cambridge, MA: MIT Press).
Hargreaves, J. (1993) "The Victorian cult of the family and the early years of female sport." in E. Dunning, J. A. Maguire, and R. E. Pearton (eds.) *The Sports Process: A Comparative and Developmental Approach* (Champaign, IL: Human Kinetics Publications).
Hargreaves, J. (1994) *Sporting Females: Critical Issues in the History and Sociology of Women's Sports* (London: Routledge).
Harmond, R. (1971/1972) "Progress and flight: an interpretation of the American cycle craze of the 1890s." *Journal of Social History* 5(2), 235–257.
Herlihy, D. V. (2004) *Bicycle: The History* (New Haven, CT: Yale University Press).
Hilborn, E. (2018) "Den eleganta cyklisten Cykling, mode och kvinnlighet i sekelskiftets svenska och danska damtidningar." *Historisk Tidskrift* 138(1), 3–32.
Hurst, R. J. (2009) *The Cyclist's Manifesto: The Case for Riding on Two Wheels Instead of Four* (Guilford, CT: Falcon Guides).
Jönsson, L. (2014) "Äventyrscyklisten." *Gränsløs: tidskrift för studier av Öresundsregionens historia, kultur och samhällsliv* 4, 39–49.
Jungnickel, K. (2018) *Bikes and Bloomers: Victorian Women Inventors and their Extraordinary Cycle Wear* (London: Goldsmiths Press).
Lykke Poulsen, A. (2005) *"Den kvindelige kvinde": kampe om kvindelighed, medborgerskab og professionalisering i dansk kvindegymnastik 1886–1940.* (Köpenhamn: Diss).
Mackintosh, P. G. & Norcliffe, G. (2007) "Men, women and the bicycle: gender and social geography of cycling in the late nineteenth-century." in D. Horton, P. Rosen & P. Cox (ed.) *Cycling and Society* (Aldershot, England: Ashgate) 153–178.
Macy, S. (2011) *Wheels of Change: How Women Rode the Bicycle to Freedom (with a few flat tires along the way).* (Washington, DC: National Geographic).
Marks, P. (1990) *Bicycles, Bangs, and Bloomers: The New Woman in the Popular Press* (Lexington, KY: University of Kentucky Press).
Meinert, C. (1998) "Wheelmen wore wool". Paper presented at the *9th International Cycle History Conference*, Ottawa, Canada.
Norcliffe, G. (2001) *The Ride to Modernity: The Bicycle in Canada, 1869–1900* (Toronto: University of Toronto Press).
Park, J. (1989) "Sport, dress reform, and the emancipation of women in Victorian England: a reappraisal." *The International Journal of the History of Sport* 6(1), 10–30.
Pratt, C. (1880) *The American Bicycler* (Boston, MA: Rockwell & Churchill).
Riegel, R. E. (1963) "Women's clothes and women's rights." *American Quarterly* 15(3), 390–401.
Rush, A. (1983) "The bicycle boom of the gay nineties: a reassessment." *Material History Review: Revue d'histoire de la culture matérielle* 18(1), 1–12.
Shapiro, S. C. (1991) "The mannish new woman: Punch and its precursors." *Review of English Studies* 42(168), 510–522.
Sims, S. (1991) "The bicycle, the bloomer, and dress reform in the 1890s." In P. A. Cunningham and S. Voso Lab (ed.), *Dress and Popular Culture* (Bowling Green, OH: Bowling Green State University Popular Press).
Strange, L. S. (2002) "The bicycle, women's rights, and Elizabeth Cady Stanton." *Women's Studies* 31(5), 609–626.
Tolvhed, H. (2008) *Nationen på spel: kropp, kön och svenskhet i populärpressens representationer av olympiska spel 1948–1972* (Umeå: h:ström - Text & Kultur).
Tolvhed, H. (2015) *På damsidan: femininitet, motstånd och makt i svensk idrott 1920–1990* (Göteborg: Makadam).
Warner, P. C. (2006) *When the Girls Came Out to Play: The Birth of American Sportswear* (Amherst, MA: University of Massachusetts Press).
Wilson, E. (2003) *Adorned in Dreams: Fashion and Modernity* (London: Tauris).

41

CYCLE POSTERS OF THE *BELLE ÉPOQUE*

Nadine Besse

Publicity posters and advertisements aimed at selling bicycles have been published since the 1870s. Many are mundane, but the posters produced in huge numbers, mainly in France at the end of the 19th century, are of important artistic interest and within the domain of graphic arts are considered to be of great originality. Cycle manufacturers, who were in rapid commercial expansion at that time, employed artists of the highest professional standing to promote their cycles. The study of these mass-produced works naturally requires a certain level of interest in social history, but also in the history of the makers who used symbols and emblems to identify their various makes by highlighting the unique characteristics of their products for cyclists.

The development of this form of advertising promotion allows us to understand the fundamental importance of this *Belle Époque* period, creating a visual social vocabulary of the cycle which, having focused on the masculine market in the first phase of cycling, turned its acquisitive eye toward the feminine, in the full awareness that advertising depicting women and seemingly aimed at them, could equally appeal to men.

Up until 1891, the bicycle was still only used by sportsmen drawn from the leisured classes. It found some use in military circles, and for postal and delivery purposes. Ladies could only admire their husband members of the Touring Club of France, which was established in 1890. In France, this date indicates the establishment of the philosophical divide between sport and touring. It was not until between 1892 and 1894 that women were first seriously courted by advertising. To attract women, the need was felt to create social relations around cycling that were not about sport and competition that typified masculine cycling. On 1 July 1893, an evening meeting was held in the Bois de Boulogne with a Festival of Lanterns as its theme. The ride caused a sensation and helped to establish bicycling as a fashionable activity suitable for ladies, while strengthening the ties between male and female cycling enthusiasts. Several posters made reference to the event. While the image of women cycling that this event offered encouraged others to take up the activity, it was also highly attractive to male cyclists and observers.

By 1893 Paris had become an immense capital city of more than three million inhabitants. Bicycle enthusiasts were city dwellers already dazed by the commercial bustle of the great city, by its department stores, by the concentration of the fashionable goods and by the desire to buy. Possessing leisure time and being financially comfortable, they found in sport a simple means of maintaining physical fitness, a distraction and a means of escape. They left for the seaside, the mountains or simply to breathe the fresh air of the Bois de Boulogne (Vignette H).

DOI: 10.4324/9781003142041-61

In Paris numerous cycle shops sprang up all along the Avenue de la Grande Armée, the 'royal route' to the Bois de Boulogne. As products, bicycles tend to anonymity, they all look rather the same and the machine is far less visible than its rider. To advertise their products cycle agencies looked for ways to visually capture the experience of cycling, rather than focusing on the bicycles themselves. This demanded creative works of imagination, of sublimation, demanding the greatest of advertising skill, where fantasy utilizes the most accomplished artistic expression to mobilize the action of buying (Chapter 39).

In France in 1890, there were 50,000 bicycles. In 1891, that had already become 150,000 and the decade saw that figure 'explode': 250,000 in 1895, 300,000 in 1895, 409,000 in 1897 and the million line was passed by 1900. This increase was at first based on the reduction in weight of the machines and the widespread adoption of the pneumatic tyre that made them easy and comfortable to ride, while their design opened bicycling to a female market. The market enjoyed a fashionable 'boom' in the mid-1890s, when the machines were relatively costly, seen as status symbols and indicators of their riders not only being wealthy, but also modern. The price of machines rapidly reduced in the final years of the 1890s which allowed a real democratization of the bicycle, a process that continued for the next three decades. With this democratization, art posters tended to become less daring and more conventional in aesthetic and imagery than those of the boom and the last years of the 19th century.

The most popular touring machines in France were English, such as Humber (usually 'Beeston Humber', their best make) Rudge, Singer, Quadrant, Whitworth and Raleigh. They were followed by French machines such as Clément, Hirondelle, Ouragan, Jussy and Dombret. In particular, the French brands Clément, Gladiator, Gauloise, Peugeot and Rochet tried to compete with the English for the top-of-the-range market. On the American side, Luthy and Peoria were top of the range, while many others, such as Columbia, Crescent and Cleveland, were active on the French market, often exploiting the gap in supply caused by comparatively slow production of English machines during the sales boom of the 1890s. However, their position was undermined by over-production and dumping in the years immediately following and then rapid contraction in the US industry. While the post-boom contraction was far less marked in the UK, many of the makers that had opened French agencies, often on some scale, economized by closing them and the 'international' nature of art posters in terms of who commissioned them was therefore brief.

The bicycle 'art poster', as it was to become known internationally, begins to appear in France at the height of, and therefore the close of, the boom. It was no wonder that as an icon of modernity, bicycle advertising should adopt the style of the most modern art practices and, as a fashionable plaything, those activities associated with the leisured and wealthy. While this could be said of bicycling in general at the time, its expression in art posters was largely and peculiarly French and metropolitan. This reflects the fact that the art poster had its roots in Parisian theatre and entertainment, it spread to alcohol, tobacco and cycling, effectively aligning the activity with a particular Parisian lifestyle that could be termed 'café culture'.

It is important to note that although both American and British makers commissioned art posters for the French market, few issued works of similar aesthetic quality on their home markets. Commonly, this is assumed to be because there was a belief in both the USA and the UK that the French were far more daring and willing to accept more painterly graphic styles and morally challenging imagery. However, it may be more that American and British makers' street advertising to the French market was more or less completely controlled by their French agencies. There was probably little correspondence between them and their head offices over the matter of what poster content should be, beyond that it advertised the maker. Essentially, then, these posters, while attracting an international audience of connoisseurs, were peculiarly French and largely commissioned locally, even if the bicycles they advertised were made abroad.

In turn, there is little evidence surviving as to how the cycle agencies commissioned the poster artists. The great variance between and individuality of each poster tends to suggest that the artists were usually given free range in terms of design and imagery. This results in poster styles that are universal, rather than associated with particular brands, with common themes developing as artists saw what others were doing.

In terms of effectiveness, it is now almost impossible to gauge what sales art posters generated over and above more conventional advertisements, but their enthusiastic reception by collectors, even at the time of their currency, has meant that disproportionate numbers survive today, giving historians a belief that they must have been very successful sales tools at the time of their currency, outstripping other forms of advertisement. Again, this is questionable. Rather, they were very desirable to a small, elite group of enthusiasts. A particular irony of the power of the art poster in longer art-collecting terms is that some makers which were short-lived and would otherwise be long forgotten, such as 'La Chaine Simpson' (a UK company set up in 1895 offering a design of chain that gave no physical advantage over others but claimed it did), remain internationally famous. But, this is hardly useful to them, given they collapsed in 1898.

The poster artists

Aiming for fame and prestige, poster art brought in the best graphic artists and thereby captured the artistic styles then in vogue. The function of the advertising poster, whether it was displayed on the street or in the shop, was to catch the eye of the viewer. Their imposing format, the play of their figures and their bright colors made for vicarious enjoyment, and was the stuff of dreams.

In Paris in the decade 1890–1900 there was a craze amongst both artists and art lovers for the posters that adorned the streets of the city during the *Belle Époque*. The entertainment industry, the theaters, the night clubs and the circus, all of which enlivened *la vie Parisienne* of the time, was the first to be attracted to, and to adopt these art works of the street, while huge chromolithographed posters for the cycle followed shortly after. Gaiety, humor, youth, liberty, creativity – seductive provocations suggesting a loosening of morals and liberty – that were captured in the spirit of the cabarets at the Moulin Rouge and the Chat Noir, was also part of the attraction to this new object of modernity, namely the bicycle.

The artist Jules Chéret (b. Paris 1836–d. Nice 1932) was a pioneer of large-format stone lithography together with the printers Chaix, who drove this innovation and drew into it numerous painters and illustrators who were part of this same world. These latter were seldom unknown and while they did not have the artistic reputation of Steinlen, Eugène Grasset or Toulouse-Lautrec, they were often painters who exhibited at the Salon. They enjoyed a certain popularity due to their enthusiastic humorous depictions of sport and leisure activities. They illustrated artistic and light reviews, such as *Frou-Frou*, *Le Rire*, *La Plume* and *Cocorico*, and songwriters' works which embodied Gallic gaiety as well as the talent of such singers as Aristide Bruant; they painted backdrops for the popular theater, providing settings in which sophisticated ladies became involved in amorous entanglements. Their skills relied upon a true professionalism in publicity art of all types, fed by the growing popularity of consumer products and they moved into the realm of mechanized sports once the cycling craze had subsided.

The techniques of stone chromolithography gave rise to problems in the production of the large-sized advertising posters which insistently enlivened the streets of Paris and of larger French towns. The width of posters, set originally at 90 cm, soon rose to 110 cm and, with a height of 160 cm, they frequently required two big sheets for a single poster. So it was the American process of using much lighter zinc plates which gave rise to many of the posters during the period 1894–1910. The masses of primary colors, blue, yellow, red, black and the white emerging from the solid colors, dominated the visual effect which caught the eye of the passer-by. Other colors obtained by superimposition were much less used, apart from the green of rural scenes with bicyclists taking a ride around a lake and of tourist scenes like those of the Phébus marque and others made by the designer Eugène Ogé before 1894. New forms of artistic inspiration decorated the graphics of *La Belle Époque*. The influence of 'Japonism' led the designer Minos, of the Chéret studios, in 1896 to place people drawn as solid black images resembling Chinese shadow puppets into scenes from the *Chalet du Cycle*.

Chéret printed the first lithographic works of Misti, a graduate of the École des Arts Décoratifs, before he established his own studio in 1894. Misti (Ferdinand Mifliez, b. Paris 1865–d. Neuilly 1922) made posters for Humber-Beeston, Gladiator, Clément, Alcyon, American Crescent and the aperitif Quinquina Dubonnet, as Chéret had done. A wholly graphic culture was presented to the consumer. Moreover, Chéret's favourite type of woman abound to such an extent that these fresh, cheerful, elegant and dynamic young girls became known as 'Chérettes,' models for a generation of active young women who loosened their corsets and adopted voluminous knickerbockers to ride diamond frame bicycles with a top tube.

Between 1889 and 1894, Henri de Toulouse-Lautrec painted numerous works, much inspired by the atmosphere of Montmartre: magazines, dancers, girls of dubious virtue. In 1891, he created the famous poster of La Goulue for the Moulin Rouge, which replaced the one by Chéret of 1889. Totally taken with lithography, it was Lautrec who elevated poster art to the nobility. Between 1892 and 1896, he produced 32 posters and three major projects, mainly for the theater, the circus and for singers. Making the advertising poster an art form in its own right, Toulouse-Lautrec drew 'La Chaîne Simpson' in 1896, simply showing with a strong drawn line and solid primary colors, the champion Constant Huret pacing behind a quad tandem. In the mid-distance and shown in grey, one sees a kindly caricatured L. B. Spoke, the French agent for the marque, in front of an enthusiastic brass band.

Personalities imposed their own styles, often followed by others. Thus, one of the most prolific of artists, PAL, from 1894, introduced scarcely veiled, naked beauties wearing just a light drapery of organdie. His best-known work is the iconic poster for Clément in 1894, in which the beautiful lady glides on a bicycle before a crescent moon. This Romanian artist Jean de Paleogolu (b.1855 Bucharest–d. Miami Beach USA 1942) studied in Paris and in London. After several visits to London, he returned to Paris and then in 1900 settled in the United States. A fervent disciple of Art Nouveau, he contributed to the may art posters commissioned for JOB cigarettes. He was the creator of a series of posters for Whitworth, spread out along a fence from where emerge the faces of a cohort of girls admiring the manliness of the happy owner of this bicycle, so light that he can hold it out at arm's length. PAL also contributed his talent to Peoria and to the beautiful poster of 1894 for Falcon Franco-American.

PAL's style inspired the painter and poster artist Tamagno (b. Turin 1862–d. 1933). He worked for drinks companies and for the railways, for Automoto and for Ravat Wonder of Saint-Etienne and made the beautiful poster with the lion for Peugeot and a large format work for the American Luthy. It was he who suggested in 1902 a major series of posters for Cycles Terrot showing an elegant cyclist being unsuccessfully chased by a train coming out of a tunnel.

One of the masters of the *Art Nouveau* poster art who is still much appreciated today was Eugène Grasset (b. 1845 Lausanne–d.1917 Sceaux). After much travelling, he settled in Paris in 1871. After a spell in the architectural studios of Viollet-le-Duc, he worked in decorative arts, including designs for postage stamps, designs for tableware, decorative sculpture, carpet design, stained glass windows, mosaics, and jewellery. He was a collector of Japanese prints who was equally inspired by the Renaissance in his typically *Art Nouveau* compositions: exotic typography, animal and vegetation motifs, strong outlines, girls with pale complexions and long hair within a busy décor. His attraction toward the Arts and Crafts movement of William Morris led him to invent a typographical form influenced by *Art Nouveau* and by Japanese style. It was Eugène Grasset who contributed to the Larousse dictionary the celebrated maxim, 'I sow to the four winds.'

The Bohemian atmosphere of Montmartre, both eclectic and picturesque, imparted to all its enthusiastic Parisian followers its *joie de vivre* and its confidence in the future.

Symbols and emblems of the cycle makes

Publicity posters provide us with much information. Although an advertising document, the poster is also a valuable industrial archive. Texts record the marque, particular graphics, makers' names, information about the locality of manufacturers and factories addresses, locality of sales and sometimes even an image of the factory.

While some brands are distinguished by the thematic continuity of their favorite iconography, all of them show an infatuation for the color yellow and symbols of the sun. The relationship of yellow with the sun by way of the analogy with the spokes of the wheel and the rays of the sun seems obvious. The wheel against a yellow background is frequently used as a manufacturer's badge. Certain firms add the initials of their founders or a distinctive emblem. Thus, the Whitworth logo shows a black hand in the middle of a wheel on a yellow background while the Georges Richard company shows a trefoil within a yellow circle – in fact a four-leaved clover to represent good luck (Figure 41.1), while Strock and Co. has a letter 'S' in place of the hub of the wheel. As another example, the Société de Vélocipèdes Clément opted for a cockerel standing on a wheel, a symbol of the dawn but also a symbol of vigilance

Figure 41.1 Georges Richard, 1897. Dessinateur Eugène Grasset. Imprimerie de Vaugirard, Paris. 116 × 153cm (Collection du musée d'Art et d'Industrie de la ville de Saint-Etienne: Cliché Yves Bresson).

during the French Revolution as well as connecting the firm with the roots of the country (the French Cockerel) clearly revived by the cycle industry.

Other solar symbols or those related to the zodiac appear among the makers' badges: the moon, the star, the lion. The moon takes us back to Venus, goddess of love (Rudge, La Déesse, Fernand Clément and Co.). Her light in the night reveals a new pathway given to man by the bicycle: the milky way. In this sense, the Rouxel et Dubois poster is very successful. Some brands play on names chosen according to their evocative power. Kosmos speaks of an earth-bound means of transport transcended, while strength, reliability and ruggedness are incarnated in Gladiator, a make represented by a horse-rider upon a world globe.

Liberator spoke of lightness (presence of wings), Phoebus related to the sun-bicycle (in mythology, Phoebus is equivalent of Apollo), Déesse (Goddess) raised up the 'little queen' to the realm of the Gods (Figure 41.2) and revealed a play on the initials of the directors of the firm, Duncan and Suberbie. Diamond recalled beauty, hardness, luxury and durability but also brilliance. Other makers preferred to name their bicycles via imaginary links to efficient natural forces, in a baroque association of themes of metamorphosis and movement. Papillon (butterfly): lightness, flight, manoeuvrability; Ouragan (storm): Irresistible force of nature; La Guèpe (wasp): lightness like a butterfly but also precision, speed and fineness; Falcon: the falcon climbs rapidly into the sky and never misses his target.

Figure 41.2 Déesse, vers 1895. Dessinateur PAL. Imprimerie Paul Dupon, Paris. 149 × 111cm (Collection du musée d'Art et d'Industrie de la ville de Saint-Etienne: Cliché Yves Bresson)

Some brands played on the names chosen for their capacity to evoke: for Kosmos, it was overtaking other terrestrial transport, rising up and performance; for Gladiator, it was a symbol of strength, of endurance, with the make represented by a horse-rider on a world-globe. Liberator spoke of lightness and of freedom (presence of wings), Phoebus evoked the sun and beauty (in mythology, Phoebus is the equivalent of Apollo).

Illustrators delineate the thematic preferences of makers

Seduction (Whitworth – Figure 41.3), lightness and universality (Rouxel et Dubois), these are examples of the characteristics of a product used to target a specific clientele. The inclusion of cyclists in poster images gives, as a leitmotif, interesting indications on dress and on different uses of the cycle (Chapter 40).

One tries to attach a national origin to the product: La Française or la Métropole. One evokes a craftsman-like and workshop competence: like Decauville with its blacksmith, but not without humor, Decauville again relates its product to the produce of the Earth with a pretty folklore peasant girl on show to the cycling market. And finally, if we examine the company

Figure 41.3 Whitworth, 1897. Dessinateur Jules Alexandre Grün. Imprimerie Bourgerie et cie, Paris. 117 × 84,5cm (Collection du musée d'Art et d'Industrie de la ville de Saint-Etienne: Cliché Yves Bresson).

letterhead, we see the lettering of the company address, the company status, the invested capital of the enterprise, all of which assert the strength of the make.

Posters made for World Fairs and major cycle exhibitions (Chapter 16) also celebrate the excitement present in the realm of cycling. Manufacturers profited from the attraction generated by the bicycle to publicize quite separately the accessories that depended on the cycle industry and benefitted from its fame. At the very least, these accessories, often prestigious names in their own right, gave their support to the bicycle brands.

There were pneumatic tires, which were the predominant novelty of 1890s (Chapter 10). The poster for Kosmos is a subtle reminder of the reasons why a crowd has gathered round a machine, under the watchful eye of a gendarme. The G and J (Gormully and Jeffery), the 'king of tires with beads', attracts the admiring gaze of the young ladies. And as for Stella pneumatic tires, they even give wings to a fleeing hobo. La Fucosine offers 'automatic sealing' as its innovation for tires. The Electroleine is a totally new lighting product for which the poster gives a long written text, in which it presents it as a 'scientific discovery.' This gives a clever impression by playing upon the craze at this time for practical scientific novelties as demonstrated at fairs and referred to in popular reviews of the 19th century for the dissemination of science. This same approach is used in a masterly way for Christy Saddles. In the centre, on a yellow background, a young woman offers an 'anatomical and hygienic' saddle surmounted by a correct profile of the female pelvic structure while on the left-hand side is shown the 'former unhygienic and harmful system' deforming the female skeleton, in much the same terms as the famous rejection of the murderous corset.

Appropriate cycle themes

We have to explore further to identify the most eye-catching themes, aimed at first toward a masculine clientele and then rapidly switching toward the feminine. Typically, the bicycle itself is not the object of a very detailed representation in a poster, and one would have difficulty in recognizing one make from another, from these simple images. The tone used to in a poster is very different from that used in a catalog. It is the whole of the graphic surroundings, all of the contextual positioning of the object offered for sale, which speaks to it, which comments upon it, and which promotes it. More than the make of the machine or the bicycle itself, the manner of treatment and the fashion of presentation determines the effectiveness of the poster and the desire to make a purchase.

Following the decisive impact of the Paris–Brest–Paris race of 1891, race victories offered an immense opportunity for the promotion of machines via the careful portrayal of the champions themselves. The first Paris–Bordeaux race, won by Rivierre in 1891, followed by three more victories including that of 1898, promoted Omega cycles, in the same way that the victory of Stéphane in the race of May 1892 enhanced the reputation of the Clément make (Figure 41.4). The opening in December 1893 of the Vélodrome d'Hiver created another social space where brand competition was pursued, including the rider Jacquelin for La Française who was celebrated in giant form in a format with four joined sheets.

The sport was directed initially to men. Military cycling quickly became a subsidiary element in the progress of publicity, with pleasure outclassing practicality. In this *fin de siècle*, the bicycle brought *joie de vivre*, festive spirit and the prospect of amorous conquests. Amorous adventures often permeate brands such as Gladiator and Whitworth. The poster exploits all forms of desire in order to promote the bicycle, with amorous conquest often cross-linked to the theme of horse riding. The reference to the horse, which had persisted for a long time, is contradictory, and often used as a foil. It presented an outdated, retrograde element

Figure 41.4 Clément Paris, 1892. Dessinateur inconnu. Imprimerie Kossuth, Paris. 150 × 99,5cm (Collection du musée d'Art et d'Industrie de la ville de Saint-Etienne: Cliché Yves Bresson.

of comparison: 'abandon the horse for the steel steed.' Aspects of worldly refinement of the equestrian sphere with its distinctive social customs and practices as well as elegance were, however, retained. The cycling vocabulary is permeated with analogous terms: the saddle, the circular track, and the velodrome relating to the hippodrome as a horse-riding ring. After having outstripped the horse in speed, the Terrot bicycle is shown faster than the train.

The man–bicycle combination, the new Sagittarius, remained barely detached from the equestrian model. Reference to the horse, enhancing the model of elegance (and thus one of the forms of seduction), is deployed above all in the staging of abductions or amorous escapes. The imagery of seduction, deeply rooted in the myth of the abduction of Sabine women, still resonates strongly in the pneumatic tread of the new steel horse.

The enjoyment and *joie de vivre* offered by the bicycle brought in its wake, via the implied erotic reference, the greater respect for woman who, as the goddess of this revolution, won her right to liberty in an epoch when social constraints were becoming less rigorous. The iconic Mother Goddess, decked with Greco-Latin or often Celtic and Nordic mythological attributes, brings to the bicycle an extraordinary will to contextualize the machine within the heritage of the human experience. The warrior goddess, she exalts her will to uphold the new modernity, with a banner in her hand like an emblematic French Marianne upon the barricades.

More to the point, if a woman rides the bicycle in order to demonstrate that she has tamed it into ease, comfort and safety, more often than not, she rides it up into the sky – the woman of the Humber and Déesse posters is the earthly support for the celestial machine. The woman personifies the bicycle as 'the little queen'. It is she who crowns the champion (Strock), she who rises upon her winged feet, she who floats in the air by the magic of her draperies, and she who embodies the flight of the aerial machine. The woman/bicycle is the muse of a new art – the art of cycling.

The discovery of the sensation of flight, brought about by the alliance of lightness and speed, is exploited by Maurice Leblanc in his novel: *Voici des Ailes* (Here are Wings), published in 1898. The illustrators of the book captured this important theme, either by the inclusion of wings themselves or by floating draperies or hair. The male cyclist is never shown with wings, and seldom is the bicycle itself. Its drawn geometric form requires the additional undraped curves of the female form, to show that it will rise upon the air and is free to float up into the sky.

Social life opening to cycling

Themes of festivity, *joie de vivre*, pleasure without ulterior motive, all are present, it would seem, in a form specific to the bicycle. It lends itself to freedom, the rediscovery of nature, physical self-expression, and brings about unimaginable joy. These themes need to make a connection with the customer. His or her reaction to what is offered by the producer will influence the product range. Bicycle use by the customer must be socially acceptable in order to create new desires as well as real needs. Sport and tourism are the bicycle's two main forms of practical use which the specialized press, backed by literature, will write about, promote and cultivate in numerous articles and illustrations.

One of the greatest merits of these posters, inspired by the psychology of the masses and the art of selling, is that they are high-level, mainstream art, accessible on the street. At a time when television did not exist as a major medium, the cycle poster also tells us about the changing world and patterns of consumption. It informs us about companies, products, as well as about consumers' dreams. It announced the emergence of a new world and prefigured the advertisements and the imaginary world that later became inhabited by the automobile and, to a lesser extent, by the airplane.

Further Reading

Besse, Nadine (2002) *Voici des Ailes: Affiches de Cycles* (St. Etienne: Musee d'Art et d'Industrie de Saint Etienne).

Dodge, Pryor (1996) *The Bicycle* (Paris: Flammarion). This book presents numerous color prints of *Belle Époque* bicycle posters.

42

ART AND THE CYCLE

Scotford Lawrence

The first form of the bicycle – the *Laufmaschine*, *draisienne* or *hobby horse* – was invented in 1817 and was in use in the following two years. This was a period of enormous national, cultural, social, technological and economic upheaval which had endured throughout Europe for the previous quarter of a century of the Napoleonic Wars. It had dominated all aspects of European life, and was to continue in the years which followed. As is so often the case during a period of warfare, it was also a springboard for inventiveness and technological innovation. The inventor, Karl von Drais, had envisaged his two-wheeled machine as a practical means of getting about more quickly and with less fatigue, inspired, no doubt, by carrying out his duties as Forest Master to the Dukes of Baden. He produced a brochure for his *Laufmaschine* with, as a frontispiece, a stylized engraving of a rider striding along, dressed in the uniform of an army staff messenger as an indicator of an obvious use to which the newly invented machine might be put. This is one of the very first images of any bicycle-type machine.

When, some months later, the machine was introduced into England, an engraved illustration of the machine and its rider appeared in the monthly magazine, *Ackermann's Repository of Arts* showing a rider travelling through an idealized country landscape. In these and related engravings of the same period, the first bicycle had established its visual identity – it was to be seen through the popular engraving and the printed image. Within a very few years, the printed image itself was to undergo a major transformation as the mechanically powered printing press allowed the rapid production of thousands of copies of a newspaper, magazine or publicity sheet. The image and its accompanying printed text was to become a commonplace of the nineteenth century.

From its inception, the bicycle was wedded to the printed image. This had the effect of inhibiting its appearance as a subject of classical art, so that there are very few paintings and even fewer sculptures of the cycle and its users. One would have hoped to find an example of the formal portrait at some point during the nineteenth century, of the proud owner standing alongside his machine in the same way as in equestrian portraits or portraits of proprietors of land or of valued objects. But there is only a single, unusual example and, because this is so, the object, the cycle itself, is rendered more commonplace and loses that first wonderment that arose from the fact of a man travelling by his own power alone, faster, further and with less effort than ever before.

DOI: 10.4324/9781003142041-62

Figure 42.1 Artist unknown. Exercising a Hobby from Wales to Hertford. Sidebethem print 1819.

In England, there had been a prolific production of political and wartime cartoons during the Napoleonic Wars and this activity was now turned toward internal politics, the mocking of post-war high fashion and the outrageous activities of the Prince Regent and his fashionable circle (Figure 42.1). In many cases these were rolled together with the appearance of the newly introduced hobby horse. During the short period of its use in England from early 1818 until it had been more or less laughed out of use in 1819, some 90 hobby horse cartoons were published in the form of hand-colored engravings (Street 2014). Thus, the image of the first form of the bicycle was allied, right from the very start, with printed illustrations. Besides this, there are just a few amateur watercolors of the hobby horse and a single, small oil painting, and that is all.

This outpouring of satirical, hobby horse cartoons, in its turn influenced the manner in which the viewer saw the machine itself. The two-wheeled machine and its balanced rider were to be a source of amusement and mockery and this element has also contributed to the nature of the cycling image and the manner in which it is viewed down the two centuries since its invention.

The German philosopher and critic Walter Benjamin (1892–1940), in his essay, 'The work of art in the age of mechanical reproduction', posits that the portrayed object is itself demystified and rendered commonplace by its image being mass-produced (Benjamin 1935). If you could buy for a few pennies a satirical cartoon of the Prince Regent riding a hobby horse with his mistress, then the machine itself becomes a part of the satire. And if you could be given *gratis* an illustrated leaflet of the *Laufmaschine*, then that famous stylized image of the messenger becomes a commonplace. Even though the hobby horse was far more expensive than could be afforded by most of those who saw the image, nevertheless it could be looked at over and over again and so become an element of daily life.

On the other hand, there are few artists' images of the machine as a subject of serious art. With only those two or three amateur watercolors and a single known oil painting, made in France, that was the one country in which the use of the Draisienne was to continue among a few enthusiasts for many years, even to the extent that, if we are to believe the story told by the Michaux family, one was still being ridden by a certain M. Brunel in Paris in the 1860s (Michaux 1893).

The hobby horse, during its short reign, had established itself as the subject for the illustrator and the cartoonist rather than for the 'serious' artist. This is not surprising. The nineteenth century was a period of innovation and development such as the western world had never experienced before, and culture and the visual arts did not keep up with the technological genie which had emerged from the lamp and was now so vigorously active amongst them.

Not only the steam engine, but also those industries which were being developed in order to engineer its components, were rapidly coming into being. But they too were hardly a 'suitable' subject for the artist and there are surprisingly few paintings showing industrial activity when it was at its most dynamic and vigorous. Joseph Wright of Derby (1734–1797) seems a precursor of the industrial image as a manifestation of the Age of Reason, but there is little industrial art thereafter.

After the disappearance of the hobby horse, for a period of over forty years, little is heard of the cycle in any form, although the quadricycle and various forms of the tricycle were developed during those four decades and technical, explanatory and satirical images of them were made. A gentleman amateur from Essex, England, produced a couple of watercolors of his home-built experimental machines in use but they are themselves 'curiosities' if only by their rarity.

In the 1860s, the invention of the pedal velocipede should have produced an outpouring of new art – but it did not. It was illustrated in printed catalogs and instruction manuals and there was even, in France, a short-lived periodical, *Le Vélocipède Illustré*, but formal, classical art eschewed the new machine which was daily visible on the boulevards of Paris and on the streets of London and many other cities. The velocipede was a French invention and the fact of its not appearing in formal art must, to some extent, be attributed to the manner in which the arts were controlled and conducted in France. Art was under national, political control. Promising students were brought to Paris to train and to work at the *Académies* and the *École des Beaux Arts* in the rue de Seine. The opportunity to make their work known to the general public came in the form of the annual Salon Exhibition which was the only official, recognized commercial outlet. There was a hierarchy of acceptable subjects: (1) History Painting; (2) Portraits; (3) Genre Painting; (4) Landscapes; (5) Still Life. Works were bought by a government fund and distributed to local art museums throughout the country for the education and cultural improvement of the masses. There was no place at all in this hierarchy of art or in its formal teaching and promotion, for the portrayal of the technical innovation of the modern world which was happening right outside the doors of the Salon. The velocipede continued to be the subject of the illustrators' engraving and the outpouring of the steam powered printing press.

There are few classical, oil-on-canvas paintings of the velocipede and its riders. The most remarkable is a complete scene of the start of a velocipede race in the small town of Gray in eastern France, which took place on 22 April 1869 and was recorded in a formal painting by the locally born artist, Joseph Roux (1832–1913) (Figure 42.2). It is a completely original work with almost no equivalent anywhere else in the canon of formal painting (Lawrence 2007). Moreover, it is extremely detailed and, from press reports and local documentation, we know the details of the event and even the names of the participating riders. It was shown at the Salon of 1870, but was not bought and remained in local ownership until the twentieth

Figure 42.2 Joseph Roux. Course de Vélocipèdes. Gray 1869. Oil on Canvas (Courtesy of National Fietsmuseum Velorama. Netherlands).

century. Although a remarkably accomplished piece of representational art, a formal painting of a cycling scene was not a subject in which the public at large was interested.

There is a single formal painting of the cyclist in classic 'me-and-my-bicycle' pose, of the famous courtesan Blanche d'Antigny by a minor artist, Henri de l'Étang (1809–1873). But the main significance of this small painting is that it shows Blanche d'Antigny wearing a demur costume but with britches. She had been a circus performer and went on to fame as the model for Nana in the eponymous novel by Émile Zola.

The only other bicycle picture of any significance is from the very end of the nineteenth century when a minor genre painter Jean Béraud (1849–1935) made a painting of the gathering of fashionable cyclists at the Châlet du Cycle in the Bois de Boulogne in ca.1900.

In the mid-century, another development came to determine the place of the cycle in the world of art – photography. By the 1860s, the various methods of photography had become rationalized into the chemical process which was to reign supreme for the next hundred years: the use of a camera to 'take' a silver chloride negative image from which a positive photograph could be made, again and again without limit. Exposure times were lengthy but they did allow the photographing of the cyclist, posed with his machine, though 'moving' images were not possible until the last decade of the nineteenth century. Photography enabled cycle owners to acquire a visual document for which an extensive market rapidly arose: the formally posed, 'me-and-my-bicycle' proud image. These came in the form of *carte de visite* (visiting card) prints of a size to be carried in the pocket, often in a neat little folding frame covered in fabric or in bookbinder's leather. Thousands of these were produced and many have survived, showing a standing rider with his machine proudly displayed. Occasionally, the owner is shown mounted, with a strategically placed block behind the velocipede to support the rider and to give something of the appearance of the rider in motion.

Photography, like the printed engraving, also served to position the bicycle within the realm of the visual image. But here again, the bicycle and its image belongs with mechanical innovation and technology, not in the world of art and the art museum.

The 1860s was the period of the velocipede and was also the decade in which the French art world was shaken to its depths by another 'French Revolution' – that of the impressionists. Here was a group of young artists who broke away completely from the tenets of the formal art of the Academies and the Salon. Their work was unacceptable to the judges of the Salon and was rejected wholesale, so they simply established a '*Salon des Refusés*' – a 'Salon of the Rejected.' These young artists believed in painting direct from the natural scene out-of-doors, the 'plein air' movement. They painted informal poses of people going about their daily activities, they painted what they found of interest and visual challenge. They were revolutionary, in subject matter, style and technique. As they worked, they met and shared their Bohemian lives together, particularly those who visited the forests around Paris to paint their pictures. In these same woodlands they certainly must have met the newly adventurous velocipedists – but they did not paint them. The impressionists may have been revolutionary but they wanted their 'plein air' to be free of the ever-encroaching world of technological innovation. If we look at the riverside paintings of Claude Monet (1840–1926) along the Seine, we see, in the far distance, the factory chimneys and the smoke of Paris encroaching as a threat to their newly-discovered, beautiful world. There were among the impressionists, artists who used the smoky activities of the new technologies for interesting light and shade effects such as Monet's studies of the light through smoke under the canopy of the Gare St. Lazare and Gustave Caillebotte's (1848–1894) street scenes and views across the steel structure of the Pont de l'Europe which spanned those same railway stations, but in all of these works and many others besides there is not a single velocipede. In fact, the only impressionist velocipede image at all was painted by Edouard Manet (1832–1883) of a head-on view of his son riding a velocipede, and that is on a strip of canvas of only 53 cm (h) × 20 cm which is believed to have been cut off from a bigger painting.

The visual record of the velocipede consists of that remarkable painting by Joseph Roux, a small portrait by Henri de l'Étang, a fragment by Manet and a host of photographs, but it causes barely a ripple in the mainstream of the visual arts. The printed image of the velocipede exists in numerous technical manuals, drawn images for press reports and as illustrations in magazines and books. Its successor, the much more widely produced and distributed high bicycle or *penny farthing* is much the same.

We must also remember that the only method by which the photograph could be mass-produced in print was by the image being hand-engraved, and the engraved block being inserted into the assembled type page. It is not uncommon to see the words, 'Engraved from an original photograph' attached to images in magazines, periodicals and books of the mid- to late-nineteenth century. It was not until the invention of the chemically etched, half-tone printing block toward the end of the nineteenth century that a direct image from a photograph could be printed. So, whatever its origin, the engraved image reigned supreme in print throughout the nineteenth century.

Because photographic exposure speeds were slow, engraved images were widely used as illustrations for press reports of cycle races and competitive events. But as touring and leisure cycling became more common, drawn and engraved illustrations of cyclists in the countryside and of the scenes through which they rode became a commonplace of the cycling magazines which proliferated in the high bicycle era and on into the era of the safety bicycle. Some remarkably talented artists came to the fore in this period and some of these continued to work long after direct photographic illustrations in magazines and periodicals became common. Many of these men did not regard themselves as 'artists' at all but as illustrators working to a schedule and in a 'job' rather than enjoying the superior status of the inspired artist in his studio. Illustrators such as Frank Patterson (1871–1952) (Figure 42.3) and George Moore in

Figure 42.3 Frank Patterson. Line drawing. My Favourite County. 1917.

Cycling in England and Joseph Pennell (1857–1926) and W. A. Rogers in *The Wheelman* in the United States produced work of remarkable talent and remain some of the best graphic artists the world has ever produced.

The bicycle is essentially a thing of tubes and wires – it is visually linear and, while it is a suitable subject for the techniques of the graphic artist, for the painter it presents an inherent difficulty, having neither mass nor surfaces in the same way as a human body or a bowl of fruit. This is an even greater problem for the sculptor and, as a result, there are few sculptures of cycling and cyclists – and even fewer that can be regarded as successful. The demand for the heavy mass of the body of the rider to be supported on the thin structure of the machine presents particular problems which are seldom satisfactorily resolved.

At the Manchester Velodrome in England, there is a dramatic sculpture by James Butler of the multiple world champion Reg Harris swooping down from the banking in his characteristic hump-backed pose. But go round to the other side of the sculpture and there is a crude network of scaffolding poles in full view which support the figure. Perhaps Jemma Pearson's statue of the English composer, Edward Elgar, standing and using the bicycle as a prop, against which to lean back and gaze up in wonder at the cathedral at Hereford is a better solution. But ultimately the bronze standing statue by Aristide Maillol (1861–1944) of Gaston Colin solves the whole 'problem.' Colin was a motor-paced cycle racer and Maillol shows him standing alone as a simple nude figure, which invokes perfectly the unprotected danger and risk in which these riders were involved as they followed the *gros motos* at high speed on the track. There is no bicycle there at all.

The cycling image, of whatever type, imposes another inherent problem. The cycle – and particularly the bicycle – is essentially an object of movement. The rider has to propel it to maintain balance and to make it 'work' and fulfil its purpose. The art image and even the photograph is essentially static, but in order to give the cycle image that life which it requires, the artist must find ways of making the image 'move.' It took almost a hundred years before artists in general overcame this fundamental visual problem. Toward the end of the nineteenth century they started to blur or even to omit the spokes in the wheel of moving cyclists. In fact, what is seen is rather more complicated in that, since the wheel and the rider are passing in front of the viewer, the spokes at the bottom of the wheel, where it is in contact with the road surface are, in fact, static to the viewer, while those at the top of the wheel are travelling at twice the speed of the machine itself. Pictures of hobby horse riders, velocipede riders and cyclists on the high bicycle and safety machine had all been shown with spokes carefully drawn or at least finely indicated. Henri de Toulouse-Lautrec, in his famous poster for *La Chaîne Simpson*, identifies this and shows only a few spokes at the bottom edge of the cyclist, Constant Huret's wheel where it is contact with the track and omits the spokes altogether at the top of the wheel (Chapter 41).

By the last decade of the nineteenth century the safety bicycle had become widespread and almost everyone had at least seen one in use and most had tried the machine. It had become a commonplace but also a symbol of the freedom to travel where and when one wanted. For the urban dweller, it also became an icon of the ability to leave the city and escape in the countryside. That modest, self-propelled minor adventure came to symbolize so much that a new age required. For a short period the bicycle became the popular emblem of wind-in-the-hair, speed. When the Futurists, that bizarre group of early twentieth-century iconoclastic disruptors, wanted an indicator of speed, they turned to attempts to analyze and resolve that inherent problem of making the static image of the bicycle 'move' on canvas or on paper. But of course, for the Futurists, the bicycle had an inherent fault. It did not make a noise! They wanted their disruption to be fast, destructive and, above all, noisy. Within a couple of images, their attention was taken by the motor car and then by the airplane. And the Futurists themselves disappeared altogether when they got all the noise, disruption and mayhem they wanted and more besides in the First World War.

In the twentieth century, the purpose of art itself underwent an irreversible change, since, if viewers wanted an accurate image of an object, they turned to photography. Drawn and painted art and sculpture had to change to fulfil a different purpose from that which they had been doing since the first images in the caves of primitive man. It was no longer required to show an object as it was, in that time-honored, triple interchange between the artist, the object and the viewer. The main function of art became an opportunity for the artist to demonstrate to the viewer the workings of the artist's own mind with the object merely as a means of doing so. As far as images of the cycle were concerned, they became an intermediary used to demonstrate, possibly freedom, liberty, leisure, and even speed again, as seen by the artist.

In the jolly, tubular cubism of Fernand Léger (1881–1955) the bicycle finds its way into his paintings as a symbol of happy times, sunlit holidays and innocent fun. *Les Loisirs* of 1948–49 celebrates the liberty of France after five years of German occupation and its figures are dressed in bathing costumes and bright, striped maillots, with bicycles as holiday objects.

But despite the camera, the fragmentation and abstraction of art, the image of the bicycle and its movement is still to be found today, and some of it of the most striking nature. There is no longer any place for those masterly illustrators and engravers of the late nineteenth century, but there are just a few artists who try to portray the movement and speed inherent in the bicycle and its rider. But in order to do so, the artist has had to make use of the tricks which

Figure 42.4 Claude le Boul. Gouache. The Champion Eddy Merckx. 1987.

were first 'discovered' over a century ago and were examined and analyzed by the photographer Eadweard Muybridge (1830–1904) and the artist Marcel Duchamp (1887–1968).

The contemporary French artist Claude le Boul (b.1947–) produced in 1987 a series of dynamic watercolors of the Belgian champion cyclist Eddy Merckx, where he uses again and again, the 'trick' of the multiple and blurred image in order to make the rider 'move' on the paper (Figure 42.4). These are striking in their immediacy and effective in communicating to the viewer the power, the effort and the speed of the cyclist in movement. But this series, *The Champion Eddy Merckx*, is an oddity, a curiosity and a byway on the path of modern art and illustration.

Everything has been tried and the artistic difficulties confronted and, in most cases, resolved. If we want a modern-day image of the cyclist, the machine and the activity, it is to be found in photography or even 'live' on real-time television. We can watch and participate in everything from a leisure ride on a sunny afternoon to an *hors catégorie* climb in the Tour de France at the click of a switch. Cycling art has always been a minor subject used to show the machine and its use, to advertise the product and to afford to the enthusiast something of the pleasure of the enjoyment of the bicycle. Throughout two centuries, thousands of images have been produced for a myriad of purposes and almost every form and style of art has attempted to show the cycle to the viewer. But it has remained the case that the bicycle and, following it, every other mechanical invention and development of the last centuries has never made it into the forefront of 'serious' classical art. Art itself had remained bound within that hierarchy of subjects so firmly defined by the French state so many years ago, despite the attempts of various modernists and movements to break free. It is amusing to note that the work of that greatest of the modernists, Pablo Picasso, still falls neatly into those same hierarchical categories laid down by the *Académies* in the eighteenth century – the history painting, the portrait, still life and so on.

Many highly competent artists have made a niche for themselves in portraying the motor car, the airplane and the inside working of technology and of modern industry but, by their choice of subject, they have precluded themselves from the world of 'serious' art and remained in that limbo, that purgatorial no-man's land which sees them dismissed by the critic and the aesthete as mere illustrators. Cycling art has been with us for two hundred years but it is a subject into which artists with a serious purpose have only, from time to time, dipped a toe but have never attempted fully to immerse themselves. But let us, the cyclists, enjoy it.

Bibliography

Benjamin, W. (1935) 'The Work of Art in the Age of Mechanical Reproduction.' *Illuminations*. Ed. H. Arendt, Trans H. Zohn 1969 (New York, NY: Schocken) 217–251.

Lawrence, S. (2007) *Joseph Roux and the Course de Vélocipèdes, 1869* (Nijmegen, Netherlands: Nationaal Fietsmuseum Velorama).

Michaux, H. (1893) *L'Éclair*, 28 March 1893.

Street, R. (2014) *Before the Bicycle. The Regency Hobby Horse Prints* (Christchurch, England: Artesius Publications).

Vignette M
THE SPACE BETWEEN

Hilary Norcliffe

Eastern design theory has long given equal value to "what is" and "what is not" – to both form and space – in the pursuit of balance and harmony. This contrasts with the West, where the emphasis in design is often weighted more toward the physical object than the surrounding space. The bicycle, however, is a linear design that calls attention to the many spaces that fill it.

Two shapes dominate bicycle design: the circle and triangle. Circles invite movement (wheels, cogs) and triangles brace for strength (spokes, frame). Additional triangulations are formed when the rider mounts. As the geometric lines of the machine connect to the organic lines of the rider, a new hybrid creature on wheels emerges. Negative spaces open and close like a kaleidoscope around the rider's legs while they pedal.

An immense (roughly) 5000-year gap in time exists between the invention of the cart (two wheels side by side) and the invention of the bicycle (two wheels, one behind the other). It seems odd that this simple rearrangement of two circles should have taken so long – but these two configurations activate space in completely different ways, and enormously change the way a human interacts with that space and form.

When hitched, the cart has the same reliable triangular balance whether stationary or in motion. The design of the bicycle is infused with tension: it requires motion to keep it balanced upright – a machine that teases the space it slices through. Its dramatic potential vastly supersedes the cart; the balancing-in-motion requires some effort and attention by the rider and is visually engaging for the spectator.

In his 1913 piece "Bicycle Wheel", Marcel Duchamp made his first readymade sculpture by attaching the front fork and wheel of a bicycle to the seat of a common kitchen stool (see https://www.moma.org/collection/works/81631). Combining these two ordinary, manufactured objects in a way that removed them from their day-to-day usefulness forced viewers to look at them differently – to think both about their social roles (conceptual) and also to look at them aesthetically as pure form in space. Lifting the wheel off the ground where it could spin freely also triggered Duchamp's interest in optical illusions. "Spinning the wheel introduces the viewer to the visual illusions of strobing spokes, which seem to rotate forward or backward, or even to disappear completely into a shimmering blur, depending on the speed of rotation" (De la Croix et al. 1991: 975). The space between now quivers.

DOI: 10.4324/9781003142041-63

Ai Weiwei's "Forever Bicycles"

These massive pieces, shown in Figure M.1, have been installed both indoors and out, most occurring between 2013 and 2019, in cities such as Toronto, Winnipeg, Austin, Taipei, Melbourne, Venice, and London.

Chinese artist and social activist Ai Weiwei is one of the most powerful voices in the art world today. A relentless critic of the Chinese government and human rights violations around the world, Ai Weiwei uses social media as a platform to advocate for political transparency, liberal thinking and individual expression. His artwork falls under the umbrellas of conceptual art and excessivism.

Ai Weiwei's decision to sculpt with hundreds of bicycles stems from several ties he has to the machine. As with much of his work, he plants one foot firmly in the language of art, and with the other stamps on Chinese politics (Siyuan Chinnery 2014). In "Forever Bicycles", he celebrates global cycling culture as well.

Ai Weiwei grew up in the poverty of a labor camp under Mao's regime where bicycles were rare and much sought after. At that time, the "Forever" company made some of the best bicycles in China. By the 1970s, bicycles had become ubiquitous in China – everyone had one. As car culture has taken over transportation in China, the bicycle has been downgraded, even to the point of being blamed for congestion and accidents (Grenier 2017). Massive skyscrapers and mass-manufacturing in China sit on the shoulders of a massive collective of humans, driving through heavily polluted air, discouraged by their leaders from straying from the crowd.

In contrast, Ai Weiwei also recognizes that, in many nations, cycling has become a symbol of health and fitness – good for the individual and good for the planet. "[Bicycles are] designated

Figure M.1 Forever Bicycles: Ai Weiwei (2015–2016) (National Gallery of Victoria, Melbourne). With artist's permission.

for the body and operated by your body. There are very few things today that are like that" (Fisher 2016). It's an honest machine.

Ai Weiwei was one of the first group of students to benefit from studying abroad after China's 1980 reform. He lived in the United States as a young adult from 1981–1993, getting exposed not only to democracy and free speech, but also to new ways of making art. Duchamp's readymades struck a chord. Now a century after Duchamp created his "Bicycle Wheel", Ai Weiwei not only embraces the found object for its material metaphor, but also furthers Duchamp's fascination with optical illusion.

Taking the bicycle as a repeated unit and using it like an architect might use a brick, Ai Weiwei connects forks to axles and draws through space with anywhere from 1100 to over 1500 bicycles. In most arrangements, the frontal composition is organic and dynamic. It resembles a giant creature, wheels cantilevered outwards, getting ready to roll itself away. The rhythms of the single cyclist have multiplied into an enormous orchestration, arching through the air. But Weiwei repeats this composition over and over again, moving backwards with machine-precision spacings. Without any actual movement, the viewer can stand still and experience a delirious blur, as if the whole collective is taking off at warp speed through a worm hole. Like the tweets Ai Weiwei blasts daily to expose injustices – a platform that is both everywhere and nowhere in space – this giant bicycle-borg blasts into a strange, liminal space of its own, somewhere between now and "forever".

Bibliography

De la Croix, H., Tansey, R.G., and Kirkpatrick, D. (1991) *Gardener's Art Through the Ages: Renaissance and Modern Art*, 9th ed. (San Diego: Harcourt, Brace, Jovanovich).

Fisher, L. (2016) "The Bicycle as Dissident Object." *Artlink*. March 1, 2016. https://www.artlink.com.au/articles/4433/the-bicycle-as-dissident-object/

Grenier, E. (2017) "Provocative Forever: Ai Weiwei Turns 60". *DW Arts*. August 28, 2017. https://www.dw.com/en/provocative-forever-ai-weiwei-turns-60/a-40201567

Klayman, Alison (2012) "Ai Weiwei: Never Sorry." https://vimeo.com/18018860

Siyuan Chinnery, C. (2014) "Ai Weiwei". *Frieze* (Reviews) Issue 165, August 13, 2014. https://www.frieze.com/article/ai-weiwei-0

Further Reading

Ai Weiwei (2013) Video: "Scotiabank Nuit Blanche 2013: Ai Weiwei on the inspiration behind Forever Bicycles." https://www.youtube.com/watch?v=k-Ba6X6Sivw

Forbes, M. (2020) "Understanding Ai Weiwei In 10 Works Of Art." *The Collector* (June 20) https://www.thecollector.com/ai-weiwei/

Huang, F. (2018) "The Rise and Fall of China's Cycling Empires." *Foreign Policy* (December 31) https://foreignpolicy.com/2018/12/31/a-billion-bicyclists-can-be-wrong-china-business-bikeshare/

43

CYCLING AND CINEMA

Revolutionary films

Bruce Bennett

The difference between flying in an airplane, walking, and riding a bicycle is the same as that between looking through a telescope, a microscope, and a movie camera. Each allows for a particular way of seeing. From an airplane the world is a distant representation of itself. On two legs, we are condemned to a plethora of microscopic detail. But the person suspended over two wheels, a meter above the ground, can see things as if through the lens of a movie camera: he can linger on minutiae and choose to pass over what is unnecessary (Luiselli 2014: 36–7).

With her *Manifesto à vélo*, Valeria Luiselli draws an analogy between cycling and cinema, observing that the bicycle and the movie camera have the capacity to change the way we see the world. The mobile gaze of the cyclist reveals new visual and conceptual perspectives on the environments through which she travels, bringing significant details into dramatically sharp focus (Chapter 32). Cycling has the potential to reintroduce us to spaces around us that have become invisible through overfamiliarity, and it can also give us a different sense of our body's physical presence in the world; as academic and former cycle courier Jon Day writes, cycling makes 'you *feel* a landscape rather than merely seeing it. By bike, your environment writes itself onto your body' (Day 2015: 15).

The similarity between cycling and cinema is more than a matter of vague correspondences but marks the complex historical relationship between the two technologies. The bicycle and the film camera are both products of industrial modernity and the expressions of a desire for movement that characterized many of the technical, social, and cultural developments of the nineteenth and twentieth centuries. They are revolutionary technologies of (virtual and literal) bodily mobility that allowed people to travel more freely than was previously possible, and the relationship between cycling and cinema is exemplified by one of the earliest surviving films, *La Sortie de l'Usine Lumière à Lyon* (*Leaving the Lumière factory in Lyon*), made in 1895 by Auguste and Louis Lumière, the owners of the photographic materials factory shown in the film. This was the opening film in a programme of ten single-shot films, each lasting around 50 seconds, shown at the Grand Café in Paris on December 28, 1895. The admission charge was one franc, and 'within weeks the Lumières were offering twenty shows a day, with long lines of spectators waiting to get in' (Bordwell & Thompson 2003: 10). This was not the first

DOI: 10.4324/9781003142041-64

film screening for a paying audience, but it was the phenomenal popularity of these shows, that secured the success of this new medium. The Paris screenings mark the beginning of cinema's emergence as a powerful force in global culture and they also coincided with the cycling boom of 1895, a conjunction that is captured by the first film in the programme.

La Sortie de l'Usine Lumière à Lyon shows workers passing through the factory gates at the end of a shift and heading off in different directions. The film, itself an artefact of new industrial technology, presents us with a self-reflexive, archetypal image of industrial modernity, epitomized by the crowd of workers, the vast majority women, spilling out of the workplace into the sunny street. Film historian Tom Gunning has observed:

> The twentieth century might be considered the century of the masses, introducing mass production, mass communication, mass culture. We could redescribe this transformation as the entrance of the working class [...] onto a new stage of visibility.
>
> *(Gunning 2004: 50)*

More than the other nine films shown in Paris, which included snapshots of, parents feeding a baby, blacksmiths at work, and a practical joke, the factory-gate film condenses the socio-cultural transformation described by Gunning into a single dynamic image of mobilized labor. There are three versions of the film with a virtually identical composition, and it's unclear which of them was the first film screened on December 28, but a significant element in each is that, alongside the pedestrians, dogs and horse-drawn carts emerging from the factory, we see a number of cyclists pushing through the crowd. In the first version, a man wheels his bicycle onto the street before mounting it and riding off, while a second cyclist follows behind a horse and cart. In the second version, two cyclists squeeze through the crowd and ride off, and another has to dismount as he tries to push his way through a group of women. In the third film, two men wheel their bicycles out of the gates with colleagues perched unsteadily on the saddles while another two head off confidently in opposite directions, one of them riding a drop-handled bike. Just before the film ends, a fifth cyclist emerges from the gates and a push from behind by one of his workmates sends him wobbling out of the frame before the gates closed behind him.

Although the cyclists, all riding modern 'safety' bicycles, appear to be an unremarkable detail in what would have seemed to Victorian audiences to be a fascinating technical novelty, for film historian Richard deCordova, they play an essential role in showing viewers how this new medium could depict the world differently from painting or photography. What made early cinema new, de Cordova argues, was not the illusion of movement, but the fact that objects and figures moved *toward* the camera and spectator, producing a powerful illusion of three-dimensional space. 'The workers do not just move: they move *in perspective*', deCordova explains, and this is shown most clearly by the cyclists who are initially hidden in the crowd, before becoming visible as they push past their co-workers (deCordova 1990: 78). The revolutionary presence of moving bicycles is what makes this film (which is, in a sense, the first film) a film. Cinema begins with a bicycle.

Bicycles were a particularly appropriate subject for early cinema since the safety bicycle and the film camera were both signs of the way that technological innovation was transforming modern life. Precision-engineered machines that were dependent on developments in industrial production, both cameras and bicycles have similar mechanisms: an assemblage of cranks, sprockets and axles, chains and gears, powered by rotary motion generated by hands and feet. Indeed, an early film projector looks very similar to an upturned bicycle with reels of film in place of wheels, and film historian Jennifer Barker's description of early hand-cranked film-viewing machines underscores the similarities between cycling and cinema:

> The situation [...] in which the spectator seeks entertainment through a medium that cannot operate without his or her continuous participation, and in which [...] the speed of the spectator's bodily movements determines the speed of the images, demonstrates the remarkable extent to which the human body is figured as an intimate and integral component of the cinema.
>
> (Barker 2009: 134)

Media theorist Marshall McLuhan, meanwhile, suggests that film cameras are developed directly from the wheel: 'The movie camera and the projector were evolved from the idea of reconstructing mechanically the movement of feet. The wheel that began as extended feet... took a great evolutionary step into the movie theater' (McLuhan 1994: 181–2).

From the first film onwards, cinema is full of bicycles and cyclists. Just as they are part of the fabric of everyday life, they are also ubiquitous components of the *mise-en-scène* of film after film, from mainstream comedies and dramas through to avant-garde experiments such as the surrealist film, *Un Chien Andalou* (Buñuel, Dalí, 1929), which features a male cyclist in a nun's habit, crashing bizarrely outside his lover's apartment, or the expressionist classic, *From Morn to Midnight* (Martin, 1920), which situates a six-day velodrome race at the centre of a nightmarish story about the destructive lure of money. While some films are concerned with depicting cycling cultures – such as *2 Seconds* (Briand, 1998), an indie drama about couriers – and others are concerned with capturing the experience of cycling – especially documentaries about riders suffering for their sport, such as *Time Trial* (Pretsell, 2017), or *Moon Rider* (Dencik, 2012) – for many filmmakers, bicycles are passing details, props or minor elements in the background of the image. Cinematic cycles are often useful narrative devices, rather than objects of fascination in themselves, and, apart from promotional films or children's fantasy films like *Pee-Wee's Big Adventure* (Burton, 1985) and *The Sky-Bike* (Frend, 1967), which feature marvellously improbable contraptions, it is rare that films pay attention to cycle design.

Nevertheless, although many films about bicycles are uninterested in detailing the history of specialized cycling cultures, they offer us a frame through which to study the social and cultural significance of cycling. In placing an image of a bicycle on screen, a film transforms it from an object into a symbol. In doing so it invites the viewer to ponder its meaning, to consider the associations that a bicycle carries (along with its riders). A bicycle is a signifying machine, a vehicle for different meanings, and cinema brings this into focus, treating bicycles not merely as functional modes of transport, but as images that are loaded with meaning and require interpretation. Cinematic bicycles are often devices through which filmmakers explore questions of identity; in *She's Gotta Have It* (Lee, 1986), for example, the immaturity and irresponsibility of Spike Lee's character Mars Blackmon is encapsulated by his job as a cycle messenger, sporting a cap with the logo of the Italian 'Brooklyn' cycling team. Cinema treats cycling as an expressive activity and, in doing so, reveals to us that riding a bike, or a related vehicle, is a performance that, in conveying a rider to a particular destination, is also conveying meanings to a spectator. The title of Lance Armstrong's autobiography (2001), *It's Not About the Bike*, describes perfectly the way that cinema treats cycling. Cycling on film is, typically, not about the bike, but is a means to explore a wide range of topics, including individual identity and the body, gender, sexuality and racism, family and community, politics and injustice, capitalism and technological progress, urbanization and environmental change.

In this chapter I will outline some of the dominant themes with which cinematic depictions of cycling are concerned: comedy, work, sport, gender and childhood.

Comedy

The spectacle of people riding – and falling off – bicycles is a rich source of comedy for the cinema. There is hilarity in the indignity of the cyclist's sweating, labouring body, backside protruding in the air, and in the mismatch between body and bicycle. Films find humor in images of large people balanced precariously on little bicycles, of too many people riding a bicycle made for one, or of bicycles that have a mind of their own (Chapter 46). If cycling is a matter of keeping one's balance, the spectacle of men and women struggling to stay upright lends itself perfectly to film comedy, but there is an additional principle of film comedy on display here where humour is derived from the hapless protagonist's unsuccessful encounter with machines. These range from the simple mechanism of a window sash or a revolving door through to escalators, locomotives and the factory machinery that swallows Charlie Chaplin in *Modern Times* (1936). Film comedy offers us images of characters who are repeatedly defeated by what Hollywood director Frank Capra called the 'intransigence of objects', and for screen comedians, the bicycle is a particularly intransigent object that rarely goes where the rider wants or is liable, as Laurel and Hardy discover when using one as a getaway vehicle in *Duck Soup* (Guiol, 1927), to disintegrate suddenly (Dale 2000: 10). These slapstick films whose characters are locked in a perpetual struggle to master even such everyday objects as an umbrella or a bicycle express the wearying experience of wrestling with the complexities of modern life far more effectively than any earnest, allegorical drama.

The most systematic exploration of the comedy of the bicycle is found in the films of Jacques Tati, which consistently examine the social costs of the preoccupation with speed and efficiency. For example, his first feature film, *Jour de Fête* (1949), is set in a sleepy French village during a visit from a travelling fair, and follows the local postman François (played by Tati) as he cycles lazily around the area on his rounds. Tricked into getting blind drunk by the unpleasant fairground workers, he wobbles home at night in a virtuoso display of trick cycling before passing out in a railway truck. The following day, inspired by a documentary about the US postal service's use of motorbikes, planes and helicopters, he races furiously around his postal route, misdelivering his letters, overtaking a peloton of racing cyclists, heading cross-country before, inevitably, crashing into a river. Like all of Tati's films, *Jour de Fête* is a gentle commentary on the absurdity and alienation of contemporary life, its argument crystallized in a single shot where a large black car screeches to a halt as François wheels his bike across the street in the centre of the village. François stops, staring at the impatient driver who is revving his engine, and then continues forward, forcing the driver to wait until he has passed. In the context of this film, the bicycle symbolizes tradition, continuity, and the nonchalant French character, while the car symbolizes speed, modernity and the violence of social change.

Work

Cycling is work. No matter how fit the cyclist or how light the machine, riding a bicycle demands physical effort. As is documented throughout the history of cinema, however, the bicycle is also a means by which people can do work in the economic sense. Indeed, among the various meanings the image of the bicycle conveys, perhaps the dominant connotation is work, and the Lumières' factory-gate film cements the association between labor and the bicycle. Tracking the movement of bicycles across cinema allows us to see both the variety of ways in which work is conceptualized and the changing nature of labor since the late 1800s. However, the bicycle has an ambiguous significance with regard to work, since, although it is firmly associated with employment, it also promises escape from work. This idea is encapsulated by the government-funded documentary, *Spare Time* (Jennings 1936), a poetic record of the varied leisure activities undertaken by laborers in heavy industry such as coal miners and cotton mill-workers. One sequence shows Sheffield steel workers riding into the countryside on their bikes and drinking in a village pub, although, as Theodor Adorno observed, reflecting upon the emergence of the expression 'free time' and 'spare time', the belief that areas of everyday life in capitalist society are 'autonomous', was misguided: 'Free time is shackled to its opposite [...] free time depends on the totality of social conditions, which continues to hold people under its spell' (Adorno 1991: 187). In offering temporary escape from wearying labor, Adorno suggests, a crucial function of free time is to allow people to work harder by offering brief periods of recovery. Like other leisure activities, cycling enables many of us to cope with work.

The most celebrated film to explore the relationship between cycling and work is undoubtedly *The Bicycle Thieves* (*Ladri di Biciclette*) (DeSica, 1948). Concerned with the desperate precarity of working-class life in impoverished post-war Rome (Chapter 19), the film has had a pervasive influence on world cinema. The film's influence is due partly to its radical embrace of realism, which involved casting non-professional actors, extensive location shooting, and eschewing formulaic narrative structures. The most well-known example of the 'neo-realist' films produced by Italian film-makers in the 1940s and 1950s, it exemplifies the movement's preoccupation with the city and urban life, with working-class experience and with cinema's political potential.

The simple narrative follows the fortunes of Antonio Ricci, who secures a job as a poster hanger and has to retrieve his bike from the pawn shop in order to take up the position. Tragically, Ricci's bike is stolen on his first morning at work and he spends the rest of the film, accompanied by his patient son, trying to recover the bike in increasing desperation before finally attempting to steal someone else's. The concluding scene makes clear that crime is not a moral failure, but a consequence of the vicious cycle of poverty wrought by a combination of fascism and years of war.

The film has been globally influential partly because it demonstrates that a formally sophisticated, affecting artwork can be made using very limited resources, and partly because this film fable of mobility, masculinity and economic precarity is so translatable. Among the significant films derived from de Sica's film fable are Satyajit Ray's *Pather Panchali* [*Song of the Road*] (1955), Ousmane Sembène's *Borrom Sarret* [*The Cart Driver*] (1963), one of the first films by a black African director which highlights the inequalities of postcolonial Senegal, and *Muerte de un Ciclista* [*Death of a Cyclist*] (Bardem, 1955), a bitter critique of bourgeois hypocrisy in Franco's Spain. *Beijing Bicycle* (2001), by sixth-generation Chinese director Wang Xiaoshuai, translates the story to modern-day Beijing and centers on a teenage migrant worker who moves from the countryside to work as a cycle courier before the brand-new mountain bike he was issued with is stolen. Like *The Bicycle Thieves*, it describes a society in transition – in this case one that is being reshaped by post-socialist China's embrace of global capitalism – foregrounding the direct impact of social transformations upon some of the poorest inhabitants: migrant workers and children.

The bicycle messenger is an ideal cinematic protagonist, visually articulating cinema's twin preoccupations – the body in movement and the city. The messenger is identified with an intriguing, romanticized subculture, and their job also lends itself to tense narrative drama. It is physically dangerous, involves fleeting contact with an array of different characters and the transgression of boundaries between separate social spheres from high finance to the criminal underworld, and, like the plot of a thriller, is organized around fixed deadlines. For all the romanticization of the messenger's maverick status, however, and an intoxicating sense of individual freedom and fluid movement, the bicycle and the rider remain intrinsic components of the economic system, consigned to what Jeffrey L. Kidder discusses as 'dirty work' or 'edgework' (Kidder 2006).

Cycle couriers skirt the edges of many films about modern cities but *Quicksilver* (Donnelly, 1988) is the film that establishes the template for this genre. It tells the story of Jack Casey, an exceptionally successful stockbroker who makes a tremendous loss in a single day, losing all his parents' savings in the process, and goes on to quit his job and start work as a cycle courier. Early in the film, Jack explains the attraction of the role to a former colleague who is trying to persuade him to return to finance: 'now all I'm responsible for is me. I pick up here, I deliver there. It's simple. Nobody worries, nobody gets hurt. I pay my rent. I got a coupla extra bucks. If I need more, I work a few more runs. Look, when I'm on the street I feel good, man. I feel exhilarated. I go as fast as I like, faster than anyone. The street sign says "One-way east", I go west.' It is an expression of transgressive individualism and the pleasures of speed but, as Lars Kristensen observes, the notion of the bicycle courier's freedom rests on the contradiction that 'the economy that these urban warriors refuse to take part in is actually upheld by their bicycle work' (Kristensen 2013: 256). Moreover, as Kristensen comments, although messengers may extol the freedom offered by their job, the bicycle courier exemplifies a troubling model of post-industrial neoliberal labor in which the boundary between work and leisure is erased as work extends beyond the confines of the working day and leisure too becomes capitalized and exploitable: 'Neoliberalism packages this as a fashionable lifestyle, a desirable living condition for a free individual. And, true to the dictum, compared to his previous life as a broker, Casey believes in his happiness' (ibid.: 258).

However, as Adorno observed presciently, 'unfreedom is gradually annexing 'free time', and the majority of unfree people are as unaware of this process as they are of the unfreedom itself' (Adorno 1991: 188). This is evident in the ways that work penetrates free time – with mobile devices that make it difficult for workers ever to leave work at the end of the working day, for example – and it is also evident in the ways that leisure activities are incorporated into work – as represented in the job of the bicycle messenger where the pleasures of cycling are co-opted and exploited through low-paid, unregulated piece-work. A New York-based messenger interviewed in *The Godmachine*, Jan Steffen's documentary about global cycle-courier culture reflects that, 'I've seen over the years, like, ten, fifteen guys die. Messengers, you know? [...] It's kinda like this big machine you know, and people get grinded up in it every once in a while.' This comment challenges assumptions that shifting from industrial production to a 'knowledge economy' or 'information economy' results in emancipation, equality and the end of work. Indeed, the cycle couriers that populate cinema remind us of the vast army of workers doing poorly paid, physically dangerous, sometimes deadly work in order to keep the modern city moving.

Sport

A rather different form of work, the high-speed, dangerous sport of competitive cycling, lends itself well to cinematic representation, and some of the earliest glimpses of bicycles on screen are in depictions of road races in films by the Lumières. Sport cycling offers the viewer the voyeuristic pleasures of scrutinizing athletic bodies in motion, and it enhances the spectatorial pleasures of viewing sport, allowing us to study details of faces and bodies in slow-motion, and from multiple angles (Vignette F). Whether it is the pretext for formal experimentation – in films like *La Course en Tête* (1974), Joël Santoni's kaleidoscopic film portrait of Eddy Merckx – or a more formally conservative document of an event or portrait of an athlete – such as *The Flying Scotsman* (Mackinnon, 2006) or *The Program* (Frears, 2015), biopics of Graham Obree and Lance Armstrong respectively – sports films are vehicles for the symbolic examination of national identity, the technological or cybernetic augmentation of the body, the delineation and policing of gender boundaries, and the strained relationship between individualism and responsibility to a team. Whether we play sports or not, as sociologist Jean-Marie Brohm contends, 'it is through the model of sport that the body is understood in practice, collectively hallucinated, fantasised and individually experienced' (Brohm 1978: 5). Cinema's treatment of cycle racing provides us with a particularly vivid examination of the disciplined, productive, machinic body in late capitalism.

The celebrated documentaries of the Danish director Jorgen Leth offer some of the most comprehensive accounts of cycle racing in documentary cinema. *Stars and Water Carriers* (1974), which follows the Giro d'Italia, *A Sunday in Hell* (1976), about the Paris-Roubaix race, and *The Impossible Hour* (1975), about an attempt on the hour record, convey the ways in which cycling is sustained both by extremes of passion and idealism and also mundane financial arrangements. They provide the spectator with intimate access to the stages in similar ways to TV coverage, with cameras in cars and on motorbikes following the racers and generating a dynamic variety of shots. In addition, the films take us inside hotel rooms and restaurants, allowing us to watch riders relaxing, receiving massages, eating and discussing tactics. However, rather than demystifying the sport, the films celebrate the way that the participants in major cycle races are all engaged not just with athletic competition, or with earning money, but with the construction of an epic mythic narrative. As Leth explains in his voiceover to *Stars and Water Carriers*, the riders are performers: 'It's their job to participate in a ritual play of great beauty and depth. They're actors on a classical stage.'

Gender

In their fascination with cyclists' bodies, Leth's films draw attention indirectly to a feature of the culture of sport cycling, which is its uncritical celebration of masculinity. This is especially striking given the popular association of the safety bicycle with Victorian women's campaigns for enhanced rights and greater political representation. The image of a well-dressed woman on a bicycle is an enduring symbol of the social transformations that were taking place in industrialized countries at the end of the nineteenth century, and the bicycle was the emblem of the suffragettes and the 'New Woman' (Chapters 2 and 40). This is captured by Cecil Hepworth's 1899 single-shot film, *Floral Parade of Lady Cyclists*, which documents a ride through London by dozens of women members of the Catford Cycling Club, all dressed in full skirts and hats. However, while the image of a woman on a bicycle is an enduring sign of the optimism of first-wave feminism, it is also an image laced with ambiguity, a symbolic threat to the established social order. As Marilyn Bonnell observes, for the Victorians, 'the bicycle became a terrible reminder of the disruptive power of women, a power which they feared existed and preferred not to witness' (Bonnell 1990: 229). A survey of images of women cyclists in cinema indicates that, depressingly, this figure remains a troubling one. For instance, the documentaries *Hardihood* (Hahn, 2002) and *Half the Road* (Bertine, 2014), which deal with women's mountain bike racing and road racing, respectively, both tackle the persistent history of the exclusion of women athletes and the absurd notion of female fragility that is used to justify this discrimination.

One of the most remarkable films to explore the significance of the bicycle as a device for challenging conventional concepts of gender is *Wadjda* (2012), the first feature film by a Saudi woman director, Haifaa al-Mansour (Vignette K). Wadjda is a 10-year-old girl who spots a beautiful bike in a toy shop and sets out to earn the money to buy it. An unscholarly non-conformist, she resolves to win a competition to recite the Quran, but her teacher confiscates the cash prize when she learns of Wadjda's intentions, telling her, 'a bike isn't a toy for girls. Especially not for well-behaved, devout girls who protect their soul and honour.' Eventually, her mother buys her the bike in a rebellious gesture when Wadjda's father marries another woman, and the final shots of the film are imbued with optimism as Wadjda races past her male friend and arrives at a junction, considering which direction to head in. It is a hopeful ending but it is also frustrating that, a century on from the Hepworth film, the image of a young woman on a bike still symbolizes a demand for urgent political change.

Children

In addition to their function as, variously, means of transport, tool for work, sports apparatus, exercise equipment, badge of social class and symbol of gender, cinema also reminds us that bicycles are associated with childhood as a means of establishing independence through greater mobility and joyful escape from the constraining gaze of parental authority. This is evident in films about, and *for*, children and young people, and perhaps the most well-known example is one of the highest-grossing films of all, Steven Spielberg's fantasy blockbuster, *E.T.: The Extra-Terrestrial* (1982). The film tells the story of a boy, Elliott, who happens upon an alien in his back garden after it is stranded on Earth when its spaceship left abruptly. Elliott hides the alien from his mother, naming it 'ET', and takes care of it, with his siblings' help, feeding and teaching it. The alien is eventually located by shadowy government forces, but before they can take ET away, Elliott and his friends rescue the creature, using their BMX bikes to escape from the police and federal agents, racing through their Californian suburban estate and into the woods where ET boards the returning spacecraft. Although brief, the cycling sequences are a principal source of pleasure in the film, producing a sense of free, transgressive movement as the children take their bikes off-road, follow shortcuts, and run rings around the adults, outpacing them and even riding across a police car as if it were a jump on a BMX track. As they navigate the sweeping topography of suburbia they follow improvised routes that betray a different knowledge and experience of their spatial environment from that of the lumbering, car-grown-ups. This is epitomized most directly in the two scenes in which, due to the alien's magical power, the bicycles and their riders literally take flight. These sequences articulate the idea of bikes as an expression of independence for children as effectively as any film, but what makes them so arresting for adult viewers, is the way that they also evoke the thrilling experience of learning to ride. Elliott and his friends are elated but terrified and the scenes evoke the moment a child first manages to balance on a bicycle. 'The bike always starts with a miracle', writes author Paul Fournel, recalling the moment at which he rode away from a parent. 'And then one morning I no longer heard the sound of someone running behind me, the sound of rhythmic breathing at my back. The miracle had taken place. I was riding. I never wanted to put my feet back down for fear that the miracle wouldn't happen again' (Fournel 2004: 26).

Writing in the 1970s, cycling historian Andrew Ritchie reflected that.

> We take the bicycle too much for granted, but at the same time, we in England do not use it enough. It is a simple machine. But it *could* have a far-reaching and revolutionary effect on the world in the next century
>
> *(Ritchie 1975: 179)*

The bicycle's revolutionary potential rests partly in its practical functionality, but also, as the cinematic history of cycling demonstrates, in the way that it allows us to see, think about, and inhabit the world differently. Over a century after the Lumières introduced their new device by screening a film in which workers cycled out of the gates of a factory, cinema continues to insist upon the power of the bicycle to turn our world upside down.

References

Adorno, T. (1991) *The Culture Industry* (London: Routledge).

Armstrong, L., with Jenkins, S. (2001) *It's Not About the Bike: My Journey Back to Life* (London: Yellow Jersey Press).

Barker, J. (2009) *The Tactile Eye: Touch and the Cinematic Experience* (Oakland, CA: University of California Press).

Bonnell, Marilyn (1990) 'The Power of the Pedal: The Bicycle and the Turn-of-the-Century Woman', *Nineteenth Century Contexts*, 14(2): 215–239.

Bordwell, D., Thompson, K. (2003) *Film History: An Introduction* (New York: McGraw-Hill).

Brohm, Jean-Marie (1978) *Sport: A Prison of Measured Time* (London: Ink Links).

Dale, A. (2000) *Comedy Is a Man in Trouble: Slapstick in American Movies* (Minneapolis: University of Minnesota Press).

Day, J. (2015) *Cyclogeography: Journeys of a London Bicycle Courier* (Honiton: Notting Hill Editions).

DeCordova, R. (1990) 'From Lumière to Pathé: The Break-Up of Perspectival Space', in *Early Cinema: Space, Frame, Narrative*, ed. Elsaesser, T. 76–85 (London: BFI).

Fournel, P. (2004) *Need for the Bike*. trans. A. Stoekl (Lincoln, NE: University of Nebraska Press).

Gunning, T. (2004) 'Pictures of Crowd Splendour: The Mitchell and Kenyon Factory Gate Films', in *The Lost World of Mitchell and Kenyon: Edwardian Britain on Film*, eds. Toulmin, V., Russell, P., Popple, S. 49–58 (London: BFI).

Kidder, J.L. (2006) '"It's the Job that I Love": Bike Messengers and Edgework', *Sociological Forum*, 21(1), 31–54.

Kristensen, L. (2013) 'Work in Bicycle Cinema: From Race Rider to City Courier', in *Work in Cinema and the Human Condition*, ed. Mazierska, E., 249–264 (New York: Palgrave Macmillan).

Luiselli, V. (2014) *Sidewalks* (Minneapolis: Coffee House Press).

McLuhan, M. (1994) *Understanding Media: The Extensions of Man* (Cambridge, MA: MIT Press).

Ritchie, A. (1975) *King of the Road: An Illustrated History of Cycling* (London: Wildwood House).

Further Reading

Bennett, Bruce (2019) *Cycling and Cinema* (London: Goldsmiths Press).

PART VIII

Cycling in literature

An introduction

Una Brogan

H. G. Wells was first compelled to turn from academic to creative writing as a result of learning to ride a bike. Wells recorded that while living at Woking in 1895:

> I learnt to ride my bicycle upon sandy tracks with none but God to help me; he chastened me considerably in the process, and after a fall one day I wrote down a description of the state of my legs which became the opening chapter of *The Wheels of Chance* [1896].
>
> *(Wells 1969 [1934]: 543)*

Mark Twain (1992 [1884]), in a similar vein, responded to the steep learning curve of learning to ride a high-wheeler by penning the humorous essay "Taming the Bicycle" (1884). Bicycles, like books, take us on journeys that require imaginative and energetic input from riders and readers alike, and the connection between these two pursuits has been established since the earliest days of the pursuit.

While the literary and cultural role played by other modes of transport such as walking, the railway, cars and planes has been widely examined (see, for instance, Beaumont and Freeman 2007; Desportes 2005; Jarvis 1997; Schivelbusch 1986 [1979]; Solnit 2000; Wallace 1994; Wenzel and Strasen 2010), the bicycle's rich contribution to culture has long been sidelined, just as cyclists still struggle to be given space on the roads. Thankfully, that balance has begun to be redressed in recent years, with this volume contributing to a growing body of academic work on the social, cultural, geographical and economic significance of cycling (see Furness 2010; Withers and Shea 2016). The chapters in this section aim to provide a rapid bird's-eye view of the vast range of literature that the bicycle has inspired since its invention, while considering texts on cycling within the wider social and cultural context of their production.

Many of the first cyclists in the late 19th century instantly recognized the rich aesthetic potential of this new means of transport, which opened up a range of urban and rural environments to a new public and offered an embodied, yet mechanized form of progression which blended multisensory stimulation with speed. As the chapters in this section highlight, this literary legacy has persisted to the present day, with novelists, poets and travel writers continuing to draw inspiration from cycling. As Justin Belmont's chapter outlines, the particular embodied experience of riding continues to fascinate and challenge us, spurring poets to

DOI: 10.4324/9781003142041-65

attempt to express their experience through verse. Belmont draws on a selection of better and lesser-known poets and establishes a compelling connection between the hard graft of cycling and that of reading and writing verse. My own chapter focuses on a specific aspect of the literary character of the bicycle: its role in provoking laughter and providing opportunities to tie the crucial element of humor into storytelling. Using Bergson's theory on humor, I show how the bicycle is uniquely well placed to be funny, due to the way in which it mingles the human and the mechanical, while combining distinct geographical and temporal spheres.

A human-powered vehicle, the bicycle was well placed to inherit aspects of the literary tradition associated with walking established by Romantic poets in the early 19th century. Edward Nye's chapter traces the genesis of French literature inspired by bicycles while highlighting the connections between early cycling literature and the Romantic aesthetic. Notably, cycling provokes a feeling of physical and psychological liberation, while encouraging both introspection and close observation of the environment. The first cyclists revived the Romantic "pedestrian" tradition associated with attentive bodily progress, while also being uniquely placed to pay attention to the increasingly mechanized, industrial character of their surroundings, just as Romantic poets bore witness to the transformation of the landscape as a result of the first Industrial Revolution.

Yet from the outset cyclists have looked both forwards and backwards. While affording an intimate connection with nature and the past, the bicycle was a thoroughly modern machine that became associated with visions of the future and the budding genre of science fiction at the start of the 20th century. Jeremy Withers's chapter takes us on a journey through historical and contemporary fiction that foregrounds the bicycle as an instrument of liberation and positive social change. As he shows, the bicycle has consistently accompanied characters on their journeys to greater autonomy. Withers explores this in a range of young adult fiction which integrates the bicycle as a means to achieve physical freedom, and in contemporary novels which foreground the bicycle as a means for gender nonconformists to liberate themselves from strictly defined categories of identity.

Dave Buchanan's chapter provides an overview of a further genre of writing that appeared at the outset of cycling and continues to flourish in the present day: cycle-travel literature. As Buchanan shows, authors of cycle-travel literature respond to much more than an impulse to simply record the facts of the journey itself. His categorization of these travel writers into "pilgrims", "ramblers" and "adventurers" provides a compelling means of conceptualizing the genre, while pointing to the literary and aesthetic sensibility shared across these categories. Whether seeking to connect with a specific person, place or event from the past, attempting to achieve a transcendental experience or engaging in extraordinary feats of endurance, authors of cycle-travel literature draw meaningful inspiration from bike journeys and establish myriad connections between the acts of writing and riding.

The stories we tell about bicycles are important. The way that we integrate this freedom-giving, accessible, levelling, environmentally friendly form of transport into our cultural narrative plays a major role in deciding the role we give the bicycle in society. Scholars on cycling literature have thus far tended to focus on the European and North American context, and the chapters in this section are unfortunately no exception to that trend. There are clear blind spots that need to be addressed in order to come to an understanding of the global significance of writing on bicycles as we move into a low-carbon future. The hegemony of the car since the mid-20th century has been bolstered through the cultural imagination of motorized vehicles as an outward symbol of freedom, power and social standing. Though the bicycle has always sent its riders' imaginations spinning, it has never yet come to dominate our cultural narratives nor our streets in the way that the car continues to do. Reconquering a cultural and imaginative space for the bicycle is as crucial as, and goes hand in hand with, defending its right to be on the road.

Bibliography and Further reading

Beaumont, M., and Freeman, M. J. (2007). *The Railway and Modernity: Time, Space, and the Machine Ensemble*. Oxford; New York: Peter Lang.

Desportes, M. (2005). *Paysages en mouvement : transports et perception de l'espace, XVIIIe–XXe siècle*. Paris: Gallimard.

Furness, Z. (2010). *One Less Car: Bicycling and the Politics of Automobility*. Philadelphia: Temple University Press.

Jarvis, R. (1997). *Romantic Writing and Pedestrian Travel*. Basingstoke: Macmillan Press.

McGonagle, S. (1969). *The Bicycle in Life, Love, War, and Literature*. South Brunswick: Pelham Books.

Nye, E. (2000). *A bicyclette: anthologie*. Paris: Sortilèges.

Pierre Thiesset, P., and Quentin Thomasset, Q. (2013). *Les bienfaits de la vélocipédie: anthologie*. Vierzon: le Pas de Côté.

Schivelbusch, W. (1986 [1979]). *The Railway Journey: The Industrialization of Time and Space in the 19th Century*. Trans. Anselm Hollo. Berkeley: University of California Press.

Solnit, R. (2000). *Wanderlust: A History of Walking*. New York: Viking.

Starrs, J. E., and Schaeffer, K. (1997). *The Literary Cyclist*. New York: Breakaway Books.

Twain, M. (1992 [1884]). "Taming the Bicycle" in *Collected Tales, Sketches, Speeches & Essays, 1852–1890*. New York: Literary Classics of the United States, 892–96.

Wallace, A. D. (1994). *Walking, Literature, and English Culture: The Origins and Uses of Peripatetic in the Nineteenth Century*. Oxford: Clarendon Press.

Watson, R., and Gray, M. (1978). *The Penguin Book of the Bicycle*. Harmondsworth: Penguin Books.

Wells, H. G. (1969 [1934]). *Experiment in Autobiography*, vol. 2. London: Jonathan Cape.

Wells, H. G. (1896 [1935]). *The Wheels of Chance, A Bicycling Idyll; The Time Machine*. London: J. M. Dent.

Wenzel, P., and Strasen, S., eds. (2010). *Discourses of Mobility, Mobility of Discourse: The Conceptualization of Trains, Cars and Planes in 19th- and 20th-Century Poetry*. Trier: WVT.

Withers, J., and Shea, D. P., eds. (2016). *Culture on Two Wheels: The Bicycle in Literature and Film*. Lincoln: University of Nebraska Press.

44

THE BICYCLE AND THE CREATIVE PURSUIT IN FRENCH LITERATURE

Edward Nye

The earliest fiction on the bicycle (or at least its ancestor, the velocipede) is likely to be French. In 1818, the young dramatist Eugène Scribe (1818), destined to be one of the most successful of his generation, staged his comedy, *Les Vélocipèdes, ou la poste au chevaux*. It was the perfect subject for his style of light-hearted contemporary social commentary: a fashionable new invention claiming to be an alternative to the horse, one which was fast, tiring, eccentric and foreign. By the end of his play, the high expectations that this new invention gave rise to are dashed when the German mechanic Monsieur Fiacrenbourg finally arrives astride his draisine, late and utterly exhausted from his short journey. The moral of the tale, sung by all the characters in the finale, is that the grand designs of great men are often child's play. Even if Scribe appears sceptical about the future of the draisine, he was clearly attracted to it for its potential on the stage and as a literary theme.

After the short-lived craze for the draisine, the next craze in the 1860s for the cranked velocipede (or 'vélo de Michaux', after one of the assumed inventors of the pedal) inspired one of the most interesting and influential French poets of the nineteenth century to write the first significant poem in French on the bicycle in July 1868. This was Théodore de Banville's 'Le Vélocipède':

> Half wheel and half brain,
> Such is the man-velocipede.
> He goes, docile as a calf,
> Half wheel and half brain.
> He, a new animal, mocks
> Buffon and Lacépède!
> Half wheel and half brain,
> Such is the man-velocipede.
> (Banville, 'Le Vélocipède', 1875: 104, my translation)

There are a number of reasons why this poem is so significant. Typical of Banville's poetry, it uses content and form to hark back to distant times but also forwards to modernity. The highly formal 'triolet' versification (which our literal translation cannot render) is a medieval invention which Banville consciously seeks to promote in many of his collections, but it also

DOI: 10.4324/9781003142041-66

gives the poem a machine-like feel which is entirely fitting for a poem on the new invention of the cranked velocipede. The reference to the eighteenth-century naturalist Buffon, who famously wrote that the horse is 'man's greatest conquest' suggests that the velocipede is the new horse, that this modern invention is an extension of past human achievements. The reference to the early evolutionist Lacépède, however, suggests that the vélocipède is something completely new, akin to the metamorphosis of one species into an entirely different species. The contrasts in this poem speak of an era of rapid change in France, both economically and politically, which contemporaries could respond to either by holding on to the past or looking forward to the future. For Banville, this tension was also palpable in poetry; his virtuoso formal versification claimed its roots in centuries of poetic precedent, but it also preceded Rimbaud's revolutionary invention of French free verse in the 1870s. In subsequent literature on the bicycle, the dichotomy between past and present becomes a familiar refrain, but Banville is the first to articulate it.

Before the cranked velocipede was banned by local ordinance from the streets of many French towns and cities in 1869 (Hadland and Lessing 2014: 64, 69), and before the Franco-Prussian War of 1870 brought a juddering halt to the cultural and technical progress of this new machine, there was a brief flowering of enthusiastic literature. The most well-known is by Richard Lesclide, the future secretary to Victor Hugo. His *Manuel du Vélocipède* (1869) is a guide to the cultural significance of the new invention as much as its practical use. *Le Tour du monde en vélocipède* (1870) is a fervent and animated fictional tale of a ride from Paris to western Siberia, first serialized in the bi-monthly magazine he founded in 1869, *Le Vélocipède illustré* (1869–72), almost the first of its kind in France.

There is very little other contemporary literature on the velocipede, until the popularity of the 'safety bicycle' in the 1890s provided a new source of inspiration. From this point on, however, the bicycle has been an evergreen subject for novelists, poets, dramatists, essayists and journalists (Chapter 45). The abundance of writing on the bicycle since the 1890s begs the question: what is it, exactly, which makes it such a good literary subject? The answer varies with the rise and fall of different literary movements, but there is one distinct constant: the intense psycho-physical sensations experienced by the rider. Since the Romantic movement, Western literary culture has considered the key characteristic and objective of literature to be the externalization of internal experience, to 'ex-press' or 'turn out' what is within, or to paraphrase the well-known treatment of the subject by M.H. Abrams in *The Mirror and the Lamp* (1953): to allow the light burning within us to shine outwards so that others can share it. Literature thus conceived seeks out experiences which stimulate reactions and changes in one's internal world, experiences which kindle the light within us and cause it to shine more brightly. Cycling turns out to be one of these experiences; it is a tool for 'ex-pression' of inner feeling. The Romantic movement *qua* movement came and went, but the principle of 'ex-pression' has remained with us in one way or another ever since. When, therefore, the development of the bicycle accelerated in the latter part of the nineteenth century after several decades of relative stasis, sensibilities were well primed with Romantic values to exploit its literary potential.

In this sense, the essence of literature on the bicycle is Romantic. Underlying much writing on the bicycle is what English readers would think of as a Wordsworthian aesthetic, and what the French would associate with Lamartine. Neither of these poets would have taken to one particular aspect of the bicycle, its speed, something which enthralled later writers. Yet both sought inspiration from the interaction between nature and their physical place in it or movement through it, much as writers on the bicycle later did (Chapter 32). The hypnotic effect of one's own repetitive physical movement cuts one's mind off from everyday concerns while facilitating concentration and inner harmony, fostering a meditative quality essential to poetry. Wordsworth's 'wise passiveness' is also the cyclist's (Wordsworth 1798: 183–5). Lamartine's intimate connection with nature which he addresses with the familiar 'tu', is also the cyclist's (Lamartine 1820: 46–9). This Romantic aesthetic, combined with the modernity of speed, is fundamental to all the texts we shall discuss.

As one might expect from these Romantic roots, writers on the bicycle at the end of the nineteenth century were poets and travellers (Chapter 48). Sometimes they were both. Edouard de Perrodil was a record-holding cyclotourist who published accounts of his epic rides in the 1890s in Algeria, from Paris to Milan, or from Paris to Vienna, and who later worked as a journalist for periodicals like *Auto-vélo* and *Le Figaro*. His first piece of bicycle writing, however, is a poem in his first published work, *Les Echos* (1886). It is a hymn to the agile, aerial, mythical grace of the bicycle ('her' in the stanza below) which concludes, as so often in contemporary poems on this theme, with the poet's feelings for the special rapport with the natural space around him:

> Amazed, across the plains,
> I ride without end
> Breathless with her.
> Visions of nymphs,
> Subtle sprights and dryads,
> Inspired by the limitless space,
> The fields and woods like a kingdom
> To be crossed at will.
> (Perrodil, 'La Bicyclette', 1886; my translation)

The first complete collection of poems in French on the bicycle was *La Bicyclette* in 1896 by Jules Riol. Very little is known about this author, other than that he was probably also a travel writer. Recurrent themes of physical and spiritual exaltation, of transcending time and space, and of the figure of the rider-poet can be found in these concluding lines:

> Thus the key to true riches
> Is within you who regenerate
> My senses, my muscles, my body;
> Space has boundaries no longer,
> The city imprisons no more,
> And the soul of the poor poet
> Ever anxious for infinity
> Sees the horizons widen:
> Long live my bicycle!
> (Riol, 'Cueillette poétique', 1896; my translation)

Time and space in 19th-century literature have been well studied; the increase in urban living, the growth of rail travel, the invention of telegraphic communication all impacted temporal and spatial experience. Among such contemporary inventions, the bicycle is the most individualistic, and consequently has an unusual propensity for subjective perception of space and time. This feature is all the more pronounced because of the unique position of the bicycle rider who, despite animated physical exertion, remains a seated passenger watching the world fly past. Passive contemplation of vivid sense impressions has echoes of a Romantic aesthetic.

It is reminiscent of Rousseau sitting for hours on the banks of Lake Bienne contemplating the gentle ebb and flow of the waves (Rousseau 1782). It leads poets like Riol to explore their internal, subjective perception of space and time.

One of the earliest cycling novels in French is by Maurice Leblanc, *Voici des ailes* in 1898, two years after the first novel in English by H.G. Wells, *The Wheels of Chance*. Leblanc later become better known for the highly successful series of novels and short stories he started writing in 1905 about the gentleman thief Arsène Lupin, inspired by E.W. Hornung's Raffles character. Before he devoted almost two decades of his career to crime fiction, however, he was an avid writer on the bicycle, and an avid cyclist. In the 1890s, he cycled long distances round his native Normandy and indeed from Paris to his home on the white cliffs of Étretat. He wrote fervent accounts of his cycling experiences for the newspaper *Gil Blas* in a column entitled 'Elle' (the feminine pronoun designating 'la bicyclette'), and *Voici des ailes* was serialized in this newspaper before publication as a novel. It is the tale of two couples who embark on a cycling tour of Normandy which is physically and mentally draining, until they realize that the configuration of couples they started with was wrong. Once they exchange partners, passionate love blossoms and, reinvigorated, their bicycles seem to fly along the roads, as the title (*Here are Wings*) suggests. Liberation from the constraints of social and moral convention comes through cycling; the characters feel like caged birds who can finally fly: 'our wings are growing [...], we will soar though the air like immense and serene birds [...], our free souls have wings!' (Leblanc 1898: 145–6). Future writers return frequently to this theme of flight, both in the sense of escape from oppression as well as aerial motion. It is worth remembering that when Leblanc wrote this novel, the Wright brothers had not yet achieved the first manned flight. Leblanc is therefore appealing to the readers' imagination and sense of mythical awe at the liberating qualities of the bicycle which defies the force of moral gravity.

French literature can lay claim to one of the most unusual pieces of writing on the bicycle in any language: Alfred Jarry's *Le Surmâle, roman moderne* (1902). A scientist invents a performance enhancing drug he calls 'perpetual-motion-food' (in English in the text) with which he hopes to produce the eponymous 'supermale' who would be capable of astonishing feats of sexual prowess. To test the new drug, he organizes a race between a high-speed train and a five-man bicycle from Paris to Siberia and back again. The cyclists are connected to each other at the knee by aluminium bars, they tow behind them a small trailer carrying a dwarf, and the whole contraption travels at up to 250 km/h in the slipstream of a driverless, torpedo-shaped pacesetting car. One of the cyclists dies during the race, but since his legs are joined to the other riders, he appears to continue pedalling. The 'quintuplette' bicycle beats the train, but both are overtaken by a flying funnel which later is found to be the 'supermale'. Jarry was an avant-garde author who always set out to shock. His most famous work, the play *Ubu roi*, was partly a parody of Shakespeare's *Macbeth* and partly a warped children's fairy tale, and caused uproar because, among other obscenities, it opens with (and subsequently repeats) the word 'merdre', the French for 'shit' with an additional 'r'. Jarry was unique in his era, but was later claimed as one of their own by the Surrealists, the avant-garde dramatists, and by the school of Absurdist writers. *Le Surmâle* defies simple interpretation, but at its heart is a battle between the human and the machine, an ambivalence toward modernity which, on the one hand, can enhance human potential but also risks leaving humanity behind. France industrialized much later than did Great Britain, so the ambivalence toward the machine which shows in the poetry of Wordsworth in the 1830s appears in French literature much later, in the last decades of the century. Jarry's novel is an extension of this ambivalence. The bicycle riders are augmented human beings who prove the worth of humanity over the machine, but do so by conjoining themselves to a very particular kind of machine, the bicycle, which remains more human than most because it requires the continual participation of the human mind and body to drive it forward.

The bicycle is an individualist machine, but it is not a socially isolated one. It brings a heightened sense of self, but it also brings a heightened sense of one's relation to others. Proust's masterwork, *À la recherche du temps perdu* (1913–1927), is renowned for its exploration of the highly subjective perception of an individual's world, but it also manages to paint a wide social panorama of *Belle Époque* society in which the bicycle makes an appearance (Chapter 41). When the narrator meets his future lover Albertine in the second volume, *À l'Ombre des jeunes filles en fleurs*, she is pushing a bicycle. She and her cycling friends burst into the narrator's consciousness like a flock of raucous seagulls crossing his path along the beach (Proust 1988a: 146). Through the bicycle, Proust evokes a particular social milieu of rising bourgeoisie, the personal confidence which shows in their physical and social ease, in the cyclist's gaze which faces resolutely ahead, 'happy and a little persistent', almost impertinent (Proust 1988a: 199). The evocation is also of an individual, Albertine, the 'Bacchant on a bicycle' (Proust 1988a: 228) with her coarse manners which manifest themselves, for example, in the way she refers to her bicycle with the term 'bécane' (Proust 1988a: 231). This slang word for 'machine' reveals a combination of affection and possessiveness for her bicycle that she will later show to Marcel and which will be the undoing of their relationship. In the narrator's imagination, she is a 'being of flight' who 'flies by on her bicycle' from one lover to the next (Proust 1988b: 576, 600). While for some in this period a woman on a bicycle was controversial, Albertine's use of 'bécane' shows how liberated she is from bourgeois conventions, and how perfectly she is in possession of herself, her bicycle, her lover, and her bisexuality, which the narrator later discovers.

Proust's contemporary, Jules Romains, also uses the bicycle as a link between the profoundly personal and the social, but on very different principles. The title of his novel *Les Copains* (1913) suggests that it is about 'friends', but is in fact about a much more esoteric notion of individuals banding together in perfect harmony, forming a 'union' of souls. For a quarter of a century and in 27 novels, Romains employs this collectivist literary aesthetic which he called 'Unanimisme' in which characters never act as individuals and narrative never dwells on individual introspection. It could hardly be more different to Proust's narrative. The challenge for Romains is to find catalysts which awaken in individuals a consciousness of their group identity. In *Les Copains*, the catalyst is the bicycle. Bénin and Broudier cycle alongside each other in the penumbra of the evening on their way to a hostel where they will spend the first night of their cycling journey. Their conversation gradually wanes and is replaced by an unspoken feeling of togetherness as they cycle alongside each other:

> Occasionally one of them touches his brake slightly so as not to overtake the other [...]. Bénin rides on the left, Broudier on the right. Soon there is no more right or left. There is Bénin's side and Broudier's side.
>
> *(Romains 1913: 92)*

Thus, group consciousness or the 'collective soul' in this novel arises through friendship, and friendship arises in turn from shared experiences like the sensations of movement, speed and the spectacle of nature provided by the bicycle.

Romains's interest in the way cycling can foster a collective spirit is more the exception than the rule in French literature. The tendency is more toward the individualism of the bicycle, even its rebelliousness. The novelist Julien Gracq's 1950 pamphlet *La Littérature à l'estomac* rails against the chattering literary classes in France who judge a book according to how much it is being talked about rather that what it is like to read. He thinks too many readers have forgotten that reading a piece of fiction is an emotional experience, not one based on rules, conventions and public opinion. The reader either bonds personally with an author or rejects him or her. Gracq compares the relationship between author and reader to that of the motorbike pacing the 'stayer' cyclist in his slipstream: if the cyclist makes the requisite effort to stay as close as possible, s/he will feel the sensation of lightness and freedom that comes from moving as one with the pacemaker; they will be a single, well-oiled block of speed, until the last page is turned and the author-pacemaker abruptly 'turns off the gas', leaving the reader-stayer dazed and drained from the passionately all-consuming experience. In this analogy, the reader-stayer follows no one but the author, and only follows an author if s/he has enough passionate commitment to him or her. Hence the title of this pamphlet: 'A gut feeling for literature'. Once again, the bicycle stands for authentic depth of personal feeling, for opposition to the force of convention. Gracq was true to his principles when, the year after he published this pamphlet, he was awarded France's highest literary prize, the Prix Goncourt, which he promptly refused.

Knowing the politics and personality of Jacques Perret, maybe his short story *Le Vélo* (1955) is also fundamentally more about a certain libertarian outlook than simply a colorful account of the wonders and then demise of his bicycle. In his typically flamboyant, touching but ironic style, he extols the exhilarating speed of the bicycle which feels much faster than any other mode of transport:

> It is not my habit to regulate my existence according to the opinion of a man with a stopwatch [...]. Up to a point, and in certain circumstances, I'm ready to agree that speed has some connection to time, sometimes even to space, [...] but I simply cannot understand cycling in the way that a technician of the theory of relativity would, or as a setter of high school examination problems would [...]. All I want is the immediate data of cycling consciousness [...] which proves the absolute velocity of my bicycle. [...] It is primarily a matter of sensation. I perceive speed through my eyes, my ears, my skin, the pit of my stomach [...]. I do not disagree that cars travel quickly, but in that case the true emotion is reserved for the car, whereas on a bike it is for me.
>
> *(Perret 1955: 77–8; my translation)*

Perret obliquely cites the title of Henri Bergson's *Essai sur les données immédiates de la conscience* (1889, translated in English as *Time and Free Will: An Essay on the immediate data of consciousness*), a highly influential philosophical treatise concerning the difference between quantifiable and non-quantifiable human experience and the way our freedom depends on it. Perret's account of the subjective sensations of cycling could quite easily figure in Bergson's treatise as an instance of non-quantifiable experience which no one can regulate for us and which is consequently a fundamental part of our freedom. The 'immediate data of cycling consciousness' cannot be 'regulated by a man with a stopwatch' because there is no way of objectively quantifying the sensation of the wind in your hair, the vision of the landscape rushing past in a fluid, almost cinematographic cadence of images, or the feeling in the pit of your stomach which is the thrill of the speed. In other words, your consciousness is your own, perhaps even the essence of your existence, and no matter how many things seek to control it, it is fundamentally your domain of freedom. Bergson's philosophy obviously appeals to Perret, as doubtless did Bergson's evocative and highly figurative style of writing which, in an odd way, is not so far from his own. Perhaps Perret's tendency to find himself seriously at odds with contemporary society is as much a product of his concept of freedom as his view of the bicycle is. He wrote articles for extreme right-wing newspapers attacking, among other things, human rights, parliamentary democracy, the Second Vatican Council, Algerian independence and De Gaulle. Such was the virulence of his writing amid the crisis, riots, bombing and attempted military putsch in 1961 that President De Gaulle stripped Perret of his war medals and his civil rights. Harmless and wholesome as the bicycle appears to be in comparison to this, it often seems to be associated with a spirit of dissent. Not every cyclist has extreme politics, but how many go through a red light?

A subdued example of this combination of the harmless bicycle and dissenting spirit is to be found in the play by the Québécois author Roch Carrier, *La Céleste bicyclette* (1980). The sole character of the play is a nameless actor who explains in monologue to his audience that he has been committed to an asylum because he stands by his explanation to doctors that he broke his leg when he fell off his bicycle as it flew him through the sky. He invites the audience to join him in his celestial flight, and the audience is never quite sure whether to interpret his story as insanity or a poetic vision. Many spectators would conclude the latter, knowing that Roch Carrer was one of a generation of writers in the 1960s and 1970s who formulated a new national literary identity in opposition to oppressive and inward-looking agrarian and Church traditions. *La Céleste bicyclette* is an invitation to launch oneself into the wide blue yonder, to leave *terra firma* behind and embrace change and new experience. As we have seen before, the bicycle evokes feelings of flying, escape, freedom; the risk of falling off is worth taking, as every young child learning to ride knows.

It is hard to write about French literature and cycling without mentioning the Tour de France. For the first fifty years of its existence, it had more readers than spectators. It was a race created in the minds of those who read the daily articles published in the magazines and newspapers like serialized stories during the three weeks of the event. Radio in the 1930s was an oral storytelling experience. Television, and more recently the internet, have made it more of a visual and less of a verbal narrative, but this does not seem to have dampened the interest from writers who continue to be inspired by it. There were those, like Albert Londres (1996), Antoine Blondin (1988) or Jacques Perret (2005), who literally followed the tour in its peregrinations around France, writing copy for the newspapers in their hotel room after the day's finish. There are others like Paul Fournel (2002, 2012) who do not necessarily follow the caravan, but who nevertheless feel the need to translate the race into a narrative. Roland Barthes wrote from a physical and intellectual distance, mulling over the extraordinary socio-cultural power of this event. In *Mythologies* (1957), he takes the Tour de France as an example of how societies create modern myths: cyclists are given epic names, their rivalry described in Homeric terms, the geography is personified and becomes a supernatural friend or foe, and the whole event becomes a kind of Odyssey. Other authors have brought out the symbolic value of an event which follows the borders of an entity which has always needed defining: the Republic. There is topological symbolism in a 'tour' of a country which every year for more than a century has linked together the social, cultural, historical and political 'landmarks' which constitute either the actual or the desired sense of national identity, the collectively imagined sense of self. The very name for the race was probably chosen to recall the practice of the journeyman artisans, the 'compagnons', who travelled France as part of their apprenticeships. The potential for civic education and patriotic edification of such a 'tour' had already been exploited in Feuillée's bestselling *Le Tour de la France de deux enfants* (1877), a didactic story of two children leaving the Alsace-Lorraine region recently annexed by Germany and travelling through France to find relatives. The cycle race is yet another 'tour' of France which promotes a sense of national identity. Even clichés are sometimes true: the Tour de France is not just a race; it is invested with so much meaning that it is not surprising that authors write about it. It is one of the driving forces behind more than two centuries of French literature on the bicycle. Like most ideologies, however, it sometimes needs a little humorous deflation, for which purpose San Antonio's *Vas-y-Béru!* (1965) and Aymé's *Le Dernier* (1934) are without equal.

References

Abrams, M.H. (1953) *The Mirror and the Lamp: Romantic Theory and the Critical Tradition*, New York: Oxford University Press.

Aymé, Marcel (1934) *Le Dernier*, in *Le Nain*, Paris: Gallimard.

Banville, Théodore de (1875) 'Le Vélocipède', *Les Occidentales*, in *Œuvres poétiques complètes*, ed. Peter J. Edwards et al., Paris: Honoré Champion, 1994–2001, 9 vols (Vol. 5).

Barthes, Roland (1957) 'Le Tour de France comme épopée', in *Mythologies*, Paris: Le Seuil.

Blondin, Antoine (1988) *L'Ironie du Sport. Chroniques de* L'Équipe *1954–1982*, Paris: Bourin.

Carrier, Roch (1980) *La Céleste bicyclette*, Montreal CA: Stanké.

Feuillée, Augustine (1877) *Le Tour de la France de deux enfants*, Paris: Belin.

Fournel, Paul (2002) *Besoin de vélo*, Paris: Points.

Fournel, Paul (2012) *Anquetil tout seul*, Paris: Seuil.

Gracq, Julien (1950) *La Littérature à l'estomac*, Paris: J. Corti

Hadland, Tony and Hans-Erhard Lessing (2014), *Bicycle Design. An Illustrated History*, Cambridge, MA: M.I.T.

Jarry, Alfred (1902) *Le Surmâle, roman moderne*, Paris: éditions de la Revue blanche.

Lamartine, Alphonse de (1820) 'Le Lac', in *Méditations poétiques*, Paris: Librairie grecque-latine-allemande

Leblanc, Maurice (1898) *Voici des ailes!*, Paris: Ollendorf.

Lesclide, Richard (1869) *Manuel du Vélocipède, publié par le Grand Jacques*, Paris: Librairie du Petit Journal.

Lesclide, Richard (1869–1872) *Le Vélocipède illustré.*

Lesclide, Richard (1870) *Le Tour du monde en vélocipède, par le Grand Jacques*, Paris: Librairie de la publication.

Londres, Albert (1996) *Les Forçats de la route*, Paris: Arléa.

Perret, Jacques (1955) 'Le Vélo', in *Le Machin*, Paris: Gallimard.

Perret, Jacques (2005) *Articles de Sports*, Paris: La Table ronde.

Perrodil, Edouard de (1886) 'La Bicyclette', in *Les Echos*, Paris: Melun, A. Beauvais.

Proust, Marcel (1988a) *À l'Ombre des jeunes filles en fleurs*, Deuxième partie, in *A la Recherche du temps perdu*, 7 volumes, Paris: Gallimard, Vol. 2.

Proust, Marcel (1988b) *La Prisonière*, in *À la Recherche du temps perdu*, 7 volumes, Paris: Gallimard, Vol. III.

Riol, Jules (1896) 'Cueillette poétique' in *La Bicyclette, monologue en vers dédié au Touring-Club de France*, Paris: Lanée.

Romains, Jules (1913) *Les Copains*, Paris: Gallimard.

Rousseau, Jean-Jacques (1782) 'Cinquième promenade' in *Les Rêveries du promeneur solitaire*, Geneva, n.p.

San Antonio [Frédéric Dard] (1965) *Vas-y Béru!* Paris: Fleuve noire.

Scribe, Eugène, (1818) *Les Vélocipèdes, ou la poste au chevaux, à-propos-vaudeville en un acte*, first performed at the Théâtre des Variétés on May 2, 1818; Œuvres complètes de Eugène Scribe, 75 vols, Paris: E. Dentu, 1874–1875 (series 2, Vol. 4).

Wordsworth, William (1798) 'Expostulation and reply', *Lyrical Ballads*, London: J.A. Arch.

45

THE LIBERATING BICYCLE IN LITERATURE

Jeremy Withers

During the Black Lives Matter demonstrations roiling the United States in the late spring and summer of 2020, the bicycle once again displayed its frequent connections to liberation. Across dozens of cities such as New York, San Francisco, and Atlanta, cyclists rode in solidarity protest rides. In Seattle, cyclists formed "Bicycle Brigades," phalanxes of bikes creating protective barriers for demonstrators protesting racially motivated violence. Of course, bicycles can be tools of oppression and marginalization too (Hoffman 2016; Cox 2019: 19, 30; Chapter 3). For example, during some of these demonstrations, police weaponized bikes as battering rams to be deployed against protesters.

But ever since the bicycle's invention in the 19th century, it has more often been associated with people's quest for more freedom. Initially, the bike liberated people from costly draft animals and train tickets. It freed people from confinement to their village or city. Perhaps most famously, the bicycle was championed by suffragists such as Susan B. Anthony, Elizabeth Cady Stanton and Frances Willard as integral to liberating women from patriarchal power (see, however, Norcliffe 2018, for a challenging of this narrow assessment). In a famous quote, Anthony (1896: 10) declared that the bicycle "has done more to emancipate women than anything else in the world." More recently, bikes have been celebrated for contributing to freedom from fossil fuels and overly sedentary, unhealthy lifestyles. This chapter focuses on some ways in which 19th-, 20th- and 21st-century English literature (with some brief forays into French literature) portrays the bicycle's relationship to liberation in terms of age, gender identity and economic class. More specifically, it demonstrates how literary genres ranging from young adult literature and science fiction, to realism and song lyrics, proclaim the emancipatory power of bikes for adolescents, gender nonconformists, and the working class.

Adolescents, bikes, and liberation

As Holdsworth (2014: 421) notes, childhood is associated "with freedom – that is, freedom to move around and play away from the surveillance of adults and freedom to develop one's own sense of self-hood in preparation for adulthood." Increasingly in the post-World War II era, countless advertisements, articles, television shows, films, and other texts have celebrated the bicycle as a machine associated with children and their desire for some form of independence

DOI: 10.4324/9781003142041-67

(see Turpin 2018: 158–92; Longhurst 2015: 152–85). Unfortunately, these repeated connections between childhood and bicycles have led some to malign adult cyclists as abnormal or immature (Parsons 2002; O'Rourke 1987). Nevertheless, while waiting to obtain what is perceived as the greater freedom granted by a driver's license, children – especially adolescents – often embrace the bicycle as a way to break free of the boundaries erected around them and seize a modicum of private space. Unsurprisingly, young adult (YA) fiction contains several notable examples of adolescents using bikes to shatter barriers and gain independence, to find solace from their confusing or painful lives.

Robert Cormier's *I Am the Cheese* (1977) and Beverley Brenna's *The White Bicycle* (2012) both portray bicycles as representing and contributing to freedom from oppressive authority figures. The former work focuses largely on the journey of teenager Adam Farmer, who is biking from his home in Massachusetts to visit his hospitalized father in Vermont. Although his journey is often rife with danger – cars, bullies, and predatory adults sometimes veer too close – Adam deliberately chooses the bike over a Greyhound bus because the latter would make him feel too "confined" (Cormier 1977: 6). While on his bike, he feels like he can "join the wind" (Cormier 1977: 8), as if he has learned to flow with the independence of elemental nature. But in a dramatic twist-ending, Cormier reveals that Adam is incarcerated in a government mental asylum, and the bike ride was only one of Adam's delusions. Although Adam does indeed own a bike, he rides it only around the hospital grounds. Near the novel's end, Adam says, "I glance outside through the gates. Someday I will ride my bike out there" (Cormier 1977: 224). Whenever Adam imagines being free from the hospital, it is atop his bicycle. This machine is a vital component of his mental world, a signifier of that still-independent part of himself that his government captors and interrogators repeatedly try and fail to access and control.

Brenna's *The White Bicycle* follows the story of Taylor Jane, a teenager with Asperger syndrome who lives with an overbearing mother. Several times in the narrative, when awake and when dreaming, Taylor Jane rides a white bicycle only to find that when the trail gets too rough, she must carry the bike, an act that signifies her stifled independence, her inability to move in the world as she pleases. Yet her dreams conclude with her "[putting] the white bicycle down… [slipping] onto the seat, and [putting her] feet to the pedals" before "[skimming] along home" (Brenna 2012: 92). This association between a rolling bicycle and liberation exists outside of Taylor Jane's dreams too as she often uses modes of transport such as bike rides (and bus rides) to demonstrate her independence from her mother. Like Cormier, Brenna presents the bicycle as antithetical to authoritarianism, as an essential tool for preserving a modicum of freedom from oppressors.

The bicycle's connection to adolescent freedom continues in other works of YA fiction such as Shirley Hughes's *Hero on a Bicycle* (2017). Set in Nazi-occupied Florence, Italy, during World War II, the book focuses on Paolo, a boy fond of sneaking out for middle-of-the-night bike rides to "escape… from the boredom and from the pinched wartime austerities of his home" (Hughes 2017: 5). Later in the book, the bicycle's initial association with mental liberation morphs into its association with more physical forms of liberation. After encountering some members of the Partisans (i.e., the Italian resistance movement) during one of his nocturnal bike rides, Paolo soon uses his bike to try to help a couple of Allied POWs escape Nazi captivity. His plan fails. But, in another scene, Partisan leader Il Volpe successfully uses the speed and nimbleness of Paolo's bike to escape Nazi execution. *Hero on a Bicycle* ends by shifting back to the bike's relationship to mental freedom as Paolo – now living in newly liberated Florence – still uses his bike "to be on his own, to try to get his thoughts in order" (Hughes 2017: 212).

In Clémentine Beauvais's *Piglettes* (2017) and Ryan Andrews's *This Was Our Pact* (2019), other adolescent characters similarly use bikes for independent mobility. But instead of exploring their own city or the hills surrounding it at night like Paolo does, the main characters in *Piglettes* and *This Was Our Pact* journey to exotic, faraway worlds. *Piglettes* follows a trio of French overweight, cyber-bullied girls as they embark on a weeklong bicycle trip from their provincial town of Bourg-en-Bresse to Paris, the opulent capital that none of them has ever visited (for more on young girls and freedom through bicycle mobility, but in a Middle Eastern context, see Al-Mansour 2015). Although the older brother of one of the girls accompanies them, for the most part these girls, while astride their bicycles, experience freedom like they have never known before. In some remarks she makes before the trip, the mother of Astrid (one the three girls) gestures toward the close connection between kids, bikes, and freedom: "I think that the journey is an excellent idea. People today are much too afraid of letting children roam freely, sleep in the countryside and explore the world" (Beauvais 2017: 78). The bicycle trip liberates the girls in several ways. The bond these new friends form while cycling together frees them from the social isolation they have endured because of their looks. Pedaling day in and day out, amid downpours and heat, they free themselves and their steadily growing audience (both virtual and in person) from any doubts that overweight bodies cannot also be robust and resilient ones. And two of the three girls – Mireille and Hakima – are eventually purged of their ignoble feelings of vengeance that motivated them to go on the trip.

As opposed to the splendid, yet ultimately somewhat commonplace locale of Paris, the locale where the two main characters in Andrews's *This Was Our Pact* journey to a more magical place. Ben and Nathaniel take off on their bicycles following the paper lanterns that their town annually floats downriver as part of its autumn festival. After the boys cross a bridge – "the barrier that all our parents made [them] promise never to cross" (Andrews 2019: 20) – on their bikes, they begin a marvelous and dreamlike journey, entering a world populated by a talking bear, a supernaturally massive crow, a magic potion maker, and other wonders. This YA graphic novel highlights how bicycles function as machines that help liberate young people from mundane realities by transporting them to strange but enchanting new worlds – worlds that provide characters with fresh opportunities to learn about themselves, form new friendships, and acquire important knowledge (a scenario similar to that found in science fiction novels such as Ridley 1926; Fischman 1979; Sirota 1991; Anthony 1991).

Gender nonconformists, bikes, and liberation

The perception that bikes destabilize gender identity in scandalous or unflattering ways has a long history. In the Victorian era, many observers expressed outrage over what they perceived as the masculinization of women who cycled (Friss 2015: 160–85). In more recent decades, American popular culture has often derided adult male cyclists for not being properly "masculinized through… motorized transportation" (Furness 2010: 111). But rather than portray this destabilization as objectionable, some texts celebrate a connection between bicycles and liberation from gender solidity. They present cycling as a preferred form of mobility for people who exhibit forms of transgenderism. As one influential gender theorist notes, a transgender body is one that is in motion, one that refuses to be defined "in relation to a destination, a final form" (Halberstam 2018: 4). The transgender body – like the cycling body – represents opposition to a "concrete land of fixity and stasis" (Halberstam 2018: 132).

In the late-Victorian, anonymously authored *Wheels: A Bicycle Romance* (1896), a bicycle tour across Europe that a group of Americans take becomes an opportunity for several main characters to experiment with gender variance. The novel opens with two upper-class sisters, Georgiana and Helena, planning to "play man" (*Wheels* 1896: 41) while cycling the Old World. But they plan to do more than just wear pants, like many women cyclists of the era were doing. Instead, "for fun" and "a little novelty" (*Wheels* 1896: 183), they adopt the appearance, names, mannerisms, and habits of men. While crossing the Atlantic, the two women – now known as George and Carl – meet Mr. and Mrs. James Montgomery Fayette Browne, who also will be cycling across Europe. The four team up for the cycling trip. At the end of *Wheels*, the reader learns that Mrs. Browne is also a gender bender: "She" is really James Montgomery, the law partner of Mr. Browne who adopted his female persona for reasons never clearly explained. Besides creating an opportunity for people to relish performing a different gender, the bike tour also allows queer desires to percolate, with some of the characters periodically expressing their attraction to the same-sex qualities of other characters. Bicycle mobility, then, liberates these characters from the rigid gender and sexuality norms of upper-class Victorian society, becoming a vehicle to explore "a transgender fantasy" and "shades of alternative sexuality" (Smethurst 2015: 95).

This Victorian era trend of aligning the bicycle with gender fluidity continues in one of the masterpieces of 20th-century fiction, Marcel Proust's multivolume *In Search of Lost Time* (1913–1927). As Una Brogan (2016: 124) argues, "Proust's narrative provides a subtle, subversive, and very modern portrait of a fluid spectrum of gender and sexual identities." In particular, the character of Albertine employs her bicycle to perform a new gender identity different from the one typically permitted women in *fin de siècle* France (the narrative's setting). Throughout much of the narrative, Albertine presents herself as a hybrid of male and female traits, an androgynous cyclist who often talks, dresses, and cycles in a way that is conventionally coded as masculine, but who at other times displays conventional feminine beauty and behavior. As Brogan (2016: 125) puts it, the link between "mobile cyclists and fluctuating gender reoccurs" across several volumes of this work. In later volumes, Albertine becomes imprisoned in the narrator's house, thus losing her vital connection to the bicycle and its liberating mobility. But for a while at least, the bike helps her escape the constrictive gender identities of society.

As Halberstam (2018: 21) observes, "transgenderism has long been situated as a site of futurity and utopian/dystopian potential." Fittingly, science fiction has recently contributed several notable works that also foreground an association between bicycle and gender mobility (in addition to the texts below, see Rogue 2020). E.L. Bangs's "From an Interview with the Famed Roller Sarah Zephyr Cain" (2014) focuses on the life of the eponymous Roller – a bicycle courier – in a futuristic, postapocalyptic America shattered by radioactive landscapes, political collapse, and energy scarcity. Sarah has transitioned to a new gender identity of female, much to the confusion of others who wonder why someone would voluntarily choose to become the "weaker sex" in such a harsh, brutal world. But Sarah believes that her way of life, cycling across the United States and having adventures, embodies something vital that has been "inherited from the world that ended before us: the idea of freedom" (Bangs 2014: 24). And Sarah's bicycle is integral to her identity as a free, transgender person for a couple of reasons.

First, a bike is one of the few functioning modes of transport left in her ravaged, post-oil world that is capable of carrying her away from "the cruel, theocratic town" (Bangs 2014: 22) where she grew up, that is, away from a place that persecutes people like her. Moreover, a bike transports her to one of the few places still performing hormone therapy for transgender people, a place Sarah describes as a "bright Oz of greenhouses where genetically engineered phyto-estrogens and androgens grow on trees, and where there are more doctors, and fewer stupid questions, than in most places" (Bangs 2014: 22). For Sarah, a bike is essential for fleeing a malevolent environment and traveling to a place that accepts and helps her. Second, as Sarah describes it, the bike carries over the "infinite mobility" from the pre-apocalypse world, with its freedom and its feeling of being "untethered from any one corner" (Bangs 2014: 24). Such language parallels the nonessentialist, nonfixed view of gender held by transgender people and other gender nonconformists. Sarah understands her gender to be just as mobile as her bike, and is most content when both can be as unhindered and nomadic as she wants.

Kim Westwood's *The Courier's New Bicycle* (2011) also employs a postapocalyptic setting and features a gender-nonconforming bike messenger: Salisbury, an androgynous courier who identifies as both male and female. Salisbury delivers illegal hormone drugs around a near-future Melbourne that is shattered by a flu pandemic, energy scarcity, ecological disruption, and plummeting fertility rates. Although the intolerant theocratic group Nation First controls Melbourne – a group that persecutes "gender transgressor[s]" (Westwood 2011: 209) like Salisbury and their transgender bike mechanic friend Albee – Salisbury still feels like their combination of male and female elements, their position between those two poles, constitutes "a freedom, albeit a dangerous one" (Westwood 2011: 19). When cycling, Salisbury feels like their body can perform beyond the limitations of the female-sexed body they were born with (a feeling also occasionally savored by Emmeline, the muscular, cross-dressing, cycling protagonist in Street 2016). Instead, because Salisbury felt like their body was masculine too, they have always embraced, not suppressed, their nonfeminine "physiology, made for sport" (Westwood 2011: 8). This athletic, "lean and lanky girlboy" (Westwood 2011: 14) body "makes cycling lots of fun and [Salisbury] a useful employee" (Westwood 2011: 8) because they are fast, nimble, and tireless when on their bike.

The liberation Salisbury feels when on their bicycle manifests itself in other ways too. Since they were a child, they have used bikes to mitigate feeling "really anxious, or really angry" (Westwood 2011: 221), a feeling often roused by their experiences with being misgendered. Also, due to Melbourne's various climate, energy, and political crises, laws now regulate transportation and movement. As Salisbury states, however, when on their bike, "I can get anywhere I want in the inner city, regardless of traffic regulations and curfews, and I can do it at speed" (Westwood 2011: 8). In sum, Salisbury's agile and resilient gender identity finds a counterpart and close ally in the agile and resilient bicycle.

An important function of these texts, then, is to help free readers' minds from a stagnant, monolithic view of what a typical or "correct" cyclist looks like. As one transgender bike mechanic has lamented, "Anyone who doesn't present as male always has to work harder to prove their worth to other mechanics and to the cyclists" (Seplavy 2020: 79). Another transgender cyclist who led a bicycle advocacy group noted of his first national meeting of bicycle advocacy leaders: "I was shocked to find that 80% or so of the groups was straight white cis-dudes… As a mixed race, gender non-conforming butch dyke, I definitely did not see myself reflected in this group of leaders at all" (Rivera 2020: 101). But texts like Bangs's "From an Interview with the Famed Roller Sarah Zephyr Cain" and Westwood's *The Courier's New Bicycle* promote an understanding that the gender identity of cyclists can be found at any point on the gender spectrum. Demolishing stereotypes about who does or does not belong on a bike is vital for getting and keeping more people on bicycles by making them feel part of an inclusive community.

The working class, bikes, and liberation

At the dawn of the bicycle age, the bike was a machine that reinforced boundaries between various economic classes. The earliest bicycle – the draisine, invented in 1817 – was an elitist plaything that enjoyed a brief popularity only among upper-class dandies and Ivy League students. Later iterations of the bike, such as midcentury velocipedes and late-century high wheelers, were also reserved for those in the more privileged classes, who could afford their high costs and had the leisure time to ride them. But, with the arrival of the safety bicycle in the mid-1880s, and with its declining price and increasing availability in second-hand markets throughout the 1890s, the middle and lower classes (collectively, the working class) finally had access to this transformative machine.

Although late-Victorian workers often used bicycles for practical purpose such as commuting from homes to work, they also used bikes for leisure activities – just like their wealthier peers. People from all classes embraced cycling for recreation during the 1890s, blurring at times class distinctions. Perhaps no other literary work captures the bicycle's social leveling power than does H.G. Wells's *The Wheels of Chance* (1896) (although it depicts such power as ephemeral). In the novel, a store clerk whose life is balanced precariously between a lower-class and middle-class existence takes a ten-day cycling holiday. The protagonist, Hoopdriver, savors how much his bicycle frees him from the drudgery of his low-paying job and how people cannot discern his class background when he is on his bike. In one memorable scene, a lower-class character refers to Hoopdriver as a "bloomin' Dook" (i.e., a bloody Duke), contemptuous words that thrill instead of offend Hoopdriver because they suggest he can pass for a more upper-class gentleman when cycling. "But now, was he not a bloomin' Dook, palpably in the sight of common men?," the narrator asks. "Then round the corner to the right – bell banged furiously – and so along the road... Whoop for Freedom and Adventure!" (Wells 2018: 32). Although at the novel's end, Hoopdriver must return to a job that barely allows him to live above poverty, when he was on his bike, he could feel and be treated like a duke (as well as a gentleman, a knight errant, and so forth). In other words, he could escape the feelings of insignificance and powerlessness that often torment those not born into wealth and privilege.

As Wells's *Wheels of Chance* also demonstrates, during the late 19th and early 20th centuries, many working-class people – confined most of the time to the cramped, polluted cities where they lived and worked – relied on bicycles to escape to the countryside whenever possible (see also Perry 1969, for how bikes helped the working class already living in the countryside to expand their world and escape rural isolation). As Hoopdriver embarks on his journey, he joyfully anticipates the "scores of miles... before him, [of] pinewood and oak forest, purple, heathery moorland and grassy down, lush meadows, where shining rivers wound their lazy ways" (Wells 2018: 39–40). Urban workers believed that these more verdant environments replenished their bodies and minds, freeing them momentarily from the enervating effects of cities and from the reality of their own limited power within the capitalist system.

In *The Type-Writer Girl* (1897), Grant Allen portrays Juliet, a young, impoverished typist, as using her bicycle to flee to the rural counties surrounding London, where she can experience feelings of ownership in a life in which she owns little but the bare essentials. "The sky overhead is mine," Juliet enthuses during one of these rides, "mine the road underfoot... the long sweep of the swift launching himself on the air from the battlements of the church-tower. All these I own, by virtue of my freehold in the saddle of my bicycle" (Allen 2003: 109). Juliet's rhapsodic remarks call to mind those of other middle-class cyclists such as those in the socialist Clarion Cycling Club (founded in 1895). The Clarions, too, composed verses in the mid-1890s about their forays into restorative nature atop the wheel: "Where thrushes sing and the busy bee hums, / Far from the stinking, stifling slums, / We'll pitch our tents by a troutling stream, / Sink all sorrow, nor think of the morrow, / But look on life as a happy dream" (qtd. in Pye 2014: 43).

Besides contributing to the present freedom of the working class, the bicycle also laid the foundations for greater liberation that might soon come. The Clarion Scouts, a subset of the larger Clarion movement, used bicycles to spread the gospel of socialism (Horton 2009: 5–7).

To do so, they often cycled "twenty to fifty miles on Saturdays and Sundays to address public meetings in towns and villages which had, as yet, no Socialist organisations. Cyclist supporters could paste walls and fences with stickers bearing Socialist slogans" (Pye 2014: 28). This zeal for a socialist utopia that would liberate workers from capitalist exploitation is captured in the "Song of the Clarion Scout" (published in the May, 1895 *Scout* journal). The song's rhyming couplets express the hard work needed to create future freedom: "What tho' the weather be cold as an icicle, / Bravely [the Scout] clings to his Clarion bicycle / Scattering leaflets, sticking up labels, / Filling a breach at old hostelry tables" (qtd. in Pye 2014: 29). Such songs and poems of the Clarions celebrate the bike as an integral tool for spreading the propaganda that might engender a critical mass of socialists ready to reject the oppressions and injustices of capitalism.

In post-Victorian literature, the intersection between bicycle mobility and working-class liberation become less pronounced. But it does appear several times in the form of that avatar of a supposedly more liberated form of labor: the bicycle messenger. Although this job is notoriously low-paying, dangerous, and physically demanding, within the cultural mythology of countries like the United States and the United Kingdom it has become synonymous with freedom (Finchman 2006, 2007; Kidder 2011). Unlike their higher-earning white-collar counterparts chained to desks in drab offices, bike messengers get paid to do what they love: ride bikes outside in the open air. The bike messenger rolls across the pages of several noteworthy works that range in genre from sentimental realism (Saroyan 1943) to science fiction (Gibson 1993), YA literature (Dorfman and Dorfman 2003), and memoir (Culley 2001). In each of these works, the bicycle helps a certain subset of the working class acquire freedom and bliss, even when their jobs do not give them financial wealth or social prestige. Regarding the freedom acquired by the bike messenger in Gibson, the narrator explains, "Sometimes, when she rode hard... Chevette got free of everything: the city, her body, even time. That was the messenger's high" (1993: 131). Similarly, the young messenger in Dorfman and Dorfman finds the grueling work of all-day cycling emancipating. Bicycle mobility frees him momentarily from being consumed with anxiety about the many crises and catastrophes unfolding in the world, such as "[melting] polar ice caps, slaughter overseas, lost votes... AIDS" (Dorfman and Dorfman 2003: 30).

Conclusion

Since the modern bicycle's arrival in the late-Victorian age, it has often functioned as a freedom machine. The bike has liberated many people from an array of constraining circumstances. Literature reflects this elevation of the bicycle as being more than a mere plaything, a fitness device or practical mode of transport. An impressive number of literary texts highlight the ways in which this deceptively simple machine can also be a tool of liberation. This chapter has focused on three groups that literature has represented as benefiting from the emancipatory power of bikes: adolescents, gender nonconformists and the working class. For these people, bikes are vital for breaking free of pernicious boundaries, accessing more wholesome and restorative spaces, destabilizing hierarchies, and providing an agile, nonimperious mobility. Rejecting the ubiquitous automobile – a machine that often divides, harms, and oppresses – these literary texts celebrate people riding to liberation at the speed of a bicycle.

References

Allen, G. (2003). *The type-writer girl*. Clarissa Suranyi (ed.) Peterborough, Ontario: Broadview.
Al-Mansour, H. (2015). *The green bicycle*. New York: Dial Books.

Andrews, R. (2019). *This was our pact*. New York: First Second.
Anthony, P. (1991). *Mercycle*. New York: Ace.
Anthony, S. B. (1896). "Champion of her sex." *New York World* (February 2), p. 10.
Bangs, E. L. (2014). "From an interview with the famed roller Sarah Zephyr Cain," in E. Blue (ed.) *Bikes in space: volume 2*. Portland: Elly Blue, pp. 21–7.
Beauvais, C. (2017). *Piglettes*. London: Pushkin Press.
Brenna, B. (2012). *The white bicycle*. Markham, ON: Red Deer Press.
Brogan, U. (2016). "Albertine the cyclist: a queer feminist bicycle ride through Proust's *In search of lost time*," in J. Withers and D. Shea (eds.) *Culture on two wheels: the bicycle in literature and film*. Lincoln: University of Nebraska Press, pp. 116–35.
Cormier, R. (1977). *I am the cheese*. New York: Alfred A. Knopf.
Cox, P. (2019). *Cycling: a sociology of vélomobility*. London: Routledge.
Culley, T. H. (2001). *The immortal class: bike messengers and the cult of human power*. New York: Villard.
Dorfman, A., and Dorfman, J. (2003). *Burning city*. New York: Random House.
Finchman, B. (2006). "Bicycle messengers and the road to freedom," *The Sociological Review* 54, no. 1, pp. 208–22.
Finchman, B. (2007). "'Generally speaking people are in it for the cycling and the beer': bicycle couriers, subculture and enjoyment," *The Sociological Review* 55, no. 2, pp. 189–202.
Fischman, B. (1979). *The man who rode his 10-speed bicycle to the moon*. New York: Richard Marek.
Friss, E. (2015). *The cycling city: bicycles and urban America in the 1890s*. Chicago: University of Chicago Press.
Furness, Z. (2010). *One less car: bicycling and the politics of automobility*. Philadelphia: Temple University Press.
Gibson, W. (1993). *Virtual light*. New York: Bantam.
Halberstam, J. (2018). *Trans*★*: a quick and quirky account of gender variability*. Oakland: University of California Press.
Hoffman, M. (2016). *Bike lanes are white lanes: bicycle advocacy and urban planning*. Lincoln: University of Nebraska Press.
Holdsworth, C. (2014). "Child," in P. Adey, D. Bissell, K. Hannam, P. Merriman, and M. Sheller (eds.) *The Routledge handbook of mobilities*. London: Routledge, pp. 421–28.
Horton, D. (2009). "Social movements and the bicycle" [Online]. Available at thinkingaboutcycling.wordpress.com (Accessed: 4 March, 2021).
Hughes, S. (2012). *Hero on a bicycle*. Somerville, MA: Candlewick Press.
Kidder, J. L. (2011). *Urban flow: bike messengers and the city*. Ithaca: Cornell University Press.
Longhurst, J. (2015). *Bike battles: a history of sharing the American road*. Seattle: University of Washington Press.
Norcliffe, G. (2018). "Women and cycling: a revisionist interpretation," in *Cycle History 28: Proceedings of the 28th international cycling history conference* (Verona, NJ: ICHC Publications), pp. 86–9.
O'Rourke, P. J. (1987). "A cool and logical analysis of the bicycle menace," in *Republican party reptile: essays and outrages*. New York: Atlantic Monthly Press, pp. 122–27.
Parsons, T. (2002). "I just don't Lycra these cycle yobs," *The Mirror* (5 August 2002) [online]. Available at mirror.co.uk (Accessed 4 March 2021).
Perry, P. J. (1969). "Working-class isolation and mobility in rural Dorset, 1837–1936: A study of marriage distances," *Transactions of the Institute of British Geographers* 46, pp. 121–41.
Pye, D. (2014). *Fellowship is life: the story of the National Clarion Cycling Club*. Bolton: National Clarion.
Ridley, F. A. (1926). *The green machine*. London: Noel Douglas.
Rivera, R. (2020). "Putting the trans in transportation," in Rogue, L. (ed.) *Trans-galactic bike ride: feminist bicycle science fiction stories of transgender and nonbinary adventurers*. Portland: Microcosm, pp. 99–109.
Rogue, L. (ed.) (2020). *Trans-galactic bike ride: feminist bicycle science fiction stories of transgender and nonbinary adventurers*. Portland: Microcosm.
Saroyan, W. (1943). *The human comedy*. New York: Harcourt.
Seplavy, T. (2020). "Finding fear, validation and community," in Rogue, L. (ed.) *Trans-galactic bike ride: feminist bicycle science fiction stories of transgender and nonbinary adventurers*. Portland: Microcosm, pp. 69–76.
Sirota, M. (1991). *Bicycling through space and time*. New York: Ace.
Smethurst, P. (2015). *The bicycle – towards a global history*. Houndmills, Basingstoke, Hampshire: Palgrave Macmillan.
Street, E. J. (2016). *The velocipede races*. Portland: Elly Blue Publishing.

Turpin, R. (2018). *First taste of freedom: a cultural history of bicycle marketing in the United States*. Syracuse: Syracuse University Press.
Wells, H. G. (2018). *The wheels of chance*. Jeremy Withers (ed.) Brighton: Sussex Academic Press.
Westwood, K. (2011). *The courier's new bicycle*. Sydney: HarperCollins.
Wheels: a bicycle romance (1896). New York: G. W. Dillingham.

46

CYCLING HUMOR IN TURN-OF-THE-CENTURY LITERATURE

Una Brogan

> The attitudes, gestures and movements of the human body are laughable insofar as the body brings to mind a simple mechanism.
>
> *(Bergson 1981 [1900]: 22–3: author's translation)*

From the earliest days of cycling, writers have found in bicycles a rich source of comic inspiration. The humor associated with bicycles relies on a blurring of the established lines between human and mechanical, animate and inanimate; themes which mirror crucial turn-of-the-century preoccupations. Critically examining humor may seem an impossible task; it escapes any definition we try to pin on it, and often a joke obstinately loses its power once it has been pulled apart for analysis. In his classic treatise *Laughter* (*Le Rire*) (1900), the French theorist Henri Bergson recognized this fact from the outset, but nonetheless insisted on the importance of coming to an understanding of *le comique*, since it can provide important insights into "the social, collective and popular imagination" (2). Following Bergson's reasoning, I analyze how and why two-wheeled transportation had the capacity to provoke mirth and provide fresh possibilities for amusement, play and creativity in texts written in Britain and the US in the early days of the technology. This will lead us to a closer understanding of the role it came to play in the popular imagination and allow us to appreciate the lasting legacy of the humor associated with cycling.

Bergson's essay provides a compelling critical framework for an examination of humor in cycling texts. It appeared in 1900, at a time when ideas around humor were evolving as a result of contact with British models. It is not French *humour* – understood as a form of satire – which interests Bergson here, but rather self-deprecating British humor. This eccentric, self-conscious laughter has been called "*l'humour 1900*" in France, and is tied to turn-of-the-century ideas of Britishness, inherited from authors such as Lewis Carroll and Edward Lear. Bergson's text thus enacts the contemporary cultural dialog between British and French concepts of humor, interestingly mirroring the to-ing and fro-ing of ideas around the social significance of the new technology of the bicycle at this time. This typically British style of humour is in evidence in *Three Men on the Bummel* (1900), Jerome K. Jerome's sequel to the bestselling comic classic *Three Men in a Boat* (1889). This novel sees the three friends from the earlier work (now older, though not much wiser) heading off on a cycling tour in Germany. As Murray Roston notes,

DOI: 10.4324/9781003142041-68

the long-lasting humor of *Three Men on a Boat* relies on Jerome's first-person narration, by means of which "he created a pseudo-self, a projection of himself seemingly unaware of the foibles, misapprehensions and illusions for which the story lampoons him" (2011: 198). The same sort of self-deprecating humor is in evidence in the memorable last lines of Mark Twain's essay "Taming the Bicycle," in which he ambiguously recommends: "Buy a bicycle. You will not regret it, if you live" (1992 [1884]: 896). Jerome's cycling novel provides a clear example of this kind of self-reflexive British humor. Throughout *Three Men on the Bummel*, cycling affords rich opportunities for self-mockery, since the characters come to realize that "human performance lags ever behind human intention" (Jerome 1994 [1900]: 139). As Roston (2011: 198) observes, the narrator is "to be laughed at whenever the gap between his illusions and the reality of his situation is perceived." Due to fatigue, clumsiness or inattention, the cycling protagonists repeatedly fail to realize their grand aims of rising early and cycling great distances, and instead suffer various minor setbacks and disasters. Indeed, after a drawn-out departure and a leisurely tour by rail around Germany, they only begin their eponymous cycling "Bummel" in the Black Forest two-thirds of the way through the novel.

Bergson focuses on analyzing comic situations at close range, pulling apart funny situations to uncover and reflect on why they make us laugh. He summarizes the essential conditions which may provoke laughter in the following terms: "it seems that humor arises when people in a group direct their attention to one of its members, silencing their sensitivity and calling only upon their intelligence" (Bergson 1981 [1900]: 6). This description brings to mind the new sight of cyclists who navigated streets at the time Bergson was writing. By adopting this individual means of transport, early cyclists singled themselves out from the group and became a fascinating object for the collective gaze. Going against the grain of the paradigm of mass public transport enshrined in trains and trams, early cyclists were certainly a spectacle, drawing the attention, fascination or ridicule of those they encountered. Despite having now become commonplace, cyclists continue to be a visible minority on the streets today, where they are also frequently the target of other road users' jokes or derision.

Bergson's first example of a humorous situation is a description of a man in the street who trips and falls, which is easily transposable onto the very visible falls and collisions of cyclists. Falling was part and parcel of riding a bicycle, especially in the early period of its adoption. The nature of the high-wheeler, or Ordinary bicycle, meant that the rider was placed directly above the front wheel some two meters above ground. While this provided an agreeable lofty sensation and increased efficiency thanks to the large wheel diameter, falls were frequent and dangerous. Testament to this is Mark Twain's essay "Taming the Bicycle," which focuses on the repeated falls suffered by the apprentice cyclist. On purchasing his bicycle, the author arms himself with "a barrel of Pond's Extract" to treat his abrasions, which is liberally applied each time the rider involuntarily "dismounts," as Twain euphemistically puts it (892). Bergson reasons that falls are funny due to people's failure to be reactive and adaptable: "Due to a lack of flexibility, distraction or stubbornness, or through stiffness or acquired speed, the muscles continued to make the same movement when the circumstances required something else. This is why the man fell, this is what makes the passers-by laugh" (1981 [1900]: 6). He goes on to insist on the mechanical aspect to this rigidity, arguing that laughter is provoked by "a certain mechanical stiffness where you would expect to find attentive suppleness and the lively flexibility of a human being" (1981 [1900]: 8). Pedaling cyclists conform very closely to this description. Their movements are not completely free, but rather mechanized by the limits the machine imposes on them; notably, the need to balance, to keep the machine in motion and to continuously turn the pedals. As Twain astutely observed when learning to steer and balance the bicycle, "my nature, habit, and breeding moved me to attempt it in one way, while some immutable and unsuspected law of physics required that it be done just the other way" (1992 [1884]: 893). The uncanny mechanization of human movement became a recurring theme in literary evocations of bicycles, from Alfred Jarry's grotesquely powerful "Supermale" (1902 [1989]) to Samuel Beckett's depiction of Molloy (1951 [1989]) attaching his crutches to his bicycle and managing to pedal in spite of his disability (Chapter 7).

Automatism in human movement is a theme that was often employed to comic effect in turn-of-the-century cycling narratives. The mechanical limits imposed upon the body and its instincts were keenly felt by early cyclists, as H. G. Wells reminds us in his iconic cycling novella *The Wheels of Chance* (1896). A novice to cycling, the hero Hoopdriver "doubted his steering so much that, for the present, he had resolved to dismount at the approach of anything else upon wheels" (Wells 1935 [1896]: 14). When he encounters a cart, he attempts to stop and instead falls off his bicycle, unable to recall quickly enough how to get off the machine. The author J. W. Allen relates a similar situation in his fictionalized account of cycle touring, *Wheel Magic* (1909):

> I remember once beholding an elderly lady riding slowly and carefully, straight at me. I was well on my proper side of the road. I rang my bell, and she looked at me and came on, as it were fascinated. If I turned out of her way, it seemed likely that she would turn also. I dismounted and stood facing her ten yards away. And still she came on, very slowly and resolutely, still she struck my front wheel and sprawled in the roadway.
>
> *(40–1)*

In the above examples, it is the mechanization of the cyclists' actions, their rigidity and lack of adaptability to the current situation, which result in comic effect. Several short stories from the collection *The Humours of Cycling* (1897) involve similar scenes, where riders do not react in the way the reader would have expected, but rather in the way the machine obliges them to, inevitably resulting in a collision (Burginthere 1897: 26; Wishaw 1897: 22).

While mechanizing human movements and instincts, the bicycle could provoke humor by requiring its rider to adopt an animalistic posture. The bodily attitude of cyclists was held up for ridicule by some of the detractors of the technology, notably in connection with the notorious, eccentric, fast-paced cyclists who were respectively termed "scorchers" and "*vélocipédards*" in the UK and France. R. J. Muir's mock-Platonic dialogue *Plato's Dream of Wheels* (1902) describes these pariahs of the cycling world in the following terms:

> Tamias – Indeed, I have seen young men [...] strangely curved as to the back, and I have felt impelled to cry onto them in the words of Persius – "Oh, crooked souls, forever bent to the earth" but they ever flashed past without stopping to listen.
>
> Eremus – That is true, for they are even as squirrels, *skiouroi*, in their rapid motion and in the clutch of their fore-paws, whence, I fancy, they have derived their names of *skiourchers* or scorchers.
>
> *(7–8)*

Here Muir paints cyclists as a hybrid of human and animal elements which is as likely to produce mirth as the combination of mechanical and human characteristics. The hunchback position described by Muir is one Bergson (1981 [1900]: 18) refers to as particularly humorous due to its capacity to mimic a facial expression or "make the body grimace." The unusual bodily position the bicycle obliged its rider to adopt was a further source of comic inspiration for authors who integrated the technology into their texts.

The bicycle also provides a context for a humorous disruption of the temporal sphere. Bergson (1981 [1900]: 8) argues that the overlapping of temporal spheres can result in comic effect:

> Let us imagine a certain naive inflexibility of the senses and intelligence, which results in us continuing to see what is no longer there, to hear sounds that have stopped, to say things that are no longer appropriate, to adapt to a past, imagined situation when we should be responding to present reality.

This description could apply to the falling cyclists described above, who seem unable to change their minds about their course of action, in spite of changed circumstances. The sensation of being out of sync with reality is evoked by J. W. Allen who, as he falls, wishes for "a half-minute back from remorseless Time – nay, ten seconds – that is all that one requires [...] And the Past is suddenly merged in the acutest of Presents. The misused machine lies prone" (1909: 39–40). In the description of Hoopdriver's fall in front of a heathkeeper, at least three different temporal spheres are invoked:

> He gripped the handles and released the brake, standing on the left pedal and waving his right foot in the air. Then – these things take so long in the telling – he found the machine was falling over to the right. While he was deciding upon a plan of action, gravitation appears to have been busy. He was still irresolute when he found the machine on the ground, himself kneeling upon it, and a vague feeling in his mind that again Providence had dealt harshly with his shin.
>
> *(Wells 1935 [1896]: 14)*

Here, narrative time ("these things take so long in the telling") is longer than subjective time ("he was deciding on a plan of action") while objective, scientific time ("gravitation") is the quickest of all, acting to place the cyclist on the ground before he or the reader expect it. We could also infer a fourth, "Providential" time, corresponding perhaps to John Urry and Phil Macnaghten's (1998: 147) "glacial" or "evolutionary" time, which is "immensely long and imperceptibly changing." At a time when thinkers such as Einstein and Bergson were revolutionizing contemporary attitudes toward time, novelists such as Wells experimented with the separation and superimposition of discrete temporal spheres for comic effect.

The humor around cycling also relied heavily on the close mingling of human and mechanical elements. In her study of comedy, *The Odd One In* (2008), Alenka Zupančič engages with Bergson's theory of the comic, reasoning that, in Bergson's terms, comedy arises because there is "something mechanical encrusted on the living" (111). Yet rather than suggesting that this is a one-way process, Zupančič argues that in comic situations there is a mutual exchange of agency between humans and objects; while the human takes on the inertia of the material world, objects gain animation in turn. In her example of a baron slipping in a puddle, not only does the baron become "mechanical," but the puddle manifests "elasticity" and "changeability." What we laugh at is humans' sense of self-importance, their desire to control the material world, and their blindness to the agency of objects. Zupančič (2008: 115) asks: "is not the comic precisely the reversal in which we come upon something rigid at the very core of life, and upon something vivid at the very core of inelasticity?" The intimacy of cyclists' connection to their machines provides a compelling opportunity for a blurring of the living and the non-living, a theme that was used to comic effect from the earliest days of cycling literature. Twain, for instance, heavily personifies the bicycle in his abovementioned essay, and later memorable incarnations of "the humanity of the bicycle" (O'Brien 1988 [1967]: 89) include novels by Beckett (1970 [1934]) and Flann O'Brien's classic *The Third Policeman* (1988 [1967]). Jerome's cycling novel mobilizes the humorous motif of self-deprecation, as we have seen, but it also relies on the comic agency of non-human actors. While in *Three Men in a Boat*, the intelligent dog Montmorency or a stubborn tin of pineapples provide examples of the agency of objects, in the sequel the bicycle allows for a rich comic exploration of the vibrancy of non-human elements. This is well illustrated by a passage in which the exasperated narrator watches an incompetent friend attempt to "overhaul" his bicycle, which puts up stiff resistance to the would-be mechanic:

> The bicycle, I was glad to see, showed spirit; and the subsequent proceedings degenerated into little else than a rough-and-tumble fight between him and the machine. One moment the bicycle would be on the gravel path, and he on top of it; the next, the position would be reversed – he on the gravel path, the bicycle on him. Now he would be standing flushed with victory, the bicycle firmly fixed between his legs. But his triumph would be short-lived. By a sudden, quick movement it would free itself, and, turning upon him, hit him sharply over the head with one of its handles.
>
> *(Jerome 1994 [1900]: 41)*

This struggle between man and machine makes us laugh since it refuses the conventional distinction between inert objects and active humans. While the man attempts to dismember it, the bicycle fights back, attempting to injure the man in turn. What is funny here is the man's undue sense of self-importance and expertise; the reader observes the struggle alongside the increasingly frustrated narrator, knowing that the man's efforts are in vain. The bicycle mocks the man's stubborn belief in his own unique agency and his inability to notice that of the non-human elements around him.

Drawing on the lively agency of the objects a cyclist encounters and the uncanny temporal space the technology opens up, Wells makes active use of the comic potential of the bicycle in *The History of Mr Polly* (1910). The bicycle figures in a humorous episode that recalls Jerome's comic technique of depicting a gap between the first-person narrator's illusions and reality. Chatting with his young female cousins, "Mr Polly struck a vein of humour in telling them how he learnt to ride the bicycle. He found the mere repetition of the word 'wabble' sufficient to produce almost inextinguishable mirth" (Wells 1963 [1910]: 81). A disjointed account of running into a pedestrian follows, giving a vivid impression of the jumbled, atemporal impressions received by the cyclist:

> Hears the bell! Wabble. Gust of wind. Off comes the hat smack into the wheel. Wabble. *Lord!* what's going to happen? Hat across the road, old gentleman after it, bell, shriek. He ran into me. Didn't ring his bell, hadn't *got* a bell – just ran into me. Over I went clinging to his venerable head. Down he went with me clinging to him. Oil can blump, blump into the road.
>
> *(Wells 1963 [1910]: 81)*

Polly's snappy sentences, which often leave out pronouns or verbs, his exclamations, interrogations, and mixing of present and past tenses provide a colorful retelling of the collision. The humor of this scene relies on the elements we discussed above: the mechanization of the cyclist's movements, the disruption of the temporal sphere and the agency of inanimate objects. Following the accident, Polly immediately attributes agency and blame to the man's hat – "I told him he oughtn't to come out wearing such a dangerous hat – flying at things. Said if he couldn't control his hat he ought to leave it at home" (81). Polly is convinced that this object willfully caused the accident, something that he claims occurs frequently to cyclists: "that's the sort of thing that's constantly happening you know – on a bicycle," he observes, "People run into you, hens and cats and dogs and things. Everything seems to have its mark on you; everything" (81). Yet what is humorous above all here is Polly's blindness to his own responsibility – and that of his bicycle – in causing the accident. When his cousin sarcastically comments "*You* never run into anything," Polly "very solemnly" replies "Never. Swelpme" (81). His cousins are laughing at, rather than with, Polly, yet his bicycle allows him to create vital human connections. It provides the means by which he is able to travel to visit his young relatives, while also giving him stories to amuse them with once there: "Mr Polly had never been such a social success before. They hung upon his every word – and laughed" (82). His popularity with his cousins eventually leads to his engagement to one of them, reminding us of the importance of humor as a basic building block of social interaction. The bicycle, although usually an individual mode of transport, is nonetheless an inherently sociable instrument that provides rich opportunities for creating connections between people.

While humor is a crucial human trait, it should not be forgotten that laughter can quickly turn into derision or ridicule. Bergson (1981 [1900]: 15–16) makes the insightful remark that laughter is society's "punishment" for those who fail to conform to what is considered normal or acceptable. In light of this, the laughter that society directed at bicycles could be seen as a form of social policing, chastisement for non-conformity or eccentricity. Writers who used humor in their sketches of cycling were, in part, reflecting the fact that cyclists stood out from other road users. Nicholas Oddy (2001: 175) examines Victorian cycling-related paraphernalia and comes to the conclusion that "the machine got off to a derision-laden beginning that characterized its subsequent public perception." Russell Mills (1995: 11) cites an example from a General Motors exhibition in which comic relief was provided in the form of a man crashing his bicycle into a pig pen, drawing the conclusion that "bicycles are not ordinarily taken seriously in twentieth-century industrial culture. Jet planes, locomotives, nuclear reactors, and other machines are taken seriously – but not bicycles." Both Oddy and Mills suggest that the derision connected to bicycles functioned to effectively rule them out as a serious transport technology. While humor plays a role in many of the texts I have been discussing, these authors also show evidence of being in earnest about the potential of this new machine to alter human capacities and transform our interaction with our environment. It is interesting to note that in Jules Romains's comic novel *Les Copains* (1913), one of the only scenes in which a serious, metaphysical and lyrical tone is adopted is that in which the friends set off on a cycle tour. Bénin sincerely declares "I love these machines. They do not carry us stupidly. They extend our limbs and let our energy reach its full potential. How silently they go! This loyal silence! This silence that respects everything" (88). This solemn passage stands in stark contrast to the irreverent, comic tone of the rest of the novel, reminding us that bicycles can provoke not only laughter, but also respect, wonder, gratitude and admiration.

A sense of humor is part of being human, and the fact that bicycles were used to comic effect in literature from their earliest days is, first and foremost, proof of the preponderant place they rapidly came to occupy in day-to-day life, and of the new opportunities they opened up for play and enjoyment. They reflected and interacted with evolving turn-of-the-century ideas around humor being formulated by writers such as Bergson. Their comic potential stems from their ability to combine elements of mechanical and organic, human and animal, past, present and future in a single artefact. The bicycle provided writers with a rich new terrain on which to experiment with humor. Yet bicycles have always been subversive and counter-cultural objects, and laughter directed at them could also be used as a means to defuse the threat they posed to the established social order. As a society, we still do not take bicycles seriously enough.

Bibliography

Allen, J. W. (1909). *Wheel Magic; Or, Revolutions of an Impressionist*. London: J. Lane.

Beckett, S. (1970 [1934]). *More Pricks than Kicks*. New York: Grove Press.

Beckett, S. (1989 [1951]). *Molloy*. Paris: Éditions de Minuit.

Burginthere, G. B. (1897). Some Emotions and – No Morals; Or, How to Learn "to Bike". In: *The Humours of Cycling*. London: James Bowden, pp. 25–28.

Bergson, H. (1981 [1900]). *Le Rire: essai sur la signification du comique*. Paris: PUF.

Jarry, A. (1996 [1902]). *Le Surmâle: roman moderne*. Paris: Mille et une nuits.

Jerome, J. K. (1990 [1889]). *Three Men in a Boat*. London: Penguin.

Jerome, J. K. (1994 [1900]). *Three Men on the Bummel*. London: Penguin.

Macnaghten, P., and Urry, J., eds. (1998). *Contested Natures*. London: Sage.

Mills, R. (1995). Thinking about Thinking About Cycles. *Cycle History*, 5, 11–18.

Muir, R. J. (1902). *Plato's Dream of Wheels; Socrates, Protagoras, and the Hegeleatic Stranger; with an Appendix by Certain Cyclic Poets*. London: T.F. Unwin.

O'Brien, F. (1988 [1967]). *The Third Policeman*. London: Paladin.

Oddy, N. (2001). Cycling in the Drawing Room. *Cycle History*, 11, 169–176.

Romains, J. (1922). *Les Copains*. Paris: Éditions Gallimard.

Roston, M. (2011). *The Comic Mode in English Literature from the Middle Ages to Today*. London: Continuum International.

Twain, M. (1992 [1884]). "Taming the Bicycle" in *Collected Tales, Sketches, Speeches & Essays, 1852–1890*. New York: Literary Classics of the United States, 892–896.

Wells, H. G. (1935 [1896]). *The Wheels of Chance, A Bicycling Idyll; The Time Machine*. London: J. M. Dent.

Wells, H. G. (1963 [1910]). *The History of Mr Polly*. London: Pan Books.

Wishaw, F. (1897). Pogeley's Ride Down Town. In: *The Humours of Cycling*. London: James Bowden, pp. 19–24.

Zupančič, A. (2008). *The Odd One In: On Comedy*. Cambridge, MA: MIT Press.

47

ON BARDS ON BICYCLES

The art of cycling poetry

Justin Daniel Belmont

Bicycling has long been esteemed for its poetic qualities. Rarely does a sport lend itself so graciously to movement; to metaphor; to style; to thought. Your mind races on the road. It *dances on the pedals*. Sometimes it's about love, loss, the meaning of life; other times, about whether the brake pads are rubbing. I don't know which is more profound.

On returning home from a ride, you begin to wonder. Where did all those great thoughts *come* from? And, perhaps just as puzzling, where did they *go*? Novelist R. K. Narayan (1976: 13–14) offers the following character portrait in *The Painter of Signs*: "While bicycling, his mind attained a certain passivity, and ideas bubbled up, lingered a while, burst and vanished."

So there you are, afterwards, back indoors, endorphins gone, left with only a vague muscle memory of the mental exercise that, at the time, seemed so utterly essential and precious and brilliant. Time to remount the saddle.

Of course, unless your last name happens to be Sagan or Froome, though you may feel you live to ride, you don't ride for a living. Thoughts about biking are squeezed around work and weekdays.

If only.

If you're reading this (which I suspect you are), even apart from scholarly interests, chances are your passion for cycling lacks an on-and-off shifter. Your workdays pass by on visualized cranksets, daydreams of derailleurs and Gu; Monday on last weekend's dirt, Friday on the smooth tarmac ahead. Before long it's time: the sun shining high, the tar sparkling, you and your slick wheels are all pumped up and ready to go. *Free at last. Breaking Away.*

Still, the philosopher nags: Away from *what*? Troubles? Responsibility? Real life? "Where are you *going*?" asks a concerned spouse or Lycra-skeptic friend, bemused by the sight of a full-grown *you* clad in clown tights racing by the Joneses at the speed of an adolescent fantasy, trying, vainly, to beat your own times.

"To Johnny's house and back?" "Just down the block?" "Dunno, back by dinner." "Out."

With an enigmatic smirk you exit the scene of domesticity, parting the side door to behold in ecstasy, astride your singsong steely steed, an untouched jungle paved in possibility, calling you to play.

Then, invariably, you ask yourself the same question: "Where *am* I going?" What the hell am I doing out here?

DOI: 10.4324/9781003142041-69

Poetry on wheels

For more than a century (that's years, not miles), inspired pedallers have inched toward answers by setting their daydreams to metered words.

That's right: poetry. *Bicycling poetry.*

Poems not just about the machine itself, in its bare essential elements, its cranks and sprocket clusters, but also the range of joys and pains and memories it inspires in the psychology, the psyche, of the cyclist.

You might be thinking, "Poetry… about bicycling?" It's an odd concept to be sure. But not a new one. Even apart from spoken wordsmiths à la Phil Liggett, traditional bike bards have been around for as long as the bike itself, from the late 19th century up through the modern era. Paul Fournel, a contemporary avant-garde French writer, introduces the connection between cycling and the poetic process (2003: 127):

> There are a lot of walker-poets who write their verses to the rhythm of their feet. Cyclist-poets are less numerous, it seems, but that's due to inattentiveness, since the bike is a good place to work for a writer. First, he can sit down; then he's surrounded by windy silence, which airs out the brain and is favorable to meditation; finally, he produces with his legs a fair number of different rhythms, which are so much music to verse.

And there's a lot of verse, in fact. In assembling an anthology of bike-themed poetry, *The Art of Bicycling* (Belmont 2005), I was struck by the scope and depth of this truly "niche" canon, which ranges from late 19th century "Anonymous" poems (often used as light fillers in periodicals) to work by some of the most recognizable names in poetry: from Dylan Thomas, Pablo Neruda, and Rita Dove to C.K. Williams, Philip Larkin, and Seamus Heaney.

Many bike poems are surprisingly good.

And alas, yes: others are bad. Like, *really* bad.

Most fall somewhere in the middle. They might be slightly uneven, like cobblestone, but still quaintly navigable, cobbling together a few striking images and ideas, a cute wordplay or two.

Some poems move us more than others, but all decent poems *move*. William Golding, penman of *Lord of the Flies*, once suggested the following analogy (1982: 178):

> Consider a man riding a bicycle. Whoever he is, we can say three things about him. We know he got on the bicycle and started to move. We know that at some point he will stop and get off. Most important of all, we know that if at any point he stops moving and does not get off the bicycle he will fall off of it. That is a metaphor for the journey through life of any living thing.

If poetry can capture anything about the way we think or the way we ride, it must find a way to articulate life in movement.

★★★

So… where does that leave us as readers, as surveyors of this "unique" landscape? One potential route is an historicist approach, viewing verse through the lens of world and bike history and the various poetic "movements." Another is to treat these lyrical pacelines to close readings. Such traditional courses of study may be instructive, but perhaps not much more so than auditing poetry or history in general.

Our goal, here, might rather be to explore how bike poems can help us see new dimensions of bicycling itself – particularly, the subjective experience of riding. In line with Fournel, we might start by recognizing that the link between cycling and verse isn't random and incidental but fruitful and deserving of unpacking. Indeed, without getting too caught up in the spokes of poststructuralist theory, it may be useful to examine not only the *products* (the poems) themselves but the *process* – the rhetorically parallel processes – of riding, writing, and reading.

We're led to wonder if there might be a common athletic imperative behind these activities. What drives a bike writer? Where's the fuel? Wherefore grunt and sweat and be made weary over such vintage vehicles, good old forms of conveyance? *Where do bike poems come from?*

A flat Wikipedia survey won't do. If we want to enter the ontology of these lyrical oddities, we might do well to engage poetry on its own playful terms. That is to say, to get in the minds of poets and bike riders, to see what makes their spokes tick, we need to speak a bit in their own speculative creative language. Ergo: we need to have fun.

Why poetry?

> Everything is bicycle.
>
> *Stephen Crane (1896: 859)*

That you enjoy riding is no secret – to anybody. It's part of your life. In a large part it defines your identity. In the words of author William Saroyan (1952: 27), "What is the bicycle? Well, my bike is himself (myself)." That is to say, by riding, you are *being yourself.*

The obvious corollary is this: By riding, *you are not someone else.* However hard you try, however many clubs you join, the truth is you can never know, viscerally, emotionally, intellectually, what anything – cycling – fully means to another person. You can get close. A brief look over your shoulder and you can tell, often, if your roadie buddies are in pain or in heaven, or if you can beat them to the town line. But never can you *be* them.

So we bridge the gap. We communicate. We share nonverbal signs like a pat on the back, a gesture, a hand, even a finger. Even the act itself of sharing a ride is an occasion for bonding (Chapter 4). Assuming we possess minimal social skills, we may even use words. There's a lot to talk about on the road. We can express deep feelings ("uh, good ride") or sincere concern ("you dead?"), talk shop ("how expensive was *that*?"), grunt (""), hypothesize ("where are we?"), even philosophize ("did Spinoza spin?"). Cyclists are smart. They like to conserve energy; they give few-word answers. They're not blowing you off; it's all you need to know. You know the rest. Don't you?

Poetry is communication, condensed. Quick. To the point. Without the small talk. Using the fewest words to say the most. The lightest gear. The leanest language. The best words. In the best order.

A claim often leveled against poets and cyclists, the creative and sometimes solitary bunch they are, is that their pursuits are a form of cheap escape, a flight from the demands of real life. But life, you know, is still there; you're just experiencing it on a higher-incline plane. Your everyday speaking part has just been translated into the language of the road: the whirring and whispers of wheels and traction; the crackling of leaves; of tires treading tar; of the private cognitive soundtracks that go off in your mind on a clear day, when you think clearly and can make sense of things.

Poetry translates it all back. It takes these unmarked lines of free-flowing thought along the road and rhymes them vaguely with sense. This is the charter of the bike bards: to put the living language of cycling into words.

They fail beautifully. Continues writer Paul Fournel (2003: 29):

> When you get on your first bike you enter a language you'll spend the rest of your life learning, and you transform every move and every event into a mystery for the pedestrian.

And on that last point, let's not leave out drivers:

> I must look funny
> to those of you in cars:
> pedaling furiously,
> going slow…
> (Kirchner 2005: 213)

Cyclists are indeed a mysterious bunch. From a pragmatic or economic or even a physical standpoint, we don't *need* to ride. We have cars and ride-share apps. And there are other aerobic exercises, like swimming and polka dancing. There is something more that drives us. Something more hidden.

An outsider can't help but think, in driving by riders, that he's observing some strange ritualistic creed, some sacred, secret society – and he's right. Bikes pass cars without interest. They pass other bikes and they perk up, able to sniff each other out from a mile away. "Bikes talk to each other like dogs," writes author Daniel Behrman (1973); "they wag their wheels and tinkle their bells." With a tacit smirk or an all-knowing nod, secret handshakes are exchanged.

Join the Bike Club. Cycling poems are a testament to and for that club. It isn't the club that plans Sunday doughnut rides and charity tours. It's the covenant, or the glorious myth of such a covenant, that is unwritten and largely unspoken for and which binds us to the wider world of riders. The language none of us knows only all too well.

The language of cycling

Fresh out of college, as an editorial greenhorn (translation: entry-level peon) at *Bicycling* magazine, one of my first tasks was reading reader mail. Letters would pour in on all topics, from cranksets to calf toning, leaky water bottles to fluid retention.

More than half were love letters.

About bikes.

Neither submissions nor formal requests, these mystery manifestos had no apparent reason for being written or sent. They simply appeared, waiting, wanting to be read. Steeped in recurrent images of thrashing wheels and fresh air and aromatic azaleas, the documents recounted both specific rides and riding philosophies, everything from the "wide array" of roadside sights to the "childlike thrill" of going downhill to the feeling of being "free as a bird."

Clichéd or not, they were about the simple joys of cycling (Vignette I). And they read like private diaries. They read… sort of like poetry.

Each letter described in more or less the same idiom that which asserted itself as a fully unique and individual quest. Freedom. Speed. Childhood. Happiness. Rhododendrons. I couldn't blame them. I began to think: With only so many words in the language, really how many ways are there to say one enjoys the sensation of manually pedaling a two-wheeled contrivance over a hard surface?

In a passage by Marcel Proust, the narrator recalls an episode from youth where, after being shut indoors for hours, he energetically takes a walk outside. After "struggl[ing] cheerfully" against rain and wind, he is dazzled by a reflection of roof tiles in a pond and finds himself confined to muttering "gosh, gosh, gosh, gosh!" (1928: 218) – registering for the first time the failure of language.

It doesn't seem a stretch to imagine the same spectator, this time on wheels, cruising up hills and feeling on top of the world as he admires in passing flashes of speckled foliage and sun.

"Dude," he might reflect at ride's end, "unreal." Um, good ride, good ride. Same time tomorrow?

Reduced to idiots, in a way, in more than the minds of motorists, riders have the habit of journeying from inspired philosophers "in the moment" to inarticulate athletes after the game. *Yeah, I guess I pulled through today, just doing my thing. Woohoo!* You know?

Sportsmen just aren't sensitive enough. They always look at a loss for words. They can't express what they "feel." Words are for airy intangibles like glory and pain; physical action is raw and real and there ain't nothin' like it. A rhyme, however persuasive and well-reasoned, can't argue the real rhyme and reason of a ride.

Reality check: when you say these things, no longer are you speaking sensibly or literally. You're thinking... like a poet. You are feeling what you feel cannot be said, at least in full. Poetry isn't the opposite of an "I can't describe it" moment. Often it's its fullest, loudest, most cultivated expression. A poem's conception presupposes that while some things are hard to say, one must try anyway – an effortful struggle to which all riders can surely relate.

> I would like to write a poem
> About how my father taught me
> To ride a bicycle one soft twilight...
>
> (Bilgere 2002: 1)

A true pro's poem doesn't start out on halcyon peaks, on Alpe d'Huez, looking down on us mortals with our puny expressions. More often it begins at an impasse, a gap or difficulty, and proceeds to grind its way up like the best of us, using the same rhetorical equipment (OK, theirs is a little better), only with far more grace and greater flair, and talent, and with more intensely fueled desire.

In the end, good bike poems elevate us to where the air is clarified. Like the central metaphor in Seamus Heaney's "Wheels Within Wheels" (1991: 46), they make ineffable *things* crystal clear, at once more sensible and more spacious – in a word, "transparent":

> The first real grip I ever got on things
> Was when I learned the art of pedaling
> (By hand) a bike turned upside down, and drove
> Its back wheel preternaturally fast.
> I loved the disappearance of the spokes,
> The way the space between the hub and rim
> Hummed with transparency.

Accidental bunch that they are, cyclists and poets revel in strange distances. With equally strange recording instruments, they actively explore and record the steps they took (figuratively) to get to where they are now, often starting out with a vague direction and ending up someplace new; poetry logs what happens on the way.

A brief history of rhyme

Bike bards are as old as the bike itself. They first surfaced in the 1880s and 1990s, a period generally known as the *belle époque* ("the golden age") – which coincided with "the cycling craze" (Chapter 41) As the *bi-cycle* became a household name, daily periodicals became flooded with verse, songs, and odes in praise of the dazzling new invention, a machine that combined recreation and utility and provided an alternative mode of transport to smelly horses:

> O the hum of the wheel, my steed of steel,
> And the rush of the welcoming wind [...]
> (Waugh 1898: 99)

> Thou and I, my fifty-four!
> Willing steed and master!
> How we skin the roadway o'er!
> Never bird went faster!
> (S. K. B. 1888: 545)

In the preface to his 1897 bike song anthology, *Lyra Cyclus*, Edmond Redmond (1897) explicates the connection:

> A new school of poesy has arisen to celebrate the tribulations and triumphs of the Bicycling world. The Bards of the Bicycle have invaded Helicon in force and have drunk deeply from the waters of its sacred rill.
>
> *(iii)*

It should be noted that until recently, much of this vintage verse was effectively lost, consigned to newsprint even too old for microfiche and hiding in out-of-print books and the occasional academic footnote. Now, with online search directories and large-scale print indexing projects, these pieces are becoming slightly more discoverable.

Even in early verse, we already see the conceptual roots for what we now take for granted as the very discourse, the *language* of cycling. Youth, love, first rides, letting go, commuting, communing with nature – these light, witty ditties already say it all, albeit in relic rhetoric. Naturally, where their poetic successors power slightly slicker equipment, experimenting with newer structures (e.g., free verse), the older bards traffic in more traditional forms.

These first poems are rhyming paeans to an infant sport. The vestigial verbiage of Romanticism abides in recurring themes of inspiration (my bike, my Muse!), often tongue-in-cheek, set against modernist themes around novelty, innovation, and progress. As an exciting new form of transport, indeed of technology, cycling was discussed in much the same spirit as we now talk about autonomous and electric vehicles, AI and blockchain.

By the "bike boom" of the 1970s, with bike ownership higher than ever, bike verse saw a parallel resurgence with its own set of themes. By now, the very idea of riding had become metaphor. Bikes were figured not just as symbolic objects but as subjects, rhetorical vehicles, for exploring such deep personal issues as memory, psychology, and an irrevocable fear of trucks:

> The bicycle, the bicycle surely, should always be the vehicle of novelists and poets.
> (Morley 1926: 35).

That freewheeling feeling

For the last half-century or so, bicycling has come to be seen as a rite of passage; the bicycle, as a symbol of childhood. In contemporary verse we encounter countless treatments of this trope. Thematically, here, we find our youth and all its affiliate themes – innocence and self-discovery, freedom and independence:

> And my hands drop from the bars
> in that quickening
> I felt long ago when Daddy let go
> and I coasted off in the lawn,
> exquisitely balanced, absolved
> from all attachment.
>
> (Kasdorf 1992: 57)

Naturally, like children themselves, poems about childhood are often less simple and one-dimensional than they first appear. When executed well, as in the excerpt above, youthful themes are textured by the orientation of the writer/narrator as a grown-up looking back. Enter themes around nostalgia, regret, change, time passing you left and right. The ablest poets deftly balance the two sides and play them against each other, intimating joy and at the same time, time's tandem partner, collapse. Here is Linda Pastan in a widely anthologized piece, "To a Daughter Leaving Home" (1998: 196):

> I kept waiting
> for the thud
> of your crash as I
> sprinted to catch up,
> while you grew
> smaller, more breakable
> with distance,
> pumping, pumping
> for your life, screaming
> with laughter,
> the hair flapping
> behind you like a
> handkerchief waving
> goodbye.

A like blend of bliss and loss caps Gregory Orr's "Lament," which projects an angelic memory onto a girl fixing her bike, "playing… a round harp / on a desolate coast" (2002: 199–200):

> There on the highway's
> edge where gusts
> from passing cars
> whipped the grass
> like wind off the sea
> and she was kneeling,
> her arms moving
> among the metal spokes

plucking from them
a music lost
in the louder
impersonal sound
of traffic (and I thought
of you
as I drove past).

Of course, the freewheeling feeling isn't limited to children. Many poems figure the bike as a vehicle for adult independence. All grown up and consumed with earthly cares – the rat race, the mortgage rates – narrators are frequently shown breaking away from work and weekday concerns:

Away from the struggle and strife and sin,--
Out of the city's roar and din,
    Cycle and I
    Swiftly fly,
Leaving behind all care, hurrah!
(Harvey 1884: 338)

Away from the office and desk at last,
The business-haunted room,
The roar of a city, hurrying past,
The heat, the worry, the gloom,
To the glorious red of the sunset sky,
The sweet, cold wine of the air...
(Allen 1888: 192)

More generally, as we see even in these early examples, cycling means freedom not just from a physical space but also a state of mind. Interestingly, *the ride* – as mental mini-journey – is rarely instantiated solely in terms of denial or cheap escapism ("Get me outta here"). More often, taking to *the open road* is a movement toward greater openness. Riding may mean parting from something, but separateness isn't the thing; narrator-riders frequently move from a head space of separation toward integration, often a fusion with nature (subject, of course, to the temporal constraints of its own natural cycle):

A victory! To leave your loneliness
panting behind you on some street corner
while you float free into a cloud of sudden azaleas,
pink petals that have never felt loneliness,
no matter how slowly they fell.
(Nye 1998: 81)

The first mover

Mood-wise, there's no complicating the obvious: Cycling is fun. Thus, many bike poems are happy. ("Oh, merry are the wheelman's days; / [...] He glides down all life's troubled ways") (Pastnor 1883: 143). Simple enough?

But many poems don't stop at mere contentment. Often, they take pleasure as the starting point and journey upwards toward ecstasy or bliss, taking on an almost visionary air en route to the Romantic sublime. In "The Cyclist," John Morgan (1986: 255) pictures the divinely inspired poet-rider:

> By the late sheen of an arctic sky
> alive with branches shimmying with
> light he comes to me: the cyclist,
> active, floating, magical, observant,
> and the poem comes from him –
> whatever he can make it: the hope
> that what he turns to will take hold.

Bike poems are rarely devotional but religious themes abound. Occasionally poems take us to church itself, as in Philip Larkin's "Church Going" (2003: 58): "Hatless, I take off / My cycle-clips in awkward reverence." More often, they play on biblical tropes in a more generalized context to convey powerful feelings. Consider an early anonymous poem, "Benedicite" (Anon. 1884: 16):

> God bless my wheel! it knows nor care nor strife,
> For one day out the ever-coming seven
> I run with it far from the hells of life,
> To find in nature's handiwork a heaven.

Or take another, newer quatrain, a cheeky revision of the nightly prayer:

> When every pedal stroke's a bore
> When bum and back and neck are sore
> When flesh is mortified for sure,
> Then, I believe, my soul will soar.
> (Ferguson 1999: 294)

Occasionally, as in the tongue-in-cheek "God on a Bike" (Winslow 2005: 233), we even meet the maker himself:

> God's a rock hopper, a weekend mud churner,
> a star on wheels. He can fix a puncture [...]

Even more frequently, bikes themselves are pictured as vehicles of divinity, not only personified but deified.

In other moving poems, like Michael S. Smith's "Sunday Morning Services," the shifting natural landscape is simply heavenly (2005: 236):

> Concrete lots of cars weigh down
> The earth, while steeples rise untaxed and tall
> As watch towers, and the dark front doors frown
> And hoard the words inside. Bright flowers nod
> And bow, and I smile at their graceful charm,

And the cows turn with my progress, their god-
Like eyes wide. I speak my peace. The moist warm
Air swaddles me, and the life-giving sun
Answers the prayers of everything green.

Just as a bike ride might spur us to think big thoughts, many bike poems are existential in nature (dirt, clouds, and all); but more specifically, as above, they're affirmational. The bike, the universal "yes"; riding, saying yes to life. Pedaling home, the narrator of William Stafford's "Maybe Alone on My Bike" (1975: 256) reflects on the "splendor of our life," intoning:

O citizens of our great amnesty:
We might have died. We live. Marvels
coast by, great veers and swoops of air
so bright the lamps waver in tears
and I hear in the chain a chuckle I like to hear.

Riding along the turning world we laugh with tears of joy.

All work and all play

Anticipating the road ahead, we wonder: What's the future of bike poetry?

Whatever course it takes, we imagine its path will run roughly parallel to larger poetic trends – as formal structures evolve (both in traditional and new online outlets) alongside changes in bicycling itself, in bikes and their role in our lives.

Reflecting on the present moment, we find ourselves returning roughly to where we started: the natural affinity between cycling and poetry. For just as we rode and read rhymes as kids, bikes and poems are in some sense throwbacks in history. Nostalgic relics from a "simpler time." In our mature broadband age of on-demand, instant gratification, they slow us down.

And, technically speaking, these pleasures are not "essential" but discretionary.

By and large we don't *need* a bike to get to work. It's often faster and easier to drive a car. (And before long, cars themselves may do all the work.)

Ditto reading prose vs. poetry. The former is easier and faster by far. (Better yet, try channel surfing; or sit back while Netflix auto-plays.)

As life gets (theoretically) simpler and simpler, and we find ourselves further and further distanced from physical and mental labor, cycling and reading poems stand out noticeably as *difficult pleasures*.

This isn't an oxymoron. These pursuits remind us of the *deep* enjoyment that comes from meaningful hard work. That joy isn't born from play alone but from the marriage of work and play; vitality from labor:

... your legs
begin to throb as if
the body communicates
in a code of pain, saying
never mind the future,
you're here
right now, alive.
(Pastan 1998: 271)

As readers, if we look at poetic meters as broken sentences, free verse as whacked-out prose on speed, we won't get far. If our measurement is *mph*, a Schwinn will never beat a Subaru. It's our perspective that needs shifting. We appreciate things on their own relative terms. Engaged in difficult pleasures, we learn to question the equation of *speed* and *good, ease* and *meaning*. If cycling is a language, our goal is not to gloss it over, like cars over a road, or to get it over with as if the road were itself a roadblock. We read. We revel in the process, in the challenge.

★★★

Passion is a strange thing. We suspect we read and write poems and songs because we feel we must, much as one might rise some wintry Sunday and leave her plush light comforter for a ride in the snow. Maybe it's a soft rebellion. A refusal to pass through experience passively, like a body asleep, without reinvigorating it in spirit. As German theorist Wolfgang Sachs suggests in *For Love of the Automobile* (1992: 199):

> To attack the pedals may be strenuous over the short run, but is an expression of trust in one's own powers, for with the bicycle everything depends on the self. Those who wish to control their own lives and move beyond existence as mere clients and consumers – those people ride a bike.

The art of bicycling is ultimately a personal art. Poems cannot begin to tell you what it means to you. At best they can do the next best thing – sharing what riding means to others, speaking on behalf of a passion that lifts you up mountains and moves you beyond words.

New climbs await, and Marco Pantani reminds us the time is now:

> My language is my bike …
> and I want to continue writing
> the last chapter of my book
> that I much too long
> have left unfinished …
> (Pantani 2005: 315)

References

Allen, W. B. (1888) "Homeward: A Twilight Song of the Wheel", *Outing: An Illustrated Monthly Magazine of Recreation*, 11 (2, October 1887–March 1888), p. 192.

Anon. (1884) "Benedicite", in *Wheel Songs: Poems of Bicycling*. Edited by S. C. Foster. New York: White, Stokes, & Allen, pp. 9–18.

Behrman, D. (1973) *The Man Who Loved Bicycles: The Memoirs of an Autophobe*. New York: Harper's Magazine Press.

Bilgere, G. (2002) "Like Riding a Bicycle," in *The Good Kiss*. Akron: The University of Akron Press, pp. 1–2.

Crane, S. (1896; 1984) "A Glittering Spectacle" (Samuel S. McClure's newspaper syndication), in *Prose and Poetry*. New York: Library of America, p. 859.

Ferguson, G. (1999) "Puritan Against the Wind", in *The Art of Bicycling*. Edited by Justin Belmont. Halcottsville, NY: Breakaway Books, p. 294.

Fournel, P. (2003) *Need for the Bike*. Translated by Allan Stoekl. Lincoln: University of Nebraska Press. (*Besoin de vélo* © Editions du Seuil, 2001.)

Golding, W. (1982) "Utopias and Antiutopias", in *A Moving Target*. London: Faber and Faber, p. 178.

Harvey, J. (1884) "Swiftly We Fly", *Outing and the Wheelman: An Illustrated Monthly Magazine of Recreation*, 4 (5, April–September), p. 338.

Heaney, S. (1991) "Wheels Within Wheels", in *Seeing Things*. London: Faber & Faber, pp. 46–47.
Kasdorf, J. (1992) "Riding Bike with No Hands", in *Sleeping Preacher*. Pittsburgh: University of Pittsburgh Press, p. 57.
Kirchner, F. W. (2005) "Reasons to Commute by Bicycle", in *The Art of Bicycling*. Edited by Justin Belmont. Halcottsville, NY: Breakaway Books, p. 213.
Larkin, P. (2003) "Church Going", in *Collected Poems*. Edited by Anthony Thwaite. London: Faber and Faber, pp. 58–59.
Morgan, J. (1986) "The Cyclist", in *The Art of Bicycling*. Edited by Justin Belmont. Halcottsville, NY: Breakaway Books, pp. 253–255.
Morley, C. (1926) *The Romany Stain*. Garden City, NY: Doubleday, Page & Company.
Narayan, R. K. (1976) *The Painter of Signs*. New York: The Viking Press.
Nye, N. S. (1998) "The Rider", in *Fuel*. Rochester, NY: BOA Editions, Ltd., p. 81.
Orr, G. (2002) "Lament", in *The Caged Owl: New and Selected Poems*. Port Townsend, Washington: Copper Canyon Press, pp. 199–200.
Pantani, M. (2005) "Sometimes We Close Our Eyes", in *The Art of Bicycling*. Edited by Justin Belmont. Halcottsville, NY: Breakaway Books, p. 315.
Pastan, L. (1998) "'To a Daughter Leaving Home' and 'Stationary Bicycle'", in *Carnival Evening: New and Selected Poems, 1968–1998*. New York: W. W. Norton, p. 196 and p. 271.
Pastnor, P. (1883) "The Wheelman's Joy", *Outing and the Wheelman, An Illustrated Monthly Magazine of Recreation*, 3 (2, October 1883–March 1884), p. 143.
Proust, M. (1928; trans. 1992) *In Search of Lost Time: Volume One*. Translated by D. Enright, from original translation by T. Kilmartin and C. Moncrieff. New York: The Modern Library.
Redmond, E. (ed.) (1897) *Lyra Cyclus: The Bards and the Bicycle*. New York: M. F. Mansfield.
Sachs, W. (1992) *For Love of the Automobile: Looking Back into the History of Our Desires*. Translated by Don Reneau. Berkeley: University of California Press.
Saroyan, W. (1952) *The Bicycle Rider in Beverly Hills*. New York: Scribner's.
S. K. B. (1888) 'Cycle and I', *Outing: An Illustrated Monthly Magazine of Recreation* (September), 12 (6, April–September), pp. 544–545.
Smith, M. (2005) "Sunday Morning Services", in *The Art of Bicycling*. Edited by Justin Belmont. Halcottsville, NY: Breakaway Books, p. 236.
Stafford, W. (1975) "Maybe Alone on My Bike", in *The Art of Bicycling*. Edited by Justin Belmont. Halcottsville, NY: Breakaway Books, p. 256.
Waugh, A. (ed.) (1898) "The Wheel and the Wind", in *Legends of the Wheel*. Bristol, UK: J.W. Arrowsmith, pp. 99–101.
Winslow, P. (2005) "God on a Bike", in *The Art of Bicycling*. Edited by Justin Belmont. Halcottsville, NY: Breakaway Books, pp. 233–234.

Further reading

Belmont, J. (ed.) (2005) *The Art of Bicycling*. Halcottsville, NY: Breakaway Books.

48

"THE STUTTER OF THE WORLD BENEATH YOU"

The literature of cycle travel

Dave Buchanan

Almost as soon as people began taking trips on cycles, they wrote about their travels. Readers were keen to hear about the possibilities of this new way of holidaying and exploring. By the 1890s, and cycling's first big boom, cycle-travel narratives abounded in both the periodical press and on booksellers' shelves in North America and the United Kingdom. Since then, the popularity of cycle-travel literature has risen and fallen with that of cycling in general. This chapter offers an overview of that literature, beginning with an important distinction: I separate *writing* about cycle travel from *literature* about it. The earliest accounts of cycle travel, from the 1870s, mostly in magazines in England and America, tended to be not so much literary as itinerary: plain-prose descriptive narratives of distances rode, places visited, and technical and logistical details about things like road conditions, supply points, and accommodation. Cycle-travel *literature*, I argue, offers something more: an aesthetic element, some kind of artful or captivating narrative, and, sometimes, a direct engagement with other literature and writers. For the purpose of this overview, I have classified producers of cycle-travel literature into three main categories – pilgrims, ramblers, and adventurers – and I will trace the evolution of the literature produced by each.

Pilgrims

In the 1880s and 1890s, Elizabeth Robins Pennell and Joseph Pennell, a husband and wife duo from Philadelphia, were pioneers of the literary cycling pilgrimage. She did the writing, he did the illustrating, and together they published five books and dozens of magazine articles about cycle travel in England and Europe, delighting readers curious about this new means of travel and inspiring others to hit the road on two or three wheels. The Pennells adapted the successful literary travel-writing style of Robert Louis Stevenson, whose breezy *Travels with a Donkey in the Cevennes* (1879) had been a success. The Pennells' version swapped out the donkey for a tandem tricycle, but maintained a Stevensonian spirit of literary nostalgia and charm (Buchanan 2015: xxxi).

Their first book, *A Canterbury Pilgrimage* (1885), which sees the Pennells follow the footsteps of Chaucer's pilgrims on a three-day tricycle trip from London to Canterbury, exploring the literary geography of Kent along the way, hit on a formula that would prove popular for cycle-travel writing. Nicola J. Watson (2006) has traced the rise of what she calls "readerly tourism"

DOI: 10.4324/9781003142041-70

in the 19th century, the impulse to travel to places associated with particular authors or books "in order to savour text, place and their intersections" (1). From Shakespeare's Stratford to Sir Walter Scott's Abbotsford, the Brontës' Haworth, and Hardy's Wessex, homes, haunts, and resting places of beloved writers became tourist destinations in the Victorian period. Watson says this phenomenon peaked in England between the 1880s and 1920s, a period that overlaps with cycle travel's flourishing, so it's no surprise to see the literary pilgrimage become such a prominent part of early cycle-travel literature (Figure 48.1). The Pennells employed this successful angle again in their third book, *Our Sentimental Journey* (1887), which recounts their tricycle trip across France, this time following the trail of Laurence Sterne's *A Sentimental Journey* (1765).

The Pennells may have invented the cycling version of the literary pilgrimage, but the idea was taken up by others, both in magazines like *The Wheelman/Outing* and in books. F. W. Bockett's *Some Literary Landmarks for Pilgrims on Wheels* (1901) is structured around a series of 1–2-day cycling trips to what Bockett calls "historic shrines" (8) of literary significance in England: locales associated with his favorite writers, including Jane Austen, Percy Shelley, Lord Alfred Tennyson, and Charles Lamb. Bockett evokes the religious connotations of pilgrimage too, though he sees the bicycle as an improvement on tradition: "staff and sandals" of yore can't compete with "rubber-tyred wheels" (vi). A serious pilgrim, Bockett travels with each beloved author's sacred text – his "dearest treasures" (v) – in his pocket, which he pulls out, once at his destinations, to read from like scripture.

A decade later, Edward Thomas's *In Pursuit of Spring* (1913) continues this literary pilgrimage tradition. He recounts his Easter weekend journey by bicycle from South London to the

Figure 48.1 Joseph Pennell, "The Hospice on the Grimsel." From *Over the Alps on a Bicycle* (1898) by Elizabeth Robins Pennell.

Quantock Hills in Somerset, specifically Nether Stowey, where, in the late 1790s, the poet Samuel Taylor Coleridge had written some of his most famous poems. As was the case in the work of the Pennells and Bockett, Thomas's writing is infused with references to writers connected to the landscapes he passes through: William Cobbett in Surrey; the naturalist W.H. Hudson in Wiltshire; Thomas Hardy in Wessex; and, of course, Coleridge in Somerset.

In the 1920s, the American Charles S. Brooks continued the cycling pilgrimage tradition in *A Thread of English Road* (1924) and *Roads to the North* (1928). Brooks offers charming accounts of his quiet spins along the backroads of England, visiting spots of literary interest – birthplaces, homes, and graveyards of famous writers, from Shakespeare to the Brontës, Hardy to Hazlitt. Stylistically, Brooks is a kind of hybrid of the American perspective of the Pennells and the nostalgic eccentricity of Bockett. The pages are studded with allusions to novelists who wrote of the English countryside – Dickens, Smollett, and Richardson – and playful imitations of Pepys and Fielding.

In general, the literary pilgrimage craze was in decline by this time, as Watson says, but one more significant contribution arrived in the form of the first cycling book by Bernard Newman, cycling's most prolific travel writer. *In the Trail of the Three Musketeers* (1934) is a classic example of the literary pilgrimage. Newman grew up savoring Alexandre Dumas' books about the adventures of d'Artagnan, Athos, Porthos, and Aramis. So Newman sets out to "follow them yard by yard" (19) across France, and he chooses the bicycle as his ideal means of transport: "it has the approximate speed of the horse [the Musketeers' method of transportation], and is not so blatantly modern as a car; further, it can go almost anywhere" (19). Like the Pennells seeking a travelling experience close to that of Laurence Sterne in the 18th century, Newman sees the bicycle as the perfect bridge between past and present. While criss-crossing France, Newman, like many pilgrims before him, experiences the "excited feet" phenomenon: "I was now literally treading the paths, maybe the very stones, which my heroes had trodden," he gushes (37).

The pilgrimage awheel faded in popularity in the middle of the 20th century, but it was revived during another flourishing of cycle-travel literature in the 1980s and 1990s, though often with a broader interpretation of pilgrimage. Bettina Selby employs a pilgrimage angle in several of her books. In *Riding the Desert Trail* (1988), she traces the route of Victorian explorer and author Amelia Edwards up the Nile; and in *Frail Dream of Timbuktu* (1991), Selby follows the footsteps of Scottish explorer Mungo Park, as outlined in his classic *Travels in the Interior Districts of Africa* (1799). Elsewhere in, for example, *Riding to Jerusalem* (1986) and *Pilgrim's Road: A Journey to Santiago de Compostela* (1994), Selby invokes and enacts the concept of religious, rather than literary, pilgrimage. Like Bockett long before her, Selby argues that the bicycle is ideal for such sacred travel, not just because its "gentle speeds" allow her to take in all the sights, sounds, and smells of the country, but also because the bicycle itself functions as the modern-day equivalent of the pilgrim's staff and cockle shell, providing assurance of safety traditionally afforded pilgrims (Selby 1988: 3). Anne Mustoe, the British former headmistress-turned-round-the-world cycle traveller, incorporates elements of pilgrimage in all seven of her books, though she is more of an historical than a religious or literary pilgrim. In *A Bike Ride: 12000 Miles Around the World* (1991), Mustoe, a Classicist by training, traces old Roman routes in Europe and follows the footsteps of Alexander the Great through the Middle East; in Pakistan and India, she traces the Moghuls and the Raj; in America, she follows the pioneer routes of the 19th century.

This pilgrimage motif pops up in numerous cycle-travel books in the late twentieth and early 21st century. Tom Vernon follows historical routes in *Fat Man on a Roman Road* (1983); Pamela Watson traces the 14th-century Moroccan explorer Ibn Batutta in *Esprit de Battuta:*

Alone Across Africa on a Bicycle (1999); Brady Fotheringham embarks *On the Trail of Marco Polo* (2000); and Edward Enfield traces the route of Lord Byron in 1809 in *Greece on My Wheels* (2003). A notable recent variation of the cycling pilgrimage can be found in Tim Moore's *French Revolutions* (2011) and *Gironimo*! (2014), humorous travelogues of trips along historical routes of professional bicycle races, the Tour de France and the Giro d'Italia, respectively.

Ramblers

Not all cycle-travel writers emphasize destinations or specific routes the way pilgrims do. In fact, a vibrant tradition of cycle-travel writing from the 1890s to the 1940s is more concerned with celebrating the experience of the ride as an end in itself. For these ramblers, as I call them, almost all of whom are British writers, cycle travel is not about moving quickly or getting to a particular spot so much as it is about philosophical or spiritual reflection and experiencing the natural world. In the rambler tradition of cycle travel, the destination is a state of mind (Chapter 32).

In the early decades of cycle travel, the literary inspiration for these philosophical cycle travellers was Izaak Walton's 17th-century classic *The Compleat Angler; or the Contemplative Man's Recreation* (1653), a celebration of the art and spirit of the gentlemanly pastime of angling. Walton's highly literary book, structured as a conversation between three voices, offers a combination of instruction, anecdotes, recipes, quotations, and poetry. The book gave some late 19th-century ramblers a kind of template for viewing a sporting pastime – even one as new as cycling – as a genteel, aesthetic activity. F. W. Bockett makes explicit the link to Walton, arguing that cycling is to Bockett's age what angling was to Walton's. Both are civilized, literary pastimes for "gentle folk" (1901: 230) – though Bockett claims that his sport is even more civilized than Walton's. Cycling is, Bockett says, "a companion to the solitary, a friend that is always exhilarating and never selfish, an aid to reflection; it gives inspiration to the poet, health and strength to the plain man, vigour to the man of science, and breadth to the philosopher" (Bockett 1901: 7).

One of the best early examples of this Walton-inspired rambleresque cycle-travel literature is William S. Beekman's *Cycle Gleanings: or, Wheels and Wheeling for Business and Pleasure and the Study of Nature* (1894). Like *The Compleat Angler*, it's a curious mashup of genres: polemic, dramatic dialog, poetry, photos, and travel narrative. Beekman argues that riding bicycles can enhance our understanding and appreciation of Nature, the cosmos, and ourselves. In particular, Beekman says that travelling by bicycle enables a unique perception of, and connection to, the natural world. The cyclist, he claims, "finds a warmth in the woodland paths" that the "ordinary tourist," afoot or in a carriage, cannot attain (13). On a bicycle, Beekman says, the rider is alive and attuned to nature in a unique way, capable of a special kind of perception that is almost mystical (14). Cycling, he argues, "allows one to view the kaleidoscopic panorama of Nature's face" (2).

J. W. Allen's *Wheel Magic: Revolutions of an Impressionist* (1909), another classic of the rambler tradition, also references Izaak Walton. Allen politely declines the mantle of the "Izaak Walton of cycling," but his book is, nevertheless, solidly in that *Compleat Angler* tradition, though with a modern twist. Allen describes a series of mostly day trips around southern England, favoring back roads and old routes, stopping at country inns and ruins. The actual destinations seem almost arbitrary; he refuses guidebooks (they're "disenchanting" (72)). Instead he wanders, sometimes with companions, other times alone, in search of a certain experience – pastoral, nostalgic, contemplative, mystical. Despite his dark view of the modern world, Allen concludes that bicycle travel can be a kind of therapy: his machine, he says, "will carry me from out the shadow of any cloud" (167).

Allen's views on the restorative powers of cycle rambling would have resonated with Edward Thomas. While *In Pursuit of Spring* is, as I've said, a literary pilgrimage, it's also firmly in the "country writing" tradition of his other books, cataloging marvelous details of English rural life and nature viewed from the saddle. Thomas's poetic inclination reveals itself in his description of clouds "hung like pudding bags all over the sky" and the "inhuman lamentation" of telegraph wires in the wind (Thomas 1913: 47, 66). Like Wordsworth, Thomas says he found in nature a regenerative presence "both sensuous and spiritual," recalling some of the mysticism of Allen's cycling experience (275). Although this was Thomas's only cycling book, it captures as few have the transcendental possibilities of bicycle travel induced by movement through nature. On his bicycle, Thomas says,

> I was a great deal nearer to being a disembodied spirit than I can often be… I fed through the senses directly, but very temperately, through the eyes chiefly, and was happier than is explicable or seems reasonable.
>
> *(210)*

Perhaps the best-known writer in the rambler tradition is W. Fitzwater Wray, who, under the pen-name "Kuklos," wrote cycling columns for London newspapers between 1894 and 1919. His essays about trips in England and Ireland, collected in *A Vagabond's Notebook* (1908), and travel pieces in *The Kuklos Papers* (1927), epitomize the wry, genteel, literary style that garnered him a loyal and large following. "A Puncture in Paradise," from the latter collection, captures his rambleresque style. Describing a flat tire on a trip through the Cotswolds with his wife, Klossie, Kuklos offers a droll blow-by-blow account of his "leisurely"(103) approach to fixing a flat, of which F. W. Bockett would have been proud. It entails a disquisition on pipe smoking, repair kits, and beech nuts (Figure 48.2).

Figure 48.2 "Kuklos" or W. Fitzwater Wray (1931). Photographer unknown. Credit: The Cycling Photographica Collection of Lorne Shields.

The 1930s and 1940s saw a few successors to Wray's elegant, if self-deprecating accounts of cycle rambling. Bernard Newman, in his many books about European travel awheel, brings a charming "light style" and "casual attitude," as he puts it (1953: 18). Never a slave to the clock or a plan, Newman rambled freely through Poland, Germany, France, Albania, and Italy, mixing insightful observations and a mellow perspective. Similarly, James Arnold's *The Joyous Wheel* (1940) offers low-key accounts of his solo rides through the rural landscapes of the Chilterns and Cotswolds, evoking an idyllic, pre-war aesthetic of country innocence, much like the charming illustrations of Frank Patterson in the Cyclists' Touring Club *Gazette*, and *Cycling Weekly* magazine.

Adventurers

Both pilgrims and ramblers tend to take a leisurely, recreational, small-scale approach to travel, one that emphasizes interactions between traveller, place, history, texts, and nature rather than distances covered and difficult terrain traversed. But a parallel tradition of what I call adventure cycling – more extreme travel, grandly ambitious, on an epic scale, full of risk and even danger – can also be traced from cycling's origins to the present. Such accounts are literary in a different way – not through direct allusion, appeals to literary nostalgia or history or philosophy but rather through the sheer narrative pull of adventure storytelling, in the tradition of Defoe, Dumas, and Stevenson.

Alfred M. Bolton, author of one of the first British cycle-travel books, *Over the Pyrenees: A Bicyclist's Adventures among the Spaniards* (1883), plays up this adventure angle in his early cycling writing. His article about cycling in Sweden and Norway appears in *The Boys Own Paper* next to harrowing narratives of polar missions, African expeditions, and wild escapades in the Oregon gold fields. In keeping with the genre, his prose is punctuated with accounts of run-ins with Spanish brigands, toughs, and highwaymen. Bolton's thesis was obvious: cycle travel was the stuff of adventure.

The most famous pioneer of the adventure-cycling genre was Thomas Stevens. Around the same time that the Pennells began publishing their literary accounts of leisurely tricycle pilgrimages, this transplanted Englishman with virtually no riding experience and a background in journalism set off on a bold excursion to circle the globe on his high-wheeler, leaving San Francisco in April 1884. After reaching New York, he struck a deal with Colonel Albert Pope, owner of *Outing Magazine*, to become a travelling correspondent. As he journeyed across Europe and Asia, Stevens sent regular dispatches from the road, which became a popular feature with readers hungry for more tales of two-wheeled adventure.

After completing his remarkable trip in December 1886, Stevens gathered these articles into a book, *Around the World on a Bicycle* (1887). Stevens' writing was more journalistic, and not as explicitly literary as his contemporaries the Pennells. But, as Duncan R. Jamieson says, Stevens "had the good sense of a travel and adventure writer" (46), and knew enough to focus on exotic locales and thrilling encounters, while omitting the day-to-day details. His remarkable yarns, such as his account of huddling against the wall of tunnel in the Sierra Nevada mountains as a train passed by or his encounters with mountain lions and bears or the time he was arrested as a Russian spy in Afghanistan, reveal an eye for detail "clearly worthy of Jules Verne's pen" (Jamieson 2015: 46).

In America, after the example of Stevens, much early adventure cycle-travel writing was closely connected to newspapers. Inspired by Stevens, New Englander George B. Thayer straddled his high wheel, picked up his pen, and set off on a grand adventure, writing a series of dispatches for the *Hartford Evening Post*, which was later published as a book, *Pedal and*

Path: Across the Continent Awheel and Afoot from Connecticut to California (1887). In 1892, Frank Lenz from Pittsburgh set out on his safety bicycle to follow Stevens' trail, writing dispatches for *Outing*, though his pieces were never collected into a book; he disappeared mysteriously somewhere in Turkey in May 1894. Tom Winder, from Indiana, rode the circumference of the United States, penning pieces for the *Buffalo Express* as he went, eventually turning those into *Tom Winder's Famous Twenty-Thousand Mile Ride* (1895). A few years after that, the American Darwin McIlrath wrote about his and his wife Hattie's around-the-globe cycle trip for the *Chicago Inter Ocean* newspaper, eventually collecting the dispatches in *Around the World on Wheels, for the Inter Ocean* (1898).

This forum of publication – newspapers – greatly influenced the content and style of early adventure cycling narratives. Adventure writers focused on action and intrigue: overcoming extreme hardship, surviving encounters with wild animals, braving moments of danger. The McIlraths were almost lynched in Nevada, when a posse of vigilantes mistook Darwin for a bandit; he was saved only by a lack of gold fillings in his teeth that proved he was not their man (McIlrath 1898: 22). When Tom Winder found himself trying to evade collision on a train trestle bridge, he tied his bicycle to a rope and lowered it off the side while the train passed. Such stories sold newspapers. Eventually, the focus on high adventure became part of the genre, and even cycle-travel books that had no affiliation with newspapers embraced the convention. The oeuvre of Englishman Robert Louis Jefferson, a former racer who wrote a series of adventure-style cycle-travel books, including *To Constantinople on a Bicycle* (1894) and *Across Siberia on a Bicycle* (1896), is a case in point. Jefferson appealed to those looking for extreme adventure. He was not interested in literature; nor was he a dawdler or ponderer of nature. He rode fast, covered enormous distances, ate whatever he could find, and slept rough. Readers liked this angle enough for him to sell five books' worth of his adventures. So popular was this strain of adventure-cycle writing in the 1890s, that even Elizabeth Robins Pennell, that ultimate leisure-cycle pilgrim, dabbled in the genre. Her *Over the Alps on a Bicycle* (1898) is a one-off, some would say ironic, experiment in the adventure cycling tradition (Chapter 33).

In what Jamieson calls the "Golden Age" of cycle-travel writing, 1895–99, adventure ruled (2015: 83). The books of American couple Fanny Bullock Workman and William Hunter Workman were less literary but more adventurous than their contemporaries the Pennells. In volumes such as *Algerian Memories: A Bicycle Tour over the Atlas to the Sahara* (1895) and *Sketches Awheel in Modern Iberia* (1897), the Workmans describe their exotic encounters with, for instance, snake charmers, surly camels, and bull fights. Despite her insistence that they travelled "at leisure, stopping where and when we pleased," her accounts of her cycle travels often seem anything but leisurely (1897: v). Fanny carried a whip in North Africa to ward off menacing dogs, but she also packed a revolver, in the tradition of Thomas Stevens, and had to use it on more than one occasion, to scare off thugs and thieves.

John Foster Fraser's *Round the World on a Wheel* (1899) is a classic of the adventure-cycling genre. Like Stevens and others before him, the Englishman Fraser and his friends Edward Lunn and F.H. Lowe embraced danger and the unknown as they set out from London on their Rover bicycles to visit 17 countries across more than 19,000 miles in a little over two years. They navigated atrocious roads (and sometimes no roads), coping with mechanical breakdowns in remote places, encountering hostile natives (they were pelted with stones so frequently that Fraser mentions it the way one might mention a rain shower), bad food (or no food for days at a time), illness, filthy lodgings, blizzards, wolves and bears and mobs. Like Stevens and the Workmans, they travelled with revolvers – which came in handy several times.

In the first half of the 20th century, adventure cycling took on some new forms, such as in *Across France in Wartime* (1916) by the rambler Fitzwater Wray. Like so many adventure cyclists before him, Wray wrote trip reports for a newspaper, in his case London's *Daily News*. But the dangers he faced on this trip were of a different magnitude than his predecessors'. Wray cycled a 520-mile loop into and back out of France's war zone, taking on levels of risk far beyond anything experienced by Fraser or the Workmans. Pedalling through the "tortured and desolate countryside"(80) of shattered trees and the "ghastly ruin" of burned buildings, the sounds of rifle fire in the distance (137), Wray on his Raleigh three-speed is arrested at one point, taken for a German in disguise on more than one occasion, and accosted by bayonet-wielding soldiers (52).

Epic cycling-adventure narratives declined in popularity in the first half of the century, but there are some notable instances, such as *Cycling Over the Roof of the World* by Framji Jamshedji Davar (1929), Fred Birchmore's *Around the World on a Bicycle* (1939), and Harold Elvin's *Ride to Chandigarh* (1959). But the tradition was revived in a big way with Dervla Murphy's *Full Tilt* (1965), her classic first book about a 3000-mile ride from Ireland to India, which hearkens back to the adventure tradition of Thomas Stevens. In fact, she follows the same general route that he did eighty years earlier (Jamieson 2015: 150). Murphy rides solo through perilous territory in Yugoslavia, Iran, Afghanistan, and Pakistan, on her bicycle Roz (a nod to Don Quixote's trusty steed). Her plucky, indefatigable diary-style account captures more of the day-to-day grind of adventure-cycle travel than did her predecessors, punctuated by moments of high drama and hardship – running out of food and water; frostbite and heatstroke; and other chilling encounters. Like Stevens and Fanny Workman before her, Murphy packs a pistol, which she uses to ward off starving wolves, menacing men, and would-be bicycle thieves. But amid the tribulations, Murphy also writes of the astonishing beauty of the Himalayas and the relentless kindness of the people she meets.

Twenty years later, epic cycle-travel adventure stories began to boom again. In *Miles from Nowhere* (1983), Barbara Savage revives the couples tradition of cycle-travel adventure writing pioneered by the Pennells and Workmans, and carried on by the likes of the McIlraths and Jim and Elisabeth Young's *Bicycle Built for Two* (1940). Savage's book, a classic of the genre, may be the best-known travel book inspired by the Bikecentennial phenomenon that saw thousands of young people ride and tramp their way across America in 1976. It features plenty of adventure, from diving into ditches to avoid crazed Floridian motorists to dodging rocks hurled by Egyptian children; but also humor (Kiwi Geoff's "pet" tapeworm has become legendary); and Savage's disarming candor, recounting petty fights in the ditch but also the profound intimacy of cycle touring with one's partner.

In the 1990s, a new variant of the adventure-cycle travelogue took off: the cycle-trip memoir. In this subgenre, an epic and often-arduous cycling adventure becomes a metaphor for personal growth. In *The Wind in My Wheels* (1992), for instance, young Josie Dew overcomes initial insecurities to embrace her independence as a solo woman traveller, roving across four continents, making insightful and often humorous observations, touching on the hazards of a woman travelling alone but also the beauty of a life devoted to two-wheeled wanderlust. Sometimes these memoir travelogues are about reconnecting with the landscape of an individual's past, as in Andrew X. Pham's *Catfish and Mandala* (1999); or working through personal loss, trauma, or spiritual crisis, as in Brian Newhouse's *A Crossing: A Cyclist's Journey Home* (1998). Yet another variant is the coming-of-middle or even old-age cycle-travel book, such as David Lamb's *Over the Hills* (1996) and Daryl Farmer's *Bicycling Beyond the Divide* (2008), both about a mid-life reckoning; and Lorraine Veisz's *Conquering the Borderlands* (2009), about a woman in her 60s, exploring both geographical and chronological borderlands.

Lands of Lost Borders: A Journey on the Silk Road (2018) by Canadian Kate Harris is a notable recent example of how some cycle-travel literature combines features of the pilgrimage, rambler, and adventurer traditions. Harris, like many before her, traces the famous trade route through Asia, but with a literary guide, Alexandra David-Néel, whose book *My Journey to Lhasa* (1924) recounts her travels in Tibet almost a century earlier. Like so many of her predecessors in cycle-travel literature, from Pennell to Bockett to Brooks, Harris treasures both books and bicycles, for where both can take her. She references a dizzying array of literary influences, from Marco Polo and Charles Darwin to Annie Dillard and Rebecca Solnit. There's adventure too: sneaking across international borders under the cover of darkness, struggling with hunger and sickness. But Harris also has a touch of the rambler-poet to her. The book itself blurs generic borders, combining travel writing, memoir, philosophy, and metaphor. Harris captures the experience – shared by many a pilgrim, rambler, and adventurer: "Travelling by bicycle is a life of simple things taken seriously: hunger, thirst, friendship, the weather, the stutter of the world beneath you" (2).

The evolution of cycle-travel literature is not unlike that of the bicycle itself: by the late 1880s, the concept had been established, and the forms really haven't changed significantly since. The three main types of cycle-travel writers that emerged in the late 19th century – pilgrims, ramblers, and adventurers – remain responsible for the vast majority of cycle-travel literature produced today. The voices are more diverse, the perspectives broader, the number of subgenres multiplied, but the essential impulse remains the same: to articulate in a compelling fashion the experience of seeing the world over the handlebars and capture on the page the "stutter" of life under one's wheels.

References

Allen, J. W. (1909) *Wheel Magic or Revolutions of an Impressionist*. London: John Lane.

Beekman, W. S. (1894) *Cycle Gleanings, or Wheels and Wheeling for Business and Pleasure and the Study of Nature*. Boston: Skinner, Bartlett & Co.

Bockett, F. W. (1901) *Some Literary Landmarks for Pilgrims on Wheels*. London: J. M. Dent & Co.

Buchanan, Dave. (2015) Introduction. *A Canterbury Pilgrimage and An Italian Pilgrimage* by E. R. Pennell and J. Pennell. Edited by D. Buchanan. Edmonton: University of Alberta Press.

Harris, Kate. (2018) *Lands of Lost Borders: Out of Bounds on the Silk Road*. Toronto: Knopf.

Jamieson, D. R. (2015) *The Self-Propelled Voyagers: How the Cycle Revolutionized Travel*. Lanham: Rowman & Littlefield.

McIlrath, H. D. (1898) *Around the World on Wheels, for the "Inter Ocean"*. Chicago: Inter Ocean.

Newman, B. (1953) *Ride to Rome*. London: Herbert Jenkins.

Selby, B. (1988). *Riding the Desert Trail*. London: Random House.

Thomas, E. (1913, 2016). *In Pursuit of Spring*. Lower Dairy: Little Toller.

Workman, F. B. and Workman, W. H. (1897) *Sketches Awheel in Modern Iberia*. New York: G. P. Putnam's.

Watson, Nicola J. (2006) *The Literary Tourist: Readers and Places in Romantic and Victorian Britain*. London: Palgrave Macmillan.

Wray, W. Fitzwater ('Kuklos.') (1916) *Across France in War-Time*. London: J. M. Dent & Sons.

Wray, W. Fitzwater. (1927) *The Kuklos Papers*. London: J.M. Dent and Sons.

INDEX

3D printing 105–106, 157, 178, 190–191

A-Team 151, 164–168, 192–193, 197–200
academic cycling studies 1, 2, 11–13, 15–16, 30, 62, 202, 237, 345
acceleration: developmental 55, 70, 165–166, 191, 232, 272, 413, 498; physical 92, 178, 264, 268, 307–308, 318, 323
accessibility 26, 33–34, 38–39, 44–45, 55–63, 65–67, 69, 70, 74–77, 80, 142–144, 204, 205, 208, 217, 230, 235, 254–259, 311, 339, 356, 373–374, 380, 386, 390, 400, 402–406, 432, 454, 494, 510; *see also* inclusivity
accessories 59, 101, 128, 131, 150, 169, 172, 174, 184, 191, 469
accidents 6, 37, 76–77, 126, 217, 265–266, 339, 346, 400, 406, 415, 424, 428, 482, 515–519
active: lifestyle 26, 60, 75; travel/transport 21, 42, 48, 237, 254, 258–261, 332–333
activism 12, 38, 216–217, 245, 247, 255, 428, 435–436; *see also* advocacy; campaigning
actor network theory (ANT) 20, 374, 376
Adams, Jay Howe 367
adolescence 28, 84, 284, 335–337, 339, 382, 450–451, 488, 505–507, 521; *see also* Youth
advanced stop line 224
adventure 4, 363–364, 371, 376–384, 436, 454, 458, 469, 476, 478, 486, 494, 508, 510, 533; adventure cycling 538–541
advertising 28, 52, 68, 149–150, 181–183, 187, 201, 208, 235–236, 288, 315, 380, 391, 440, 443–452, 453, 456–458, 462–471, 479, 505
advocacy 4, 12, 29, 44, 46, 55, 62, 66, 201–202, 207–208, 215–217, 229, 243–246, 251–252, 272, 310–311, 317, 374, 386, 421, 436, 509; *see also* activism; campaigning
aerodynamics 68, 95, 97, 101–104, 106, 108, 110–111, 312, 318, 324–325; *see also* air drag
aesthetics 4, 104, 106, 152, 156, 284, 289–290, 310, 338, 355, 365, 439, 443–452, 463, 481, 493–494, 498–499, 501, 533, 536, 538
affluent cyclists 400–401, 405–406
affordability (cost) 2, 7, 33–34, 45, 56, 59, 66–67, 83, 85, 95, 128–129, 202, 290, 292, 318, 320, 392, 399–405, 449, 454, 460, 473, 510
affordances 20, 26, 203–204, 206–209, 479, 494
Africa 4, 7, 27, 44, 59, 108, 124, 137–139, 181, 201, 255, 302, 355–356, 373, 381, 388–395, 488, 535–539
agency (includes capacity) 2, 18, 20, 39, 43, 62, 140, 237, 245, 363, 366, 404, 484, 514, 517–518
aggression 6, 25, 29–30, 278, 310
air drag 92, 94, 95; *see also* aerodynamics
air quality 232, 250, 332, 339
Allen, Grant 510
Allen, J. W.: *Wheel Magic* 516–517, 536–537
alley cat 29, 207
alps *see* mountains
aluminium 84–85, 104–105, 109–111, 125, 133, 157–158, 160, 318
amateur cycling 3, 16, 183–185, 283–285, 287–297, 301, 315, 328, 331, 347–348, 428, 474
Amaury Sports Organisation (ASO) 301, 303–304
ambivalence 357, 394, 500
Amsterdam 44, 215–216, 232, 244, 255, 272, 278, 380, 453
Andrews, Ryan 507
anthropology 36, 89, 356
anti-cycling 498; *see also* traduced bicycle
anti-doping policies 285, 347–351
appearance 27, 318, 325, 347, 416, 443, 446, 453, 460, 472–475, 508
appropriation of technology and culture 388, 391

Arnold, Schwinn & Company *see* Schwinn
assemblage 2, 20, 43, 49, 144, 485
associations *see* collectives
Australia 24, 25, 27, 30, 59, 152, 181, 208, 265–266, 272, 288, 292, 295, 302, 316, 378, 427–429
authenticity 290, 293, 502
automation 158, 161, 165, 260
automobility as a system 3, 21, 34, 42, 47, 49, 211, 217, 233, 235, 245–247, 267, 269, 276–279, 401
automobiles 33, 55, 65, 68–70, 84, 90, 150, 153, 156, 181, 201–202, 245–247, 276–279, 356– 357, 364–366, 369, 377, 380, 382, 388–389, 403, 430–436, 471, 511, 531; *pico y placa* rules (Bogotá) 434; *see also* cars
autonomy 21, 44, 73, 76, 389, 494; *see also* independence
axles 93, 97, 102, 108–109, 119, 142–143, 145, 416, 445

balance (equilibrium) 8, 86, 89, 93, 95, 131, 142, 145, 410, 473, 478, 481, 487, 491, 515, 527
Bangs, E. L. 508–509
Banville, Théodore de 497–498
barriers to cycling: physical/infrastructural 73–80, 133, 264, 316; social 20, 24, 25, 55–58, 61–62, 73–80, 140, 217, 234, 236, 264, 265, 267, 272–275, 316, 374
Barkman, Albert 370
Barthes, Roland 440, 503
Bassett, Abbott 371
Bassett, Helen Drew (pseud. Daisie) 363–364, 371
bearings 92, 93, 100, 108, 109, 116, 119, 122, 123, 131, 157, 211
Beauvais, Clémentine 507
Beckett, Samuel: *Molloy* 516, 518
Beekman, William S. 363–364, 536
behaviour 19–20, 30, 60, 67, 78, 91, 98, 102–106, 168, 216, 227, 236, 255, 258, 273, 335, 345, 456, 508
behaviour change 45, 219–220, 233, 251, 334, 500
Beijing 5, 152, 211–213, 288, 410, 414–415
Beijing Bicycle (Wang, 2001) 488
Benefit: risk ratio 333, 339–340
Benjamin, Walter 473
Bennett, Jane 376
Bergson, Henri: *Laughter* 494, 502, 514–519
Berhman, Daniel 380, 524
The Bicycle Thieves (De Sica, 1948) 488
Bicycle Booms 26, 27, 31, 90, 96, 129, 131, 145, 153, 181, 233, 241, 263–264, 268, 447, 463, 485, 526, 533, 540
bicycles for development (BFD) 373–374
bike kitchens 59; *see also* repairs
bike-share/bikesharing 3, 7, 16, 58, 59, 139, 146, 161, 168, 175, 212, 230, 233–235, 252, 254–262, 267, 269, 357, 380–381, 405, 416
bike to work 3, 251, 267
bikespace 2, 12, 42–49
biopolitics 232, 234–236
Birchmore, Fred 377, 540
Blondin, Antoine 503
Bloomer, Amelia 91, 456
bloomers 26, 455–456, 458, 460
Boardman, Chris 85, 325
bodaboda 391
body (human) 20, 24–26, 30, 33, 37, 91, 94, 102, 114, 130, 140–142, 178, 203–204, 207, 275, 279, 284, 290–294, 307–308, 322–323, 328–329, 334, 346, 363, 424, 454–455, 477, 483, 484, 486–490, 499–500, 507–511, 514, 516, 530–531
Bogotá (Colombia) 28–30, 37, 43–44, 47–48, 216, 243–247, 251, 357, 430–437; *Ciclopaseos de los Miércoles* (Bogotá) 436; Pro-Cicla (Bogotá) 434, 435
Bogotá Bicycle Users Assembly 435
Bois de Boulogne 4, 359–362, 462–463, 475
bone health 145, 335–336
le Boul, Claude 479
boys 28, 182, 336, 491, 506, 507, 509, 538; *see also* children
brakes and brake levers 86, 93, 94, 109–112, 122–127, 307, 318–319, 329, 392
braking 101, 133, 501, 517
brand (ing) 90, 130, 149–151, 153–162, 163–169, 173–179, 181–190, 192–199, 204, 236, 290, 293, 413, 416, 435, 445, 463–469
Brenna, Beverly 506
British Cycling 288–289
Broadcasting rights 302–303
Brooks, Charles 535
built environment 42, 258, 273, 368, 371, 431
Buffon, Georges-Louis Leclerc, Comte de 498
Burston, George and Harry Stokes 378–379
Business models 38, 112, 118–190, 284, 298, 301, 303–304, 351
buzz 172–180

campaigning: cycle 3, 11, 12, 26, 42, 44, 48, 56, 61, 73, 80, 139, 251, 261, 267, 391, 406, 432, 490; *see also* activism; advocacy
capacity 56, 73, 244–245, 335, 404, 468, 484, 514, 516; *see also* agency
Carrier, Roch 503
carrying capacity *see* load
capital: cultural 208, 295; financial 34, 35, 38, 151–152, 153, 158–159, 173, 199, 234–235, 242, 256, 405, 413, 416, 469
carbon fibre 85, 104–106, 108, 110–111, 115, 116, 129, 142, 157, 158, 160, 161, 197, 292–293, 313
cargo bicycles and tricycles 75, 78, 106, 122, 128, 133, 137–139, 175, 203–207, 221, 227–228, 263, 269, 393, 421, 422, 430, 432

caring 29, 30, 35, 37, 38, 230, 236, 246
cars 5, 6, 21, 58, 65–72, 74, 90, 129, 135, 156, 211, 213, 215, 221, 229, 237, 264, 266, 269, 272, 276–279, 295, 308, 386, 402, 403, 411, 414, 419–425, 433–434, 450, 502, 506, 524, 527, 529–531; *see also* automobiles
Cavendish, Mark 288, 292
Census of India 401
Chéret, Jules 443, 464–465
children 4, 8, 28, 29, 35, 55–60, 62, 78, 83, 86, 90, 92, 94, 128–134, 140, 143, 146, 150, 153, 156, 181–183, 205, 212, 216, 236, 246, 257, 264–267, 275, 284, 289, 295, 335–339, 376, 381, 392, 413–414, 419, 421–424, 441, 486–489, 491, 503, 505, 509, 524, 527, 528, 540
China 4, 7, 66, 90, 124, 137, 144, 149, 151, 154–162, 164, 175, 182–183, 192–199, 211–213, 263, 266, 268, 356–357, 377–381, 392, 409–417, 440, 482–483, 488; bicycle exports 416–417; bicycles under Mao 411–414
Christie-Robin, Julia 455, 456
Ciclovía 4, 47–49, 216–218, 243–244, 247, 357, 430–432, 434–435
ciclorutas 431, 433, 434, 435; *see also* cycle tracks
cinema *see* film
citizenship 4, 28, 62, 140, 199, 216, 233, 237, 241, 243, 403–405, 411, 415, 428, 434–436
class (social) 1, 12, 19–21, 25–26, 33–34, 36, 44–45, 136, 146, 151, 173, 179, 201, 203, 207, 236, 277–279, 283, 289, 295, 310, 346, 356, 374, 389, 390, 397, 401–402, 405, 410–411, 419, 430, 433–436, 439–440, 445–451, 454–460, 462, 485, 488, 491, 501, 505, 508–511
clean cycling 345, 349, 351
climate change/climate crisis 1, 21, 65–66, 80, 232, 276, 332, 339, 440, 509
clothing 4, 25, 26, 28, 30, 83, 91, 142, 150, 177, 283, 287, 289, 293, 308, 313, 314, 325, 328, 360, 378, 382, 385, 401, 413, 419, 421, 439–440, 448–449, 453–461, 468, 472, 478, 490, 508–509; *see also* dress
clubs 26, 27, 42, 44, 53, 59, 86, 89, 128, 216, 283, 291, 295, 317, 338, 359–360, 367, 370, 377–379, 406, 427–429, 453, 462, 464, 490, 510, 523–524, 538; *see also* collectives
collectives 29, 37, 40, 48, 61, 95, 152, 216, 251, 277, 301, 311, 317, 435; *see also* clubs
collectivity 39, 46, 49, 207, 247, 290, 356, 414, 482–483, 490, 501–503, 514
colonialism 4, 8, 35, 40, 153, 203, 356, 388–393
comedy and cycling *see* humour
community 8, 20, 42, 44–46, 58, 61–62, 71, 74, 146, 151, 169, 188, 203, 207, 216–218, 234, 241–247, 250, 254, 258, 277, 287, 305, 311, 313, 315, 317, 355, 380, 392, 397, 509
commuting 1, 6, 7, 34, 44, 74–75, 89, 90, 150, 165, 168–169, 175, 188, 201, 203, 208, 211, 215, 229, 233–237, 241, 247, 250–252, 267, 276–278, 332–338, 377, 380, 391, 400–404, 414, 420–424, 433, 510, 526
competences 20, 28, 37, 56–61, 277, 518; *see also* skills
composites *see* carbon fibre
configuration of cycles 87–96, 115, 142, 158, 222, 284
Congo 389, 391
congestion 1, 5, 6, 68, 135, 205, 211, 213, 215, 232–233, 235, 250–251, 257–258, 300, 360, 380, 402–403, 415, 423, 482
consumers/consumer societies 2, 3, 26, 67, 131, 132, 158, 172–180, 182, 188–189, 238, 261, 290, 457, 465
consumer goods 6, 90, 391–393, 412, 439, 464
consumption 34, 69, 136, 151, 157, 173, 182, 190, 202, 237, 266, 290, 295, 310, 347, 389–393, 413, 439–440, 456, 471
contingency 1, 434
cost *see* affordability
cool *see* social acceptability
Copenhagen 44, 83, 177, 215, 232, 234, 244, 255, 272, 278, 357, 419–426, 453
Copenhagenize 216, 421, 432
Cormier, Robert 506
corsets 454, 458, 465, 469
counterculture 47, 48, 520
couriers 34–39, 45, 47, 201, 205, 206–207, 472–473, 484, 486, 488, 489, 508–509, 511
critical mass 44, 47–49, 216, 245, 436
cycle deliveries 5, 33–41, 45, 70, 139, 151, 201, 204–208, 382, 390, 400, 411, 413, 415, 450, 462, 487, 489
cycle lanes 5–7, 24, 36, 42–44, 49, 56, 58, 70, 76, 94, 133, 139, 216–217, 230, 235, 241–242, 251–252, 258–259, 263, 269, 274, 300, 367, 380, 385–386, 400–403, 413, 419–423, 429, 425, 432
cycle logistics 201
cycle paths *see* cycle lanes, cycleways
cycle racing 3, 4, 25–27, 42, 52, 88–89, 97, 101, 103, 104, 106, 110–113, 120, 122, 125, 127, 139, 150, 181–185, 196, 283–284, 288–293, 301, 310, 314, 315, 318–326, 327–331, 351, 428, 431, 432, 449, 453, 487, 490, 491, 521
cycle retailing 3, 90, 150, 157, 159, 173, 176–179, 181–191, 206, 287, 380; mass v. specialty 183; new-wave and mobile 188
cycle routing 217, 219, 221–223, 230, 235, 250, 274, 369–371; *see also* cycle track; cycle lanes
cycle sales: online 3, 145, 151, 176, 183–185, 188–189, 290; specialty independent (IBD) 90, 183, 187–188, 447

cycle sport 3, 16, 27, 30, 52–54, 60, 88, 109, 120, 142, 149, 165, 169, 267–268, 361, 374, 428, 432, 434, 447–449, 453–455, 457–458, 460, 462, 464, 469, 471, 486–490, 521, 526, 536
cycle superhighway 233–234
cycle taxi 7, 128, 133, 135–136, 138, 139, 144, 204–205, 213, 257, 391–393, 411
cycle touring, cycle tourists 59, 93–94, 120, 122, 126, 363–364, 370, 377, 382, 447, 449, 462, 476, 500, 516, 538, 540
cycle track 224, 226–227, 250, 357, 433; *see also* cycleway; cycle route
Cycle4Change 404
cycleways 3, 217, 220, 222–228; *see also* cycle routing; cycle track
cycling initiatives 55–64, 216, 229, 237, 242, 245–247, 400, 404–406, 431, 434
cycling mobility system 55–56
cycling press/magazines 52–54, 89, 315, 455, 471, 476, 498, 524, 533, 538
cycling programs 55–64, 334
cycling teams 138, 167, 188, 288, 299–304, 328, 348, 486
cycling trajectories (restorative and resilient) 267–268
Cyclists' Touring Club *see* clubs
Cyclocross 4, 284, 310, 312, 316

data privacy 260, 261
Day, Jon 484
Deliveroo *see* food delivery
Desgrange, Henri 325
demand (for cycles/cycling) 2, 42, 43, 45, 61, 85, 112, 132, 149–150, 153, 155–156, 158, 163, 167–169, 197–199, 220, 224, 254, 268, 391, 405–406, 411–412, 423
democratization 463
demographics 39, 73, 83, 202, 217, 254, 256–257, 273, 274, 311, 313–314, 332, 423
device caps 258
dirt 122, 305, 310–312, 521, 530; dirt jumping 314–317
disability 3, 8, 29, 59, 62, 69, 73–81, 84, 86, 94, 128, 133, 140–147, 159, 221, 227, 236, 259, 516
disciplinarity (academic) 1, 2, 8, 11, 12, 15, 18, 24, 152, 273, 275, 369, 374
doping 4, 284–285, 288, 298, 303, 305, 345–354; definitions of 345
downhill 101, 104, 312–317, 320, 524
dress *see* clothing
Drais, Karl 87–88
draisine *see* hobby horse
Duchamp, Marcel 440, 479, 481, 483
duties (import and export) 153, 174, 182, 183

e-bike 3, 5, 18, 45, 67, 95–96, 106, 115, 139, 141, 144, 151–161, 168, 175–176, 178, 184, 191, 197, 199, 211–213, 215–217, 255, 263–271, 275, 317, 335, 416
e-bike benefit 335
early cinema and cycling 484–485
ecology 266, 316, 509; *see also* environmental concerns
economic growth 236, 238
economy (cycling) 45, 149–152, 202–203, 241
efficiency: of mobility 5, 69, 236, 238, 250, 267, 403, 487; economic/labour 5, 37, 172, 235, 411; energy-efficiency 68, 69, 83–84, 202, 447; mechanical 98–99, 106, 115, 118–127, 203, 312–313, 446, 450, 515; space-efficiency 5, 66, 68, 215, 220, 227
elderly cycling *see* older riders
electric vehicles (EV) 66, 135, 191, 260, 264, 526
elites: social 203, 208, 247, 397, 409, 435, 436, 464, 509; sport 283, 284, 288–293, 295, 315, 324, 325, 331, 346–347, 350–351
emancipation 12, 36, 208, 456, 489, 505, 511
embodiment 106, 277, 356, 357, 419, 424–425, 429, 431, 494, 537
emotions 29, 43, 68, 79, 178, 244, 357, 365, 373, 378, 382, 424–425, 502, 523
empowerment 33, 44, 59, 388
enforcement 4, 28, 156, 222, 255, 258, 259, 261, 278
entrepreneurship 33, 35, 197–200, 211, 236, 391, 401
endurance 25, 27, 284, 305, 312–313, 318–319, 321–324, 345, 428–429, 468, 494
environment/nature, interaction with 2, 7, 76, 310, 317, 338, 498, 501, 510, 519, 526, 528, 530, 536–539
environmental concern/impacts/benefits 2–4, 6–8, 29, 44, 48, 83, 106, 136, 137, 139, 150, 152, 156, 165, 167, 168, 197, 199, 202, 205, 215–218, 231, 233–236, 238, 242–246, 251, 254, 257, 266, 291, 311, 316–317, 364, 381, 391, 519
epidemiology 273, 333–334, 337–338
equilibrium *see* balance
E.T: The Extra-Terrestrial (Spielberg, 1982) 491
ethnicity 16, 33, 35, 38–39, 45, 236, 275, 374
ethnography 37, 216, 357, 392, 431
Eurobike 175–177, 179
everyday life 243, 340, 388, 399, 419, 423–425, 486, 488
everyday cycling 24, 27, 89, 220–222, 291, 295, 399–407, 419, 424, 434, 456
exclusion 12, 16, 21, 29, 45, 49, 55, 61–62, 77, 89, 279, 401, 405–406, 419, 433, 450, 460, 490
exercise 1, 2, 7, 26, 27, 30, 47, 73, 75, 77–80, 141, 143, 145, 151, 173, 199, 250–251, 263, 317, 334, 397, 424, 453–455, 458, 473, 491, 524
exhibitions 161, 164, 172–180, 185, 359–362, 464, 469, 474, 519; *see also* trade shows
experiences (of cycling) 3, 8, 16, 24, 29, 36, 49, 60, 73–74, 79, 85, 140, 143, 146, 201, 215–216, 243, 247, 264, 275, 277–279, 284,

288, 290, 291, 294, 307–309, 338, 363–365, 371, 376–384, 400, 419, 424, 432, 440, 463, 486, 491, 493, 498, 500, 507, 523, 536–541
experiential learning 244; knowledge 174, 177, 178
export-oriented development 163–164, 196

face-to-face (F2F) communication 172, 178, 179
fairings 95, 101
family cycling 59, 60, 75, 78, 129, 204, 380
fashion 4, 8, 26, 42, 84–85, 150, 289, 432, 446, 453–461, 462–463, 473, 475, 489, 497; *see also* clothing
fat biking 313
Fatma Aliye Topuz 396
fear 18, 27, 34, 37, 60, 84, 246, 277, 357, 360, 380, 396, 430–432, 436, 455, 490–491, 526
fees 6, 258–259, 300, 303, 405, 416
femininity 26, 27, 364, 382, 440, 454–455, 456, 457–458, 460, 462, 469, 508, 509
feminism 18, 19, 38, 61, 91, 242, 245–247, 382, 396, 435, 456, 490
female cyclists *see* women cycling
Festina affair 348, 350
film 4, 8, 53, 397, 440–441, 450, 484–492, 502
filtered permeability 220
fitness 28, 42, 45, 130, 145–146, 159, 216, 233, 265, 289, 292, 307, 334–336, 371, 380, 401, 427, 449, 462, 482, 511
flight 68, 467, 471, 491, 500, 501, 503
food delivery 33, 34, 37, 39, 206, 208, 415; *see also* couriers
folding bike *see* portable cycle
Fournel, Paul 491, 503, 522–524
frame (cycle) 84–86, 88, 91, 97–106, 108, 109, 114–116, 119, 121–122, 124–125, 129–131, 133–134, 145, 157–160, 165, 185, 188, 269, 292, 313, 318, 329, 377–378, 392, 445, 447, 449, 454, 465, 481
freedom 8, 21, 25, 26, 73–81, 145, 288, 295, 355–356, 376–378, 383, 388, 397, 400, 411, 424, 431, 441, 456, 458, 468, 471, 478, 489, 494, 502–503, 505–513, 524, 527–528
France 44–47, 52–54, 59, 84, 89–93, 95, 108, 129, 149, 153, 173, 183–185, 255, 265, 279, 301, 303, 305, 324–325, 348, 361, 377–378, 397, 440, 443–446, 453, 462–471, 474–480, 487, 507, 514, 522, 536
French literature 494, 497–504, 505, 516, 522, 524–525

gearing *see* transmission; hubs
gender 1, 2, 12, 16, 18–21, 24–32, 33, 35–38, 44, 61, 62, 91, 146, 203, 236, 246, 265, 267, 275–279, 294, 291, 294–295, 316, 330, 340, 373–374, 389, 393, 396, 401, 419, 439, 446, 448, 456, 460, 486–487, 490–494, 505–509
geofence 256, 258–259
Giant Bicycle Company 3, 90, 151, 155, 156, 158–161, 164–169, 176, 182, 187, 192–200, 316
gig economy 33–41
girls 27–28, 61, 84, 246, 316, 336, 374, 397, 399, 455, 458, 465, 466, 468, 491, 507, 509, 510, 527
Giro d'Italia 89, 298, 349, 490, 536
global production network 150, 154, 173, 192, 196, 209
Global Cycling Network (GCN) 371
Global South 3, 8, 35, 45, 49, 151, 156, 201, 203, 208, 242–243
globalization 150, 153, 161, 173–175, 192, 199, 302, 351, 355, 409, 414
Glovo *see* food delivery
GPS 2, 150, 256, 371, 416
governance 48, 62, 220, 233, 275, 283, 284, 400
gravel 4, 156, 169, 185, 284, 299, 310, 312, 518
Gracq, Julien 501–502
Great Exhibition 172–174

Hampden Cycle Co. 444
Harvey, David 232
Hassam, F. Childe 367–368
health 2–4, 16, 17, 25, 27–29, 37, 42, 55, 56, 60, 73–80, 130, 141, 146, 156, 168, 176, 202, 205, 215–218, 233–237, 250, 256–258, 264–266, 272–274, 283, 353, 373, 382, 391, 399, 401, 404, 422, 424, 432, 437, 449, 482, 536
health benefits adults 332–335
health benefits children 335–336
health risks 273–274, 338–339, 348
Heaney, Seamus 522
hegemony 47, 278, 381, 494
Hering, Carl 371
heritage 27, 293, 366, 470
Herrera, Lucho 432
high-wheel bicycle 26, 84, 87, 89, 91, 93, 94, 118, 123, 319, 325, 361, 363–366, 377–379, 447, 453–454, 494, 515, 538
hills 119, 120, 135, 137–139, 263–267, 314, 335, 363, 368, 370, 379, 425, 432, 507, 525, 535, 540; *see also* mountains
hobby horse (also draisine) 86, 88, 89, 114, 118, 123, 203, 361, 409, 445, 472–474, 478, 497, 509
horses, equestrianism 26, 89, 93, 160, 232, 304, 311, 316, 319, 324, 327, 331, 360, 365, 411, 428, 453, 467–470, 472, 485, 497–498, 526, 535
Hour Record 284, 324–325, 490
Hoy, Chris 288, 323
hub brakes 110, 124–127
hub gears 84, 95, 109, 119, 121–123
hubs (wheel) 108, 109, 111, 115, 116, 118, 119, 121, 122, 124, 466, 525

Hughes, Shirley 506
human power 2, 84, 87, 93–95, 118, 129, 135, 181, 191, 324, 440, 445, 451, 472, 494
humour 4, 207, 293, 356, 456, 464, 468, 487, 493–494, 514–520, 524, 529, 536, 540
Huntington, Charles 370
hypermobility 37

identity 1, 33, 38, 44, 55, 60, 76, 150, 152, 156, 172, 235, 278, 284, 290, 295, 299, 350, 374, 382, 435, 445, 460, 482, 486, 494, 501, 505, 507–509, 523; national identity 139, 490, 503; visual identity 443–452, 472; *see also* collectivity
ideology 95, 310, 357, 431, 504, 526, 531
imagery 20, 73, 315, 365, 367, 431, 436, 463–464, 470
imaginary (political) 26, 207, 242, 245, 355, 431, 432
immigration *see* migration
impairment *see* disability
imperialism 356, 376, 381
import substitution 163
inclusivity 2–3, 13, 55–64, 73, 77–80, 201, 208, 236, 237, 242, 246, 311, 374, 404, 430, 509
independence 25, 29, 60, 76, 79, 266, 355, 364, 378, 382, 396–398, 491, 505–507, 527–528, 540; *see also* autonomy
independent cycle dealers (IBD) 90, 183, 187–188, 447
inertia 89, 92–94, 318, 517
India 4, 59, 124, 137, 149, 156–160, 181–182, 255, 356, 373, 381, 399–408, 535, 540
industrial clusters 158, 165–166, 176, 192, 196
inequalities 12–13, 19, 21, 27, 33–34, 37, 45, 151, 209, 238, 246, 275, 357, 431–432, 488
infrastructure 2, 3, 20, 21, 29, 38, 42–45, 49, 55–56, 58, 60, 62, 67, 69, 71, 75–76, 79–80, 83, 86, 139, 140–141, 146, 152, 159, 169, 201–202, 215–219, 228–231, 233–234, 236–237, 241–249, 252, 258–261, 267, 269, 273–275, 300, 339, 355, 374, 385, 388–389, 400–408, 413, 419–426, 428, 434–435
informal sector economy 35, 137, 151–152, 201, 203–205, 207–208, 211, 391, 393, 399–401, 404, 430
information economy 177, 489
injury 4, 28, 47, 68, 75, 77, 143, 144, 258, 272–276, 335, 338–339, 356, 360, 382, 399, 428, 429, 430, 518
innovation 18, 19, 44, 66–71, 84–85, 93, 121–122, 129, 133–134, 141, 144, 150–151, 154, 159, 165–167, 173–175, 183, 192–200, 207–208, 215–217, 220, 224, 226, 237, 265, 305, 351, 360, 391, 393, 396, 416, 431, 434, 469, 472, 476, 485, 526
intellectual property *see* patent
interdisciplinarity 1, 2, 8, 11, 24, 152, 374; *see also* disciplinarity
intersectionality 33–39, 44, 61, 272, 275–279
introspective journeying 376, 382, 494, 501

Jakarta 3, 6, 244, 250–253
Jäeger, Gustav 454
Jackson, J. B. (John Brinckerhoff) 364
Jackson, Tim 238
Jarry, Alfred: *The Supermale* 500–501, 516
Jenkins, Jedediah 376
Jerome, Jerome K: *Three Men on the Bummel* 514–515, 518
joy 30, 55, 337, 376–378, 382, 410, 431, 451, 471, 491, 510, 522, 524–526, 528–530, 538; *see also* play
Just-in-time (JIT) 150, 158, 159, 167, 198
justice 2, 3, 12, 43, 46, 49, 65–72, 140, 208, 217, 236–237, 245, 272–281, 381, 397, 432

keirin 94, 284, 288, 318–319, 321, 327–331
Kennard, Mrs. Edward 382
kinopolitical constellations 43, 44, 48
Kron, Karl (pseud. of Lyman Hotchkiss Bagg) 365, 379

labour: employment/waged 1, 2, 33–38, 151, 153, 154, 158–161, 164, 173, 194–196, 199, 205–208, 234–235, 246, 310, 393, 411, 485, 487–489, 511; physical effort 135, 530
Lacépède, Bernard-Germain de 497–498
Lamartine, Alphonse de 498–499
landscape 4, 21, 137, 291, 294, 355, 363–372, 377, 472, 474, 484, 494, 502, 508, 522, 529, 535, 538, 540
Larkin, Philip 522
laws *see* legal
League of American Wheelmen (LAW) 52, 89, 363; road books and maps 365, 369–371
Leblanc, Maurice 471, 500
legal 18, 25, 34, 36, 87, 106, 212, 221, 228, 255, 263, 347, 349, 351, 382, 398, 406, 429, 509; *see also* regulations
leisure 7, 8, 16, 21, 27, 34, 35, 59, 75, 83, 89, 120, 129–132, 141, 145, 165, 168–169, 175, 179, 215, 241, 247, 268, 331, 334–338, 378, 406, 450, 460, 462–464, 476, 478–479, 488–489, 510, 515, 537–539
Lesclide, Richard 498
Leth, Jorgen 490
liberalization 402
liberation 12, 33, 396, 494, 500, 505–513
light segregation 222
liminality 4, 46, 140, 177, 219, 483
liveable neighbourhoods, livability 12, 66, 71, 215, 220, 235, 423
livelihoods 35–37, 201, 203, 208, 399

load (mechanical) 97–99, 101, 104, 111, 115, 125, 336
load carrying 3, 7, 29, 35, 36, 85–86, 101–102, 106, 129–130, 133, 135, 137–138, 203–208, 263, 264, 269, 308, 377, 392, 413
London: Ontario 319; UK 219, 232–235, 238–240, 278, 288–289, 295, 331, 382, 454, 465, 474, 482, 490, 510, 534, 537, 539
Londres, Albert 503
López-Gydosh, Dilia 455, 456
low carbon: economy 42; footprint 167, 401; futures 494; mobility 21, 44, 266, 415, 494
low traffic neighbourhoods 58, 220, 259
luggage *see* load carrying
Luiselli, Valeria 484
Lumière, Auguste and Louis, *La Sortie de l'Usine Lumière à Lyon* (1895) 484, 485
Lund 229–231

machinery 19, 166, 195, 292, 412, 487
madison 321–324
Madison Square Garden 284, 322–324
magazines, ladies' 455, 456, 457–459; *see also* cycling press; periodicals
maintenance: cycle 2, 45, 59, 106, 116–117, 119, 122, 124, 127, 128, 134, 137, 167, 212, 227, 269, 404; infrastructure 5, 6, 58, 71, 311, 314–315, 222, 254, 386, 387, 435
MAMILS 73, 84, 283, 285, 287–297, 453, 460; *see also* masculinities
manufacturing 6, 18, 106, 151–152, 154, 156, 163–169, 174–176, 181–182, 185, 192–200, 203, 401, 413, 440, 482
mapping *see* maps
maps 24, 234, 369–371, 376, 379–380
marginalization 5, 37, 201–205, 208, 237, 276–277, 355, 373, 381, 450, 505
marketing 26, 114, 154, 164, 165, 169, 192, 196, 233, 235
masculinities 26–30, 61, 246, 277–278, 283, 287–297, 315, 448, 454–455, 457, 460, 462, 469, 488, 490, 507–509; *see also* MAMILS
mass production 34, 95, 108, 131, 151, 166, 199, 264, 462, 473, 476, 485
mechanics (persons) 45–47, 61, 84, 92, 181, 246, 308, 497, 509, 518
mental health 16, 73–76, 266, 284, 332, 336–338, 339–340
Melbourne 4, 46, 59, 427–429, 482, 509
Merida 151, 156, 158, 160, 164–169, 182, 195–198
Merckx, Eddy 325, 479, 490
messengers *see* couriers
Mexico 37–38, 44, 162, 203, 377
Mexico City 37, 45, 202, 204–208, 244–245
Michaux, Henri 359
micromobility 67–70, 137–139, 175, 208, 215, 217, 254–262
Middle East 255, 396–398, 507, 535
migrants, migration 28–30, 33, 35–39, 45, 356, 393, 399, 401, 428, 488
Mills, Russell 519
mobility as a service (MaaS) 232, 234–235, 255, 261
mobility data specification (MDS) 261
mobility on demand (MOD) 261
mobilities 11–12, 21, 43, 48–49, 229, 230, 235, 242, 244, 246, 275, 423
Mockus, Antanas 434
modal split/modal share 12, 42, 216, 218, 229, 233, 403, 413
modal shift 16, 266, 399, 403
modernism 229, 231, 433, 439, 448–450, 479, 526
modernity 150, 355, 388, 391, 401, 439–440, 443, 445, 448, 463–464, 470, 484–485, 487, 497, 499–500
mopeds 44, 88, 254–256, 263–264, 402
motility 56–60
motivations for cycling 62, 78, 216–218, 264, 265, 267, 283, 291–292, 437
mountain bike 4, 8, 77, 85–86, 99, 102, 104, 106, 115–117, 120–122, 125, 126–127, 130, 137, 144, 156, 159, 169, 173–174, 182–183, 199, 284, 289, 310–317, 432, 450, 451, 488, 490
mountains 8, 119, 125, 137, 288, 305, 307, 314, 377, 382, 432, 450, 462, 525, 531, 534, 538, 539; *see also* hills
MTB *see* mountain bike
Multimodality 257
Murphy, Dervla 378, 381, 540

National Urban Transport Policy (NUTP) 403
neoliberalism 33, 40, 153, 489
Newman, Bernard 381, 535, 538
Nigeria 390, 392
normalization 348, 350
nostalgia 365–366, 450, 525, 527, 530, 533, 535–538
Nutting, Wallace 365–366

O'Brien, Flann: *The Third Policeman* 518
obesity/overweight 141, 144, 233, 235–236, 333–336, 507
objectification 27
off-road cycling 4, 100, 101, 114, 120, 132, 135, 142–144, 159, 283–284, 289–290, 310–317, 385, 491
older riders 29, 35, 59, 60, 69, 73, 83, 84, 144–146, 168, 199, 205, 215, 236, 264–265, 274, 313, 314, 335, 337–338, 419, 422, 450, 459, 516
Olympic games 183, 185, 288, 291, 295, 299, 301, 312, 315, 31–24, 331, 348, 350
ontology 17, 18, 21, 522–523, 530

oppression 35, 39–40, 276–277, 500, 505; ableism 140; sexism 84
Original Brand Manufacture (OBM) 164
Original Design Manufacture (ODM) 156, 157–159, 161, 164, 196, 199
Original Equipment Manufacturer (OEM) 151, 154, 157–161, 163–164, 166, 192–194, 196, 199
Orzada, Belina T. 455, 456
Ottoman empire 396–397

pain 76, 142, 339, 346, 365, 404, 523, 525, 530
paint 104, 106, 222, 224, 386, 446–447, 476, 291
Paleogolu, Jean de (PAL) 444, 465–466
Pantani, Marco 531
Paris 1, 3, 4, 52, 234, 359–361, 380, 446, 455, 463–466, 498–500, 507
Paris-Roubaix 298–299, 304, 490
pandemic 1, 34–35, 37–38, 45, 48–49, 55, 68, 70, 80, 156, 168–169, 172, 175, 185, 191, 197, 199, 213, 252, 254, 256, 260–261, 268–269, 272, 301, 316, 318, 385, 399, 404, 436, 439, 509
parking: car 58, 65, 68, 219, 222, 227, 257, 274, 386, 402, 420–424, 428–429, 434; cycle 42, 65, 76, 93, 94, 216, 217, 220, 227, 235, 241, 416, 428, 430, 435; micromobility 256, 258–261
participation 16, 20, 26–30, 43–44, 56–58, 75, 80, 216, 237, 246, 267, 272, 274, 283, 288, 310, 316, 333, 336, 397, 432, 436, 455
participation ladder 56–58
passenger carrying 3, 86, 128–136, 138, 143, 144, 201, 203–205, 413
patent (intellectual property) 18, 26, 89, 92, 95, 108, 144, 159, 177
peace 1, 4, 7, 44, 92, 373–375, 382, 397–398, 530
pedaling 91, 94, 97, 115, 123, 143, 207, 222, 241, 263–264, 318, 323, 373, 378, 500, 507, 515, 524, 525, 530, 540
pedelec 263; Speed pedelec/s-pedelec 269; *see also* e-bike
pedestrian crossing 226
peloton 3, 284, 290, 300, 302, 305, 307–309, 487
Peñalosa, Enrique 48, 432, 434, 436
Pennell, Elizabeth and Joseph 367, 378, 381–382, 477, 533–535, 538–540
people of colour 61, 62, 73, 277
perceived safety 265–266, 274–275
performance (action) 26, 27, 30, 90, 106, 141, 172–180, 236, 269, 293, 318, 336, 337, 345– 354, 371, 443, 449, 486, 490, 500–501, 515
periodicals 498, 499, 500; *see also* cycling press; ladies magazines
permit fee 259
Perret, Jacques 502–503
Perrodil, Édouard de 499
phenomenology 523, 530
philosophy, philosophers 16, 17, 20, 189, 241, 292, 312, 388, 434, 462, 473, 502, 521, 523–525, 536, 538, 541
photography 53, 290, 355, 360, 365–367, 369, 371, 381, 389, 397, 398, 440–441, 475–476, 478, 479, 484, 485
physiology 68, 131, 284, 322–323, 332, 334, 509
pilgrimage: literary 533–535; historical 535–536
place marketing 235
planners 3, 42–44, 49, 67, 71, 90, 146, 174, 215, 217, 220, 241, 243, 247, 269, 421
planning 21, 42, 48–49, 55, 65–71, 86, 140, 215–218, 219–228, 229, 237, 241–247, 260, 272–273, 279, 284, 346, 356–357, 400–405, 419–423, 425, 431, 433; *The Bicycle Account* 420
play, cycling as 29, 34–36, 44, 237, 245, 414, 463, 509, 511, 519, 521
platform capitalism 34
platform economy 35, 37, 38
platform work 33–41
pleasure 35, 42, 141, 149, 152, 159, 216, 284, 293, 315, 338, 355, 357, 363, 378, 382, 430, 440, 447, 449, 455, 460, 469, 471, 479, 489, 529–531
poetry 1, 497–500, 521–532, 536
policy: cycle/transport 1, 3, 12, 16, 17, 20, 29, 45, 66, 70, 73, 77, 83, 146, 215–217, 219–220, 227, 230, 232–240, 241–245, 247, 254–262, 264, 272, 356, 385, 402–406, 434–436; industrial/economic 151, 154, 163, 169, 192–196, 199, 414
policy makers 3, 4, 27, 55, 80, 146, 168, 219, 227, 233, 241, 247, 414
pollution 5, 6, 130, 135, 197, 211, 215, 233, 250–251, 266, 276, 339, 380, 391, 401, 415, 434
portable cycle 96, 97, 100, 116–117, 134, 156, 197, 267
posters 440–441, 443–444, 462–471, 478, 488
postmodernity 450, 451
poverty 38, 60, 76, 137, 138, 356, 373, 381, 401, 431, 482, 488, 510
Pratt, Charles 367, 369, 450
Pré-Catelan 359–361
presence 3, 27, 42, 43, 246, 537
precarious work 33–41, 45, 56, 206, 211–212, 488, 510
priority junctions 222–223
prize money 299, 302–303, 305, 327, 346, 428
professional cycling 285, 298–305, 346, 348, 351, 352, 408
protectionism 152, 160, 174
Proust, Marcel 501, 508, 525
public space 12, 26–29, 37, 48–49, 79, 141, 206, 277, 357, 359–361, 397, 416, 431, 433, 434
public transport/transit 7, 45, 55, 74, 77, 135, 168, 199, 211, 213, 215, 229, 230, 232,

234–237, 251–252, 254–257, 261, 266, 268, 333, 400, 402, 405, 411, 423, 515
public bike systems 139, 161, 233, 234, 403–405, 415–417; *see also* shared mobility/micromobility
Pune Bicycle Plan (PBP) 403

Quicksilver (Donnelly, 1988) 489
Quintana, Nairo 432

racing *see* cycle racing
racism 45, 53, 217, 277–279, 381, 486
realism (depiction) 440, 443, 488, 505, 511; as social theory 17–18
rebelliousness 398, 458, 491, 501–503, 531
recreational cycling 400–401, 405–406
recumbent 73, 75–79, 84–85, 89, 94, 95, 100–102, 142–144, 156, 324
recycling 20, 59, 106, 212, 266
regulations 18, 36, 67, 70, 84, 87, 89, 90, 101, 106, 182, 204, 213, 222, 227, 261, 269, 325, 345, 347–349, 374, 415, 509; *see also* legal
religion 396, 529–530, 535
repair (cycle) 6, 13, 45–47, 56, 59, 90, 96, 112, 113, 149, 150, 152, 181, 191, 203–204, 211–213, 374, 392, 537; *see also* maintenance
resistance: political 27, 38, 39, 42, 205, 245, 410, 432, 435, 506, 518; mechanical 98, 99, 104, 142, 308, 318, 323–324; rolling 92–94, 101, 103, 112–114, 116, 134–135
respectability 397, 448, 449, 456, 459–460
retailing *see* cycle retailing
rhythm 141, 143, 294, 308, 421, 483, 491, 522
rickshaw *see* cycle taxi
risk 3, 8, 18, 25, 27–30, 36–39, 95, 112, 114, 215–217, 224, 226, 230, 258, 261, 273, 276, 278, 284, 295, 300, 310, 313, 315, 333–339, 346, 348, 350, 356, 382, 400, 405, 406, 445, 454, 460, 477, 503, 538, 540
ritual 207, 283, 435, 490, 524
Riol, Jules 499–500
road cycling 3, 4, 263, 284–285, 287–297, 298–309, 312, 345, 427–429
road racing 3, 104, 183, 185, 284–285, 288, 290, 328, 361, 427, 449, 489, 490
roadster 448, 449–450
Rogers, Cline 367–369
Romains, Jules 501, 519
romanticism (literary movement) 494, 498–499, 526, 529
roundabouts 217, 219, 222, 225–226, 228, 300
Roux, Joseph 474–475
rhythm 143, 150, 177, 284, 294, 308, 522
Rwanda 3, 137–139

safety bicycle 19, 26, 30, 84, 93, 123, 128, 181, 264, 275, 363–365, 447, 453, 454, 458, 476, 478, 485, 490, 498, 510, 539
sales *see* retail, supply chains, trade shows
salon art in France 464, 474, 476
Sauer, Carl 364, 369, 371
schools 8, 29, 43, 58, 60, 75, 134, 136, 150, 246, 294, 328, 336, 339, 346, 376, 380, 391–393, 397, 413, 420, 422, 424, 450
Schwinn 151, 153, 164, 174, 181, 185–187, 193, 531
scooters 67, 69, 78, 80, 87, 95, 118, 156, 163, 168, 213, 217, 237, 254–262, 263, 269
Scribe, Eugène 497
sculpture 8, 388, 440, 465, 472, 477, 478, 481–483
secondhand bikes 45
sensation, sensory experience 8, 43, 80, 177, 338, 356, 363–364, 368, 371, 378, 431–432, 462, 471, 498–502, 515, 517, 524–525, 537
Shanghai 232, 234–235, 239–240
shared mobility/micromobility 45, 168, 193, 212, 215, 217, 234–237, 254–262, 400, 402, 415, 416; *see also* public bike systems
shared (road)space 5, 66, 141, 219, 222, 230, 277, 386, 420
sharing economy 168
shock absorption *see* suspension
sidepaths 5, 365; sidepath movement 367, 369
signification 44, 208, 245, 249, 419, 506
signifier *see* signification
Simpson (lever chain) 448, 449–450
skills 20, 28–29, 56–62, 138, 156, 212, 246, 299, 309, 314, 323, 328, 432; *see also* competencies
skirts 26, 91, 93, 95, 440, 447, 454–456, 458–459, 490; skirt guards 454, 458
Smart City Mission 404
social construction 18, 19, 140, 450
social inclusion *see* inclusivity
social interaction 1, 28, 43–45, 56, 241, 338, 519
social practice 20
sociotechnology 19, 203, 207, 216
spectacle 176–177, 298, 324, 359, 487, 501, 515
speed: high speed 102–103, 133, 134, 258, 321, 323, 328, 422, 450, 454, 455, 477, 489; low speed 56, 58, 103, 221, 429
spokes 86, 108–111, 125, 446, 466, 478, 481, 523, 525, 527
sponsorship 150, 164, 182–183, 235, 289, 300, 303–304, 361, 371, 378, 380
sport (cycle) *see* amateur cycling, professional cycling
sportives 283, 288
steel 84, 85, 104–105, 108–111, 113, 115–117, 125, 127, 129, 130, 132, 137, 156, 157, 158, 318, 329, 470, 526
steering 93, 94, 97, 100, 102–104, 133, 142, 144, 516
Stevens, Thomas 364, 378–379, 381–382, 538–540
Stilgoe, John 366

Stone, Alison 382–383
storage of cycles 448
storytelling (as strategy) 423
street trade, traders 151, 201–210, 413
strength: material 98, 99, 101, 104–106, 110–111, 125, 318, 328, 392, 405, 467, 481; physical 30, 141, 299, 310, 328, 345, 382, 410, 432, 455, 468, 536
sublimity 525, 529–530
subjectivity 221, 237, 273, 274, 393, 499–502, 517, 523, 527, 531
supply chain 3, 149, 154, 157, 159, 164, 181–182, 185, 199, 202
surveillance capitalism 235
suspension 86, 100–101, 109, 112, 114–117, 121, 130, 134, 142, 144, 169, 310, 312–314, 374, 450
sustainability 2, 3, 16, 20, 21, 33, 48, 69, 75, 106, 201, 217, 232, 235, 237–238, 242, 247, 291, 301, 399, 432
sustainable cities 202, 242, 244, 252
sustainable mobility 12, 19, 27, 49, 139, 168, 201–203, 216, 229–231, 233, 317, 356, 389, 399–405, 416, 422, 425
sustainable transport *see* sustainable mobility
Sylvester, H. E. 367–368
systematic reviews 284, 333–337

tacit knowledge 57, 174, 177, 178, 198, 305
tactical urbanism 243, 247, 423
tandem 93–94, 101, 128, 129, 143–144, 156, 168, 185, 221, 227, 367, 377, 465, 527, 533
tariffs 153, 160, 162, 174, 182, 191
Taipei International Cycle Show (TICS) 159, 173–175, 178
Taiwan 3, 90, 144, 149, 151, 153–162, 163–171, 173–176, 182, 192–200, 415
Taiwan Bicycle Association 164, 168, 175, 199; *see also* A-Team
Tati, Jacques, *Jour de Fête* (1949) 487
Taylor, Frank Hamilton 367, 371
taxes 2, 93, 152, 259, 367, 434
teenage *see* Youth
theatre 360, 463, 490, 497
Thomas, Edward 534–535
titanium 104–106, 115, 142, 157, 293
Toulouse-Lautrec, Henri de 444–445, 464–465
Tour de France 89, 123, 183, 283, 288–229, 291, 295, 298–306, 324, 325, 348, 449, 479, 503, 536
touring bikes 94, 120, 122, 125, 203, 371, 377, 382, 463
track racing and track bikes 3, 52–54, 89, 99, 110, 112, 115, 156, 185, 283–284, 288–289, 300, 318–326
trade associations 164, 168, 175, 188, 191, 194, 199, 417
trade show 3, 150, 151, 154, 157, 159, 172–180, 360
trail (in cycle design) 103; *see also* steering
trails, trail riding 5, 8, 159, 199, 261, 284, 310, 317, 373, 506
trailers 59, 86, 128, 129, 132–135, 143, 144, 201, 221, 265, 377, 421, 500
trailer bikes 86, 131
transactional economy 172–180, 255
transmission 3, 84–86, 94, 96, 118–127, 416
travel writing 493–495, 497–504, 533–541
tricycle 8, 19, 59, 84, 91, 93–95, 97, 128, 133, 136, 141–146, 156, 173, 175, 199, 201, 203–208, 211, 221, 228, 363, 364, 367, 396, 409, 413, 447, 448, 474, 533–534, 538
trophies 448
trust 164, 174, 177–178, 198, 236, 245, 307, 392, 404, 531
Twain, Mark: "Taming the Bicycle" 494, 515–517
tyres/tires 65, 85, 86, 92, 96, 101, 108, 111–114, 116, 123, 125, 134, 158, 165, 174, 211, 292, 310, 312–314, 319, 329, 359, 363, 392, 416, 444, 447, 450, 454, 463, 469, 523, 534

Uber Eats *see* food delivery
Uganda 391, 392
Union Cycliste Internationale (UCI) 84, 87–90, 95–96, 300–302, 304, 315, 319–320, 324–325, 328
urban design 2, 3, 12, 231, 237, 243
utility cycles and cycling 25, 27, 35, 85, 102, 104–106, 216, 218, 267–268, 295, 447, 450, 526; *see also* everyday cycling
utopia 508, 511

value chains 162–171, 192, 195
velocipede 89, 91, 93, 144, 173, 359–361, 409–410, 428, 440, 445–446, 453, 466, 474–476, 478, 497–498, 509, 516
velodrome 43, 52, 53, 113, 242, 284, 318–326, 327–331, 469, 470, 477, 486; *see also* track racing
velomobiles 95, 201, 269
vélomobility 1, 6, 21, 43, 48, 269
Vietnam 157, 160, 16, 164, 196, 377
vehicular cycling 30
Vuelta a Colombia 432
Vuelta a España 298, 304

Wadjda (al-Mansour, 2012) 491
walking 42, 69–70, 73–75, 77–79, 91, 139, 145, 202, 229, 237, 244, 257–258, 275, 333–335, 338, 400–404, 411, 423, 432, 484, 493–494
Warmbrunn, Erica 377, 381
Weiwei, Ai 8, 440, 482–483
Wells, H. G. 510; *The Wheels of Chance* 516; *The History of Mr Polly* 518–519
Westwood, Kim 509
wheelchairs 70, 73–74, 78, 144, 145

wheels 3, 84, 85, 87–94, 100–103, 108–117, 132–133, 142, 144, 158, 165, 204, 259, 308, 318, 325, 351, 360, 363, 365, 376–377, 398, 445, 448, 481–483, 484–487; disc 108, 318, 325
Wheels: A Bicycle Romance 507–508
Wiggins, Bradley 290, 349; 'The Wiggo Effect' 288–289
winter cycling 385–387
women and cycling 18, 24–32, 37, 38, 359, 363–364, 382, 396–399, 401, 440, 454–460, 462–463, 469, 490, 508
Wood, Henry 369–370
Wordsworth 498–499, 500, 537
work: employment 2, 6, 33–42, 46, 62, 66, 73–75, 84, 87, 135, 150–152, 201–210, 211–213, 235, 237, 250–253, 267, 294–295, 352, 373, 377, 378, 380–381, 385, 391, 397, 399–405, 411–414, 419–420, 424, 433–436, 450, 485, 487–489, 509–511; physical effort 114, 128, 138, 299, 308, 313–315, 322–324, 487–489, 509, 530
working bicycles, tricycles 57, 211
workshops 45–47, 49, 59, 130, 137, 244, 246, 359; *see also* shared mobility
"World Bike Capital," (Bogotá) 430–437
Wray, W. Fitzwalter (Kuklos) 537–538, 540

youth/young people 7, 26–28, 45, 61, 84, 93, 135, 137–138, 143, 182, 205, 236–237, 264–265, 277, 284, 293, 316–317, 335–339, 374, 464, 390, 392–393, 397, 412, 419, 428, 432, 436, 444, 446–447, 451, 453, 455, 459, 465, 469, 491, 511, 516, 518–519, 525–527, 540; *see also* adolescence

zoning 154, 205, 258, 259, 423
Zupančič, Alenka 517